Concrete

About the Authors

P. Kumar Mehta, Ph.D., is Professor Emeritus, Civil and Environmental Engineering at the University of California at Berkeley. The recipient of the Berkeley Citation—the highest honor for contributions to his field and to the University—Dr. Mehta is also an Honorary Member of the American Concrete Institute (ACI). He is the author or coauthor of numerous scientific papers and four books in the field of concrete technology.

Paulo J. M. Monteiro, Ph.D., is Roy W. Carlson Distinguished Professor, Department of Civil and Environmental Engineering, University of California at Berkeley and Faculty Scientist, Department of Materials, Lawrence Berkeley National Laboratory. Dr. Monteiro has authored or coauthored over 190 archival articles, and numerous conference proceedings, and reports.

Concrete

Microstructure, Properties, and Materials

P. Kumar Mehta

Paulo J. M. Monteiro

Fourth Edition

McGraw Hill Education (India) Private Limited

McGraw Hill Education (India) Reprint 2014

Twelfth reprint 2026

Reprinted in India by arrangement with McGraw-Hill LLC, New York

Sales territories: India, Pakistan, Nepal, Bangladesh, Sri Lanka and Bhutan

Cataloging-in-Publication Data is on file with the Library of Congress

ISBN (13): 978-93-392-0476-1
ISBN (10): 93-392-0476-X

12 13 14 15 16 SAU 30 29 28 27 26

Published by McGraw Hill Education (India) Private Limited,
Anjana Complex No: 5/90 A, Butt Road, St. Thomas Mount, Chennai – 600016, Tamil Nadu, India, and
Printed and bound in India at Saurabh Printers Pvt. Ltd., 67A-68, Ecotech Ext-I, Kasna, Greater Noida, Gautam Budh Nagar (U.P.), 201308

Visit us at: www.mheducation.co.in; Phone (Toll free in India): **1800 103 5875**
Write to us at: professional_india@mheducation.com

CIN: U80302TN2010PTC111532

This book is dedicated to students, researchers, and practicing engineers in the concrete community who are faced with the challenges of extending the uses of the material to new frontiers of human civilization and to make it more durable, sustainable, and environment friendly.

Contents

Part III Advances in Concrete Technology

Foreword

The authors of this outstanding textbook, Professors Mehta and Monteiro, are leading researchers who have made lasting and seminal contributions in the field of nano- and microstructure of concrete, cement chemistry, concrete technology and mechanics. Apart from their distinguished academic career, they have also served as consultants for challenging real-world projects. The perceptive case studies described in the textbook will inspire engineers to find creative and sustainable solutions for concrete materials and structures. The text material has been presented in a remarkably clear and intuitive manner; it is greatly appreciated by students and professional engineers. The quality of the textbook has been recognized worldwide; previous editions have been translated into Japanese, Chinese, Greek, Spanish, Portuguese, and Persian.

For this fourth edition, every chapter has been carefully revised or rewritten to incorporate new developments in the field, including recent construction projects worldwide, shrinkage-reducing admixtures, self-consolidating concrete, shotcrete, pervious concrete, internal curing, electrical tomography of concrete, and other cutting-edge innovations. A strong feature of the textbook is the synergy between concrete materials, concrete technology, and concrete mechanics. For the new edition, the chapter on concrete mechanics includes the ground-breaking poromechanics formulation to model concrete exposed to freezing temperatures, analysis of the length scale in nanomechanics and an insightful review of the modeling of the alkali-aggregate reaction in concrete.

The authors made a profound analysis of the environmental impacts caused by the global concrete industry. They estimated that in 2010, 33 billion tonnes of concrete were produced globally, consuming nearly 3.7 billion tonnes of portland cement clinker and 27 billion tonnes of aggregate, in addition to 2.7 billion tonnes of mixing water and a small amount of chemical admixtures. The mining, processing, and transport of such huge quantities of materials for making concrete require considerable energy, and adversely affect the ecology of virgin lands. Also, the world's yearly cement production of mostly portland cement is responsible for nearly 7 percent of the global anthropogenic CO_2 emissions. Professors Mehta and Monteiro provided solid guidelines for the development of green concrete by using large amounts of supplementary cementing materials, particularly coal fly ash or blast-furnace slag. Chemical and mineral admixtures are fully described in Chapter 8 and their applications to the development of high-performance green concrete are given in Chapter 12. Professors Mehta and Monteiro rounded off the textbook with a discussion of concrete in the era of global warming and sustainability. This should be required reading for everyone committed to sustainable concrete construction.

Victor Li, FASCE, FASME, FWIF, FACI
E. Benjamin Wylie Collegiate Professor
Department of Civil and Environmental Engineering
Department of Materials Science and Engineering
University of Michigan

Preface

There is a direct relationship between population and urbanization. During the last 100 years, the world population has grown from 1.5 to 7 billion and nearly 3 billion people now live in and around the cities. Seventeen of the 20 megacities, each with a population of 10 million or more, happen to be situated in developing countries where enormous quantities of materials are required for the construction of housing, factories, commercial buildings, drinking water and sanitation facilities, dams and canals, roads, bridges, tunnels, and other parts of the infrastructure. And the principal material of construction is portland cement concrete. *By volume, the largest manufactured product in the world today is concrete.* Naturally, design and construction engineers need to know more about concrete than about other materials of construction.

Written primarily for the use of students in civil engineering, it covers a wide *spectrum of topics in modern concrete technology* that should be of considerable interest to practicing engineers. For instance, to reduce the environmental impact of concrete, roles of pozzolanic and cementitious by-products as well as superplasticizing admixtures in producing highly durable products are thoroughly covered.

One of the objectives of this book is to present the *art and science of concrete in a simple, clear, and scientific manner*. Properties of engineering materials are governed by their microstructure. Therefore, it is highly desirable that structural designers and engineers interested in properties of concrete become familiar with microstructure of the material. In spite of apparent simplicity of the technology of producing concrete, the *microstructure of the product is highly complex*. Concrete contains a heterogeneous distribution of many solid compounds as well as voids of varying shapes and sizes that may be completely or partially filled with alkaline solution.

Compared to other engineering materials like steel, plastics, and ceramics, the microstructure of concrete is not a static property of the material. This is because two of the three components of the microstructure, namely, the bulk cement paste and the interfacial transition zone between aggregate and cement paste change with time. In fact, the word *concrete* comes from the Latin term *concretus*, which means to grow together. The strength of concrete depends on the volume of the cement hydration products that continue to form for several years, resulting in a gradual enhancement of strength. Depending on the exposure to environment, solutions penetrating from the surface into the interior of concrete sometimes dissolve the cement hydration products causing an increase in porosity which reduces the strength and durability of concrete; conversely, when the products of interaction recrystallize in the voids and microcracks, it may enhance the strength and durability of the material. This explains why analytical methods of material science that work well in modeling and predicting the behavior of microstructurally stable and homogeneous materials do not seem to be satisfactory in the case of concrete.

In regard to organization of the subject matter, the *first part* of this three-part book is devoted to hardened concrete microstructure and properties, such as strength, modulus of elasticity, drying shrinkage, thermal shrinkage, creep, tensile strain capacity, permeability, and durability to various processes of degradation. Definition of each property, its significance and origin, and factors controlling it are set forth in a clear manner. The *second part* of the book deals with concrete-making materials and concrete processing. Separate chapters contain state-of-the-art reviews on composition and properties of cements, aggregates, and admixtures. There are also separate chapters on proportioning of concrete mixtures, properties of concrete at early ages, and nondestructive test methods. The *third part* covers special topics in concrete technology. One chapter is devoted to composition, properties, and applications of special types of concrete, such as lightweight concrete, high-strength concrete, high-performance concrete, self-consolidating concrete, shrinkage-compensating concrete, fiber-reinforced concrete, pervious concrete, shotcrete, concretes containing polymers, and mass concrete. A separate chapter deals with advances of concrete mechanics covering composite models, creep and shrinkage, thermal stresses, fracture of concrete, and durability of concrete. The final chapter contains some reflections on current challenges to concrete as the most widely used building material, with special emphasis on ecological considerations.

A special feature of the book is the inclusion of numerous unique diagrams, photographs, and summary tables intended to serve as teaching aids. New terms are indicated in italics and are clearly defined. Each chapter begins with a preview of the contents, and ends with a self-test and a guide for further reading.

Acknowledgments

This thoroughly revised fourth edition of the book would not have been possible without the help and cooperation of many friends and professional colleagues. The authors thank all of them most sincerely.

Grigory Isaakovich Barenblatt for his insightful comments on the nanomechanics of concrete.

Fabbri Antonin, Teddy Fen-Chong, and Patrick Dangla for their critical review of the poromechanics of freezing concrete. Olivier Coussy would be proud of his team.

John Bolander for his valuable contribution to fiber reinforced concrete in Chapter 12.

Omkar Deo for the new material describing previous concrete.

Maria Garci for the perceptive review of Chapters 2 and 10.

Guoging Geng for the superb work during the final proofreading.

Claire Johnson for the careful editing of the manuscript.

Mônica Batista Leite for the discussions on recycled aggregate.

Maryanne McDarby for the continuous support with the editing process.

Joclyn Norris for dedicated work with the illustrations.

Jae-Eun Oh for the photographs of modern concrete construction.

Rossella Pignatelli for discussions on the poromechanics of alkali-aggregate reaction.

John Provis for his comprehensive review of Chapters 6 and 8.
Silvia Vieira for the detailed review of the chapters on aggregates and cements.
George W. Scherer for discussion on the mechanism for salt scaling.
Gaurav Sant for the description of shrinkage reducing admixtures.
Rae Taylor for the many discussions on the microstructure of the hydration products and the beautiful TEM images.
Mason Walters for the photographs of the Cal Stadium.

P. Kumar Mehta
Paulo J. M. Monteiro
University of California at Berkeley

Concrete

PART I

Microstructure and Properties of Hardened Concrete

CHAPTER 1
Introduction

Preview

This chapter describes important applications of concrete, and examines the reasons that made concrete the most widely used structural material in the world today. The principal components of modern concrete are identified and defined. A brief description of the major concrete types is given.

For the benefit of beginning students, an introduction to important properties of engineering materials, with special reference to concrete, is also included in this chapter. The properties discussed are strength, elastic modulus, toughness, dimensional stability, and durability.

1.1 Concrete as a Structural Material

In an article published by the *Scientific American* in April 1964, S. Brunauer and L.E. Copeland, two eminent scientists in the field of cement and concrete, wrote:

> The most widely used construction material is concrete, commonly made by mixing portland cement with sand, crushed rock, and water. Last year in the U.S. 63 million tons of portland cement were converted into 500 million tons of concrete, five times the consumption by weight of steel. In many countries the ratio of concrete consumption to steel consumption exceeds ten to one. The total world consumption of concrete last year is estimated at three billion tons, or one ton for every living human being. Man consumes no material except water in such tremendous quantities.

Today, the rate at which concrete is used is much higher than it was 50 years ago. It is estimated that the present consumption of concrete in the world is of the order of 33 billion metric tonnes every year.

Concrete is neither as strong nor as tough as steel, so why is it the most widely used engineering material? There are at least three primary reasons. First, concrete* possesses excellent resistance to water. Unlike wood and ordinary steel, the ability of concrete to withstand the action of water without serious deterioration makes it an ideal material for building structures to control, store, and transport water. In fact, some of the earliest known applications of the material consisted of aqueducts and waterfront retaining walls constructed by the Romans. The use of *plain concrete* for

*In this book, the term *concrete* refers to portland cement concrete unless stated otherwise.

FIGURE 1-1 Itaipu Dam, Brazil. (*Photograph courtesy of Itaipu Binacional, Brazil.*)

This spectacular 12,600 MW hydroelectric project at Itaipu, estimated cost $18.5 billion, includes a 180-m high hollow-gravity concrete dam at the Paraná River on the Brazil-Paraguay border. By 1982 twelve types of concrete, totaling 12.5 million cubic meters, had been used in the construction of the dam, piers of diversion structure, and the precast beams, slabs, and other structural elements for the power plant.

The designed compressive strengths of concrete ranged from as low as 14 MPa at 1 year for mass concrete for the dam to as high as 35 MPa at 28 days for precast concrete members. All coarse aggregate and about 70 percent of the fine aggregate was obtained by crushing basalt rock available at the site. The coarse aggregates were separately stockpiled into gradations of 150, 75, 38, and 19 mm maximum size. A combination of several aggregates containing different size fractions was necessary to reduce the void content and, therefore, the cement content of the mass concrete mixtures. As a result, the cement content of the mass concrete was limited to as low as 108 kg/m^3, and the adiabatic temperature rise to 19°C at 28 days. Furthermore, to prevent thermal cracking, it was specified that the temperature of freshly cooled concrete would be limited to 7°C by precooling the constituent materials.

dams, canal linings, and pavements is now a common sight almost everywhere in the world (Fig. 1-1). Structural elements exposed to moisture, such as piles, foundations, footings, floors, beams, columns, roofs, exterior walls, and pipes, are frequently built with reinforced and prestressed concrete (Fig. 1-2). *Reinforced concrete* is a concrete usually containing steel bars, which is designed on the assumption that the two materials act together in resisting tensile forces. With *prestressed concrete* by tensioning the steel tendons, a precompression is introduced such that the tensile stresses during service are counteracted to prevent cracking. Large amounts of concrete find their way into reinforced or prestressed structural elements. The durability of concrete to aggressive waters is responsible for the fact that its use has been extended to severe industrial and natural environments (Fig. 1-3).

Figure 1-2 Precast concrete girders under installation for the Skyway Segment of the eastern span crossing the San Francisco Bay. (*Photograph courtesy of Joseph A. Blum.*)

The Loma Pietra earthquake caused damage in the eastern span of the San Francisco Bay Bridge. After years of studying the seismic performance of the bridge, the engineers decided that the best solution was to construct a new span connecting Oakland to the Yerba Buena Island. The two new twin precast segmental bridges accommodate five lanes of traffic in each direction and a bike path on one side. The superstructure, constructed using the segmental cantilever method, will require 452 precast girders, each weighting as much as 750 tons.

The second reason for the widespread use of concrete is the ease with which structural concrete elements can be formed into a variety of shapes and sizes (Figs. 1-4 to 1-10). This is because freshly made concrete is of a plastic consistency, which enables the material to flow into prefabricated formwork. After a number of hours when the concrete has solidified and hardened to a strong mass, the formwork can be removed for reuse.

The third reason for the popularity of concrete with engineers is that it is usually the cheapest and most readily available material on the job. The principal components for making concrete, namely aggregate, water, and portland cement are relatively inexpensive and are commonly available in most parts of the world. Depending on the components' transportation cost, in certain geographical locations the price of concrete may be as high as U.S. $75 to $100 per cubic meter, at others it may be as low as U.S. $60 to $70 per cubic meter.

Some of the considerations that favor the use of concrete over steel as the construction material of choice are as follows:

Maintenance. Concrete does not corrode, needs no surface treatment, and its strength increases with time; therefore, concrete structures require much less maintenance.

Figure 1-3 Statfjord B offshore concrete platform, Norway. (*Photograph courtesy of Norwegian Contractors, Inc.*)

Since 1971, twenty concrete platform requiring about 1.3 million cubic meters of concrete have been installed in the British and Norwegian sectors of the North Sea. Statfjord B, the largest concrete platform, built in 1981, has a base area of 18,000 m^2, 24 oil storage cells with about 2 million barrels of storage capacity, four prestressed concrete shafts between the storage cells and the deck frame, and 42 drilling slots on the deck. The structure was built and assembled at a dry dock in Stavanger; then the entire assembly, weighing about 40,000 tonnes, was towed to the site of the oil well, where it was submerged to a water depth of about 145 m. The prestressed and heavily reinforced concrete elements of the structure are exposed to the corrosive action of seawater and are designed to withstand 31-m-high waves. Therefore, the selection and proportioning of materials for the concrete mixture was governed primarily by consideration of the speed of construction by slip-forming and durability of hardened concrete to the hostile environment. A free-flowing concrete mixture (220-mm slump), containing 380 kg/m^3 of finely ground portland cement, 20 mm of maximum-size coarse aggregate, a 0.42 water-cement ratio, and a superplasticizing admixture was found satisfactory for the job. The tapered shafts under slip-forming operation are shown in the figure.

Steel structures, on the other hand, are susceptible to rather heavy corrosion in offshore environments, require costly surface treatment and other methods of protection, and entail considerable maintenance and repair costs.

Fire resistance. The fire resistance of concrete is perhaps the most important single aspect of offshore safety and, at the same time, the area in which the advantages of concrete are most evident. Since an adequate concrete cover on reinforcement or tendons is required for structural integrity in reinforced and prestressed concrete

Figure 1-4 The tallest self-supporting tower in the world, Tokyo, Japan, 2012. (*Photograph courtesy of Tobu Tower Skytree Co. and Obayashi Co.*)

TOKYO SKYTREE™ is a 634-m-high tower that has become a landmark structure. The architecture of the tower was selected to symbolize the integration between the past and the future. As one example of the symbolism that characterized the design process, the designers originally had specified a 610-m tower. In the final design; however, they insisted that the building be 634 m high because in old Japanese numbers it sounds as "mu-sa-shi," which is reminiscent of the old Musahi Province, a large area including Tokyo and Saitama. The tower was built with a central column similar to a traditional Japanese five-story pagoda, which is another link to the past. Advanced slip form construction was successfully used to place the 54-MPa high-strength concrete during the winter. The observation decks—at 350 and 450 m—offer a remarkable view of the Tokyo skyline and were made with lightweight concrete. Special care was taken to minimize the thermal stresses in the mat foundation, which was cast with 60 MPa high-strength concrete using a low-heat portland cement.

FIGURE 1-5 Interior of the Sports Palace in Rome, Italy, designed by Pier Luigi Nervi, for Olympic games in 1960. (*Photograph from Ediciones Dolmen.*)

Nervi was a creative engineer with full appreciation of structural concept, practical constructability, and new materials. He was a pioneer of "ferro-cement" technology, which involves embedding a thin metallic mesh in a rich cement mortar to form structural elements with high ductility and crack-resistance. The above photograph shows the Palazzo dello Sport Dome built with a 100-m span, for a seating capacity of 16,000. Thin-walled precast elements with higher flexibility, elasticity, and strength capacity were created.

FIGURE 1-6 Seismic retrofit of California Memorial Stadium, University of California, Berkeley. (*Photograph courtesy of Forell Elsesser Engineers.*)

The California Memorial Stadium, a landmark structure in the Berkeley campus, was designed in 1923 by the renowned architect John Galen Howard. Because the active Hayward fault bisects the stadium, in 2000 it was decided to perform a major retrofit to the structure. A series of creative retrofit schemes accommodates the concentrated ground deformations of an estimated 1.8 m of horizontal movement and 0.6 m of vertical offset generated by the surface fault rupture. Part of the retrofit strategy involved using a green, sustainable concrete. A high-volume fly concrete mixture was utilized that was easy to pump, place, and consolidate in heavily reinforced structural members. During 2009-2011, the repair project consumed over 20,000 m^3 of concrete in constructing mat foundations, footings, grade beams, columns, and pile caps. Although specified to produce 27 MPa compressive strength, the concrete consistently measured 41 MPa or more at 56 days. The concrete mixture proportions were as follows (in kg/m^3): portland cement Type II/V: 196, ASTM Class F fly ash: 196, water content: 148, fine aggregate: 857, coarse aggregate: 1038. A superplasticizer was used with a dosage of 1-3 L/m^3.

FIGURE 1-7 Concrete frieze at the Colegio Official de Arquitectos in Barcelona, Spain. Picasso designed the drawings, and the Norwegian artist Carl Nesjar transferred them to concrete via sandblasting. (*Photograph by P.J.M. Monteiro.*)

Pablo Picasso was born in Andalusia, Spain, but he always maintained a strong emotional attachment to Barcelona, the city that nurtured his early artistic career. When invited to design a frieze for the new building of the Colegio de Arquitectos, he enthusiastically accepted the commission, but only if it would be done in sandblasted concrete by Carl Nesjar, the famed Norwegian painter and sculptor. Picasso's drawings were reminiscent of his early days in Barcelona, which depicts the joyful Corpus Christi procession with young children waving palms. The artistic and technical challenges entailed transferring the spontaneity of Picasso's drawings with their different lines intensities to concrete. A vibrant black basalt from the Pyrenees was selected as aggregate for the concrete in the frieze. Picasso's drawings were scaled up to their final size onto large sheets of paper, and the layout was then traced to the surface of the concrete. The last step was to carefully sandblast the concrete surface to reveal the black basalt aggregate and recreate the vitality of Picasso's drawings. Even though the concrete has been exposed to ambient condition for over fifty years, the quality of the lines, texture, and contrast remain fresh and alluring.

structures, the protection against failure due to excessive heat is provided at the same time.

Resistance to cyclic loading. The fatigue strength of steel structures is greatly influenced by local stress fields in welded joints, corrosion pitting, and sudden changes in geometry, such as from thin web to thick frame connections. In most codes of practice, the allowable concrete stresses are limited to about 50 percent of the ultimate strength; thus the fatigue strength of concrete is generally not a problem.

1.2 Components of Modern Concrete

Although composition and properties of materials used for making concrete are discussed in Part II, here it is useful to define concrete and the principal concrete-making components. The following definitions are adapted from ASTM C 125* (*Standard Definition of Terms Relating to Concrete and Concrete Aggregates*), and ACI Committee 116 (A Glossary of Terms in the Field of Cement and Concrete Technology):

*The ACI committee reports and the ASTM (American Society for Testing and Materials) standards are updated from time to time. The definitions given here are from the ASTM standard approved in the year 2004.

Figure 1-8 Fountain of Time: a sculpture in concrete. (*Photograph courtesy of David Solzman.*)

"Time goes, you say? Ah, no. Alas, time stays; we go." Concrete is an extraordinary material because it can be not only cast into a variety of complex shapes, but also given special surface effects. Aesthetically pleasing sculpture, murals, and architecture ornaments can be created by suitable choice of concrete-making materials, formwork, and texturing techniques. Fountain of Time is a massive 120 by 36 by 5 by 4 m (120 by 18 by 14 ft) work of art in concrete on the south side of the University of Chicago campus. The sculpture is a larger-than-life representation of 100 individual human figures, all cast in place in the exposed aggregate finish. In the words of Steiger, the central figure is Time the conqueror, seated on an armored horse and surrounded by young and old, soldiers, lovers, religious practitioners, and many more participants in the diversity of human life, finally embracing death with outstretched arms. Lorado Taft made the model for this sculpture in 1920 after 7 years of work. About the choice of concrete as a medium of art, the builder of the sculpture, John J. Earley, had this to say: "Concrete as an artistic medium becomes doubly interesting when we realize that in addition to its economy it possesses those properties which are the most desirable of both metal and stone. Metal is cast, it is an exact mechanical reproduction of the artist's work, as in concrete . . . Stone (sculpture) is an interpretation of an original work and more often than not is carried out by another artist. But stone has the advantage of color and texture which enable it to fit easily into varied surroundings, a capability lacking in metal. Concrete, treated as in the Foundation of Time, presents a surface almost entirely of stone with all its visual advantages while at the same time offering the precision of casting that would otherwise only be attained in metal."

Concrete is a composite material that consists essentially of a binding medium within which are embedded particles or fragments of aggregate. In hydraulic-cement concrete, the binder is formed from a mixture of hydraulic cement and water.

Aggregate is the granular material, such as sand, gravel, crushed stone, crushed blast-furnace slag, or construction and demolition waste that is used with a cementing medium to produce either concrete or mortar. The term *coarse aggregate* refers to the aggregate particles larger than 4.75 mm (No. 4 sieve), and the term *fine aggregate* refers

FIGURE 1-9 Baha'i Temple, Wilmette, Illinois. (*Photograph courtesy of David Solzman.*)

The Baha'i Temple is an example of the exceedingly beautiful, ornamental architecture that can be created in concrete. Describing the concrete materials and the temple, F. W. Cron (Concrete Construction, Vol. 28, No. 2, 1983) wrote: "The architect had wanted the building and specially the great dome, 27-m diameter, to be as white as possible, but not with a dull and chalky appearance. To achieve the desired effect Earley proposed an opaque white quartz found in South Carolina to reflect light from its broken face. This would be combined with a small amount of translucent quartz to provide brilliance and life. Puerto Rican sand and white portland cement were used to create a combination that reflected light and imparted a bright glow to the exposed-aggregate concrete surface. On a visit to the Temple of Light one can marvel at its brilliance in sunlight. If one returns at night, the lights from within and the floodlights that play on its surface turn the building into a shimmering jewel. The creativity of Louis Bourgeois and the superbly crafted concrete from Earley Studios have acted in concert to produce this great performance."

to the aggregate particles smaller than 4.75 mm but larger than 75 μm (No. 200 sieve). *Gravel* is the coarse aggregate resulting from natural disintegration by weathering of rock. The term *sand* is commonly used for fine aggregate resulting from either natural weathering or crushing of stone. *Crushed stone* is the product resulting from industrial crushing of rocks, boulders, or large cobblestones. *Iron blast-furnace slag,* a by-product of the iron industry, is the material obtained by crushing blast-furnace slag that solidified by slow cooling under atmospheric conditions. Aggregate from construction and demolition waste refers to the product obtained from recycling of concrete, brick, or stone rubble.

Mortar is a mixture of sand, cement, and water. It is like concrete without a coarse aggregate. *Grout* is a mixture of cementitious material and aggregate, usually fine aggregate, to which sufficient water is added to produce a pouring consistency

FIGURE 1-10 Burj Khalifa building in the United Arab Emirates. (*Photograph courtesy of Samsung C&T Corporation.*)

The Burj Khalifa building, 828 m high and with 162 floors, is presently the world's tallest building. The Burj Khalifa surpassed Taipei 101 building in Taiwan (see Fig. 12-6), the previous record holder, by 320 m, which is a remarkable increase of 61%. Historically, tall buildings have used steel as a construction material, but nowadays reinforced concrete and composite structures are dominant. For instance, in 1990, 57 of the 100 tallest buildings only had steel in the structural framework; presently, the number of purely steel building has been reduced to 24. Because the Burj Khalifa can accommodate up to 35,000 people, transportation and evacuation of the building were carefully studied. Fifty-seven elevators and 8 escalators were strategically installed, and some of the elevators were designed to permit evacuation during fire. In the event of an emergency, pressurized and air-conditioned refuge areas are available every 25 floors.

The design of the Burj Khalifa building combined reinforced concrete with a steel framework; the high-performance concrete mixtures were specified to have low permeability and high durability. Concrete mixtures with 80 MPa and 60 MPa compressive strength were composed of portland cement, fly ash, silica fume, and local aggregates. The concrete mix was designed to have a high stiffness, which significantly reduced the displacement; the 80 MPa concrete had an impressive high elastic modulus of 43.8 GPa at 90 days. Two notable achievements during the construction of this building were as follows: the concrete was pumped up to 600 m, and an efficient system of automatic self-climbing formwork was developed. Three specialized satellites for GPS measurements were used to monitor the verticality of the structure; therefore, the contractors were able to maintain the margin of error within 5 mm.

without segregation of the constituents. *Shotcrete* refers to a mortar or concrete that is pneumatically transported through a hose and projected onto a surface at high velocity.

Cement is a finely pulverized, dry material that by itself is not a binder but develops the binding property as a result of hydration (i.e., from chemical reactions between cement minerals and water). A cement is called *hydraulic* when the hydration products are stable in an aqueous environment. The most commonly used hydraulic cement for making concrete is *portland cement*, which consists essentially of reactive calcium silicates; the calcium silicate hydrates formed during the hydration of portland cement are primarily responsible for its adhesive characteristic, and are stable in aqueous environment.

The foregoing definition of concrete as a mixture of hydraulic cement, aggregates, and water does not include a fourth component, namely admixtures that are frequently used in modern concrete mixtures.

Admixtures are defined as materials other than aggregates, cement, and water, which are added to the concrete batch immediately before or during mixing. The use of admixtures in concrete is now widespread due to many benefits which are possible by their application. For instance, chemical admixtures can modify the setting and hardening characteristic of the cement paste by influencing the rate of cement hydration. Water-reducing admixtures can plasticize fresh concrete mixtures by reducing the surface tension of water; air-entraining admixtures can improve the durability of concrete exposed to cold weather; and mineral admixtures such as pozzolans (materials containing reactive silica) can reduce thermal cracking in mass concrete. Chapter 8 contains a detailed description of the types of admixtures, their composition, and mechanism of action.

1.3 Types of Concrete

Based on unit weight, concrete can be classified into three broad categories. Concrete containing natural sand and gravel or crushed-rock aggregates, generally weighing about 2400 kg/m^3 (4000 lb/yd^3), is called *normal-weight concrete*, and it is the most commonly used concrete for structural purposes. For applications where a higher strength-to-weight ratio is desired, it is possible to reduce the unit weight of concrete by using natural or pyro-processed aggregates with lower bulk density. The term *lightweight concrete* is used for concrete that weighs less than about 1800 kg/m^3 (3000 lb/yd^3). *Heavyweight concrete*, used for radiation shielding, is a concrete produced from high-density aggregates and generally weighs more than 3200 kg/m^3 (5300 lb/yd^3).

Strength grading of cements and concrete is prevalent in Europe and many other countries but is not practiced in the United States. However, from standpoint of distinct differences in the microstructure-property relationships, which will be discussed later, it is useful to divide concrete into three general categories based on compressive strength:

- *Low-strength concrete:* less than 20 MPa (3000 psi)
- *Moderate-strength concrete:* 20 to 40 MPa (3000 to 6000 psi)
- *High-strength concrete:* more than 40 MPa (6000 psi)

	Low-Strength (kg/m³)	Moderate-Strength (kg/m³)	High-Strength (kg/m³)
Cement	255	356	510
Water	178	178	178
Fine aggregate	801	848	890
Coarse aggregate	1169	1032	872
Cement paste proportion			
percent by mass	18	22.1	28.1
percent by volume	26	29.3	34.3
Water/cement by mass	0.70	0.50	0.35
Strength, MPa	18	30	60

TABLE 1-1 Typical Proportions of Materials in Concrete Mixtures of Different Strength

Moderate-strength concrete, also referred to as ordinary or normal concrete, is used for most structural work. High-strength concrete is used for special applications. It is not possible here to list all concrete types. There are numerous modified concretes which are appropriately named: for example, fiber-reinforced concrete, expansive-cement concrete, and latex-modified concrete. The composition and properties of special concretes are described in Chap. 12.

Typical proportions of materials for producing low-strength, moderate-strength, and high-strength concrete mixtures with normal-weight aggregate are shown in Table 1-1. The influence of the cement paste content and water-cement ratio on the strength of concrete is obvious.

1.4 Properties of Hardened Concrete and Their Significance

The selection of an engineering material for a particular application has to take into account its ability to withstand the applied force. Traditionally, the deformation occurring as a result of applied load is expressed as *strain*, which is defined as the change in length per unit length; the load is expressed as *stress*, which is defined as the force per unit area. Depending on how the stress is acting on the material, the stresses are further distinguished from each other: for example, compression, tension, flexure, shear, and torsion. The stress-strain relationships in materials are generally expressed in terms of strength, elastic modulus, ductility, and toughness.

Strength is a measure of the amount of stress required to fail a material. The working stress theory for concrete design considers concrete as mostly suitable for bearing compressive load; this is why it is the compressive strength of the material that is generally specified. Since the strength of concrete is a function of the cement hydration process, which is relatively slow, traditionally the specifications and tests for concrete strength are based on specimens cured under standard temperature-humidity conditions for a period of 28 days. Typically, the tensile and flexural strengths of concrete are of the order of 10 and 15 percent, respectively, of the compressive strength. The reason for

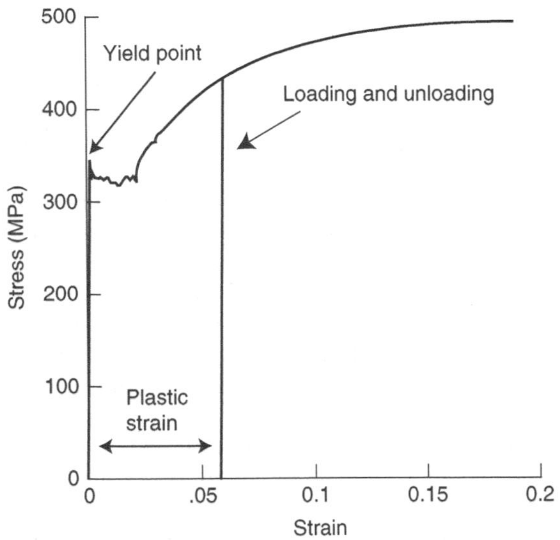

FIGURE 1-11 Stress-strain behavior of a steel specimen subjected to incremental loads.

such a large difference between the tensile and compressive strength is attributed to the heterogeneous and complex microstructure of concrete.

With many engineering materials, such as steel, the observed stress-strain behavior when a specimen is subjected to incremental loads can be divided into two parts (Fig. 1-11). Initially, when the strain is proportional to the applied stress and is reversible on unloading the specimen, it is called the *elastic strain*. The *modulus of elasticity* is defined as the ratio between the stress and the reversible strain. In homogeneous materials, the elastic modulus is a measure of the interatomic bonding forces and is unaffected by microstructural changes. This is not true of the heterogeneous multiphase materials like concrete. The elastic modulus of concrete in compression varies from 14×10^3 to 40×10^3 MPa (2×10^6 to 6×10^6 psi). The significance of the elastic limit in structural design lies in the fact that it represents the maximum allowable stress before the material undergoes permanent deformation. Therefore, the engineer must know the elastic modulus of the material because it influences the rigidity of a design.

At a high stress level (Fig. 1-11), the strain no longer remains proportional to the applied stress, and also becomes permanent (i.e., it will not be reversed if the specimen is unloaded). This strain is called the *plastic* or *inelastic strain*. The amount of inelastic strain that can occur before failure is a measure of the *ductility* of the material. The energy required to break the material, the product of force times distance, is represented by the area under the stress-strain curve. The term *toughness* is used as a measure of this energy. The contrast between toughness and strength should be noted; the former is a measure of energy, whereas the latter is a measure of the stress required to fracture the material. Thus, two materials may have identical strength but different values of toughness. In general, however, when the strength of a material goes up, the ductility and the toughness go down; also, very high-strength materials usually fail in a brittle manner (i.e., without undergoing any significant plastic strain).

Although under compression concrete appears to show some inelastic strain before failure, typically the strain at fracture is of the order of 2000×10^{-6}, which is considerably lower than the failure strain in structural metals. For practical purposes, therefore, designers do not treat concrete as a ductile material and do not recommend it for structures that are subject to heavy impact loading unless reinforced with steel.

Concrete is a composite material, however, many of its characteristics do not follow the laws of mixtures. For instance, under compressive loading both the aggregate and the hydrated cement paste, if separately tested, would fail elastically, whereas concrete itself shows inelastic behavior before fracture. Also, the strength of concrete is usually much lower than the individual strength of the two components. Such anomalies in the behavior of concrete can be explained on the basis of its microstructure, specially the important role of the interfacial transition zone between coarse aggregate and cement paste.

The stress-strain behavior of the material shown in Fig. 1-11 is typical of specimens loaded to failure in a short time in the laboratory. For some materials the relationship between stress and strain is independent of the loading time; for others it is not. Concrete belongs to the latter category. If a concrete specimen is held for a long period under a constant stress, for instance 50 percent of the ultimate strength of the material, it will exhibit plastic strain. The phenomenon of gradual increase in strain with time under a sustained stress is called *creep*. When creep in concrete is restrained, it manifests itself as a progressive decrease of stress with time. The stress relief associated with creep has important implications for the behavior of plain, reinforced, and prestressed concrete structures.

Strains can arise even in unloaded concrete as a result of changes in the environmental humidity and temperature. Freshly formed concrete is moist; it undergoes *drying shrinkage* when exposed to the ambient humidity. Similarly, shrinkage strains result when, due to the heat generated by cement hydration, hot concrete is cooled to the ambient temperature. Massive concrete elements register considerable rise in temperature because of poor dissipation of heat, therefore significant *thermal shrinkage* occurs on cooling. Shrinkage strains can be detrimental to concrete because, when restrained, they manifest into tensile stress. As the tensile strength of concrete is low, concrete structures often crack as a result of restrained shrinkage caused by humidity and temperature changes. In fact, the cracking tendency of the material is one of the serious disadvantages in structures built with concrete.

Professional judgment in the selection of construction materials should take into consideration not only the strength, dimensional stability, and elastic properties of the material but also its durability, which has serious implications for the life-cycle cost of a structure. *Durability* is defined as the service life of a material under given environmental condition. Generally, watertight concrete structures endure for a long time. The excellent conditions of the 2700-year-old concrete lining of a water storage tank on the Rodos Island in Greece and several aqueducts built in Europe built by the Romans nearly 2000 years ago, are a living testimony to the long-term durability of concrete in moist environments. In general, there is a relationship between strength and durability when low strength is associated with high porosity and high permeability. Permeable concretes are, of course, less durable. The permeability of concrete depends not only on mix proportions, compaction, and curing, but also on microcracks caused by the ambient temperature and humidity cycles. Finally, as discussed in Chap. 14, ecological and sustainability considerations are beginning to play an important role in the choice of materials for construction.

1.5 Units of Measurement

The metric system of measurement, which is prevalent in most countries of the world, uses millimeters and meters for length; grams, kilograms, and tonnes for mass; liters for volume; kilogram force per unit area for stress; and degrees Celsius for temperature. The United States is the only country in the world that uses old English units of measurement such as inches, feet, and yards for length; pounds or tons for mass; gallons for volume; pounds per square inch (psi) for stress; and degree Fahrenheit for temperature. Multinational activity in the design and construction of large engineering projects is commonplace in the modern world. Therefore, it is becoming increasingly important that scientists and engineers throughout the world speak the same language of measurement.

The metric system is simpler than the old English system and has recently been modernized in an effort to make it universally acceptable. The modern version of the metric system, called the *International System of Units* (Systѐme International d'Unités), abbreviated as SI, was approved in 1960 by many participating nations in the General Conference on Weights and Measures.

In SI measurements, meter and kilogram are the only units permitted for length and mass, respectively. A series of approved prefixes, shown in Table 1-2, are used for the formation of multiples and submultiples of various units. The force required to accelerate a mass of 1 kilogram (kg) at the rate of 1 meter per square second (m/s^2) is expressed as 1 newton (N), and a stress of 1 newton per square meter (N/m^2) is expressed as 1 pascal (Pa). The ASTM Standard E 380-70 contains a comprehensive guide to the use of SI units.

Multiplication Factor	Prefix	SI Symbol
1 000 000 000 = 10^9	giga	G
1 000 000 = 10^6	mega	M
1 000 = 10^3	kilo	k
100 = 10^2	hecto*	h
10 = 10^1	deka*	da
0.1 = 10^{-1}	deci*	d
0.01 = 10^{-2}	centi*	c
0.001 = 10^{-3}	milli	m
0.000 001 = 10^{-6}	micro	μ
0.000 000 001 = 10^{-9}	nano†	n

*Not recommended but occasionally used.

†0.1 nanometer (nm) = 1 angstrom (Å) is a non-SI unit which is commonly used.

Table 1-2 Multiple and Submultiple SI Units and Symbols

To Convert From:	To:	Multiply By:
yards (yd)	meters (m)	0.9144
feet (ft)	meters (m)	0.3048
inches (in.)	millimeter (mm)	25.4
cubic yards (yd^3)	cubic meters (m^3)	0.7646
U.S. gallons (gal)	cubic meters (m^3)	0.003785
U.S. gallons (gal)	liters	3.785
pounds, mass (lb)	kilograms (kg)	0.4536
U.S. tons (t)	tonnes (T)	0.9072
pounds/cubic yard (lb/yd^3)	kilograms/cubic meter (kg/m^3)	0.5933
kilogram force (kgf)	newtons (N)	9.807
pounds force (lbf)	newtons (N)	4.448
kips per square inch (ksi)	megapascal (MPa or N/mm^2)	6.895
degrees Fahrenheit (°F)	degrees Celsius (°C)	(°F – 32)/1.8

TABLE 1-3 Conversion Factors from the U.S. to SI Units

In 1975, the U.S. Congress passed the Metric Conversion Act, which declares that it will be the policy of the United States to coordinate and plan the increasing use of the metric system of measurement (SI units). Meanwhile, a bilinguality in the units of measurement is being practiced so that engineers should become fully conversant with both systems. To aid quick conversion from the U.S. customary units to SI units, a list of the commonly needed multiplication factors is given in Table 1-3.

Test Your Knowledge

1.1 Why is concrete the most widely used engineering material?

1.2 Compared to steel, what are the engineering benefits of using concrete for structures?

1.3 Define the following terms: fine aggregate, coarse aggregate, gravel, grout, shotcrete, hydraulic cement.

1.4 What are the typical unit weights for normal-weight, lightweight, and heavyweight concretes? How would you define high-strength concrete?

1.5 What is the significance of elastic limit in structural design?

1.6 What is the difference between strength and toughness? Why is the 28-day compressive strength of concrete generally specified?

1.7 Discuss the significance of drying shrinkage, thermal shrinkage, and creep in concrete.

1.8 How would you define durability? In general, what concrete types are expected to show better long-time durability?

Suggestions for Further Study

ACI, 2010, "ACI Concrete Terminology," American Concrete Institute, Farmington Hills, MI, http://terminology.concrete.org.

Ashby, M.F., and D.R.H. Jones, *Engineering Materials 1*, Butterworth-Heinemann, Oxford, 1996.

ASTM International, *Annual Book of ASTM Standards*, Vol. 04.01 (Cement, Lime, and Gypsum), Philadelphia, PA, 2012.

ASTM International, *Annual Book of ASTM Standards*, Vol. 04.02 (Concrete and Mineral Aggregates), Philadelphia, PA, 2012.

Callister, W.D. Jr., D.G. Rethwisch, *Materials Science and Engineering: An Introduction*, 8th ed. John Wiley & Sons, Hoboken, NJ, 2010.

Mindess, S., R.J. Gray, and A. Bentur, *The Science and Technology of Civil Engineering Materials*, Prentice Hall, Upper Saddle River, NJ, p. 384, 1998.

Smith, W.F., and J. Hashemi, *Foundations of Materials Science and Engineering*, 5th ed. McGraw-Hill, New York, 2009.

CHAPTER 2

Microstructure of Concrete

Preview

Microstructure-property relationships are at the heart of modern material science. Concrete has a highly heterogeneous and complex microstructure. Therefore, it is very difficult to constitute realistic models of its microstructure from which the behavior of the material can be reliably predicted. However, knowledge of the microstructure and properties of the individual components of concrete and their relationship to each other is useful for exercising control on the properties. This chapter describes the three components of the concrete microstructure, namely, hydrated cement paste, aggregate, and interfacial transition zone between the cement paste and aggregate. Finally, microstructure-property relationships are discussed with respect to their influence on strength, dimensional stability, and durability of concrete.

2.1 Definition

The type, amount, size, shape, and distribution of phases present in a solid constitute its *microstructure*. The gross elements of the microstructure of a material can readily be seen from a cross section of the material, whereas the finer elements are usually resolved with the help of a microscope. The term *macrostructure* is generally used for the gross microstructure visible to the human eye; the limit of resolution of the unaided human eye is approximately one-fifth of a millimeter (200 μm). The term *microstructure* is used for the microscopically magnified portion of a macrostructure. The magnification capability of modern electron microscopes is of the order of 10^6 times. Therefore, application of transmission and scanning electron microscopy techniques has made it possible to resolve the microstructure of materials to a fraction of one micrometer.

2.2 Significance

Progress in the field of materials has resulted primarily from recognition of the principle that the properties originate from the internal microstructure; in other words, properties can be modified by making suitable changes in the microstructure of a material. Although concrete is the most widely used structural material, its microstructure is heterogeneous and highly complex. The microstructure-property relationships in concrete are not yet fully developed; however, some understanding of the essential elements of the microstructure would be helpful before discussing the factors influencing the important engineering properties of concrete, such as strength (Chap. 3), elasticity, shrinkage, creep, and cracking (Chap. 4), and durability (Chap. 5).

2.3 Complexities

From examination of a cross section of concrete (Fig. 2-1), the two phases that can easily be distinguished are aggregate particles of varying size and shape, and the binding medium, composed of a hydrated cement paste made up of numerous different phases. At the macroscopic level, therefore, concrete may be considered as a two-phase material, consisting of aggregate particles dispersed in a matrix of cement paste.

At the microscopic level, the complexities of the concrete microstructure are evident. It becomes obvious that the two phases of the microstructure are neither homogeneously distributed with respect to each other, nor are they themselves homogeneous. For instance, in some areas the *hydrated cement paste* mass appears to be as dense as the aggregate, while in others it is highly porous (Fig. 2-2). Also, if several specimens of concrete containing the same amount of cement but different amounts of water are examined at various time intervals, it will be seen that, in general, the volume of capillary voids in the *hydrated cement paste* decrease with decreasing water-cement ratio or with increasing age of hydration. For a mature cement paste, the inhomogeneous distribution of solids and

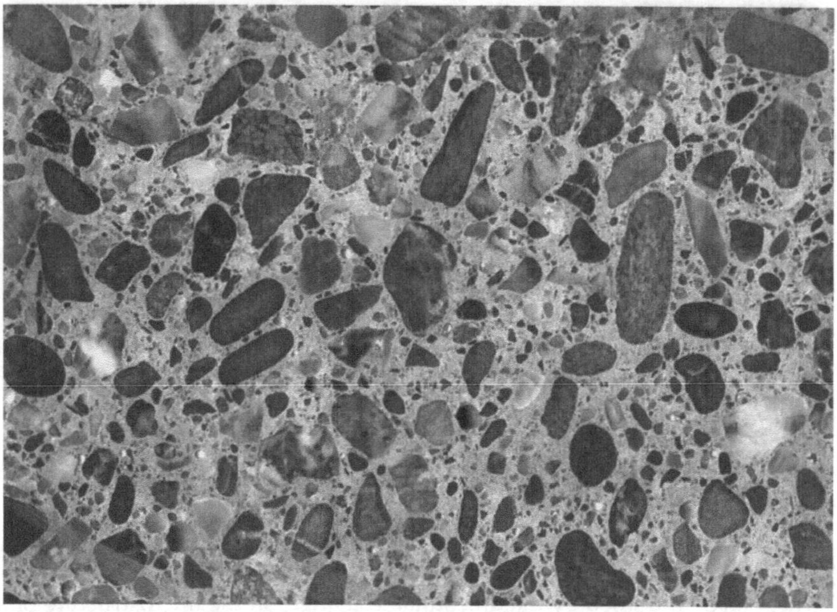

FIGURE 2-1 Polished section from a concrete specimen. (*Photograph courtesy of Gordon Vrdoljak.*)

Macrostructure is the gross structure of a material that is visible to the unaided human eye. In the macrostructure of concrete two phases are readily distinguished: aggregate of varying shapes and size, and the binding medium, which consists of an incoherent mass of the hydrated cement paste.

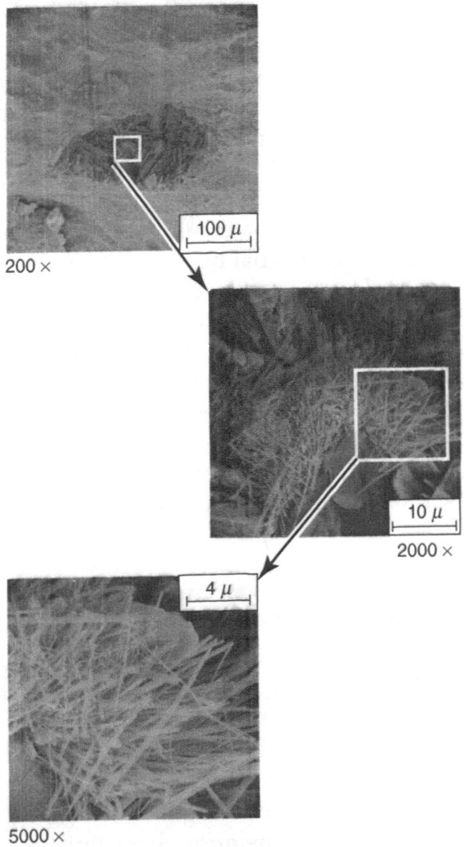

FIGURE 2-2 Microstructure of a hydrated cement paste.

Microstructure is the subtle structure of a material that is resolved with the help of a microscope. A low-magnification (200×) electron micrograph of a hydrated cement paste shows that the structure is not homogeneous; while some areas are dense, the others are highly porous. In the porous area, it is possible to resolve the individual hydrated phases by using higher magnifications. For example, massive crystals of calcium hydroxide, long and slender needles of ettringite, and aggregation of small fibrous crystals of calcium silicate hydrate can be seen at 2000 × and 5000 × magnifications.

voids alone can perhaps be ignored when modeling the behavior of the material. However, microstructural studies have shown that this cannot be done for the *hydrated cement paste* present in concrete. In the presence of aggregate, the microstructure of *hydrated cement paste* in the vicinity of large aggregate particles is usually very different from the microstructure of bulk paste or mortar in the system. In fact, many aspects of concrete behavior under stress can be explained only when the cement paste-aggregate interface is treated as a third phase of the concrete microstructure.

Thus the *unique features of the concrete microstructure* can be summarized as follows: First, there is the *interfacial transition zone*, which represents a small region next to the particles of coarse aggregate. Existing as a thin shell, typically 10 to 50 μm thick around large aggregate, the interfacial transition zone is generally weaker than either of the two main components of concrete, namely, the aggregate and the bulk hydrated cement paste; therefore, it exercises a far greater influence on the mechanical behavior of concrete than is reflected by its size. Second, each of the three phases is itself a multiphase in character. For instance, each aggregate particle may contain several minerals in addition to microcracks and voids. Similarly, both the bulk *hydrated cement paste* and the interfacial transition zone generally contain a heterogeneous distribution of different types and amounts of solid phases, pores, and microcracks, as will be described later. Third, unlike other engineering materials, the microstructure of concrete is not an intrinsic characteristic of the material because the two components of the microstructure, namely, the *hydrated cement paste* and the interfacial transition zone, are subject to change with time, environmental humidity, and temperature.

The highly heterogeneous and dynamic nature of the microstructure of concrete are the primary reasons why the theoretical microstructure-property relationship models, that are generally so helpful for predicting the behavior of engineering materials, are not of much practical use in the case of concrete. A broad knowledge of the important features of the microstructure of each of the three phases of concrete, as provided below, is nevertheless essential for understanding and control of properties of the composite material.

2.4 Microstructure of the Aggregate Phase

The composition and properties of different types of aggregates are described in detail in Chap. 7. Given here is only a brief description of the elements that exercise a major influence on properties of concrete.

The aggregate phase is predominantly responsible for the unit weight, elastic modulus, and dimensional stability of concrete. These properties of concrete depend to a large extent on the bulk density and strength of the aggregate, which in turn are determined by physical rather than chemical characteristics of the aggregate. In other words, the chemical or the mineralogical composition of the solid phases in aggregate is arguably less important than the physical characteristics, such as volume, size, and distribution of pores.

In addition to porosity, the shape and texture of the coarse aggregate also affect the properties of concrete. Some aggregate particles are shown in Fig. 2-3. Generally, natural gravel has a rounded shape and a smooth surface texture. Crushed rocks have a rough texture; depending on the rock type and the choice of crushing equipment, the crushed aggregate may contain a considerable proportion of flat or elongated particles that adversely affect many properties of concrete. Lightweight aggregate particles from pumice, which is highly cellular, are also angular and have a rough texture, but those from expanded clay or shale are generally rounded and smooth.

Being stronger than the other two phases of concrete, the aggregate phase has usually no direct influence on the strength of normal concrete except in the case of some highly porous and weak aggregates, such as pumice. The size and the shape of coarse aggregate can, however, affect the strength of concrete in an indirect way. It is obvious from Fig. 2-4 that the larger the size of aggregate in concrete and the higher the proportion of elongated and flat particles, the greater will be the tendency for water films to accumulate next to the aggregate surface, thus weakening the interfacial transition zone. This phenomenon, known as *bleeding*, is discussed in detail in Chap. 10.

FIGURE 2-3 Shape and surface texture of a coarse aggregate particles: (*a*) gravel, rounded and smooth; (*b*) crushed rock, equidimensional; (*c*) crushed rock, elongated; (*d*) crushed rock, flat; (*e*) lightweight, angular and rough; (*f*) lightweight, rounded and smooth.

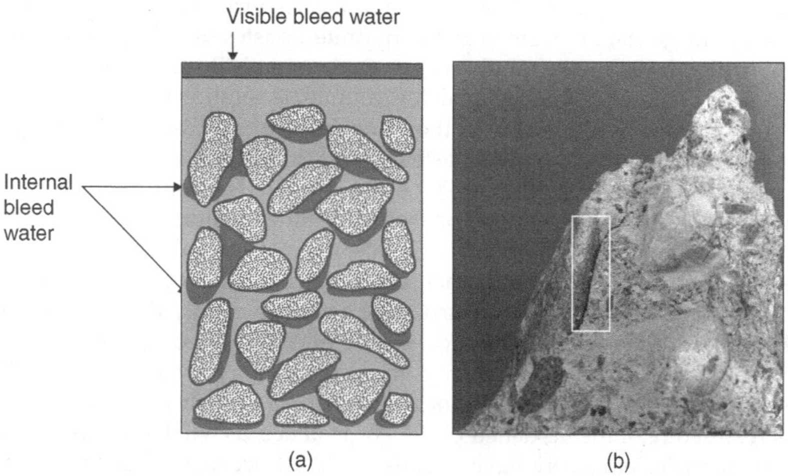

FIGURE 2-4 (*a*) Diagrammatic representation of bleeding in freshly deposited concrete; (*b*) shear-bond failure in a concrete specimen tested in uniaxial compression.

Internal bleed water tends to accumulate in the vicinity of elongated, flat, and large pieces of aggregate. In these locations, the aggregate-cement paste interfacial transition zone tends to be weak and easily prone to microcracking. This phenomenon is responsible for the shear-bond failure at the surface of the aggregate particle marked in the photograph.

2.5 Microstructure of the Hydrated Cement Paste

The term hydrated cement paste as used here refers to pastes made from portland cement. Although the composition and properties of portland cement are discussed in detail in Chap. 6, a summary of the composition will be helpful before discussing how the microstructure of the hydrated cement paste develops as a result of chemical reactions between portland-cement compounds and water.

Anhydrous portland cement is a gray powder composed of angular particles typically in the size range from 1 to 50 μm. It is produced by pulverizing a clinker with a small amount of calcium sulfate, the clinker being a heterogeneous mixture of several compounds produced by high-temperature reactions between calcium oxide and silica, alumina, and iron oxide. The chemical composition of the principal clinker compounds corresponds approximately to C_3S,* C_2S, C_3A, and C_4AF. In ordinary portland cement their respective amounts usually range between 45 and 60, 15 and 30, 6 and 12, and 6 and 8 percent.

When portland cement is dispersed in water, the calcium sulfate and the high-temperature compounds of calcium begin to go into solution, and the liquid phase gets rapidly saturated with various ionic species. As a result of interaction between calcium, sulfate, aluminate, and hydroxyl ions within a few minutes of cement hydration, the needle-shaped crystals of calcium trisulfoaluminate hydrate, called ettringite, first make their appearance. A few hours later, large prismatic crystals of calcium hydroxide and very small fibrous calcium silicate hydrates begin to fill the empty space formerly occupied by water and the dissolving cement particles. After some days, depending on the alumina-to-sulfate ratio of the portland cement, ettringite may become unstable and will decompose to form monosulfoaluminate hydrate, which has a hexagonal-plate morphology. Hexagonal-plate morphology is also the characteristic of calcium aluminate hydrates that are formed in the hydrated pastes of either undersulfated or high-C_3A portland cements. A scanning electron micrograph illustrating the typical morphology of phases prepared by mixing a calcium aluminate solution with calcium sulfate solution is shown in Fig. 2-5. A model of the essential phases present in the microstructure of a well-hydrated portland cement paste is shown in Fig. 2-6.

From the microstructural model of the hydrated cement paste shown in Fig. 2-6, it may be noted that the various phases are neither uniformly distributed nor are they uniform in size and morphology. In solids, microstructural inhomogeneities can lead to serious effects on strength and other related mechanical properties because these properties are controlled by the microstructural extremes, not by the average microstructure. Thus, in addition to the evolution of the microstructure as a result of the chemical changes, which occur after cement comes in contact with water, attention has to be paid to certain rheological properties of freshly mixed cement paste that also influence the microstructure of the hardened paste. For instance, as will be discussed later, the anhydrous particles of cement have a tendency to attract each other and form flocs, which entrap large quantities of mixing water. Obviously, local variations in water-cement ratio would be the primary source of evolution of the heterogeneous microstructure. With a highly flocculated cement paste system, not only the size and shape of pores but also the crystalline products of hydration would be different when compared to a well-dispersed

*Cement chemists use the following abbreviations: C = CaO; S = SiO_2; A = Al_2O_3; F = Fe_2O_3; $\overline{S}$ = SO_3; H = H_2O.

FIGURE 2-5 Scanning electron micrograph of typical hexagonal crystals of monosulfate hydrate and needlelike crystals of ettringite formed by mixing calcium aluminate and calcium sulfate solutions. (*Courtesy of Locher, F.W., Research Institute of Cement Industry, Dusseldorf, Federal Republic of Germany.*)

system. A modern description of the mechanisms of cement hydration can be found in Bullard et al.[1]

2.5.1 Solids in the Hydrated Cement Paste

The types, amounts, and characteristics of the four principal solid phases in the hydrated cement paste that can be resolved by an electron microscope are as follows:

Calcium Silicate Hydrate

The calcium silicate hydrate phase, abbreviated as **C-S-H**, makes up 50 to 60 percent of the volume of solids in a completely hydrated portland cement paste and provides the majority of the long-term strength and durability, making it the most important phase in determining the properties of the paste. The fact that the term C-S-H is hyphenated signifies that C-S-H is not a well-defined compound; the C/S ratio varies between 1.2 and 2.3 and the structural water content varies even more. The morphology of C-S-H also varies from poorly crystalline fibers to reticular network. Due to their colloidal

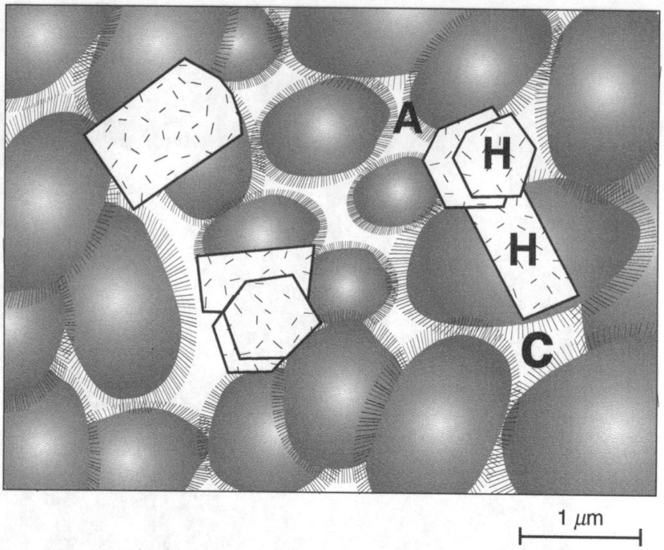

FIGURE 2-6 Model of a well-hydrated portland cement paste. "**A**" represents aggregation of C-S-H particles which have at least one colloidal dimension (1 to 100 nm). Inter-particle spacing within an aggregation is 0.5 to 3.0 nm (avg. 1.5 nm). **H** represents hexagonal crystalline products such as CH=C_4AH_{19}=$C_4A\bar{S}H_{18}$. They form large crystals, typically 1 μm wide. **C** represents capillary cavities or voids which exist when the spaces originally occupied with water do not get completely filed with the hydration products of cement. The size of capillary voids ranges from 10 nm to 1 μm, but in well-hydrated pastes with low water/cement, they are less than 100 nm.

dimensions and a tendency to cluster, C-S-H crystals could only be resolved with the advent of electron microscopy. In older literature, the material is often referred to as *C-S-H gel*. The internal crystal structure of C-S-H also remains unresolved; previously it was assumed to resemble the natural mineral tobermorite, which is why C-S-H was sometimes called *tobermorite gel*.

Although the exact structure of C-S-H is not known, several models have been proposed to explain the properties of the materials. According to the *Powers-Brunauer model*,[2] the material has a layer structure with a very high surface area. Depending on the measurement technique, surface areas on the order of 100 to 700 m^2/g have been proposed for C-S-H, and the strength of the material is attributed mainly to van der Waals' forces. The size of *gel pores*, or the solid-to-solid distance, is reported to be about 18 Å. The *Feldman-Sereda model*[3] visualizes the C-S-H structure as being composed of an irregular or kinked array of layers which are randomly arranged to create interlayer spaces of different shapes and sizes (5 to 25 Å). Richardson[4] gives a detailed description of the nature of C-S-H in hardened cement paste.

Calcium Hydroxide

Calcium hydroxide crystals (also called portlandite) constitute 20 to 25 percent of the volume of solids in the hydrated paste. In contrast to the C-S-H, calcium hydroxide is a compound with a definite stoichiometry, $Ca(OH)_2$. It tends to form large crystals with a distinctive hexagonal-prism morphology. The morphology usually varies from nondescript to stacks of large plates, and is affected by the available space, temperature of hydration, and impurities present in the system. Compared with C-S-H, the strength-contributing potential of calcium hydroxide is limited as a result of considerably lower surface area.

Calcium Sulfoaluminates Hydrates

Calcium sulfoaluminate hydrates occupy 15 to 20 percent of the solid volume in the hydrated paste and, therefore, play only a minor role in the microstructure-property relationships. It has already been stated that during the early stages of hydration the sulfate/alumina ionic ratio of the solution phase generally favors the formation of trisulfate hydrate, $C_6A\bar{S}_3H_{32}$, also called *ettringite*, which forms needle-shaped prismatic crystals. In pastes of ordinary portland cement, ettringite eventually transforms to the monosulfate hydrate, $C_4A\bar{S}H_{18}$, which forms hexagonal-plate crystals. The presence of the monosulfate hydrate in portland cement concrete makes the concrete vulnerable to sulfate attack. It should be noted that both ettringite and the monosulfate contain small amounts of iron, which can substitute for the aluminum ions in the crystal structure.

Unhydrated Clinker Grains

Depending on the particle size distribution of the anhydrous cement and the degree of hydration, some unhydrated clinker grains may be found in the microstructure of hydrated cement pastes, even long after hydration. As stated earlier, the clinker particles in modern portland cement generally conform to the size range 1 to 50 μm. With the progress of the hydration process, the smaller particles dissolve first and disappear from the system, then the larger particles become smaller. Because of the limited available space between the particles, the hydration products tend to precipitate in close proximity to the hydrating clinker particles, which gives the appearance of a coating formation around them. At later ages, due to the lack of available space, *in situ* hydration of clinker particles results in the formation of a very dense hydration product, the morphology of which may resemble the original clinker particle.

2.5.2 Voids in the Hydrated Cement Paste

In addition to solids, the hydrated cement paste contains several types of voids which have an important influence on its properties. The typical sizes of both the solid phases and the voids in hydrated cement paste are illustrated in Fig. 2-7*a*. The various types of voids and their amount and significance are discussed next. Just for information the size range of several objects ranging from human height to Mars' diameter is shown in Fig. 2.7*b*.

Interlayer Space in C-S-H

Powers assumed the width of the interlayer space within the C-S-H structure to be 18 Å and determined that it accounts for 28 percent porosity in solid C-S-H; however, Feldman and Sereda suggested that the space may vary from 5 to 25 Å. This void size is too small to have an adverse effect on the strength and permeability of the hydrated cement paste. However, as discussed below, water in these small voids can be held by hydrogen bonding, and its removal under certain conditions may contribute to drying shrinkage and creep.

Capillary Voids

Capillary voids represent the space not filled by the solid components of the hydrated cement paste. The total volume of a typical cement-water mixture remains essentially unchanged during the hydration process. The average bulk density of the hydration

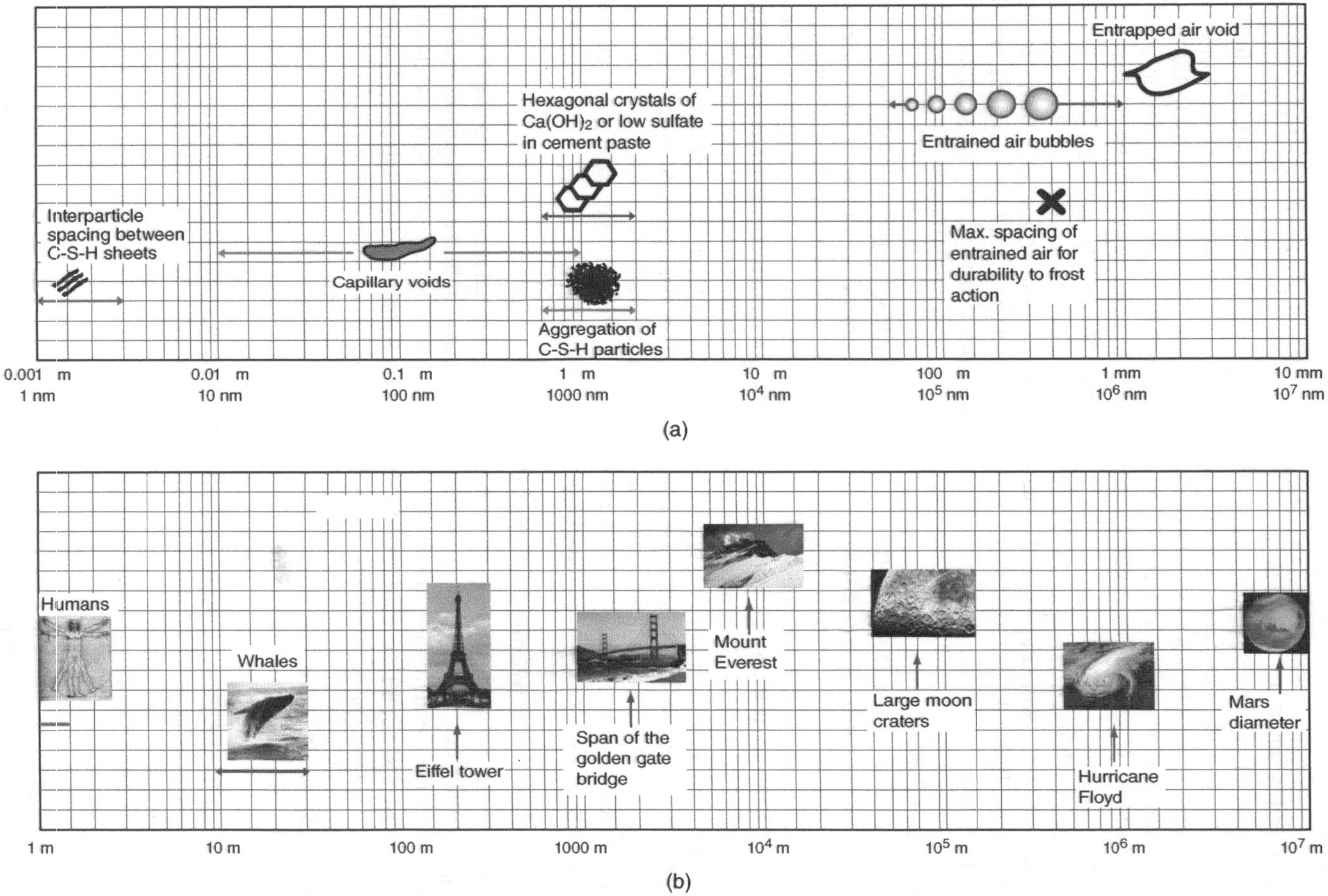

Figure 2-7 (*a*) Dimensional range of solids and pores in a hydrated cement paste. (*b*) In Fig. 2-7*a*, the dimensional range covers seven orders of magnitude. To illustrate how wide the range is, Fig. 2-7*b* illustrates a similar range using the height of a human being as a starting point and planet Mars as the ending point.

products* is considerably lower than the density of anhydrous portland cement; it is estimated that 1 cm^3 of cement, on complete hydration, requires about 2 cm^3 of space to accommodate the products of hydration. Thus, cement hydration may be looked upon as a process during which the space originally occupied by cement and water is being replaced more and more by the space filled by hydration products. The space not taken up by the cement or the hydration products consists of capillary voids, the volume and size of the capillary voids being determined by the original distance between the anhydrous cement particles in the freshly mixed cement paste (i.e., water-cement ratio), and the degree of cement hydration. A method of calculating the total volume of capillary voids, popularly known as *porosity*, in portland cement pastes having either different water-cement ratios or different degrees of hydration will be described later.

In well-hydrated, low water-cement ratio pastes, the capillary voids may range from 10 to 50 nm; in high water-cement ratio pastes, at early ages of hydration, the capillary voids may be as large as 3 to 5 μm. Typical pore size distribution plots of several hydrated cement paste specimens tested by the mercury intrusion technique are shown in Fig. 2-8. It has been suggested that the pore size distribution, not the total capillary porosity, is a better criterion for evaluating the characteristics of a hydrated cement paste. Capillary voids larger than 50 nm, referred to as *macropores* in modern literature, are probably more influential in determining the strength and impermeability characteristics, whereas voids smaller than 50 nm, referred to as *micropores*, play an important part in drying shrinkage and creep.

Air Voids

Whereas capillary voids are irregular in shape, air voids are generally spherical. A small amount of air usually gets trapped in the cement paste during concrete mixing. For various reasons, as discussed in Chap. 8, admixtures may be added to concrete to entrain purposely tiny air voids. Entrapped air voids may be as large as 3 mm; entrained air voids usually range from 50 to 200 μm. Therefore, both the entrapped and entrained air voids in the hydrated cement paste are much bigger than the capillary voids, and are capable of adversely affecting the strength.

2.5.3 Water in the Hydrated Cement Paste

Under electron microscopic examination, voids in the hydrated cement paste appear to be empty. This is because the specimen preparation technique calls for drying the specimen under high vacuum. Actually, depending on the environmental humidity and the porosity of the paste, the untreated cement paste is capable of holding a large amount of water. Like the solid and the void phases discussed above, water can exist in the hydrated cement paste in many forms. The classification of water into several types is based on the degree of difficulty or ease with which it can be removed from the hydrated cement paste. As there is a continuous loss of water from a saturated cement paste when the relative humidity of the environment is reduced, the dividing line between the different states of water is not rigid. In spite of this, the classification is useful for understanding the properties of the hydrated cement paste. In addition to vapor in empty or partially water-filled voids, water exists in the hydrated cement paste in the following states:

*Note that the interlayer space within the C-S-H phase is considered as a part of the solids in the hydrated cement paste.

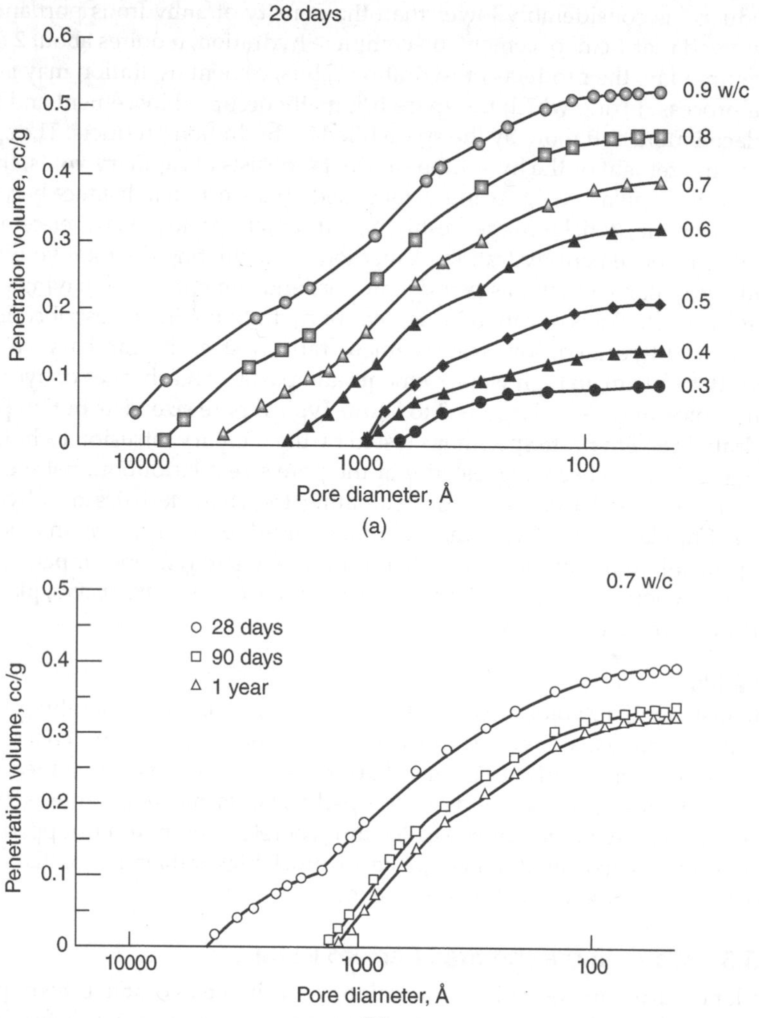

FIGURE 2-8 Pore size distribution in hydrated cement pastes. (From Mehta P.K., and D. Manmohan, *Proceedings of the Seventh International Congress on the Chemistry of Cements*, Editions Septima, Vol. III, Paris, 1980.)

It is not the total porosity but the pore size distribution that actually controls the strength, permeability, and volume changes in a hardened cement paste. Pore size distributions are affected by water-cement ratio, and the age (degree) of cement hydration. Large pores influence mostly the compressive strength and permeability; small pore influence mostly the drying shrinkage and creep.

Capillary Water

This is the water present in voids larger than about 50 Å. It may be pictured as the bulk water that is free from the influence of the attractive forces exerted by the solid surface. Actually, from the standpoint of the behavior of capillary water in the hydrated cement paste, it is desirable to divide the capillary water into two categories: the water in large voids of the order of >50 nm (0.05 μm), which may be called *free water* (because its removal does not cause any volume change), and the water held by capillary tension in small capillaries (5 to 50 nm), the removal of which may cause shrinkage of the system.

Adsorbed Water

This is the water that is close to the solid surface of the hydration products. Under the influence of attractive forces, water molecules are physically adsorbed onto the surface of solids in the hydrated cement paste. It has been suggested that up to six molecular layers of water (15 Å) can be physically held by hydrogen bonding. Because the bond energies of the individual water molecules decrease with distance from the solid surface, a major portion of the adsorbed water can be lost when hydrated cement paste is dried to 30 percent relative humidity. The loss of adsorbed water is responsible for the shrinkage of the hydrated cement paste.

Interlayer Water

This is the water associated with the C-S-H structure. It has been suggested that a monomolecular water layer between the layers of C-S-H is strongly held by hydrogen bonding. The interlayer water is lost only on strong drying (i.e., below 11 percent relative humidity). The C-S-H structure shrinks considerably when the interlayer water is lost.

Chemically Combined Water

This is the water that is an integral part of the microstructure of various cement hydration products. This water is not lost on drying; it is evolved when the hydrates decompose on heating. Based on the Feldman-Sereda model, different types of water associated with the C-S-H are illustrated in Fig. 2-9.

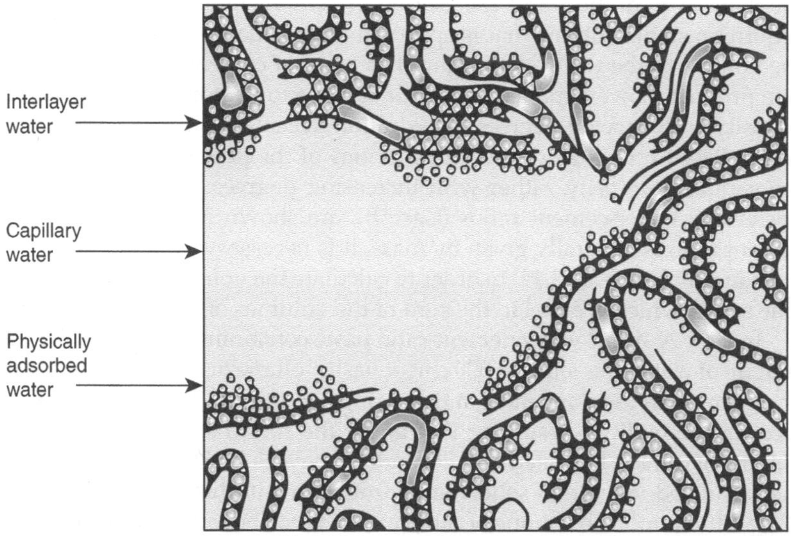

FIGURE 2-9 Diagrammatic model of the types of water associated with the calcium silicate hydrate. [Based on Feldman, R.F., and P.J. Sereda, *Eng. J. (Canada)*, Vol. 53, No. 8/9, 1970.]

In the hydrated cement paste, water can exist in many forms; these can be classified depending on the degree of ease with which water can be removed. This classification is useful in understanding the volume changes that are associated with water held by small pores.

2.5.4 Microstructure-Property Relationships in the Hydrated Cement Paste

The desirable engineering characteristics of hardened concrete—strength, dimensional stability, and durability—are influenced not only by the proportion but also by the properties of the hydrated cement paste, which, in turn, depend on the microstructural features (i.e., the type, amount, and distribution of solids and voids). The microstructure-property relationships of the hydrated cement paste are discussed next.

Strength

It should be noted that the principal source of strength in the solid products of the hydrated cement paste is the existence of the van der Waals forces of attraction. Adhesion between two solid surfaces can be attributed to these physical forces, the degree of the adhesive action being dependent on the extent and the nature of the surfaces involved. The small crystals of C-S-H, calcium sulfoaluminate hydrates, and hexagonal calcium aluminate hydrates possess enormous surface areas and adhesive capability. These hydration products of portland cement tend to adhere strongly not only to each other, but also to low surface-area solids, such as calcium hydroxide, anhydrous clinker grains, and fine and coarse aggregate particles.

It is a well-known fact that there is an inverse relationship between porosity and strength in solids. Strength resides in the solid part of a material; therefore, voids are detrimental to strength. In hydrated cement paste, the interlayer space with the C-S-H structure and the small voids, which are within the influence of the van der Waals forces of attraction, are not considered detrimental to strength because stress concentration and subsequent rupture on application of load begin at large capillary voids and microcracks that are invariably present. As stated earlier, the volume of capillary voids in a hydrated cement paste depends on the amount of water mixed with the cement at the start of hydration and the degree of cement hydration. When the paste sets, it acquires a stable volume that is approximately equal to the volume of the cement plus the volume of the water. Assuming that 1 cm^3 of cement produces 2 cm^3 of the hydration product, Powers made simple calculations to demonstrate the changes in capillary porosity with varying degrees of hydration in cement pastes of different water-cement ratios. Based on his work, two illustrations of the process of progressive reduction in the capillary porosity, either with increasing degrees of hydration (Case A) or with decreasing water-cement ratios (Case B), are shown in Fig. 2-10. Because the water-cement ratio is generally given by mass, it is necessary to know the specific gravity of portland cement (e.g., 3.14) in order to calculate the volume of water and the total available space, which is equal to the sum of the volumes of water and cement.

In Case A, a 0.63 water-cement-ratio paste containing 100 cm^3 of the cement requires 200 cm^3 of water; this sums to 300 cm^3 of paste volume or total available space. The degree of cement hydration depends on the curing conditions (duration of hydration, temperature, and humidity). Assuming that under the ASTM standard curing conditions,* the volume of cement hydrated at 7, 28, and 365 days is 50, 75, and 100 percent, respectively, the calculated volume of solids (anhydrous cement plus the hydration product) is 150, 175, and 200 cm^3. The volume of capillary voids can be found from the difference between the total available space and the total volume of solids. This turns out to be 50, 42, and 33 percent, respectively, at 7, 28, and 365 days of hydration.

*ASTM C31 requires moist curing at 23 ± 1°C until the age of testing.

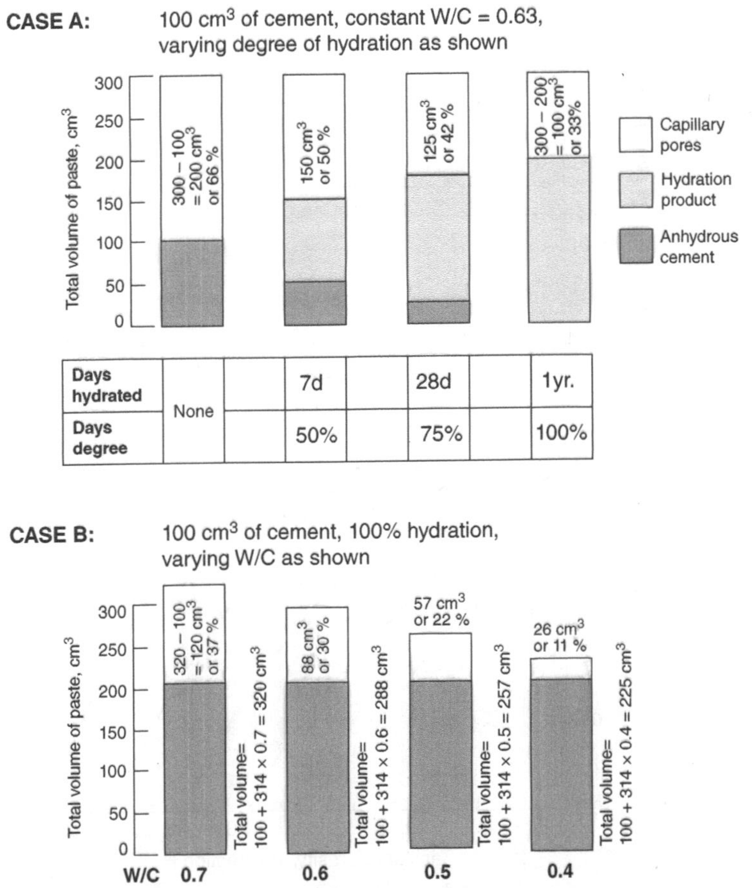

FIGURE 2-10 Changes in the capillary porosity with varying water-cement ratio and degree of hydration.

By making certain assumptions, calculations can be made to show how, with a given water-cement ratio, the capillary porosity of a hydrated cement paste would vary with varying degrees of hydration. Alternatively, capillary porosity variations, for a given degree of hydration but variable water-cement ratios, can be determined.

In Case B, a 100 percent degree of hydration is assumed for four cement pastes made with different amounts of water corresponding to water-cement ratios of 0.7, 0.6, 0.5, or 0.4. For a given volume of cement, the paste with the largest amount of water will have the greatest total volume of available space. However, after complete hydration, all the pastes would contain the same quantity of the solid hydration product. Therefore, the paste with the greatest total space would end up with a correspondingly larger volume of capillary voids. Thus 100 cm^3 of cement at full hydration would produce 200 cm^3 of solid hydration products in every case; however, because the total available space in the 0.7, 0.6, 0.5, or 0.4 water-cement-ratio pastes was 320, 288, 257, and 225 cm^3, the calculated capillary voids are 37, 30, 22, and 11 percent, respectively. Under the assumptions made here, with a 0.32 water-cement-ratio paste, there would be no capillary porosity when the cement had completely hydrated.

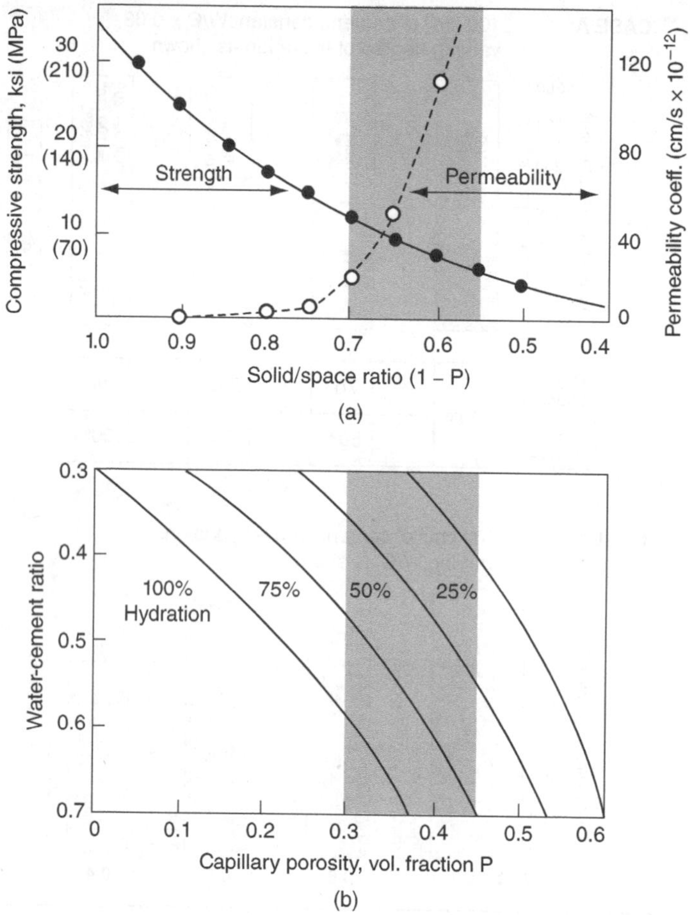

FIGURE 2-11 Influence of water-cement ratio and degree of hydration on strength and permeability.

A combination of water-cement ratio and degree of hydration determines the porosity of hydrated cement paste. The porosity and the opposite of porosity (solid-space ratio) are exponentially related to both the strength and permeability of the material. The shaded area shows the typical capillary porosity range in hydrated cement pastes.

For normally hydrated portland cement mortars, Powers showed that there is an exponential relationship of the type $f_c = ax^3$ between the compressive strength f_c and the solids-to-space ratio (x), where a is a constant equal to 234 MPa (34,000 psi). Assuming a given degree of hydration, such as 25, 50, 75, and 100 percent, it is possible to calculate the effect of increasing the water-cement ratio, first on the porosity and subsequently on the strength by using Powers' formula. The results are plotted in Fig. 2-11*a*. The permeability curve of this figure will be discussed later.

Dimensional Stability

Saturated hydrated cement paste is not dimensionally stable. As long as it is held at 100 percent relative humidity (RH), practically no dimensional change will occur.

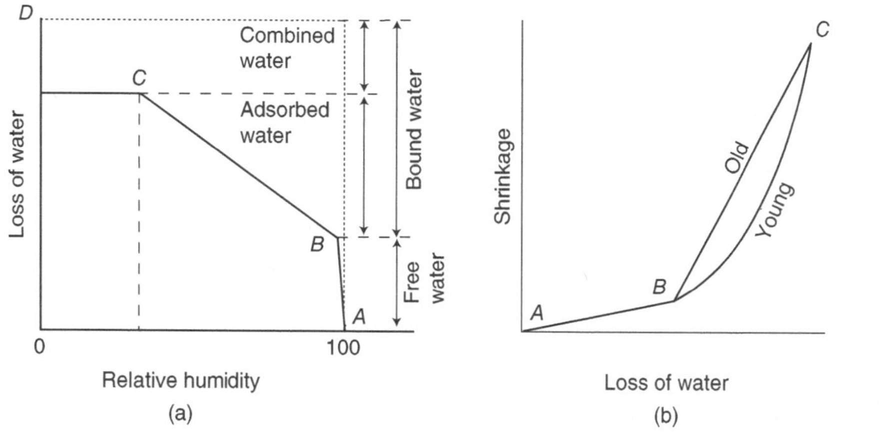

FIGURE 2-12 (*a*) Loss of water as a function of the relative humidity and (*b*) shrinkage of a cement mortar as a function of the water loss. (From Hermite, R. L., *Proceedings of the Fourth International Symposium on Chemistry of Cements*, Washington, D.C., 1960.)

From a saturated cement paste, it is the loss of adsorbed water that is mainly responsible for the drying shrinkage.

However, when exposed to environmental humidity, which normally is much lower than 100 percent, the material begins to lose water and shrink. How the water loss from saturated hydrated cement paste is related to RH on one hand, and to the drying shrinkage on the other, is described by L'Hermite (Fig. 2-12). As soon as the RH drops below 100 percent, the free water held in large cavities (e.g., >50 nm) begins to escape to the environment. Because the free water is not attached to the microstructure of the hydration products by any physical-chemical bonds, its loss would not be accompanied by shrinkage. This is shown by curve '*A* – *B*' in Fig. 2-12. Thus, a saturated hydrated cement paste exposed to slightly less than 100 percent RH can lose a considerable amount of total evaporable water before undergoing any shrinkage.

When most of the free water has been lost, it is found on continued drying that further loss of water results in considerable shrinkage. This phenomenon, shown by curve '*B* – *C*' in Fig. 2-12, is attributed mainly to the loss of adsorbed water and the water held in small capillaries (see Fig. 2-9). It has been suggested that when confined to narrow spaces between two solid surfaces, the adsorbed water causes *disjoining pressure*. The removal of the adsorbed water reduces the disjoining pressure and brings about shrinkage of the system. The interlayer water, present as a mono-molecular water film within the C-S-H layer structure, can also be removed by severe drying conditions. This is because the closer contact of the interlayer water with the solid surface, and the tortuosity of the transport path through the capillary network call for a stronger driving force. Because the water in small capillaries (5 to 50 nm) exerts *hydrostatic tension*, its removal tends to induce a compressive stress on the solid walls of the capillary pore, thus also causing contraction of the system.

Note that the mechanisms that are responsible for drying shrinkage are also responsible for creep of hydrated cement paste. In the case of creep, a sustained external stress becomes the driving force for the movement of the physically adsorbed water and the water held in small capillaries. Thus, creep strain can occur even at 100 percent RH.

Durability

Hydrated cement paste is alkaline; therefore, exposure to acidic waters is detrimental to the material. Under these conditions, *impermeability*, or *watertightness*, becomes a primary factor in determining the durability. The impermeability of hydrated cement paste is a highly prized characteristic because it is assumed that an impermeable hydrated cement paste would result in an impermeable concrete (the aggregate in concrete is generally assumed to be impermeable). *Permeability* is defined as the ease with which a fluid under pressure can flow through a solid. It should be obvious that the size and continuity of the pores in the microstructure of the solid would determine its permeability. Strength and permeability of the hydrated cement paste are two sides of the same coin in the sense that both are closely related to the capillary porosity or the solid-space ratio. This is evident from the permeability curve shown in Fig. 2-11, which is based on the experimentally determined values of permeability by Powers.

The exponential relationship between permeability and porosity shown in Fig. 2-11 can be understood from the influence that various pore types exert on permeability. As hydration proceeds, the void space between the originally discrete cement particles gradually begins to fill up with the hydration products. It has been shown (Fig. 2-10) that the water-cement ratio (i.e., original capillary space between cement particles) and the degree of hydration determine the total capillary porosity, which decreases with the decreasing water-cement ratio and/or increasing degree of hydration. Mercury-intrusion porosimetric studies[5] on the cement pastes shown in Fig. 2-8, hydrated with different water-cement ratios and to various ages, demonstrate that the decrease in total capillary porosity was associated with reduction of large pores in the hydrated cement paste (Fig. 2-13). From the data in Fig. 2-11 it is obvious that the coefficient of permeability registered an exponential drop when the fractional volume of capillary pores was reduced from 0.4 to 0.3. This range of capillary porosity, therefore, seems to correspond to the point when both the volume and the size of capillary pores in a hydrated cement paste are reduced such that the interconnections between them no longer exist. As a result, the permeability of a fully hydrated cement paste may be of the order of 10^6 times less than that of a young paste. Powers showed that even on complete hydration a 0.6-water-cement-ratio paste can become as impermeable as a dense rock such as basalt or marble.

Note that the porosities represented by the C-S-H interlayer space and small capillaries do not contribute to the permeability of hydrated cement paste. On the contrary, with increasing degree of hydration, although there is a considerable increase in the volume of pores due to the C-S-H interlayer space and small capillaries, the permeability is greatly reduced. In hydrated cement paste a direct relationship was noted between the permeability and the volume of pores larger than about 100 nm.[3] This is probably because the pore systems, comprised mainly of small pores, tend to become discontinuous.

2.6 Interfacial Transition Zone in Concrete

2.6.1 Significance of the Interfacial Transition Zone

Have you ever wondered why:

- Concrete is brittle in tension but relatively tough in compression?
- The components of concrete when tested separately under uniaxial compression remain elastic until fracture, whereas concrete itself shows inelastic behavior?

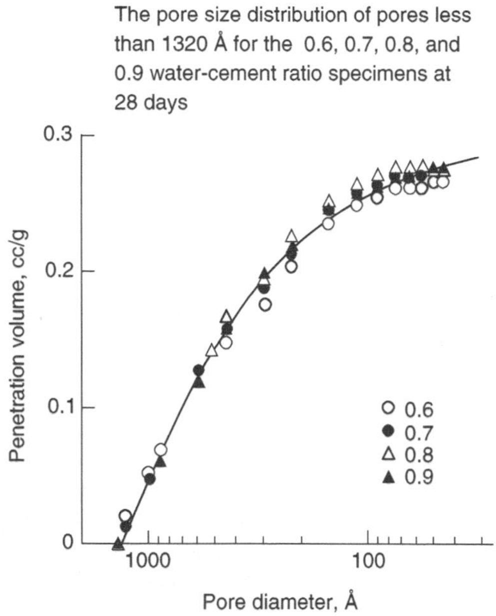

FIGURE 2-13 Distribution plots of small pores in cement pastes of varying water-cement ratios. (From Mehta, P.K., and D. Manmohan, *Proceedings of the Seventh International Congress on the Chemistry of Cement*, Paris, 1980.)

When the data of Fig. 2-8 are replotted after omitting the large pores (i.e., >1320 Å), it was found that a single curve could fit the pore distributions in the 28-day-old pastes made with four different water-cement ratios. This shows that in hardened cement pastes, the increase in total porosity resulting from increasing water-cement ratios manifests itself in the form of large pores only. This observation has great significance from the standpoint of the effect of water-cement ratio on strength and permeability, which are controlled by large pores.

- The compressive strength of a concrete is higher than its tensile strength by an order of magnitude?
- At a given cement content, water-cement ratio, and age of hydration, cement mortar will always be stronger than the corresponding concrete? Also, the strength of concrete goes down as the coarse aggregate size is increased.
- The permeability of a concrete containing even a very dense aggregate will be higher by an order of magnitude than the permeability of the corresponding cement paste?
- On exposure to fire, the elastic modulus of a concrete drops more rapidly than its compressive strength?

The answers to the above and many other enigmatic questions on concrete behavior lie in the interfacial transition zone that exists between large particles of aggregate and the hydrated cement paste. Although composed of the same elements as hydrated cement paste, the microstructure and properties of the interfacial transition zone are different from bulk hydrated cement paste. It is, therefore, treated as a separate phase of the concrete microstructure.

2.6.2 Microstructure

Because of experimental difficulties, information about the interfacial transition zone in concrete is scarce; however, based on a description given by Maso,[6] some understanding of its microstructural characteristics can be obtained by following the sequence of its development from the time concrete is placed.

> *First,* in freshly compacted concrete, water films form around the large aggregate particles. This would account for a higher water-cement ratio closer to the larger aggregate than away from it (i.e., in the bulk mortar).
>
> *Next,* as in the bulk paste, calcium, sulfate, hydroxyl, and aluminate ions, produced by the dissolution of calcium sulfate and calcium aluminate compounds, combine to form ettringite and calcium hydroxide. Owing to the high water-cement ratio, these crystalline products in the vicinity of the coarse aggregate consist of relatively larger crystals, and therefore form a more porous framework than in the bulk cement paste or mortar matrix. The platelike calcium hydroxide crystals tend to form in oriented layers, for instance, with the *c*-axis perpendicular to the aggregate surface.
>
> *Finally,* with the progress of hydration, poorly crystalline C-S-H and a second generation of smaller crystals of ettringite and calcium hydroxide start filling the empty space that exists between the framework created by the large ettringite and calcium hydroxide crystals. This helps to improve the density and hence the strength of the interfacial transition zone.

A scanning electron micrograph and diagrammatic representation of the interfacial transition zone in concrete are shown in Fig. 2-14.

2.6.3 Strength

As in the case of hydrated cement paste, the cause of adhesion between hydration products and the aggregate particle is van der Waals force of attraction; therefore, the strength of the interfacial transition zone at any point depends on the volume and size of voids present. Even for low water-cement ratio concrete, at early ages the volume and size of voids in the interfacial transition zone will be larger than in bulk mortar; consequently, the former is weaker in strength. However, with increasing age the strength of the interfacial transition zone may become equal to or even greater than the strength of the bulk mortar. This may occur as a result of crystallization of new products in the voids of the interfacial transition zone by slow chemical reactions between the cement paste constituents and the aggregate, formation of calcium silicate hydrates in the case of siliceous aggregates, or formation of carboaluminate hydrates in the case of limestone. Such interactions are strength contributing because they also tend to reduce the concentration of the calcium hydroxide in the interfacial transition zone. Large calcium hydroxide crystals possess less adhesion capacity, not only because of the lower surface area and correspondingly weak van der Waals forces of attraction, but also because they serve as preferred cleavage sites owing to their tendency to form an oriented structure.

In addition to the large volume of capillary voids and oriented calcium hydroxide crystals, a major factor responsible for the poor strength of the interfacial transition zone in concrete is the presence of microcracks. The amount of microcracks depends on numerous parameters, including aggregate size and grading, cement content, water-cement ratio, degree of consolidation of fresh concrete, curing conditions, environmental humidity, and thermal history of concrete. For instance, a concrete mixture containing poorly graded aggregate is more prone to segregation during consolidation; thus, thick

(a)

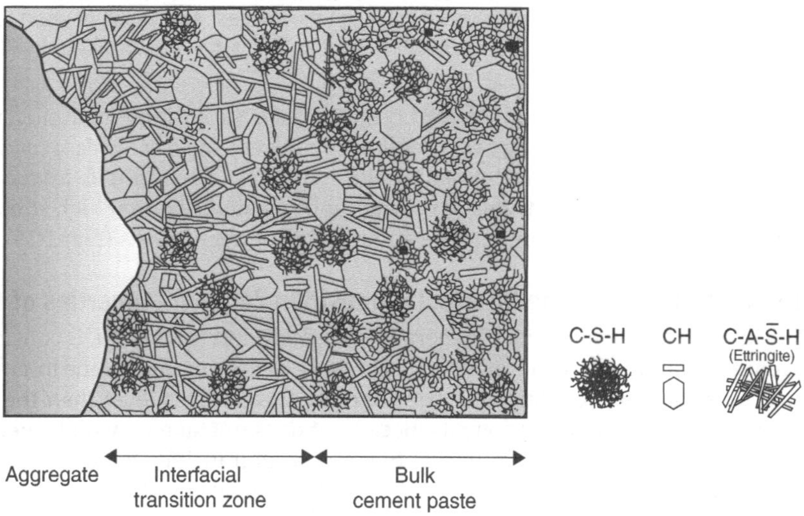

(b)

FIGURE 2-14 (*a*) Scanning electron micrograph of the calcium hydroxide crystals in the interfacial transition zone. (*b*) Diagrammatic representation of the interfacial transition zone and bulk cement paste in concrete.

At early ages, especially when a considerable internal bleeding has occurred, the volume and size of voids in the transition zone are larger than in the bulk cement paste or mortar. The size and concentration of crystalline compounds such as calcium hydroxide and ettringite are also larger in interfacial transition zone. The cracks are formed easily in the direction perpendicular to the c-axis. Such effects account for the lower strength of the transition zone than the bulk cement paste in concrete.

water films can form around the coarse aggregate, especially beneath the particle. Under identical conditions, the larger the aggregate size the thicker the water film. The interfacial transition zone formed under these conditions will be susceptible to cracking when subjected to the influence of tensile stresses induced by differential movements

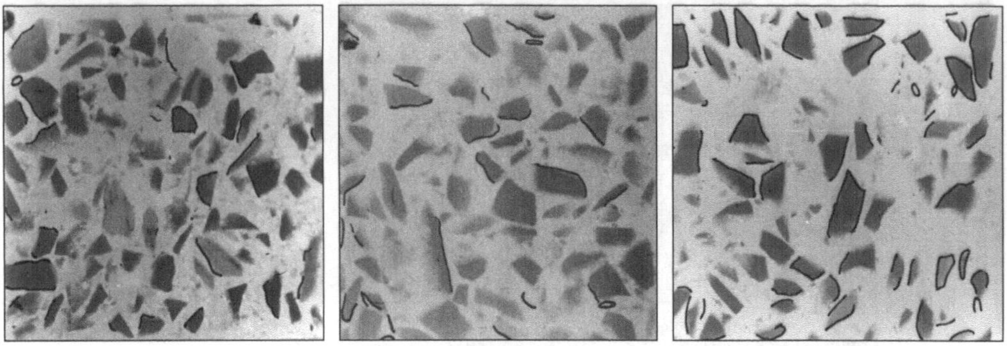

(a) (b) (c)

Figure 2-15 Typical cracking maps for normal (medium-strength) concrete: (*a*) after drying shrinkage; (*b*) after short-term loading; (*c*) for sustained loading for 60 days at 65 percent of the 28-day compressive strength. (From Ngab, A.J., F.O. Slate, and A.M. Nilson, *J. ACI, Proc.*, Vol. 78, No. 4, 1981.)

The light areas in the image corresponds to cement paste/mortar and the darker areas to aggregates. Black lines were drawn to accentuate the visual cracks in the images. As a result of short-term loading, drying shrinkage, and creep, the interfacial transition zone in concrete contains microcracks.

between the aggregate and hydrated cement paste. Such differential movements commonly arise either on drying or on cooling of concrete. *In other words, a concrete can have microcracks in the interfacial transition zone even before a structure is loaded.* Obviously, short-term impact loads, drying shrinkage, and sustained loads at high stress levels will have the effect of increasing the size and number of microcracks (Fig. 2-15).

2.6.4 Influence of the Interfacial Transition Zone on Properties of Concrete

The interfacial transition zone, generally the *weakest link of the chain*, is considered as the strength-limiting phase in concrete. It is because of the presence of the interfacial transition zone that concrete fails at a considerably lower stress level than the strength of either of the two main components. Because it does not take very high energy levels to extend the cracks already existing in the interfacial transition zone, even at 50 percent of the ultimate strength, higher incremental strains may be obtained per unit of applied stress. This explains the phenomenon that the components of concrete (i.e., aggregate and hydrated cement paste or mortar) usually remain elastic until fracture in a uniaxial compression test, whereas concrete itself shows inelastic behavior.

At stress levels higher than about 70 percent of the ultimate strength, the stress concentrations at large voids in the mortar matrix become large enough to initiate cracking. With increasing stress, the matrix cracks gradually spread until they join the cracks originating from the interfacial transition zone. When the crack system becomes continuous, the material ruptures. Considerable energy is needed for the formation and extension of matrix cracks under a compressive load. On the other hand, under tensile loading, cracks propagate rapidly and at a much lower stress level. This is why concrete fails in a brittle manner in tension but is relatively tough in compression. This is also the reason why the tensile strength is much lower than the compressive strength of concrete. This subject is discussed in greater detail in Chaps. 3 and 4.

The microstructure of the interfacial transition zone, especially the volume of voids and microcracks present, has a great influence on the stiffness or the elastic modulus of concrete. In the composite material, the interfacial transition zone serves as a bridge between the two components: the mortar matrix and the coarse aggregate particles. Even when the individual components are of high stiffness, the stiffness of the composite is reduced because of the *broken bridges* (i.e., voids and microcracks in the interfacial transition zone), which do not permit stress transfer. Thus, due to microcracking on exposure to fire, the elastic modulus of concrete drops faster than the compressive strength.

The characteristics of the interfacial transition zone also influence the durability of concrete. Prestressed and reinforced concrete elements often fail due to corrosion of embedded steel. The rate of corrosion of steel is greatly influenced by the permeability of concrete. The existence of microcracks in the interfacial transition zone at the interface with steel and coarse aggregate is the primary reason that concrete is more permeable than the corresponding hydrated cement paste or mortar. It should be noted that the penetration of air and water is a necessary prerequisite to corrosion of the embedded steel in concrete.

The effect of the water-cement ratio on the permeability and strength of concrete is generally attributed to the relationship that exists between the water-cement ratio and the porosity of hydrated cement paste in concrete. The foregoing discussion on the influence of microstructure and properties of the interfacial transition zone on concrete shows that, in fact, it is more appropriate to think in terms of the effect of the water-cement ratio on the concrete mixture as a whole. This is because, depending on the aggregate characteristics, such as the maximum size and grading, it is possible to have large differences in the water-cement ratio between the mortar matrix and the interfacial transition zone. In general, everything else remaining the same, the larger the aggregate size the higher the local water-cement ratio in the interfacial transition zone and, consequently, the weaker and more permeable would be the concrete.

Test Your Knowledge

2.1 What is the significance of the microstructure of a material? How do you define microstructure?

2.2 Describe some of the unique features of the concrete microstructure that make it difficult to predict the behavior of the material from its microstructure.

2.3 Discuss the physical-chemical characteristics of the C-S-H, calcium hydroxide, and calcium sulfoaluminates present in a well-hydrated portland cement paste.

2.4 How many types of voids are present in a hydrated cement paste? What are their typical dimensions? Discuss the significance of the C-S-H interlayer space with respect to properties of the hydrated cement paste.

2.5 How many types of water are associated with a saturated cement paste? Discuss the significance of each. Why is it desirable to distinguish between the free water in large capillaries and the water held in small capillaries?

2.6 What would be the volume of capillary voids in an 0.2-water-cement ratio paste that is only 50 percent hydrated? Also calculate the water-cement ratio needed to obtain zero porosity in a fully hydrated cement paste.

2.7 When a saturated cement paste is dried, the loss of water is not directly proportional to the drying shrinkage. Explain why.

2.8 In a hydrating cement paste the relationship between porosity and impermeability is exponential. Explain why.

2.9 Draw a typical sketch showing how the microstructure of hydration products in the aggregate-cement paste interfacial transition zone is different from the bulk cement paste in concrete.

2.10 Discuss why the strength of the interfacial transition zone is generally lower than the strength of the bulk hydrated cement paste. Explain why concrete fails in a brittle manner in tension but not in compression.

2.11 Everything else remaining the same, the strength and impermeability of a mortar will decrease as coarse aggregate of increasing size is introduced. Explain why.

2.12 When concrete is exposed to fire, why the elastic modulus shows a relatively higher drop than the compressive strength?

References

1. Bullard, J.W., H.M. Jennings, R.A. Livingston, A. Nonat, G.W. Scherer, J.S. Schweitzer, K.L. Scrivener, J.J. Thomas, Mechanisms of cement hydration, *Cem. Concr. Res.*, Vol. 41, pp. 1208–1223, December 2011.
2. Powers, T.C., *J. Am. Ceram. Soc.*, Vol. 61, No. 1, pp. 1–5, 1958; and Brunauer, S., *Am. Sci.*, Vol. 50, No. 1, pp. 210–229, 1962.
3. Feldman, R.F., and P.J. Sereda, *Eng. J. (Canada)*, Vol. 53, No. 8/9, pp. 53–59, 1970.
4. Richardson, I.G., The nature of C-S-H in hardened cements, *Cem. Concr. Res.*, Vol. 29, pp. 1131–1147, 1999.
5. Mehta, P.K., and D. Manmohan, *Proceedings of the Seventh International Congress on the Chemistry of Cements*, Editions Septima, Vol. III, Paris, 1980.
6. Maso, J.C., *Proceedings of the Seventh International Congress on the Chemistry of Cements*, Editions Septima, Paris, 1980.

Suggestions for Further Study

Aitcin, P.C., *Binders for Durable and Sustainable Concrete*, Taylor and Francis, New York, 2008.

Hewlett, P.C., ed., *Lea's Chemistry of Cement and Concrete*, 4th ed. Arnold, London; p. 1053, 1998.

Jennings, H.M., and Bullard, J.W., From electrons to infrastructure: Engineering concrete from the bottom up, *Cem. Concr. Res.*, Vol. 41, pp. 727–735, 2011.

Klieger, P., and J.F. Lamond, eds., *Concrete and Concrete Making Materials*, ASTM STP, 169, American Society for Testing and Materials, Philadelphia, PA, Chap. 2, 1994.

Lea, F.M., *The Chemistry of Cement and Concrete*, Chemical Publishing Company, New York, Chap. 10, 1971, The Setting and Hardening of Portland Cement.

Locher, F.W., *Cement: Principles of Production and Use*, Verlag Bau + Technik GmbH, Dusseldorf, 2006.

Maso, J.C., ed., *Interfacial Transition Zone in Concrete*, E & FN SPON, London, 1996.

Monteiro, P.J.M., Maso, J.C., Ollivier, J.P., The aggregate-mortar interface, *Cem. Concr. Res.*, Vol. 15, pp. 953–958, 1985.

Powers, T.C., *Properties of Fresh Concrete*, Wiley, New York, Chaps. 2, 9, and 11, 1968.

Proceedings of the 7th International Congress on the Chemistry of Cement (Paris, 1980); 8th Congress (Rio de Janeiro, 1986); 9th Congress (New Delhi, 1992); 10th Congress (Gothenburg, 1998); 11th Congress (Durban, 2003); 12th Congress (Montreal, 2007); 13th Congress (*Cem. Concr. Res.*, Vol. 41, No. 7, pp. 615–798, 2011).

Struble, L., P. Livesey, P. del Strother, and G.C. Bye, *Portland Cement: Composition, Production and Properties*, 3d ed., ICE Publishing, Westminster, London, 2011.

Taylor, H.F.W., *Cement Chemistry*, 2d ed., T. Telford, London, p. 459, 1997.

CHAPTER 3

Strength

Preview

The strength of concrete is the property most valued by designers and quality control engineers. In solids, there exists a fundamental inverse relationship between porosity (volume fraction of voids) and strength. Consequently, in multiphase materials such as concrete, the porosity of each component of the microstructure can become strength-limiting. Natural aggregates are generally dense and strong; therefore, it is the porosity of the cement paste matrix as well as the interfacial transition zone between the matrix and coarse aggregate, which usually determines the strength characteristic of normal-weight concrete.

Although the water-cement ratio is important in determining the porosity of both the matrix and the interfacial transition zone and hence the strength of concrete, factors such as compaction and curing conditions (degree of cement hydration), aggregate size and mineralogy, admixtures types, specimen geometry and moisture condition, type of stress, and rate of loading can also have an important effect on strength. In this chapter, the influence of various factors on concrete strength is examined in detail. Since the uniaxial strength in compression is commonly accepted as a general index of the concrete strength, the relationships between the uniaxial compressive strength and other strength types like tensile, flexural, shear, and biaxial strength are discussed.

3.1 Definition

The *strength* of a material is defined as the ability to resist stress without failure. Failure is sometimes identified with the appearance of cracks. However, as described in Chap. 2, microstructural investigations of ordinary concrete show that unlike most structural materials concrete contains many fine cracks even before it is subjected to external stresses. In concrete, therefore, strength is related to the stress required to cause failure and it is defined as the maximum stress the concrete sample can withstand. In tension testing, the fracture of the test piece usually signifies failure. In compression the test piece is considered to have failed even when no signs of external fracture are visible; however, the internal cracking has reached such an advanced state that the specimen is unable to carry a higher load.

3.2 Significance

In concrete design and quality control, strength is the property generally specified. This is because, compared to most other properties, testing of strength is relatively easy. Furthermore, many properties of concrete, such as elastic modulus, watertightness or impermeability, and resistance to weathering agents including aggressive waters, are believed to be dependent on strength and may therefore be deduced from the strength data. As pointed out earlier (Chap. 1) the compressive strength of concrete is several times greater than other types of strength, therefore a majority of concrete elements are designed to take advantage of the higher compressive strength of the material. Although in practice most concrete is subjected simultaneously to a combination of compressive, shearing, and tensile stresses in two or more directions, the uniaxial compression tests are the easiest to perform in laboratory, and the 28-day compressive strength of concrete determined by a standard uniaxial compression test is accepted universally as a general index of the concrete strength.

3.3 Strength-Porosity Relationship

In general, there exists a *fundamental inverse relationship between porosity and strength* of solids. For simple homogeneous materials, it can be described by the expression

$$S = S_0 e^{-kp} \tag{3-1}$$

where S = strength of the material which has a given porosity p
S_0 = intrinsic strength at zero porosity
k = constant

For many materials the ratio S/S_0 plotted against porosity follows the same curve. For instance, the data in Fig. 3-1*a* represent normally cured cements, autoclaved cements, and a variety of aggregates. Actually, the same strength-porosity relationship is applicable to a very wide range of materials, such as iron, plaster of Paris, sintered alumina, and zirconia (Fig. 3-1*b*).

Powers[1] found that the 28-day compressive strength f_c of three different mortar mixtures was related to the gel/space ratio, or the ratio between the solid hydration products in the system and the total space:

$$f_c = ax^3 \tag{3-2}$$

where a is the intrinsic strength of the material at zero porosity p, and x the solid/space ratio or the amount of solid fraction in the system, which is therefore equal to $1 - p$. Powers data are shown in Fig. 3-1*c*; he found the value of a to be 234 MPa (34,000 psi). The similarity of the three curves in Fig. 3-1 confirms the general validity of the strength-porosity relationship in solids.

Whereas in hardened cement paste or mortar the porosity can be related to strength, with concrete the situation is not simple. The presence of microcracks in the interfacial transition zone between the coarse aggregate and the matrix makes concrete too complex a material for prediction of strength by precise strength-porosity relations. The general validity of strength-porosity relation, however, must be respected because porosities of the component phases of concrete, including the interfacial transition zone, indeed become strength-limiting. With concrete containing the conventional low-porosity or high-strength aggregates, the strength of the material will be governed both by the strength of the matrix and the strength of the interfacial transition zone.

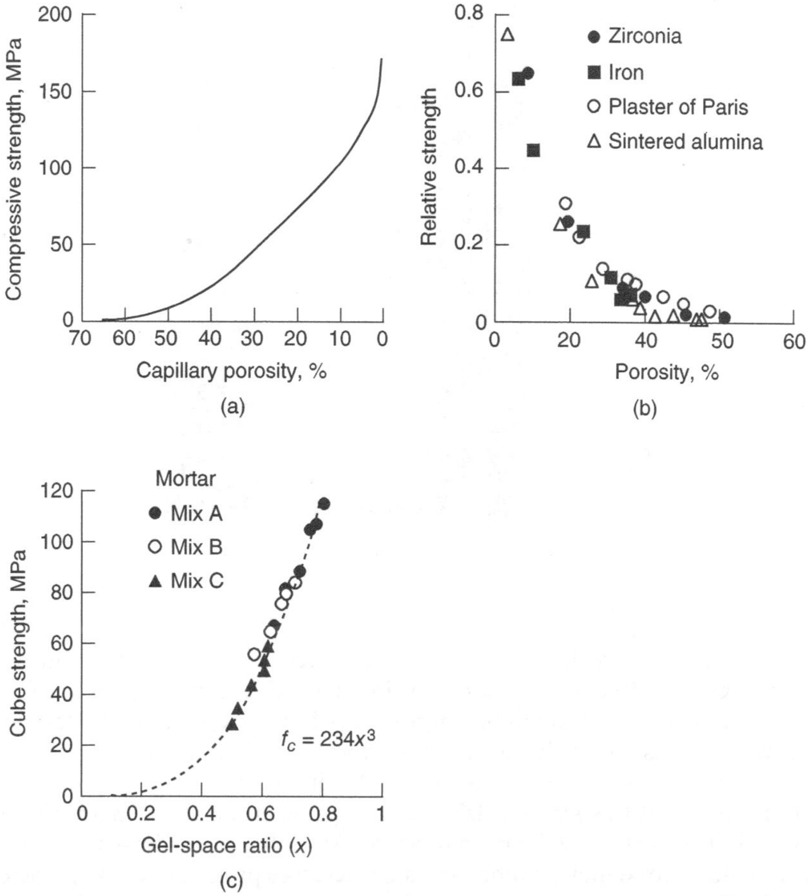

FIGURE 3-1 Porosity-strength relation in solids: (*a*) normally cured cements, autoclaved cements, and aggregates; (*b*) iron, plaster of Paris, sintered alumina, and zirconia; (*c*) portland cement mortars with different mix proportions. [(*a*) From Verbeck, G.J., and R.A. Helmuth, *Proceedings of Fifth International Symposium on Chemistry of Cements*, Tokyo, Vol. 3, pp.1–32, 1968; (*b*) from Neville, A.M., *Properties of Concrete*, Pitman Publishing, Marshfield, MA, p. 271, 1981; (*c*) from Powers, T.C., *J. Am. Ceram. Soc.*, Vol. 41, No. 1, pp. 1–6, 1958.]

The inverse relationship between porosity and strength is not limited to cementitious products; it is generally applicable to a very wide variety of materials.

3.4 Failure Modes in Concrete

With a material such as concrete, which contains void spaces of various size and shape in the matrix and microcracks at the interfacial transition zone, the failure modes under stress are very complex and vary with the type of stress. A brief review of the failure modes, however, will be useful in understanding and control of the factors that influence concrete strength.

Under uniaxial tension, relatively less energy is needed for the initiation and growth of cracks in the matrix. Rapid propagation and interlinkage of the crack system,

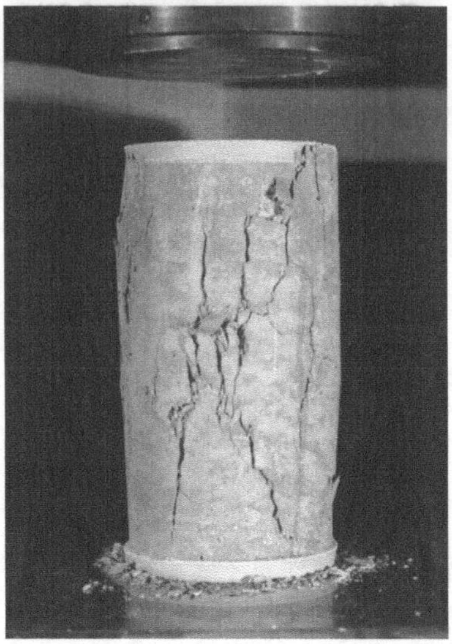

FIGURE 3-2 Typical failure mode of concrete in compression.

consisting of preexisting cracks at the interfacial transition zone and newly formed cracks in the matrix, account for the brittle failure. In compression, the failure mode is less brittle because considerably more energy is needed to form and to extend cracks in the matrix. It is generally agreed that, in a uniaxial compression test on medium- or low-strength concrete, no cracks are initiated in the matrix up to about 50 percent of the failure stress; at this stage a stable system of cracks, called *shear-bond cracks*, already exists in the vicinity of coarse aggregate. At higher stress levels, cracks are initiated within the matrix; their number and size increases progressively with increasing stress levels. The cracks in the matrix and the interfacial transition zone (shear-bond cracks) eventually join up, and generally a failure surface develops at about 20° to 30° from the direction of the load, as shown in Fig. 3-2.

3.5 Compressive Strength and Factors Affecting It

The response of concrete to applied stress depends not only on the stress type but also on how a combination of various factors affects porosity of the different structural components of concrete. The factors include properties and proportions of materials that make up the concrete mixture, degree of compaction, and conditions of curing. From the standpoint of strength, the relationship between *water-cement ratio and porosity* is undoubtedly the most important factor because, independent of other factors, it affects the porosity of both the cement mortar matrix and the interfacial transition zone between the matrix and the coarse aggregate.

Direct determination of porosity of the individual structural components of concrete—the matrix and the interfacial transition zone—is impractical, and therefore precise models of predicting concrete strength cannot be developed. However, over a period of time many useful empirical relations have been found, which, for practical use, provide enough indirect information about the influence of numerous factors on compressive strength (compressive strength being widely used as an index of all other types of strength). Although the actual response of concrete to applied stress is a result of complex interactions between various factors, to facilitate a clear understanding of these factors they can be separately discussed under three categories: (1) characteristics and proportions of materials, (2) curing conditions, and (3) testing parameters.

3.5.1 Characteristics and Proportions of Materials

Before making a concrete mixture, the selection of proper component materials and their proportions is the first step toward obtaining a product that would meet the specified strength. The composition and properties of concrete-making materials are discussed in detail in Chaps. 6, 7, and 8; however, some of the aspects that are important from the standpoint of concrete strength are considered here. It should be emphasized again that, in practice, many mixture design parameters are interdependent, and therefore their influences cannot really be separated.

Water-Cement Ratio

In 1918, as a result of extensive testing at the Lewis Institute, University of Illinois, Duff Abrams found that a relation existed between water-cement ratio and concrete strength. Popularly known as *Abrams' water-cement ratio rule*, this inverse relation is represented by the expression

$$f_c = \frac{k_1}{k_2^{w/c}} \tag{3-3}$$

where w/c represents the water-cement ratio of the concrete mixture and k_1 and k_2 are empirical constants. Typical curves illustrating the relationship between water-cement ratio and strength at a given moist-curing age are shown in Fig. 3-3.

From an understanding of the factors responsible for the strength of hydrated cement paste and the effect of increasing the water-cement ratio on porosity at a given degree of cement hydration (Fig. 2-10, case B), the w/c-strength relationship in concrete can easily be explained as the natural consequence of a progressive weakening of the matrix caused by increasing porosity with increase in the water-cement ratio. This explanation, however, does not consider the influence of the water-cement ratio on the strength of the interfacial transition zone. In low- and medium-strength concrete made with normal aggregate, both the interfacial transition zone porosity and the matrix porosity determine the strength, and a direct relation between the water-cement ratio and the concrete strength holds. This seems no longer to be the case in high-strength (i.e., very low water-cement ratio) concrete mixtures. For water-cement ratios under 0.3, disproportionately high increases in the compressive strength can be achieved with very small reductions in water-cement ratio. The phenomenon is attributed mainly to a significant improvement in the strength of the interfacial transition zone at very low water-cement ratios. Furthermore, with low water-cement ratio the crystal size of the hydration products is much smaller and the surface area is correspondingly higher.

Air Entrainment

For the most part, it is the water-cement ratio that determines the porosity of the cement paste matrix at a given degree of hydration; however, when air voids are incorporated into the system, either as a result of inadequate compaction or through the use of an

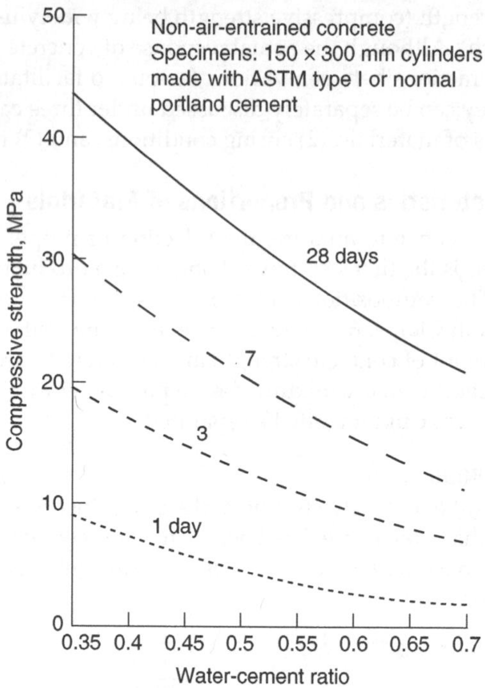

FIGURE 3-3 Influence of the water-cement ratio and moist curing age on concrete strength. (From *Design and Control of Concrete Mixtures*, 13th ed., Portland Cement Association, Skokie, Ill., p. 6, 1988.)

Compressive strength of concrete is a function of the water-cement ratio and degree of cement hydration. At a given temperature of hydration, the degree of hydration is time dependent and so is the strength.

air-entraining admixture, they also have the effect of increasing the porosity and decreasing the strength of the system. At a given water-cement ratio, the effect on the compressive strength of concrete of increasing the volume of entrained air is shown by the curves in Fig. 3-4*a*.

It has been observed that the extent of strength loss as a result of entrained air depends not only on the water-cement ratio of the concrete mixture (Fig. 3-4*a*), but also on the cement content. In short, as a first approximation, the strength loss due to air entrainment can be related to the general level of concrete strength. The data in Fig. 3-4*b* show that at a given water-cement ratio, high-strength concretes (containing a high cement content) suffer a considerable strength loss with increasing amounts of entrained air, whereas low-strength concretes (containing a low cement content) tend to suffer only a little strength loss or may actually gain some strength as a result of air entrainment. This point is of great significance in the design of mass-concrete mixtures (Chap. 12).

The influence of the water-cement ratio and cement content on the response of concrete to applied stress can be explained from the two opposing effects caused by incorporation of air into concrete. By increasing the porosity of the matrix, entrained air will have an adverse effect on the strength of the composite material. On the other hand, by improving the workability and compactibility of the mixture, entrained air tends to

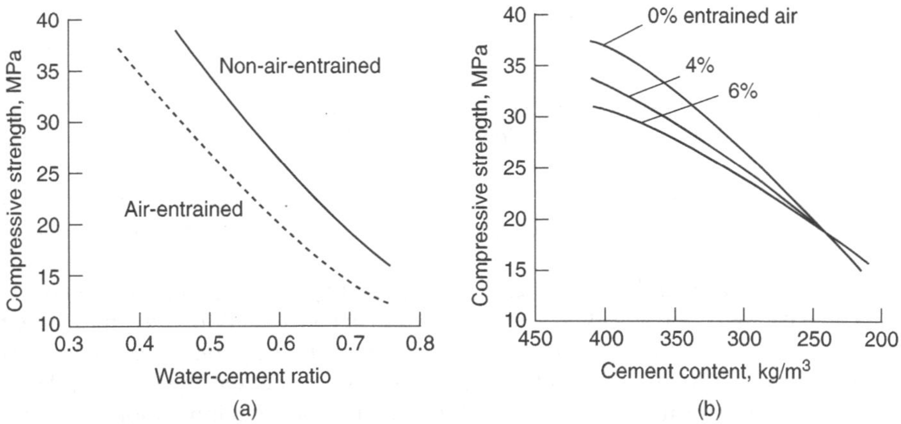

FIGURE 3-4 Influence of the water-cement ratio, entrained air, and cement content on concrete strength. (From *Concrete Manual*, U.S. Bureau of Reclamation, 1981, and Cordon, W.A., *Properties, Evaluation, and Control of Engineering Materials*, McGraw-Hill, New York, 1979.)

At a given water-cement ratio or cement content, entrained air generally reduces the strength of concrete. For very low cement contents, entrained air may actually increase the strength.

improve the strength of the interfacial transition zone (especially in mixtures with very low water and cement contents) and thus improves the strength of concrete. It seems that with concrete mixtures of low cement content, when air entrainment is accompanied by a significant reduction in the water content, the adverse effect of air entrainment on the strength of the matrix is more than compensated by the beneficial effect on the interfacial transition zone.

Cement Type

It may be recalled from Fig. 2-10 that the degree of cement hydration has a direct effect on porosity and consequently on strength. At ordinary temperature ASTM Type III portland cement, which has a higher fineness, hydrates more rapidly than other types; therefore, at early ages of hydration (e.g., 1, 3, and 7 days) and a given water-cement ratio, a concrete containing Type III portland cement will have a lower porosity and correspondingly a higher strength. On the other hand, compared to ASTM Type I, Type II, and Type III portland cements, the rates of hydration and strength development with Type IV and Type V cements (Chap. 6), and with portland-slag and portland-pozzolan cements are slower up to 28 days; however, the differences usually disappear thereafter when they have achieved a similar degree of hydration.

Aggregate

In concrete technology, an overemphasis on the relationship between water-cement ratio and strength has caused some problems. For instance, the influence of aggregate on concrete strength is not generally appreciated. It is true that aggregate strength is usually not a factor in normal strength concrete because, with the exception of lightweight aggregates, the aggregate particle is several times stronger than the matrix and the interfacial transition zone in concrete. In other words, with most natural aggregates the strength of the aggregate is hardly utilized because the failure is determined by the other two phases.

There are, however, aggregate characteristics other than strength, such as the size, shape, surface texture, grading (particle size distribution), and mineralogy, which are known to affect concrete strength in varying degrees. Frequently the effect of aggregate characteristics on concrete strength can be traced to a change of water-cement ratio. But there is sufficient evidence in the published literature that this is not always the case. Also, from theoretical considerations it may be anticipated that, independent of the water-cement ratio, the size, shape, surface texture, and mineralogy of aggregate particles would influence the characteristics of the interfacial transition zone and therefore affect concrete strength.

A change in the *maximum size* of well-graded coarse aggregate of a given mineralogy can have two opposing effects on the strength of concrete. With the same cement content and consistency, concrete mixtures containing larger aggregate particles require less mixing water than those containing smaller aggregate. On the contrary, larger aggregates tend to form weaker interfacial transition zone containing more microcracks. The net effect will vary with the water-cement ratio of the concrete and the type of applied stress. Cordon and Gillispie[2] (Fig. 3-5) showed that, in the No. 4 mesh to 3 in. range (5 to 75 mm) the effect of increasing maximum aggregate size on the 28-day compressive strengths of the concrete was more pronounced with a high-strength (0.4 water-cement ratio) and a moderate-strength (0.55 water-cement ratio) concrete than with a low-strength concrete (0.7 water-cement ratio). This is because at lower water-cement ratios the reduced porosity of the interfacial transition zone begins to play an important role in the concrete strength. Furthermore, since the interfacial transition zone characteristics have more effect on the tensile strength of concrete compared to the compressive strength, it is to be expected that with a given concrete mixture any changes in the coarse aggregate properties would influence the tensile-compressive strength ratio of the material. For instance, a decrease in the size of coarse aggregate, at a given water-cement ratio, will increase the tensile-compressive strength ratio.

A change in the *aggregate grading* without any change in the maximum size of coarse aggregate, and with water-cement ratio held constant, can influence the concrete

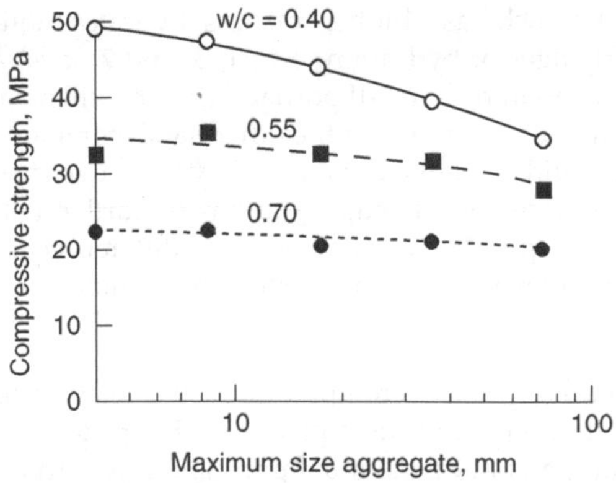

Figure 3-5 Influence of the aggregate size and the water-cement ratio on concrete strength. (From Cordon, W.A., and H.A. Gillespie, *J. ACI, Proc.*, Vol. 60, No. 8, 1963.)

Generally, the compressive strength of high strength (i.e., low water-cement ratio) concrete is adversely affected by increasing the size of aggregate. The aggregate size does not seem to have much effect on the strength in the case of low-strength or high water-cement ratio concrete.

strength when this change causes a corresponding change in the consistency and bleeding characteristics of the concrete mixture. In a laboratory experiment, with a constant water-cement ratio of 0.6, when the coarse/fine aggregate proportion and the cement content of a concrete mixture were progressively raised to increase the consistency from 2 to 6 in. (50 to 150 mm) of slump, there was about 12 percent decrease in the average 7-day compressive strength. The effects of increased consistency on the strength and the cost of concrete mixtures are shown in Fig. 3-6. The data demonstrate the economic significance of making concrete mixtures at the stiffest possible consistency that is acceptable from the standpoint of constructibility.

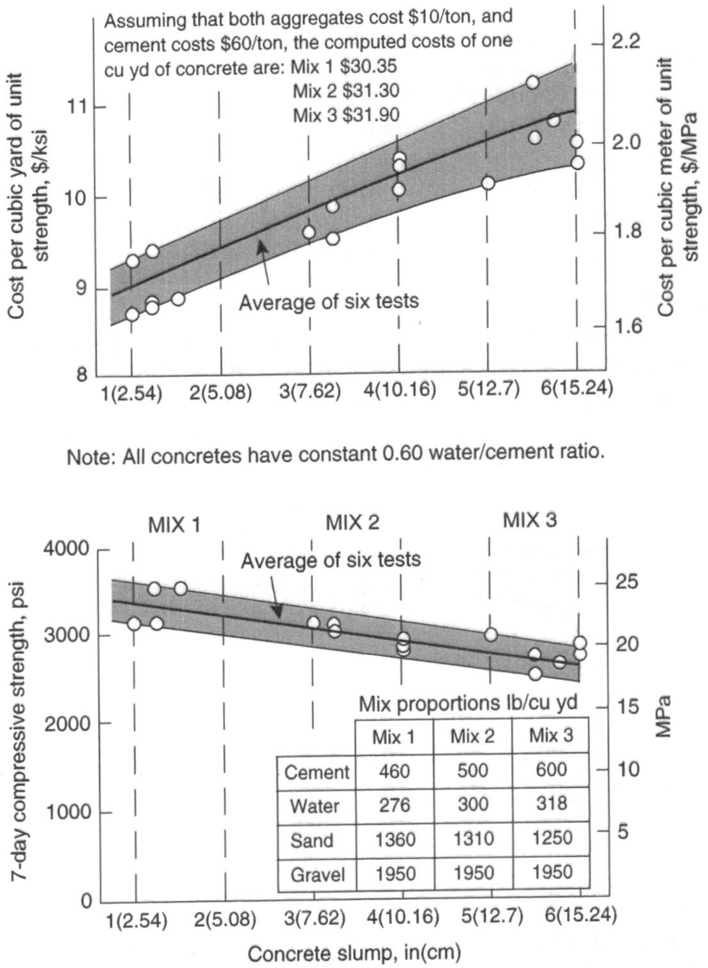

	Mix 1	Mix 2	Mix 3
Cement	460	500	600
Water	276	300	318
Sand	1360	1310	1250
Gravel	1950	1950	1950

FIGURE 3-6 Influence of the concrete slump on compressive strength and cost. (Data from student experiments, University of California at Berkeley.)

For a given water-cement ratio, concrete mixtures with higher slumps tend to bleed and therefore give lower strength. It is not cost-effective to produce concrete mixtures with slumps higher than needed.

It has been observed that a concrete mixture containing a *rough-textured* or crushed aggregate would show somewhat higher strength (especially tensile strength) at early ages than a corresponding concrete containing smooth or naturally weathered aggregate of similar mineralogy. A stronger physical bond between the aggregate and the hydrated cement paste is assumed to be responsible for this. At later ages, when chemical interaction between the aggregate and the cement paste begins to take effect, the influence of the surface texture of aggregate on strength may be reduced. From the standpoint of the physical bond with cement paste, it may be noted that a smooth-looking particle of weathered gravel, when observed under a microscope would appear to possess adequate roughness and surface area. Also, with a given cement content, somewhat more mixing water is usually needed to obtain the desired workability in a concrete mixture containing rough-textured aggregates; thus the small advantage due to a better physical bonding may be lost as far as the overall strength is concerned.

Differences in the *mineralogical composition* of aggregates are also known to affect the concrete strength. Reports show that, with identical mix proportions, the substitution of a calcareous for a siliceous aggregate can result in strength improvement. For instance, according to Fig. 3-7 not only a decrease in the maximum size of coarse aggregate (Fig. 3-7*a*), but also a substitution of limestone for sandstone (Fig. 3-7*b*), improved the 56-day strength of concrete significantly. This may be due to the higher interfacial bond strength with limestone aggregate at late ages.

Mixing Water

Impurities in water used for mixing concrete, when excessive, may affect not only the concrete strength but also setting time, *efflorescence* (deposits of white salts on the surface of concrete), and the corrosion of reinforcing and prestressing steel. In general, mixing water is rarely a factor in concrete strength, because many specifications for making concrete mixtures require that the quality of water used should be fit for

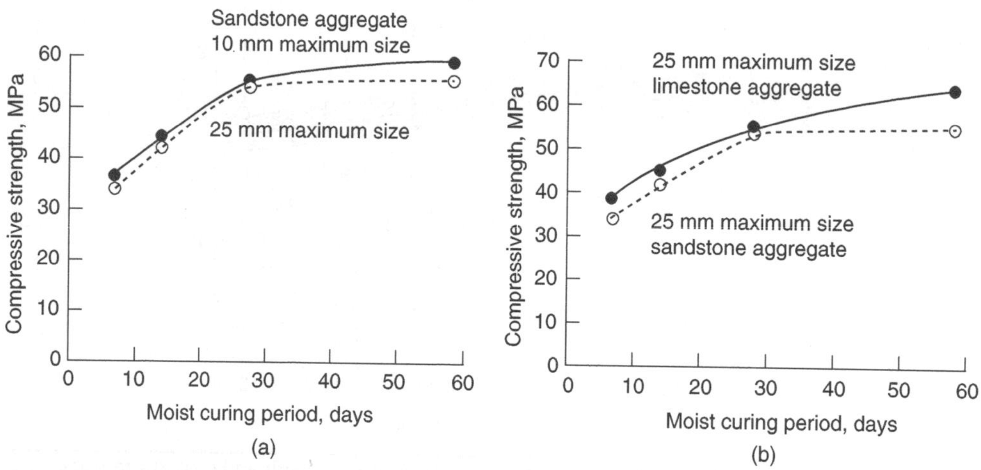

FIGURE 3-7 Influence of the aggregate size and mineralogy on compressive strength of concrete. (Data from students experiments, University of California at Berkeley.)

For a given water-cement ratio and cement content, the strength of concrete can be significantly affected by the choice of aggregate size and type.

drinking, and municipal drinking waters seldom contain dissolved solids in excess of 1000 ppm (parts per million).

As a rule, a water that is unsuitable for drinking may not necessarily be unfit for mixing concrete. Slightly acidic, alkaline, salty, brackish, colored, or foul-smelling water should not be rejected outright. This is important because of the water shortage in many areas of the world. Also, recycled waters from cities, mining, and many industrial operations can be safely used as mixing waters for concrete. The best way to determine the suitability of a water of unknown performance for making concrete is to compare the setting time of cement and the strength of mortar cubes made with the unknown water with reference water that is clean. The cubes made with the questionable water should have 7- and 28-day compressive strengths equal to or at least 90 percent of the strength of reference specimens made with clean water; also, the quality of mixing water should not affect the setting time of cement to an unacceptable degree.

Seawater, which contains about 35,000 ppm dissolved salts, is not harmful to the strength of plain concrete. However, with reinforced and prestressed concrete it increases the risk of steel corrosion; therefore, the use of seawater as concrete-mixing water should be avoided under these circumstances. As a general guideline, from standpoint of the concrete strength, the presence of excessive amounts of algae, oil, salt, or sugar in the mixing water should send a warning signal.

Admixtures

The adverse influence of air-entraining admixtures on concrete strength has already been discussed. By their ability to reduce the water content of a concrete mixture, *at a given consistency*, the water-reducing admixtures can enhance both the early and the ultimate strength of concrete. *At a given water-cement ratio*, the presence of water-reducing admixtures in concrete generally has a positive influence on the rates of cement hydration and early strength development. Admixtures capable of accelerating or retarding cement hydration obviously would have a great influence on the rate of strength gain; however, the ultimate strengths may not be significantly affected. Many researchers have pointed out the tendency toward a higher ultimate strength of concrete when the rate of strength gain at early ages was retarded.

For ecological and economic reasons, the use of pozzolanic and cementitious byproducts as mineral admixtures in concrete is gradually increasing. When used as a partial replacement for portland cement, mineral admixtures usually have a retarding effect on the strength at early ages. However, the ability of a mineral admixture to react at normal temperatures with calcium hydroxide (present in the hydrated portland cement paste) and to form additional calcium silicate hydrate can lead to significant reduction in porosity of both the matrix and the interfacial transition zone. Consequently, considerable improvements in the ultimate strength and watertightness of concrete are achievable by incorporation of mineral admixtures. It should be noted that mineral admixtures are especially effective in increasing the tensile strength of concrete.

3.5.2 Curing Conditions

The term *curing of concrete* involves a combination of conditions that promote the cement hydration, namely time, temperature, and humidity conditions immediately after the placement of a concrete mixture into formwork.

At a given water-cement ratio, the porosity of a hydrated cement paste is determined by the degree of cement hydration (Fig. 2-10, case A). Under normal temperature conditions some of the constituent compounds of portland cement begin to hydrate as soon as water is added, but the hydration reactions slow down considerably when the products of hydration coat the anhydrous cement grains. This is because hydration can proceed satisfactorily only under conditions of saturation; it almost stops when the vapor pressure of water in capillaries falls below 80 percent of the saturation humidity. Time and humidity are therefore important factors in the hydration process controlled by water diffusion. Also, like all chemical reactions, temperature has an accelerating effect on the hydration reactions.

Time

It should be noted that the time-strength relations in concrete technology generally assume moist-curing conditions and normal temperatures. At a given water-cement ratio, the longer the moist curing period the higher the strength (Fig. 3-3), assuming that the hydration of anhydrous cement particles is still going on. In thin concrete elements, if water is lost by evaporation from the capillaries, air-curing conditions prevail, and strength will not increase with time (Fig. 3-8).

The evaluation of compressive strength with time is of great concern to structural engineers. ACI Committee 209 recommends the following relationship for moist-cured concrete made with normal portland cement (ASTM Type I):

$$f_{cm}(t) = f_{c28}\left(\frac{t}{4+0.85t}\right) \tag{3-4}$$

For concrete specimens cured at 20°C, the FIP Model Code (2010) suggests the following relationship:

$$f_{cm}(t) = \exp\left(s\left(1-\sqrt{\frac{28}{t}}\right)\right)f_{cm} \tag{3-5}$$

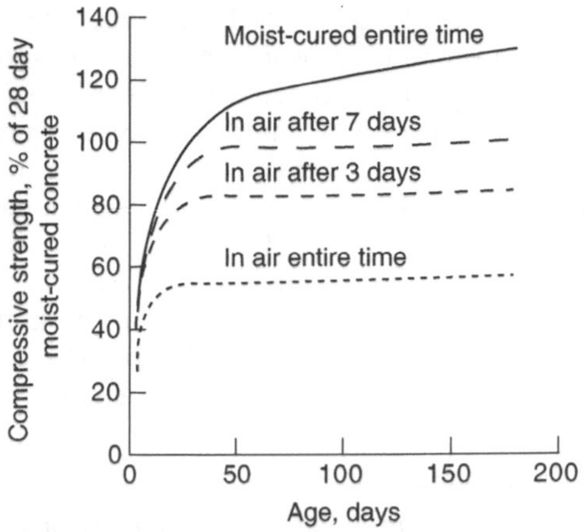

Figure 3-8 Influence of curing conditions on strength. (From *Concrete Manual*, 8th ed., U.S. Bureau of Reclamation, 1981.)

The curing age would not have any beneficial effect on the concrete strength unless curing is carried out in the presence of moisture.

where $f_{cm}(t)$ = mean compressive strength in MPa at age t in days
f_{cm} = mean 28-day compressive strength in MPa
s = coefficient depending on the cement type, such as $s = 0.20$ for high early strength cements, $s = 0.25$ for normal hardening cements; $s = 0.38$ for slow hardening cements. For $f_{cm} > 60$ MPa, $s = 0.20$ for all classes of cements.
t = concrete age in days

Humidity

The influence of the curing humidity on concrete strength is obvious from the data in Fig. 3-8, which show that after 180 days at a given water-cement ratio, the strength of the continuously moist-cured concrete was three times greater than the strength of the continuously air-cured concrete. Furthermore, probably as a result of microcracking in the interfacial transition zone caused by drying shrinkage, a slight retrogression of strength occurs in thin members of moist-cured concrete when they are subjected to air drying. The rate of water loss from concrete soon after the placement depends not only on the surface/volume ratio of the concrete element but also on temperature, relative humidity, and velocity of the surrounding air.

A minimum period of 7 days of moist-curing is generally recommended with concrete containing normal portland cement; obviously, with concrete mixtures containing either a blended portland cement or a mineral admixture, longer curing period is desirable to ensure strength contribution from the pozzolanic reaction. Moist curing is provided by spraying or ponding or by covering the concrete surface with wet sand, sawdust, or cotton mats. Since the amount of mixing water used in a concrete mixture is usually more than needed for portland cement hydration (estimated to be about 30 percent by weight of cement), proper application of an impermeable membrane soon after the concrete placement provides an acceptable way to maintain the strength development at a satisfactory rate. However, moist-curing should be the preferred method when control of cracking due to autogenous shrinkage or thermal shrinkage is important.

Temperature

With moist-cured concrete the influence of temperature on strength depends on the time-temperature history of *casting and curing*. This can be illustrated with the help of three cases: concrete cast and cured at the same temperature, concrete cast at different temperatures but cured at a normal temperature, and concrete cast at a normal temperature but cured at different temperatures.

In the temperature range 5 to 46°C, when concrete is cast and cured at a specific constant temperature, it is generally observed that up to 28 days, the higher the temperature the more rapid the cement hydration and the strength gain. From the data in Fig. 3-9, it is evident that the 28-day strength of specimens cast and cured at 5°C was about 80 percent of those cast and cured at 21 to 46°C. At later ages, when the differences in the degree of cement hydration disappear, so do the differences in the concrete strength. On the other hand, as explained below, it has been observed that the higher the casting and curing temperature, the lower will be the ultimate strength.

The data in Fig. 3-9*b* represent a different time-temperature history of casting and curing. The casting temperature (i.e., the temperature during the first 2 h after making concrete) was varied between 10 and 46°C; thereafter, all concrete mixtures were moist-cured at a constant temperature of 21°C. The data show that ultimate strengths (180 days)

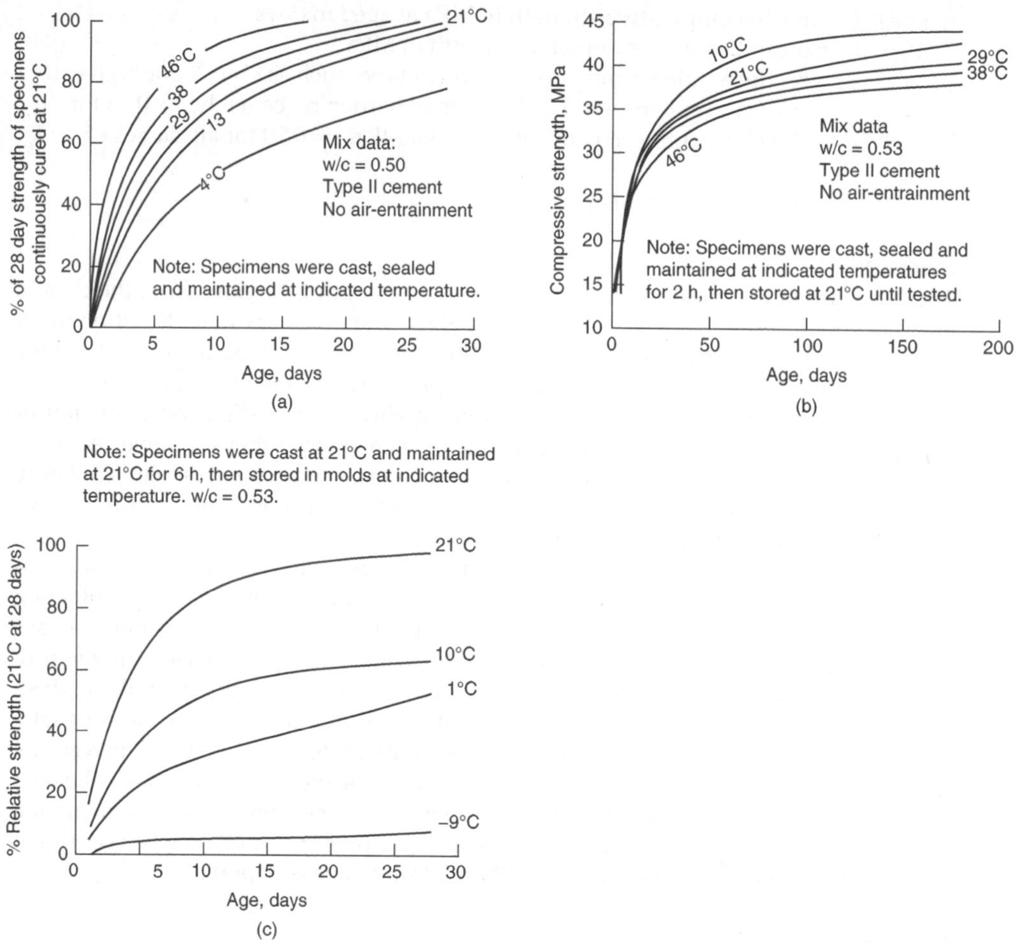

FIGURE 3-9 Influence of casting and curing temperatures on concrete strength. (From *Concrete Manual*, U.S. Bureau of Reclamation, 1975.)

Concrete casting and curing temperatures control the degree of cement hydration and thus have a profound influence on the rate of strength development as well as the ultimate strength.

of the concrete cast at 5 or 13°C were higher than those cast at 21, 30, 38, or 46°C. From microscopic studies many researchers have concluded that, with low temperature casting, a relatively more uniform microstructure of the hydrated cement paste (especially the pore size distribution) accounts for the higher strength.

With concrete mixtures cast at 21°C and subsequently cured at different temperatures from below freezing to 21°C, the effect of the curing temperature on strength is shown in Fig. 3-9*c*. In general, the lower the curing temperature, the lower would be the strength up to 28 days. At a curing temperature near freezing, the 28-day strength was about one-half of the strength of the concrete cured at 21°C; hardly any strength developed at the below-freezing curing temperature. Since the hydration reactions of portland cement compounds are slow, it appears that adequate temperature levels must be maintained for a sufficient time to provide the needed activation energy for the reactions to begin. This enables the strength development process that is associated with progressive filling of voids with hydration products, to proceed unhindered.

The influence of time-temperature history on concrete strength has several important applications in the concrete construction practice. Since the curing temperature is far more important to the strength than the placement temperature, ordinary concrete mixtures that are placed in cold weather must be maintained above a certain minimum temperature for a sufficient length of time. Concrete cured in summer or in a tropical climate can be expected to have a higher early strength but a lower ultimate strength than the same concrete cured in winter or in a colder climate. In the precast concrete products industry, steam curing is used to accelerate strength development to achieve quicker mold release. In massive elements, when no measures for temperature control are taken, for a long time the temperature of concrete will remain at a much higher level than the environmental temperature. Therefore, compared to the strength of the specimens cured at normal laboratory temperature, the in situ concrete strength will be higher at early ages and lower at later ages.

3.5.3 Testing Parameters

It is not always appreciated that the results of concrete strength tests are significantly affected by parameters involving the test specimen and loading conditions. Specimen parameters include the influence of size, geometry, and the moisture state of concrete; loading parameters include stress level and duration, and the rate at which stress is applied.

Specimen Parameters

In the United States, the standard specimen for testing the compressive strength of concrete is a 15- by 30-cm cylinder. While maintaining the height/diameter ratio equal to 2, if a concrete mixture is tested in compression with cylindrical specimens of varying diameter, the larger the diameter the lower will be the strength. The data in Fig. 3-10 show that, compared to the standard specimens, the average strength of 5- by 10-cm and 7.5- by 15-cm cylindrical specimens was 106 and 108 percent, respectively. When the diameter is increased beyond 45 cm (18 in.), a much smaller reduction in

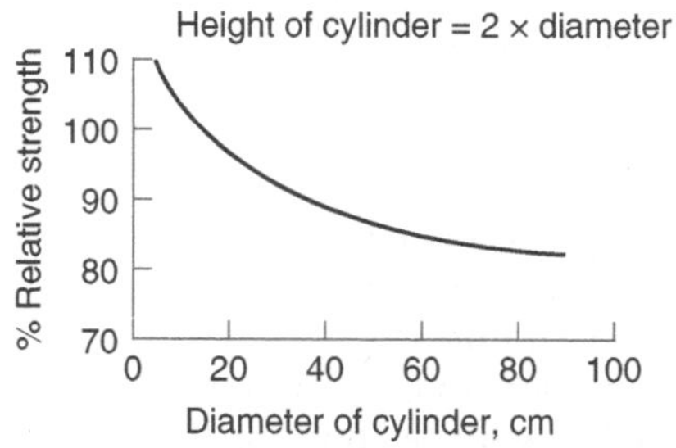

FIGURE 3-10 Influence of the specimen diameter on concrete strength when the height-diameter ratio is equal to 2. (From *Concrete Manual*, U.S. Bureau of Reclamation, pp. 574–575, 1975.)

Specimen geometry can affect the laboratory test data on concrete strength. The strength of cylindrical specimens with a slenderness ratio (H/D) above 2 or a diameter above 30 cm is not much influenced by the size effects.

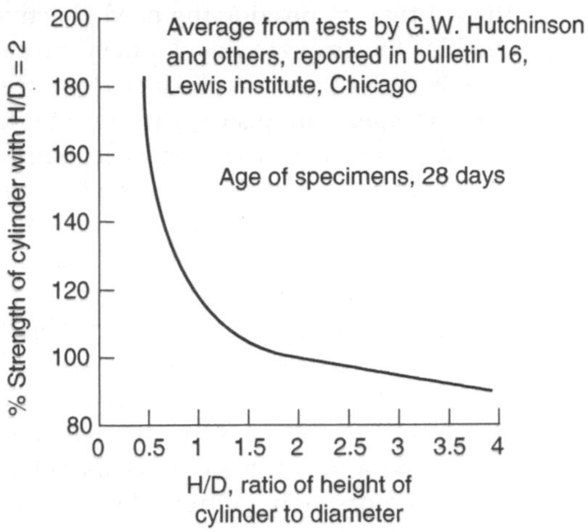

FIGURE 3-11 Influence of varying the length/diameter ratio on concrete strength. (From *Concrete Manual*, U.S. Bureau of Reclamation, pp. 574–575, 1975.)

strength is observed. Chapter 13 describes this phenomenon in greater details and presents mathematical equations for the scaling law.

The effect of change in the specimen geometry (height/diameter ratio) on the compressive strength of concrete is shown in Fig. 3-11. In general, the greater the ratio of the specimen height to diameter, the lower will be the strength. For instance, compared to the strength of the standard specimens (height/diameter ratio equal to 2), the specimens with the height/diameter ratio of 1 showed about 15 percent higher strength. It may be of interest to point out that the concrete strength testing based on 15-cm (6-in.) standard cube, which is prevalent in Europe, is reported to give 10 to 15 percent higher strength than the same concrete mixture tested in accordance with the standard U.S. practice.

Because of the effect of moisture state on the concrete strength, the standard procedure requires that the specimens continue to be in a moist condition at the time of testing. In compression tests it has been observed that air-dried specimens show 20 to 25 percent higher strength than corresponding specimens tested in a saturated condition. The lower strength of the saturated concrete is attributed to the disjoining pressure within the cement paste.

Loading Conditions

The compressive strength of concrete is measured in the laboratory by a uniaxial compression test (ASTM C 469) in which the load is progressively increased to fail the specimen within 2 to 3 min. In practice, most structural elements are subjected to a dead load for an indefinite period and, at times, to repeated loads or to impact loads. It is, therefore, desirable to know the relationship between the concrete strength under laboratory testing conditions and actual loading conditions. The behavior of concrete under various stress states is described in the next section. From this description it can be concluded that the loading condition has an important influence on the strength.

To appreciate at a glance the complex web of numerous variables that influence the strength of concrete, a summary is presented in Fig. 3-12.

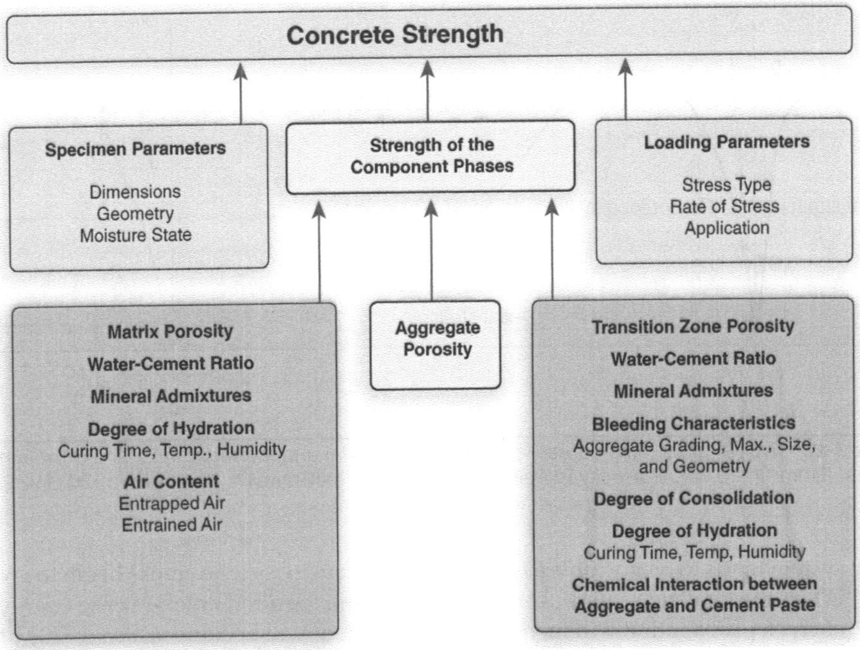

FIGURE 3-12 Interplay of factors influencing the concrete strength.

3.6 Behavior of Concrete under Various Stress States

It was described in Chap. 2 that, even before any load has been applied, a large number of microcracks exist in the interfacial transition zone (i.e., the region between the cement paste matrix and coarse aggregate). This characteristic of the structure of concrete plays a decisive role in determining the behavior of the material under various stress states that are discussed next.

3.6.1 Behavior of Concrete under Uniaxial Compression

Stress-strain behavior of concrete subjected to uniaxial compression will be discussed in detail in Chap. 4; only a summary is presented here. The stress-strain curve (Fig. 3-13*a*) shows a linear-elastic behavior up to about 30 percent of the ultimate strength f'_c, because under *short-term loading* the microcracks in the interfacial transition zone remain undisturbed. For stresses above this point, the curve shows a gradual increase in curvature up to about $0.75f'_c$ to $0.9f'_c$, then it bends sharply (almost becoming flat at the top) and, finally, descends until the specimen is fractured.

From the shape of the stress-strain curve it seems that, with a stress level that is between 30 and 50 percent of f'_c, the microcracks in the interfacial transition zone show some extension due to stress concentration at the crack tips; however, no cracking occurs in the mortar matrix. Until this point, crack propagation is assumed to be *stable* in the sense that crack lengths rapidly reach their final values if the applied stress is held constant. With a stress level between 50 and 75 percent of f'_c, increasingly the crack

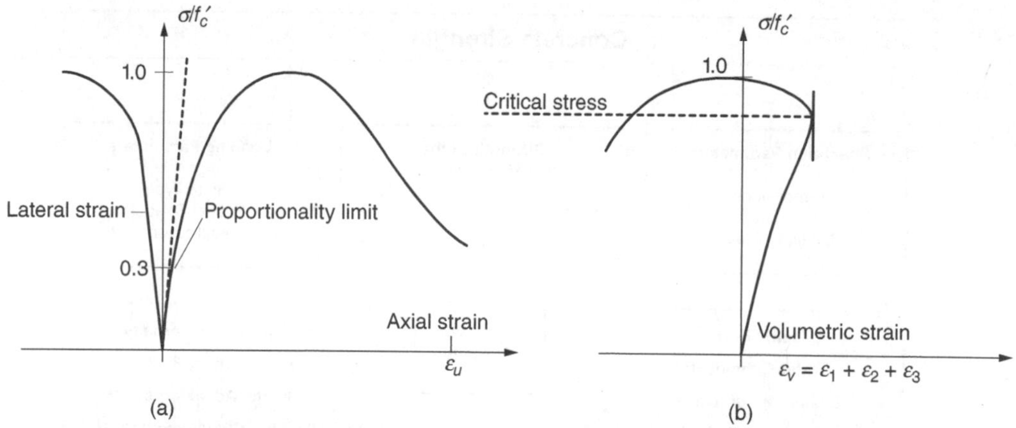

FIGURE 3-13 Typical plots of compressive stress vs. (*a*) axial and lateral strains, and (*b*) volumetric strains. (From Chen, W.F., *Plasticity in Reinforced Concrete*, McGraw-Hill, New York, p. 20, 1982.)

system tends to be unstable as the interfacial transition zone cracks begin to grow again. When the available internal energy exceeds the required crack-release energy, the rate of crack propagation will increase and the system will become *unstable*. This happens at the compressive stress levels above 75 percent of f'_c, when complete fracture of the test specimen can occur by bridging of the cracks between the matrix and the interfacial transition zone.

The stress level of 75 percent of f'_c, which represents the onset of unstable crack propagation, is called *critical stress*[3]; critical stress also corresponds to the maximum value of volumetric strain (Fig. 3-13*b*). From the figure it may be noted that when volumetric strain $\varepsilon_v = \varepsilon_1 + \varepsilon_2 + \varepsilon_3$ is plotted against stress, the initial change in volume is almost linear up to about $0.75f'_c$; at this point the direction of the volume change is reversed, resulting in a volumetric expansion near or at f'_c.

Above the critical stress level, concrete shows a time-dependent fracture; that is, under *sustained stress conditions* the crack bridging between the interfacial transition zone and the matrix would lead to failure at a stress that is lower than the short-term loading strength f'_c. In an investigation by Price[4] when the sustained stress was 90 percent of the ultimate short-time stress, the failure occurred in 1 h; however, when the sustained stress was about 75 percent of the ultimate short-time stress, it took 30 years. As the value of the sustained stress approaches that of the ultimate short-time stress, the time to failure decreases. Rusch[5] confirmed this in his tests on 56-day-old, 34 MPa (5000 psi) compressive-strength specimens. The long-time failure limit was found to be about 80 percent of the ultimate short-time stress (Fig. 3-14).

In regard to the *effect of loading rate* on concrete strength, it is generally agreed that the more rapid the rate of loading, the higher the observed strength. However, Jones and Richart[6] found that within the range of customary testing, the effect of rate of loading on strength is not large. For example, compared with the data from the ASTM C469 standard test, which requires the rate of uniaxial compression loading to be 0.25 MPa/s, a loading rate of 0.007 MPa/s reduced the indicated strength of concrete cylinders by about 12 percent; on the other hand, a loading rate of 6.9 MPa/s increased the indicated strength by a similar amount.

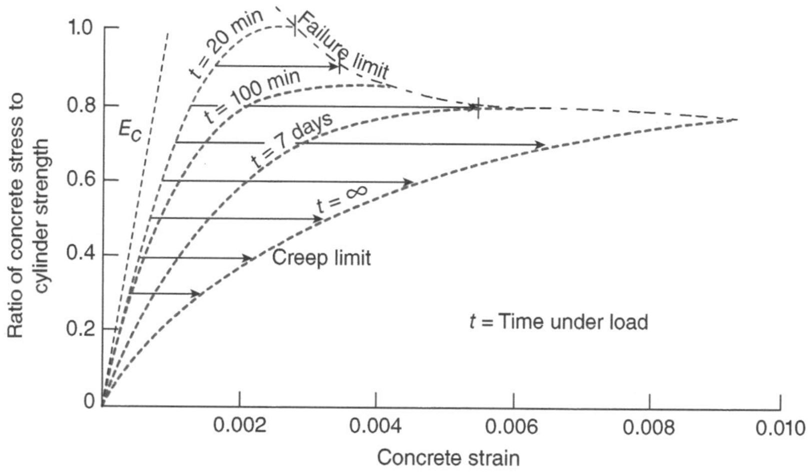

FIGURE 3-14 Relationship between the short-term and long-term loading strengths. (From Rusch, H., *J. ACI, Proc.*, Vol. 57, No. 1, 1960.)

The ultimate strength of concrete is also affected by the rate of loading. Due to progressive microcracking at sustained loads, a concrete will fail at a lower stress than that induced by instantaneous or short-time loading normally used in the laboratory.

It is interesting to point out here that the *impact strength* of concrete increases greatly with the rate at which the impact stress is applied. It is generally assumed that the impact strength is directly related to the compressive strength, as both are adversely affected by the presence of microcracks and voids. This assumption is not completely correct; for the same compressive strength, Green[7] found that the impact strength increased substantially with the angularity and surface roughness of coarse aggregate, and decreased with the increasing size of aggregate. It seems that the impact strength is more influenced by the interfacial transition zone characteristics than by the compressive strength. Therefore, the impact strength is more closely related to the tensile strength.

The FIP Model Code (2010) recommends that the increase in compressive strength due to impact, with rates of loading less than 10^6 MPa/s, can be computed using the relationship:

$$\frac{f_{c,imp}}{f_{cm}} = \left(\frac{\dot{\sigma}_c}{\dot{\sigma}_{c0}}\right)_s^{0.014} \tag{3-6}$$

where $f_{c,imp}$ = impact compressive strength
f_{cm} = compressive strength of concrete
$\dot{\sigma}_{c0}$ = 1.0 MPa/s
$\dot{\sigma}_c$ = impact stress rate

Ople and Hulsbos[8] reported that, *repeated* or *cyclic loading* has an adverse effect on concrete strength at stress levels greater than 50 percent of f_c'. For instance, in 5000 cycles of repeated loading, concrete failed at 70 percent of the ultimate monotonic loading strength.

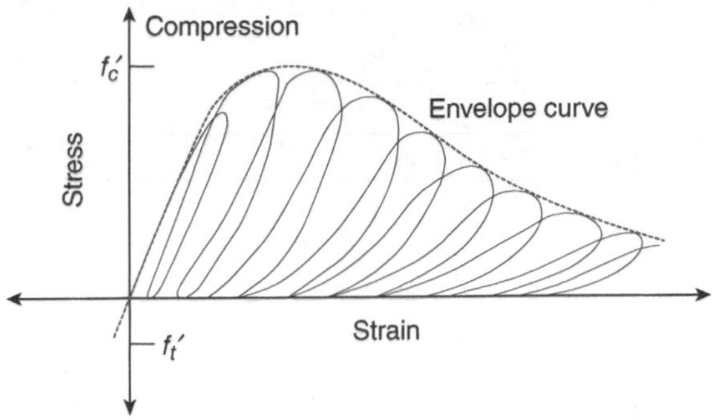

FIGURE 3-15 Response of concrete to repeated uniaxial loading. (Adapted from Karson, P., and J.O. Jirsa, *ASCE Jour. Str. Div.*, Vol. 95, No. ST12, Paper 6935, 1969.)

Progressive microcracking in the interfacial transition zone and the matrix are responsible for this phenomenon.

Typical behavior of plain concrete subjected to cyclic compressive loading is shown in Fig. 3-15. For stress levels between 50 and 75 percent of f_c', a gradual degradation occurs in both the elastic modulus and the compressive strength. As the number of loading cycles increases, the unloading curves show nonlinearity and a characteristic hysteresis loop is formed on reloading. For stress levels at about 75 percent of f_c', the unloading-reloading curves exhibit strong nonlinearity (i.e., the elastic property of the material has greatly deteriorated). In the beginning, the area of the hysteresis loop decreases with each successive cycle but eventually increases before fatigue failure. Figure 3-15 shows that the stress-strain curve for monotonic loading serves as a reasonable envelope for the peak values of stress for concrete under cyclic loading.

3.6.2 Behavior of Concrete under Uniaxial Tension

The shape of the stress-strain curve, the elastic modulus, and the Poisson ratio of concrete under uniaxial tension are similar to those under uniaxial compression. However, there are some important differences in the behavior. As the uniaxial tension state of stress tends to arrest cracks much less frequently than the compressive states of stress, the interval of stable crack propagation is expected to be short. Explaining the relatively brittle fracture behavior of concrete in tension tests, Chen[3] states:

> The direction of crack propagation in uniaxial tension is transverse to the stress direction. The initiation and growth of every new crack will reduce the available load-carrying area, and this reduction causes an increase in the stresses at critical crack tips. The decreased frequency of crack arrests means that the failure in tension is caused by a few bridging cracks rather than by numerous cracks, as it is for compressive states of stress. As a consequence of rapid crack propagation, it is difficult to follow the descending part of the stress-strain curve in an experimental test.

The ratio between uniaxial tensile and compressive strengths is generally in the range of 0.07 to 0.11. Owing to the ease with which cracks can propagate under a tensile stress, this is not surprising. Most concrete elements are therefore designed under the assumption that the concrete would resist the compressive but not the tensile stresses. However, tensile stresses cannot be ignored altogether because cracking of concrete is frequently the outcome of a tensile failure caused by restrained shrinkage; the shrinkage is usually due either to lowering of the concrete temperature or to drying of moist concrete. Also, a combination of tensile, compressive, and shear stresses usually determines the strength when concrete is subjected to flexural or bending loads, such as in highway pavements.

In the preceding discussion on factors affecting the compressive strength of concrete, it was assumed that the compressive strength is an adequate index for all types of strength, and therefore a direct relationship ought to exist between the compressive and the tensile or flexural strength of a given concrete. As a first approximation, the assumption is valid; however, this may not always be the case. It has been observed that the relationship among various types of strength is influenced by factors like the methods by which the tensile strength is measured (i.e., direct tension test, splitting test, or flexure test), the quality of concrete (i.e., low-, moderate-, or high-strength), the aggregate characteristics (e.g., surface texture and mineralogy), and admixtures (e.g., air-entraining and mineral admixtures).

Testing Methods for Tensile Strength

Direct tension tests of concrete are seldom carried out, mainly because the specimen holding devices introduce secondary stresses that cannot be ignored. The most commonly used tests for estimating the tensile strength of concrete are the ASTM C 496 splitting tension test and the ASTM C 78 third-point flexural loading test (Fig. 3-16).

In the splitting tension test a 15- by 30-cm concrete cylinder is subjected to compression loads along two axial lines which are diametrically opposite. The load is applied continuously at a constant rate within the splitting tension stress range of 0.7 to 1.3 MPa until the specimen fails. The compressive stress produces a transverse tensile stress, which is uniform along the vertical diameter. The splitting tension strength is computed from the formula

$$T = \frac{2P}{\pi l d} \tag{3-7}$$

where T = tensile strength
P = failure load
l = length
d = diameter of the specimen

Compared to direct tension, the splitting tension test is known to overestimate the tensile strength of concrete by 10 to 15 percent (see box).

In the third-point flexural loading test, a 150- by 150- by 500 mm concrete beam is loaded at a rate of 0.8 to 1.2 MPa/min (125 to 175 psi/min). Flexural strength is expressed in terms of the modulus of rupture, which is the maximum stress at rupture computed from the flexure formula

$$R = \frac{PL}{bd^2} \tag{3-8}$$

where R = modulus of rupture
P = maximum indicated load
L = span length
b = width
d = depth of the specimen

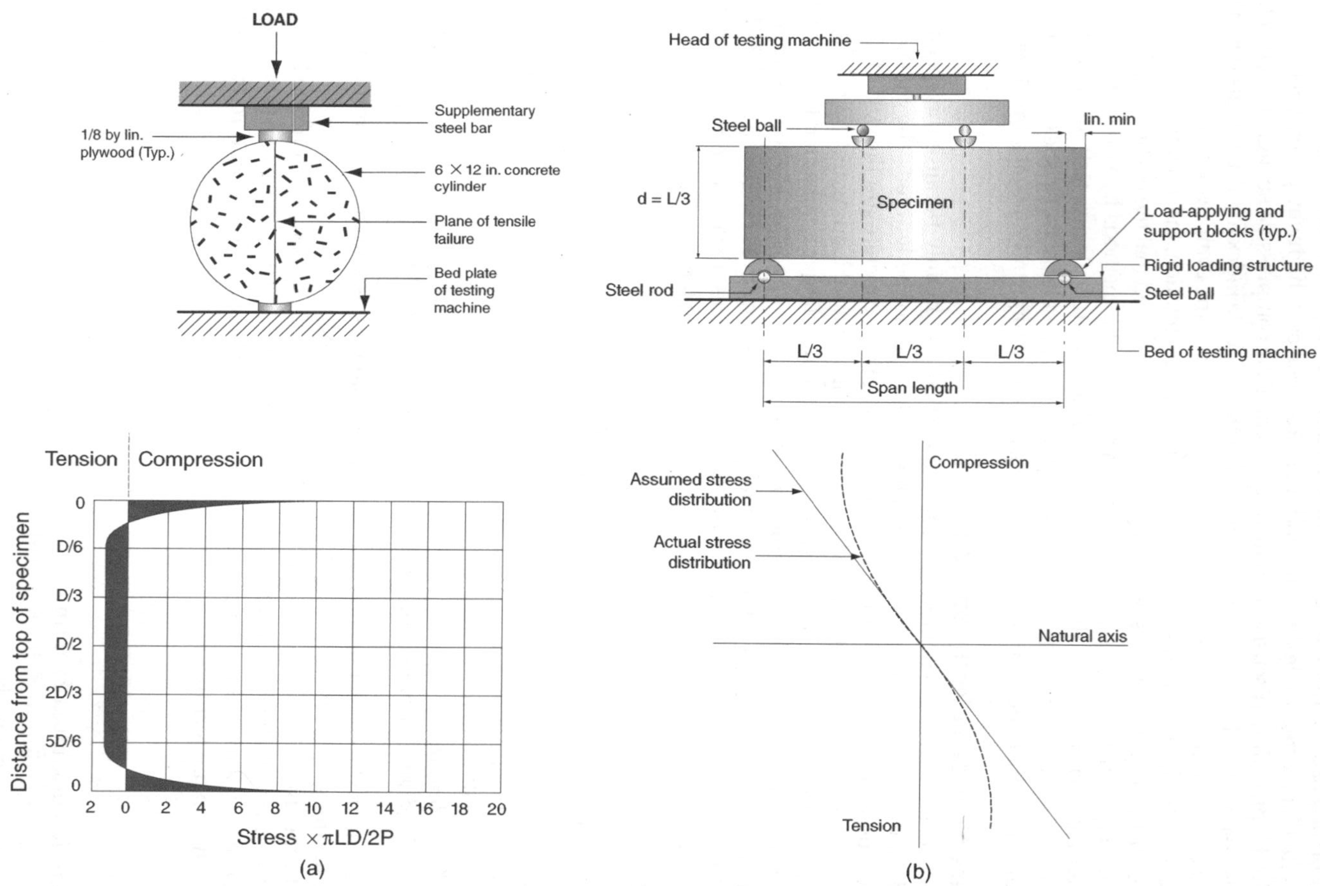

Figure 3-16 (*a*) Splitting tension test (ASTM C 496): top, diagrammatic arrangement of the test; bottom, stress distribution across the loaded diameter of a cylinder compressed between two plates. (*b*) Flexural test by third-point loading (ASTC C 78): top, diagrammatic arrangement of the test; bottom, stress distribution across the depth of a concrete beam under flexure.

Origin of the Splitting Tension Test

Behind the origin of the "splitting tension test," the method of determining the tensile strength of concrete by applying diametrically opposite compressive forces on a plane passing through the center of a cylinder, is an interesting story. During World War II, the Brazilian city of Rio de Janeiro expanded very fast, necessitating enlargement and redesign of the avenues along the Guanabara Bay. The small church of São Pedro, built in 1740, occupied a section of the redesigned roadway system and therefore plans were made for its relocation. Because of the war, steel rollers were in short supply, therefore, concrete cylinders (0.3 m diameter and 1.2 m long) covered by 9 mm thick steel plates were investigated for use as rollers to transport the church. Lobo Carneiro, the young engineer in charge of testing the load-bearing capacity of the concrete cylinders when loaded diametrically (without the steel plates), noticed that the cylinders had a uniform and consistent splitting failure in all the tests. Intrigued, he studied the work of Hertz, who had performed theoretical analysis of stress distribution for concentrated loads applied to cylinders. Carneiro noticed that the tensile stresses normal to the plane of the load were uniform and, therefore, concluded that this configuration would be appropriate for measuring the indirect tensile strength of concrete. Unfortunately, the plans for relocating the church were abandoned when studies indicated that the masonry was weak and there was a risk of collapse during transport. However, the splitting test proposed by Carneiro for measuring the tensile strength of brittle materials became popular. In rock mechanics, this test is often referred to as the "Brazilian test," but in concrete technology it is called the splitting tension test.

View of the Guanabara Bay in Rio de Janeiro, Brazil. (*Photograph courtesy of Luis Arouche.*)

The formula is valid only if the fracture in the tension surface is within the middle third of the span length. If the fracture is outside by not more than 5 percent of the span length, a modified formula is used:

$$R = \frac{3Pa}{bd^2} \tag{3-9}$$

where a is equal to the average distance between the line of fracture and the nearest support measured on the tension surface of the beam. When the fracture is outside by more than 5 percent of the span length, the results of the test are rejected.

The results from the modulus of rupture test tend to overestimate the tensile strength of concrete by 50 to 100 percent, mainly because the flexure formula assumes a linear stress-strain relationship in concrete throughout the cross section of the beam. Additionally, in direct tension tests the entire volume of the specimen is under applied stress, whereas in the flexure test only a small volume of concrete near the bottom of the specimen is subjected to high stresses. The data in Table 3-1 show that with low-strength concrete the modulus of rupture can be as high as twice the strength in direct tension; for moderate or high-strength concrete the values are about 70 percent and 50 to 60 percent higher, respectively. Nevertheless, the flexure test is usually preferred for quality control of concrete for highway and airport pavements, where the concrete is loaded in bending rather than in axial tension.

The FIP Model Code (2010) suggests the following relationship between the mean axial tensile strength (f_{ctm}) and mean flexural strength ($f_{ctm,fl}$)

$$f_{ctm} = f_{ctm,fl}\,\frac{0.06 \cdot h_b^{0.7}}{1+0.06 \cdot h_b^{0.7}} \tag{3-10}$$

where h_b is the depth of the beam in mm and strengths are expressed in MPa units.

Strength of Concrete (MPa)			Ratio (%)		
Compressive	Modulus of Rupture	Tensile	Modulus of Rupture to Compressive Strength	Tensile Strength to Compressive Strength	Tensile Strength to Modulus of Rupture
7	2	1	23.0	11.0	48
14	3	1	18.8	10.0	53
21	3	2	16.2	9.2	57
28	4	2	14.5	8.5	59
34	5	3	13.5	8.0	59
41	5	3	12.8	7.7	60
48	6	4	12.2	7.4	61
55	6	4	11.6	7.2	62
62	7	4	11.2	7.0	63

Source: Price, W.H., *J. ACI, Proc.*, Vol. 47, p. 429, 1951.

TABLE 3-1 Relation between Compressive, Flexural, and Tensile Strength of Concrete

3.6.3 Relationship between the Compressive and the Tensile Strength

It has been pointed out before that the compressive and tensile strengths are closely related; however, there is no direct proportionality. As the compressive strength of concrete increases, the tensile strength also increases but at a decreasing rate (Fig. 3-17). In other words, the tensile-compressive strength ratio depends on the general level of the compressive strength; the higher the compressive strength, the lower the ratio. Relationship between the compressive and tensile strengths in the f_c range 7.0 to 62 MPa is also shown in Table 3-1. It appears that the tensile-to-compressive strength ratio is approximately 10 to 11 percent for low-strength, 8 to 9 percent for moderate-strength, and 7 percent for high-strength concrete.

The CEP-FIP Model Code (2010) recommends that the upper and lower bound values of the characteristic tensile strength, $f_{ctk,\max}$ and $f_{ctk,\min}$ may be estimated from the characteristic strength f_{ck} (in MPa units) for concrete grades less or equal to C50:

$$f_{ctk,\min} = 0.21(f_{ck})^{2/3} \quad \text{and} \quad f_{ctk,\max} = 0.39(f_{ck})^{2/3} \tag{3-11}$$

The mean value of the tensile strength is given by the relationship:

$$f_{ctm} = 0.3\,(f_{ck})^{2/3} \tag{3-12}$$

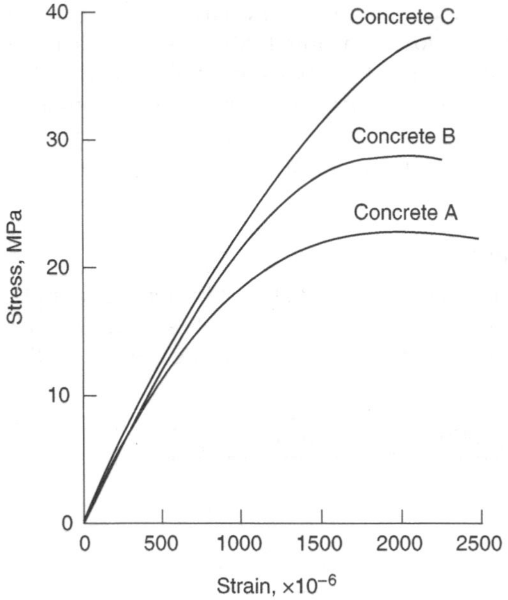

Mix No. →	A	B	C
Water-cement ratio	0.68	0.57	0.48
Slump, mm	165	180	170
f'_c, MPa	22.4	29.0	40.0
f'_{st}, MPa	2.6	2.9	3.5
f'_{st}/f'_c	0.11	0.10	0.09

*Unpublished data from students experiments, University of California at Berkeley.

Mix Proportions and Properties of Non-Air-Entrained Concrete*

Figure 3-17 Influence of the water-cement ratio on tensile and compressive strengths.

The relationship between the compressive strength and the tensile-to-compressive strength ratio seems to be determined by the combined effect of various factors on properties of both the matrix and the interfacial transition zone in concrete. It is observed that not only the curing age but also the characteristics of the concrete mixture, such as water-cement ratio, type of aggregate, and admixtures, affect the tensile-to-compressive strength ratio to varying degrees. For example, after about 1 month of curing the tensile strength of concrete is known to increase more slowly than the compressive strength; that is, the tensile-compressive strength ratio decreases with the curing age. At a given curing age, the tensile-compressive ratio also decreases with decrease in the water-cement ratio.

With concrete containing calcareous aggregate or mineral admixtures it is possible to obtain, after adequate curing, a relatively high tensile-compressive strength ratio even at high levels of compressive strength. From Table 3-1 it may be observed that with ordinary concrete, in the high compressive strength range (55 to 62 MPa), the direct-tensile-compressive strength ratio is about 7 percent (the splitting tensile-compressive strength ratio will be slightly higher). Splitting tension data for the high-strength concrete mixtures of Fig. 3-7 are shown in Table 3-2. The beneficial effect on the f_{st}/f_c ratio by reducing the maximum size of coarse aggregate, or by changing the aggregate type is clear from the data. Also, it has been found that compared to a typical 7 to 8 percent splitting tension/compressive strength ratio (f_{st}/f_c) for a high-strength concrete with no fly ash, the ratio was considerably higher when fly ash was present in the concrete mixtures.

Whereas factors causing a decrease in the porosity of the matrix and the interfacial transition zone lead to a general improvement of both the compressive and the tensile strengths of concrete, it seems that the magnitude of increase in the tensile strength of concrete remains relatively small unless the intrinsic strength of hydration products comprising the interfacial transition zone is improved at the same time. That is, the tensile strength of concrete with a low-porosity interfacial transition zone will continue to be weak as long as large numbers of oriented crystals of calcium hydroxide are present there (see Fig. 2-14). The size and concentration of calcium hydroxide crystals in the interfacial transition zone can be reduced by chemical reactions when either a pozzolanic admixture (see Fig. 6-14) or a reactive aggregate is present. For example, a possible chemical interaction between calcium hydroxide and the calcareous aggregate is probably the reason for the relatively large increase in the tensile strength of concrete, as shown by the data in Table 3-2.

3.6.4 Tensile Strength of Mass Concrete

Engineers working with reinforced concrete ignore the low value of the tensile strength of concrete and use steel to pick up tensile loads. With massive concrete structures, such as dams, it is impractical to use steel reinforcement. Therefore, a reliable estimate of the

	f_c (MPa)	f_{st} (MPa)	f_{st}/f_c
Sandstone, 25 mm max.	55.8	5.2	0.09
Limestone, 25 mm max.	63.9	7.0	0.11
Sandstone, 10 mm max.	58.9	5.9	0.10

TABLE 3-2 Effect of Aggregate Mineralogy and Size on Tensile-Compressive Strength Relations in High-Strength Concretes (60-Days Moist Cured)

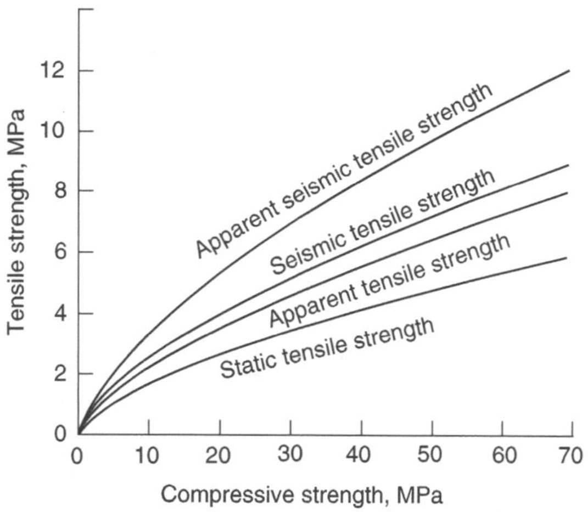

FIGURE 3-18 Design chart for tensile strength. (From Raphael, J., *J. ACI, Proc.*, Vol. 81, No. 2, pp. 158–164, 1984.)

tensile strength of concrete is necessary, especially for judging the safety of a dam under seismic loading. Raphael[9] recommends the values obtained by the splitting test or the modulus of rupture test, augmented by the multiplier found appropriate by dynamic tensile tests, or about 1.5. Alternatively, depending on the loading conditions, the plots of tensile strength as a function of compressive strength (Fig. 3-18) may be used for this purpose. The lowest plot $f_t = 1.7f_c^{2/3}$ represents actual tensile strength under long-time or static loading. The second plot $f_t = 2.3f_c^{2/3}$ is also for static loading but takes into account the nonlinearity of concrete and is to be used with finite element analyses. The third plot $f_t = 2.6f_c^{2/3}$ is the actual tensile strength of concrete under seismic loading, and the highest plot $f_t = 3.4f_c^{3/2}$ is the apparent tensile strength under seismic loading that should be used with linear finite element analyses.

3.6.5 Behavior of Concrete under Shearing Stress

Pure shear is not encountered in concrete structures, however, an element may be subject to the simultaneous action of compressive, tensile, and shearing stresses. Therefore, the failure analysis under multiaxial stresses is carried out from a phenomenological rather than a material standpoint. Although the Coulomb-Mohr theory is not exactly applicable to concrete, the Mohr rupture diagram (Fig. 3-19) offers a way of representing the failure under combined stress states from which an estimate of the shear strength can be obtained.

In Fig. 3-19, the strength of concrete in pure shear is represented by the point at which the failure envelope intersects the vertical axis, τ_0. By this method it has been found that the shear strength is approximately 20 percent of the uniaxial compressive strength.

3.6.6 Behavior of Concrete under Biaxial and Multiaxial Stresses

Biaxial compressive stresses $\sigma_1 = \sigma_2$ can be generated by subjecting a cylindrical specimen to hydrostatic pressure in radial directions. To develop a truly biaxial stress state, the friction between the concrete cylinder and the steel plates must be avoided.

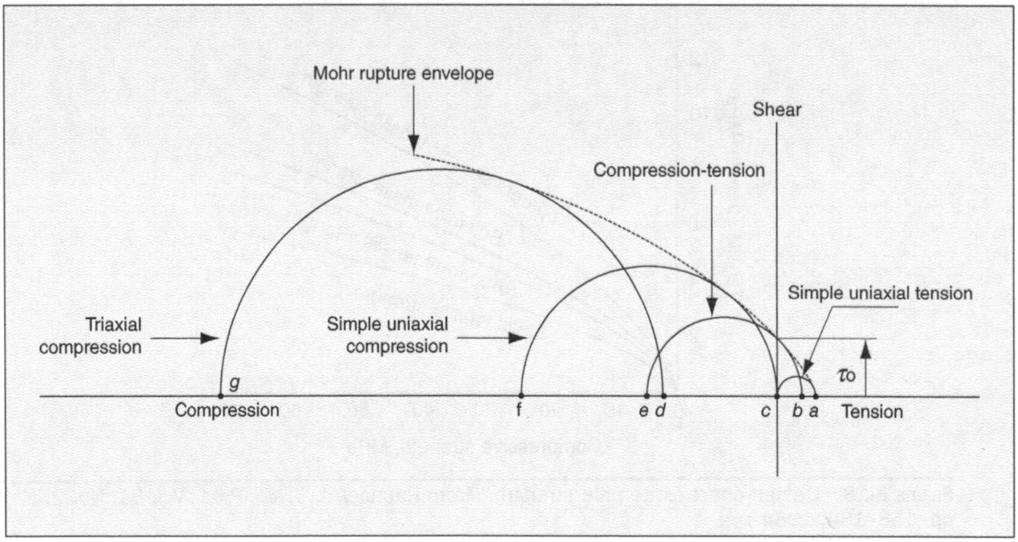

Figure 3-19 Typical Mohr rupture diagram for concrete. (From Mindess, S., and J.F. Young, *Concrete*, p. 401, 1981. Reprinted by permission of Prentice Hall, Englewood Cliffs, NJ.)

Also penetration of the pressure fluid into microcracks and pores on the surface of concrete must be prevented by placing the specimen into a suitable membrane.

Kupfer, Hilsdorf, and Rusch[10] investigated the biaxial strength of three types of concrete (18.6, 30.7, and 57.6 MPa unconfined uniaxial compressive strengths), when the specimens were loaded without longitudinal restraint by replacing the solid bearing platens of a conventional testing machine with *brush bearing platens*. These platens consisted of a series of closely spaced small steel bars that were flexible enough to follow the concrete deformations without generating appreciable restraint of the test piece. Figure 3-20 shows the typical stress-strain curves for concrete under (a) biaxial compression, (b) combined tension-compression, and (c) biaxial tension. Biaxial stress interaction curves are shown in Fig. 3-21.

The test data show that the strength of concrete subjected to biaxial compression (Fig. 3-20*a*) may be up to 27 percent higher than the uniaxial strength. For equal compressive stresses in two principal directions, the strength increase is approximately 16 percent. Under biaxial compression-tension (Fig. 3-20*b*), the compressive strength decreased almost linearly as the applied tensile strength increased. From the biaxial strength envelope of concrete (Fig. 3-21*a*) it can be seen that the strength of concrete under biaxial tension is approximately equal to the uniaxial tensile strength.

Chen[3] points out that concrete *ductility* under biaxial stresses has different values depending on whether the stress states are compressive or tensile. For instance, in biaxial compression (Fig. 3-20*a*) the average maximum compressive microstrain is about 3000 and the average maximum tensile microstrain varies from 2000 to 4000. The tensile ductility is greater in biaxial compression than in uniaxial compression. In biaxial tension-compression (Fig. 3-20*b*), the magnitude at failure of both the principal compressive and

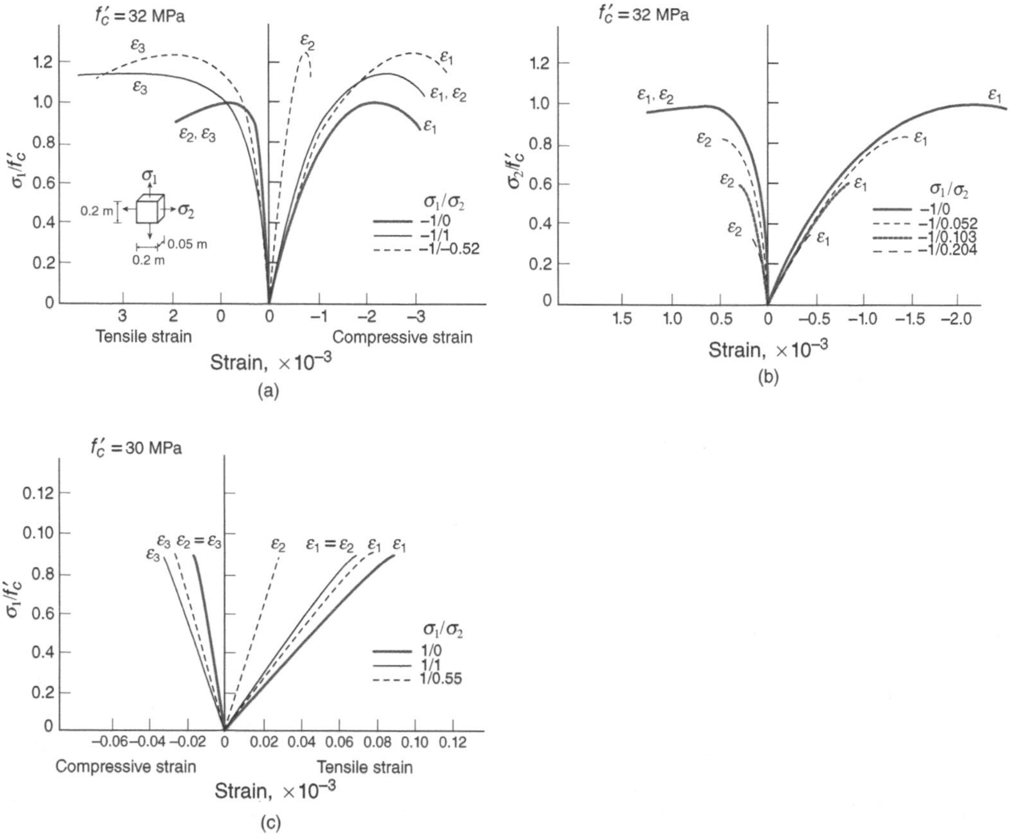

Figure 3-20 Experimental stress–strain curves for concrete under (*a*) biaxial compression, (*b*) combined tension and compression, and (c) biaxial tension. (From Kupfel, H., H.K. Hilsdorf, and H. Rush, *J. ACI, Proc.*, Vol. 66, No. 8, pp. 622–663, 1969.)

tensile strains decreases as the tensile stress increases. In biaxial tension (Fig. 3-20*c*), the average value of the maximum principal tensile microstrain is only about 80.

The data in Fig. 3-21*a* show that the level of uniaxial compressive strength of concrete virtually does not affect the shape of the biaxial stress interaction curves or the magnitude of values (the uniaxial compressive strength of concretes tested was in the range 18.6 to 57.6 MPa). However, in compression-tension and in biaxial tension (Fig. 3-21*b*), it is observed that the relative strength at any particular biaxial stress combination decreases as the level of uniaxial compressive strength increases. Neville[11] suggests that this is in accord with the general observation that the ratio of uniaxial resilient strength to compressive strength decreases as the compressive strength level rises (see Table 3-2).

The behavior of concrete under multiaxial stresses is very complex and, as was explained in Fig. 3-19, it is generally described from a phenomenological point of view. Unlike the laboratory tests for determining the behavior of concrete under uniaxial compression, splitting tension, flexure, and biaxial loading, there are no standard tests for concrete subjected to multiaxial stresses.

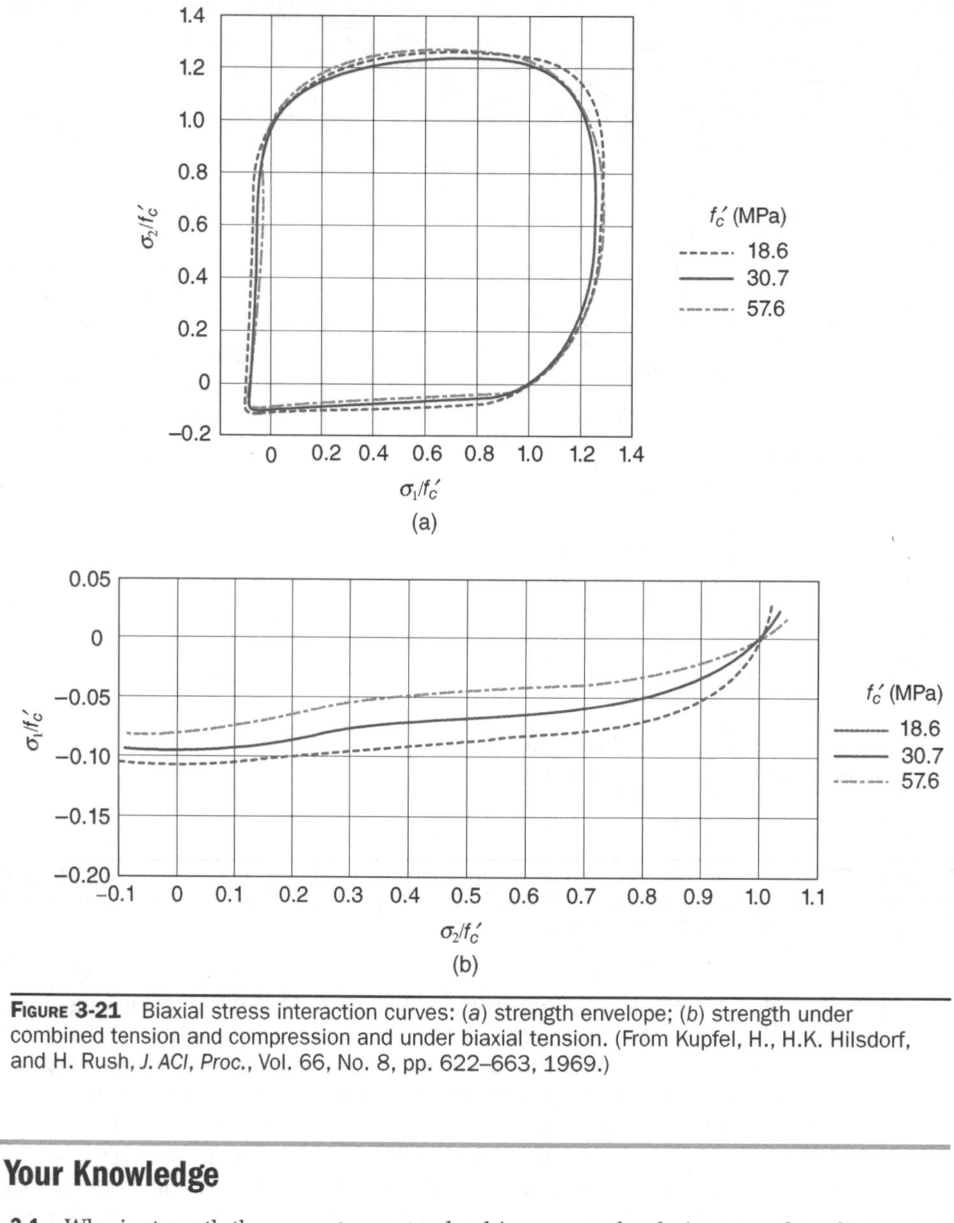

FIGURE 3-21 Biaxial stress interaction curves: (*a*) strength envelope; (*b*) strength under combined tension and compression and under biaxial tension. (From Kupfel, H., H.K. Hilsdorf, and H. Rush, *J. ACI, Proc.*, Vol. 66, No. 8, pp. 622–663, 1969.)

Test Your Knowledge

3.1 Why is strength the property most valued in concrete by designers and quality control engineers?

3.2 In general, discuss how strength and porosity are related to each other.

3.3 Abrams established a rule that relates the water-cement ratio to strength of concrete. List two additional factors that have a significant influence on the concrete strength.

3.4 Explain how water-cement ratio influences the strength of the cement paste matrix and the interfacial transition zone in concrete.

3.5 Why does air entrainment reduce the strength of moderate- and high-strength concrete mixtures but may increase the strength of low-strength concrete mixtures?

3.6 For the ASTM Types I, III, and V of portland cements, at a given water-cement ratio would the ultimate strength values be different? Would the early-age strength values be different? Explain your answer.

3.7 In regard to concrete strength, discuss the two opposing effects that are caused by an increase in the maximum size of aggregate in a concrete mixture.

3.8 At a given water-cement ratio, either a change in the cement content or aggregate grading can be made to increase the consistency of a concrete mixture. Which one of the two options would you recommend? Why is it not desirable to produce concrete mixtures of a higher consistency than necessary?

3.9 Can we use recycled water from industrial operations as mixing water in concrete? What about the use of seawater for this purpose?

3.10 What do you understand by the term *curing of concrete*? What is the significance of curing?

3.11 From the standpoint of concrete strength, which of the two options is undesirable, and why?

(*a*) Concrete cast at 5°C and cured at 21°C.
(*b*) Concrete cast at 21°C and cured at 5°C.

3.12 Many factors have an influence on the compressive strength of concrete. Briefly explain which one of the two options listed below will result in higher strength at 28 days:

(*a*) Water-cement ratio of 0.5 vs. 0.4.
(*b*) Moist curing temperature of 25°C vs. 10°C.
(*c*) Using test cylinder of size 150 by 300 mm vs. 75 by 150 mm.
(*d*) Using a compression test loading rate of 3 MPa/s vs. 0.3 MPa/s.
(*e*) Testing the specimens in a saturated condition vs. air-dry condition.

3.13 The temperature during the placement of concrete is known to have an effect on later age strength. What would be the effect on the 6-month strength when a concrete mixture is placed at (a) 10°C and (b) 35°C.

3.14 In general, how are the compressive and tensile strengths of concrete related? Is this relationship independent of concrete strength? If not, why? Discuss how admixtures and aggregate mineralogy can affect the relationship.

References

1. Powers, T.C., *J. Am. Ceram. Soc.*, Vol. 41, No. 1, pp. 1–6, 1958.
2. Cordon, W.A., and H.A. Gillispie, *J. ACI, Proc.*, Vol. 60, No. 8, pp. 1029–1050, 1963.
3. Chen, W.F., *Plasticity in Reinforced Concrete*, McGraw-Hill, New York, pp. 20–21, 1982.
4. Price, W.H., *J. ACI, Proc.*, Vol. 47, pp. 417–432, 1951.
5. Rusch, H., *J. ACI, Proc.*, Vol. 57, pp. 1–28, 1960.
6. Jones, P.G., and F.E. Richart, *ASTM Proc.*, Vol. 36, pp. 380–391, 1936.
7. Green, H., *Proceedings, Institute of Civil Engineers (London)*, Vol. 28, No. 3, pp. 383–396, 1964.
8. Ople, F.S., and C.L. Hulsbos, *J. ACI, Proc.*, Vol. 63, pp. 59–81, 1966.
9. Raphael, J., *J. ACI, Proc.*, Vol. 81, No. 2, pp. 158–164, 1984.

10. Kupfer, H., H.K. Hilsdorf, and H. Rusch, *J. ACI, Proc.*, Vol. 66, pp. 656–666, 1969.
11. Neville, A., *Hardened Concrete: Physical and Mechanical Aspects*, ACI Monograph No. 6, pp. 48–53, 1971.

Suggestions for Further Study

Brooks, A.E., and K. Newman, eds., The Structure of Concrete, *Proceedings of International Conference*, London, Cement and Concrete Association, Wesham Springs, Slough, U.K., p. 49318, 1968.

Collepardi, M., *The New Concrete*, Grafiche Tintoretto, 2006.

FIB Bulletin No. 65, *Model Code 2010 - Final draft*, Volume 1, International Federation for Structural Concrete, Lausanne, Switzerland, 2012.

Klieger, P., and J.F., Lamond, eds., *Concrete and Concrete Making Materials*, ASTM STP 169, American Society for Testing and Materials, Philadelphia, Chaps. 14 and 15, 1994.

Li, Z., *Advanced Concrete Technology*, Wiley, Hoboken, 2011.

Neville, A.M., *Properties of Concrete*, 5th edition, Pearson, Harlow, 2011.

Newman, J., and B.S. Choo, eds., *Advanced Concrete Technology: Concrete Properties*, Oxford, England; Burlington, MA: Butterworth-Heinemann, 2003.

Popovics, S., *Strength and Related Properties of Concrete: A Quantitative Approach*, New York: Wiley, 535 p., Vol. 92, No. 1, 1998.

Ross, C.A., J.W. Tedesco, and S.T. Kuennen, Effects of strain rate on concrete strength, *ACI Materials J.*, pp. 37–47, 1995.

CHAPTER 4

Dimensional Stability

Preview

Concrete shows elastic as well as inelastic strains on loading, and shrinkage strains on drying or cooling. When restrained, shrinkage strains result in complex stress patterns that often lead to cracking.

In this chapter, causes of nonlinearity in the stress-strain relation of concrete are discussed, and different types of elastic moduli and the methods of determining them are described. Explanations are provided as to why and how the aggregate, the cement paste, the interfacial transition zone, and the testing parameters affect the modulus of elasticity.

The stress effects resulting from the drying shrinkage and the viscoelastic strains in concrete are not the same; however, with both phenomena the underlying causes and the controlling factors have much in common. Important parameters that influence the drying shrinkage and creep are discussed, such as aggregate content, stiffness, water content, cement content, time of exposure, relative humidity, and size and shape of the concrete member.

Thermal shrinkage is of great importance in massive concrete elements. Its magnitude can be controlled by controlling the coefficient of thermal expansion of aggregate, cement content and type, and temperature of concrete-making materials. The concepts of extensibility, tensile strain capacity, and their significance to concrete cracking are also discussed.

4.1 Types of Deformations and Their Significance

Deformations in concrete, which often lead to cracking, occur as a result of the material's response to external load and environment. When freshly hardened concrete (whether loaded or unloaded) is exposed to the ambient temperature and humidity, it generally undergoes *thermal shrinkage* (shrinkage strain associated with cooling)* and *drying shrinkage* (shrinkage strain associated with the moisture loss). Which one of the two shrinkage strains will be dominant under a given condition depends, among other factors, on the size of the member, characteristics of concrete-making materials, and mix proportions. Generally, with massive structures (e.g., nearly 1 m or more in thickness), the drying shrinkage is less important a factor than the thermal shrinkage.

---

*Exothermic reactions between cement compounds and water tend to raise the temperature of concrete (see Chap. 6).

It should be noted that concrete members are almost always under restraint, sometimes from subgrade friction and end members, but usually from reinforcing steel and from differential strains that develop between the exterior and the interior of concrete. When the shrinkage strain in an elastic material is fully restrained, it results in elastic tensile stress; the magnitude of the induced stress σ is determined by the product of the strain ε and the elastic modulus E of the material ($\sigma = E\varepsilon$). The elastic modulus of concrete is also dependent on the characteristics of concrete-making materials and mix proportions, but not necessarily to the same degree as the shrinkage strains. The material is expected to crack when a combination of the elastic modulus and the shrinkage strain induces a stress level that exceeds its tensile strength (Fig. 4-1). Given the low tensile strength of concrete, this does happen in practice but, fortunately, the magnitude of the stress is not as high as predicted by the elastic model.

To understand the reason why a concrete element may not crack at all or may crack but not soon after exposure to the environment, we have to consider how concrete would respond to sustained stress or to sustained strain. The phenomenon of a gradual increase in strain with time under a given level of sustained stress is called *creep*. The phenomenon of gradual decrease in stress with time under a given level of sustained strain is called *stress relaxation*. Both manifestations are typical of viscoelastic materials. When a concrete element is restrained, the viscoelasticity of concrete will manifest into a progressive decrease of stress with time (Fig. 4-1, curve b). Thus, under the restraining conditions present in concrete, the interplay between the elastic tensile stresses induced by shrinkage strains and the stress relief due to viscoelastic behavior is at the heart of deformations and cracking in most structures.

In practice, the stress-strain relations in concrete are much more complex than indicated by Fig. 4-1. First, concrete is not a truly elastic material; second, neither the strains

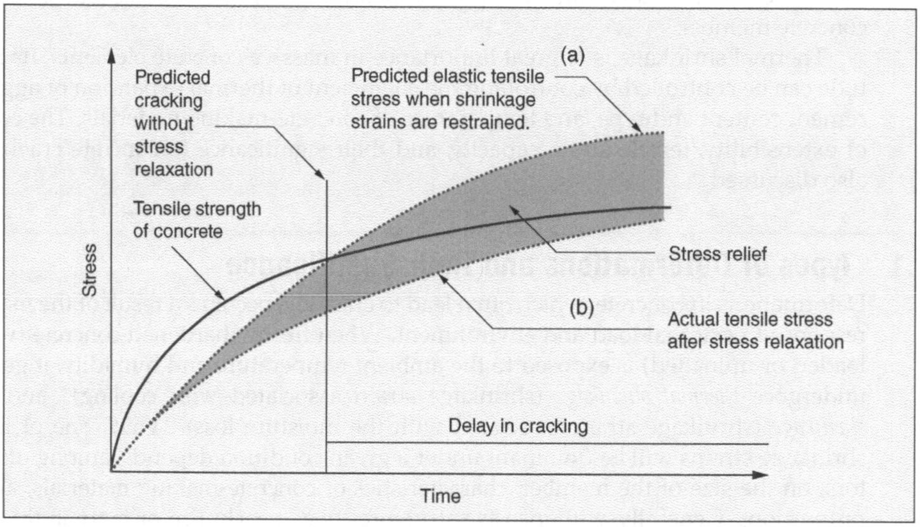

FIGURE 4-1 Influence of shrinkage and creep on concrete cracking. (Troxell, G.E., H.E. Davis, and J.W. Kelly, *Composition and Properties of Concrete*, McGraw-Hill, New York, p. 342, 1968.)

Under restraining conditions in concrete, the interplay between the elastic tensile stresses induced by shrinkage strains and the stress relief due to the viscoelastic behavior is at the heart of deformations and cracking in most structures.

nor the restraints are uniform throughout a concrete member; therefore, the resulting stress distributions tend to vary from point to point. Nevertheless, it is important to know the elastic, drying shrinkage, thermal shrinkage, and viscoelastic properties of concrete and the factors affecting them.

4.2 Elastic Behavior

The elastic characteristics of a material are a measure of its stiffness. In spite of the nonlinear behavior of concrete, an estimate of the elastic modulus (the ratio between the applied stress and instantaneous strain within an assumed proportional limit) is necessary for determining the stresses induced by strains associated with environmental effects. It is also needed for computing the design stresses under load in simple elements, and moments and deflections in complicated structures.

4.2.1 Nonlinearity of the Stress-Strain Relationship

From typical $\sigma - \varepsilon$ curves for aggregate, hardened cement paste, and concrete loaded in uniaxial compression (Fig. 4-2), it becomes immediately apparent that unlike the aggregate and the cement paste, concrete is not an elastic material. Neither is the strain on instantaneous loading of a concrete specimen found to be directly proportional to the applied stress, nor is it fully recovered upon unloading. The cause for nonlinearity of the stress-strain relationship is explained from studies on progressive microcracking of concrete under load by researchers, from the Cornell University[1] (Fig. 4-3 and a review of their work by Glucklich[2]).

In regard to the relationship between stress level (expressed as percent of the ultimate load) and microcracking in concrete, Fig. 4-3 shows that concrete behavior can be divided into four distinct stages. Under normal atmospheric exposure conditions (when

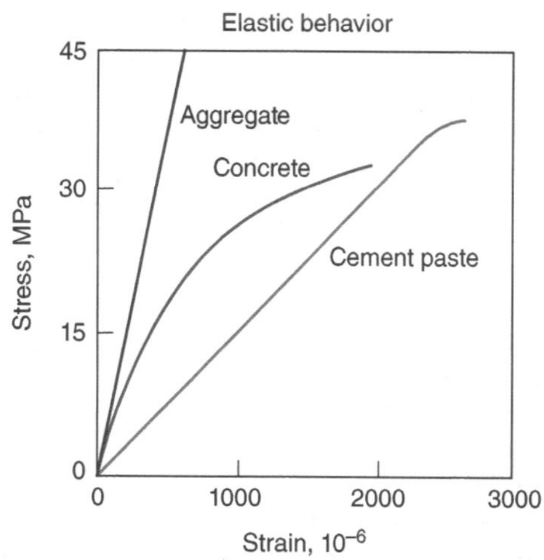

FIGURE 4-2 Typical stress-strain behaviors of cement paste, aggregate, and concrete. (Based on Hsu, T.C., ACI Monograph 6, p. 100, 1971.)

The properties of complex composite materials need not to be equal to the sum of the properties of their components. Thus, both hydrated cement paste and aggregates show linear elastic properties, whereas concrete does not.

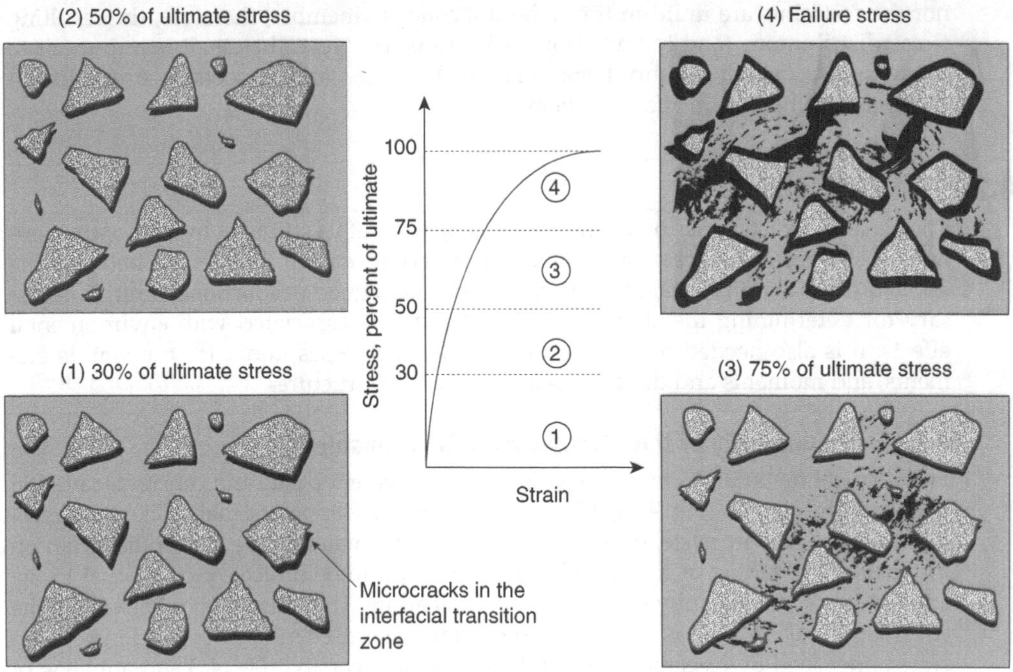

Figure 4-3 Diagrammatic representation of the stress-strain behavior of concrete under uniaxial compression. (Based on Glucklich, J., *Proceedings of International Conference on the Structure of Concrete*, Cement and Concrete Association, Wexham Springs, Slough, U.K., pp. 176–185, 1968.)

The progress of internal microcracking in concrete goes through various stages, which depend on the level of applied stress.

a concrete element is subjected to drying or thermal shrinkage effects) due to the differences in their elastic moduli differential strains are set up between the matrix and the coarse aggregate, causing cracks in the interfacial transition zone. Therefore, even before the application an external load, microcracks already exist in the interfacial transition zone between the matrix mortar and coarse aggregate. The number and width of these cracks in a concrete specimen depend, among other factors, on the bleeding characteristics, and the curing history of concrete. Below about 30 percent of the ultimate load, the interfacial transition zone cracks remain stable; therefore, the $\sigma-\varepsilon$ curve remains linear. This is Stage 1 in Fig. 4-3.

Above 30 percent of the ultimate load, with increasing stress, the interfacial transition zone microcracks begin to increase in length, width, and number. Thus, the ε/σ ratio increases and the curve begins to deviate appreciably from a straight line. However, until about 50 percent of the ultimate stress, *a stable system of microcracks* appears to exist in the interfacial transition zone. This is Stage 2 and at this stage the matrix cracking is negligible. At 50 to 60 percent of the ultimate load, cracks begin to form in the matrix. With further increase in stress level up to about 75 percent of the ultimate load, not only does the crack system in the interfacial transition zone becomes *unstable* but also the proliferation and propagation of cracks in the matrix increases, causing the $\sigma-\varepsilon$ curve to bend considerably toward the horizontal. This is Stage 3. At 75 to 80 percent of the ultimate load, the rate of strain energy release seems to reach the critical level necessary for spontaneous crack growth under sustained stress, and the material strains to failure. In short, above 75 percent of the ultimate load, with increasing stress very high strains are developed, indicating that the crack system is becoming continuous due to the rapid propagation of cracks in both the matrix and the interfacial transition zone. This is the final stage (Stage 4).

4.2.2 Types of Elastic Moduli

The *static modulus of elasticity* for a material under tension or compression is given by the slope of the $\sigma - \varepsilon$ curve for concrete under uniaxial loading. Since the curve for concrete is nonlinear, three methods for computing the modulus are used. This has given rise to the three types of elastic moduli, as illustrated by Fig. 4-4:

1. The *tangent modulus* is given by the slope of a line drawn tangent to the stress-strain curve at any point on the curve.
2. The *secant modulus* is given by the slope of a line drawn from the origin to a point on the curve corresponding to a 40 percent stress of the failure load.
3. The *chord modulus* is given by the slope of a line drawn between two points on the stress-strain curve. Compared to the secant modulus, instead of the origin the line is drawn from a point representing a longitudinal strain of 50 μm/m to the point that corresponds to 40 percent of the ultimate load. Shifting the base line by 50 microstrain is recommended to correct for the slight concavity that is often observed at the beginning of the stress-strain curve.

The *dynamic modulus of elasticity*, corresponding to a very small instantaneous strain, is approximately given by the *initial tangent modulus*, which is the tangent modulus for a line drawn at the origin. It is generally 20, 30, and 40 percent higher than the static modulus of elasticity for high-, medium-, and low-strength concretes, respectively. For stress analysis of structures subjected to earthquake or impact loading it is more appropriate to

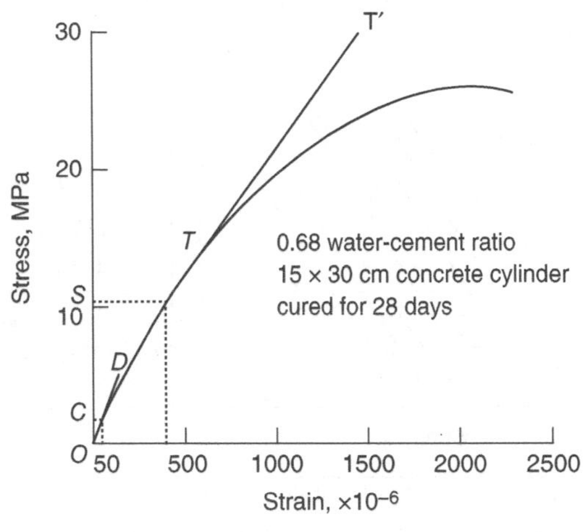

Calculating the Elastic Moduli

f_t = 26 MPa
40% f_t = 10.4 MPa = SO

Secant Modulus: Slope of the line corresponding to stress SO
$10.4/(417 \times 10^{-6}) = 24.9$ GPa

Chord Modulus: Slope of the line corresponding to stress SC $(10.4 - 1.6)/(417 \times 10^{-6} - 50 \times 10^{-6}) = 24.0$ GPa

Tangent Modulus: Slope of the line TT′ drawn tangent to any point on the $\sigma - \varepsilon$ curve
$(30 - 14.6)/(1445 \times 10^{-6} - 625 \times 10^{-6}) =$ 18.8 GPa

Dynamic Modulus (Initial Tangent Modulus): Slope of the line OD from the origin $5/143 \times 10^{-6} = 34.9$ GPa

Figure 4-4 Different types of elastic moduli and the method by which these are determined.

use the dynamic modulus of elasticity, which can be determined more accurately by a sonic test.

The *flexural modulus of elasticity* may be determined from the deflection test on a loaded beam. For a beam simply supported at the ends and loaded at midspan, ignoring the shear deflection, the approximate value of the modulus is calculated from:

$$E = \frac{PL^3}{48I\Delta}$$

where Δ = midspan deflection due to load P
L = span length
I = moment of inertia

The flexural modulus is commonly used for design and analysis of pavements.

4.2.3 Determination of the Static Elastic Modulus

ASTM C 469 describes a standard test method for measurement of the static modulus of elasticity (the chord modulus) and Poisson's ratio of 150 by 300 mm concrete cylinders loaded in longitudinal compression at a constant loading rate within the range 0.24 ± 0.03 MPa/s. Normally, the deformations are measured by a linear variable differential transformer. Typical $\sigma - \varepsilon$ curves, with sample computations for the secant elastic moduli of the three concrete mixtures of Fig. 3-17, are shown in Fig. 4-5.

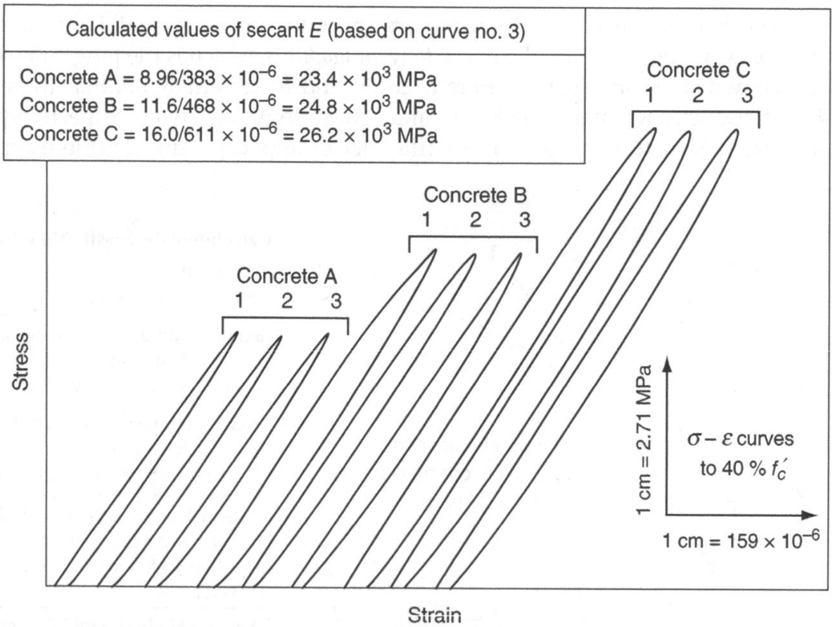

FIGURE 4-5 Determination of the secant modulus in the laboratory (ASTM C 469).
See Fig. 3-17 for the composition and strength characteristics of concrete mixtures. (*Unpublished data from student experiments, University of California at Berkeley.*)

The elastic modulus values used in concrete design computations are usually estimated from empirical expressions that assume direct dependence of the elastic modulus on the strength and density of concrete. As a first approximation this makes sense because the stress-strain behavior of the three components of concrete, namely the aggregate, the cement paste matrix, and the interfacial transition zone, would indeed be determined by their individual strengths, which in turn are related to the ultimate strength of the concrete. Furthermore, it may be noted that the elastic modulus of the aggregate (which controls the aggregate's ability to restrain volume changes in the matrix) is directly related to its porosity, and the measurement of the unit weight of concrete happens to be the easiest way of obtaining an estimate of the aggregate porosity.

According to ACI Building Code 318, with a concrete unit weight between 1500 and 2500 kg/m^3, the modulus of elasticity can be determined from

$$E_c = w_c^{1.5} \times 0.043 f_c'^{1/2}$$

where E_c = static modulus of elasticity (MPa)
w_c = unit weight (kg/m^3)
f_c' = 28-day compressive strength of standard cylinders (MPa)

In the FIP Model Code (2010), the modulus of elasticity of normal-weight concrete can be estimated from

$$E_c = 2.15 \times 10^4 (f_{cm}/10)^{1/3}$$

where E_c is the 28-day modulus of elasticity of concrete (MPa) and f_{cm} the average 28-day compressive strength. If the actual compressive strength is not known, f_{cm} should be replaced by $f_{ck} + 8$, where f_{ck} is the characteristic compressive strength. The elastic modulus-strength relationship was developed for quartzitic aggregate concrete. For other types of aggregates, the modulus of elasticity can be obtained by multiplying E_c with factors α_e from Table 4-1. It should be mentioned that the FIP expression is valid for characteristic strengths up to 120 MPa, whereas the ACI equation is valid up to 41 MPa only. Extensions to the ACI formulation are presented in Chap. 12 (see high-strength concrete). Assuming concrete density to be 2320 kg/m^3, the computed values of the modulus of elasticity for normal-weight concrete according to both the ACI Building Code and FIP Model Code (2010) are shown in Table 4-2.

From the following discussion of the factors affecting the modulus of elasticity of concrete, it will be apparent that the computed values shown in Table 4-2, which are based on strength and density of concrete, should be treated as approximate only. This

Aggregate Type	α_e
Basalt, dense limestone	1.2
Quartzitic	1.0
Limestone	0.9
Sandstone	0.7

TABLE 4-1 Effect of Type of Aggregate on Modulus of Elasticity

ACI Building Code		FIP Model Code	
Compressive Strength	Modulus of Elasticity	Compressive Strength	Modulus of Elasticity
psi (MPa)	$\times 10^6$ psi (GPa)	psi (MPa)	$\times 10^6$ psi (GPa)
3000 (21)	3.1 (21)	3000 (21)	4.0 (28)
4000 (27)	3.6 (25)	4000 (27)	4.3 (30)
5000 (34)	4.1 (28)	5000 (34)	4.7 (32)
6000 (41)	4.4 (30)	6000 (41)	5.0 (34)

Table 4-2 Elastic Moduli for Normal-Weight Concretes (Quartzitic Aggregate)

is because the transition-zone characteristics and the moisture state of the specimen at the time of testing do not have a similar effect on the strength and elastic modulus.

4.2.4 Poisson's Ratio

For a material subjected to simple axial load, the ratio of the lateral strain to axial strain *within the elastic range* is called *Poisson's ratio*. Poisson's ratio is not generally needed for most concrete design computations; however, it is needed for structural analysis of tunnels, arch dams, and other statically indeterminate structures.

With concrete the values of Poisson's ratio generally vary between 0.15 and 0.20. There appears to be no consistent relationship between Poisson's ratio and concrete characteristics such as water-cement ratio, curing age, and aggregate gradation. However, Poisson's ratio is generally lower in high-strength concrete, and higher for saturated concrete and for dynamically loaded concrete.

4.2.5 Factors Affecting Modulus of Elasticity

In homogeneous materials, a direct relationship exists between density and modulus of elasticity. In heterogeneous, multiphase materials such as concrete, the volume fraction, the density and the modulus of elasticity of the principal constituents, and the characteristics of the interfacial transition zone, determine the elastic behavior of the composite. Since density is oppositely related to porosity, obviously the factors that affect the porosity of aggregate, cement paste matrix, and the interfacial transition zone would be important. For concrete, the direct relation between strength and elastic modulus arises from the fact that both are affected by the porosity of the constituent phases, although not to the same degree.

Aggregate

Among the coarse aggregate characteristics that affect the elastic modulus of concrete, porosity seems to be the most important. This is because aggregate porosity determines its stiffness, which in turn controls the ability of aggregate to restrain the matrix strain. Dense aggregates have a high elastic modulus. In general, the larger the amount of coarse aggregate with a high elastic modulus in a concrete mixture, the greater would be the modulus of elasticity of concrete. Because with low- or medium-strength concrete, the strength is not affected by normal variations in the aggregate porosity, this shows that all variables may not control the strength and the elastic modulus in the same way.

Rock core tests have shown that the elastic modulus of natural aggregates of low porosity such as granite, trap rock, and basalt is in the range 70 to 140 GPa (10 to 20×10^6 psi), while with sandstones, limestones, and gravels of the porous variety it varies from 21 to 49 GPa (3 to 7×10^6 psi). Lightweight aggregates are highly porous; depending on the porosity, the elastic modulus of a lightweight aggregate may be as low as 7 GPa (1×10^6 psi) or as high as 28 GPa (4×10^6 psi). Generally, the elastic modulus of lightweight-aggregate concrete ranges from 14 to 21 GPa (2.0 to 3.0×10^6 psi), which is between 50 and 75 percent of the modulus for normal-weight concrete of the same strength.

Other properties of aggregate also influence the modulus of elasticity of concrete. For example, aggregate size, shape, surface texture, grading, and mineralogical composition can influence the microcracking in the interfacial transition zone and thus affect the shape of the stress-strain curve.

Cement Paste Matrix

The elastic modulus of the cement paste matrix is determined by its porosity. The factors controlling the porosity of the cement paste matrix, such as water-cement ratio, air content, mineral admixtures, and degree of cement hydration, are listed in Fig. 3-12. Values in the range 7 to 28 GPa (1 to 4×10^6 psi) as the elastic moduli of hydrated portland cement pastes of varying porosity have been reported. It should be noted that these values are similar to the elastic moduli of lightweight aggregates.

Transition Zone

In general, capillary voids, microcracks, and oriented calcium hydroxide crystals are relatively more common in the interfacial transition zone than in the bulk matrix; therefore, they play an important part in determining the stress-strain relations in concrete. The factors controlling the porosity of the interfacial transition zone are listed in Fig. 3-12.

It has been reported that the strength and elastic modulus of concrete are not influenced to the same degree by curing age. With different concrete mixtures of varying strength, it was found that at later ages (i.e., 3 months to 1 year) the elastic modulus increased at a higher rate than the compressive strength (Fig. 4-6). It is possible that the beneficial effect of

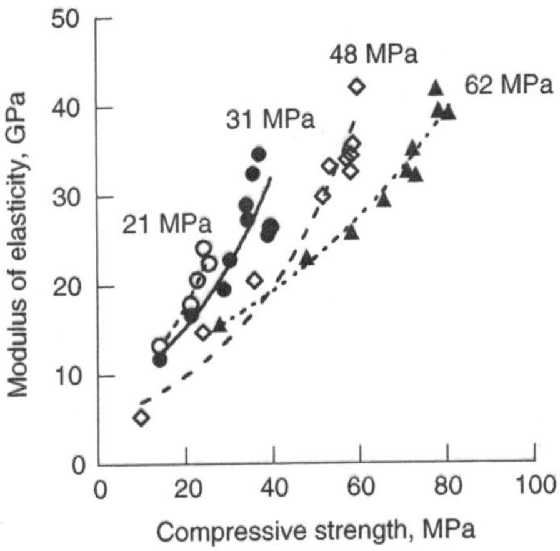

FIGURE 4-6 Relationship between the compressive strength and elastic modulus. (Based on Shideler, J.J., *J. ACI*, *Proc.*, Vol. 54, No. 4, 1957.)

The upward tendency of the $E - f'_c$ curves from different-strength concrete mixtures tested at regular intervals up to 1 year shows that, at later ages, the elastic modulus increases at a faster rate than the compressive strength.

improvement in the density of the interfacial transition zone, as a result of slow chemical interaction between the alkaline cement paste and aggregate, is more pronounced for the stress-strain relationship than for the compressive strength of concrete.

Testing Parameters

It is observed that regardless of mix proportions or curing age, concrete specimens that are tested in wet conditions show about 15 percent higher elastic modulus than the corresponding specimens tested in a dry condition. Interestingly, the compressive strength of the specimen behaves in the opposite manner; that is, the strength is higher by about 15 percent when the specimens are tested in dry condition. It seems that drying of concrete produces a different effect on the cement paste matrix than on the interfacial transition zone; while the former gains in strength owing to an increase in the van der Waals force of attraction in the hydration products, the latter loses strength due to microcracking. The compressive strength of the concrete increases when the matrix is strength-determining; however, the elastic modulus is reduced because increases in the transition-zone microcracking greatly affects the stress-strain behavior. There is yet another explanation for the phenomenon. In a saturated cement paste the adsorbed water in the C-S-H is load-bearing, therefore its presence contributes to the elastic modulus; on the other hand, the disjoining pressure in the C-S-H (see Chap. 2) tends to reduce the van der Waals force of attraction, thus lowering the strength.

The advent and degree of nonlinearity in the stress-strain curve obviously would depend on the rate of application of load. At a given stress level the rate of crack propagation, and hence the modulus of elasticity, is dependent on the rate at which load is applied. Under instantaneous loading, only a little strain can occur prior to failure, and the elastic modulus is very high. In the time range normally required to test the specimens (2 to 5 min), the strain is increased by 15 to 20 percent, hence the elastic modulus decreases correspondingly. For very slow loading rates, the elastic and the creep strains would be superimposed, thus lowering the elastic modulus further.

Figure 4-7 presents a summary showing all the factors discussed above, which affect the modulus of elasticity of concrete.

4.3 Drying Shrinkage and Creep

For a variety of reasons it is desirable to discuss the drying shrinkage and the viscoelastic phenomena (creep and stress relaxation) together. First, both the drying shrinkage and creep originate from the same source, that is, the hydrated cement paste; second, the strain-time curves are very similar; third, the factors that influence the drying shrinkage also influence the creep generally in the same way; fourth, in concrete the microstrain of each phenomenon, 400 to 1000×10^{-6}, is large and it cannot be ignored in structural design; and fifth, both drying shrinkage and creep are partially reversible.

4.3.1 Causes

As described in Chap. 2, a saturated cement paste will not remain dimensionally stable when exposed to ambient humidities that are below saturation, mainly because the loss of physically adsorbed water from C-S-H results in a shrinkage strain. Similarly, when a hydrated cement paste is subjected to a sustained stress, depending on the magnitude and duration of applied stress, the C-S-H will lose a large amount of the physically adsorbed water, and the paste will show a creep strain. This is not to suggest that there

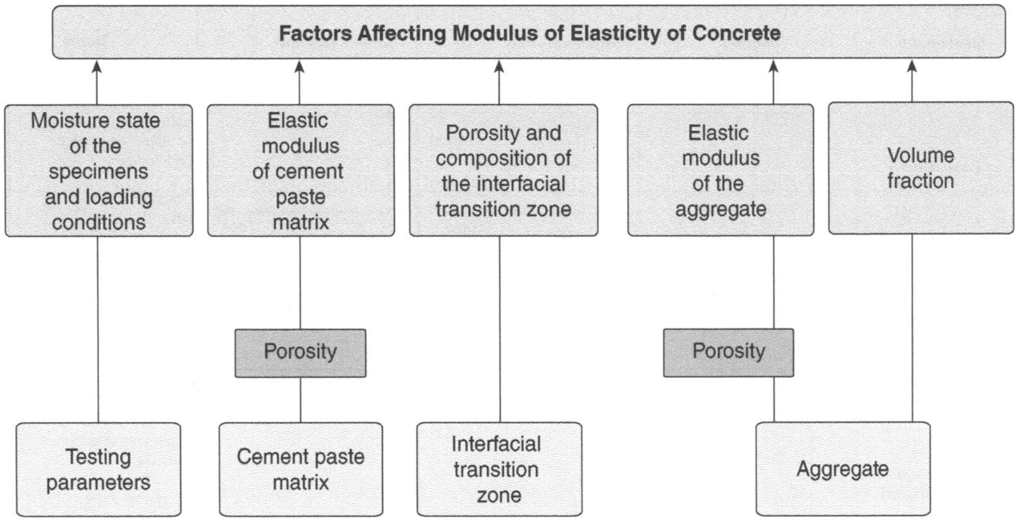

FIGURE 4-7 Various parameters that influence the modulus of elasticity of concrete.

are no other causes contributing to creep in concrete; however, the loss of adsorbed water under sustained pressure appears to be the most important cause. In short, both the drying shrinkage and creep strains in concrete are assumed to be related mainly to the *removal of adsorbed water* from the hydrated cement paste. The difference is that in one case the differential relative humidity between concrete and the environment is the driving force, while in the other it is the sustained applied stress. Again, as stated in Chap. 2, a minor cause of the contraction of the system, either as a result of drying or applied stress is the removal of water held by hydrostatic tension in small capillaries (<50 nm) of the hydrated cement paste.

The causes of creep in concrete are more complex. It is generally agreed that in addition to moisture movements there are other causes that contribute to the creep phenomenon. The nonlinearity of the stress-strain relation in concrete, especially at stress levels greater than 30 to 40 percent of the ultimate stress, clearly shows the contribution of the interfacial transition zone microcracks to creep. Increase in creep strain, which invariably occurs when concrete is simultaneously exposed to the drying condition, is caused by additional microcracking in the interfacial transition zone owing to drying shrinkage.

The occurrence of delayed elastic response in aggregate is yet another cause of creep in concrete. Since the cement paste and the aggregate are bonded together, the stress on the former gradually declines as load is transferred to the latter, which with increasing load transfer deforms elastically. Thus the delayed elastic strain in aggregate contributes to total creep.

4.3.2 Effect of Loading and Humidity Conditions on Drying Shrinkage and Viscoelastic Behavior

In practice, drying shrinkage and viscoelastic phenomena usually take place simultaneously. Consider the various combinations of loading, restraining, and humidity conditions presented in Table 4-3. Application of a constant stress on a concrete specimen

Mechanism	Diagram	Strain vs. Time	Stress vs. Time	Notes
Basic Creep	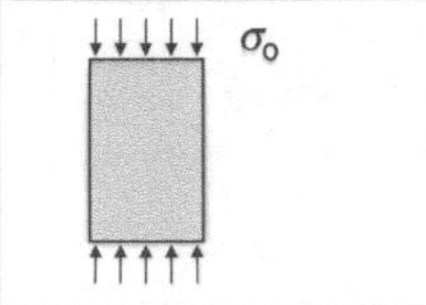	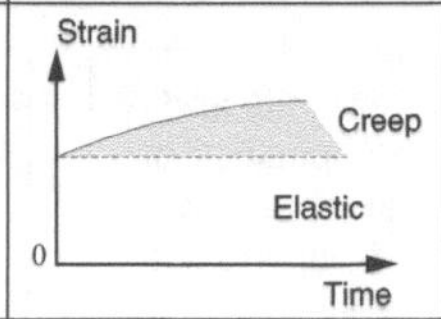	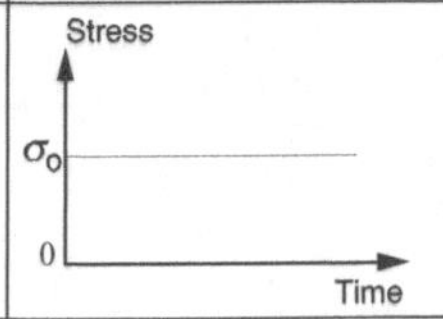	No moisture movement between concrete and ambient (no drying shrinkage) Constant stress over time
Stress Relaxation	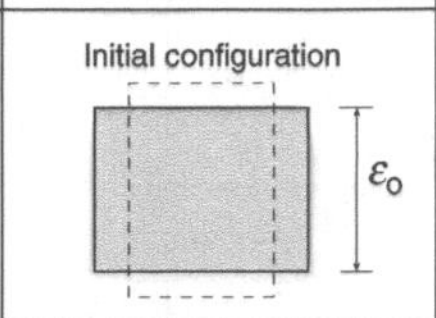	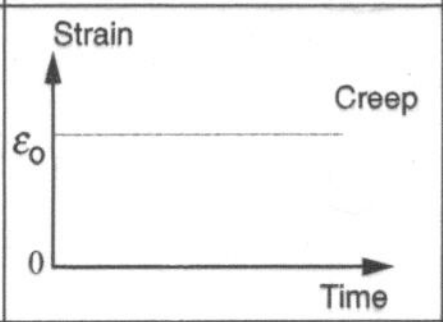	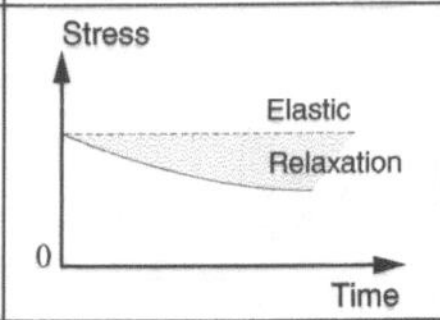	Constant strain over time
Drying Shrinkage (Unrestrained)	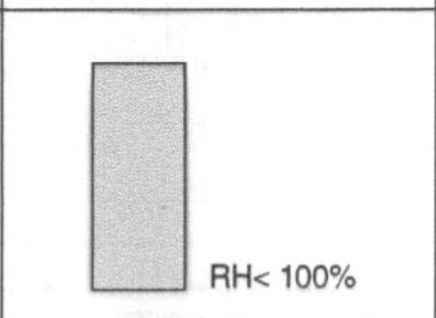	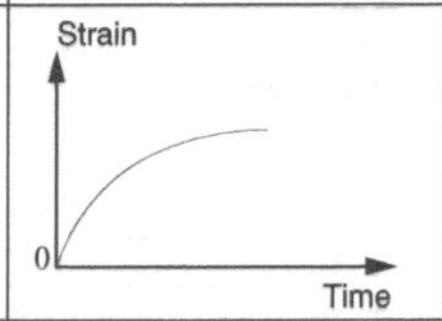	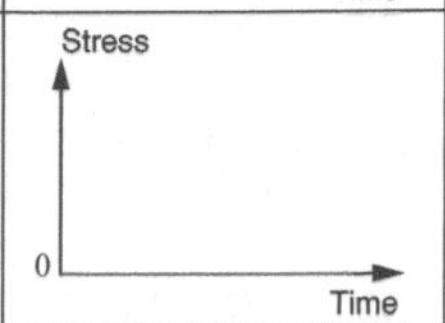	The member is free to move. No stresses are generated.
Drying Shrinkage (Restrained)	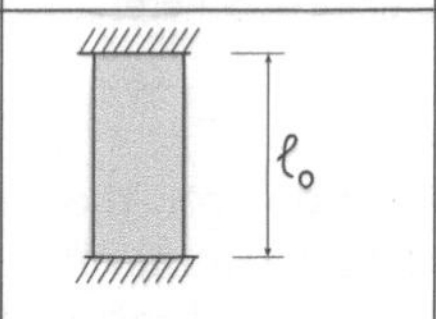	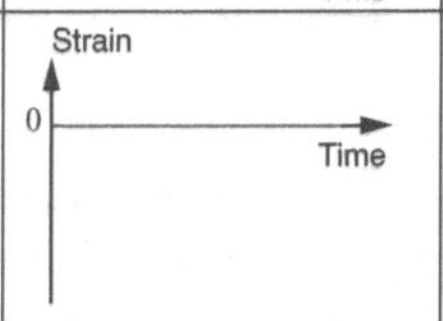	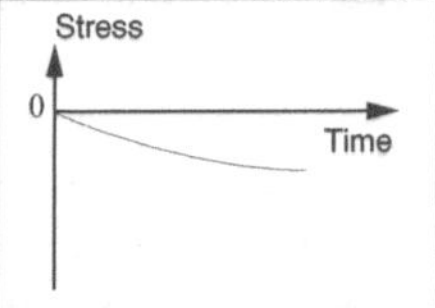	Development of tensile stress
Drying Shrinkage (Under constant strain)	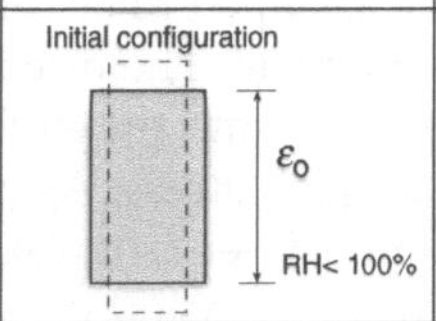	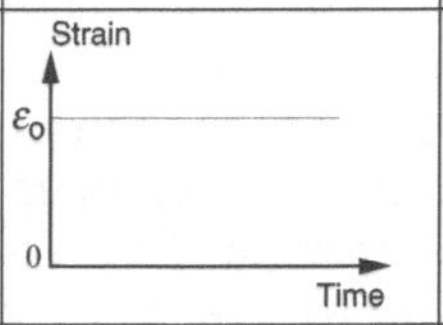	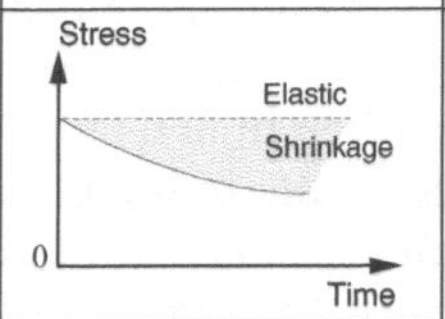	The previous example is a particular case with $\varepsilon = 0$.
Creep + Drying Shrinkage	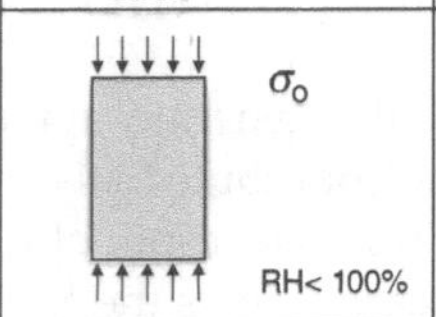	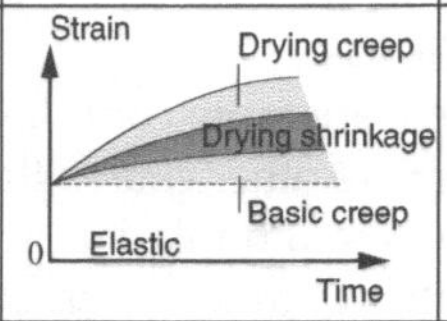	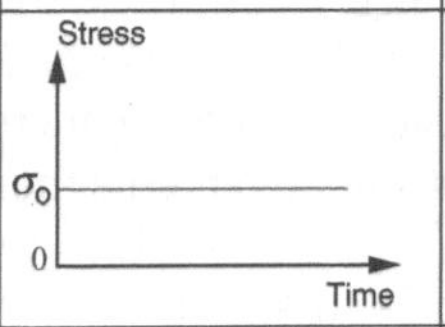	The total strain is not the sum of the elastic, basic creep, and drying shrinkage strain. The strain due to drying creep should be included.
Drying Shrinkage + Stress Relaxation (Restrained)	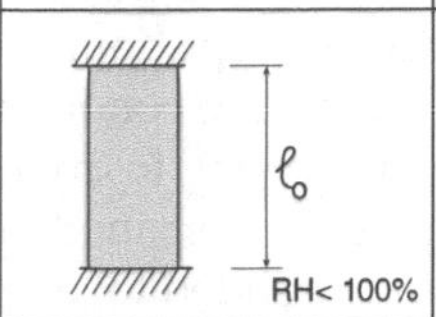	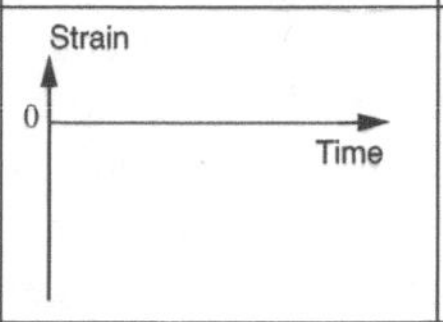	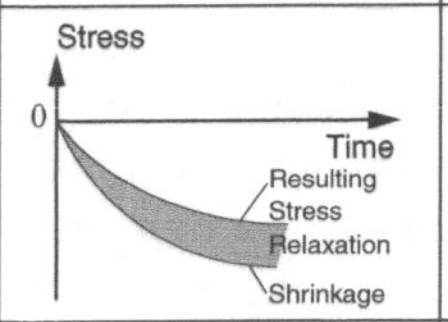	The relaxation stress opposed the stress due to drying shrinkage.
Drying Shrinkage + Stress Relaxation (Under constant strain)	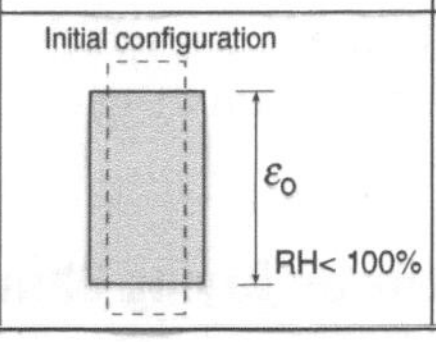	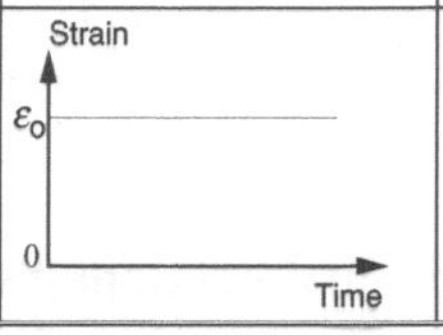	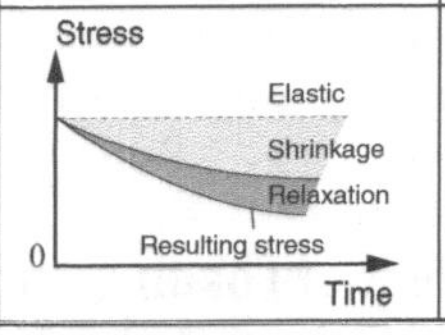	Shrinkage and relaxation stress act in the same direction.

TABLE 4-3 Combination of Loading, Restraining and Humidity Conditions

under conditions of 100 percent relative humidity leads to an increase of strain over time, this is called *basic creep*. This condition often arises in massive concrete structures where drying shrinkage can be neglected. Now, instead of applying a constant stress let us analyze the case where a constant strain is imposed on the concrete specimen. When the strain is applied, the concrete specimen will have an instantaneous elastic stress; however, the stress will decrease over time by the phenomenon of stress relaxation. Both creep and stress relaxation can be visualized as resulting from the application of stress to a classical spring and dashpot model (springs and dashpots connected either in series or in parallel are discussed in Chap. 13).

Exposure of an unrestrained concrete specimen to low relative humidity conditions causes drying shrinkage, which increases over time. However, if the specimen is restrained, that is, if it is not free to move, the strain will be zero but tensile stresses will develop over time. This is the reason for cracking due to the drying shrinkage.

It has been observed that when a concrete is under load and is simultaneously exposed to low relative humidity environment, the total strain is higher than the sum of elastic strain, free shrinkage strain (drying shrinkage strain of unloaded concrete), and basic creep strain (without drying). The additional creep that occurs when the specimen under load is also drying is called *drying creep*. Total creep is the sum of basic and drying creep; however, it is a common practice to ignore the distinction between the basic and the drying creep, and creep is simply considered as the deformation under load in excess of the sum of the elastic strain and free drying shrinkage strain.

The interaction between the restrained drying shrinkage strain and stress relaxation due to the viscoelastic behavior of concrete was illustrated in Fig. 4-1, and is also shown in Table 4-3. Because of the boundary conditions, the strain is zero and the magnitude of tensile stresses caused by drying shrinkage is reduced by stress relaxation. Note that presentation of creep data can be done in different ways, which have given rise to special terminology. For instance, *specific creep* is the creep strain per unit of applied stress and *creep coefficient* is the ratio of creep strain to elastic strain.

4.3.3 Reversibility

Typical behavior of concrete on drying and rewetting or on loading and unloading is shown in Fig. 4-8. Both the drying shrinkage and the creep phenomena in concrete exhibit a degree of irreversibility that has practical significance. Figure 4-8 shows that after the first drying, concrete did not return to the original dimension on rewetting. Drying shrinkage has therefore been categorized into *reversible shrinkage* (which is the part of total shrinkage that is reproducible on wet-dry cycles) and *irreversible shrinkage* (which is the part of total shrinkage on first drying that cannot be reproduced on subsequent wet-dry cycles). The irreversible drying shrinkage is probably due to development of chemical bonds within the C-S-H structure as a consequence of drying. The improvement in the dimensional stability of concrete as a result of first drying has been used to advantage in the manufacture of precast concrete products.

The creep curve for a plain concrete specimen subjected to sustained uniaxial compression for 90 days and thereafter unloaded is shown in Fig. 4-8*b*. When the specimen is unloaded the instantaneous or elastic recovery is approximately of the same order as the elastic strain on first application of the load. The instantaneous recovery is followed by a gradual decrease in strain called *creep recovery*. Although the creep recovery occurs more rapidly than creep, the reversal of creep strain is not total. Similar to the drying

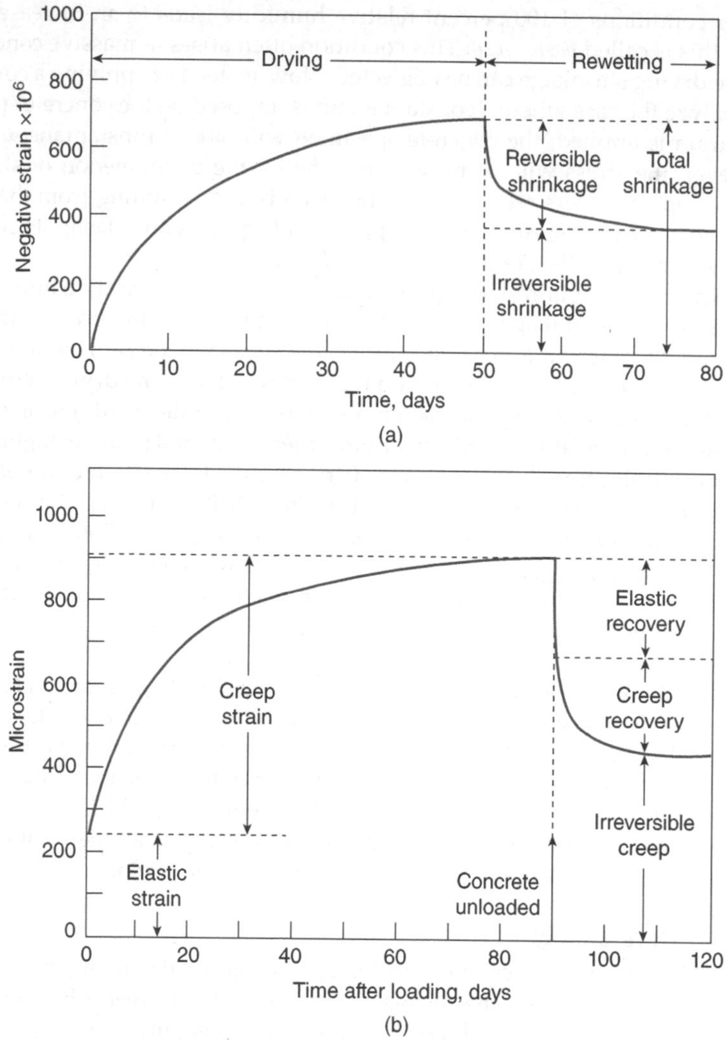

FIGURE 4-8 Reversibility of drying shrinkage and creep. (From Mindess, S., and J.F. Young, *Concrete*, 1981, pp. 486, 501. Reprinted by permission of Prentice-Hall, Englewood Cliffs, NJ.)

There is a remarkable similarity of concrete behavior on: (*a*) drying and rewetting and (*b*) loading and unloading (uniaxial compression).

shrinkage (Fig. 4-8*a*), this phenomenon is defined by the corresponding terms, *reversible* and *irreversible creep*. A part of the reversible creep may be attributed to the delayed elastic strain in aggregate, which is fully recoverable.

4.3.4 Factors Affecting Drying Shrinkage and Creep

In practice, moisture movements in the hydrated cement paste, which essentially control the drying shrinkage and creep strains in concrete, are influenced by numerous simultaneously interacting factors. The interrelationships among these factors are quite complex and not easily understood. The factors are categorized and discussed below individually, solely for the purpose of understanding their relative significance.

Materials and Mix-Proportions

The main source of moisture-related deformations in concrete is the hydrated cement paste. Therefore, many attempts have been made to obtain expressions relating the drying shrinkage or creep strain to the volume fraction of the hydrated cement paste in concrete (as determined by the cement content and the degree of hydration). Although both the drying shrinkage and the creep strain are a function of the hydrated cement paste content, a direct proportionality does not exist because the restraint against deformation has a major influence on the magnitude of deformation.

Most theoretical expressions for predicting the drying shrinkage or creep of concrete assume that the elastic modulus of concrete can provide an adequate measure of the degree of restraint against deformation and that, as a first approximation, the elastic modulus of aggregate determines the elastic modulus of concrete. When the elastic modulus of the aggregate becomes a part of the mathematical expression, it is convenient to relate the drying shrinkage or the creep strain to the aggregate fraction rather than to the cement paste fraction in concrete. This is easily done because the sum of the two is constant.

Powers[3] investigated the drying shrinkage of concrete specimens containing two different aggregates and water-cement ratios of 0.35 or 0.50. From the data shown in Fig. 4-9*a* the ratio of the shrinkage of concrete (S_c) to the shrinkage of the cement paste (S_p) can be related exponentially to the volume fraction of aggregate (g) in concrete

$$\frac{S_c}{S_p} = (1-g)^n \tag{4-1}$$

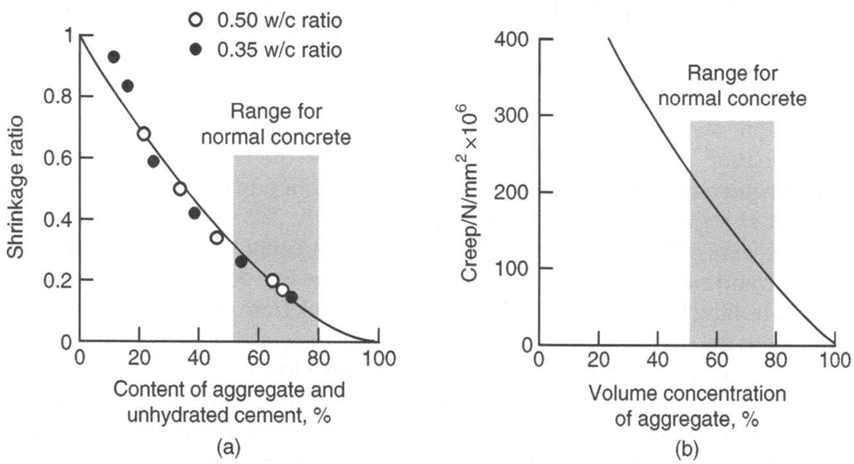

FIGURE 4-9 Influence of aggregate content on (*a*) drying shrinkage and (*b*) creep. [(*a*) From ACI Monograph 6, 1971, p. 126; from Concrete Society (London), Technical Paper 101, 1973.]

Aggregate content in concrete is the most important factor affecting drying shrinkage and creep. Unhydrated cement does not shrink and therefore may be included in the aggregate.

L'Hermite[4] found that values of n varied between 1.2 and 1.7 depending on the elastic modulus of aggregate. From the standpoint of shrinkage-causing and shrinkage-restraining constituents in concrete, Powers suggested that any unhydrated cement present may be considered a part of the aggregate (Fig. 4-9*a*).

Figure 4-9 shows that a similar relationship exists between the volume concentration of aggregate and creep of concrete. Neville[5] suggested that creep of concrete (C_c) and cement paste (C_p) can be related to the sum of the aggregate (g) and unhydrated cement (μ) contents:

$$\log \frac{C_p}{C_c} = \alpha \log \frac{1}{1-g-\mu} \tag{4-2}$$

In well-cured concrete, neglecting the small fraction of unhydrated cement (μ), the expression can be rewritten as

$$\frac{C_c}{C_p} = (1-g)^{\alpha} \tag{4-3}$$

Thus the expressions for creep and drying shrinkage are similar.

The grading, maximum size, shape, and texture of aggregate have also been suggested as factors influencing the drying shrinkage and creep. It is generally agreed that the *modulus of elasticity of the aggregate* is the most important factor; the influence of other aggregate characteristics may be indirect, that is, through their effect on the *aggregate content* of concrete or on the compactibility of the concrete mixture.

The influence of aggregate characteristics, primarily the elastic modulus, was confirmed by Troxell et al.'s study[6] (Fig. 4-10) of creep and shrinkage of concrete over a period of 23 years. As the modulus of elasticity of aggregate affects the elastic deformation of concrete, a good correlation exists between the elastic deformation of concrete and the drying shrinkage or creep values. Using fixed mix proportions, it was found that the 23-year drying shrinkage values of concrete mixture containing quartz and limestone aggregate were 550 and 650 $\times 10^{-6}$, respectively; concrete mixtures containing gravel and sandstone showed 1140 and 1260 $\times 10^{-6}$ drying shrinkage strains, respectively. The elastic deformation of the concretes containing either quartz or limestone aggregate was approximately 220 $\times 10^{-6}$ and the elastic deformation of the concretes containing either gravel or sandstone aggregate was approximately 280 $\times 10^{-6}$. The corresponding creep values were 600, 800, 1070, and 1500 $\times 10^{-6}$ for the concrete containing limestone, quartz, gravel, and sandstone aggregate, respectively. The importance of the aggregate modulus in controlling the concrete deformations is obvious from Troxell et al.'s data, which show that both the drying shrinkage and the creep of concrete increased 2.5 times when a high elastic modulus aggregate was substituted with a low elastic modulus aggregate.

Although the influence of aggregate type on creep and drying shrinkage is similar, a closer examination of the curves in Fig. 4-10 shows subtle differences. For example, compared to the drying shrinkage strain, the creep of concrete containing basalt or quartz aggregate was relatively higher. A plausible explanation is the higher degree of microcracking in the interfacial transition zone that is possible when a relatively nonreactive aggregate is used for concrete making. This underscores the point that creep in concrete is controlled by more than one mechanism.

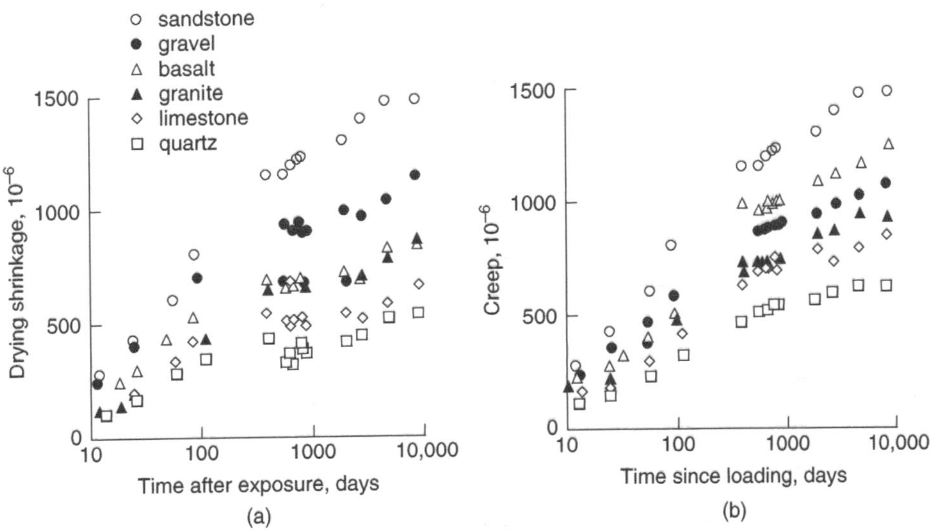

FIGURE 4-10 Influence of aggregate type on drying shrinkage and creep. (From Troxell, G.E. et al., *Proc. ASTM*, Vol. 58, 1958; and ACI Monograph 6, 1971, pp. 128, 151, Reprinted with permission from ASTM Copyright, ASTM, 1916 Race Street, Philadelphia. PA.)

The modulus of elasticity of aggregate can affect the magnitude of ultimate drying shrinkage and creep up to 2-1/2 times. Generally, dense limestones and quartz have higher elastic moduli than sandstones and gravel.

Within limits, variations in the *fineness and composition of portland cement* affect the rate of hydration, but not the volume and the characteristics of hydration products. Therefore, many researchers have observed that normal changes in cement fineness or composition, which tend to influence the drying shrinkage behavior of small specimens of cement paste or mortar, have a negligible effect on concrete. Obviously, with a given aggregate and mix proportions, if the type of cement influences the strength of concrete at the time of application of load, the creep of concrete will be affected. When loaded at early ages, concrete containing an ordinary portland cement generally shows higher creep than the corresponding concrete containing a high-early-strength cement (Fig. 4-11*b*). Due to their low early-strength concrete mixtures made with portland blast-furnace slag cement and portland-pozzolan cement also show higher creep at early age than the corresponding Type I cement concrete.

In general, the influence of *cement content and water-cement ratio* of concrete on the drying shrinkage and creep is not direct, because an increase in the cement paste volume means a decrease in the aggregate fraction and, consequently a corresponding increase in the moisture-dependent deformations in concrete. For a given cement content, with increasing water-cement ratio, both the drying shrinkage and creep are known to increase. A decrease in the strength (therefore, the elastic modulus) and an increase in the permeability of the system are probably responsible for this behavior. The data in Fig. 4-11*a* show that, for a given water-cement ratio, both the drying shrinkage and the creep increased with increasing cement content. This is expected due to an increase in the volume of the cement paste; however, in actual practice this does not always happen.

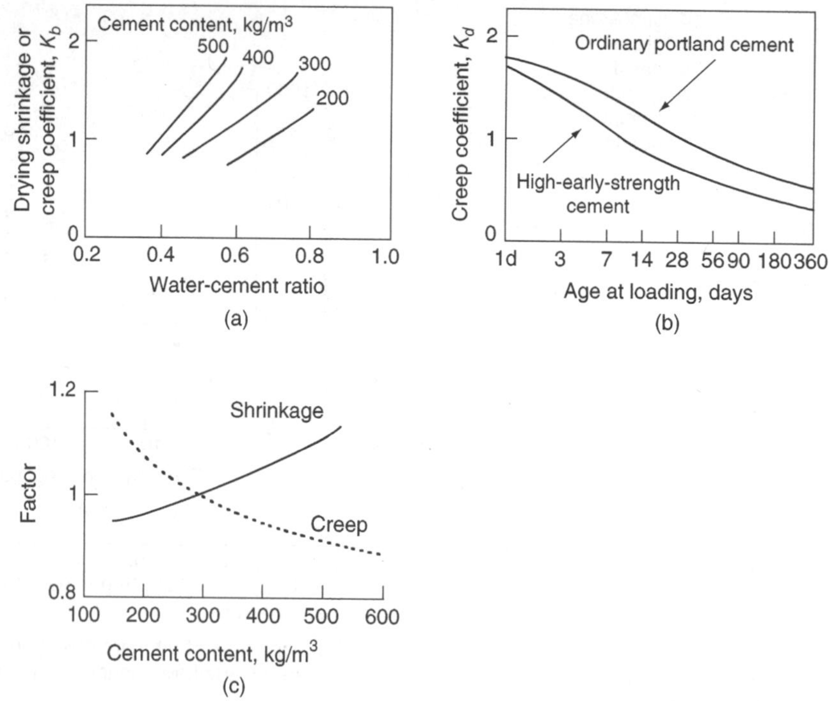

FIGURE 4-11 (*a*) Effect of water-cement ratio on shrinkage or creep; (*b*) effect of cement type on creep; (*c*) influence of cement content on drying shrinkage and creep. [(*a*) to (*b*), From International Recommendations for the Design and Construction of Concrete Structures, CEB/FIP, 1970; (*c*) From Jones, T.R., T.J. Hirsch, and H.K. Stephenson, Texas Transportation Institute Report E52, 1959; and ACI Monograph 6, p. 178, 1971.]

The results from many experimental investigations have shown that the foregoing theoretical analysis holds good for drying shrinkage but not always for creep. Experimental data show that within a wide range of concrete strengths, creep is inversely proportional to the *strength of concrete* at the time of application of load. It appears, therefore, that any increase in creep as a result of lowering the aggregate content is over-compensated by a reduction in creep that is associated with the increase in concrete strength. Curves illustrating the effect of cement content on both drying shrinkage and creep at a constant water-cement ratio are shown in Fig. 4-11*c*.

Concrete admixtures, such as calcium chloride, granulated slag, and pozzolans, tend to increase the volume of fine pores in the cement hydration product. Since drying shrinkage and creep in concrete are directly associated with the water held by small pores in the size range 3 to 20 nm, concretes containing admixtures causing pore refinement usually show higher drying shrinkage and creep. Water-reducing and set-retarding admixtures, which are capable of effecting better dispersion of anhydrous cement particles in water, also lead to pore refinement in the hydration product. Furthermore, in general, the admixtures which increase the drying shrinkage also tend to increase the creep.

Time and Humidity

Diffusion of the adsorbed water and the water held by capillary tension in small pores (under 50 nm) of the hydrated cement paste into large capillary voids within the system or to the atmosphere is a time-dependent process that takes place over long periods. From long-time creep and drying shrinkage tests lasting more than 20 years, Troxell et al. found that with a wide range of concrete mixture proportions, aggregate types, and environmental and loading conditions only 20 to 25 percent of the 10-year drying shrinkage was realized in 2 weeks, 50 to 60 percent in 3 months, and 75 to 80 percent in 1 year (Fig. 4-12*a*). Surprisingly, similar results were found for the creep strains (Fig. 4-12*b*).

An increase in the atmospheric humidity is expected to slow down the relative rate of moisture flow from the interior to the outer surfaces of concrete. With a given condition of exposure, the effects of relative humidity (RH) of air on the drying shrinkage strain (Fig. 4-13*a*) and creep coefficient (Fig. 4-13*b*) are illustrated in the charts published by the Comité Euro-International du Béton (CEB). At 100 percent RH, the drying shrinkage (ε_c) is assumed to be zero, rising to about 200 microstrain at 80 percent RH, and 400 microstrain at 45 percent RH. Similarly, the creep coefficient, which is one of the five partial coefficients contributing to total creep, is assumed to be 1 at 100 percent RH, it rises to about 2 at 80 percent RH, and 3 at 45 percent RH. The data showing the effect of humidity conditions and thickness of concrete structure on the ultimate drying shrinkage and creep are presented in Fig. 4-14.

Geometry of the Concrete Element

Because of the resistance to water transport from the interior of concrete to the atmosphere, the rate of water loss would obviously be controlled by the length of the path traveled by the water, which is being expelled during the drying shrinkage and/or creep. At a constant RH, both the size and the shape of a concrete element

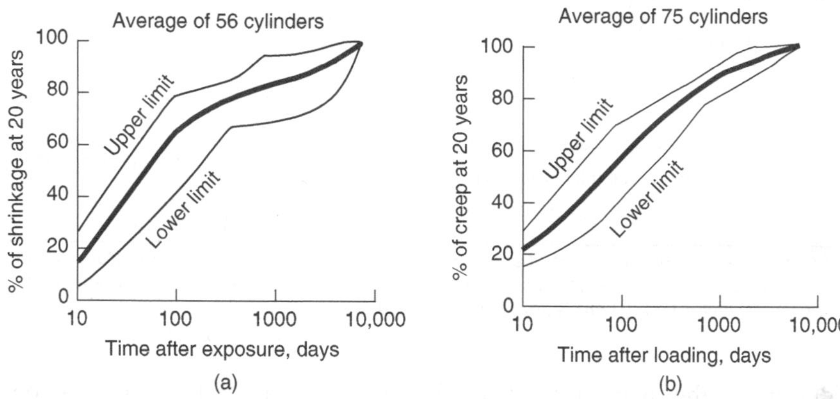

FIGURE 4-12 The time dependency of (*a*) drying shrinkage and (*b*) creep. (From Troxell, G.E. et al., *Proc. ASTM*, Vol. 58, 1958. Reprinted with permission from ASTM. Copyright ASTM, 1916 Race Street, Philadelphia, PA.)

For a wide range of concrete mixtures, drying shrinkage and creep show a similar time dependency.

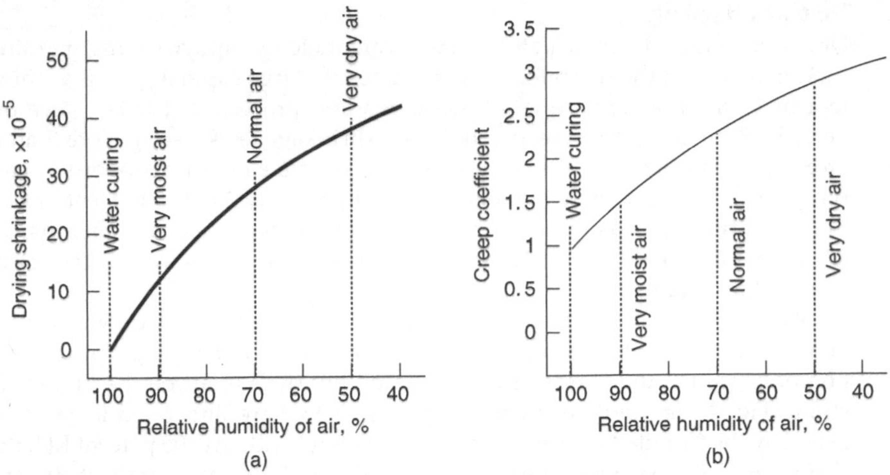

FIGURE 4-13 Influence of the relative humidity on (*a*) drying shrinkage and (*b*) creep. (From International Recommendations for the Design and Construction of Concrete Structures, CEB/FIP, 1970.)

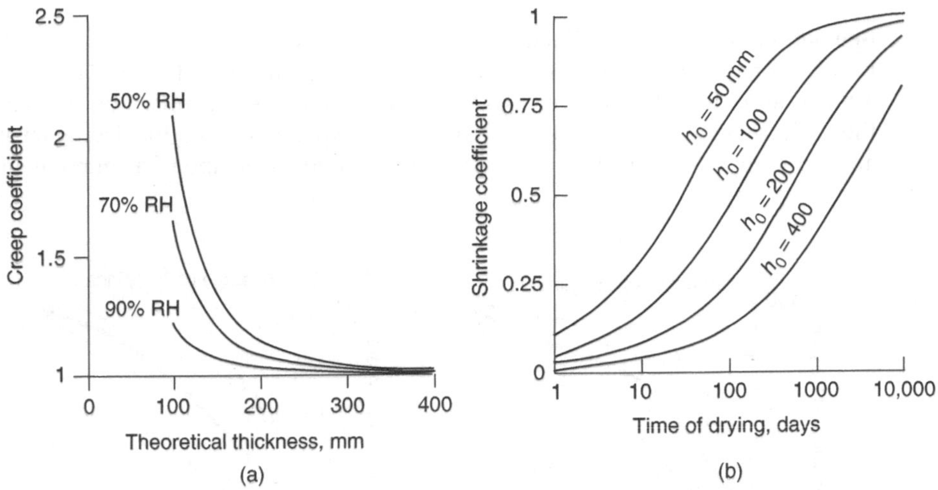

FIGURE 4-14 (*a*) Influence of specimen size and relative humidity on the creep coefficient, (*b*) influence of exposure time and specimen size on the drying shrinkage coefficient. (Data from equations given by CEB-FIP Model Code, 1990.)

determine the magnitude of the drying shrinkage and creep. It is convenient to express the size and shape parameters by a single quantity expressed in terms of effective or *theoretical thickness*, which is equal to the area of the section divided by the semiperimeter in contact with the atmosphere. The relations between theoretical thickness and drying shrinkage or creep coefficient, as given in the CEB charts, are illustrated by Fig. 4-14.

Additional Factors Affecting Creep

The curing history of concrete, temperature of exposure, and magnitude of applied stress are known to affect drying creep more than drying shrinkage, probably because of a greater influence of these factors on the interfacial transition zone characteristics (i.e., porosity, microcracking, and strength). Depending on the *curing history* of a concrete element, creep strains in practice may be significantly different from those in laboratory tests that are carried out at a constant humidity. For instance, drying cycles can enhance microcracking in the interfacial transition zone and thus increase the creep. For the same reason, it has often been observed that alternating the environmental humidity between two limits would result in a higher creep than that obtained at a constant humidity (within those limits).

The *temperature* to which concrete is exposed can have two counteracting effects on the creep. If a concrete member is exposed to higher than normal temperature as a part of the curing process before it is loaded, the strength will increase and the creep strain would be less than that of a corresponding concrete stored at a lower temperature. On the other hand, exposure to high temperature during the period under load can increase the creep. Nasser and Neville[7] found that, in the temperature range 21 to 71°C, the 350-day creep increased approximately 3.5 times with temperature (Fig. 4-15). The influence of temperature on the creep is of considerable interest to nuclear PCRV (prestressed concrete reactor vessel) structures because neutron attenuation and gamma-ray absorption causes the concrete temperature to rise (see Chap. 12).

In regard to the *intensity of the applied stress*, Troxell et al. found a direct proportionality between the magnitude of sustained stress and the creep of concrete made with 0.69-water-cement ratio (20 MPa nominal compressive strength). Specimens cured for 90 days and then loaded for 21 years showed 680, 1000, and 1450 × 10^{-6} creep strain, corresponding to sustained stress levels of 4, 6, and 8 MPa, respectively (Fig. 4-16).

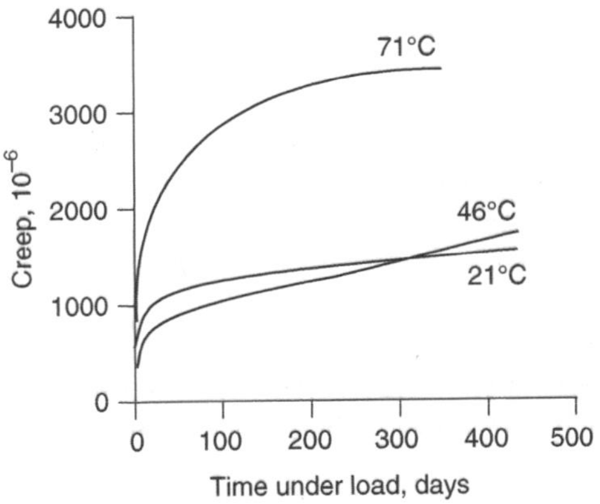

FIGURE 4-15 Effect of concrete temperature on creep. (From Nasser, K.W., and A.M., Neville, *J. ACI, Proc.*, Vol. 64, No. 2, 1967; and ACI Monograph 6, p. 162, 1971.)

At a stress-strength ratio of 70 percent, the creep can increase 3 to 5 times if the surrounding temperature is raised from 21 to 71°C.

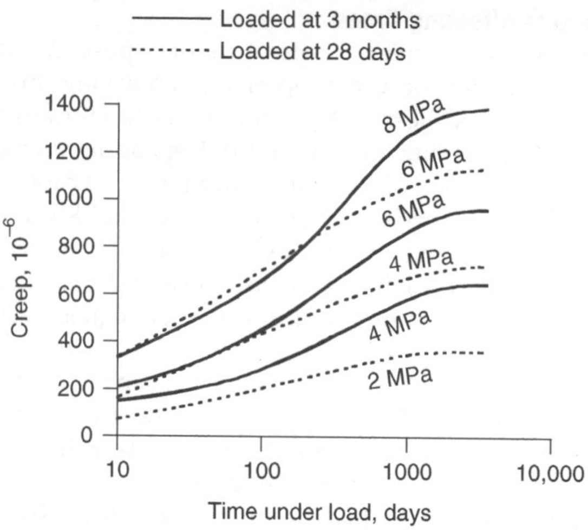

FIGURE 4-16 Effect of magnitude of sustained stress on creep. (From Troxell, G.E. et al., *Proc. ASTM*, Vol. 58, 1958. Reprinted with permission from ASTM, Philadelphia, PA.)

Creep is directly proportional to the magnitude of sustained stress. With 90-day-old concrete specimens, the amount of ultimate creep doubled when the loading stress was increased from 4 to 8 MPa. Because of the effect of strength on creep, the figure shows that, at a given stress level, lower creep values were obtained for the longer period of curing before application of the load.

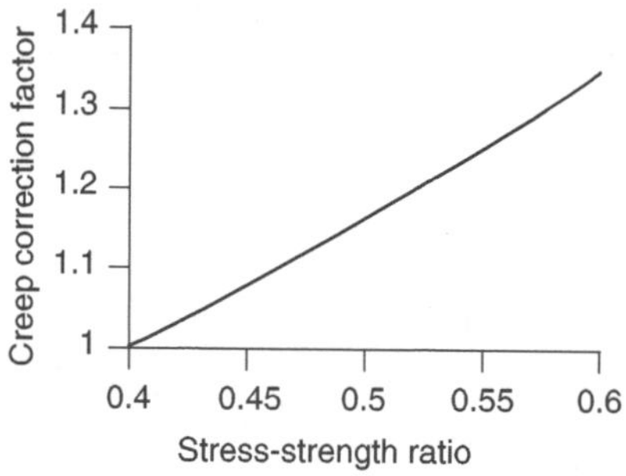

FIGURE 4-17 Correction factor for computing the creep coefficient at high stress levels. (Data from equations given by CEB-FIP Model Code, 1990.)

The proportionality is valid as long as the applied stress is in the linear domain of stress-to-strain relationship (e.g., 0.4 stress-strength ratio for a compressive stress). At high stress-strength ratios a correction factor should be used from the data in Fig. 4-17.

4.4 Thermal Shrinkage

In general, solids expand on heating and contract on cooling. The strain associated with change in temperature will depend on the coefficient of thermal expansion of the material and the magnitude of temperature drop or rise. Except under extreme climatic conditions, ordinary concrete structures suffer little or no distress from changes in ambient temperature. However, in massive structures, the combination of heat produced by cement hydration and relatively poor heat dissipation conditions results in a large rise in concrete temperature within a few days after placement. Subsequently, cooling to the ambient temperature often causes the concrete to crack. Since the primary concern in the design and construction of mass concrete structures is that the completed structure remains a monolith, free of cracks, every effort to control the temperature rise is made through selection of proper materials, mix proportions, curing conditions, and construction practice (see Chap. 12).

With low tensile strength materials, such as concrete, it is the shrinkage strain from cooling that is more important than the expansion from heat generated by cement hydration. This is because, depending on the elastic modulus, degree of restraint, and stress relaxation due to creep, the resulting tensile stress can be large enough to cause cracking. For instance, assuming that the coefficient of thermal expansion α of concrete is 10×10^{-6} per °C, and the temperature rise above the ambient ΔT from heat of hydration is 15°C, then the thermal shrinkage ε caused by the 15°C temperature drop will be $\alpha\Delta T$ or 150×10^{-6}. The elastic modulus E of ordinary concrete may be assumed as 20 GPa. If the concrete member is fully restrained ($K_r = 1$), the cooling would produce a tensile stress of $\varepsilon E = 3$ MPa. Since the tensile strength of ordinary concrete is usually less than 3 MPa, it is likely to crack if there is no relief due to stress relaxation (Fig. 4-1).

However, there is always some relaxation of stress due to creep. When the creep coefficient is known, the resulting tensile stress σ_t can be calculated from the expression:

$$\sigma_t = K_r \frac{E}{1+\varphi} \alpha \, \Delta T \tag{4-4}$$

where σ_t = tensile stress
K_r = degree of restraint
E = elastic modulus
α = coefficient of thermal expansion
ΔT = temperature change
φ = creep coefficient

The factors influencing the modulus of elasticity and the creep of concrete are described in the previous sections. An analysis of other factors in Eq. (4-4) which affect thermal stresses is presented next.

4.4.1 Factors Affecting Thermal Stresses

Degree of Restraint (K_r)

A concrete element, if free to move, would have no stress development associated with thermal deformation on cooling. However, in practice, the concrete mass will be restrained either externally by the rock foundation or internally by differential deformations within different areas of concrete due to the presence of temperature gradients. For example, assuming a rigid foundation, there will be full restraint at the concrete-rock interface ($K_r = 1.0$), however, as the distance from the interface increases, the restraint will decrease, as shown in Fig. 4-18. The same reasoning can be applied to determine the restraint between different concrete lifts. If the foundation is not rigid,

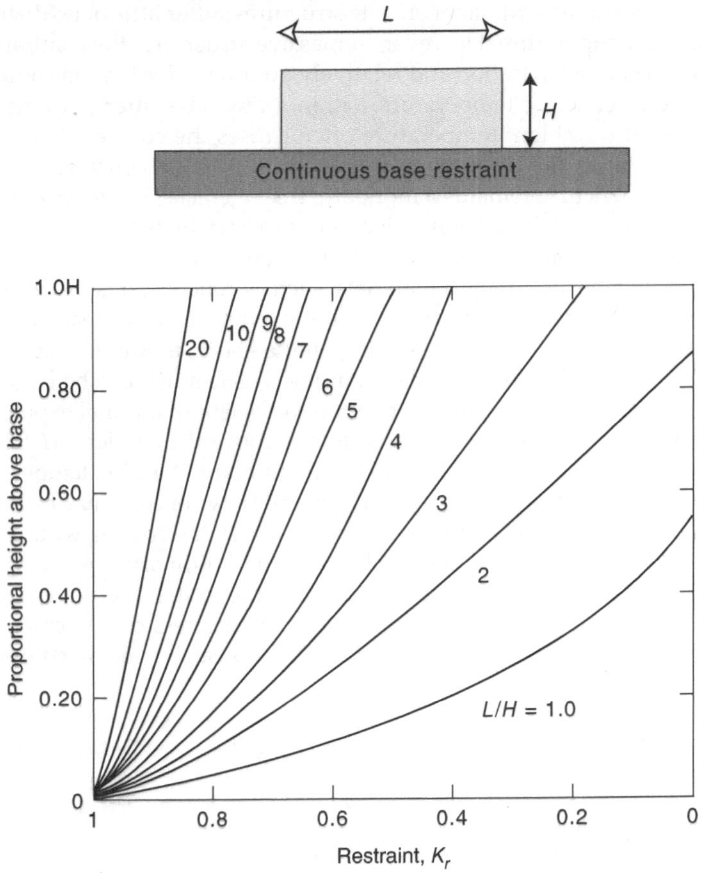

FIGURE 4-18 Degree of tensile restraint at center section. (*Source: ACI Committee 207, Cooling Mass Concrete, 1998.*)

the degree of restraint will decrease. When dealing with a nonrigid foundation, ACI-207.2R recommends the following multipliers for K_r:

$$\text{multiplier} = \frac{1}{1 + \frac{A_g E}{A_f E_f}} \tag{4-5}$$

where A_g = gross area of concrete cross section

A_f = area of foundation or other restraining element (For mass rock, A_f can be assumed as 2.5 A_g.)

E_f = modulus of elasticity of foundation or restraining element

E = modulus of elasticity of concrete

Temperature Change (ΔT)

The hydration of cement compounds involves exothermic reactions, which generate heat and increase the temperature of concrete mass. Heating causes expansion, and expansion under restraint results in compressive stress. However, at early ages, the elastic

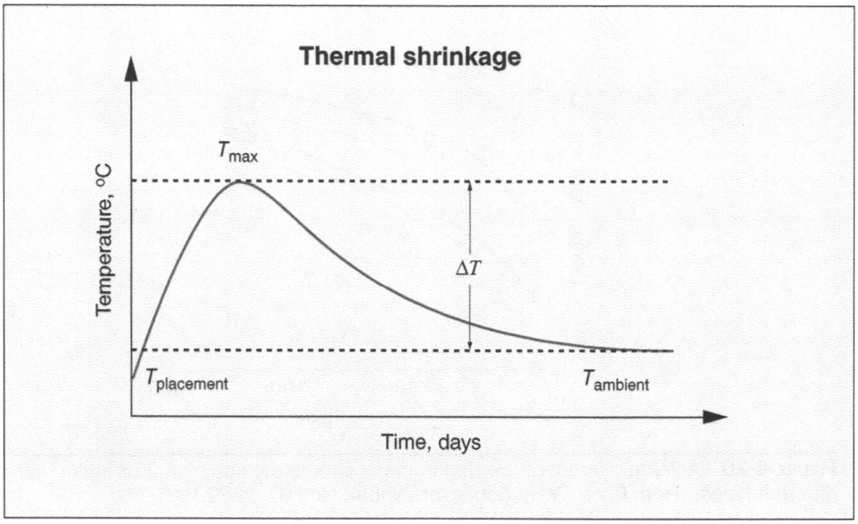

FIGURE 4-19 Temperature change with time.

modulus of concrete is low and the stress relaxation is high, therefore, the compressive stress will be very small, even in areas of full restraint. In design, to be conservative, it is assumed that a condition of no initial compression exists. The change of temperature ΔT in Eq. (4-4) is the difference between the peak temperature of concrete and the service temperature of the structure, as shown in Fig. 4-19. The change of temperature can be calculated as follows:

$$\Delta T = \text{placement temperature of fresh concrete} + \text{adiabatic temperature rise} - \text{ambient or service temperature} - \text{temperature drop due to heat losses}$$

Controlling the *placement temperature* of concrete is one of the best ways to avoid thermal cracks. *Precooling of fresh concrete* is a commonly used method of controlling the subsequent temperature drop. Often, chilled aggregates and/or ice shavings are specified for making mass concrete mixtures in which the temperature of fresh concrete is limited to 10°C or less. During the mixing operation the latent heat needed for fusion of ice is withdrawn from other components of the concrete mixture, providing a very effective way to lower the temperature. ACI 207.4R suggests a placement temperature such that the tensile strain caused by the temperature drop should not exceed the tensile strain capacity of the concrete. This is expressed by the relationship:

$$T_i = T_f + \frac{C}{\alpha K_r} - T_r \tag{4-6}$$

where T_i = placement temperature of concrete
T_f = final stable temperature of concrete
C = tensile strain capacity of concrete
K_r = degree of restraint
α = coefficient of thermal expansion
T_r = initial temperature rise of concrete

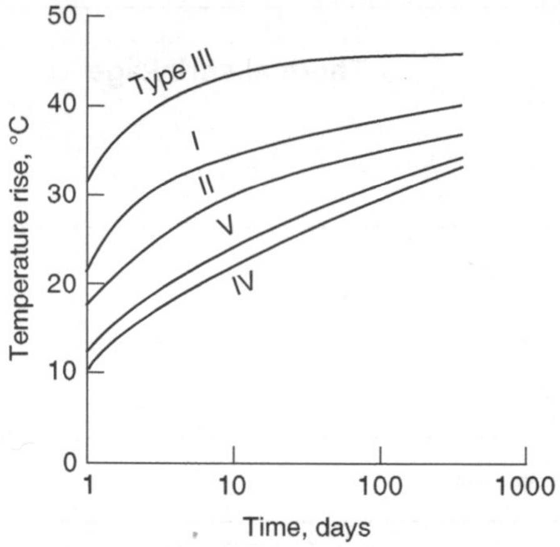

FIGURE 4-20 Adiabatic temperature rise in mass concrete containing 223 kg/m^3 cement of different types. (From Price, W.H., *Concr. Int.*, Vol. 4, No. 10, 1982.)

The rate and magnitude of the *adiabatic temperature rise* is a function of the amount, composition and fineness of cement, and the temperature during hydration. Finely ground portland cements, cements with relatively high C_3A and C_3S content show higher heats of hydration than coarser cements or cements with low C_3A and C_3S (see Chap. 6). The adiabatic temperature rise curves for a concrete containing 223 kg/m^3 cement and one of the five types of portland cements are shown in Fig. 4-20. It can be seen that between a normal portland cement (Type I) and a low-heat cement (Type IV) the difference in temperature rise is 13°C in 7 days. At this cement content the total adiabatic temperature rise was above 30°C even with the ASTM Type IV, low-heat cement. Also, as shown in Figs. 4-20 and 4-21, the composition of cement and the placement temperature appear to affect mainly the rate of heat generation rather than the

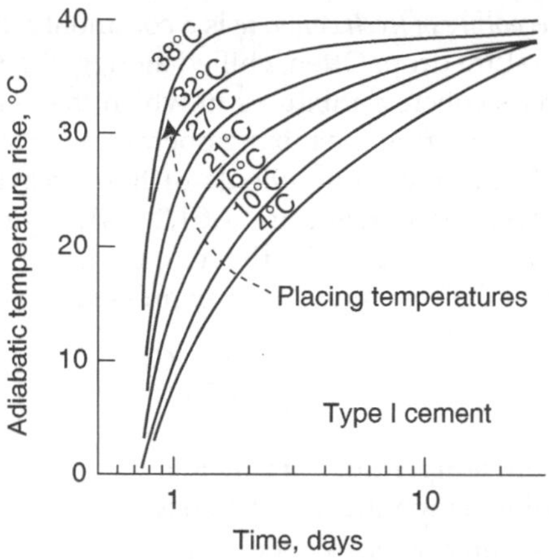

FIGURE 4-21 Effect of placing temperature on temperature rise of mass concrete containing 223 kg/m^3 on Type I cement. (ACI Committee 207. Effect of restraint, volume changes, and reinforcement on cracking of massive concrete, 1995.)

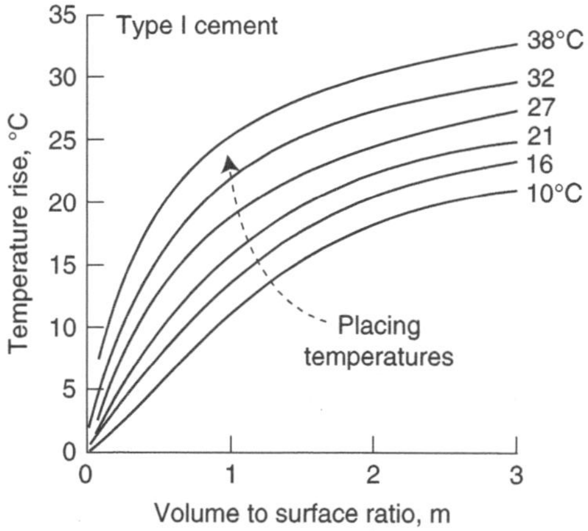

FIGURE 4-22 Temperature rise of concrete members containing 223 kg/m³ of cement. (ACI Committee 207. Effect of restraint, volume changes, and reinforcement on cracking of massive concrete, 1995.)

total heat produced. Figure 4-22 shows the effect of volume-to-surface ratio of concrete on the adiabatic temperature rise at different placement temperatures.

Another effective means of reducing the magnitude of the adiabatic temperature rise is the inclusion of a pozzolan as a partial replacement for cement. Typical data given by Carlson et al.[8] on adiabatic temperature rise in mass concrete containing different types and amounts of cementitious materials are shown in Fig. 4-23. In a concrete containing 223 kg/m³ cement, the replacement of ASTM Type I cement by Type II

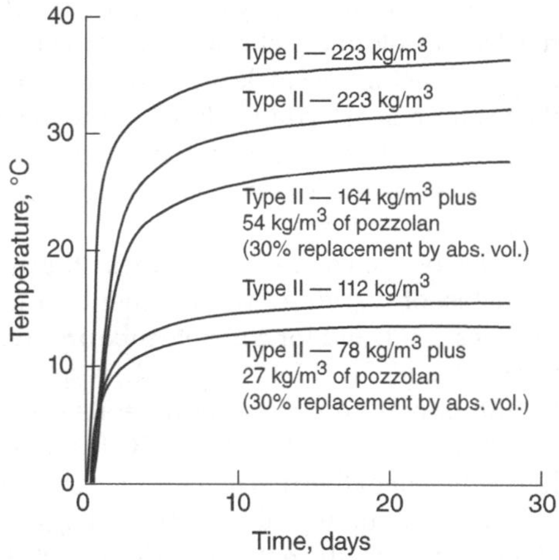

FIGURE 4-23 Effect of cement and pozzolan contents on temperature rise in concrete. (From Carlson, R.W. et al., *J. ACI, Proc.*, Vol. 76, No. 7, 1979.)

The use of a low cement content, an ASTM Type II moderate-heat portland cement instead of Type I, and a partial substitution of the portland cement by a pozzolan, are effective means by which the adiabatic temperature rise in mass concrete can be significantly reduced.

Type of Isolation	Surface Transmission Coefficient (kcal/m^2·h·C)
Concrete-air	11.6
Concrete-curing water	300.0
Concrete-wood-air	2.6
Concrete-metal-air	11.6
Concrete-isolant-air	2.0

TABLE 4-4 Coefficient of Heat Transmission of Different Isolation Environments

cement reduced the 28-day adiabatic temperature rise from 37 to 32°C; a partial replacement of the Type II cement by 30 volume percent of pozzolan (25 weight percent) further reduced the temperature rise to 28°C.

Heat losses depend on the thermal properties of concrete, and the construction technology adopted. A concrete structure can lose heat through its surface, and the magnitude of heat loss is a function of the type of environment in immediate contact with the concrete surface. Table 4-4 shows surface transmission coefficients for different isolation environments. Numerical methods of computing the temperature distribution in mass of concrete are presented in Chap. 13.

4.5 Thermal Properties of Concrete

Coefficient of thermal expansion (α) is defined as the change in unit length per degree of temperature change. Selecting an *aggregate with a low coefficient of thermal expansion* when it is economically feasible and technologically acceptable, may, under certain conditions, become a critical factor for crack prevention in mass concrete. This is because the thermal shrinkage strain is determined both by the magnitude of temperature drop and the coefficient of linear thermal expansion of concrete; the latter, in turn, is controlled primarily by the coefficient of linear thermal expansion of the aggregate which is the primary constituent of concrete.

The reported values of the coefficient of linear thermal expansion for saturated portland cement pastes of varying water-cement ratios, for mortars containing 1:6 cement/natural silica sand, and for concrete mixtures with different aggregate types are approximately 18, 12, and 6 to 12×10^{-6} per °C, respectively. The coefficient of thermal expansion of commonly used rocks and minerals varies from about 5×10^{-6} per °C for limestones and gabbros to 11 to 12×10^{-6} per °C for sandstones, natural gravels, and quartzite. Since the coefficient of thermal expansion can be estimated from the weighted average of the components, assuming 70 to 80 percent aggregate in the concrete mixture, the calculated values of the coefficient for various rock types (both coarse and fine aggregate from the same rock) are shown in Fig. 4-24. The data in the figure are fairly close to the experimentally measured values of thermal coefficients reported in the published literature for concrete tested in moist condition, which is representative of the condition of typical mass concrete.

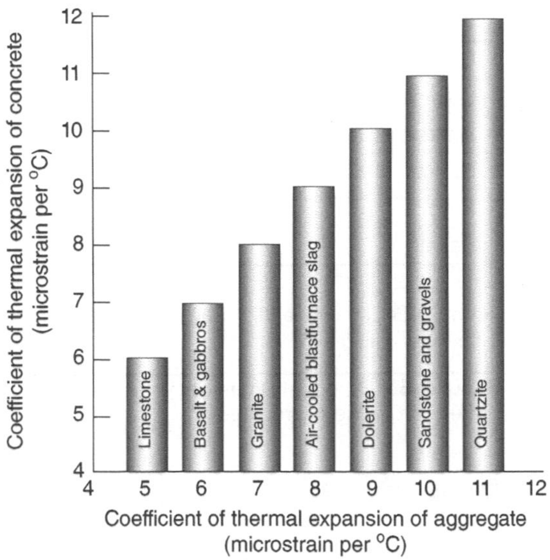

FIGURE 4-24 Influence of the aggregate type on the coefficient of thermal expansion of concrete. Since the coefficient of thermal expansion of concrete is directly related to the coefficient of expansion of the aggregate present, in mass concrete the selection of an aggregate with a lower coefficient provides another approach toward lowering the thermal strain.

Specific heat is defined as the quantity of heat needed to raise the temperature of a unit mass of a material by one degree. The specific heat of normal weight concrete is not very much affected by the type of aggregate, temperature, and other parameters. Typically the values of specific heat are in the range of 0.9 to 1.0 kJ/kg·C.

Thermal conductivity gives the heat flux transmitted through a unit area of a material under a unit temperature gradient. The thermal conductivity of concrete is influenced by the mineralogical characteristics of aggregate, and by the moisture content, density, and temperature of concrete. Table 4-5*a* shows typical values of thermal conductivity for concretes containing different aggregate types.

Aggregate Type	Thermal Conductivity	
	Btu in./h·ft²·F	W/m·K
Quartzite	24	3.5
Dolomite	22	3.2
Limestone	18–23	2.6–3.3
Granite	18–19	2.6–2.7
Rhyolite	15	2.2
Basalt	13–15	1.9–2.2

TABLE 4-5*a* Thermal Conductivity Values for Concrete with Different Aggregate Types

Coarse Aggregate	ft²/h	m²/h
Quartzite	0.058	0.0054
Dolomite	0.051	0.0047
Limestone	0.050	0.0046
Granite	0.043	0.0040
Rhyolite	0.035	0.0033
Basalt	0.032	0.0030

Source: ACI Committee 207—Cooling and Insulating Systems for mass concrete, 1998.

Table 4-5*b* Thermal Diffusivity Values for Concrete with Different Coarse Aggregates

Thermal diffusivity is defined as:

$$\kappa = \frac{K}{c\rho} \tag{4-7}$$

where κ = diffusivity
K = conductivity
c = specific heat
ρ = density of concrete

Heat will move more readily through a concrete with higher thermal diffusivity. For normal-weight concrete, the conductivity usually controls the thermal diffusivity because the density and specific heat do not vary much. Table 4-5*b* shows typical values of thermal diffusivity for concretes made with different types of coarse aggregate.

4.6 Extensibility and Cracking

As stated earlier, the primary significance of deformations caused by applied stress and by thermal and moisture-related effects in concrete is whether or not their interaction would lead to cracking. Thus the magnitude of the shrinkage strain is only one of the factors governing the cracking of concrete. From Fig. 4-1 it is clear that the other factors are:

- *Modulus of elasticity.* The lower the modulus of elasticity, the lower will be the amount of the induced elastic tensile stress for a given magnitude of shrinkage.
- *Creep.* The higher the creep, the higher is the amount of stress relaxation and lower the net tensile stress.
- *Tensile strength.* The higher the tensile strength, the lower is the risk that the tensile stress will exceed the strength and crack the material.

The combination of factors that are desirable to reduce the advent of cracking in concrete can be described by a single term called *extensibility*. Concrete is said to have a high degree of extensibility when it can be subjected to large deformations without cracking.

Obviously, for minimum risk of cracking, the concrete should undergo not only less shrinkage but also should have a high degree of extensibility (i.e., low elastic modulus, high creep, and high tensile strength). In general, high-strength concrete is more prone to cracking because of greater thermal shrinkage and lower stress relaxation. On the other hand, low-strength concrete tends to crack less, due to lower thermal shrinkage and higher stress relaxation. The preceding statement is applicable to massive concrete members; with thin sections the effect of drying shrinkage strain would be more important.

It may be of interest to point out that many factors that reduce the drying shrinkage of concrete will also tend to reduce the extensibility. For instance, an increase in the aggregate content or aggregate stiffness will reduce the drying shrinkage but at the same time reduce stress relaxation by creep, and extensibility. This example demonstrates the difficulty of practicing concrete technology from purely theoretical considerations.

The cracking behavior of concrete in the field can be more complex than indicated by Fig. 4-1, that is, the rates at which the shrinkage and stress relaxation develop may not be similar to those shown in the figure. For example, with mass concrete compressive stresses are developed during the very early period when temperatures are rising, and the tensile stresses do not develop until at a later age when the temperature begins to decline. However, due to the low strength of concrete at early ages, most of the stress relaxation takes place during the first week after the placement. It means that concrete loses most of the stress-relaxing capacity before it is needed for the prevention of cracking induced by tensile stresses.

With thermal-shrinkage cracking, whether related to internal temperature effects in mass concrete or to external temperature effects in extreme climates, the significance of *tensile strain capacity*, which is defined as the failure strain under tension, is noteworthy. It is generally agreed that the failure of concrete loaded in uniaxial compression is mainly a tensile failure. Also, there are indications that it is not a limiting tensile strength, but a limiting tensile strain that determines the fracture strength of concrete under static loading. Accordingly, Houghton[9] has described a simple method to determine the ultimate tensile strain for quick loading by taking a ratio of the modulus of rupture to the elastic modulus in compression. As the modulus of rupture is 20 to 40 percent higher than the true tensile strength, and the modulus of elasticity in compression is higher than the stress-strain ratio by a similar order of magnitude, it is claimed that the method gives a true value of the ultimate elastic strain for quick loading. By adding to this strain the creep strain due to slow loading, an estimate of the tensile strain capacity can be obtained. For the purpose of risk analysis against thermal cracking, it is suggested that the determination of the tensile strain capacity is a better criterion than the practice of converting the thermal strain to induced elastic stress. A general method of computing stress in viscoelastic materials is presented in Chap. 13, which also contains a finite element method to compute temperature distributions in mass concrete.

Test Your Knowledge

4.1 What is a truly elastic material? Is concrete truly elastic? If not, why? Describe the various stages of microcracking when a concrete specimen is loaded to failure.

4.2 Draw a typical stress-strain curve for concrete. From this, how would you determine the dynamic modulus of elasticity and the different types of the static elastic moduli? Typically, what are their magnitudes for a medium-strength concrete?

4.3 What are the assumptions underlying the formulas used by the ACI Building Code and the FIP Model Code for predicting the static elastic modulus of concrete? Can you point out any limitations of these formulas?

4.4 How does the moisture state of a concrete test specimen affect the elastic modulus and strength values? Explain why both properties are not affected in the same manner.

4.5 What is the significance of adiabatic temperature in concrete? How much adiabatic temperature rise can occur in a typical low-strength concrete containing ASTM Type II cement? How can this be reduced?

4.6 Can we control the coefficient of thermal expansion of concrete? If so, how?

4.7 What are the typical ranges of drying shrinkage strain and creep strain in concrete; what is their significance? How are the two phenomena similar to each other?

4.8 What do you understand by the terms *basic creep, specific creep, drying creep*, and *creep coefficient*?

4.9 List the most important factors that affect drying shrinkage and creep, and discuss when the effects are similar or opposite.

4.10 Which factors affect creep only, and why?

4.11 What is the significance of the term *theoretical thickness*?

4.12 Beside the magnitude of shrinkage strain, which other factors determine the risk of cracking in a concrete element?

4.13 What is the usefulness of the *extensibility* concept? Why would high-strength concrete be more prone to cracking than low-strength concrete?

4.14 Ideally, from the standpoint of crack resistance, a concrete should have low shrinkage and high extensibility. Give examples to show why this may not be possible to achieve in practice.

4.15 What is the significance of tensile strain capacity? How can you determine it?

References

1. Hsu, T.C., F.O. Slate, G.M. Sturman, and G. Winter, *J. ACI, Proc.*, Vol. 60, No. 2, pp. 209–223, 1963; Shah, S.P., and F.O. Slate, *Proceedings of a Conference on Structure of Concrete*, Brooks, A.E., and K. Newman, eds., Cement and Concrete Association, Wexham Springs, Slough, U.K., pp. 82–92, 1968.
2. J. Glucklich, ibid, pp. 176–189, 1968.
3. Powers, T.C., *Rev. Mater. Construct.* (*Paris*), Vol. No. 545, pp. 79–85, 1961.
4. Hermite, R.L., *Proceedings of Fourth International Symposium on Chemistry of Cements*, National Bureau of Standards, Washington, D.C., pp. 659–694, 1962.
5. Neville, A.M., *Mag. Concr. Res.* (*London*), Vol. 16, No. 46, pp. 21–30, 1964.
6. Troxell, G.E., J.M. Raphael, and R.E. Davis, *Proc. ASTM*, Vol. 58, pp. 1101–1120, 1958.
7. Nasser, K.W., and A.M. Neville, *J. ACI, Proc.*, Vol. 64, No. 2, pp. 97–103, 1967.
8. Carlson, R.W., D.L. Houghton, and M. Polivka, *J. ACI, Proc.*, Vol. 76, No. 7, pp. 821–837, 1979.
9. Houghton, D.L., *J. ACI, Proc.*, Vol. 73, No. 12, pp. 691–700, 1976.

Suggestions for Further Study

ACI Committee Report 207-1R-05, Guide to Mass Concrete, *ACI Manual of Concrete Practice*, American Concrete Institute, Farmington Hills, MI, 2012.

ACI Committee Report 207.2R-07, Thermal and Volume Change Effects on Cracking of Mass Concrete, *ACI Manual of Concrete Practice*, American Concrete Institute, Farmington Hills, MI, 2012.

ACI Committee Report 209R-92, Prediction of Creep, Shrinkage, and Temperature Effects in Concrete Structures, *ACI Manual of Concrete Practice*, American Concrete Institute, Farmington Hills, MI, 2012.

FIB Bulletin No. 65, Model Code 2010—*Final draft*, Volume 1, International Federation for Structural Concrete, Lausanne, Switzerland, 2012. Neville, A.M., W.H. Dilger, and J.J. Brooks, *Creep of Plain and Structural Concrete*, Longman Inc., New York, 1983.

Neville, A.M., *Properties of Concrete*, 5th ed., Pearson, Harlow, 2011.

Pijaudier-Cabot G., B. Gérard, and P. Acker (editors), *Creep, Shrinkage and Durability of Concrete and Concrete Structures*, Proceedings of the CONCREEP 7 conference held at Ecole Centrale de Nantes, France, Wiley, September 12th–September 14th, 2005, John & Sons, Incorporated, 2008.

Rusch H., D. Jungwirth, H. K. Hilsdorf, *Creep and Shrinkage: Their Effect on the Behavior of Concrete Structures*, Springer-Verlag New York, LLC, 2012 (reprint of the original 1983 edition).

Tanabe T., K. Sakata, H. Mihashi, R. Sato, K. Maekawa, H. Nakamura (editors), *Creep, Shrinkage and Durability Mechanics of Concrete and Concrete Structures*, Two Volume Set: Proceedings of the CONCREEP 8 conference held in Ise-Shima, Japan, 30 September-2 October 2008, Taylor & Francis Group, London UK, 2009.

Ulm F.J., Z.P. Bazant, and F.H. Wittman (editors), *Creep, Shrinkage, and Durability Mechanics of Concrete and other Quasi-Brittle Materials*, Proceedings of the CONCREEP-6 conference held in Cambridge (MA), USA, 2001, Elsevier Science Limited, 2001.

CHAPTER 5

Durability

Preview

Designers of concrete structures have been mostly interested in the strength characteristics of the material; for a variety of reasons, they must now become durability-conscious. Whereas properly constituted, placed, and cured concrete can be durable under most natural and industrial environments, cases of premature deterioration of concrete structures do occur and they provide valuable lessons for control of factors responsible for the lack of durability.

Water is generally involved in every form of deterioration and, with porous solids the ease of penetration of water into the solid usually determines its rate of deterioration. Therefore, at the beginning of this chapter, the structure and properties of water are described with special reference to its destructive effect on porous materials. Next, the factors controlling the permeability of cement paste, aggregate, and concrete are presented.

Physical effects that adversely influence the durability of concrete include surface wear, cracking due to crystallization of salts in pores, and exposure to temperature extremes such as during frost action or fire. Deleterious chemical effects include leaching of the cement paste by acidic solutions, and expansive reactions involving sulfate attack, alkali-aggregate reaction, and corrosion of the embedded steel in concrete. The significance, physical manifestations, mechanisms, and control of various causes of concrete deterioration are discussed in detail. A holistic model of concrete deterioration is also presented.

Special attention is given to performance of concrete in seawater. As numerous physical and chemical causes of deterioration occur simultaneously when a concrete structure is exposed to seawater, the study of the behavior of concrete in seawater provides an excellent opportunity to appreciate the complexity of durability problems that usually occur with concrete in field practice.

5.1 Definition

A long service life is considered synonymous with durability. As durability under one set of conditions does not necessarily mean durability under another, it is customary to include a general reference to the environment when defining durability. According to ACI Concrete Terminology, *durability* of portland cement concrete is defined as its ability to resist weathering action, chemical attack, abrasion, and other conditions of service. In other words, a durable concrete will retain its original form, quality, and serviceability when exposed to its intended service environment.

No material is inherently durable. As a result of environmental interactions the microstructure and consequently the properties change with time. A material is assumed to reach the end of service life when its properties, under given conditions of use, have deteriorated to an extent that its continued use is ruled either unsafe or uneconomical.

5.2 Significance

For a variety of reasons, there is a general awareness now that designers of structures must evaluate the durability characteristics of the construction materials under consideration as carefully as other aspects, such as mechanical properties and initial cost. First, there is a better appreciation of the socioeconomic implications of durability. Increasingly, repair and replacement costs of structures arising from materials failure have become a substantial portion of the total construction budget. For example, it is estimated that, in industrially developed countries, about 40 percent of the total resources of the construction industry are being applied to repair and maintenance of existing structures and only 60 percent to new installations. The escalation in replacement costs of structures and the growing emphasis on the life-cycle cost rather than the first cost are forcing engineers to pay serious attention to durability issues. Next, there is a realization that a close relation exists between durability of materials and ecology. Conservation of natural resources by making the construction materials last longer is therefore an ecological step. Failure of offshore steel structures in Norway, Newfoundland, and other parts of the world has shown that both the human and economic costs associated with sudden failure of the material of construction can be very high.* Therefore, the uses of concrete are being extended increasingly to severe environments, such as offshore platforms in the North Sea, and concrete containers for handling liquefied gases at cryogenic temperatures.

5.3 General Observations

Before a discussion of important aspects of durability of concrete, a few general remarks on the subject will be helpful. First, *water*, which is the primary agent of both creation and destruction of many natural materials, happens to be central to most durability problems in concrete. In porous solids, water is known to be the cause of many types of *physical processes of degradation*. As a vehicle for transport of aggressive ions, water can also be a source of *chemical processes of degradation*. Second, the physical-chemical phenomena associated with water transport in porous solids are controlled by the *permeability* of the solid. For instance, the rate of chemical deterioration is dependent on whether chemical attack is confined to the surface of concrete, or whether it is also happening inside the material. Third, the rate of deterioration is affected by the type and the concentration of ions present in water, and by the chemical composition of the solid. Unlike natural rocks and minerals, *concrete is essentially an alkaline material* because all of the calcium compounds that constitute the hydration product of portland cement are alkaline. Therefore, acidic waters are particularly harmful to concrete.

---

*On March 27, 1980, Alexander Kjeland, an oil-drilling *steel platform* off the coast of Stavanger (North Sea) failed suddenly, resulting in the death of 123 persons. Shortly after this incident another offshore oil-drilling steel structure collapsed near Newfoundland, causing the death of 64 persons.

Most of our knowledge of physical-chemical processes responsible for concrete deterioration is derived from case histories of structures in the field; it is difficult in the laboratory to simulate the combination of long-term conditions normally present in real life. In practice, deterioration of concrete is seldom due to a single cause. Usually, at an advanced stage of a material's degradation more than one deleterious phenomenon is at work. In general, the physical and chemical causes of deterioration are so closely intertwined and mutually reinforcing that separation of the causes from their effects often becomes impossible. Therefore, a classification of concrete deterioration processes into neat categories should be treated with some caution. The purpose of such a classification is to explain systematically and individually the various phenomena. However, one must not overlook the interactions that occur when several phenomena are present simultaneously.

5.4 Water as an Agent of Deterioration

Concrete is not the only material vulnerable to physical and chemical processes of deterioration associated with water. Therefore it is desirable to review, in general, the characteristics of water that make it the principal agent of destruction of solid materials.

Water in its various forms, such as seawater, groundwater, river water, lake water, snow, ice, and vapor, is undoubtedly the most abundant fluid in nature. Water molecules are very small and, therefore, are able to penetrate into extremely fine pores or cavities. As a solvent, water is noted for its ability to dissolve more substances than any other known liquid. This property accounts for the presence of many ions and gases in some waters which, in turn, become instrumental in causing chemical decomposition of solid materials. Also, water has the highest heat of vaporization among the common liquids; therefore, at ordinary temperatures it has a tendency to exist in the liquid state in a porous material, rather than vaporizing and leaving the material dry. Furthermore, with porous solids, internal moisture movements and structural transformations of water are known to cause disruptive volume changes of many types. For example, freezing of water into ice, formation of an ordered structure of water inside the fine pores, development of osmotic pressure due to differences in ionic concentration, and hydrostatic pressure buildup by differential vapor pressures can lead to high internal stresses. A brief review of the structure of water molecules will be useful here for understanding these phenomena.

5.4.1 The Structure of Water

The H-O-H molecule is covalent bonded. Due to differences in the charge centers of hydrogen and oxygen, the positively charged proton of the hydrogen ion belonging to a water molecule attracts the negatively charged electrons of the neighboring water molecules. This relatively weak force of attraction, called the *hydrogen bond*, is responsible for the *ordered structure of water*.

The highest manifestation of the long-range order in the structure of water due to hydrogen bonding is seen in ice (Fig. 5-1*a*). In ice, each molecule of water is surrounded by four molecules such that the group has one molecule at the center and the other four at the corners of a tetrahedron. In all three directions the molecules and groups of molecules are held together by hydrogen bonding. Ice melts at 0°C when approximately 15 percent of the hydrogen bonds break up. As a result of the partial breakdown in

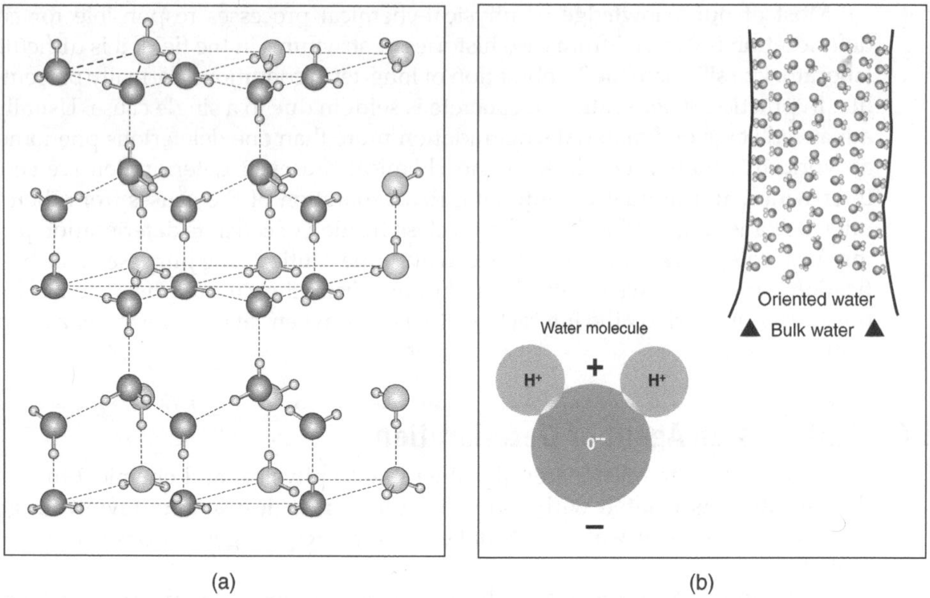

Figure 5-1 (*a*) Structure of ice. [Reprinted from Pauling, L., *The Nature of Chemical Bond, 3d.* Cornell University, 1960. Used by permission of the publisher, Cornell University Press]. (*b*) Structure of oriented water molecules in a micropore. (From Winkler, E.M., *Stone: Properties, Durability in Man's Environment*, Springer-Verlag, New York, 1973.)

The structure and properties of water are affected by temperature and by the size of pores in a solid.

directionality of the tetrahedral bond, each water molecule can acquire more than four nearest neighbors, the density thus rising from 0.917 to 1. The reversibility of the process accounts for the phenomenon that liquid water, on solidification, expands rather than shrinks.

Compared to the structure of ice, water at room temperature has approximately 50 percent of the hydrogen bonds broken. Materials in the broken-bond state have unsatisfied surface charges that give rise to surface energy. The surface energy in liquids causes surface tension, which accounts for the tendency of a large number of molecules to adhere together. It is the *high surface tension of water* (defined as the force required to pull the water molecules apart) that prevents it from acting as an efficient plasticizing agent in concrete mixtures unless certain chemical admixtures are added to reduce the surface tension.

Formation of the *oriented structure* of water by hydrogen bonding in micropores is known to cause expansion in many systems. In solids, the surface energy due to unsatisfied charges depends on the surface area; therefore, the surface energy is high when numerous fine pores are present. If water is able to permeate such micropores, and if the forces of attraction at the surface of the pores are strong enough to break down the surface tension of bulk water and orient the molecules to an ordered structure (analogous to the structure of ice), this oriented or ordered water, being less dense than the bulk water, will require more space and will therefore tend to cause expansion (Fig. 5-1*b*).

5.5 Permeability

In concrete, the role of water has to be seen in a proper perspective because, as a necessary ingredient for the cement hydration reactions and as an agent that facilitates the mixing of the components of concrete, water is present from the beginning. Gradually, depending on the ambient conditions and the thickness of a concrete element, most of the evaporable water in concrete (all the capillary water and a part of the adsorbed water) is lost, leaving the pores empty or unsaturated. As it is the evaporable water that is freezable and also free for internal movement, a concrete will not be vulnerable to water-related destructive phenomena if there is a little or no evaporable water left after drying, and if subsequent exposure of that concrete to the environment does not cause resaturation of the pores. The latter, to a large extent, depends on the *hydraulic conductivity*, which is also known as the *coefficient of permeability* (K). Note that, in concrete technology, it is a common practice to drop the adjective and refer to K simply as the permeability.

Garboczi[1] reviewed several theories that attempt to relate the microstructural parameters of cement products with either diffusivity (the rate of diffusion of ions through water-filled pores) or permeability (the rate of viscous flow of fluids under pressure through the pore structure). For materials like concrete with numerous microcracks, a satisfactory *fluid transport property factor* is difficult to determine because of the effect of unpredictable changes in the pore structure upon penetration of a fluid from outside. Note that the fluid transport property of the material is changing continuously because of cycles of narrowing and widening of the pores and microcracks due to ongoing physical-chemical interactions between the penetrating fluid and the minerals of the cement paste. According to Garboczi, *the diffusivity predictions need more development and validation before their practical usefulness can be proven*. Therefore, the discussion in this text will be limited to permeability of concrete. However, it is implied that the term, in a crude sense, refers to the overall fluid transport property of the material.

Permeability is defined as the property that governs the rate of flow of a fluid into a porous solid. For *steady-state flow*, the coefficient of permeability (K) is determined from Darcy's expression:

$$\frac{dq}{dt} = K\frac{\Delta HA}{L\mu}$$

where dq/dt = rate of fluid flow
μ = viscosity of the fluid
ΔH = pressure gradient
A = surface area
L = thickness of the solid

The coefficient of permeability of a concrete to gases and water vapor is much lower than the coefficient for liquid water; therefore, tests for measurement of permeability are generally carried out using water that has no dissolved air. Unless otherwise stated, the data in this chapter pertain to permeability of concrete to pure water. Due to their interaction with cement paste, the permeability values for solutions containing ions would be different from the water permeability.

5.5.1 Permeability of Hardened Cement Paste

In hardened cement paste, at any point in time during the hydration process the size and continuity of the pores would control the coefficient of permeability. As discussed in Chap. 2, the mixing water is indirectly responsible for the permeability of hydrated cement paste because its content determines, first, the total space and, subsequently, the unfilled space after the water has been consumed either by cement hydration reactions or by evaporation to the environment. The coefficient of permeability of freshly mixed cement paste is of the order of 10^{-4} to 10^{-5} cm/s; with the progress of hydration, as the capillary porosity decreases, so does the coefficient of permeability (Table 5-1), but there is no direct proportionality between the two. For instance, when the capillary porosity decreases from 40 to 30 percent (Fig. 2-11), the coefficient of permeability drops by a much greater amount (i.e., from about 110 to 20 $\times$ 10^{12} cm/s). However, a further decrease in the porosity from 30 to 20 percent brings about only a small drop in permeability. This is because, in the beginning, as the cement hydration process progresses, even a small decrease in the total capillary porosity is associated with considerable segmentation of large pores, thus greatly reducing the size and number of channels of flow in the cement paste. Typically, about 30 percent capillary porosity represent a point when the interconnections between the pores have already become so tortuous that a further decrease in the porosity of the paste is not accompanied by a substantial decrease in the permeability coefficient.

In general, when the water-cement ratio is high and the degree of hydration is low, the cement paste will have a high capillary porosity. It will contain a relatively large number of big and well-connected pores and, therefore, its coefficient of permeability will be high. As hydration progresses, most of the pores will be reduced in size (e.g., 100 nm or less) and will lose their interconnections, and the permeability will drop. The coefficient of permeability of cement paste when most of the capillary voids are small and not interconnected is of the order of 10^{-12} cm/s. As already stated, with conventional cement pastes the discontinuity in the capillary network is generally reached when the capillary porosity is about 30 percent. With 0.4, 0.5, 0.6, and 0.7 water-cement ratio cement pastes, this generally happens in 3, 14, 180, and 365 days of moist curing, respectively. As the

Age Days	Coefficient of Permeability (cm/s $\times$ 10^{-11})
Fresh	20,000,000
5	4,000
6	1,000
8	400
13	50
24	10
Ultimate	6

Source: Powers, T.C., L.E. Copeland, J.C. Hayes, and H.M. Mann, *J. ACI, Proc.*, Vol. 5, pp. 285–298, 1954.

TABLE 5-1 Reduction in Permeability of a 0.7 w/c Cement Paste with the Progress of Hydration water-cement ratio in most concrete mixtures seldom exceeds 0.7, theoretically, with most well-cured concrete mixtures, cement paste should not be the principal contributing factor to the coefficient of permeability of concrete.

5.5.2 Permeability of Aggregate

Compared to 30 to 40 percent capillary porosity of ordinary cement paste present in hardened concrete, the volume of pores in most natural aggregates is usually under 3 percent and rarely exceeds 10 percent. It is expected, therefore, that the permeability of aggregate would be much lower than that of the typical cement paste. However, this may not necessarily be the case. From the permeability data of some natural rocks and cement pastes (Table 5-2), the coefficient of permeability of aggregates are as variable as those of hydrated cement pastes of water-cement ratios in the range 0.38 to 0.71.

Although the coefficient of permeability of marble, traprock, diorite, basalt, and dense granite is generally of the order of 1 to 10×10^{-12} cm/s, some varieties of granite, limestone, sandstones, and cherts show values that are higher by two orders of magnitude. The reason why some aggregates with only 10 percent porosity show much higher permeability than the cement paste is that the size of capillary pores in aggregate is usually much larger. Most of the capillary porosity in a mature cement paste lies in the 10 to 100 nm range, whereas the pore sizes in aggregate are, on the average, larger than 10 μm. With some cherts and limestones the pore size distribution involves a considerable content of fine pores. Their permeability may be low, but such aggregates are vulnerable to expansion and cracking associated with sluggish moisture movements and the resulting hydrostatic pressure.

5.5.3 Permeability of Concrete

Theoretically, the introduction of low-permeability aggregate particles into a high-permeability cement paste (especially with high water-cement ratio pastes at early ages when the capillary porosity is high) is expected to reduce the permeability of the system because the aggregate particles should intercept the channels of flow within the cement paste matrix. Compared to a neat cement paste, therefore, a mortar or a concrete

Type of Rock	Coefficient of Permeability (cm/s)	Water-Cement Ratio of Mature Paste with the Same Coefficient of Permeability
Dense trap	2.47×10^{-12}	0.38
Quartz diorite	8.24×10^{-12}	0.42
Marble	2.39×10^{-11}	0.48
Marble	5.77×10^{-10}	0.66
Granite	5.35×10^{-9}	0.70
Sandstone	1.23×10^{-8}	0.71
Granite	1.56×10^{-8}	0.71

Source: Powers, T.C., *J. Am. Ceram. Soc.*, Vol. 4, No. 1, pp. 1–5, 1958.

TABLE 5-2 Comparison between the Permeability of Rocks and Cement Pastes

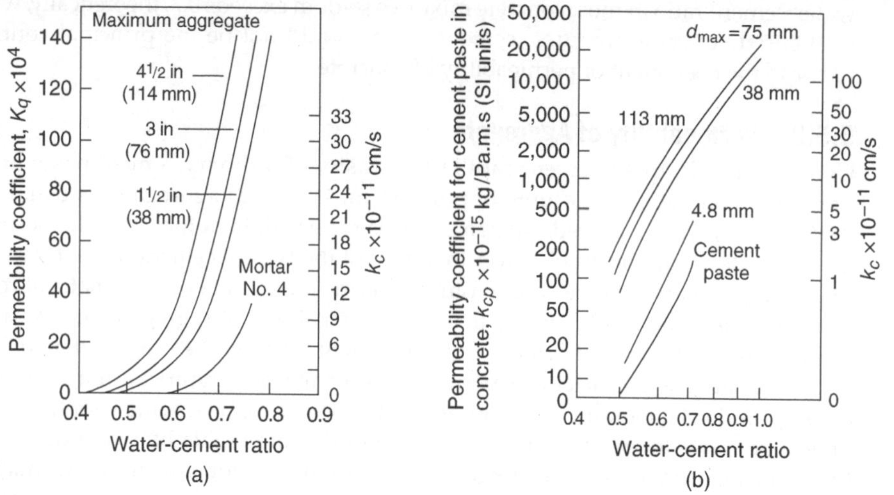

FIGURE 5-2 Influence of water-cement ratio and maximum aggregate size on concrete permeability: (*a*) *Kq* is a relative measure of the flow of water through concrete in a cubic feet per year per square foot of area for a unit hydraulic gradient. [(*a*) From *Concrete Manual*, 8th ed., U.S. Bureau of Reclamation, 1975, p. 37, (*b*) adapted from Beton-Bogen, Aalborg Cement, Aalborg, Denmark, 1979.]

The permeability of concrete to water depends mainly on the water-cement ratio (which determines the size, volume, and continuity of capillary voids) and maximum aggregate size (which influences the microcracks in the transition zone between the coarse aggregate and the cement paste).

with the same water-cement ratio and degree of maturity should give a lower coefficient of permeability. Test data indicate that, in practice, this does not happen. The two sets of data* in Fig. 5-2 clearly show that the addition of aggregate to a cement paste or a mortar increased the permeability considerably; in fact, the larger the aggregate size, the greater the coefficient of permeability. Typically, the permeability coefficients for moderate-strength concrete (containing 38 mm aggregate, 356 kg/m^3 cement, and an 0.5 water-cement ratio), and low-strength concrete used in dams (75 to 150 mm aggregate, 148 kg/m^3 cement, and an 0.75 water-cement ratio) are of the order of 1×10^{-10} and 30×10^{-10} cm/s, respectively.

The explanation as to *why the permeability of mortar or concrete is higher than the permeability of the corresponding cement paste* lies in the microcracks normally present in the interfacial transition zone between aggregate and the cement paste. As described in Chap. 2, the aggregate size and grading affect the bleeding characteristic of a concrete mixture that, in turn, influences the interfacial transition zone. During the early hydration period the interfacial transition zone is weak and vulnerable to cracking from differential strains between the cement paste and the aggregate particles that are induced by drying shrinkage, thermal shrinkage, and externally applied load. The cracks in the interfacial transition zone are too small to be seen by the naked eye, but are larger than

*The permeability coefficient in SI units is expressed as kg/Pa.m.s, which is approximately 10^{-3} times smaller than the coefficient expressed in cm/s.

most capillary cavities present in the cement paste matrix. Later, the propagation of microcracks establish the interconnections that become instrumental in increasing the permeability of the system.

Due to the significance of the permeability to physical and chemical processes of deterioration of concrete, which will be described below, a brief review of the *factors controlling the permeability* of concrete should be useful. Because strength and permeability are related to each other through the capillary porosity (Fig. 2-11), as a first approximation the factors that influence the strength of concrete (Fig. 3-14) also influence the permeability. A reduction in the volume of large (e.g., >100 nm) capillary voids in the paste matrix will reduce the permeability. This should be possible by using a low water-cement ratio, adequate cement content, and proper compaction and curing. Similarly, proper attention to the aggregate size and grading, thermal and drying shrinkage strains, and premature or excessive loading are necessary steps to reduce microcracking in the interfacial transition zone, which is the major cause of high permeability of concrete in field practice. Finally, it should also be noted that tortuosity of the path of fluid flow that determines the permeability also depends on the thickness of the concrete member.

5.6 Classification of the Causes of Concrete Deterioration

Mehta and Gerwick[2] grouped the physical causes of concrete deterioration (Fig. 5-3) into two categories: (*a*) surface wear or loss of mass due to abrasion, erosion, and cavitation; (*b*) cracking due to normal temperature and humidity gradients, crystallization of salts in pores, structural loading, and exposure to temperature extremes such as freezing or fire. Similarly, as will be discussed later in this chapter, the authors grouped the chemical causes of deterioration into three categories: (1) hydrolysis of the cement paste components by soft water; (2) cation-exchange reactions between aggressive fluids and the cement paste; and (3) reactions leading to formation of expansive products, such as in the case of sulfate attack, alkali-aggregate reaction, and corrosion of reinforcing steel in concrete.

It should be emphasized again that the distinction between the physical and chemical causes of deterioration is purely arbitrary; in practice, the two are frequently superimposed on each other. For example, loss of mass by surface wear and cracking increases the permeability of concrete, which then becomes the primary cause of one or more processes of chemical deterioration. Similarly, the detrimental effects of the chemical phenomena are physical; for instance, leaching of the components of hardened cement paste by soft water or acidic fluids would increase the porosity of concrete, thus making the material more vulnerable to abrasion and erosion.

Cracking of concrete due to normal temperature and humidity gradients was discussed in Chap. 4. A comprehensive report on the causes, mechanisms, and control of cracking in concrete is also published by the ACI Committee 224.[3] Deterioration of concrete by surface wear, crystallization of salts in pores, freeze-thaw cycles, fire, and a number of chemical processes mentioned above are discussed in this chapter.

5.7 Surface Wear

Progressive loss of mass from a concrete surface can occur due to abrasion, erosion, and cavitation. The term *abrasion* generally refers to dry attrition, such as in the case of wear on pavements and industrial floors by vehicular traffic. The term *erosion* is normally

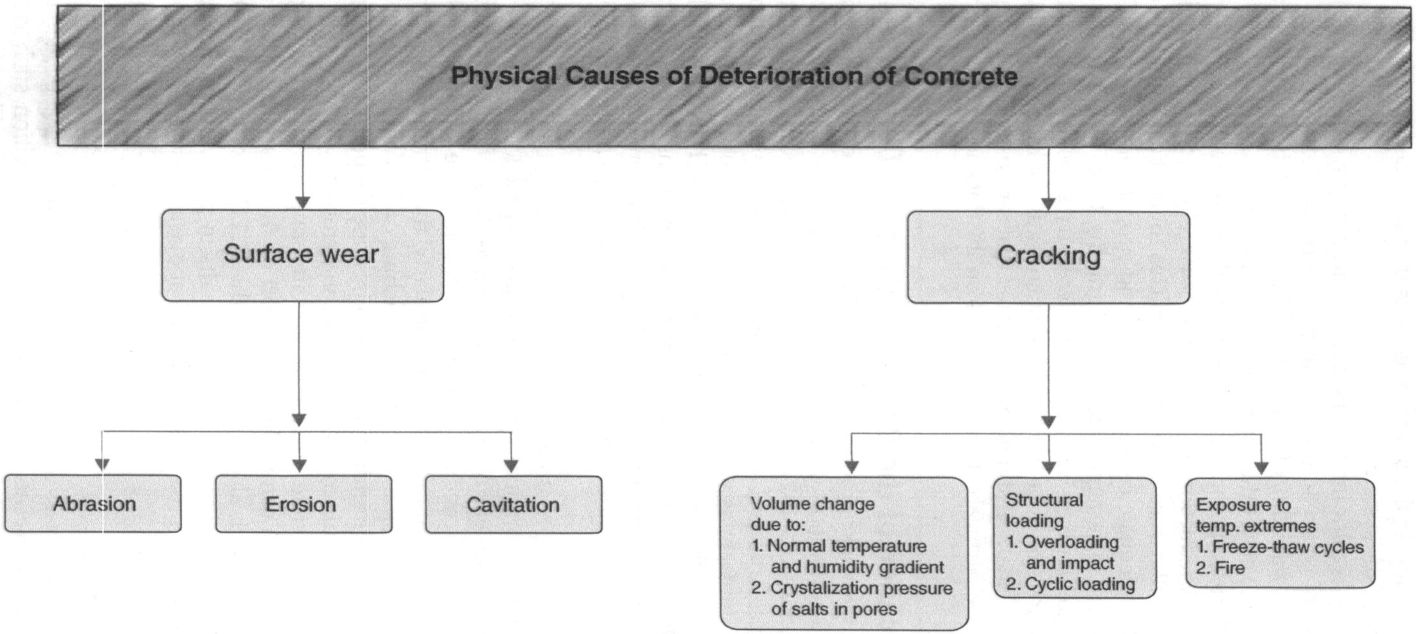

Figure 5-3 Physical causes of concrete deterioration. (From Mehta, P.K., and B.C. Gerwick, Jr., *Concr. Int.*, Vol. 4, pp. 45–51, 1982.)

used to describe wear by the abrasive action of fluids containing solid particles in suspension. Erosion takes place in hydraulic structures, for instance canal linings, spillways, and concrete pipes for water or sewage transport. Another possibility of damage to hydraulic structures is by *cavitation*, which relates to loss of mass by formation of vapor bubbles and their subsequent collapse due to sudden change of direction in a rapidly flowing water.

Hardened cement paste does not possess a high resistance to attrition. Service life of concrete can be shortened under conditions of repeated attrition cycles, especially when the cement paste in concrete is of high porosity or low strength, and is inadequately protected by an aggregate which itself lacks wear resistance. Using a special test method, Liu[4] found a good correlation between the water-cement ratio and abrasion resistance of concrete (Fig. 5-4*a*). Accordingly, *for obtaining abrasion resistance concrete surfaces*, ACI Committee 201 recommends that the compressive strength of concrete should not be less than 4000 psi (28 MPa). Suitable strength may be attained by low water-cement ratio, proper grading of fine and coarse aggregate (limit the maximum size to 25 mm), lowest consistency (e.g., 75 mm max. slump) needed for proper placing and consolidation, and a minimum air content consistent with exposure conditions.

When a fluid containing suspended solid particles is in contact with concrete, the impinging, sliding, or rolling action of particles will cause surface wear. The rate of surface erosion depends on the porosity or the strength of concrete, and on the amount, size, shape, density, hardness, and velocity of the moving particles. If the quantity and size of solids is small, such as, silt in an irrigation canal, the erosion loss will be negligible at bottom velocities up to 1.8 m/s (velocity at or above which a given particle can be transported). When *severe erosion or abrasion* conditions exist, it is recommended that, in addition to the use of hard aggregates, the concrete should be proportioned to develop at least 41 MPa compressive strength at 28 days and adequately cured before exposure to the aggressive environment. ACI Committee 201 recommends at least 7 days of continuous moist curing after the finishing of concrete.

Where *additional measures for improving the durability of concrete to abrasion or erosion* are needed it is worth remembering that the process of physical attrition of concrete occurs at the surface; therefore, particular attention should be paid to ensure that, at least, the concrete at the surface is of high quality. To reduce the formation of a weak surface, called *laitance* (the term is used for a layer of fine particles, removed from the cement paste and aggregate), it is recommended to delay the floating and trowelling operations until the concrete has lost its surface bleedwater. Heavy-duty industrial floors or pavements may be designed to have a 25- to 75-mm-thick topping, consisting of a low water-cement ratio concrete mixture and a hard aggregate of 12.5 mm maximum size. Because of their very low water-cement ratio, concrete toppings containing admixtures or superplasticizing admixtures are becoming increasingly popular for use against abrasion or erosion. Mineral admixtures, such as silica fume, are also being used to obtain high strength and impermeability. Besides enabling hardened concrete to become less permeable after moist curing, fresh concrete mixtures containing mineral admixtures are less prone to bleeding. Resistance to deterioration by permeating fluids and reduction in dusting due to attrition can also be achieved by the application of surface-hardening solutions to well-cured new floors or abraded old floors. The solutions most commonly used for this purpose are magnesium and zinc fluosilicate, or sodium silicate, which react with calcium hydroxide present in the portland cement paste to form insoluble reaction products, thus sealing the capillary pores at or near the surface.

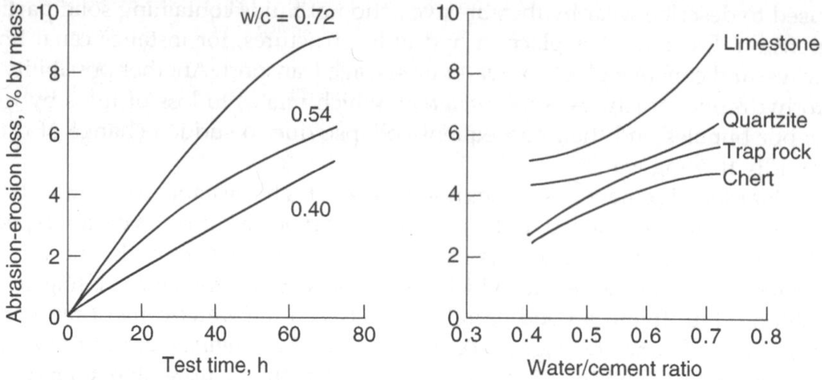

(a)

(b)

Figure 5-4 (*a*) Influence of water-cement ratio and aggregate type on abrasion-erosion damage in concrete; (*b*) cavitation damage to concrete lining in a 41-ft-diameter (12.5 m) tunnel of the Glen Canyon Dam. [(*a*) From Liu, T.C., *J. ACI, Proc.*, Vol. 78, No. 5, p. 346, 1981; (*b*) photograph courtesy of U.S. Bureau of Reclamation and William Scharf of Guy F. Atkinson Construction Co.]

While good-quality concrete shows excellent resistance to steady flow of clear water, nonlinear flow at velocities exceeding 12 m/s (7 m/s in closed conduits) may cause severe damage to concrete through *cavitation*. In flowing water, vapor bubbles form when the local absolute pressure at a given point is reduced to ambient vapor pressure of water corresponding to the ambient temperature. As the vapor bubbles flowing downstream with water enter a region of high pressure, they implode with great impact because of the entry of high-velocity water into the previously vapor-occupied space, causing severe local pitting. Therefore, the concrete surface affected by cavitation is irregular or pitted, in contrast to the smoothly worn surface by erosion from suspended solids. Also, in contrast to erosion or abrasion, a strong concrete may not necessarily be effective in preventing cavitation damage. The best solution lies in removal of the causes of cavitation, such as surface misalignments or abrupt changes of slope. In 1984, extensive repairs were needed for the concrete lining of a tunnel of the Glen Canyon Dam (Fig. 5-4*b*); the damage was caused by cavitation attributable to surface irregularities in the lining.

Test methods for the evaluation of wear resistance of concrete are not always satisfactory because simulation of the field conditions of wear is not easy in the laboratory. Therefore, laboratory methods are not intended to provide a quantitative measurement of the length of service that may be expected from a given concrete surface; they can be used for a qualitative evaluation of the effects of concrete materials and curing and finishing procedures on the abrasion resistance of concrete.

ASTM C 779 describes three optional methods for testing the relative abrasion resistance of horizontal concrete surfaces. In the steel-ball abrasion test, load is applied to a rotating head containing steel balls while the abraded material is removed by water circulation. In the dressing wheel test, load is applied through rotating dressing wheels of steel. In the revolving-disk test, revolving disks of steel are used in conjunction with a silicon carbide abrasive. In each of the tests, the degree of wear can be measured in terms of weight loss after a specified time. ASTM C 418 describes the sandblasting test, which covers the abrasion resistance characteristics of concrete by subjecting it to the impingement of pneumatically driven silica sand. There are no satisfactory tests for the erosion resistance. Due to a direct relationship between the abrasion and erosion resistance, the abrasion resistance data can be used as a rough guide for the erosion resistance.

5.8 Crystallization of Salts in Pores

Under certain environmental conditions, for example, when one side of a retaining wall or slab of a permeable solid is in contact with a salt solution and the other sides are subject to loss of moisture by evaporation, the material can deteriorate by stresses caused by crystallization of salts in the pores. Winkler[5] lists a number of salts that are known to cause cracking and spalling type of damage to rocks and stone monuments. This phenomenon was attributed to the large pressures produced by crystallization of salts from their supersaturated solutions.

From investigations of masonry damage due to salt crystallization, Binda and Baronio[6] discussed the microclimatic conditions that influence whether or not any serious damage would occur. According to the authors, the extent of damage depends on the *site* of the salt crystallization, which is determined by a dynamic balance between the rate of evaporation of water from the exposed surface of the material and the rate of supply of the salt solution to that site. When the rate of evaporation is lower than the rate of supply of water from inside the masonry, the salt crystallization takes place on the external surface, without causing any damage. Only when the rate of migration of the salt solution through the interconnected pores of the material is slower than the rate of replenishment, the drying zone occurs substantially beneath the surface. Salt crystallization under such conditions may result in sufficient expansion to cause flaking or spalling.

In the literature,[7] the terms *salt scaling, salt weathering,* and *salt hydration attack* have been used to describe the physical manifestation of a phenomenon that has been observed

Figure 5-5 Salt scaling in mortar prisms partially submerged in solutions of (*a*) sodium sulfate and (*b*) sodium carbonate. (*Photographs courtesy of Harvey Haynes.*)

with masonry and porous concrete exposed to *hydratable* salts such as sodium sulfate and sodium carbonate. Thenardite (Na_2SO_4) converts into its hydrated form, Mirabalite ($Na_2SO_4 \cdot 10H_2O$) at 20°C when the relative humidity is more than 72 percent, and at 32°C when the relative humidity is 81percent or more. Interestingly, the transformation of Thermonatrite ($Na_2CO_3 \cdot H_2O$) into Natron ($Na_2CO_3 \cdot 10H_2O$) occurs at similar temperature and humidity conditions, which happen to be within the range of everyday environmental changes in many parts of the world. Due to large differences in the density, considerable volumetric expansion is associated with the transformation of the anhydrous form of these salts into their hydrated form. As a consequence of numerous cycles of ambient relative humidity and temperature changes, a progressive deterioration of concrete at the surface occurs (Fig. 5-5).[8] This type of purely physical salt attack from a penetrating salt solution, as distinguished from the attacks involving chemical interactions with the cement hydration products, is not known to cause structural damage.[9]

5.9 Frost Action

In cold climates, damage to concrete pavements, retaining walls, bridge decks, and railings, attributable to *frost action* (freezing and thawing cycles), is one of the major problems requiring heavy expenditures for the repair and replacement of structures. The causes of deterioration of hardened concrete by frost action can be related to the complex microstructure of the material; however, the deleterious effect depends not only on characteristics of the concrete but also on the specific environmental conditions. Thus a concrete that is frost resistant under given freeze-thaw conditions can be destroyed under a different set of conditions.

Frost damage in concrete can take several forms. The most common is *cracking and spalling* of concrete caused by progressive expansion of the cement paste matrix from repeated freezing and thawing cycles. Concrete slabs exposed to freezing and thawing cycles in the presence of moisture and deicing salts are susceptible to *scaling* (i.e., the finished surface flakes or peels off). Also some coarse aggregates in concrete slabs are known to cause cracking, usually parallel to joints and edges, which eventually acquires a pattern resembling a large capital letter D (cracks curving around two of the four corners of the slab). This type of cracking is described by the term *D-cracking*. The different types of concrete deterioration due to frost action are shown by the photographs in Fig. 5-6.

Air entrainment has proved to be an effective means of reducing the risk of damage to concrete by frost action. The mechanisms by which frost damage occurs in the cement paste and how air-entrainment prevents the damage are described next.

5.9.1 Frost Action on Hardened Cement Paste

Powers aptly described the mechanisms of frost action in cement paste, and also explained why air entrainment is effective in reducing the expansion associated with this phenomenon:

> When water begins to freeze in a capillary cavity, the increase in volume accompanying the freezing of the water requires a dilation of the cavity equal to 9 percent of the volume of frozen water or forcing of the amount of excess water out through the boundaries of the specimen, or some of both effects. During this process, hydraulic pressure is generated and the magnitude of that pressure depends on the distance to an "escape boundary," the permeability of the intervening material, and the rate at which ice is formed. Experience shows that disruptive pressures will be developed in a saturated specimen of paste unless every capillary cavity in the paste is not farther than three or four thousandths of an inch from the nearest escape boundary. Such closely spaced boundaries are provided by the correct use of a suitable air-entraining agent.[10]

Powers' data and a diagrammatic representation of his hypothesis are shown in Fig. 5-7. During freezing to –24°C, the saturated cement paste specimen containing no entrained air elongated about 1600 millionths, and on thawing to the original temperature about 500 millionths permanent elongation was observed (Fig. 5-7*a*). The specimen containing 2 percent entrained air showed about 800 millionths elongation on freezing, and a residual elongation of less than 300 millionths on thawing (Fig. 5-7*b*). The specimen containing 10 percent entrained air showed no appreciable dilation during freezing and no residual dilation at the end of the thawing cycle. On the contrary, this air-entrained paste showed contraction during freezing (Fig. 5-7*c*). A diagrammatic illustration of Powers' hypothesis is shown in Fig. 5-7*d*. Figure 5-8 indicates how the presence of air-voids can reduce the stresses caused by ice formation in the concrete.

Powers also proposed that, in addition to the *hydraulic pressure* caused by water freezing in large cavities, the *osmotic pressure* resulting from partial freezing of solutions in capillaries can be another source of destructive expansion in cement paste. Water in the capillaries is not pure; it contains several soluble substances, such as alkalies,

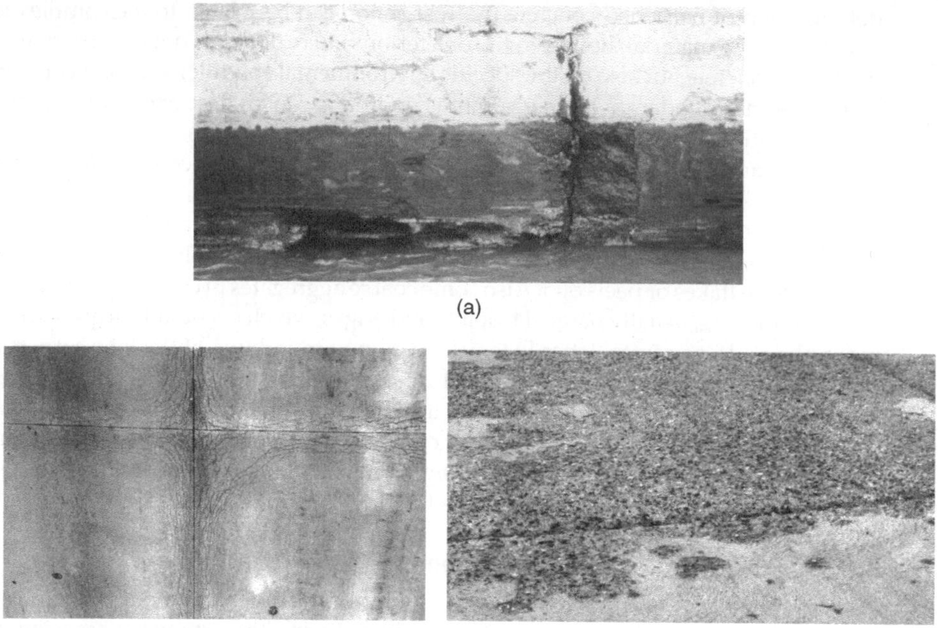

(a) (b) (c)

Figure 5-6 Types of frost action damage in concrete: (*a*) deterioration of a non-air-entrained concrete-retaining wall along the saturation line (Lock and Dam No. 3, Monongahela River, Pittsburg, PA); (*b*) severe D-cracking along longitudinal and transverse joints of a 9-year-old pavement; (*c*) scaling of a concrete surface. [(*a*) Photograph courtesy of J.M. Scanlon, U.S. Army Corps of Engineers, Vicksburg, MS; (*b*) photograph courtesy of D. Stark, from *Report RD* 023.01P, Portland Cement Association, Skokie, IL,1974; (*c*) photograph courtesy of R.C. Meininger, from *Concrete in Practice*, Publ. 2, National Ready Mixed Concrete Association, Silver Springs, MD.]

(a) Progressive expansion of unprotected (non-air-entrained) cement paste by repeated freeze-thaw cycles leads to deterioration of concrete by cracking and spalling. Many Corps of Engineers lock walls which were built prior to the use of air entrainment in concrete suffer from freezing and thawing deterioration in moist environment. Standard operating procedures normally require the water in the locks to remain at upper pool level during the winter so that the concrete is protected from free-thaw cycles. All hydraulic projects of the Corps built since 1940s have been constructed with air-entrained concrete.

(b) D-cracking in highway and airfield pavement refers to a D-shaped pattern of closely spaced cracks which occur parallel to longitudinal transverse joints. This type of cracking is associated with coarse aggregates which contain a proportionately greater pore volume in the narrow pore size range (0.1 to 1 μm).

(c) Concrete scaling or flaking of the finished surface from freezing and thawing generally starts as localized small patches which later on may merge and extend to expose large areas. Light scaling does not expose the coarse aggregate. Moderate scaling exposes the coarse aggregate and may involve loss of up to 3 to 9 mm of the surface mortar. In severe scaling, more surface has been lost and the aggregate is clearly exposed and stands out. Most scaling is caused by (i) inadequate air entrainment, (ii) application of calcium and sodium chloride deicing salts, (iii) performing finishing operations while bleed water is still on the surface, and (iv) insufficient curing before exposure of the concrete to frost action in the presence of moisture and deicing salts.

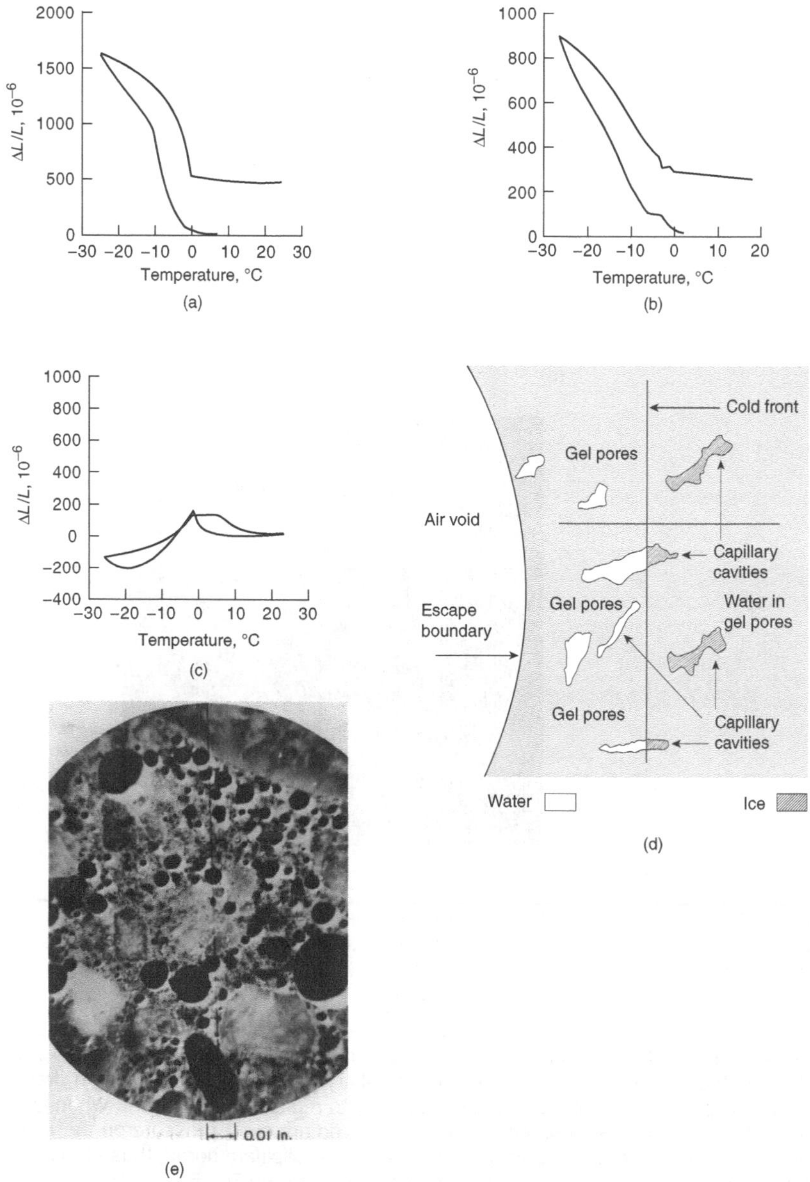

FIGURE 5-7 Response of saturated cement paste to freezing and thawing with and without entrained air. [(*a*)–(*c*), From Powers, T.C., *The Physical Structure and Engineering Properties of Concrete,* Bulletin 90, Portland Cement Association, Skokie, IL,1958, (*d*) From Cordon, W.A., *Freezing and Thawing of Concrete—Mechanism and Control,* ACI Monograph 3, 1967; (*e*) From PCA, *Design and Control of Concrete Mixtures*, 1979.]

According to Powers, a saturated cement paste containing no entrained air expands on freezing due to the generation of hydraulic pressure (a). With increasing air entrainment, the tendency to expand decreases because the entrained air voids provide escape boundaries for the hydraulic pressure [(b), (c), and (d)]. (e) Polished section of air- entrained concrete as seen through a microscope.

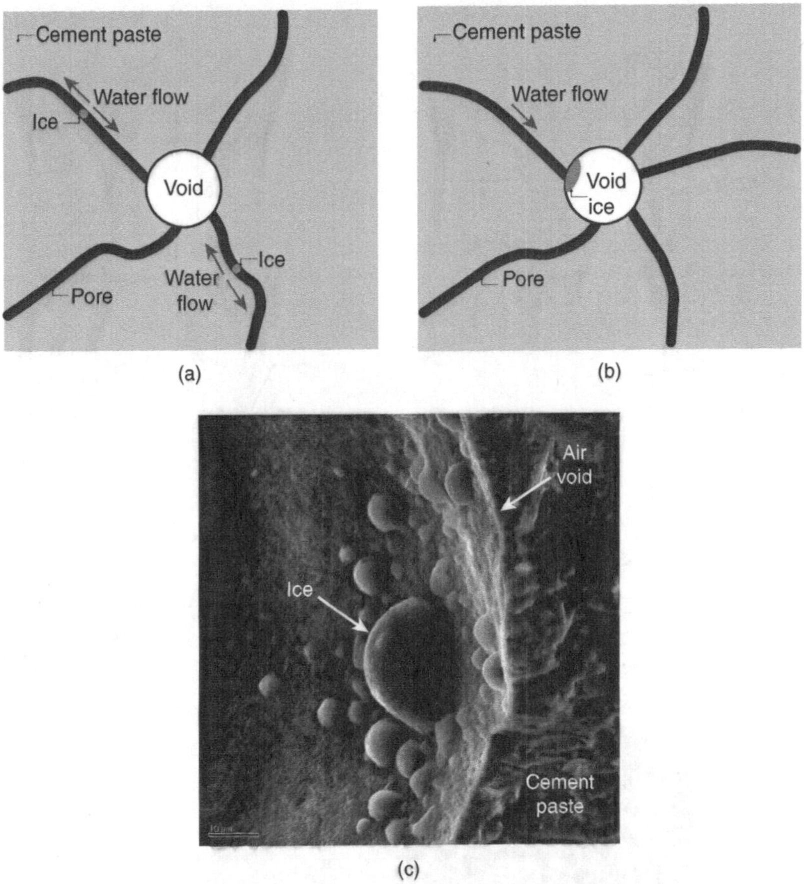

FIGURE 5-8 (*a*) Schematic diagram of ice forming in capillary voids; (*b*) ice forming in an air void; and (*c*) scanning electron micrograph of ice crystals growing in an air void [(*a*) and (*b*) courtesy of George W. Scherer, (*c*) micrograph from Corr, D.J., P.J.M. Monteiro, J. Bastacky, *ACI Mat. J.*, Vol. 99, No. 2, pp. 190–195, Mar–Apr, 2002].

The transformation of ice from liquid water generates a volumetric dilation of 9 percent. As shown in Fig. 5-8, if the transformation occurs in small capillary pores, the ice crystals can damage the cement paste by pushing the capillary walls and by generating hydraulic pressure. Air voids can provide an effective escape boundary to reduce this pressure. When ice is formed in an empty air void (Figs. 5-8b and c), the crystals do not exert pressure on the walls. The growth of ice crystals in the air void attracts water from the capillary pores, thus reducing the hydraulic pressure and inducing shrinkage in the cement paste (see Fig. 5-9).

Experimentally, it is difficult to see the ice crystals inside an air void because the traditional scanning electron microscopy requires that the sample be dried. In addition, it is not easy to maintain the low temperature required to stabilize the ice in the sample. These limitations are overcome by using a special low-temperature scanning electron microscope that is able to maintain the sample frozen for a long period of time. In Fig. 5-8c, ice crystals can be seen forming inside an air void, providing an important open space for the crystals to develop. Had these crystals formed in the cement paste, the matrix would have expanded, leading to cracking and loss of stiffness.

chlorides, and calcium hydroxide. Solutions freeze at lower temperatures than pure water; generally, the higher the concentration of salts in a solution, the lower the freezing point. The existence of local salt concentration gradients between capillaries is envisaged as the source of osmotic pressure.

Hydraulic pressure (due to an increase in the specific volume of water on freezing in large cavities), and osmotic pressure (due to salt concentration differences in the pore fluid) do not appear to be the only causes of expansion of cement pastes exposed to frost action. Expansion of cement paste specimens was observed[11] even when benzene, which contracts on freezing, was used as a pore fluid instead of water.

Analogous to the formation of ice lenses in soil, a *capillary effect*,[12] involving large-scale migration of water from small pores to large cavities, is believed to be the primary cause of expansion in porous bodies. According to the theory advanced by Litvan,[13] the rigidly held water by the C-S-H (both interlayer and adsorbed in gel pores) in cement paste cannot rearrange itself to form ice at the normal freezing point of water because the mobility of water existing in an ordered state is rather limited. Generally, the more rigidly a water is held, the lower will be the freezing point. It may be recalled that three types of water are physically held in cement paste; in order of increasing rigidity these are the capillary water in small capillaries (10 to 50 nm), the adsorbed water in gel pores, and the innerlayer water in the C-S-H structure.

It is estimated that water in the gel pores does not freeze above −78°C. Therefore, when a saturated cement paste is subjected to freezing conditions, the water in large cavities turns into ice while the gel pore water continues to exist as liquid water in a supercooled state. This creates a thermodynamic disequilibrium between the frozen water in capillaries, which acquires a low-energy state, and the supercooled water in gel pores, which is in a high-energy state. The difference in the entropy of ice and supercooled water forces the latter to migrate to the lower-energy sites (large cavities) where it can freeze. This fresh supply of water from the gel pores to the capillary pores increases the volume of ice in the capillary pores steadily until there is no room to accommodate more ice. Any subsequent tendency for the supercooled water to flow toward the ice-bearing regions would obviously cause internal pressure and expansion of the system. Further, according to Litvan, the moisture transport associated with cooling of saturated porous bodies may not necessarily lead to mechanical damage. Mechanical damage occurs when the rate of moisture transport is considerably less than demanded by the conditions (e.g., a large temperature gradient, a low permeability, and a high degree of saturation).

It may be noted that during frost action on cement paste, the tendency for certain regions to expand is balanced by other regions that undergo contraction (e.g., loss of adsorbed water from C-S-H). The net effect on a specimen is, obviously, the result of the two opposite tendencies. This explains satisfactorily why cement paste containing no entrained air showed a large elongation (Fig. 5-7*a*) while the cement paste containing 10 percent entrained air showed contraction during freezing (Fig. 5-7*c*). Microscopic observations confirmed that when ice forms inside an air-void, there is shrinkage in the cement paste (Fig. 5-9).

5.9.2 Frost Action on Aggregate

Depending on how the aggregate responds to frost action, a concrete containing entrained air in the cement paste matrix can still be damaged. The mechanism underlying the development of internal pressure on freezing a saturated cement paste is also applicable to other porous bodies; this includes aggregates produced from porous

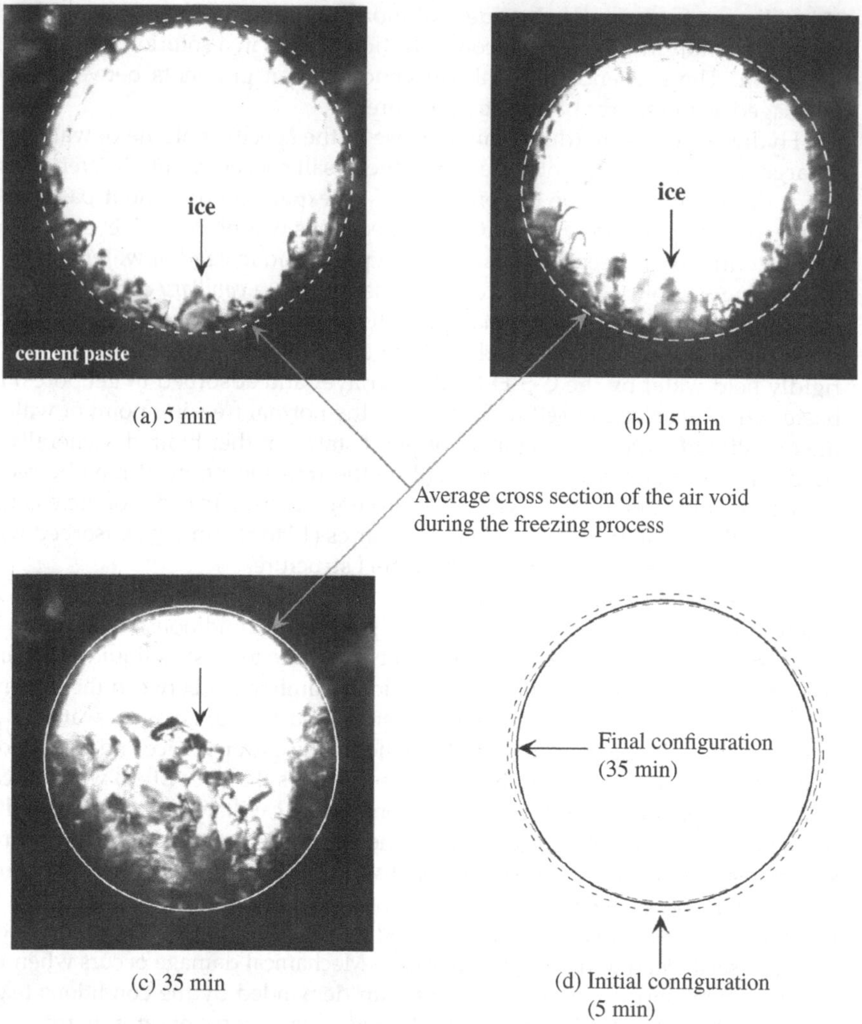

FIGURE 5-9 Sequence of ice propagation in an air-entrained void.

The images were obtained using the directional solidification method, which permits the controlled cooling and warming of a relatively large sample. The amount of time after the freezing front passed is indicated in each of the images. The external diameter of the air void is outlined to determine the change in its dimension during freezing of concrete. Note the decrease of air void diameter as freezing continues in the matrix, indicating shrinkage of the matrix. [From Piltner, R., and P.J.M. Monteiro, Cem. Concr. Res., Vol. 30, p. 847, 2000.]

rocks, such as certain cherts, sandstones, limestones, and shales. Not all porous aggregates are susceptible to frost damage; the behavior of an aggregate particle when exposed to freeze-thaw cycles depends primarily on the size, number, and continuity of pores (i.e., on the pore size distribution) and permeability.

To explain the frost damage to concrete that is attributable to aggregate, Verbeck and Landgren[14] proposed three classes of aggregate. In the first category are the aggregates of *low permeability* and high strength. On freezing of water in the pores, the elastic strain in the particle is accommodated without causing fracture. In the second category are the aggregates of *intermediate permeability*, that is, those having a significant proportion of the total porosity represented by small pores of the order of 500 nm and smaller. Capillary forces in such small pores cause the aggregate to get easily saturated and to hold water. On freezing, the magnitude of pressure depends primarily on the rate of temperature drop and the distance that water under pressure must travel to find an escape boundary to relieve the pressure. Pressure relief may be available either in the form of any empty pore within the aggregate (analogous to entrained air in cement paste) or at the aggregate surface. The critical distance for pressure relief in a hardened cement paste is of the order of 0.2 mm; it is much greater for most rocks because of their higher permeability than cement paste.

These considerations have given rise to the concept of *critical aggregate size* with respect to frost damage. With a given pore size distribution, permeability, degree of saturation, and freezing rate the large particles of an aggregate may cause damage but smaller particles of the same aggregate would not. For example, when 14-day-old concrete specimens containing a 50:50 mixture of varying sizes of quartz and chert aggregate were exposed to freeze-thaw cycles, those containing 25- to 12-mm chert required 183 cycles to show a 50 percent reduction in the modulus of elasticity, compared to 448 cycles for similarly cured concretes containing 12- to 5-mm chert.[15]

There is no single critical size for an aggregate type because this will depend on the freezing rate, degree of saturation, and permeability of the aggregate. Permeability plays a dual role: first, it determines the degree of saturation or the rate at which water will be absorbed in a given period of time; and second, it determines the rate at which water will be expelled from the aggregate on freezing (and thus development of hydraulic pressure). Generally, when aggregates larger than the critical size are present in a concrete, freezing is accompanied by *pop-outs*, that is, failure of the aggregate in which a part of the aggregate particle remains in the concrete and the other part pops out with the mortar flake.

Aggregates of high permeability, which generally contain a large number of big pores, belong to the third category. Although they permit easy entry and egress of water, they are also capable of causing durability problems. This is because the interfacial transition zone between the aggregate surface and the cement paste matrix may be damaged when water under pressure is expelled from an aggregate particle. In such cases, the aggregate particles themselves are not damaged as a result of frost action. Incidentally, this shows why the results from freeze-thaw and soundness tests on aggregate alone are not always reliable in predicting its behavior in concrete.

It is believed that with concrete pavements exposed to frost action, some sandstone or limestone aggregates are responsible for the *D-cracking phenomenon*. The aggregates that are likely to cause D-cracking seem to have a specific *pore-size distribution* characterized by a large volume of very fine pores less than <1 μm.

5.9.3 Factors Controlling the Frost Resistance of Concrete

By now it should be obvious that the ability of a concrete to resist damage due to frost action depends on the characteristics of both the cement paste and the aggregate. In each case, however, the outcome is controlled actually by the interaction of several factors, such as the location of escape boundaries (the distance by which water has to travel for pressure relief), the pore structure of the system (size, number, and continuity of pores), the degree of saturation (amount of freezable water present), the rate of cooling, and the tensile strength of the material that must be exceeded to cause rupture. As discussed later, the provision of escape boundaries in the cement paste matrix and modification of its pore structure are the two parameters that are relatively easy to control; the former can be controlled by means of air entrainment in concrete and the latter by the use of proper mix proportions and curing.

Air Entrainment

It is not the total air but the void spacing of the order of 0.1 to 0.2 mm within every point in the hardened cement that is necessary for protection of concrete against frost damage. By adding small amount of certain air-entraining agents to the cement paste (e.g., 0.05 percent by weight of the cement) it is possible to incorporate 0.05- to 1-mm bubbles. Thus, for a given volume of air, *depending on the size of air bubbles*, the number of voids, void spacing, and degree of protection against frost action can vary a great deal. In one experiment,[16] 5 to 6 percent air was incorporated into concrete by using one of five different air-entraining agents. Agents A, B, D, E, and F produced 24,000, 49,000, 55,000, 170,000, and 800,000 air voids per cubic centimeter of hardened cement paste, and the corresponding concrete specimens required 29, 39, 82, 100, and 550 freeze-thaw cycles to show 0.1 percent expansion, respectively.

Although the volume of entrained air is not a sufficient measure for the protection of concrete against frost action, assuming that mostly small air bubbles are present, it is the easiest criterion for the purpose of quality control of concrete mixtures. Because the cement paste content is generally related to the maximum aggregate size, lean concretes with large aggregates have less cement paste than rich concretes with small aggregates; therefore, the latter would need more air entrainment for an equivalent degree of frost resistance. Total air contents specified for frost resistance, according to the ACI Building Code 318, are shown in Table 5-3.

The aggregate grading also affects the volume of entrained air, which is decreased by an excess of very fine sand particles. Addition of mineral admixtures such as fly ash,

Nominal Maximum Aggregate Size (mm)*	Air Content (%)	
	Severe Exposure	Moderate Exposure
9	7½	6
12.5	7	5½
19	6	5
25	6	4½
37.5	5½	4½
50†	5	4
76†	4½	3½

*See ASTM C 33 for tolerances on oversize for various nominal maximum size designations.

†These air contents apply to total mix, as for the preceding aggregate sizes. When testing these concretes, however, aggregate larger than 37.5 mm is removed by handpicking or sieving and air content is determined on the minus 37.5 mm fraction of mix. (Tolerance on air content as delivered applies to this value.) Air content of total mix is computed from value determined on the minus 37.5 mm fraction.

Source: ACI Building Code 318.

TABLE 5-3 Total Air Content for Frost—Resistant Concrete or the use of very finely ground cements, has a similar effect. In general, a more cohesive concrete mixture is able to hold more air than either a very wet or a very stiff mixture. Also, insufficient mixing or overmixing, excessive time of handling or transportation of fresh concrete, and overvibration tend to reduce the air content. For these reasons, it is recommended that air content should be determined on concrete as placed, and the adequacy of void spacing be estimated by microscopical determination as described by the ASTM Standard Method C 457.

Water-Cement Ratio and Curing

Earlier it was explained how the pore structure of a hardened cement paste is determined by the water-cement ratio and degree of hydration. In general, the higher the water-cement ratio for a given degree of hydration or the lower the degree of hydration for a given water-cement ratio, the higher will be the volume of large pores in the hydrated cement paste (Fig. 2-8). Because the readily freezable water resides in large pores, it can therefore be assumed that, at a given temperature of freezing, the amount of freezable water will be more with high water-cement ratios and at early ages of curing. Experimental data of Verbeck and Klieger confirmed the validity of this assumption (Fig. 5-10*a*). The influence of water-cement ratio on frost resistance of concrete is shown in Fig. 5-10*b*.

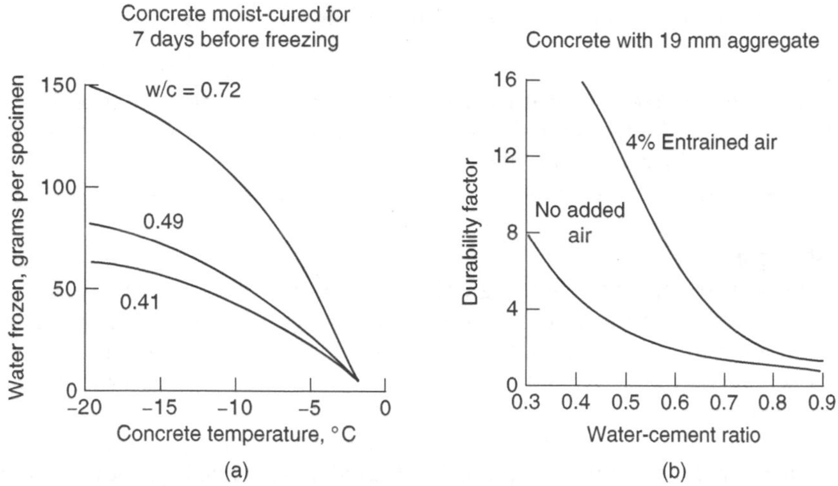

FIGURE 5-10 Influence of water-cement ratio and air content on durability of concrete to frost action. [(*a*) From Verbeck, G., and P. Klieger, *Highway Research Board Bulletin 176*, Transportation Research Board, National Research Council, Washington, D.C., pp. 9–22,1958; (*b*) From *Concrete Manual*, 8th ed., U.S. Bureau of Reclamation, p. 35, 1975.]

The figure on the left shows that the amount of water that can be frozen in concrete with a given water-cement ratio increases with decreasing temperature. It also shows that the amount of water that will freeze at a given temperature increases with water-cement ratio. The observed effect of water-cement ratio is simply that higher ratios result in larger and a greater number of capillaries in which more freezable water can be present. The figure on the right shows that a combination of low water-cement ratios and entrained air ensures a high durability factor to frost action. With ASTM Method 666 it is required to continue freezing and thawing for 300 cycles or until the dynamic modulus of elasticity is reduced to 60 percent of the original value (whichever occurs first). The durability is then assessed by the formula: durability factor = percentage of original modulus × number of cycles at the end of the test ÷ 300.

The importance of the water-cement ratio on the frost resistance of concrete is recognized by building codes. For example, ACI 318-83 requires that normal-weight concrete subject to freezing and thawing in a moist condition should have a maximum 0.45 water-cement ratio in the case of curbs, gutters, and guardrails, and 0.50 for other elements. Obviously, these water-cement ratio limits assume normal cement hydration; therefore, at least 7 days of moist curing at normal temperature is recommended prior to frost exposure.

Degree of Saturation

It is well known that dry or partially dry substances do not suffer frost damage (see box). There is a critical degree of saturation above which concrete is likely to crack and spall when exposed to very low temperature. In fact, it is the difference between the critical and the actual degree of saturation that determines the frost resistance of concrete, as explained in Fig. 5-11. A concrete may fail below the critical degree of saturation after adequate curing, but depending on the permeability it may again reach or exceed the critical degree of saturation if exposed to a moist environment. The role of the permeability of concrete is therefore important in frost action because it controls not only the hydraulic pressure associated with internal water movement on freezing but also the critical degree of saturation prior to freezing. From the standpoint of frost damage the effect of increase in the permeability of a concrete, as a result of cracking due to any physical or chemical causes, should therefore be apparent.

> Bugs do not freeze to death in winter. Some bugs are able to reduce the water content in their bodies so that they can hibernate without freezing; others contain a natural antifreeze in their blood.

Strength

Contrary to popular belief, high strength concrete does not always guarantee high durability. For example, let us consider frost damage. When comparing non-air-entrained with air-entrained concrete, the former may be of higher strength, but the latter will have a better durability to frost action because of the protection provided against the development of high hydraulic pressure during exposure to cycles of freezing and thawing. As a rule of thumb, with medium- and high-strength concretes, each 1 percent increase in the air content reduces the strength of concrete by about 5 percent. Without any change in the water-cement ratio, a 5 percent air entrainment would, therefore, lower the concrete strength by 25 percent. Due to the improved workability as a result of entrained air, it is possible to recover a part of the strength loss through a slight reduction of the water-cement ratio while maintaining the desired level of workability. Nevertheless, air-entrained concrete is generally lower in strength than the corresponding non-air-entrained concrete.

5.9.4 Freezing and Salt Scaling

According to a careful review of the existing data by Valenza and Scherer,[17] the main characteristics of salt scaling include the following: (1) the damage consists of small flakes removed from the concrete surface; (2) the pessimum concentration of the solute concentration is approximately 3 percent, which is nearly independent of the type of

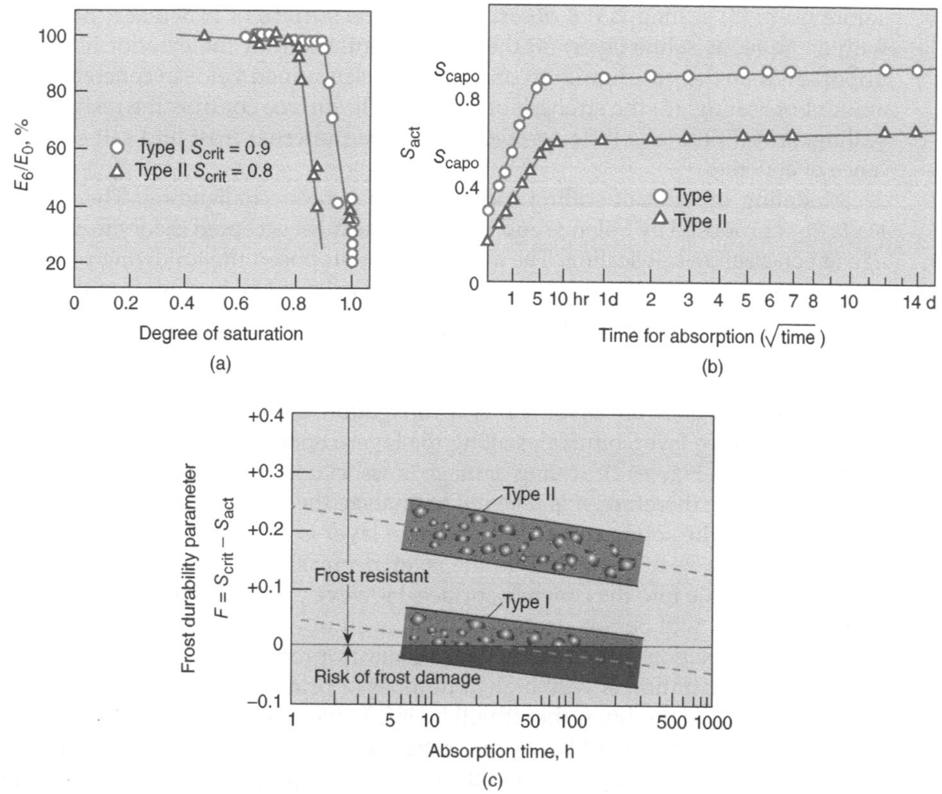

Figure 5-11 Method of predicting the frost resistance of concrete. (From *Betong-handboken*, Svensk Byggjanst, Stockholm, 430–433, 1980.)

G. Fagerlund of the Swedish Cement and Concrete Research Institute proposed a method of predicting the frost resistance of concrete which emphasizes the importance of critical degree of saturation. The frost resistance F is evaluated as a difference between the critical degree of saturation S_{crit} and the actual degree of saturation S_{act}. The degree of water saturation is defined as a ratio between the total volume of evaporable water at 105°C and the total open pore volume of available space before freezing. The micropoint on a plot of water saturation, S vs. E6/Eo (i.e., the residual dynamic elastic modulus after six freeze-thaw cycles) gives S_{crit}. Examples of determining S_{crit} for a non-air-entrained concrete (Type I) and an air-entrained concrete containing 7.1 percent air are shown in Fig. 5-11(a). An estimate of S_{act} can be obtained through a moisture absorption or a simple capillary suction test. As shown in Fig. 5-11(b), the nick point on a plot of degree saturation vs. square root of water uptake time corresponds to the capillary degree of saturation, S_{cap}. At the nick point all gel and capillary pores are filled with water; the larger air pores are the last to be filled and at a very slow rate. Frost resistance, expressed as a difference between S_{crit} and S_{act} for concrete exposed to water uptake for long periods, can be graphically determined.

For example, as shown in Fig. 5-11(c), concrete without entrained air (Type I) will be damaged by frost (F ≥ 0) after about 200 h of continuous water absorption, whereas the air-entrained concrete (Type II) will not be damaged even after very long exposure to water absorption.

solute used; (3) scaling is not observed when the surface of concrete is not in contact with an aqueous saline layer; (4) the salt concentration for the exterior liquid is more important than the pore liquid in concrete; (5) air-entrained voids in concrete reduce the amount of scaling; (6) the strength of the concrete surface controls the resistance of salt scaling; and (7) there is little correlation between internal frost and salt-scaling resistance of concrete.

Modeling of the salt-scaling mechanism has been challenging. The *glue-spalling mechanism* proposed by Valenza and Scherer is the most accepted theory to explain damage of concrete by salt scaling. The authors have proposed the following process: (1) in order to induce scaling damage, an aqueous layer should be in contact with the surface of concrete; (2) as the temperature decreases, the layer freezes; and (3) as the temperature continues to decrease, an increasing thermo-mechanical mismatch will occur between the frozen layer ($\alpha \approx 50 \times 10^{-6}$ °C^{-1}) and concrete ($\alpha \approx 10 \times 10^{-6}$ °C^{-1}), leading to eventual cracking in the layer. Crack propagation can have the following paths: is arrested at the ice layer, bifurcates along the layer-concrete interface, or penetrates into the concrete (see Fig. 5-12). Scaling damage is associated with the cracks penetrating the concrete surface; therefore, it is relevant to analyze the conditions that lead to the crack propagation. If the solute concentration in the layer is about 3 percent (pessimum concentration), pockets of brine form in the layer, decreasing its strength and allowing the crack to propagate into the concrete surface. However, if there is little or no salt present, the frozen layer will be too strong to crack, even upon cooling to −20°C, in part due to the stress relaxation generated by the viscoelastic properties of ice. On the other hand, if the salt concentration is very high, a number of brine pockets will form, weakening the frozen layer to the point that it will be incapable of exerting stress on the concrete surface. Note that the bond between concrete and the frozen layer is normally strong enough that it is highly unlikely that the crack would propagate along its interface.

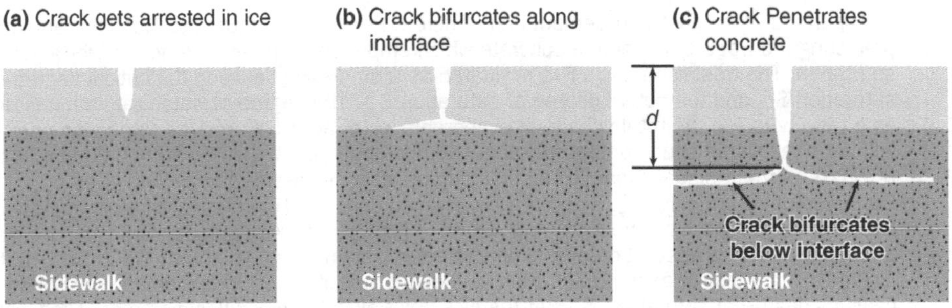

FIGURE 5-12 Illustration of the *glue-spalling mechanism* to explain salt scaling in concrete. (Figure after Valenza II, J.J., and Scherer G.W., Mechanism for Salt Scaling, *J. Am. Ceramics Soc.*, Vol. 89, No. 4, pp. 1161–1179, 2006.)

A crack is formed in the ice layer in contact with the concrete surface due to the stresses generated by the mismatch in thermal expansion between ice and concrete. The crack may (a) get arrested in the ice layer, (b) bifurcate along the ice/concrete interface, or (c) penetrate the concrete surface. The term glue-spalling comes from a technique used to decorate glass where a thin layer of epoxy is placed onto the surface. The composite is then exposed to low temperatures, which generate thermal stresses due to the thermal mismatch. Cracks form in the epoxy and then propagate, according to condition (c), penetrating into the glass surface and flaking the surface. As a result of this process, the glass surface becomes aesthetically pleasing.

The model explains many of the main observation associated to salt scaling: (1) the crack pattern, as illustrated in Fig. 5-12, generates small flakes at the concrete surface, which is a typical manifestation of salt scaling; (2) the pessimum concentration of the solute develops naturally from the crack propagation in the ice layer; (3) if there is no liquid in contact with concrete, there is no frozen layer to crack; (4) the salt concentration in the layer will affect the crack propagation while the pore solution of the concrete will not play a significant role; (5) air-entrained voids in concrete cause shrinkage in the matrix (see Fig. 5-9), thus reducing the strain mismatch with the frozen layer; (6) the strength and toughness of the concrete will control the crack propagation; and (7) the model does not require modeling of the internal ice formation in concrete.

5.10 Effect of Fire

Human safety in the event of fire is one of the considerations in the design of residential, public, and industrial buildings. Concrete has a good service record in this respect. Unlike wood and plastics, concrete is incombustible and does not emit toxic fumes on exposure to high temperature. Unlike steel, when subjected to temperatures of the order of 700 to 800°C, concrete is able to retain sufficient strength for reasonably long periods, thus permitting rescue operations by reducing the risk of structural collapse. For example, in 1972 when a 31-story reinforced concrete building in São Paulo (Brazil) was exposed to a high-intensity fire for over 4 h, more than 500 persons were rescued because the building maintained its structural integrity during the fire. It may be noted that from the standpoint of fire safety of steel structures, a 50- to 100-mm coating of concrete or any other fire-resisting material is routinely specified by building codes.

As with other phenomena, many factors control the response of concrete to fire. Composition of concrete is important because both the cement paste and the aggregate consist of components that decompose on heating. The permeability of concrete, the size of the element, and the rate of temperature rise are important because they govern the development of internal pressures from the gaseous decomposition products. Fire tests have shown that the degree of microcracking, and, therefore, the strength of concrete, is also influenced by test conditions (i.e., whether the specimens are tested hot and under load, or after cooling to the ambient humidity and temperature).

Again, the actual behavior of a concrete exposed to high temperature is the result of many simultaneously interacting factors that are too complex for precise analysis. However, for the purpose of understanding their significance, some of the factors are discussed below.

5.10.1 Effect of High Temperature on Hydrated Cement Paste

The effect of increasing temperature on hydrated cement paste depends on the degree of hydration and moisture state. A well-hydrated portland cement paste, as described before, consists mainly of calcium silicate hydrates (C-S-H), calcium hydroxide, and calcium sulfoaluminate hydrates. A saturated paste contains large amounts of free water and capillary water, in addition to adsorbed water. The various types of water are readily lost on raising the temperature of concrete. However, from the standpoint of fire protection, it may be noted that, due to the considerable heat of vaporization needed for the conversion of water into steam, the temperature of concrete will not rise until all the evaporable water has been removed.

The presence of large quantities of evaporable water can cause one problem. If the rate of heating is high and the permeability of the cement paste is low, damage to concrete may take place in the form of surface spalling. Spalling occurs when the vapor pressure of steam inside the material increases at a rate faster than the pressure relief by the release of steam into the atmosphere.

By the time the temperature reaches about 300°C, the interlayer C-S-H water and some of the chemically combined water from the C-S-H and sulfoaluminate hydrates would also be lost. Further dehydration of the cement paste due to decomposition of calcium hydroxide begins at about 500°C, but temperatures on the order of 900°C are required for complete decomposition of the C-S-H.

5.10.2 Effect of High Temperature on Aggregate

The porosity and mineralogy of the aggregate seem to exercise an important influence on the behavior of concrete exposed to fire. Depending on the rate of heating and the size, permeability, and moisture state of the aggregate, the porous aggregate may themselves be susceptible to disruptive expansion leading to pop-outs of the type described in the case of frost attack. Low-porosity aggregates should, however, be free of problems related to internal moisture movement.

Siliceous aggregates containing quartz (e.g., granite and sandstone) can cause distress in concrete at a temperature of about 573°C, because at this temperature the transformation of quartz from α to β form is associated with a sudden expansion of the order of 0.85 percent. In the case of carbonate rocks, a similar distress can begin above 700°C as a result of the decarbonation reaction. In addition to possible phase transformations and thermal decomposition of the aggregate, mineralogy of the aggregate determines the response of concrete to fire in other ways also. For instance, the aggregate mineralogy determines the differential thermal expansions between the aggregate and the cement paste and the ultimate strength of the interfacial transition zone.

5.10.3 Effect of High Temperature on Concrete

In Fig. 5-13 Abrams' data[18] illustrate the effect of short-duration exposure up to 870°C on the compressive strength of concrete specimens with an average 27 MPa f_c before the exposure. The variables included *aggregate type* (carbonate, siliceous, or lightweight expanded shale) and *testing conditions* (heated without load and tested hot; heated with load at a stress level that is 40 percent of the original strength and tested hot; and tested without load after cooling to ambient temperature).

When heated without load and tested hot (Fig. 5-13*a*), the specimens made with the carbonate aggregate and the sanded lightweight aggregate (60 percent of the fine aggregate was replaced by natural sand) retained more than 75 percent of the strengths at temperatures up to 650°C. At this temperature, concrete specimens containing the siliceous aggregate retained only 25 percent of the original strength; they had retained 75 percent of the original strength up to about 427°C. The superior performance of the concretes containing either the carbonate or the lightweight aggregate at the higher temperature of heat exposure could be due both to stronger interfacial transition zone and a smaller difference in the coefficients of thermal expansion between the matrix mortar and the coarse aggregate.

Strength of specimens tested hot but loaded in compression (Fig. 5-13*b*) was up to 25 percent higher than those of unloaded companion specimens, but the superior performance of carbonate and lightweight aggregate concretes was reaffirmed. However,

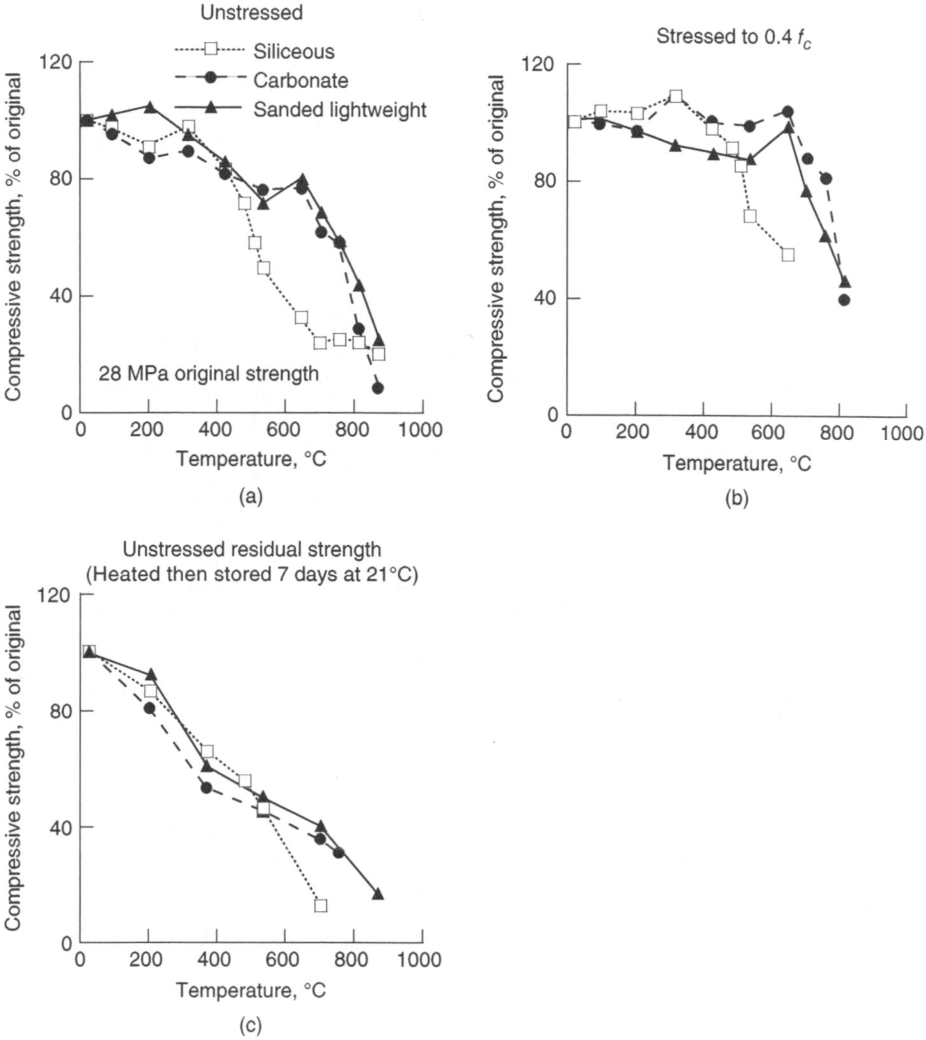

Figure 5-13 Effect of aggregate type and testing conditions on fire resistance. (From Abrams, M.S., *Temperature and Concrete*, ACI SP-25, pp. 33–58, 1973.)

Unloaded concrete specimens heated to 650°C and tested hot (a), show that the concrete containing limestone or lightweight aggregate retained 75 percent of the original strength, while the concrete containing a siliceous aggregate retained only 25 percent of the original strength. When loaded to 40 percent of the original strength (b), a similar trend was observed, although all strengths were higher by about 25 percent. However, according to Fig. 5-13c irrespective of the aggregate type, all concretes showed considerable strength loss on cooling.

the effect of aggregate mineralogy on concrete strength (Fig. 5-13*c*) was significantly reduced when the specimens were tested after cooling to 21°C, most likely as a result of microcracking in the interfacial transition zone, associated with thermal shrinkage.

In the strength range 23 to 45 MPa, Abrams found that the original strength of the concrete had little effect on the percentage of compressive strength retained after the

Figure 5-14 Fire damage to the concrete lining of the Channel Tunnel. (*Photograph courtesy of Paul Acker.*)

Opened in 1994, the English Channel Tunnel was an impressive engineering achievement and culminated the old dream of connecting England and France through underwater transportation. Over $15 billion was spent on the project which used state-of-the art construction methods and materials, including 50 MPa high-strength concrete liners. On November 18, 1996, a fire broke out.[20] Fortunately, there was no fatality but the damage was extensive. Approximately 50 m of tunnel lining was damaged and in some areas its thickness was reduced from 40 to 17 cm. The spalling of concrete, caused by the fire, led to local buckling in some sections of the reinforcing grid. It is interesting to note that the post-fire study on the undamaged concrete showed that the actual compressive strength was about 100 MPa.[21] The lining was repaired with fiber-reinforced concrete.

high-temperature exposure. In a subsequent study,[19] it was observed that when, compared to the compressive strength of heated specimens the *elastic moduli* of concrete made with the three types of aggregate dropped more rapidly as the temperature was increased. For example, at 304 and 427°C, the moduli were 70 to 80 percent and 40 to 50 percent of the original value, respectively. This can be attributed to the interfacial transition zone microcracking, which has a more damaging effect on the flexural strength and the elastic modulus than the compressive strength of concrete (Fig. 5-14).

5.10.4 Behavior of High-Strength Concrete Exposed to Fire

Compared to normal-strength concrete, laboratory research and field performance showed that high-strength concrete behaves differently when exposed to similar heat exposures. High-strength concrete has different strength loss when subjected to thermal loads and shows a higher tendency to spall in an explosive manner. Existing fire design codes were based on normal-strength concrete and extrapolation of the guidelines for high-strength concrete may not be appropriate. These codes also do not address the possibility of explosive spalling failure in high-strength concrete. Even though intensive research has been done in this field in the last decade, the results are very sensitive to the loading conditions, concrete mixture proportions, original compressive strength, and moisture content. Phan and Carino[22] have analyzed many of these variables.

As an example of the behavior of high-strength concrete exposed to high temperatures let us examine the results of an experimental study by Phan and Carino.[23] In this study, 100 mm × 200 mm high-strength concrete (HSC) cylinders were heated to different temperatures and loaded to failure while hot, or after returning to ambient temperature. The maximum temperature was 600°C and the heating rate was 5°C/min. High-strength concrete was prepared both with and without silica fume, at a water-cement ratios ranging from 0.22 to 0.57, and compressive strengths between 51 and 93 MPa. The major conclusions of the study are as follows:

1. For concrete exposed to temperatures in the range of 100 and 300°C, the strength was higher for the samples tested after cooling than for those tested hot. For concrete exposed to temperatures higher than 400°C, the trend was reversed.
2. Preloading the samples up to 40 percent of the compressive strength at room temperature had no effect on the strength reduction.
3. High-strength mixtures made with 0.22 water/cementitious material (w/cm) showed less strength loss than with 0.33 w/cm. For mixtures made with w/cm in the range of 0.33 to 0.57, the behavior was more complex and depended on the test method used, making it difficult to draw definitive conclusions.
4. For preloaded specimens, silica fume had no effect on the behavior of high-strength concrete exposed to elevated temperatures. For unstressed specimens, silica fume had no effect on the strength up to 300°C, however, when tested at higher temperature, the mixtures containing silica fume showed greater strength loss.
5. Silica-fume concrete samples showed higher residual strength (strength after cooling) than samples without the silica fume when exposed to temperatures in the range of 150 to 250°C.

High-strength concrete has a greater tendency to spall than normal-strength concrete. Spalling can compromise the structural integrity of the member and jeopardize the efficiency of the rescue and fire fighting activities. Anderberg[24] concluded that surface spalling increases with increasing moisture content, impermeability of concrete, compressive stress from external load, temperature rise, unsymmetrical temperature distribution, smaller cross sections, and higher concentration of reinforcement. The mechanisms for spalling are related to the development of vapor pressure, increase of stress due to the thermal loads, and volume change due to phase transformation in the aggregate.

In a more recent research investigation, the following conclusions were reached by Phan and Carino[22] regarding the spalling of high-strength concrete:

1. Explosive spalling was observed when the temperature of the specimen center was in the range of 200 and 325°C.
2. Preloading seems to have a mitigating effect on the development of explosive spalling.
3. Concrete samples cast with 0.22 w/cm had a greater potential for spalling under unrestrained condition than samples cast with 0.33 w/cm. However, when

the test was conducted under restrained conditions, explosive spalling only occurred with samples cast with 0.33 w/cm. Samples made with 0.57 w/cm did not have explosive spalling under any testing condition.

4. Silica fume did not have any significant influence on the explosive spalling for tests performed under restrained and unrestrained conditions.

Two mechanisms have been proposed by Bazant[25] to explain the explosive thermal spalling: (*a*) development of high pore pressure caused by oversaturation at the heating front; and (*b*) propagation of brittle fracture. The creation of a high pore pressure seems to be important to trigger explosive spalling since this failure mechanism has not been observed in dry concrete. However, as cracks open, a greater volume becomes available for the liquid and the water vapor, causing the pore pressure to decrease significantly. The energy to propagate the crack can be provided by the strain energy generated by the thermal stresses. High-strength concrete is more brittle than conventional concrete, so it is more sensitive to brittle crack growth and, consequently, to explosive spalling when exposed to high temperatures.

5.11 Deterioration of Concrete by Chemical Reactions

In concrete the deterioration processes triggered by chemical reactions involve generally, but not necessarily, *chemical interactions between the aggressive agents in the environment and the constituents of the cement paste.* The exceptions include alkali-aggregate reactions, which occur between the alkalies present in the cement paste and certain reactive materials in the aggregate, delayed hydration of crystalline CaO and MgO when present in excessive amounts in portland cement, and delayed ettringite formation.

With a well-hydrated portland cement paste, the solid phase, composed of relatively insoluble hydrates of calcium (e.g., C-S-H, CH, and C-A-$\bar{S}$-H), exists in a state of stable equilibrium with a high-pH pore fluid. Depending on the concentration of Na^+, K^+, and OH^- ions the pH value ranges from 12.5 to 13.5. Clearly, portland-cement concrete would be in a state of chemical disequilibrium when it comes into contact with acidic environmental conditions.

Theoretically, *any environment with less than 12.5 pH may be branded aggressive* because a reduction of the alkalinity of the pore fluid would, eventually, lead to destabilization of the cementitious products of hydration. This means that most industrial and natural waters will be aggressive to portland-cement concrete. However, the rate of chemical attack will be a function of the pH of the aggressive fluid and the permeability of concrete. When the permeability of concrete is low and the pH of the aggressive fluid is above 6, the rate of chemical attack is too slow to be taken seriously. Free CO_2 in soft water and stagnant waters, acidic ions such as SO_4^{2-} and Cl^- in groundwater and seawater, and H^+ ions in some industrial waters are frequently responsible for lowering the pH below 6, which would be detrimental to concrete.

Again, it should be noted that chemical attacks on concrete manifest themselves into detrimental physical effects, such as increase in the porosity and permeability, decrease in the strength, and cracking and spalling. In practice, several chemical and physical processes of deterioration act at the same time and may even reinforce each other. For the purpose of developing a clear understanding, the chemical processes can be divided into three subgroups discussed individually and shown in Fig. 5-15. Special attention will be given to sulfate attack, alkali-aggregate attack, and corrosion of the embedded

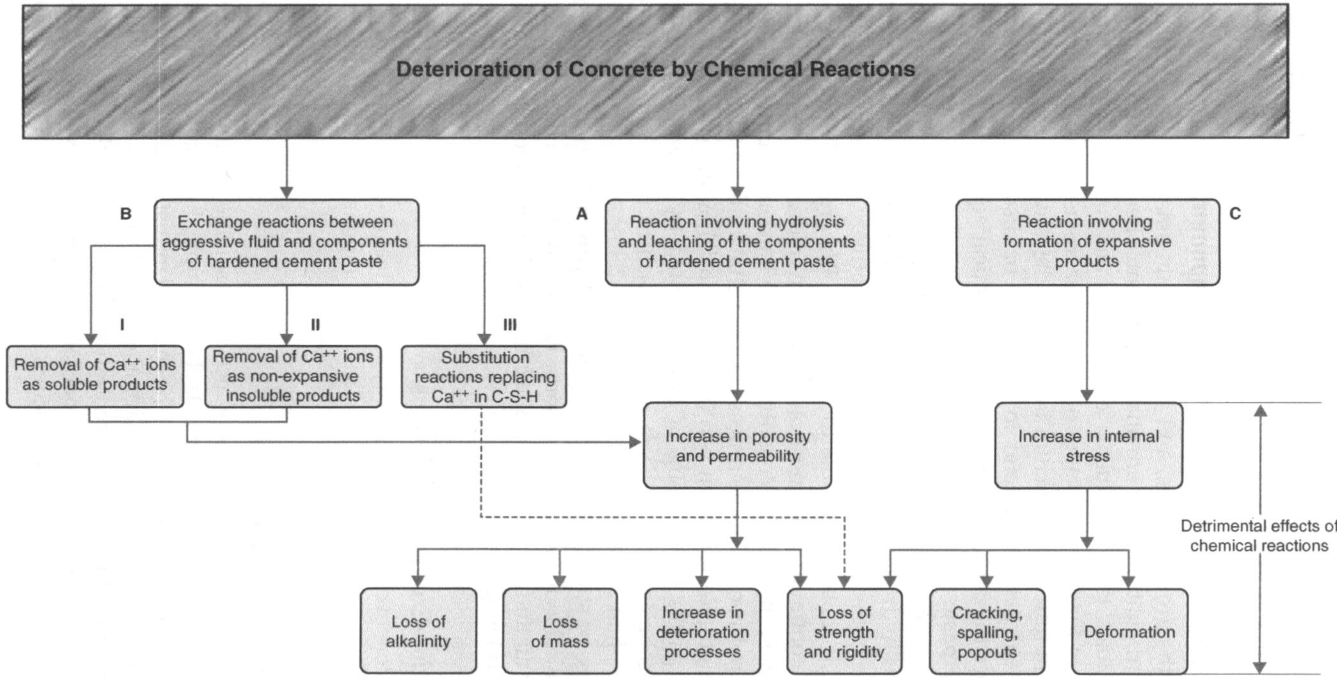

Figure 5-15 Types of chemical reactions responsible for concrete deterioration. A: soft-water attack on calcium hydroxide and C-S-H present in hydrated portland cements; B (I): acidic solution forming soluble calcium compounds such as calcium chloride, calcium sulfate, calcium acetate, or calcium bicarbonate; B (II): solutions of oxalic acid and its salts, forming calcium oxalate; B(III): long-term seawater attack weakening the C-S-H by substitution of Mg^{2+} for Ca^{2+}; C: sulfate attack forming ettringite and gypsum, alkali-aggregate attack, corrosion of steel in concrete, hydration of crystalline MgO and CaO. (From Mehta, P.K., and B.C. Gerwick, Jr., *Concr. Int.*, Vol. 4, pp. 45–51, 1982.)

steel, as these phenomena are responsible for deterioration of most concrete structures. Finally, the last section of this chapter is devoted to durability of concrete in the marine environment because coastal and offshore structures are exposed to a maze of interrelated chemical and physical processes of deterioration, which aptly demonstrate the complexities of concrete durability problems in field practice.

5.11.1 Hydrolysis of the Cement Paste Components

Waters from the ground, lakes, and rivers contain small amounts of chlorides, sulfates, and bicarbonates of calcium and magnesium. These so-called *hard waters* generally do not attack the constituents of the portland cement paste. *Pure water* from condensation of fog or water vapor and *soft water* from rain or from melting of snow and ice contain little or no calcium ions. When these waters come into contact with portland cement paste they tend to hydrolyze or dissolve the calcium-containing products. Once the contact solution attains chemical equilibrium, further hydrolysis of the cement paste stops. However, in the case of flowing water or seepage under pressure, dilution of the contact solution will take place, thus providing the condition for continuous hydrolysis. Calcium hydroxide is one of the constituent of hydrated portland cement pastes which, because of its relatively high solubility in pure water (1230 mg/L), is most susceptible to hydrolysis. Theoretically, the hydrolysis of the cement paste continues until most of the calcium hydroxide has been leached away; this exposes the cementitious constituents of the hardened cement paste to chemical decomposition. Eventually, the process leaves behind silica and alumina gels with little or no strength. Results from two investigations showing strength loss from portland cement pastes by leaching of lime are cited by Biczok.[26] Also according to Terzaghi,[27] a concrete that had lost about a fourth of its original lime content was reduced to one-half of the original strength.

Besides loss of strength, leaching of calcium hydroxide from concrete may be considered undesirable for aesthetic reasons. Frequently, the leachate interacts with CO_2 present in the air and forms an off-white crust of calcium carbonate on the surface. The phenomenon is known as *efflorescence*.

5.11.2 Cation-Exchange Reactions

Based on cation exchange, the three types of deleterious reactions that can occur between aggressive chemical solutions and the components of portland cement paste, are as follows:

Formation of Soluble Calcium Salts

Acidic solutions containing anions, which form soluble calcium salts, are frequently encountered in industrial environments. For example, hydrochloric, sulfuric, or nitric acid may be present in the effluents of the chemical industry. Acetic, formic, or lactic acid are found in many food products. Carbonic acid, H_2CO_3, is present in soft drinks and natural waters with high CO_2 concentration. The cation-exchange reaction between acidic solutions and the constituent of portland cement paste gives rise to soluble salts of calcium, such as calcium chloride, calcium acetate, and calcium bicarbonate which are removed by leaching.

Through the cation-exchange reaction, the solutions of *ammonium chloride* and *ammonium sulfate*, which are commonly found in the fertilizer industry and in agriculture, are able to transform the cement paste components into highly soluble products, for example:

$$2NH_4Cl + Ca(OH)_2 \rightarrow CaCl_2 + 2NH_4OH \tag{5-1}$$

It should be noted that since both the reaction products are soluble, the effects of the attack are more severe than, for instance, with a $MgCl_2$ solution which produces $CaCl_2$ and $Mg(OH)_2$. Because the latter is insoluble, its formation would not increase the porosity and permeability of the system.

Due to certain features of the *carbonic acid attack* on cement paste, it is desirable to discuss it in some detail. The typical cation-exchange reactions between carbonic acid and calcium hydroxide present in hydrated portland cement paste can be shown as follows:

$$Ca(OH)_2 + H_2CO_3 \rightarrow CaCO_3 + 2H_2O \tag{5-2}$$

$$CaCO_3 + CO_2 + H_2O \rightarrow Ca(HCO_3)_2 \tag{5-3}$$

After the precipitation of calcium carbonate, which is insoluble, the first reaction stops unless some free CO_2 is present in the water. By transforming calcium carbonate into soluble bicarbonate, in accordance with second reaction, the free CO_2 aids the hydrolysis of calcium hydroxide. Because this reaction is reversible, a certain amount of free CO_2 referred to as the *balancing* CO_2, is needed to maintain the reaction equilibrium. Any free CO_2 over and above the balancing CO_2 would be aggressive to the cement paste because by driving the second reaction to the right it would accelerate the process of transformation of calcium hydroxide present in the cement paste into the soluble bicarbonate of calcium. The balancing CO_2 content of water depends on its hardness (which is related to the amount of calcium and magnesium present in the water).

It should be noted that the acidity of naturally occurring water is generally due to the dissolved CO_2, which is found in significant concentration in mineral waters, seawater, and groundwater that might have been in contact with decaying vegetable or animal wastes. Normal groundwater contains 15 to 40 mg/L CO_2; however, concentrations of the order of 150 mg/L are not uncommon. Typically seawater contains 35 to 60 mg/L CO_2. As a rule, when the pH of the groundwater or seawater is 8 or above, the free CO_2 concentration is generally negligible; when the pH is below 7, harmful concentration of free CO_2 may be present.

Formation of Insoluble and Nonexpansive Calcium Salts

Certain anions when present in aggressive water may react with cement paste to form insoluble salts of calcium; their formation may not cause damage to concrete unless the reaction product is either expansive (see below) or removed by flowing water or by seepage or vehicular traffic. The products of reaction between calcium hydroxide and oxalic, tartaric, tannic, humic, hydrofluoric, or phosphoric acid belong to the category of insoluble, nonexpansive, calcium salts. When concrete is exposed to decaying animal waste or vegetable matter, the presence of humic acid is usually responsible for chemical deterioration.

Chemical Attack by Solutions Containing Magnesium Salts

Chloride, sulfate, or bicarbonate of magnesium are frequently found in groundwaters, seawater, and some industrial effluents. The magnesium solutions readily react with the calcium hydroxide present in portland cement paste to form soluble salts of calcium. As discussed in the next section, $MgSO_4$ solution is very aggressive because of sulfate attack on alumina-bearing hydrates present in the portland cement paste.

A characteristic feature of the *magnesium ion attack* on portland cement paste is that the attack eventually is extended to the calcium silicate hydrate which is the principal cementitious constituent. Upon prolonged contact with a magnesium solution, the C-S-H in the hydrated portland cement paste gradually loses calcium ions, which are partially or sometimes completely replaced by the magnesium ions. The ultimate product of this substitution reaction is a magnesium silicate hydrate, the formation of which is associated with loss of the cementitious characteristic.

5.12 Reactions Involving the Formation of Expansive Products

Chemical reactions that involve the formation of expansive products in hardened concrete can lead to certain harmful effects. Expansion may, at first, take place without any damage to concrete, but the increasing buildup of the internal stress eventually manifests itself by closing of the expansion joints, deformations, and displacements in different parts of the structure, cracking, spalling, and pop-outs. The four phenomena associated with expansive chemical reactions are: sulfate attack, alkali-aggregate attack, delayed hydration of free CaO and MgO, and corrosion of steel in concrete.

5.13 Sulfate Attack

Most soils contain some sulfate in the form of gypsum $CaSO_4 \cdot 2H_2O$ (typically 0.01 to 0.05 percent expressed as SO_4); this amount is considered harmless to concrete. The solubility of gypsum in water at normal temperatures is rather limited (approximately 1400 mg/L SO_4). Higher concentrations of sulfate in groundwaters are generally due to the presence of magnesium, sodium, and potassium sulfates. Ammonium sulfate is frequently present in agricultural soil and water. Effluents from furnaces (that use high-sulfur fuels) and from chemical industry may contain sulfuric acid. Decay of organic matter in marshes, shallow lakes, mining pits, and sewer pipes often leads to the formation of H_2S gas which is transformed into sulfuric acid by bacterial action. The water used in concrete cooling towers may also contain a high concentration of sulfate due to evaporation. Thus, it is not uncommon to find potentially deleterious concentrations of sulfate in natural and industrial waters.

Degradation of concrete as a result of chemical reactions between hydrated portland cement and sulfate ions from an outside source is known to take *two forms* that are distinctly different from each other. Which one of the deterioration processes is predominant in a given case depends on the concentration and source of sulfate ions (i.e., the associated cation) in the contact water, and composition of the cement paste in concrete.

Sulfate attack can manifest in the form of *expansion and cracking* of concrete. When concrete cracks, its permeability increases and the aggressive water penetrates more easily into the interior, thus accelerating the process of deterioration. Sometimes the expansion of concrete may cause serious structural problems, such as the displacement of building walls due to horizontal thrust by an expanding slab. Sulfate attack can also take the form of a *progressive decrease in the strength and loss of mass* due to loss of cohesiveness of the cement hydration products. A brief review of some theoretical aspects of sulfate-generated failures, selected case histories, and control of sulfate attack follows.

5.13.1 Chemical Reactions in Sulfate Attack

Calcium hydroxide and alumina-bearing phases of hydrated portland cement are more vulnerable to attack by sulfate ions. On hydration, portland cements with more than 5 percent potential C_3A will contain most of the alumina in the form of monosulfate hydrate, $C_3A \cdot C\bar{S} \cdot H_{18}$. If the C_3A content of the cement is more than 8 percent, the hydration products will also contain $C_3A \cdot CH \cdot H_{18}$. Due to the presence of calcium hydroxide in hydrated portland cement paste, when a cement paste comes in contact with sulfate ions, both the alumina-containing hydrates are converted to the high-sulfate form (ettringite, $C_3A \cdot 3C\bar{S} \cdot H_{32}$) as shown by the following equations:

$$C_3A \cdot CH \cdot H_{18} + 2CH + 3\bar{S} + 11H \rightarrow C_3A \cdot 3C\bar{S} \cdot H_{32} \quad (5\text{-}4)$$

$$C_3A \cdot C\bar{S} \cdot H_{18} + 2CH + 2\bar{S} + 12H \rightarrow C_3A \cdot 3C\bar{S} \cdot H_{32} \quad (5\text{-}5)$$

There is general agreement that the sulfate-related expansions in concrete are associated with ettringite; however, the *mechanisms* by which ettringite formation causes expansion is still a controversial subject.[28] Exertion of pressure by growing ettringite crystals, and swelling due to adsorption of water in alkaline environment by poorly crystalline ettringite, are two of the several hypotheses supported by most researchers.

Gypsum formation as a result of cation-exchange reactions is also capable of causing expansion. However, it has been observed[29] that deterioration of hardened portland cement paste by gypsum formation goes through a process that first leads to reduction of pH of the system and loss in the stiffness and strength, followed by expansion and cracking, and eventually transformation of the concrete into a mushy or noncohesive mass.

Depending on the cation type associated with the sulfate solution (i.e., Na^+, K^+, or Mg^{2+}), both calcium hydroxide and C-S-H present in the hydrated portland cement paste may be converted to gypsum by sulfate attack:

$$Na_2SO_4 + Ca(OH)_2 + 2H_2O \rightarrow CaSO_4 \cdot 2H_2O + 2NaOH \quad (5\text{-}6)$$

$$MgSO_4 + Ca(OH)_2 + 2H_2O \rightarrow CaSO_4 \cdot 2H_2O + Mg(OH)_2 \quad (5\text{-}7)$$

$$\begin{aligned} &3MgSO_4 + 3CaO \cdot 2SiO_2 \cdot 3H_2O + 8H_2O \\ &\qquad \rightarrow 3(CaSO_4 \cdot 2H_2O) + 3Mg(OH)_2 + 2SiO_2 \cdot H_2O \end{aligned} \quad (5\text{-}8)$$

In the case of sodium sulfate attack, the formation of sodium hydroxide as a by-product of the reaction ensures continuation of high alkalinity in the system, which is essential for the stability of the cementitious hydration product, C-S-H. On the other hand, in the case of magnesium sulfate attack, the conversion of calcium hydroxide to gypsum is accompanied by the simultaneous formation of magnesium hydroxide, which is insoluble and reduces the alkalinity of the system. In the absence of hydroxyl ions in the solution, C-S-H is no longer stable and is also attacked by the sulfate solution [Eq. (5-8)]. The magnesium sulfate attack is, therefore, more severe on concrete.

5.13.2 Delayed Ettringite Formation

This is a case of chemical sulfate attack when the source of sulfate ions happens to be internal (within the concrete) rather than external. The phenomenon is not new; it is known to occur when either a gypsum-contaminated aggregate or a cement containing unusually high sulfate content has been used in the concrete production. Cases of delayed ettringite formation have been reported with steam-cured concrete products. Ettringite is not a stable phase above 65°C; it decomposes to form the monosulfate hydrate if steam-curing temperatures higher than 65°C are used in the manufacturing process. The sulfate ions released by the decomposition of ettringite are adsorbed by calcium-silicate hydrate. Later, during the service, when sulfate ions are desorbed, the re-formation of ettringite causes expansion and cracking (see Fig. 5-16).

There is a general agreement among the researchers that DEF-related expansion is associated with internally available sources of sulfate, and that the ettringite formed is poorly crystalline. Some researchers believe that the decomposition of primary ettringite by high-temperature steam curing, followed by adsorption of sulfate ions by C-S-H and their subsequent desorption to reform a secondary ettringite within the early cement hydration products, are the necessary conditions for the DEF phenomenon. Others, including Collepardi,[30] have observed that DEF is not limited to heat-cured products, and that adsorption-desorption of sulfate by C-S-H is not essential to the phenomenon. He has proposed the following hypothesis:

- Microcracks resulting from the concrete manufacturing process, or chemical reactions such as the alkali-silica reaction, or loading conditions in service, increase the permeability of concrete.
- Sulfate ions are released from the cement hydration products or derived from other sources.

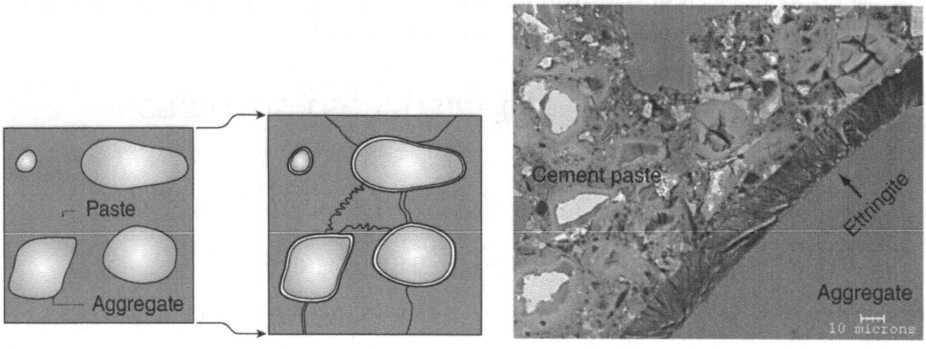

(a) (b)

FIGURE 5-16 (*a*) Diagrammatic representation of the expansion of a mortar or concrete caused by delayed ettringite formation (DEF). (After Taylor, H.F.W., C. Famy, and K.L. Scrivener, Delayed ettringite formation. *Cem. Concr. Res.*, Vol. 31, No. 5, pp. 683–693, 2001) (*b*) Scanning electron micrograph of a mortar affected by DEF. The sample was stored for 4 h at 20°C, heated at 90°C for 12 h and subsequently stored in water for 600 days. (Micrograph courtesy of C. Famy.)

Expansion in the paste caused by DEF originates cracks in the paste and at the aggregate-cement paste interface. Subsequently, ettringite recrystallizes in the cracks from sub-microscopic crystals dispersed throughout the cement paste.

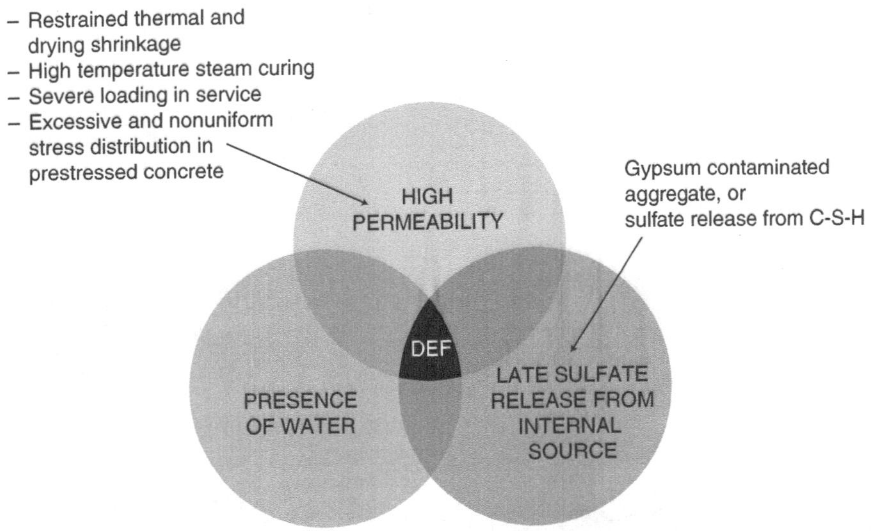

FIGURE 5-17 Holistic approach for expansion and cracking by delayed ettringite formation.

- The presence of water is necessary for the ionic migration within the concrete.
- Ettringite deposition occurs inside the existing microcracks, which propagate either by swelling or by growth of ettringite crystals.

A diagrammatic representation of Collepardi's holistic approach to DEF-related expansion and cracking is shown in Fig. 5-17.

5.13.3 Selected Cases Histories

An interesting case history of sulfate attack by spring water on Elbe River bridge piers in Magdeburg, Germany, is reported by Biczok.[31] The pier-sinking operation in a closed caisson opened up a spring. The spring water contained 2040 mg/L SO_4. The expansion of the concrete lifted the piers by 8 cm in 4 years and caused extensive cracking, which made it necessary to demolish and rebuild the piers. Obviously, such occurrences of sulfate expansion can be avoided by a thorough survey of the environmental conditions and by providing suitable protection against sulfate attack when necessary.

Bellport[32] described the experience of the U.S. Bureau of Reclamation in regard to sulfate attack on hydraulic structures located in Wyoming, Montana, South Dakota, Colorado, and California. In some cases, the soluble sulfate content of soil was as high as 4.55 percent, and the sulfate concentration of water was up to 9900 mg/L. Many cases of serious deterioration of concrete structures, 5 to 30 years old, were reported. Research studies showed that sulfate-resisting cements containing low C_3A performed better than zero-C_3A cements, which contained unusually large amounts of tricalcium silicate (58 to 76 percent).

As a result of sulfate exposure for 20 years, loss of strength and mass was reported from concrete structures of the Ft. Peck Dam in Montana (Fig. 5-18). The sulfate content

FIGURE 5-18 Sulfate attack on concrete in Fort Peck Dam, 1971. (Photographs courtesy of T.J. Reading, formerly Materials Engineer of the Missouri River Division, U.S. Corps of Engineers.)

In the northern Great Plains states (the Dakotas and Montana), and extending up into the prairie provinces of Canada, groundwater may contain 1000 to 10,000 mg/L SO_4 in areas of poor drainage. During 1935–1966, the U.S. Corps of Engineers constructed six earth-filled dams across the upper Missouri River; however, there are large auxiliary concrete structures such as tunnels, a stilling basin, a power-house, and a spillway. Four of the six projects, including the Ft. Peck (Montana) Dam contain more than 1 million cubic yards concrete each. Judged from the compressive strength (48 to 60 MPa) of cores on 20-year-old specimens, the Ft. Peck concrete made with Type I Portland cement (7 to 9 C_3A) an 0.49 water-cement ratio, and 335 kg/m^3 cement content, is of good quality (low permeability).

Inspections of concrete structures in 1957–1958 after 20 years of use showed that the overall condition of the concrete at Ft. Peck was very good; however, appreciable sulfate attacks was found in two areas; slabs in the penstock floor and the downstream end of Tunnel 1, and a tailrace training wall (shown in the photograph). The deteriorated concrete was mushy and disintegrated easily. The sulfate concentration of the groundwater, due almost entirely to sodium sulfate, was found to be about 10,000 mg/L. Between 1958 and 1971, the deteriorated area in the trailrace training wall enlarged and increased in depth to about 200 mm. Mineralogical analysis of the cement paste from deteriorated concrete specimens showed that large amounts of gypsum had formed at the expense of C-S-H and calcium hydroxide.

of groundwater, due entirely to alkali sulfates, was up to 10,000 mg/L. An investigation of the deteriorated concrete specimens (Fig. 5-19) showed large amounts of gypsum formed at the expense of the cementitious constituents normally present in hydrated portland cement pastes.[33] Similar cases of sulfate deterioration are reported from the prairie soils in western Canada which contain as high as 1½ percent alkali sulfates (groundwaters frequently contain 4000 to 9000 mg/L sulfate). Typically, as a consequence of the sulfate attack, concrete was rendered relatively porous or weak and, eventually, reduced to a mushy (noncohesive) mass.

Verbeck[34] reported the results of a long-time investigation on concrete performance in sulfate soils in Sacramento, California. Concrete specimens made with different types of portland cement and three cement contents were used. The soil contained

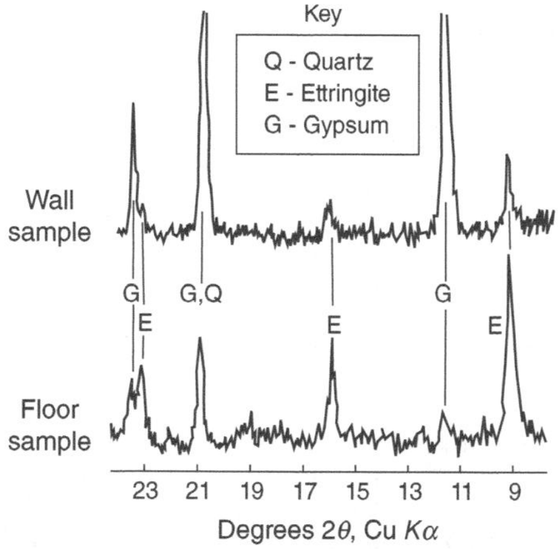

FIGURE 5-19 X-ray diffraction analysis of deteriorated concrete from Fort Peck Dam.

The X-ray diffraction (XRD) technique offers a convenient way to determine the mineralogical analysis of crystalline solids. If a crystalline mineral is exposed to X-rays of a particular wavelength, the layers of atoms diffract the rays and produce a pattern of peaks, which is characteristic of the mineral. The horizontal scale (diffraction angle) of a typical XRD pattern gives the crystal lattice spacing, and the vertical scale (peak height) gives the intensity of the diffracted ray. When the specimen being X-rayed contains more than one mineral, the intensity of characteristics peaks from the individual minerals are proportional to their amount.

Using copper K_α radiation, this XRD pattern was obtained from cement paste specimens taken from the deteriorated concrete of Ft. Peck Dam. Large amounts of ettringite and gypsum are found in the specimens instead of C-S-H, $Ca(OH)_2$, and monosulfate hydrate, which are normally present in mature portland cement concretes. This is an unmistakable evidence of strong sulfate attack on concrete. Contamination of the cement paste by aggregate is responsible for the presence of quartz peak in the XRD pattern.

approximately 10 percent sodium sulfate. The deterioration of the concrete specimens was evaluated by visual inspection and by measurement of strength and dynamic modulus of elasticity after various periods of exposure. Verbeck's data regarding the effect of C_3A content of portland cement and the cement content of concrete on the average rate of deterioration are shown in Fig. 5-20. The results clearly demonstrate that the cement content (which has a direct influence on the permeability of concrete) had more effect on the sulfate resistance than the composition of cement. For example, the performance of concrete containing 390 kg/m^3 of 10 percent-C_3A cement was reported to be two to three times better than the concrete containing 310 kg/m^3 of cement with 4 percent-C_3A (Fig. 5-20*a*). With a high-C_3A cement (11 percent C_3A), the effective C_3A content in the cementitious mixture can be reduced by adding a pozzolanic admixture such as fly ash (Fig. 5-20*b*), thus improving the sulfate resistance.

An interesting case of sulfate attack was brought to the attention of the authors, which showed that the soil, groundwaters, seawater, and industrial waters are not the only sources of sulfate. Deterioration of the drypack grout between the cantilevered precast concrete girders and the cast-in-place concrete bleacher girders was reported from Candlestick Park Stadium in San Francisco, California.[35] Apparently, the grout was not compacted properly during construction; therefore, leaching of the cementitious

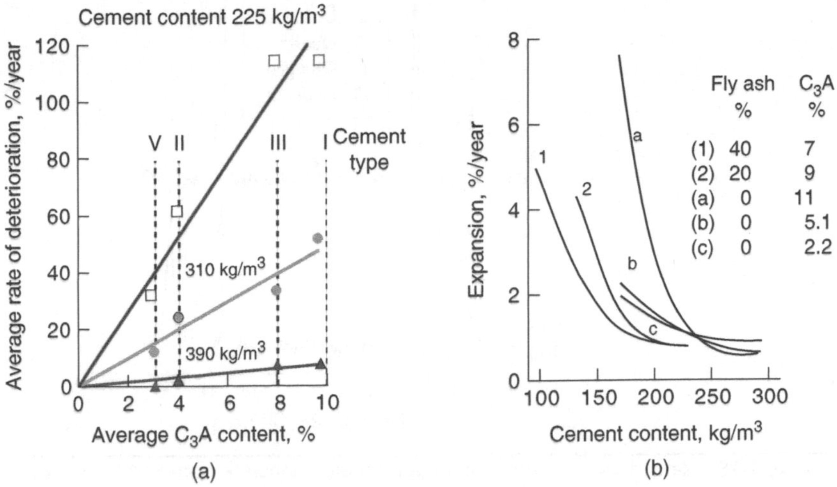

FIGURE 5-20 Effects of cement type and content and fly ash addition on sulfate attack in concrete. [(*a*) From Verbeck, G.J., in *Performance of Concrete*, Swenson, E.G., ed., University of Toronto Press, Toronto, pp. 113–124, 1968; (*b*) From Brown, G.E., and D.B. Oates, *Concr. Int.*, Vol. 5, pp. 36–39, 1983.]

The deterioration of concrete due to sulfate attack can be controlled by the cement content (w/c), cement type, and mineral admixtures. The results from a long-time study of concrete specimens exposed to a sulfate soil (containing 10 percent Na_2SO_4) at Sacramento, California, showed (figure on the left) that the low permeability of concrete (high cement content) was more important in reducing the rate of deterioration than the C_3A content of the cement. The figure on the right shows that in the case of a high-C_3A portland cement, addition of mineral admixtures (fly ash) offers another way of controlling sulfate attack, by reducing the effective C_3A content in the total cementitious material.

material resulted in a high strength loss* and caused the formation of calcium carbonate stalactites in the vicinity. X-ray diffraction analysis of the deteriorated material showed the presence of considerable amounts of ettringite and gypsum as a result of sulfate attack. It may be noted that the joint containing the grout is located 18 to 30 m above ground level. As a result of poor drainage, it was found that rain water had accumulated in the vicinity of the mortar. It seems that, due to air pollution, the sulfates present in rain water (see box below) can cause the deterioration of mortar or concrete even at elevated levels. This is likely to happen when the material is permeable and when, during the design and construction, adequate provisions are not made for proper drainage.

5.13.4 Control of Sulfate Attack

According to *BRE Digest*,[36] *factors influencing sulfate attack* are: (1) the amount and nature of the sulfate present, (2) the level of the water table and its seasonal variation, (3) the flow of groundwater and soil porosity, (4) the form of construction, and (5) the quality of concrete. If the sulfate water cannot be prevented from reaching the concrete, the

*Grout cores showed 4 to 7 MPa compressive strength against the normal 25 to 30 MPa compressive strength.

Acid Rain and Durability of Concrete

Samples collected for the Air Resources Board have shown that the average pH value of rain in Northern California ranged from pH 4.4 at San Jose . . . [to] pH 5.2 at Davis. . . Nor the occurrence of *acid rain* confined to the state's urban centers . . . In Sequoia National Park and in the Mammoth Lakes region, the average pH value of rainfall during 1980 and 1981 was 4.9 and one week averaged 3.5.

Yet those readings pale in comparison with the *disturbing levels of acidity found in fog*. In December 1982, the fog blanketing Orange County reached an all-time low reading . . . of pH 1.7 at Corona del Mar. [Even the coastal fog that rolls through at San Francisco's Golden Gate has registered as low as pH 3.5.] According to Dr. Michael Hoffman at California Institute of Pasadena, fog near urban areas routinely registers between 2.5 and 3 on the pH scale and is laden with pollutants such as *sulfate*, nitrate, ammonium ions, lead, copper, nickel, vanadium, and aldehydes.

Source: Report by K. Patrick Conner,
Published in the *San Francisco Chronicle*, June 3, 1984

Acid rain is manmade, not a natural phenomenon, with 90 percent of such pollution in the [Northeastern part of the United States] coming from industrial and automotive combustion of fossil fuels. These pollutants [the chief component is sulfur dioxide, with nitrogen oxide also playing an important role] are transported through the atmosphere long distances from their sources. . . Several thousand lakes and streams . . . have been acidified, with life in them killed or reduced. [Among other factors] acid rain may be a contributor to the decline of the forests. *Buildings, monuments, and other man-made structures are being eroded by air pollution and could be the "sleeper" issue of the acid rain problem* said Dr. J. Christopher Bernabo, executive director of the national assessment program.

Source: Report by Philip Shabecoff,
Published in the *San Francisco Chronicle*, February 24, 1985

only defense against sulfate attack lies in the control of factor (5), as discussed below. The rate of attack on a concrete structure with all faces exposed to sulfate water is less than if moisture can be lost by evaporation from one or more surfaces. Therefore, basements, culverts, retaining walls, and slabs on ground are more vulnerable than foundations and piles.

The *quality of concrete*, specifically a low permeability, is the best protection against sulfate attack. Adequate concrete thickness, high cement content, low water-cement ratio and proper compaction and curing of fresh concrete are among the important factors that contribute to low permeability. To mitigate the effect of cracking due to drying shrinkage, frost action, corrosion of reinforcement, or other causes, additional safety can be provided by the *use of sulfate-resisting portland or blended cements.*

Portland cement containing less than 5 percent C_3A (ASTM Type V) is sufficiently sulfate resisting under moderate conditions of sulfate attack (i.e., when ettringite forming reactions are the only consideration). However, when high sulfate concentrations of the order of 1500 mg/L or more are involved (which are normally associated with the presence of magnesium and alkali cations), then Type V portland cement may not be effective against the cation-exchange reactions that result in gypsum formation, especially if the C_3S content of the cement is high.[32] Under these conditions, experience shows that cements potentially containing little or no calcium hydroxide on hydration perform much better: for instance, high-alumina cements, portland blast-furnace slag cements with more than 50 percent slag, and portland pozzolan cements with at least 25 percent pozzolan (natural pozzolan, calcined clay, or low-calcium fly ash).

Based on the standards originally developed by the U.S. Bureau of Reclamation, *sulfate exposure is classified into four degrees of severity* in the ACI 201.2R-08, which contains the following requirements:

- *Negligible attack (Class 0).* When the sulfate content is under 0.1 percent in soil, or under 150 ppm (mg/L) in water, there shall be no restriction on the cement type and water-cement ratio.
- *Moderate attack (Class 1).* When the sulfate content is 0.1 to 0.2 percent in soil, or 150 to 1500 ppm in water, ASTM Type II portland cement or equivalent shall be used, with less than an 0.5 water-cementitious materials ratio for normal-weight concrete.
- *Severe attack (Class 2).* When the sulfate content is 0.2 to 2.00 percent in soil or 1500 to 10,000 ppm in water, ASTM Type V portland cement or equivalent, with less than an 0.45 water-cementitious materials ratio, shall be used.
- *Very severe attack (Class 3).* When the sulfate content is over 2 percent in soil, or over 10,000 ppm in water, ASTM Type V cement plus a pozzolan or slag shall be used, with less than an 0.40 water-cementitious materials ratio.

It is suggested that with normal-weight concrete a lower water-cement ratio (or higher strength in the case of lightweight concrete) may be required for watertightness or for protection against corrosion of embedded items. For very severe attack conditions, *BRE Digest 250* requires the use of sulfate-resisting portland cement, a maximum 0.45 water-cement ratio, a minimum 370 kg/m^3 cement content, and a protective coating on concrete. Concrete coatings are not a substitute for high-quality or low-permeability concrete because it is difficult to ensure that a thin coating will remain unpunctured or that thick coating will not crack. ACI Committee 515 recommendations should be considered for barrier coatings to protect concrete from external chemical attacks.

5.14 Alkali-Aggregate Reaction

Expansion and cracking of concrete leading to loss of strength and elastic modulus can also result from chemical reaction involving alkali and hydroxyls ions from portland cement paste and certain *reactive siliceous minerals* that are often present in the aggregate. In recent literature, the phenomenon is referred to as *alkali-silica reaction* (ASR). Pop-outs and exudation of a viscous alkali-silicate fluid are other manifestations of the phenomenon, a description of which was first published in 1940 by Stanton[37] from his investigations of cracked concrete structures in California (Fig. 5-21). Since then, numerous examples of concrete deterioration from other parts of the world have been reported to show that the alkali-silica reaction may become one of the causes of distress in structures located in humid environments, such as dams, bridge piers, and sea walls. Characteristics of cements and aggregates that contribute to this reaction, mechanisms

(a) (b)

FIGURE 5-21 Cracks caused by the alkali-silica reaction. [Photographs courtesy of the California Department of Transportation.]

Thomas E. Stanton (shown in the Fig. 5-21*a*) was the first to provide a comprehensive explanation for the damages that occurred in the California Highway system in the late '30s. He proposed that the deterioration was caused by the expansion of a gel generated by the reactive silica from the aggregate and the alkalies from the cement. His explanation caused consternation in the portland cement industry.[38] For a while, the cement companies tried to defend their product, but compelling evidence from many distressed highway structures and large concrete dams forced the technical community and the cement producers to develop methods and material for preventing the alkali-silica reaction in concrete structures.

Stanton's research team (Fig. 5-21*b*) investigated several cases of damaged concrete by alkali-silica reaction (ASR). In plain concrete, ASR generates cracks in form of "maps" or "alligator skin" (Fig. 5-21*a*), however in reinforced concrete the cracks tend to form parallel to the reinforcing bars (Fig. 5-21*b*).

associated with expansion, selected case histories, and methods of controlling the phenomenon, are discussed below.

5.14.1 Cements and the Aggregate Types Contributing to ASR

Raw materials used for the portland-cement clinker manufacture are the source of alkalies in cement which typically range from 0.2 to 1.5 percent Na_2O equivalent.* Depending on the alkali content of a cement, the pH of the pore fluid in normal concretes is generally 12.5 to 13.5. This pH represents a strongly alkaline fluid in which contain acidic rocks, composed of silica and siliceous minerals do not remain stable on long exposure.

---

*Both sodium and potassium compounds are usually present in portland cements. However, it is customary to express the alkali content of cement as acid-soluble sodium oxide equivalent, which is equal to $Na_2O + 0.658K_2O$.

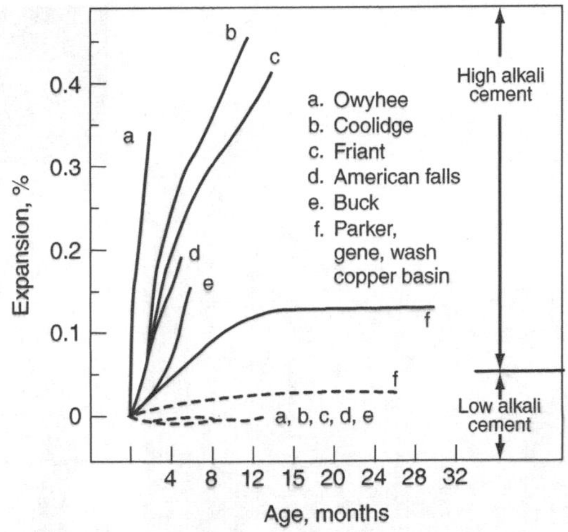

FIGURE 5-22 Alkali-reactive rocks in portland cement concrete. [Based on Blanks, R.F., and H.L. Kennedy, *The Technology of Cement and Concrete*, Vol. 1, Wiley, New York, 1955.]

Combinations of high-alkali portland cement (>0.6 percent equivalent Na_2O) and certain siliceous aggregates used for making concrete for several U.S. dams showed undesirably large expansions in a mortar prism test. The same aggregates showed only small expansions when a low-alkali cement was used in the test. Table 5-4 gives a comprehensive list of the alkali-reactive aggregate types.

Both laboratory and field data from several studies in the United States showed that portland cements containing more than 0.6 percent equivalent Na_2O, when used in combination with an alkali-reactive aggregate, can cause significant expansion due to the alkali-aggregate reaction (Fig. 5-22). ASTM C 150, therefore, designated the cements with less than 0.6 percent equivalent Na_2O as *low-alkali cements* and with more than 0.6 percent equivalent Na_2O as *high-alkali cements*. In practice, with ordinary concrete, alkali content of 0.6 percent or less is usually found insufficient to cause damage due to the alkali-aggregate reaction, irrespective of the type of reactive aggregate. With concrete mixtures containing very high cement content, even less than 0.6 percent alkali in cement may prove harmful. Investigations in Germany and England have shown that if the total alkali content of the concrete from all sources is below 3 kg/m^3, deleterious expansion does not occur.

As discussed below, the presence of both hydroxyl ions and alkali-metal ions appears to be necessary for the expansive phenomenon. Due to large amount of calcium hydroxide in a hydrated portland cement paste, the concentration of hydroxyl ions in the pore fluid remains high even with low-alkali cements; in this case, the expansive phenomenon will therefore be limited by the short supply of the alkali-metal ions unless these ions are furnished by any other source, such as alkali-containing admixtures, salt-contaminated aggregate, and seawater, or a deicing salt solution containing sodium chloride that may have penetrated into the concrete.

In regard to *alkali-reactive aggregates*, depending on the time, temperature, and particle size, all silicate or silica minerals, as well as silica in hydrous (opal) or amorphous form (obsidian, silica glass), can react with alkaline solutions, although a large number of minerals react only to an insignificant degree. Feldspars, pyroxenes, amphiboles, micas, and quartz, which are the constituent minerals of granite, gneiss, schist, sandstone, and basalt, are classified as innocuous minerals. Opal, obsidian, cristobalite, tridymite, chalcedony, chert, andesite, rhyolites, and strained or metamorphic quartz have been found to be alkali-reactive in the decreasing order of reactivity. A comprehensive list of substances responsible for deterioration of concrete by the alkali-aggregate reaction is shown in Table 5-4. A few cases of reaction between alkali and carbonate rocks are also reported in the literature, and they will not be discussed here.

5.14.2 Mechanisms of Expansion

Depending on the degree of disorder in the crystal structure of the aggregate, the porosity and the particle size, alkali-silicate gels of variable chemical composition are formed in the presence of hydroxyl and alkali-metal ions. The mode of attack in concrete

Reactive Substance	Chemical Composition	Physical Character
Opal	$SiO_2 \cdot nH_2O$	Amorphous
Chalcedony	SiO_2	Microcrystalline to cryptocrystalline; commonly fibrous
Certain forms of quartz	SiO_2	Microcrystalline to cryptocristalline; Crystalline, but intensely fractured, strained, and/or inclusion-filled
Cristobalite	SiO_2	Crystalline
Tridymite	SiO_2	Crystalline
Rhyolitic, dacitic, latitic, or andesitic glass or cryptocrystalline devitrification products	Siliceous, with lesser proportions of Al_2O_3, Fe_2O_3, alkaline earths, and alkalies	Glass or cryptocrystalline material as the matrix of volcanic rocks or fragments in tuffs
Synthetic siliceous glasses	Siliceous, with less proportions of alkalies, alumina, and/or other substances	Glass
The most important deleteriously alkali-reactive rocks (i.e., rocks containing excessive amounts of one or more of the substances listed above) are as follows:		
Opaline cherts	Andesites and tuffs	
Chalcedonic cherts	Siliceous shales	
Quartzose cherts	Phyllites	
Siliceous limestones	Opaline concretions	
Siliceous dolomites	Fractured, strained, and	
Rhyolites and tuffs	inclusion-filled quartz and	
Dacites and tuffs	quartzites	

Note: A rock may be classified as, for example, a "siliceous limestone" and be innocuous if its siliceous constituents are other than those indicated above. [From ACI Committee 201, *ACI Mat. J.*, Vol. 88, No. 5, p. 565, 1991.]

TABLE 5-4 Deleteriously Reactive Rocks, Minerals, and Synthetic Substances involves depolymerization or breakdown of the silica structure* of the aggregate by hydroxyl ions followed by adsorption of the alkali-metal ions on newly created surface of the reaction products. Like marine soils with surface-adsorbed sodium or potassium, when an alkali-silicate gel comes into contact with water, it swells by imbibing a large amount of water through osmosis. If the degree of restraint on the system is low, the hydraulic pressure developed may be sufficient to cause expansion and cracking of the affected aggregate particles, and also the cement paste matrix surrounding the aggregate.

Solubility of the alkali silicate gels in water accounts for their mobility from the interior of aggregate particles to the microcracked regions both within the aggregate and the concrete. Continued availability of water to the concrete causes enlargement and extension of the microcracks, which eventually reach the outer surface of the concrete. The crack pattern is irregular and is referred to as *map cracking*.

The presence of water is fundamental for (1) the silica dissolution process, (2) the transport of alkali inside concrete, and (3) the swelling mechanism of the alkali silica gel. The water-to-cement ratio of normal concrete lies in a range between 0.35 to 0.60; however, the amount of water required for full hydration of the cement amounts to about 0.32 water-cement ratio. Therefore, some water is always present in the pores of well-cured concrete, an amount that is sufficient to trigger the alkali silica reaction (ASR). The consensus among researchers is that at only at high levels of moisture is swelling of the reaction product gel dangerous for the integrity of concrete. Larive[39] reported axial expansions of concrete specimens over 1 percent only for relative humidity higher than 80 percent. Poyet et al.[40] performed a series of tests to assess the effect of moisture conditions on expansion due to ASR. Figure 5-23 plots the free axial expansion reached by reactive concrete specimens for different constant humidity conditions in accelerated 60°C free expansion tests. The lower level reached in the case of 100 percent relative humidity with respect to 96 percent relative humidity is due to the leaching of alkalis in consequence to the condensation of pure water on the external surface of the samples; this leads to the reduction of ASR expansion.

It should be noted that the evidence of alkali-aggregate reaction in a cracked concrete does not necessarily prove that this reaction is the principal cause of cracking. Among other factors, development of internal stress depends on the amount, size, and type of the reactive aggregate present and the chemical composition of the alkali-silicate gel formed. When a large amount of the reactive material is present in a finely divided form (i.e., under 75 μm), there may be considerable petrographic evidence of the alkali-silica reaction yet no significant expansion. On the other hand, most case histories of expansion and cracking of concrete attributable to the alkali-aggregate reaction are associated with the sand-size alkali-reactive particles, especially in the size range 1 to 5 mm. Satisfactory explanations for these observations are not available due to simultaneous interplay of many complex factors; however, a lower water adsorption tendency of alkali-silica gels with a higher silica/alkali ratio, and relief of the hydraulic pressure at the surface of the reactive particle when its size is very small may partially explain these observations.

*In the case of sedimentary rocks composed of clay minerals such as phyllites, graywackes, and argillites, *exfoliation* of the sheet structure due to hydroxyl ion attack and water adsorption is the principal cause of expansion. In the case of dense particles of glass and flint, reaction rims form around the particles with the *onion-ring type of progressive cracking and peeling*.

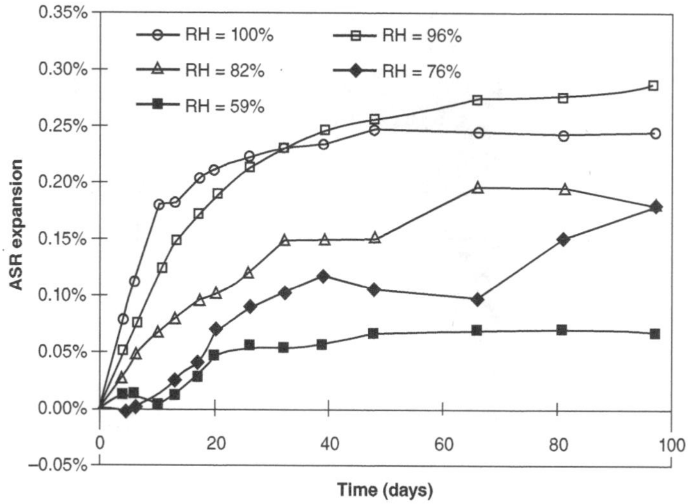

FIGURE 5-23 ASR expansion as a function of time for samples exposed to different constant external relative humidity. (From Poyet, S., A. Sellier, B. Capra, G. Thevenin-Foray, J.M. Torrenti, H. Tournier-Cognon, and E. Bourdarot, Influence of water on alkali-silica reaction. Experimental study and numerical simulation. *J. Mater. Civ. Eng.*, Vol. 18, No. 4, pp. 588–596, 2006.)

5.14.3 Selected Case Histories

From published reports of concrete deterioration due to alkali-aggregate reaction, it is apparent that availability of alkali-reactive aggregates is widespread in the United States, eastern Canada, Australia, Brazil, New Zealand, South Africa, Denmark, Germany, England, and Iceland. Blanks and Kennedy[41] describe some of the earlier cases in the United States. According to the authors, ten years after construction, deterioration was first observed in 1922 at the Buck hydroelectric plant on the New River, Virginia. As early as 1935, R.J. Holden had concluded from petrographic studies of the concrete that the expansion and cracking were caused by chemical reaction between the cement and the phyllite rock, which had been used as an aggregate. Linear expansion in excess of 0.5 percent, caused by the alkali-aggregate reaction, was reported. In another case, the crown of an arch dam in California deflected upstream by about 127 mm in 9 years after the construction. Also, measurements at Parker Dam (California-Arizona) showed that expansion of the concrete increased from the surface to a depth of 3 m, and linear expansions in excess of 0.1 percent were detected.

Because chemical reactions are a function of temperature, it was first thought that the alkali-silica reaction may not be a problem in colder countries, such as Denmark, Germany, and England. Subsequent experience with certain alkali-reactive rocks has shown that this assumption was incorrect. For example, in 1971,[42] it was discovered that concrete of the Val de la Mare dam in the United Kingdom (Fig. 5-24*a*) was suffering from alkali-silica reaction, possibly as a result of the use of a crushed diorite rock containing veins of amorphous silica. Extensive remedial measures were needed to ensure the safety of the dam. By 1981,[43] evidence of concrete deterioration attributable to alkali-silica reaction was found in 23 structures, 6 to 17 years old, located in Scotland, the Midlands, Wales, and other parts of southwestern England. Many of the structures contained concrete made with inadequately washed, sea-dredged sand.

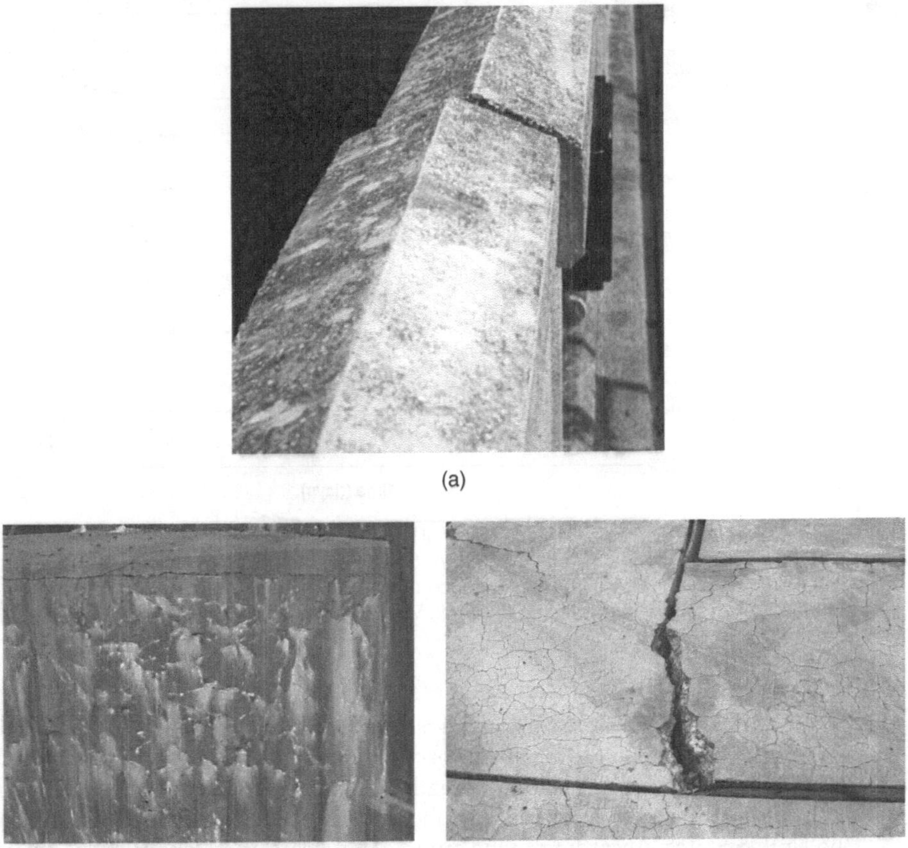

(a)

(b) (c)

FIGURE 5-24 Alkali-aggregate expansion in concrete. [Photographs courtesy of (*a*) J. Figg, Ove Arup Partnership, U.K., (*b*) Mark Desrosiers, California Department of Transportation, and (*c*) U.S. Navy, NFESC.]

(a) Parapet of the Val-de-la-Mare dam (Jersey Island, U.K.) showing misalignment caused by differential movement of adjacent blocks resulting from expansion due to alkali-aggregate reactivity; (b) The girder pedestals and abutments of a bridge built on the eastern slope of the Sierra Nevada were seriously damaged by the alkali-silica reaction; (c) Airfield parking apron at Naval Air Station Point Mugu, California. The lowest part of the apron collects rainfall and as a consequence the ASR has been more pronounced there than in adjacent rows of slabs, resulting in large differential horizontal movements, and very large cracks.

5.14.4 Control of Expansion

From the foregoing description of case histories and mechanisms underlying expansion associated with the alkali-aggregate reaction, it may be concluded that the most important factors influencing the phenomenon are: (1) the alkali content of the cement and the cement content of concrete; (2) the alkali-ion contribution from sources other than portland cement, such as admixtures, salt-contaminated aggregates, and penetration of seawater or deicing salt solution into concrete; (3) the amount, size, and, reactivity of the alkali-reactive constituent present in the aggregate; (4) the availability of moisture to the concrete structure; and (5) the ambient temperature.

When cement is the only source of alkali ions in concrete and alkali-reactive constituents are suspected to be present in the aggregate, experience shows that the use of low-alkali portland cement (less than 0.6 percent equivalent Na_2O) offers the best protection against the alkali attack. If beach sand or sea-dredged sand and gravel are to be used, they should be washed with sweet water to ensure that the *total alkali content* from the cement and aggregates in concrete does not exceed 3 kg/m^3. If a low-alkali portland cement is not available, the total alkali content in concrete can be reduced by replacing a part of the high-alkali cement with cementitious or pozzolanic admixtures such as granulated blast-furnace slag, volcanic glass (ground pumice), calcined clay, fly ash, or silica fume. It should be noted that, similar to the well-bound alkalies in most feldspar minerals, the alkalies present in slags and natural pozzolans are acid-insoluble and probably are not available for reaction with aggregate.

In addition to reducing the effective alkali content, the use of pozzolanic admixtures results in the formation of less expansive alkali-silicate products with a high silica/alkali ratio. In Iceland, only alkali-reactive volcanic rocks are available for use as aggregate, and the cement raw materials are such that only high-alkali portland cement is produced. The problem has been satisfactorily resolved by blending all portland cement with approximately eight percent silica fume, a highly reactive pozzolan (see Chap. 8).

With mildly reactive aggregates, another approach for reducing the concrete expansion is to *sweeten* the reactive aggregate with 25 to 30 percent limestone or any other nonreactive aggregate, when this is economically feasible. Finally, it should be remembered that subsequent to or simultaneously with the progress of the reaction, the availability of moisture to the structure is essential for the expansion to occur. Consequently if the access of water to concrete is prevented by prompt repair of any leaking joints, deleterious expansion may never occur.

According to Swamy[44]:

> Exclude water – and one can almost have a trouble-free concrete even if it contains reactive aggregates and mobile alkalies. Marked deterioration due to the alkali-silica reaction occurs under continuous moist exposure, and in field practice, under wet environmental conditions. . . Funny things can happen in real life–the interior columns of an exposed bridge, sheltered from sunshine and rain, showed no cracking whilst the exterior columns developed extensive cracking. The same structural member, partly sheltered and partially exposed by the nature of the structure, may show extensive cracking on the exposed faces and little or no cracking in the sheltered parts.

5.15 Hydration of Crystalline MgO and CaO

Numerous reports including a review by Mehta,[45] indicate that crystalline MgO or CaO, when present in substantial amounts in cement, hydrate and cause expansion and cracking in concrete. The expansive effect of high MgO in cement was first recognized in 1884 when a number of concrete bridges and viaducts in France failed two years after the construction. About the same time, the town hall of Kassel in Germany had to be rebuilt as a result of expansion and cracking attributed to crystalline MgO in cement. The French and the German cements contained 16 to 30 percent and 27 percent MgO, respectively. This led to restrictions on the maximum permissible MgO in cement.

For example, the current ASTM Standard Specification for Portland Cement (ASTM C 150-83) requires that the MgO content in cement shall not exceed 6 percent.

Although expansion due to hydration of crystalline CaO has been known for a long time in the United States, the deleterious effect associated with the phenomenon was recognized in the 1930s when certain 2- to 5-year-old concrete pavements cracked. Initially suspected to be due to MgO, the expansion and cracking were attributed later to the presence of hard-burnt CaO in the cement used for the construction of the pavements.* Laboratory tests showed that the cement pastes made with a low-MgO portland cement, which contained 2.8 percent hard-burnt CaO, showed considerable expansion. However, with concrete mixtures, due to the restraining effect of the aggregate, relatively large amounts of hard-burnt CaO are needed to obtain a significant expansion. The phenomenon is virtually unknown with modern concrete because better manufacturing controls on the quality of portland-cement clinker have assured that the content of uncombined or free CaO in clinker seldom exceeds 1 percent.

The crystalline MgO, periclase, in a portland cement clinker that has been exposed to 1400 to 1500°C is essentially inert to moisture at room temperature because the reactivity of periclase drops sharply when it is heated above 900°C. No cases of structural distress due to the presence of periclase in modern portland cements are reported from countries such as Brazil, where raw material limitations compel some cement producers to manufacture portland cements containing more than 6 percent MgO. Several cases of expansion and cracking of concrete structures were reported from Oakland, California where the aggregate used for making concrete was found to have been accidentally contaminated with crushed dolomite bricks containing large amounts of MgO and CaO, calcined at temperatures much lower than 1400°C.

5.16 Corrosion of Embedded Steel in Concrete

Deterioration of concrete containing embedded metals, such as conduits, pipes, and reinforcing and prestressing steel, is generally attributable to the combined effect of more than one cause; however, the corrosion of the embedded metal is invariably one of the principal causes. A survey[46] of collapsed buildings in England showed that from 1974 to 1978, the immediate cause of failure of at least eight concrete structures was the corrosion of reinforcing or prestressing steel. These structures were 12 to 40 years old at the time of collapse, except for one that was only 2 years old.

It is to be expected that when the embedded steel is protected from air by an adequately thick cover of a low-permeability concrete, the corrosion of steel and other problems associated with it would not arise. That this may not be entirely true in practice is evident from the high frequency with which even some properly built reinforced and prestressed concrete structures begin to show premature deterioration due to steel corrosion. The incidence of damage is especially large in the structures exposed to deicing chemicals or marine environment. For example, a 1991 report from the Federal Highway Administration to the U.S. Congress said that 134,000 reinforced concrete bridges in the United States (23 percent of the total) required immediate repair and

*Conversion of $CaCO_3$ to CaO can occur at 900 to 1000°C. The CaO thus formed can hydrate rapidly and is called *soft-burnt lime*. Since portland cement clinker is heat-treated to 1400 to 1500°C, any uncombined CaO present is called *hard-burnt*, and it hydrates slowly. It is the slow hydration of hard-burnt CaO in a hardened cement paste that causes expansion.

226,000 (39 percent of the total) were also deficient. Corrosion of the reinforcing steel was implicated as one of the causes of damage in the majority of cases, and the total repair cost was estimated at $90 billion dollars.[47]

The damage to concrete resulting from the corrosion of embedded steel manifests in the form of expansion, cracking, and eventual spalling of the cover (Fig. 5-25*a*). In addition to loss of cover, a reinforced-concrete member may suffer structural damage due to loss of bond between steel and concrete and loss of rebar cross-sectional area—sometimes to the extent that structural failure becomes inevitable.[48] A review of the mechanisms involved in concrete deterioration due to corrosion of embedded steel, selected case histories, and measures for control of the phenomenon are given here.

5.16.1 Mechanisms Involved in Concrete Deterioration by Corrosion of Embedded Steel

Corrosion of steel in concrete is an *electrochemical process*. The electrochemical potentials to form the *corrosion cells* may be generated in two ways:

1. Composition cells may be formed when two dissimilar metals are embedded in concrete, such as steel rebars and aluminum conduit pipes, or when significant variations exist in surface characteristics of the steel.
2. In the vicinity of reinforcing steel concentration cells may be formed due to differences in the concentration of dissolved ions, such as alkalies, and chlorides.

As a result, one of the two metals (or some parts of the metal when only one type of metal is present) becomes anodic and the other cathodic. The fundamental chemical changes occurring at the anodic and cathodic areas[49] are as follows (see also Fig. 5-25*b*).

$$\text{Anode: } \underset{\text{(metallic iron)}}{Fe} \longrightarrow 2e^- + Fe^{2+} \searrow$$

$$\underset{\text{(rust)}}{FeO \cdot (H_2O)_x} \qquad (5\text{-}9)$$

$$\text{Cathode: } \underset{\text{(air)}}{\tfrac{1}{2}O_2} + \underset{\text{(water)}}{H_2O} + 2e^- \longrightarrow 2(OH)^- \nearrow$$

The transformation of metallic iron to rust is accompanied by an increase in volume that, depending on the state of oxidation, may be as large as 600 percent of the original metal (Fig. 5-25*c*). This volume increase is believed to be the principal cause of concrete expansion and cracking. Also, like the swelling of poorly crystalline ettringite, the poorly crystalline iron hydroxides may have a tendency to imbibe water and expand. Another point worth noting is that the anodic reaction involving ionization of metallic iron will not progress far unless the electron flow to the cathode is maintained by the consumption of electrons. For the cathode process, therefore *the presence of both air and water at the surface of the cathode is absolutely necessary*. When steel is embedded in concrete, a protective passive film forms on its surface. This occurs because of existing

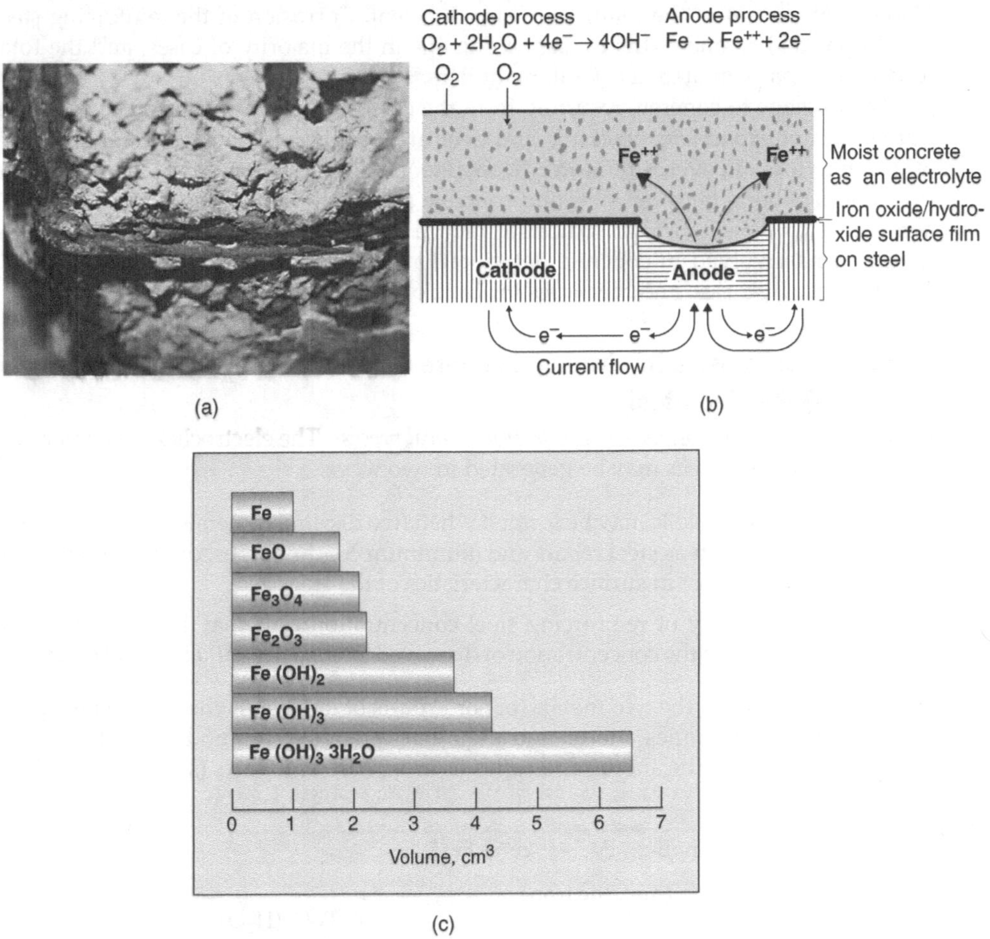

Figure 5-25 Expansion and cracking of concrete due to corrosion of the embedded steel. [(*b*), (*c*), Beton-Bogen, Aalborg Denmark, 1981.]

Figure (a) shows that deterioration of concrete due to corrosion of embedded steel manifests in the form of expansion, cracking, and loss of cover. Loss of steel-concrete bond and reduction of rebar cross section may lead to structural failure. Figure (b) illustrates the electrochemical process of steel corrosion in moist and permeable concrete. The galvanic cell constitutes an anode process and a cathode process. The anode process cannot occur until the protective or the passive iron oxide film is either removed in an acidic environment (e.g., carbonation of concrete) or made permeable by the action of Cl^- ions. The cathode process cannot occur until a sufficient supply of oxygen and water is available at the steel surface. The electrical resistivity of concrete is also reduced in the presence of moisture and salts. Part (c) shows that, depending on the oxidation state, the corrosion of metallic iron can result in up to six times increase in the solid volume.

alkaline conditions in the pore solution of the cement paste, where large concentrations of hydroxides create an environment with a pH between 12 and 14. A double-layered oxide film is formed, consisting of an inner iron II layer (mainly Fe_3O_4) and an outer iron (III) oxide layer (mainly γ-Fe_2O_3) (Fig. 5-26). This means that metallic iron is not available for the anodic reaction until the passivity of steel has been destroyed.

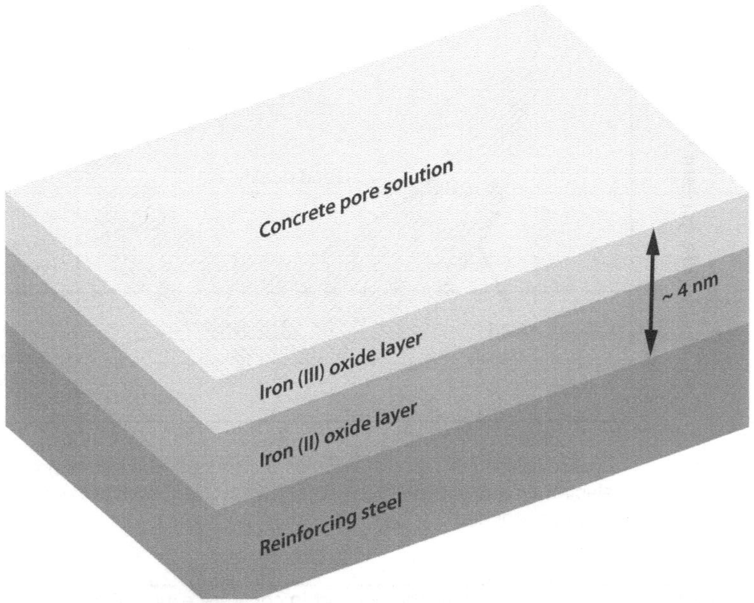

FIGURE 5-26 Schematic representation of the passive film in reinforced concrete.

In the absence of chloride ions in solution, the protective film on steel is reported to be stable as long as the pH of the solution stays above 11.5. As hydrated portland cement contains alkalies in the pore fluid and about 20 weight percent solid calcium hydroxide, normally there is sufficient alkalinity in the system to maintain the pH above 12. Under some conditions (e.g., when concrete has high permeability and alkalies and most of the calcium hydroxide have either been carbonated or leached away), the pH of concrete in the vicinity of steel may have been reduced to less than 11.5. This would destroy the passivity of steel and set the stage for the corrosion process.

In the *presence of chloride ions*, depending on the Cl^-/OH^- ratio, it is reported that the protective film is destroyed even at pH values considerably above 11.5. It seems that when Cl^-/OH^- molar ratio is higher than 0.6, steel is no longer protected against corrosion probably because the iron-oxide film becomes either permeable or unstable under these conditions. For the typical concrete mixtures normally used in practice, the critical chloride content to initiate corrosion is in the order of 0.4 percent of the cement content. However, as shown in Fig. 5-27, this critical value depends on (1) degree of carbonation, (2) quality of the concrete, and (3) relative humidity. Once the passivity of the embedded steel is destroyed, electrical resistivity and the availability of oxygen control the rate of corrosion. In fact, significant corrosion is not observed as long as the electrical resistivity of concrete is above 50 to 70 × 10^3 Ω·cm. Among the common sources of chloride in concrete are admixtures, salt-contaminated aggregate, and penetration of deicing salt solutions or seawater.

5.16.2 Selected Case Histories

A survey of the collapsed buildings and their immediate causes by the British Building Research Establishment[50] showed that, in 1974, a sudden collapse of one main beam of

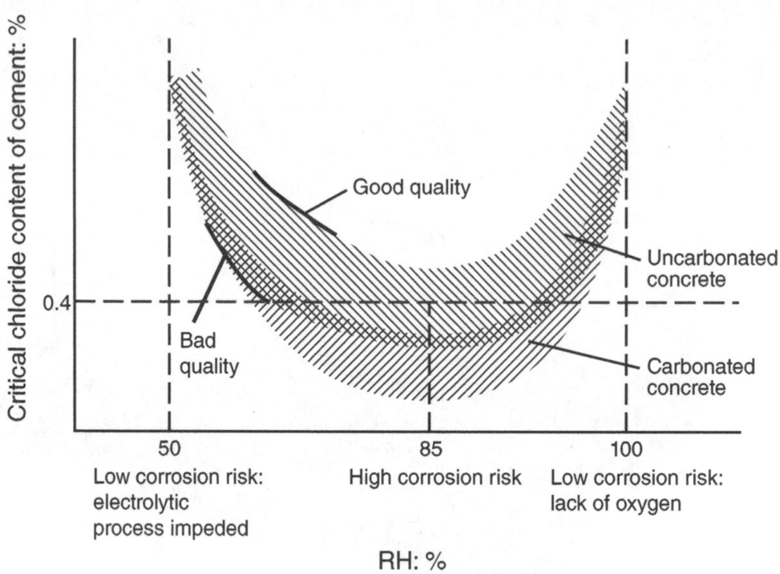

FIGURE 5-27 Variation of critical chloride content in concrete exposed to different environmental conditions. (From Durable Concrete Structures: Design Guide, CEB Task Group 20, Thomas Telford Services, Ltd., 1992.)

a 12-year-old roof with post-tensioned prestressed concrete beams was due to the corrosion of tendons. Poor grouting of ducts and the use of 2 to 4 percent calcium chloride by weight of cement as an accelerating admixture for concrete were diagnosed as the factors responsible for the corrosion of steel. A number of similar mishaps in Britain provided support for the 1979 amendment to the British Code of Practice 110 that *calcium chloride should never be added to prestressed concrete, reinforced concrete, and concrete containing embedded metal.*

A survey by the Kansas State Transportation Department showed that with bridge decks exposed to deicing salt treatment there was a strong relation between the depth of the cover and concrete deterioration by delaminations or horizontal cracking. Generally, good protection to steel was provided when the cover thickness was 50 mm or more (at least thrice the nominal diameter of the rebar, which was 15 mm); however, the normal distribution of variation in cover depth was such that about 8 percent of the steel was 37.5 mm or less deep. With the shallower cover, corrosion of steel is believed to be responsible for the horizontal cracks or delaminations in concrete. On one bridge deck, a combination of the freeze-thaw cracking and corrosion of steel extended the area of concrete delamination about eightfold in 5 years so that 45 percent of the deck surface had spalled by the time the bridge was only 16 years old. Similar case histories of bridge deck damage on numerous highways, including those in Pennsylvania have been reported (Fig. 5-28*a*).

The Kansas survey also reported that the corrosion of the reinforcing steel produced vertical cracks in the concrete deck that contributed to corrosion of the steel girders supporting the deck. Carl Crumpton's humorous observation regarding bridge deck corrosion problems due to deicing salt applications should be noted:

(a) (b)

FIGURE 5-28 Damage to reinforced concrete structures due to corrosion of steel. [(*a*) Photograph courtesy of P.D. Cady, The Pennsylvania State University, University Park, Pennsylvania; (*b*) photograph from Mehta, P.K., and B.C. Gerwick, Jr., *Concr. Int.*, Vol. 4, pp. 45–51, 1982.]

When the $Cl^-/(OH)^-$ ratio of the moist environment in contact with the reinforcing steel in concrete exceeds a certain threshold value, the passivity of steel is broken. This is the first step necessary for the onset of the anodic and cathodic reactions in a corrosion cell. In cold climates, reinforced concrete bridge decks are frequently exposed to the application of deicing chemicals containing chlorides. Progressive penetration of chlorides in permeable concrete leads to scaling, potholes, and delaminations at the concrete surface, finally rendering it unfit for use. Part (a) shows a typical concrete failure (scaling and potholes in the surface of a concrete pavement in Pennsylvania) due to a combination of frost action, corrosion of the embedded reinforcement, and other causes. Part (b) shows deterioration of concrete due to corrosion of the reinforcing steel in the spandrel beams of the San Mateo-Hayward Bridge after 17 years of service. In this case, seawater was the source of chloride ions.

> The wedding of concrete and steel was an ideal union and we used lots of reinforced concrete for bridge decks. Unfortunately, we began tossing salt to melt snow and ice instead of rice for good fertility. That brought irritation, tensions, and erosion of previously good marital relations. No longer could the two exist in blissful union; the seeds of destruction had been planted and the stage had been set for today's bridge deck cracking and corrosion problems.[51]

Mehta and Gerwick[2] reported that many heavily reinforced, 8 by 3.7 by 1.8-m spandrel beams of the San Mateo-Hayward bridge at the San Francisco Bay in California had to undergo expensive repairs due to serious cracking of concrete associated with the corrosion of embedded steel (Fig. 5-28*b*). The beams were made in 1963 with a high-quality concrete (370 kg/m^3 cement, 0.45 water-cement ratio). The damage was confined to the underside and to the windward faces exposed to seawater spray, and occurred only in the precast, steam-cured beams. No cracking and corrosion were in evidence in the naturally-cured, cast-in-place beams made at the same time with a similar concrete mixture. It was suggested that a combination of heavy reinforcement and differential cooling rates immediately following the steam-curing operation, in the massive beams might have resulted in the formation of microcracks in concrete, which were later enlarged due to severe weathering conditions on the windward side of the beams. Thereafter, penetration of the salt water promoted the corrosion-cracking corrosion type of chain reaction, eventually leading to the serious damage. More discussion of cracking-corrosion interaction and case histories of seawater attack on concrete are presented later.

5.16.3 Control of Corrosion

Because water, oxygen, and chloride ions play important roles in the corrosion of embedded steel and cracking of concrete, it is obvious that *permeability of concrete* is the key to control the various processes involved in the phenomena. Concrete-mixture parameters to ensure low permeability, such as low water-cement ratio, adequate cement content, control of aggregate size and grading, and use of mineral admixtures have been discussed earlier. Accordingly, ACI Building Code 318 specifies a maximum 0.4 water-cement ratio for reinforced normal-weight concrete exposed to deicing chemicals and seawater. Proper consolidation and curing of concrete are equally essential. Design of concrete mixtures should also take into account the possibility of increase in the permeability of concrete under service conditions due to various physical-chemical causes, such as thermal gradients, frost action, sulfate attack, and alkali-aggregate expansion.

For the corrosion protection, *maximum permissible chloride content* of concrete mixtures is also specified by ACI Building Code 318. For instance, maximum water-soluble Cl^- ion concentration in hardened concrete, at an age of 28 days, from all ingredients (including aggregates, cementitious materials, and admixtures) should not exceed 0.06, 0.15, and 0.30 percent by mass of cement, for prestressed concrete, reinforced concrete exposed to chloride in service, and other reinforced concretes, respectively. Reinforced concrete members that remain dry or protected from moisture in service are permitted to contain up to 1.00 percent Cl^- by mass of the cementitious material in concrete.

Certain design parameters also influence permeability. That is why, with concrete structures exposed to corrosive environment, Sec. 7.7 of the ACI Building Code 318 specifies *minimum concrete cover* requirements. A minimum concrete cover of 50 mm for walls and slabs, and 63 mm for other members is recommended. Current practice for coastal structures in the North Sea requires a minimum 50 mm of cover on conventional reinforcement, and 70 mm on prestressing steel. Also, ACI 224R specifies 0.15 mm as the *maximum permissible crack width* at the tensile face of reinforced concrete structures subject to wetting-drying or seawater spray. The CEB Model Code recommends limiting the crack widths to 0.1 mm at the steel surface for concrete members exposed to frequent flexural loads, and 0.2 mm to others. Many researchers find no direct relation between crack width and corrosion; however, it is obvious that by increasing the permeability of concrete to water and harmful ions and gases, the presence of a network of interconnected macrocracks and internal microcracks would expose the structure to numerous physical-chemical processes of deterioration.

The repair and replacement costs associated with concrete bridge decks damaged by the corrosion of reinforcing steel have become a major maintenance expense. Many highway agencies now prefer the extra initial cost of providing a waterproof membrane, or a thick overlay of an impervious concrete mixture on newly constructed, or thoroughly repaired surfaces of reinforced and prestressed concrete elements that are large and have flat configuration. *Waterproof membranes*, usually preformed and of the sheet-type variety, are used when they are protected from physical damage by asphaltic concrete wearing surfaces; therefore, their surface life is limited to the life of the asphaltic concrete, which is about 15 years. *Overlay of watertight concrete*, 37.5 to 63 mm thick, provides a more durable protection to the penetration of aggressive fluids into reinforced or prestressed concrete members. Typically, concrete mixtures used for overlay are of low slump, very low water-cement ratio (made possible by adding a superplasticizing admixture), and high cement content. Portland cement mortars containing polymer emulsion (latex) also show excellent impermeability and have been used for overlay purposes; however, vinylidene chloride type latex emulsions are suspected to be the cause of corrosion problems in some cases, and it is now preferred that styrene butadiene type products be used.

Reinforcing bar coatings and cathodic protection provide other approaches to prevent corrosion; however, they are costlier than producing a low-permeability concrete through quality, design, and construction controls. *Protective coatings for reinforcing steel* are of two types: anodic coatings (e.g., zinc-coated steel) and barrier coatings (e.g., epoxy-coated steel). Due to the concern for long-term durability of zinc-coated rebars in concrete, in 1976 the U.S. Federal Highway Administration placed a temporary moratorium on its use in bridge decks. Long-time performance of epoxy-coated rebars is still under investigation in many countries. *Cathodic protection* techniques involve suppression of current flow in the corrosion cell, either by supplying externally a current flow in the opposite direction or by using sacrificial anodes. Due to its complexity and high cost, the system is finding limited applications.

5.17 Development of a Holistic Model of Concrete Deterioration

Field experience shows that, *in order of decreasing importance, the principal causes for deterioration of concrete structures are the corrosion of reinforced steel, exposure to cycles of freezing and thawing, alkali-silica reaction, and sulfate attack.* With each of these four causes of concrete deterioration, the permeability and the presence of water are implicated in the mechanisms of expansion and cracking. Properly constituted, placed, consolidated, and cured concrete is essentially watertight and should therefore have a long service life under most conditions. However, as a result of environmental exposure, cracks as well as microcracks occur and propagate. When they interconnect, a concrete structure loses its water tightness, and becomes vulnerable to one or more processes of deterioration.

Mehta and Gerwick[2] gave a diagrammatic presentation of concrete cracking process due to the reinforcement corrosion (Fig. 5-29*a*). A similar illustration of cracking process due to freezing and thawing cycles was presented by Moukuwa[52] (Fig. 5-29*b*). Generally, the capillary voids in a well-cured concrete structure exposed to air are not saturated. Therefore, a normal (non-air-entrained) concrete should not expand and crack when exposed to freezing and thawing cycles. Concrete expands because weathering and other environmental effects produce cracks and microcracks, which increases the permeability of concrete and the degree of saturation of capillary voids.

Based on a report by Swamy,[53] a diagrammatic presentation of expansion and cracking of concrete due to alkali-aggregate reaction is shown in Fig. 5-29*c*. According to the author, portland cements contain some soluble alkalies and many aggregates contain alkali-reactive minerals, therefore alkali-aggregate reaction can be found in most concretes. He writes:

> In spite of the alkali-aggregate reaction occurring in a concrete, the expansion and deleterious cracking would not take place unless the environment is highly saturated. With properly selected materials, mixture proportions, processing, and curing conditions, it is possible to produce concrete structures that will remain sufficiently dry in the interior during service. Microcracking during weathering and loading effects sometimes destroys the watertightness and makes the concrete permeable.

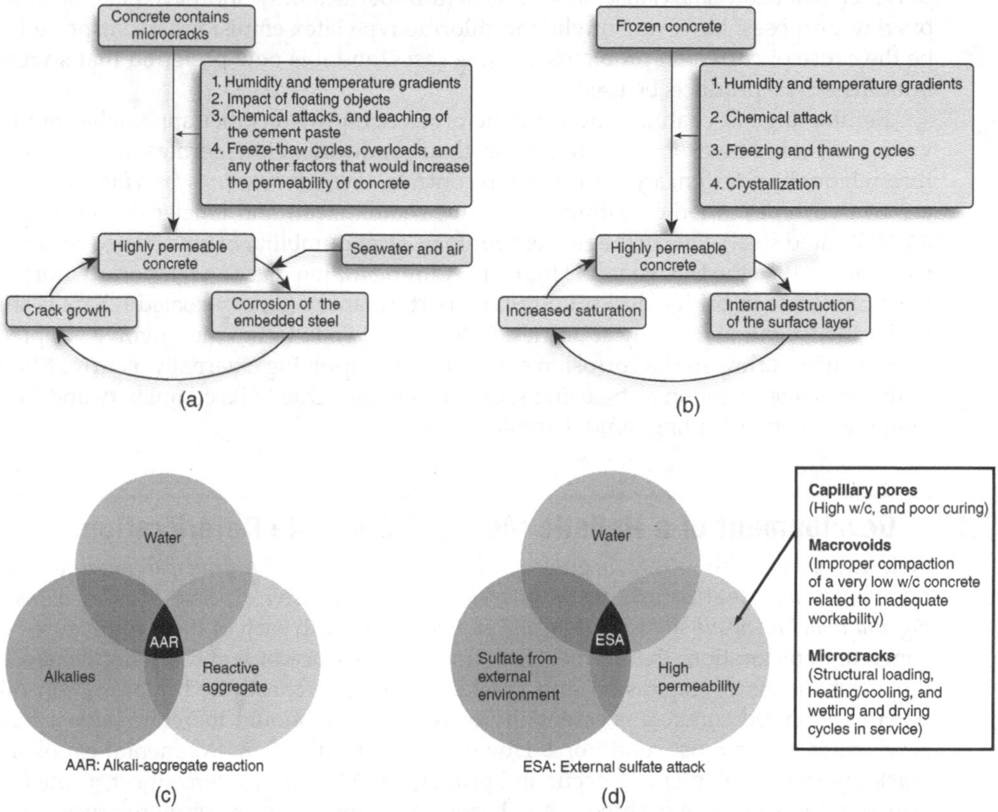

FIGURE 5-29 Diagrammatic presentation of damage to concrete from (*a*) corrosion of reinforced concrete, (*b*) cycles of freezing and thawing, (*c*) alkali-silica reaction, (*d*) external sulfate attack. [(*a*) From Mehta, P.K., and B.C. Gerwick, Jr., *Concr. Int.*, Vol. 4, pp. 45–51, 1982; (*b*) from Moukwa, M., Moukwa, *Cem. Concr. Res.*, Vol. 20, No. 3, pp. 439–446, 1990; (*c*) from Swamy, R.N., *ACI, SP 144*, pp. 105–139, 1994; (*d*) from Collepardi, M., *Concr. Int.*, Vol. 21, No. 1, pp. 69–74, 1999.]

According to the diagrammatic representation of sulfate attack by Collepardi (Fig. 5-29*d*),[30] deterioration of the hydrated cement paste as a result of interaction with sulfate ions from external source requires high permeability and presence of water. Typical causes of high permeability of concrete are high water-cement, inadequate consolidation, and cracking due to adverse weathering and loading conditions.

Integrating the concepts presented by Figs. 5-29*a*, *b*, *c*, and *d*, Mehta[54] has proposed a holistic model of concrete deterioration from the commonly encountered environmental effects (Fig. 5-30). According to this model, a well-constituted, properly consolidated, and cured concrete remains essentially watertight as long as the microcracks and pores within the interior do not form an interconnected network of pathways leading to the surface of concrete. Structural loading as well as weathering effects, such as exposure to cycles of heating-cooling and wetting-drying, facilitate the propagation of microcracks that normally preexist in the interfacial transition zone between the cement mortar and the coarse aggregate particles. This happens during *Stage 1 of the structure-environmental interaction.*

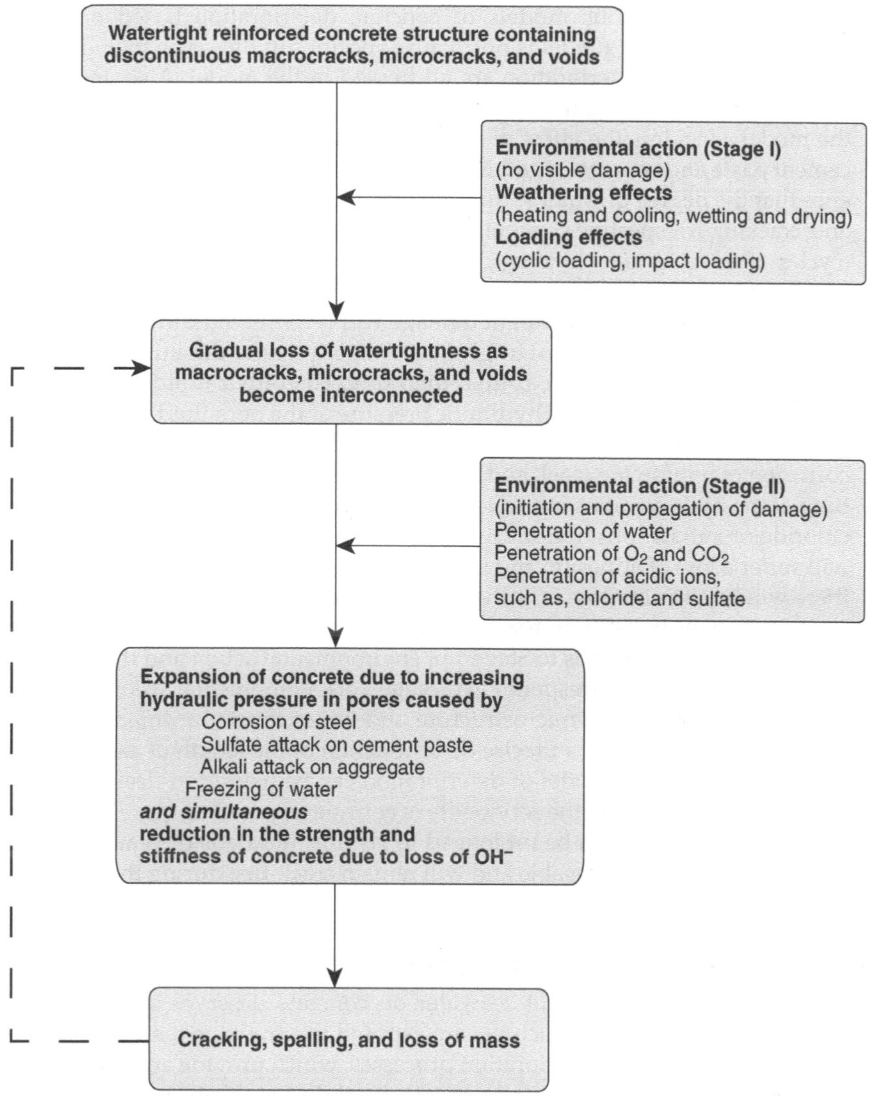

FIGURE 5-30 A holistic model of deterioration of concrete from commonly encountered environmental effects. (Mehta, P.K., *ACI, SP-144*, pp. 1–34, 1994; *Concr. Int.*, Vol. 19, No. 7, pp. 69–76, 1997.)

Radical enhancements in the durability of concrete to commonly known causes of deterioration can be achieved by preventing the loss of watertightness during service through control in the growth of microcracks that interlink the surface cracks with the interior voids and microcracks.

Once the watertightness of concrete is lost, the interior of concrete can become saturated. Consequently, water and ions which play an active role in the processes of deterioration, can now be transported readily into the interior. This marks the beginning of *Stage 2 of the "structure-environmental interaction"* during which the deterioration of concrete takes place through successive cycles of expansion, cracking, loss of mass, and increased permeability.

Unlike the previous models of concrete deterioration based on a reductionist approach, the holistic model is not "cause specific" in the sense that all of the primary causes of concrete deterioration are addressed in the model. Also, instead of holding only one of the components of the cement paste or concrete responsible for the damage, the model considers the effect of agents of deterioration on all the components of the cement paste and concrete together. Furthermore, the model recognizes the field experience that the degree of water saturation of concrete plays a dominant role in expansion and cracking irrespective of whether the primary cause of deterioration is frost action (cycles of freezing and thawing), corrosion of reinforcing steel, alkali-aggregate reaction, or sulfate attack.

Note that little or no apparent damage will be observed during Stage 1, which represents a gradual loss of watertightness. Stage 2 marks the initiation of the damage, which occurs at a slow rate at first, then proceeds rather rapidly. It is suggested that during the second stage, the hydraulic pressure of the pore fluid in a saturated concrete will rise due to one or more phenomena of volumetric expansion (e.g., freezing of water, corrosion of reinforcing steel, and swelling of ettringite or alkali-silica gel). At the same time, if the hydroxyl ions in the cement paste are being leached away and replaced by chloride or sulfate ions, the calcium silicate hydrate will decompose and the concrete will suffer a loss of adhesion and strength. As a result of these two damaging processes there will be a further loss of watertightness and acceleration of the damage.

Based on the holistic approach of concrete deterioration, it is obvious that the period of no-damage corresponds to Stage 1 of environmental action and the gradually escalating period of damage corresponding to Stage 2 of environmental action shown in Fig. 5-30. Due to variations in the microstructure and microclimate at different points within a given concrete structure, a precise determination of the length of each stage is difficult. However, the holistic model of deterioration can be helpful in designing cost-effective strategies for prolonging the service life of concrete exposed to aggressive environments. For example, Stage 1 can be prolonged to last for hundreds of years by using concrete mixtures that are impermeable and will remain crack-free during the service.

5.18 Concrete in the Marine Environment

For several reasons, effect of seawater on concrete deserves special attention. First, coastal and offshore sea structures are exposed to simultaneous attack by a number of physical and chemical deterioration processes, which provide an excellent opportunity to understand the complexity of concrete durability problems in the field practice. Second, oceans make up 80 percent of the surface of the earth; therefore, a large number of structures are exposed to seawater either directly or indirectly (e.g., winds can carry seawater spray for a few miles inland from the coast). Concrete piers, decks, breakwater, and retaining walls are widely used in the construction of harbors and docks. To relieve land from pressures of urban congestion and pollution, floating offshore platforms made of concrete are being considered for location of new airports, power plants, and waste disposal facilities. Many offshore concrete drilling platforms and oil storage tanks have been installed during the last 30 years.

Most seawaters are fairly uniform in chemical composition, which is characterized by the presence of about 3.5 percent soluble salts by mass. The ionic concentrations of Na^+ and Cl^- are the highest, typically 11,000 and 20,000 mg/L, respectively. However, from standpoint of aggressive action to cement hydration products, sufficient amounts of Mg^{2+} and SO_4^{2-} are present, typically 1400 and 2700 mg/L, respectively. The pH of seawater varies between 7.5 and 8.4; the average value in equilibrium with the atmospheric CO_2 is 8.2. Under certain conditions, such as sheltered bays and estuaries, pH values lower than 7.5 may be encountered due to high concentration of dissolved CO_2, which would make the seawater more aggressive to portland-cement concrete.

Concrete exposed to marine environment may deteriorate as a result of combined effects of chemical action of seawater constituents on the cement hydration products, alkali-aggregate expansion (when reactive aggregates are present), crystallization pressure of salts within concrete if one face of the structure is subject to wetting and others to drying conditions, frost action in cold climates, corrosion of the embedded steel in reinforced or prestressed members, and physical erosion due to wave action and floating objects. Attack on concrete due to any one of these causes tends to increase the permeability; not only would this make the material progressively more susceptible to further action by the same destructive agent but also by other types of attack. Thus a maze of interwoven chemical and physical causes of deterioration is at work when a concrete structure exposed to seawater is an advanced stage of degradation. Theoretical aspects of concrete deterioration by seawater, selected case histories, and recommendations for construction of durable concrete structures in the marine environment are discussed by Mehta,[55] and are summarized here.

5.18.1 Theoretical Aspects

In regard to chemical attack on the constituents of the hydrated cement paste, it may be expected that sulfate and magnesium ions are the harmful constituents in seawater. Note that with groundwaters, sulfate attack is classified as *severe* when the sulfate ion concentration is higher than 1500 mg/L; similarly, portland cement paste can deteriorate by cation-exchange reactions when magnesium ion concentration exceeds, for instance, 500 mg/L.

Interestingly, in spite of undesirably high sulfate content of seawater, field experience shows that even when a high-C_3A portland cement has been used and significant amounts of ettringite present as a result of sulfate attack on the cement paste, the deterioration of concrete did not happen by expansion and cracking; instead, it usually took the form of erosion or loss of solid constituents from the mass. It appears that ettringite expansion is suppressed in the environments where $(OH)^-$ ions have essentially been replaced by Cl^- ions. This is consistent with the hypothesis that an alkaline environment is necessary for the swelling of ettringite by water adsorption. Irrespective of the mechanism by which the sulfate expansion associated with ettringite is suppressed in high-C_3A portland cement concrete exposed to seawater, the influence of chloride on the sulfate expansion clearly demonstrates the error too often made in modeling the behavior of materials when, for the sake of simplicity, the effect of an individual factor on a phenomenon is predicted without sufficient regard to the other factors that may be present, and may modify the effect significantly.

According to ACI Building Code 318, the sulfate exposure in seawater is classified as *moderate* for which the use of ASTM Type II portland cement (maximum 8 percent C_3A) with a 0.50 maximum water-cement ratio in normal-weight concrete is permitted. In fact, it is stated in the ACI 318R-21, *Building Code Commentary*, that cements with C_3A up to 10 percent may be used if the maximum water-cement ratio is further reduced to 0.40.

The fact that uncombined calcium hydroxide in a mortar or concrete can cause deterioration by an exchange reaction involving magnesium ions was known as early as

1818 from investigations on the disintegration of lime-pozzolan concretes by Vicat, who undoubtedly is regarded as one of the founders of the technology of modern cement and concrete. Vicat made the profound observation:

> On being submitted to examination, the deteriorated parts exhibit much less lime than the others; what is deficient then, has been dissolved and carried off; it was in excess in the compound. Nature, we see, labors to arrive at exact proportions, and to attain them, corrects the errors of the hand which has adjusted the doses. Thus the effects that we have just described, and in the case alluded to, become the more marked, the further we deviate from these exact proportions.[56]

State-of-the-art reviews[57,58] on the performance of structures in marine environment confirm that Vicat's observation is equally valid for portland cement concrete. From long-term studies of portland cement mortars and concrete mixtures exposed to seawater, the evidence of magnesium ion attack is well established by the presence of white deposits of *brucite* or $Mg(OH)_2$ and by magnesium silicate hydrate which can be detected by mineralogical analysis. In seawater exposure, a well-cured concrete containing a large amount of slag or a pozzolan in the cementitious materials usually outperforms concrete containing only portland cement.[59] This happens, in part, because the former contains less uncombined calcium hydroxide after curing. The implication of the loss of calcium hydroxide by the hydrated cement paste, whether it has occurred by magnesium ion attack or by CO_2 attack, is obvious from Fig. 5-31*c*.

Because seawater analyses seldom include the dissolved CO_2 content, the potential for the loss of concrete mass by leaching away of solid calcium hydroxide from the hydrated cement paste due to carbonic acid attack is often overlooked. According to Feld,[60] in 1955, after 21 years of use, the concrete piles and caps of the trestle bends of the James River Bridge at Newport News, Virginia, required a $1.4 million repair and replacement cost involving 70 percent of the 2500 piles. Similarly, 750 precast concrete piles driven in 1932 near Ocean City, New Jersey, had to be repaired in 1957 after 25 years of service; some of the piles had been reduced from the original 550 mm diameter to 300 mm. In both cases, the loss of material was associated with higher than normal concentration of dissolved CO_2 in the seawater.

It should be noted that with *permeable concrete* the normal amount of CO_2 present in seawater is sufficient to decompose the cementitious products eventually. The presence of *thaumasite* (calcium silicocarbonate), *hydrocalumite* (calcium carboaluminate hydrate), and *aragonite* (calcium carbonate) has been reported in the cement pastes samples obtained from deteriorated concrete structures exposed to seawater for long periods.

5.18.2 Case Histories of Deteriorated Concrete

Compared to other structural materials, generally, concrete has a satisfactory record of performance in seawater. However, published literature contains reports on large number of both plain and reinforced concrete that has suffered serious deterioration in the marine environment. For the purpose of drawing useful lessons for construction of concrete sea structures, several case histories of deterioration of concrete as a result of long-term exposure to seawater are summarized in Table 5-5, and are discussed next.

In the *mild climates* of southern France and southern California, *unreinforced mortar and concrete specimens* remained in excellent condition after more than 60 years of seawater exposure, except when the permeability of concrete was high. Permeable specimens

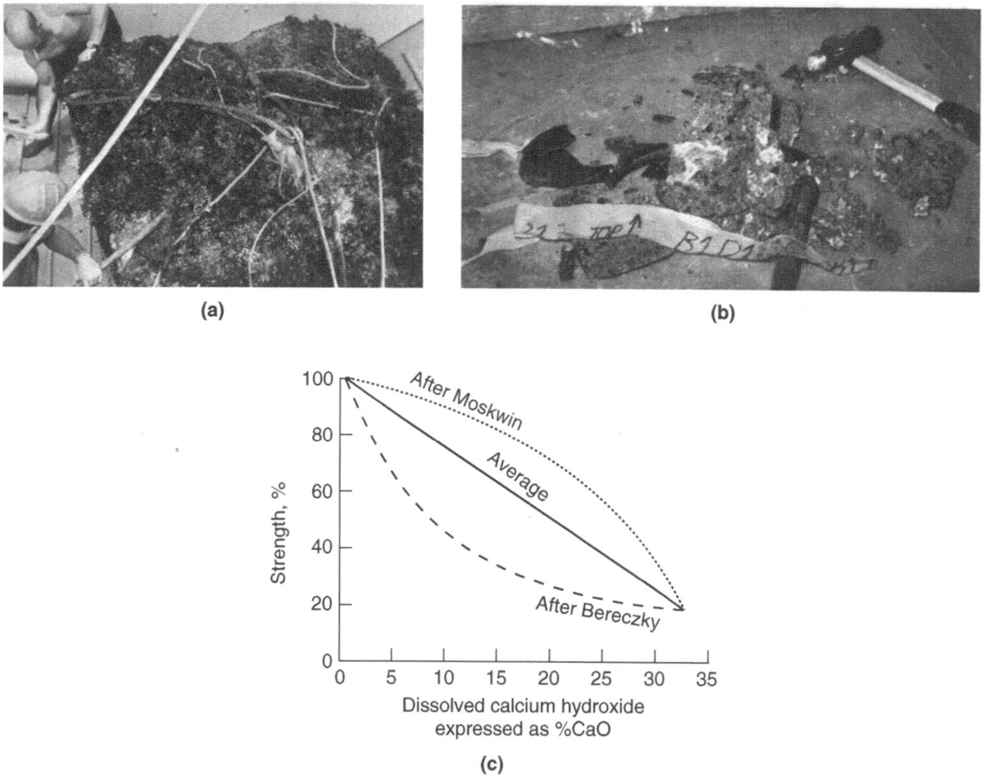

FIGURE 5-31 Strength loss in permeable concrete due to lime leaching. [(*a*), (*b*) Photographs from Mehta, P.K., and H. Haynes, *J. ASCE, Structure Div.*, Vol. 101, No. ST-8, pp. 1679–1686, 1975; (*c*) adapted from Biczok, I., *Concrete Corrosion and Concrete Protection*, Chemical Publishing Company, New York, p. 291, 1967.]

Unreinforced concrete tests blocks (1.75 by 1.75 by 1.07 m) partially submerged in seawater at San Pedro harbor in Los Angeles, California, were examined after 67 years of continuous exposure. Low-permeability concretes, irrespective of the portland cement composition were found to be in excellent condition. Concretes containing a low cement content (high permeability) showed so much reduction in the surface hardness that deep grooves were made by a wire rope on the test blocks when they were being hauled up with the help of an amphibious crane [part (a)]. Test cores showed that concrete was very porous and weak. With large pores containing deposits of a white precipitate [part (b)] which was identified as $Mg(OH)_2$ by X-ray diffraction analysis. The original products of portland cement hydration, C-S-H and $Ca(OH)_2$ were no longer present.

Numerous researchers have found that portland cement pastes, mortars, and concretes undergo strength loss when the cementitious products are decomposed and leached out as a result of attack by acidic or magnesium-containing solutions. The severity of leaching can be evaluated from the content of dissolved CaO. On the average, the compressive strength drops by about 2 percent when 1 percent CaO is removed from the portland cement paste [part (c)].

showed considerable loss of mass associated with magnesium ion attack, CO_2 attack, and calcium leaching. In spite of the use of high-C_3A portland cements, expansion and cracking of concrete due to ettringite was not observed in low-permeability concretes. Therefore, the effect of cement composition on durability to seawater appears to be less significant than effect of the permeability of concrete.

History of Structures	Results of Examination
Mild Climate	
Forty-centimeter mortar cubes made with different cements and three different cement contents, 300, 450, and 600 kg/m^3, were exposed to seawater at La Rochell, southern France, in 1904–1908.*	After 66 years of exposure to seawater, the cubes made with 600 kg/m^3 cement were in good condition even when they contained a high-C_3A (14.9 percent) portland cement. Those containing 300 kg/m^3 were destroyed; therefore, chemical composition the cement was of major importance for the low-cement-content cubes. In general, pozzolan and slag cements showed the better resistance to seawater than portland cement. Electron microscopy studies of deteriorated specimens showed the presence of aragonite, brucite, ettringite, magnesium silicate hydrate, and thaumasite.
Eighteen 1.75 × 1.75 × 1.07-m unreinforced concrete blocks made with six different portland cements and three different concrete mixtures, partially submerged in seawater in the Los Angeles harbor in 1905.†	After 67 years of exposure, the dense concrete (1:2:4) blocks, some made with 14 percent C_3A portland cement, were still in excellent condition. Lean concrete (1:3:6) blocks lost some material and were much softer (Fig. 5-31*a*). X-ray diffraction analyses of the weakened concrete showed the presence of brucite, gypsum, ettringite, and hydrocalumite.The cementing constituents, C-S-H gel and $Ca(OH)_2$, were not detected.
Concrete structures of the San Francisco Ferry Building, built in 1912. Type I portland cement with 14 to 17 percent C_3A was used. 1:5 concrete mixture contained 658 lb/yd^3 (390 kg/m^3) cement.	After 46 years of service (a) was found in excellent condition, and 90 percent of piles in (b) were in good condition. In (c), 20 to 30 percent of piles were attacked in tidal zone, and about 35 percent of the deep transverse girders had their underside and part of the vertical face cracked or spalled due to corrosion of reinforcement. Presence of microcracks due to deflection under load might have exposed the reinforcing steel to corrosion by seawater. Poor workmanship was held responsible for differences in behavior of concrete, which was of the same quality in all the structures.
Precast concrete cylinders jacket for Pier 17.	
Cast-in-place concrete cylinders for Piers 30 and 39.	
Cast-in-place concrete cylinders and transverse girders for Piers 26 and 28.†	
Cold Climate	
Many 20- to 50-year coastal structures were included in a 1953–1955 survey of 431 concrete structures in Denmark.§ Among the severely deteriorated structures were the following in Jutland.	Of the coastal structures, about 40 percent showed overall deterioration, and about 35 percent showed from severe surface damage to slight deterioration.
Oddesund bridge, Pier 7: History of the structure indicated initial cracking of caissons due to thermal stresses. This permitted considerable percolation of water through the caisson walls and the interior mass concrete filling. General repairs commenced after 8 years of service.	Examination of deteriorated concrete from the Oddesund Bridge indicated decomposition of cement and loss of strength due to sulfate attack below low-tide level and cracking due to freezing and thawing as well as alkali-aggregate reaction above hightide level. Reaction products from cement decomposition were aragonite, ettringite, gypsum, brucite, and alkali-silica gel.

Table 5-5 Performance of Concrete Exposed to Seawater

History of Structures	Results of Examination
Cold Climate (*Continued*)	
Highway Bridge, North Jutland: Severe cracking and spalling of concrete at the mean water level provided a typical hourglass shape to the piers. Concrete in this area was very weak. Corrosion of reinforcement was everywhere and pronounced in longitudinal girders.	Examination of concrete piers of the highway bridge showed evidence of poor concrete quality (high w/c). Symptoms of general decomposition of cement and severe corrosion of the reinforcement were superimposed on the evidence for the primary deleterious agents, such as freezing-thawing and alkali-aggregate reaction.
Groin 71, north barrier, Lim Fjord: Lean concrete blocks (220 kg/m^3 cement) exposed to windy weather, repeated wetting and drying, high salinity, freezing and thawing, and severe impact of gravel and sand in the surf. Some blocks disappeared in the sea in the course of 20 years.	Examination of the severely deteriorated concrete blocks from Groin 71 showed very weak, soapy matrix with loose aggregate pebbles. In addition to the alkali-silica gel, the presence of gypsum and brucite was confirmed.
Along the Norwegian seaboard, 716 concrete structures were surveyed in 1962–1964. About 60 percent of the structures were reinforced concrete wharves of the slender-pillar type containing tremie-poured underwater concrete. Most wharves had decks of the beam slabs and slab type. At the time of survey, about two thirds of the structures were 20–50 years old.¶	Below the low-tide level and above the high-tide level concrete pillars were generally in good condition. In the splashing zone, about 50 percent of the surveyed pillars were in good condition; 14 percent had their cross sectional area reduced by 30 percent or more, and 24 percent had 10–30 percent reduction in area of cross section. Deck were generally in good condition but 20 percent deck beams needed repair work because of major damage due to corrosion of reinforcement. Deterioration of pillars in the tidal zone was ascribed mainly due to frost action on poor-quality concrete.

*Regourd, M., Annales de l'Institute Technique du Bâtiment et des Travaux Publics, No. 329, June 1975, and No. 358, Feb. 1978.

†Mehta, P.K., and H. Haynes, *J. Struct. Div ASCE*, Vol. 101, No. ST-8, Aug. 1975.

‡Fluss, P.J., and S.S. Gorman, *J. ACI, Proc.*, Vol. 54, 1958.

§Idorn, G.M., Durability of Concrete Structures in Denmark, *D. Sc. dissertation*, Tech. Univ., Copenhagen, Denmark, 1967.

¶Gjorv, O.E., *Durability of Reinforced Concrete Wharves in Norwegian Harbors*, The Norwegian Committee on Concrete in Sea Water, 1968.

TABLE 5-5 Performance of Concrete Exposed to Seawater (*Continued*)

Reinforced concrete members in a mild climate (Piers 26 and 28 of the San Francisco Ferry Building in California). In spite of a low-permeability concrete mixture (390 kg/m^3 cement content), the structures showed cracking due to corrosion of the reinforcing steel after 46 years of service. Because corrosion requires permeation of seawater and air to the embedded steel, poor consolidation of concrete and structural microcracking were diagnosed to be the probable causes of the increase in the permeability which made the corrosion of steel possible.

In the *cold climates* of Denmark and Norway, concrete mixtures unprotected by entrained air were subject to expansion and cracking by frost action. (It may be noted that air entrainment was not prevalent before the 1950s.) Therefore, cracking due to freeze-thaw cycles was probably responsible for increase in the permeability, followed by other destructive processes, such as alkali-aggregate attack and corrosion of the reinforcing steel.

Investigations of reinforced concrete structures have shown that, generally, concrete fully immersed in seawater suffered only a little or no deterioration; concrete exposed to salts in air or water spray suffered some deterioration, especially when permeable; and concrete subject to tidal action suffered the most.

5.18.3 Lessons from the Case Histories

For the future construction of concrete sea structures, the following lessons from the case histories of concrete deteriorated by seawater can be drawn. These lessons confirm the validity of the holistic model of concrete deterioration already discussed:

1. Permeability is the key to durability Deleterious interactions of serious consequence between constituents of hydrated portland cement and seawater take place when seawater is not prevented from penetrating into the interior of a concrete. Typical causes of insufficient watertightness are poorly proportioned concrete mixtures, absence of properly entrained air if the structure is located in a cold climate, inadequate consolidation and curing, insufficient concrete cover on the reinforcing steel, badly designed or constructed joints, and microcracking in hardened concrete attributable to the loading conditions and other factors, such as thermal shrinkage, drying shrinkage, and alkali-aggregate reaction.

It is interesting to point out that engineers on the forefront of concrete technology are becoming increasingly conscious of the significance of the permeability of concrete to durability of structures exposed to aggressive waters. For example, concrete mixtures for offshore structures in Norway are now specified to meet a maximum permissible permeability requirement ($k \leq 10^{-13}$ kg/Pa·m·s). In the United States, concrete mixtures for the construction of decks and parking garages exposed to deicer salts are being specified to meet 2000 Coulombs or less chloride penetration rating according to the ASTM Standard Test Method C1202.

2. Type and severity of deterioration may not be uniform throughout the structure As illustrated by the diagrammatic representation of a reinforced concrete cylinder exposed to seawater (Fig. 5-32), the section that always remains above the high-tide line will be more susceptible to frost action and corrosion of embedded steel. The section that is between high- and low-tide lines will be vulnerable to cracking and spalling, not only from frost action and steel corrosion but also from wet-dry cycles. Chemical attacks due to alkali-aggregate reaction and seawater-cement paste interaction will also be at work here. Concrete weakened by microcracking and chemical attacks will eventually disintegrate by eroding action and the impact of sand, gravel, and ice; thus maximum deterioration occurs in the tidal zone. On the other hand, the fully submerged part of the structure will only be subject to chemical attack by seawater. Because it is not exposed to subfreezing temperatures, there will be no risk of frost damage. There will be little or no corrosion of the reinforcing steel due to lack of oxygen.

It appears that progressive chemical deterioration of cement paste by seawater from the surface to the interior of the concrete follows a general pattern.[61] The formation of aragonite and bicarbonate by CO_2 attack is usually confined to the surface of concrete,

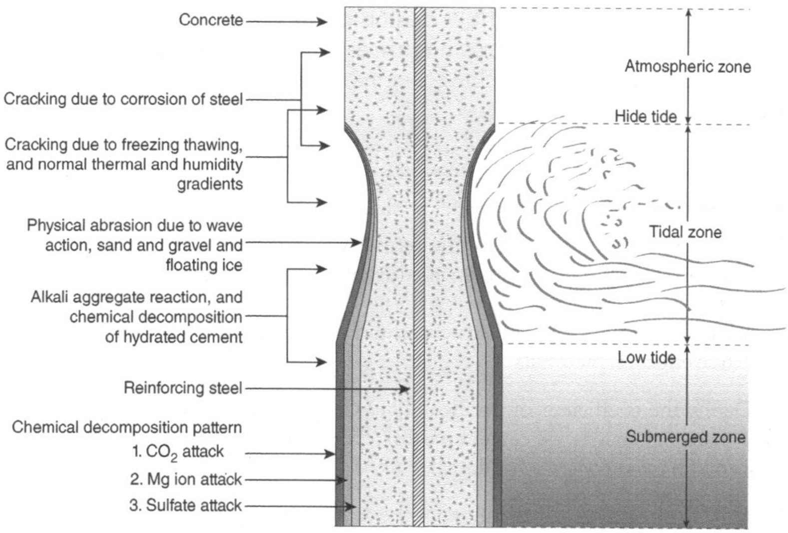

FIGURE 5-32 Diagrammatic representation of a reinforced concrete cylinder exposed to seawater. (From Mehta, P. K., *Performance of Concrete in Marine Environment*, ACI SP 65, pp. 1–20, 1980.)

The type and severity of attack on a concrete sea structure depend on the conditions of exposure. The sections of the structure that remain fully submerged are rarely subjected to frost action or corrosion of the embedded steel. Concrete at this exposure condition will be susceptible to chemical attacks. The general pattern of chemical attack from the concrete to the interior is shown. The section above the high-tide mark will be vulnerable to both frost action and corrosion of the embedded steel. The most severe deterioration is likely to take place in the tidal zone because here the structure is exposed to all kinds of physical and chemical attacks.

the formation of brucite by magnesium ion attack is found below the surface of concrete, and the evidence of some ettringite formation in the interior shows that sulfate ions are able to penetrate even deeper. Unless concrete is very permeable, no damage results from the chemical action of seawater on cement paste because the reaction products (aragonite, brucite, and ettringite), being insoluble, tend to reduce the permeability and stop further ingress of seawater into the interior of concrete. This kind of protective action would not be available under dynamic loading conditions in the tidal zone, where the reaction products would be washed away by wave action as soon as they are formed.

3. Corrosion of embedded steel is, generally, the major cause of concrete deterioration in reinforced and prestressed concrete structures exposed to seawater, but in low-permeability concrete this does not appear to be the first cause of cracking Based on numerous case histories, it appears that cracking-corrosion interactions probably follow the route diagrammatically illustrated in Fig. 5-29*a*. Because the corrosion rate depends on the cathode/anode area, significant expansion accompanying the corrosion of steel should not occur until there is sufficient supply of oxygen at the surface of the reinforcing steel (i.e., an increase in the cathode area). This will not occur as long as the concrete cover surrounding the of steel-cement paste interfacial zone remains impermeable.

Pores and microcracks already exist in the interfacial zone, but their enlargement through a variety of phenomena other than corrosion seems to be necessary before conditions exist for significant corrosion of the embedded steel in concrete. Once the conditions for significant corrosions are established, a progressively escalating cycle of cracking-corrosion-more-cracking begins, eventually leading to considerable structural damage.

Test Your Knowledge

5.1 What do you understand by the term *durability*? Compared to other considerations, how much importance should be given to durability in the design and construction of concrete structures?

5.2 Write a short note on the structure and properties of water, with special reference to its destructive effect on materials.

5.3 Define the coefficient of permeability? Give typical values of the coefficient for (a) fresh cement pastes; (b) hardened cement pastes; (c) commonly used aggregates; (d) high-strength concretes; and (e) mass concrete for dams.

5.4 How does aggregate size influence the coefficient of permeability of concrete? List other factors that determine the permeability of concrete in a structure.

5.5 What is the difference between erosion and abrasion? From the standpoint of durability to severe abrasion, what recommendations would you make in the design of concrete and construction of an industrial floor?

5.6 Under what conditions may salt solutions damage concrete without involving chemical attack on the portland cement paste? Which salt solutions commonly occur in natural environments?

5.7 Briefly explain the causes and control of scaling and D-cracking in concrete. What is the origin of laitance; what is its significance?

5.8 Discuss Powers' hypothesis of expansion on freezing of a saturated cement paste containing no air. What modifications have been made to this hypothesis? Why is entrainment of air effective in reducing the expansion due to freezing?

5.9 With respect to frost damage, what do you understand by the term *critical aggregate size*? What factors govern it?

5.10 Discuss the significance of *critical degree of saturation* from the standpoint of predicting frost resistance of a concrete.

5.11 Discuss the factors that influence the compressive strength of concrete exposed to a fire of medium intensity (650°C, short-duration exposure). Compared to the compressive strength, how would the elastic modulus be affected, and why?

5.12 What is the effect of pure water on hydrated portland cement paste? With respect to carbonic acid attack on concrete, what is the significance of *balancing* CO_2?

5.13 List some of the common sources of sulfate ions in natural and industrial environments. For a given sulfate concentration, explain which of the following solutions would be the most deleterious and which would be the least deleterious to a permeable concrete containing a high-C_3A portland cement: Na_2SO_4, $MgSO_4$, $CaSO_4$.

5.14 What chemical reactions are generally involved in sulfate attack on concrete? What are the physical manifestations of these reactions?

5.15 Critically review the *BRE Digest 250* and the ACI Building Code 318 requirements for control of sulfate attack on concrete.

5.16 What is the alkali-aggregate reaction? List some of the rock types that are vulnerable to attack by alkaline solutions. Discuss the effect of aggregate size on the phenomenon.

5.17 With respect to the corrosion of steel in concrete, explain the significance of the following terms: carbonation of concrete, passivity of steel, Cl^-/OH^- molar ratio of the contact solution, electrical resistivity of concrete, state of oxidation of iron.

5.18 Briefly describe the measures that should be considered for the control of corrosion of embedded steel in concrete.

5.19 With coastal and offshore concrete structures directly exposed to seawater, why does most of the deterioration occur in the tidal zone? From the surface to the interior of concrete, what is the typical pattern of chemical attack in sea structures?

5.20 A heavily reinforced and massive concrete structure is to be designed for a coastal location in Alaska. As a consultant to the primary contractor, write a report explaining the state-of-the-art on the choice of cement type, aggregate size, admixtures, mix proportions, concrete placement, and concrete curing procedures.

References

1. Garboczi, E.J., *Cem. Concr. Res.*, Vol. 20, No. 4, pp. 591–601, 1990.
2. Mehta, P.K., and B.C. Gerwick, Jr., *Concr. Int.*, Vol. 4, No. 10, pp. 45–51, 1982.
3. ACI Report 224R-00, *Manual of Concrete Practice*, Part 3, 2013.
4. Liu, T.C., *J. Aci. Proc.*, Vol. 78, No. 5, p. 346, 1981.
5. Winkler, E.M., *Stone: Properties, Durability in Man's Environment*, Springer-Verlag, New York, 1975.
6. Binda, L., and G. Baronio, *ACI SP 145*, pp. 933–946, 1994.
7. Goude and Vilas, *Salt Weathering Hazards*, Wiley, New York, 1997.
8. Haynes, H., R. O'Neill, and P.K. Mehta, *Concr. Int.*, Vol. 18, No. 1, pp. 63–69, 1996.
9. Mehta, P.K., *Concr. Int.*, Vol. 22, No. 8, pp. 57–61, 2000.
10. Power, T.C., *The Physical Structure and Engineering Properties of Concrete, Bulletin 90*, Portland Cement Association, Skokie, IL, 1958.
11. Beaudoin, J.J., and C. McInnis, *Cem. Concr. Res.*, Vol. 4, pp. 139–148, 1974.
12. Meier, U., and A.B. Harnik, *Cem. Concr. Res.*, Vol. 8, pp. 545–551, 1978.
13. Litvan, G.G., *Cem. Concr. Res.*, Vol. 6, pp. 351–356, 1976.
14. Verbeck, G.J., and R. Landgren, *Proc. ASTM*, Vol. 60, pp. 1063–1079, 1960.
15. Bloem, D.L., *Highway Res. Rec.*, Vol. 18, pp. 48–60, 1963.
16. Woods, H., *Durability of Concrete*, ACI Monograph 4, p. 20, 1968.
17. Valenza II, J.J., and G.W. Scherer, Mechanism for Salt Scaling, *J. Am. Ceramics Soc.*, Vol. 89, No. 4, pp. 1161–1179, 2006.
18. Abrams, M.S., *Temperature and Concrete, ACI SP-25*, pp. 33–50, 1973.
19. Cruz, C.R., *J. Res. & Dev.*, Portland Cement Association, Skokie, IL, No. 1, pp. 37–45, 1966.
20. Comeau, E., *Chunnel Vision, NFPA Journal*, pp. 75–77, March/April, 2002.
21. Ulm, F.J., Fire Damage in the Eurotunnel, *International Workshop on Fire Performance of High-Strength Concrete*, NIST Special Publication 919, National Institute of Standards and Technology, Gaithersburg, MD, 1997.

22. Phan, L.T., and J.N. Carino, Review of Mechanical Properties of HSC at Elevated Temperature, *Journal of Materials in Civil Engineering*, American Society of Civil Engineers, Vol. 10, No. 1, pp. 58–64, 1998.
23. Phan, L.T., and J.N. Carino, Mechanical Properties of High-Strength Concrete at Elevated Temperatures, *NISTIR 6726*, National Institute of Standards and Technology, Washington, D.C., 2001.
24. Anderberg, Y., *International Workshop on Fire Performance of High-Strength Concrete*, NIST Special Publication 919, National Institute of Standard and Technology, Gaithersburg, MD, 1997.
25. Bazant, Z.P., *International Workshop on Fire Performance of High-Strength Concrete*, NIST Special Publication 919, National Institute of Standards and Technology, Gaithersburg, MD, 1997.
26. Biczok, I., *Concrete Corrosion and Concrete Protection*, Chemical Publishing Company, New York, p. 291, 1967.
27. Terzaghi, R.D., *J. ACI, Proc.*, Vol. 44, p. 977, 1948.
28. Cohen, M.D., and B. Mather, *ACI Mat. J.*, Vol. 88, No. 1, pp. 62–69, 1991.
29. Mehta, P.K., *Cem. Concr. Res.*, Vol. 13, No. 3, pp. 401–406, 1983.
30. Collepardi, M., *Concr. Int.*, Vol. 21, No. 1, pp. 69–74, 1999.
31. Biczok, I., *Concrete Corrosion and Concrete Protection*, Chemical Publishing Company, New York, p. 543, 1967.
32. Bellport, B.P., in *Performance of Concrete*, Swenson, E.G., ed., University of Toronto Press, Toronto, pp. 77–92, 1968.
33. Reading, T.E., *ACI-SP-47*, pp. 343–366, 1975; and Mehta, P.K., *J. ACI, Proc.*, Vol. 73, No. 4, pp. 237–238, 1976.
34. Verbeck, G.J., in *Performance of Concrete*, Swenson, E.G., ed., University of Toronto Press, Toronto, 1968.
35. *Engineering New Record*, p. 32, January 5, 1984.
36. *Building Research Establishment Digest 250*, 1981.
37. Stanton, T.E., *Proc. ASCE*, Vol. 66, pp. 1781–1812, 1940.
38. Lepps, T.M. *Second International Conference on Alkali-Aggregate Reactions in Hydroelectric Plants and Dams*, USCOLD, Chattanooga, Tennessee, 1995.
39. Larive, C., Apports combines de l'experimentationet de la modelisation la comprehension de l'alcali-reaction et de ses effets mechaniques. Ph.D. thesis, LCPC, 1998.
40. Poyet, S., A. Sellier, B. Capra, G. Thevenin-Foray, J.M. Torrenti, H. Tournier-Cognon, and E. Bourdarot, Influence of water on alkali-silica reaction. Experimental study and numerical simulation. *J. Mat. Civ. Eng.*, Vol. 18, No. 4, pp. 588–596, 2006.
41. Blanks, R.F., and H.L. Kennedy, *The Technology of Cement and Concrete*, Vol. 1, Wiley, New York, pp. 316–341, 1955.
42. Figg, J.W., *Concrete, Cement and Concrete Association*, Grosvenor Crescent, London, Vol. 15, No. 7, pp. 18–22, 1981.
43. Palmer, D., *Concrete, Cement and Concrete Association*, Vol. 15, No. 3, pp. 24–27, 1981.
44. Swamy, R.N., *ACI SP-144*, pp. 105–131 1994.
45. Mehta, P.K., *ASTM STP 663*, pp. 35–60, 1978.
46. *Building Research Establishment News*, Her Majesty's Stationery Office, London, Winter 1979.
47. 1991 Status of the Nations Highways and Bridges: Conditions, Performance, and Capital Investment Requirements, *Federal Highway Administration*, July 2, 1991.
48. Cady, P.D., *ASTM STP 169B*, pp. 275–299, 1978.

49. Erlin, B., and G.J. Verbeck, *ACI SP-49*, pp. 39–46, 1978.
50. *Building Research Establishment News*, see Ref. 36.
51. Crumpton, C.F., ACI Convention Paper, Dallas, 1981.
52. Moukwa, M., *Cem. Concr. Res.*, Vol. 20, No. 3, pp. 439–446, 1990.
53. Swamy, R.N., *ACI, SP-144*, pp. 105–139, 1994.
54. Mehta, P.K., *ACI, SP-144*, pp. 1–34, 1994; *Concr. Int.*, Vol. 19, No. 7, pp. 69–76, 1997.
55. Mehta, P.K., *Concrete in the Marine Environment*, Elsevier, London, pp. 214, 1991.
56. Vicat, L.J., *A Practical and Scientific Treatise on Calcareous Mortars and Cements*, 1837 (translated by J.T. Smith, London).
57. Atwood, W.G., and A.A. Johnson, *Trans. ASCE*, Vol. 87, Paper 1533, pp. 204–275, 1924.
58. Lea, F.M., *The Chemistry of Cement and Concrete*, Chemical Publishing Co., New York, pp. 623–638, 1971.
59. Gjorv, O.E., *J. ACI, Proc.*, Vol. 68, pp. 67–70, 1971.
60. Feld, J., *Construction Failures*, Wiley, New York, pp. 251–255, 1968.
61. Biczok, I., *Concrete Corrosion and Concrete Protection*, Chemical Publishing Co., New York, pp. 357–358, 1967.

Suggestions for Further Study

General

ACI Committee 201, Guide to Durable Concrete, *ACI Manual of Concrete Practice*, 2008.

Gjorv, O.E., *Durability Design of Concrete Structures in Severe Environments*, Taylor & Francis, New York, 2009.

Hall, C., and W. Hoff, *Water Transport in Brick, Stone, and Concrete*, Spon Press, New York, 2002.

Maekawa, K., R. Chaube, and T. Kishi, *Modelling of Concrete Performance: Hydration, Microstructure Formation and Mass Transport*, E & FN Spon, New York, 1999.

Proceedings of Katherine and Bryant Mather Conference on Concrete Durability, Scanlon, J.M., ed. ACI SP 100, 1987.

Proceedings of CANMET/ACI International Conferences on Durability of Concrete, Malhotra, V.M., ed., *ACI Special Publications*, SP 126, 1991; SP 145, 1994; SP 170, 1997; SP 192, 2000; and SP 212, 2004.

Concrete Exposed to Elevated Temperatures

Bazant, Z.P., and M.F. Kaplan, *Concrete at High Temperatures*, Longman Group, Essex, 1996.

Chemical Aspects of Durability

Biczok, I., *Concrete Corrosion and Concrete Protection*, Chemical Publishing Company, New York, 1967.

Glasser, F.P., J. Marchand, E. Samson, Durability of concrete—Degradation phenomena involving detrimental chemical reactions, *Cem. Concr. Res.*, Vol. 38, pp. 226–246, 2008.

Lea, F.M., *The Chemistry of Cement and Concrete*, Chemical Publishing Company, New York, pp. 338–359, 623–676, 1971.

Sulfate Attack

Famy, C., K.L. Scrivener, and H.F.W. Taylor, *Delayed Ettringite Formation, Structure and Performance of Cements*, Bensted, J., and P. Barnes, ed., Spon Press, London, 2002.

Kurtis, E., P.J.M. Monteiro, and S. Madanat, *Empirical models to predict concrete expansion caused by sulfate attack. ACI Materials Journal*, V97, 156, 2000; errata V97, 173, 2000.

Marchand J., and J.P. Skalny ed, *Materials Science of Concrete: Sulfate Attack Mechanisms*, The American Ceramic Society, Westerville, 1999.

Monteiro P. J. M., Scaling and saturation laws for the expansion of concrete exposed to sulfate attack, *Proceeding of the Academy of Sciences*, 103, pp. 11467-11472, 2006.

Skalny, J., J. Marchand, and I. Odler, *Sulfate Attack on Concrete*, Spon Press, London, 2002.

Alkali-Aggregate Expansion

Blank, R.F., and H.O. Kennedy, *The Technology of Cement and Concrete*, Vol. 1, Wiley, New York, pp. 318–342, 1955.

Blight, G.E., and M.G. Alexander, *Alkali-Aggregate Reaction and Structural Damage to Concrete: Engineering Assessment*, Repair and Management, Taylor & Francis Group, London, 2011.

Diamond, S., *Cem. Concr. Res.*, Vol. 5, pp. 329–346, 1975; Vol. 6, pp. 549–560, 1976.

Gratten-Belleue, P.E., ed., *Proceedings of 7th International Conference on Alkali-Aggregate Reactions*, National Research Council, Ottawa, Canada, 1987.

Hobbs, D.W., *Alkali-Silica Reaction in Concrete*, Thomas Telford Publishing, London, 1988.

Idorn, G., *Concrete Progress: From Antiquity to the Third Millennium*, Thomas Telford, London, 1997.

Corrosion of Embedded Steel

Ahmad, S., Reinforcement corrosion in concrete structures, its monitoring and service life prediction—a review, *Cem. Concr. Composites*, Vol. 25, pp. 459–471, 2003.

Alekseev, S.N., F.M. Ivanov, S. Modry, and P. Schiessel, *Durability of Reinforced Concrete in Aggressive Media*, Balkema Publishers, Brookfield, 1993.

Bentur, A., S. Diamond, and N.S. Berke, *Steel Corrosion in Concrete: Fundamentals and Civil Engineering Practice*, E & FN Spon, London, 1997.

Broomfield, J.P., *Corrosion of Steel in Concrete: Understanding, Investigation, and Repair*, E & FN Spon, London, 1997.

Crane, A.P., ed., *Corrosion of Reinforcement in Concrete Construction*, Ellis Horwood Chichester, West Sussex, U.K., 1983.

Page, C.L., Initiation of chloride-induced corrosion of steel in concrete: role of the interfacial zone, *Materials and Corrosion*, Vol. 60, pp. 586–592, 2009.

Poulsen, E., and L. Mejlbro, *Diffusion of Chloride in Concrete: Theory and Application*, Taylor & Francis, New York, 2006.

Schiessl, P., ed., *Report of the Technical Committee 60-CSC RILEM*, Chapman and Hall, London, pp. 79–95, 1988.

Seawater Attack

Malhotra, V.M., ed., *Performance of Concrete in Marine Environment*, ACI SP-65, Concrete Institute, Detroit, 1980.

Malhotra, V.M., ed., *Performance of Concrete in Marine Environment*, ACI SP198, Concrete Institute, Detroit, 1988.

Mehta, P.K., *Concrete in the Marine Environment*, Elsevier Science Publishing Co., New York, NY, 1991.

Frost Action and Fire

ACI, *Behavior of Concrete under Temperature Extremes*, SP-39, 1973.

Betonghandboken (in Swedish), Svensk Byggtjanst, Stockholm, 1980; and Report of RILEM Committee 4 CDC, *Materials and Structures*, Vol. 10, No. 58, 1977.

Litvan, G.G., and P.J. Sereda, eds., *Durability of Building Materials and Components*, ASTM STP 691, American Society for Testing and Materials, Philadelphia, PA, 1980.

Pigeon, M., and R. Pleau, *Durability of Concrete in Cold Climates*, E & FN Spon, London, 1995.

Valenza II, J.J., and Scherer G.W., Mechanism for Salt Scaling, *J. Am. Ceramics Soc.*, Vol. 89, No. 4, pp. 1161–1179, 2006.)

A Simple Code for Builders

Hammurabi, a king of Babylon, who lived four thousand years ago, had the following rule about the responsibility of builders enforced:

"If a building falls down causing the death of the owner or his son, whichever may be the case, the builder or his son will be put to death. If the slave of the home owner dies, he shall be given a slave of the same value. If other possessions are destroyed, these shall be restored, and the damaged parts of the home shall be reconstructed at builder's cost."

To those engaged on the concrete construction industry, Hammurabi's code should be a reminder of the individual's responsibility toward durability of structures.

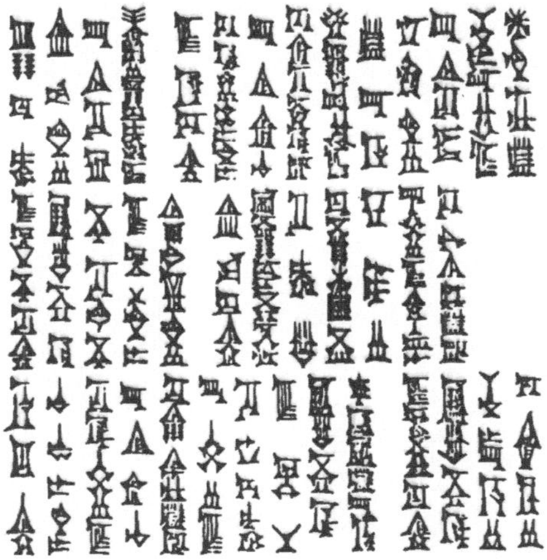

PART II

Concrete Materials, Mix Proportioning, and Early-Age Properties

CHAPTER 6

Hydraulic Cements

Preview

Hydraulic or water-resisting cements for structural applications consist essentially of portland cement and its several modifications. To understand the properties of portland cement, it is helpful to acquire some familiarity with its manufacturing process, chemical and mineralogical composition, and reactivity of the constituent compounds such as calcium silicates and calcium aluminates. Furthermore, properties of concrete containing portland cement develop as a result of chemical reactions between the portland cement compounds and water, because the hydration reactions are accompanied by changes in matter and energy.

In this chapter the composition and characteristics of the principal compounds of portland cement are described. Hydration reactions of the aluminate compounds with their influence on setting behavior of cement, and of silicate compounds with their influence on strength development, are fully discussed. The relationship between the chemistry of reactions and physical aspects of setting and hardening of portland cements is explained. Classification of portland cement types and cement specifications are also reviewed.

Portland cements do not completely satisfy the needs of the concrete construction industry. Special cements have been developed to fill those needs. The compositions, hydration characteristics, and important properties of pozzolan cements, blast-furnace slag cements, portland limestone cements, expansive cements, rapid setting and hardening cements, white and colored cements, oil-well cements, and calcium aluminate cements are described. Finally, trends in cement specifications in Europe and North America are reviewed.

6.1 Hydraulic and Nonhydraulic Cements

6.1.1 Chemistry of Gypsum and Lime Cements

Cements that not only harden by reacting with water but also form a water-resistant product are called *hydraulic cements*. The cements derived from the calcination of gypsum or calcium carbonates are *nonhydraulic* because their products of hydration are not resistant to water. The chemistry underlying gypsum and lime cements is illustrated in Fig. 6-1. Lime mortars that were used in ancient structures built by Greeks and Romans were rendered hydraulic by the addition of pozzolanic materials, which reacted with lime to produce a water-resistant, cementitious product.

Compared to gypsum and lime cements, portland cement and its various modifications are the principal cements used today for making structural concrete. Portland cement

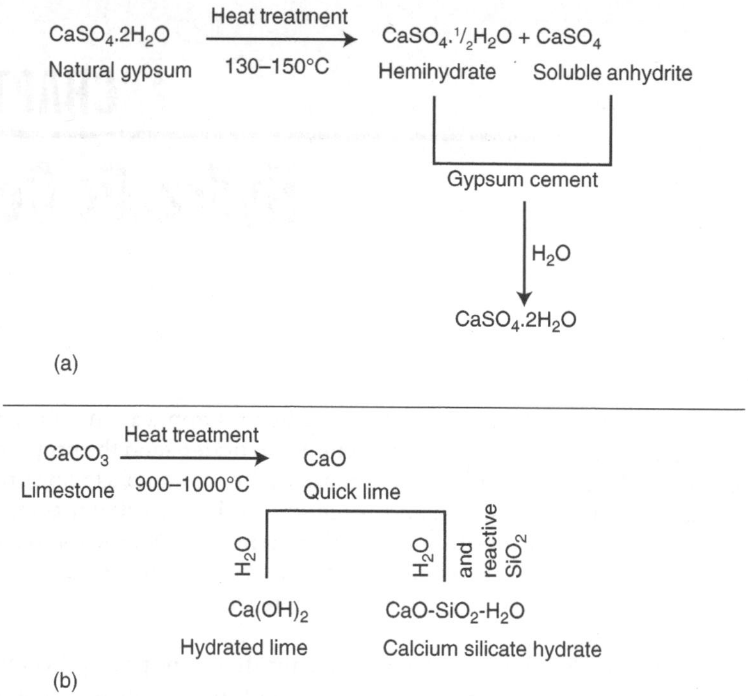

FIGURE 6-1 Chemistry of gypsum and lime cements: (*a*) production of gypsum cement, and hydration reaction; (*b*) production of lime cements, and hydration reactions both with and without pozzolans.

Crystallization of gypsum needles from a hydrated gypsum-cement is the cause of setting and hardening. Gypsum is not stable in water; therefore, the gypsum cement is nonhydraulic. Hydrated lime, $Ca(OH)_2$, is also not stable in water. However, it can slowly carbonate in air to form a stable product ($CaCO_3$). When a pozzolan (reactive silica) is present in the system, the calcium silicate hydrates formed as a result of the reaction between lime and pozzolan are stable in water.

and modified portland cements are hydraulic cements because they do not require the addition of a pozzolanic material to develop water-resisting properties.

6.2 Portland Cement

Definition. Portland cement is a hydraulic cement produced by pulverizing clinkers consisting essentially of crystalline hydraulic calcium silicates, and a small amount of one or more forms of calcium sulfate and up to 5% limestone as an interground addition. Clinkers are 5- to 25-mm-diameter nodules of a sintered material that is produced when a suitable composition and proportions of raw materials, as discussed below, is heated to high temperatures.

6.2.1 Manufacturing Process

Since calcium silicates are the primary constituents of portland cement, the raw material for the production of cement must provide calcium and silica in suitable forms and proportions. Naturally occurring calcium carbonate materials such as limestone, chalk, marl, and sea-shells are the common industrial sources of calcium, but clay or dolomite ($CaCO_3 \cdot MgCO_3$) are usually present as impurities. Clays and shales, rather than quartz, are the preferred sources of additional silica in the raw-mix for making calcium silicates because quartzitic silica is difficult to grind.

Clay minerals contain alumina (Al_2O_3), iron oxide (Fe_2O_3), and alkalies. The presence of aluminum, iron and magnesium ions, and alkalies, in the raw mix has a mineralizing effect on the formation of calcium silicates; that is, they facilitate the formation of the calcium silicate at considerably lower temperatures than would otherwise be possible. Therefore, when sufficient amounts of iron and alumina minerals are not present in the primary raw materials, these are intentionally incorporated into the raw mix through addition of secondary materials such as bauxite and iron ore. As a result, beside the calcium silicate compounds, the portland cement clinker also contains aluminates and aluminoferrites of calcium.

To facilitate the formation of the desired compounds in portland cement clinker, it is necessary to homogenize the raw-mix before heat treatment. That is why the materials are subjected to a series of crushing, grinding, and blending operations. From chemical analyses of the stockpiled materials, their individual proportions are determined by the compound composition desired in the clinker; the proportioned raw materials are usually interground in ball or roller mills to particles below 75 μm.

In the wet process of cement manufacture, the grinding and homogenization of the raw mix is carried out in the form of a slurry containing 30 to 40 percent water. Modern cement plants favor the dry process, which is more energy efficient than the wet process because the water in the slurry must be evaporated before clinkering. For the clinkering operation, the dry-process kilns equipped with multistage suspension preheaters, which permit efficient heat exchange between hot gases and the raw-mix, require a fossil-fuel energy input on the order of 800 kcal/kg of clinker compared to about 1400 kcal/kg for the wet-process kilns. Figure 6-2 shows a simplified flow diagram of the dry process for portland cement manufacture; an aerial view of a modern cement plant is shown in Fig. 6-3.

The chemical reactions taking place in the cement kiln may be expressed as follows:

$$\text{Limestone} \rightarrow CaO + CO_2$$
$$\text{Clay} \rightarrow SiO_2 + Al_2O_3 + Fe_2O_3$$
$$\text{Clay + Limestone} \rightarrow \begin{cases} 3CaO \cdot SiO_2 \\ 2CaO \cdot SiO_2 \\ 3CaO \cdot Al_2O_3 \\ 4CaO \cdot Al_2O_3 \cdot Fe_2O_3 \end{cases}$$

The final operation in the portland cement manufacturing process consists of pulverizing the clinker to an average particle between 10 and 15 μm. The operation is carried out in ball mills, also called *finish mills*. Approximately 5 percent gypsum or calcium sulfate is usually interground with clinker in order to control the early setting and hardening behavior of the cement, as will be discussed.

6.2.2 Chemical Composition

Although portland cement consists essentially of various compounds of calcium, the results of routine chemical analysis are reported in terms of oxides of the elements present.

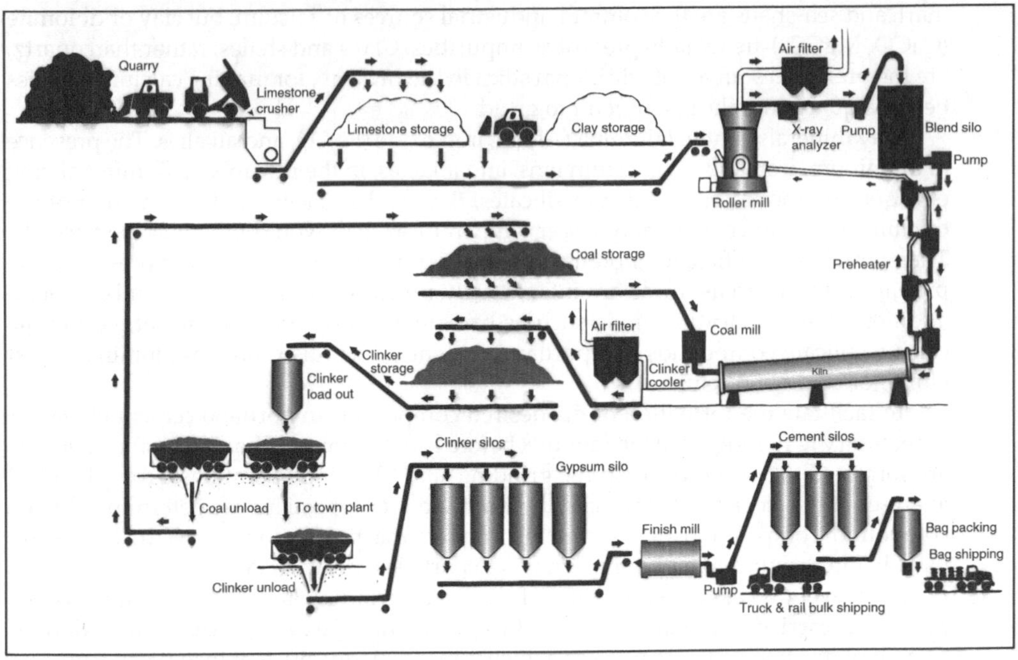

Figure 6-2 Flow diagram of the dry process for portland cement manufacture.

A major step in the process is the clinkering operation carried out in a rotary kiln, which consists of an inclined steel cylinder lined with refractory bricks. The preheated and partially calcined raw mix enters at the higher end of the continuously rotating kiln and is transported to the lower end at a rate controlled by the slope and the speed of the kiln rotation. Pulverized coal, oil, or a fuel gas is injected at the lower end in the burning zone, where temperatures on the order of 1450 to 1550°C may be reached and the chemical reactions involving the formation of portland cement compounds are completed.

Also, it is customary to express the individual oxides and cement compounds by using the following *abbreviations*:

Oxide	Abbreviation	Compound	Abbreviation
CaO	C	$3CaO{\cdot}SiO_2$	C_3S
SiO_2	S	$2CaO{\cdot}SiO_2$	C_2S
Al_2O_3	A	$3CaO{\cdot}Al_2O_3$	C_3A
Fe_2O_3	F	$4CaO{\cdot}Al_2O_3{\cdot}Fe_2O_3$	C_4AF
MgO	M	$4CaO{\cdot}3Al_2O_3{\cdot}SO_3$	$C_4A_3\bar{S}$
SO_3	$\bar{S}$	$3CaO{\cdot}2SiO_2{\cdot}3H_2O$	$C_3S_2H_3$
H_2O	H	$CaSO_4{\cdot}2H_2O$	$C\bar{S}H_2$

Since properties of portland cement are related to the compound composition, it is difficult to draw any conclusions from the cement oxide analyses, such as those shown in Table 6-1. It is a common practice in the cement industry to compute the compound

FIGURE 6-3 Aerial view of the Ash Grove Cement (West) portland cement plant at Durkee, Oregon. (Photograph courtesy of Vagn Johansen, F.L. Smidth, Copenhagen, Denmark.)

An aerial photograph of the Ash Grove Cement (West) dry process plant located near Durkee, Oregon, is shown. This 500,000 tonne/year plant, which in 1979 replaced a 200,000 tonne/year wet process plant, contains a 4.35 by 66-m-long rotary kiln equipped with a four-stage suspension preheater. The preheater exhaust gases go to an electrostatic precipitator designed for an emission efficiency of 99.93 percent. All process loops are monitored and controlled with a 2000 microprocessor-based distributed control system utilizing fuzzy logic.

composition of portland cement from the oxide analysis by using a set of equations which were originally developed by R.H. Bogue. Direct determination of the compound composition, which requires special equipment and skill (Fig. 6-4), is not necessary for routine quality control.

Oxide	Cement No. 1	Cement No. 2	Cement No. 3	Cement No. 4	Cement No. 5
S	21.1	21.1	21.1	20.1	21.1
A	6.2	5.2	4.2	7.2	7.2
F	2.9	3.9	4.9	2.9	2.9
C	65.0	65.0	65.0	65.0	64.0
S–	2.0	2.0	2.0	2.0	2.0
Rest	2.8	2.8	2.8	2.8	2.8

TABLE 6-1 Oxide Analyses of Portland Cements (%)

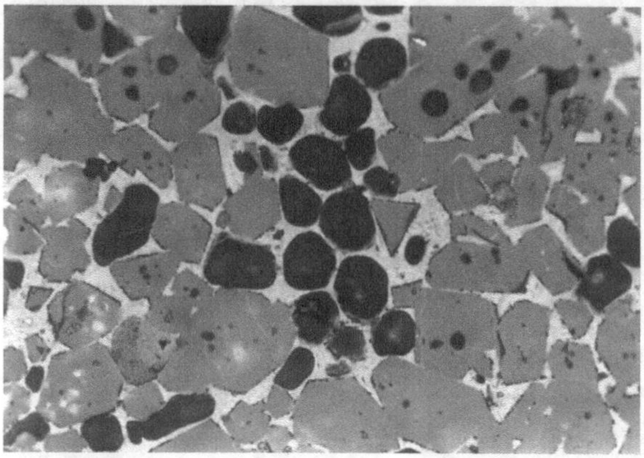

(a)

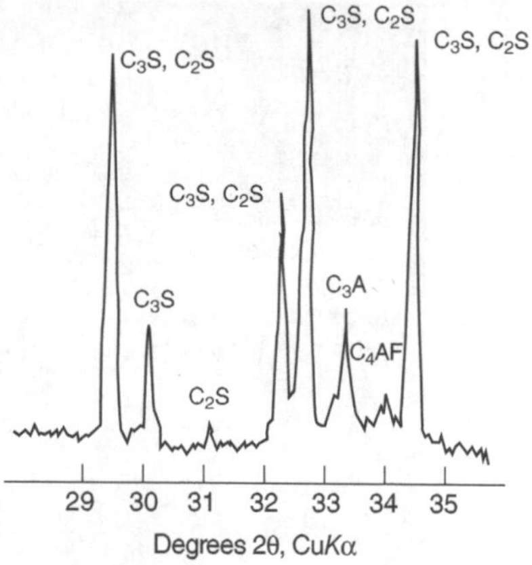

(b)

Figure 6-4 (*a*) Photomicrograph of a polished clinker specimen by reflected light microscopy; (*b*) X-ray diffraction pattern of a powdered clinker specimen.

Two methods are commonly used for direct quantitative analysis of portland cement clinker. The first method involves reflected-light microscopy of polished and etched sections, followed by a point count of areas occupied by the various compounds. Typically, C_3S appears as hexagonal-plate crystals, C_2S as rounded grains with twinning bands, and C_3A and C_4AF as interstitial phases. The second method which is also applicable to pulverized cements, involves X-ray diffraction of powder specimens. The ground-breaking Rietveld method[1] opened a new, exciting chapter in refinement of crystalline structures. The method uses the whole profile of the powder diffraction pattern, and utilizes the unit-cell dimensions for determining the peak positions, and the atomic positional and thermal displacement parameters as a model for the peak intensities. The details of using the Rietveld method to characterize anhydrous cements and hydrated cement paste can be found in Refs. 2 and 3.

6.2.3 Determination of the Compound Composition from Chemical Analysis

The *Bogue equations* for estimating the theoretical or the potential compound composition of portland cement are as follows:

$$\%C_3S = 4.071C - 7.600S - 6.718A - 1.430F - 2.852\overline{S} \tag{6-1}$$

$$\%C_2S = 2.867S - 0.7544C_3S \tag{6-2}$$

$$\%C_3A = 2.650A - 1.692F \tag{6-3}$$

$$\%C_4AF = 3.043F \tag{6-4}$$

The equations are applicable to portland cements with an A/F ratio 0.64 or higher; should the ratio be less than 0.64 another set of equations apply, which are included in ASTM C 150.

Even small differences in the oxide analyses of two cements can result in large differences in their compound composition. This is illustrated by comparing the computed compound composition (Table 6-2) of five samples of portland cements, the oxide analyses of which are shown in Table 6-1. Comparison between Cement no. 1 and Cement no. 2 shows that a 1 percent decrease in Al_2O_3 with a corresponding increase in Fe_2O_3

Compound Composition	Cement No. 1	Cement No. 2	Cement No. 3	Cement No. 4	Cement No. 5
C_3S	52.8	58.0	63.3	53.6	42.0
C_2S	20.7	16.7	12.7	17.2	28.8
C_3A	11.5	7.2	2.8	14.2	14.2
C_4AF	8.8	11.9	14.9	8.8	8.8

TABLE 6-2 Compound Composition of Portland Cements (%)

lowered the C_3A and C_2S contents by 4.3 and 4.0 percent, respectively; this change also caused an increase in the C_4AF and C_3S contents by 3.1 and 5.2 percent, respectively. Similarly, comparison between Cement no. 4 and Cement no. 5 shows that a 1 percent decrease in CaO with a corresponding increase in SiO_2 caused the C_3S to drop 11.6 percent, and the C_2S to rise by the same amount. Furthermore, as discussed next, some of the assumptions underlying the Bogue equations must be noted.

The Bogue equations assume that the chemical reactions of formation of clinker compounds have proceeded to completion, and that the presence of impurities such as MgO and alkalies can be ignored. Neither assumption is strictly valid; hence in some cases the computed compound compositions, especially the amounts of C_3A and C_4AF in cement, are known to deviate considerably from the actual compound composition determined directly. This is why the computed compound composition is also referred to as the *potential compound composition*. However, because the properties of portland cement are influenced by the proportion and the type of the compounds present, the Bogue equations serve a useful purpose by offering an easy method of providing a first estimate of the compound composition of portland cement from oxide analysis.

6.2.4 Crystal Structure and Reactivity of the Compounds

The chemical composition of the compounds present in industrial portland cements is not exactly what is expressed by the commonly used formulas, C_3S, C_2S, C_3A, and C_4AF. This is because at the high temperatures prevalent during clinker formation the elements present in the system, including the impurities such as magnesium, sodium, potassium, and sulfur, are able to become incorporated into each of the major compounds of the cement in clinker, in a "solid solution." Very small amounts of impurities in solid solution may not significantly alter the crystal structure and reactivity of a compound, but larger amounts can do so.

Beside the particle size and the temperature of hydration, the reactivity of the portland cement compounds with water is influenced by their crystal structure. Under the high-temperature and nonequilibrium conditions of the cement kiln, and with a variety of cations present, the crystal structures formed are far from perfect. The structural imperfections thus produced explain why the cement compounds are unstable in an aqueous environment. In fact, differences between the reactivity of two compounds having a similar chemical composition can only be explained from the degree of their structural instability. It is beyond the scope of this book to discuss in detail the highly complex crystal structures of cement compounds; however, essential features that account for differences in the reactivity are described next.

Calcium Silicates

Tricalcium silicate (C_3S) and beta-dicalcium silicate (β-C_2S) are the two hydraulic silicates commonly found in industrial portland cement clinkers. Both invariably contain small amounts of magnesium, aluminum, iron, potassium, sodium, and sulfur ions. The impure forms of C_3S and β-C_2S are known as *alite* and *belite*, respectively.

Three main crystalline forms of alite—triclinic, monoclinic, and trigonal—have been detected in industrial cements, and these forms are each a slight distortion of an ideal C_3S pseudostructure built from SiO_4 tetrahedra, calcium ions, and oxygen ions (Fig. 6-5*a*). According to Lea,[4] a notable feature of the ionic packing is that the coordination of oxygen ions around the calcium is irregular so that the oxygen ions are concentrated on one side of each of the calcium ion. This arrangement leaves large structural holes, which account for the high lattice energy and reactivity.

Similarly, the structure of belite in industrial cements is irregular, but the interstitial holes thus formed are much smaller, and this makes belite far less reactive than alite. By way of contrast, another crystallographic form of dicalcium silicate, namely, γ-C_2S, has a regularly coordinated structure (Fig. 6-5*b*) thus rendering this compound nonreactive.

Calcium Aluminate and Ferroaluminate

Several hydraulic calcium aluminates can occur in the CaO-Al_2O_3 system; however, the tricalcium aluminate (C_3A) is the principal aluminate compound in portland cement clinker. Calcium ferrites are not found in normal portland cement clinker; instead,

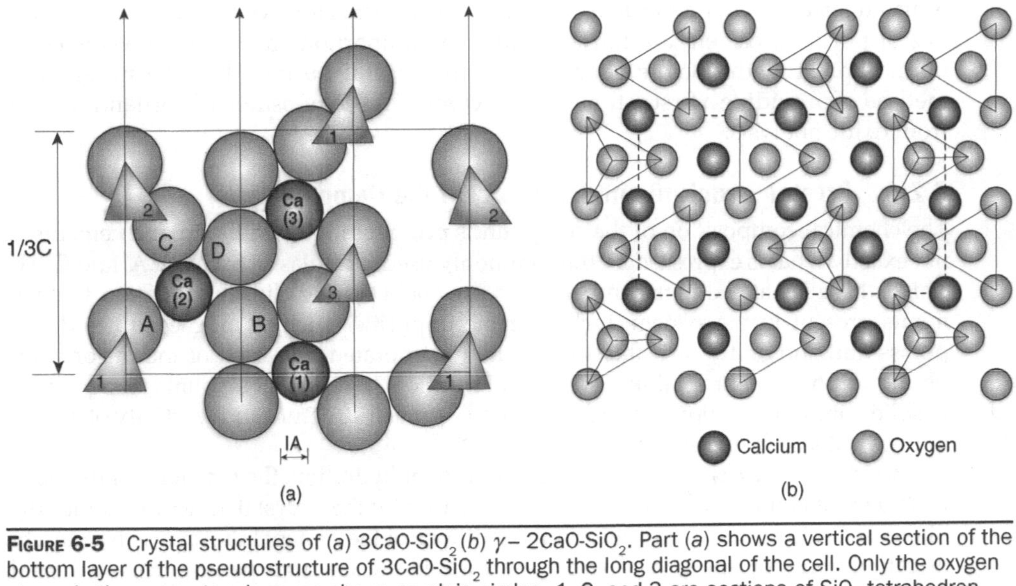

FIGURE 6-5 Crystal structures of (*a*) $3CaO$-SiO_2 (*b*) $\gamma-2CaO$-SiO_2. Part (*a*) shows a vertical section of the bottom layer of the pseudostructure of $3CaO$-SiO_2 through the long diagonal of the cell. Only the oxygen atoms in the symmetry plane are shown as plain circles. 1, 2, and 3 are sections of SiO_4 tetrahedron. Calcium atoms are labeled. In (*b*), silicon atoms are not shown; they occur at the center of the silica tetrahedra. [Lea, F.M., *The Chemistry of Cement and Concrete*, Chemical Publishing Company, New York, 1971, by permission of Edward Arnold (Publishers).]

The irregular coordination of the oxygen ions around calcium leaves large voids, which account for the high reactivity of C_3S. On the other hand, γ-C_2S has a regularly coordinated structure and is, therefore, nonreactive.

calcium aluminoferrites which belong to the C_2A-C_2F ferrite solid solution (F_{ss}) series are formed, and the most common compound corresponds approximately to the equimolecular composition, C_4AF.

Similar to the calcium silicates, in industrial clinkers both C_3A and C_4AF contain significant amounts of magnesium, sodium, potassium, and silica in their crystal structure. The crystal structure of pure C_3A is cubic; however, both C_4AF and C_3A contain large amounts of alkalies and are therefore orthorhombic. The crystal structures are very complex but are characterized by large structural holes that account for high reactivity.

Magnesium Oxide and Calcium Oxide

The source of magnesium oxide in cement is usually dolomite, which is present as an impurity in most limestones. A part of the total magnesium oxide in portland cement clinker (i.e., up to 2 percent) may enter into solid solution with the various compounds described above; however, the rest occurs as crystalline MgO, also called *periclase*. Hydration of periclase to magnesium hydroxide is a slow and expansive reaction that, under certain conditions, can cause *unsoundness* (i.e., cracking and pop-outs in cement-based products).

Uncombined or *free calcium oxide* is rarely present in significant amounts in modern portland cements. Improper proportioning of raw materials, inadequate grinding and homogenization of the raw mix, and insufficient temperature or retention time in the kiln burning zone are among the principal factors that account for the presence of free or crystalline calcium oxide in portland cement clinker. Like MgO, the crystalline CaO that has been exposed to high temperature in the cement kiln hydrates slowly and the hydration reaction is capable of causing unsoundness in cement-based products.

Both MgO and CaO form cubic structures, with each magnesium or calcium ion surrounded by six oxygens in a regular octahedron. The size of the Mg^{2+} ion is such that, in the MgO structure, the oxygen ions are in close contact and the Mg^{2+} ions are well packed in the interstices. However, in the case of the CaO structure, due to the much larger size of the Ca^{2+} ion, the oxygen ions are forced apart so that the Ca^{2+} ions are not well packed. Consequently, the crystalline MgO formed from a high-temperature (>1400°C) melt in a portland cement kiln is much less reactive with water than the crystalline CaO formed under the same temperature conditions. This is the reason why *under ordinary curing temperatures* the presence of a significant quantity of crystalline CaO in portland cement may cause unsoundness in cement-based products, whereas a similar amount of crystalline MgO may prove harmless.

Alkali and Sulfate Compounds

The alkalies (sodium and potassium) in portland cement clinker are derived mainly from the clay components present in the raw mix and coal; their total amount, expressed as Na_2O equivalent ($Na_2O + 0.64K_2O$), may range from 0.3 to 1.5 percent. The sulfates in a cement kiln generally originate from fuel. Depending on the amount of sulfate available, soluble double-sulfates of alkalies such as langbeinite and aphthitalite are known to be present in portland cement clinker. Their presence has a significant influence on the early hydration reactions of the cement.

When sufficient sulfate is not present in the kiln system, the alkalies are preferentially taken up by C_3A and C_2S, which may then be modified to compositions of the type NC_8A_3 and $KC_{23}S_{12}$, respectively. Sometimes large amounts of sulfate in the form of gypsum are purposely added to the raw mix either for lowering the burning temperature or for modification of the C_3A phase to $C_4A_3\bar{S}$, which is an important constituent of certain types of cements that will be described later.

In ordinary portland cement the source of most of the sulfate (expressed as SO_3) is calcium sulfate in one of its several possible forms, added to the clinker during grinding. The main purpose of this additive is to retard the quick-setting tendency of ground portland clinker, attributable to the highly reactive C_3A phase. Calcium sulfate can occur as gypsum ($CaSO_4{\cdot}2H_2O$), hemihydrate (plaster of paris) ($CaSO_4{\cdot}1/2H_2O$), or anhydrite ($CaSO_4$). Compared to clinker compounds, gypsum, the principal form of calcium sulfate, dissolves rather quickly in water. Hemihydrate is even more soluble than gypsum and is invariably present in cements due to decomposition of gypsum during the finish grinding operation.

6.2.5 Fineness

In addition to the compound composition, the fineness of cement also affects its reactivity with water. Generally, the finer the cement, the more rapidly it will react. For a given compound composition the rate of reactivity and hence the strength development can be enhanced by finer grinding of cement; however, the cost of grinding and the heat evolved on hydration set some limits on the fineness.

For quality control purposes in the cement industry, the fineness is easily determined as the residue on standard sieves such as No. 200 mesh (75 μm) and No. 325 mesh (45 μm). It is generally agreed that cement particles larger than 45 μm are slow to hydrate and those larger than 75 μm may never hydrate completely. However, an estimate of the relative rates of reactivity of cements with similar compound composition cannot be made without knowing the complete particle size distribution. As the determination of particle size distribution is either cumbersome or requires expensive equipment, it is a common practice in the industry to obtain a relative measure of the particle size distribution from surface area analysis of the cement by the Blaine Air Permeability Method (ASTM C 204). Typical data on particle size distribution and Blaine surface area for two samples of industrially produced portland cements are shown in Fig. 6-6.

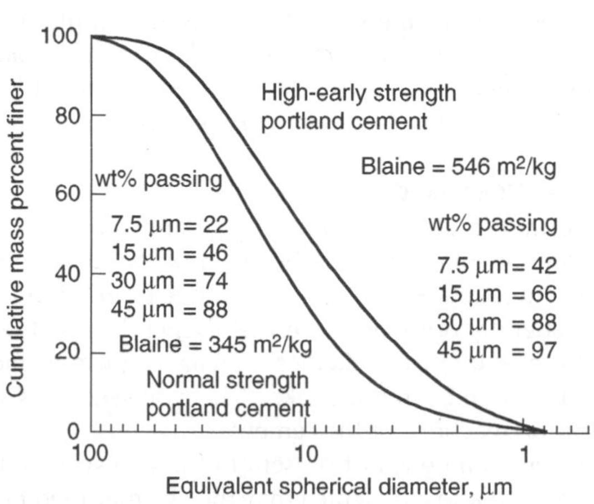

FIGURE 6-6 Typical particle size distribution data from ASTM Types I and III portland cement samples.

6.3 Hydration of Portland Cement

6.3.1 Significance

Anhydrous portland cement cannot bind sand and rock; it acquires the adhesive property only when mixed with water. This is because the chemical reaction of cement with water, commonly referred to as the *hydration of cement*, yields products that possess setting and hardening characteristics. Brunauer and Copeland aptly described the significance of portland cement hydration to concrete technology:

> The chemistry of concrete is essentially the chemistry of the reaction between portland cement and water. . . . In any chemical reaction the main features of interest are the changes in matter, the changes in energy, and the speed of the reaction. These three aspects of a reaction have great practical importance for the user of portland cement. Knowledge of the substances formed when portland cement reacts is important because the cement itself is not a cementing material; its hydration products have the cementing action. Knowledge of the amount of heat released is important because the heat is sometimes a help and sometimes a hindrance. . . . Knowledge of reaction speed is important because it determines the time of setting and hardening. The initial reaction must be slow enough to enable the concrete to be poured into place. On the other hand, after the concrete has been placed rapid hardening is often desirable.[5]

6.3.2 Mechanism of Hydration

Two mechanisms of hydration of portland cement have been proposed. The *through-solution hydration* involves dissolution of anhydrous compounds into their ionic constituents, formation of hydrates in the solution and, due to their low solubility, eventual precipitation of the hydrates from the supersaturated solution. Thus, the through-solution mechanism envisages complete reorganization of the constituents of the original compounds during the hydration of cement. According to the other proposed mechanism, called the *solid-state hydration or topochemical* of cement, the reactions take place directly at the surface of the anhydrous cement compounds without the compounds going into solution. From electron microscopic studies of hydrating cement pastes (Fig. 6-7), it appears that the through-solution mechanism is dominant in the early stages of cement hydration. At later ages, when the ionic mobility in the solution becomes restricted, the hydration of residual cement particle may occur by solid-state reactions.

Since portland cement is composed of a heterogeneous mixture of several compounds, the hydration process consists of simultaneously occurring reactions of the anhydrous compounds with water. However, the various clinker compounds do not all hydrate at the same rate. The aluminates are known to hydrate at a much faster rate than the silicates. In fact, the *stiffening* (loss of consistency) and *setting* (solidification) characteristics of a portland cement paste are largely determined by the hydration reactions involving the aluminates.

The silicates, which make up about 75 percent of ordinary portland cement, play a dominant role in determining the *hardening* (rate of strength development) characteristics. For the purpose of obtaining a clear understanding of the chemical and physical changes during the hydration of portland cement, it is desirable to discuss separately the hydration reactions of aluminates and silicates.

(a) (b) (c)

Figure 6-7 Scanning electron micrograph of a fractured specimen of a 3-day-old portland cement paste.

Calcium hydroxide crystals are massive, C-S-H crystals are poorly crystalline and show a fibrous morphology.

When a cement grain undergoes hydration, two forms of C-S-H product are formed: the outer product (Op) and inner product (Ip). Op forms in the original water-filled space and Ip in the original grain (or anhydrous clinker-filled space). When C-S-H is formed some minerals do not migrate readily from the original cement grain to the water-filled space; as such Ip and Op vary in chemical composition. The most common minerals that do migrate are Mg and Fe; and to a lesser extent Si. However, Ca migrates readily, meaning that Ip tends to have a lower Ca-Si ratio than that of Op. For hydrated portland cement paste, the morphology of the Op tends to be fibrillar. The middle-bottom left of Fig. 6-8a shows a finely fibrilled Op C-S-H. At such an early age the Op is expectedly not as dense as would be found in older pastes. The nature of the fibrils is however, variable and dependent on the space available during formation. Coarse fibrils form when the pore space is larger, whereas if the space is minimal and restricted, the fibrils will be of a finer nature. This distinction between finer and coarse

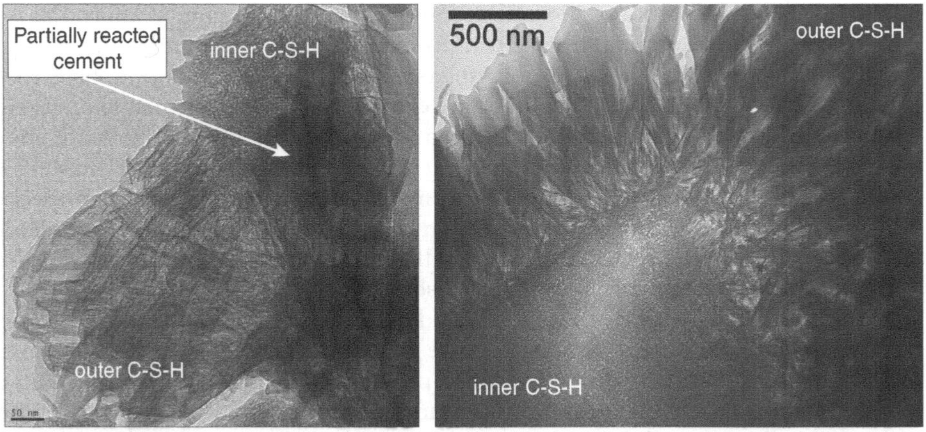

FIGURE 6-8 Transmission electron micrograph: (*a*) six-month-old portland cement paste cast with w/c 0.45, and (*b*) 40-year-old C_3S paste cast with 0.5 w/c. Note the clear boundary between Inner and Outer C-S-H. (Micrographs courtesy of Rae Taylor—unpublished work.)

fibrils in C-S-H is important because, as highlighted by Richardson,[6] the fibrillar morphology is responsible for the capillary porosity. Ip C-S-H can also been seen in this figure (see the center-top right image); at 3 months it is relatively featureless and the boundary between this and Op is discrete. The morphology of the hydrated C_3S grain is somewhat different from the portland cement paste. Figure 6-8b shows a well-defined grain boundary between Ip and Op C-S-H, where the Op C-S-H is extremely fine and very dense in many areas. In Fig. 6-8b, the Op C-S-H in the C_3S paste has no foil-like morphology whatsoever and the fibrils are long and linear. In this figure, the areas closest to the reacted grain the regions of Op C-S-H seem truncated, with a high variation in direction, in contrast to the long unidirectional fibril Op predominantly seen throughout the sample. While this Op morphology is most prominent, it is not restricted to smaller grains and, as highlighted, occasionally forms unevenly around larger grains.

6.3.3 Hydration of the Aluminates

The reaction of C_3A with water is immediate. Crystalline hydrates, such as C_3AH_6, C_4AH_{19}, and C_2AH_8, are formed quickly, with liberation of a large amount of heat of hydration. Unless the rapid hydration of C_3A is slowed down by some method, portland cement cannot be used for most construction applications. This task is generally accomplished by the addition of gypsum. Therefore, for practical purposes, it is not the hydration reactions of C_3A alone but the hydration reactions of C_3A in the presence of gypsum which are important.

From the standpoint of hydration of portland cement, it is also convenient to discuss the hydration reactions of C_3A and ferroaluminate together because when the latter reacts with water in the presence of sulfate, the products formed are structurally similar to those formed from the hydration of C_3A. For instance, depending on the sulfate concentration, the hydration of C_4AF produces either $C_6A(F)\overline{S}_3H_{32}$ or

$C_4A(F)\bar{S}H_{18}$,* which, in spite of differences in chemical composition, have crystal structures that are similar to ettringite and calcium monosulfoaluminate, respectively. However, the part played by the ferroaluminate compound in the early setting and hardening reactions of the portland cement paste depends mainly on its chemical composition and temperature of formation. Generally, the reactivity of the ferrite phase is somewhat slower than C_3A, but it increases with increasing alumina content and with decreasing temperature of formation during the clinkering process. In any case, it may be noted that the hydration reaction of the aluminates described below are applicable to both the C_3A phase and the ferrite phase in portland cement although, for the sake of simplicity, only C_3A is discussed.

Several theories have been postulated to explain the *mechanism of retardation of C_3A by gypsum*. According to one theory, since gypsum and alkalies go into solution quickly, the solubility of C_3A is depressed in the presence of hydroxyl, alkali, and sulfate ions. Depending on the concentration of aluminate and sulfate ions in the solution, the precipitating crystalline product is either calcium aluminate trisulfate hydrate or the calcium aluminate monosulfate hydrate. In solutions saturated with calcium and hydroxyl ions, the former crystallizes as short prismatic needles and is also referred by its mineralogical name, *ettringite*. The monosulfate crystallizes as thin hexagonal plates. The relevant chemical reactions may be expressed as follows:

Ettringite

$$[Al(OH)_4]^- + 3[SO_4]^{2-} + 6[Ca]^{2+} + aq. \rightarrow C_6A\bar{S}_3H_{32} \quad (6\text{-}5)$$

Monosulfate

$$[Al(OH)_4]^- + [SO_4]^{2-} + 4[Ca]^{2+} + aq. \rightarrow C_4A\bar{S}H_{18} \quad (6\text{-}6)$$

Ettringite is usually the first hydrate to crystallize because of the high sulfate/aluminate ratio in the solution phase during the first hour of hydration. In normally retarded portland cements, which contain 5 to 6 percent gypsum, the precipitation of ettringite contributes to stiffening (loss of consistency), setting (solidification of the paste), and early strength development. Later, after the depletion of sulfate when the concentration of aluminate ions in the solution increases again due to renewed hydration of C_3A and C_4AF, ettringite becomes unstable and is gradually converted into the monosulfate phase, which is the final product of hydration of C_3A and C_4AF in portland cements containing more than 5 percent C_3A:

$$C_6A\bar{S}_3H_{32} + 2C_3A + 22H \rightarrow 3C_4A\bar{S}H_{18} \quad (6\text{-}7)$$

Since the aluminate-to-sulfate balance in the solution phase of a hydrated portland cement paste primarily determines whether the setting behavior is normal or not, various setting phenomena affected by an imbalance in the $A/\bar{S}$ ratio, which have practical significance in the concrete construction practice, are illustrated by Fig. 6-9 and are discussed below:

*In recent literature the terms AFt and AFm are employed to designate the products which may have variable chemical compositions but are structurally similar to trisulfate and monosulfate hydrate, respectively.

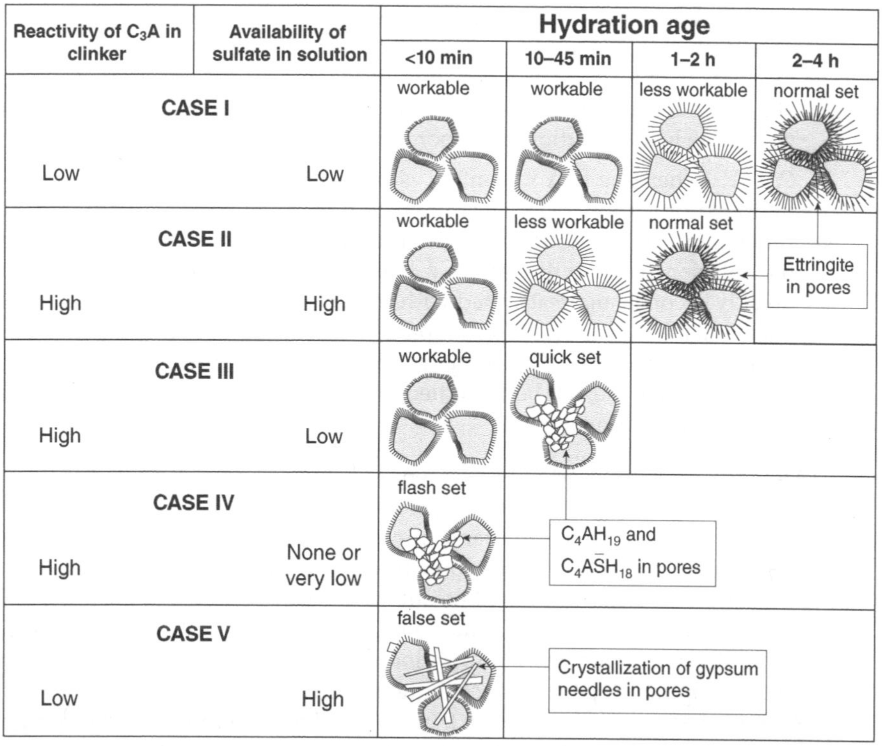

FIGURE 6-9 Influence of the aluminate/sulfate ratio in the solution phase on setting characteristics of portland cement pastes. (From Locher, F.W., W. Richartz, and S. Sprung, *Zement-Kalk Gips*, No. 6, pp. 271–277, 1980.)

Case I. When the rates of availability of the aluminate ions and the sulfate ions to the solution phase are low, the cement paste will remain workable for about 45 min; thereafter it will start stiffening as the water-filled space begins to be filled with ettringite crystals. Most so-called *normal-setting* portland cements belong to this category. The paste becomes less workable between 1 and 2 h after the addition of water, and may begin to solidify within 2 to 3 h.

Case II. When the rates of availability of the aluminate ions and the sulfate ions to the solution phase are high, large amounts of ettringite form rapidly and cause a considerable loss of consistency in 10 to 45 min, with solidification of the paste between 1 and 2 h. Freshly produced high-C_3A cements containing more than normal amounts of alkali sulfates or calcium sulfate hemihydrate are generally characterized by this type of behavior.

Case III. When the amount of reactive C_3A is high but the soluble sulfate present is less than required for normal retardation, hexagonal-plate crystals of monosulfate and calcium aluminate hydrates form quickly and in large amounts causing the cement paste to set in less than 45 min after the addition of water. This phenomenon is known as *quick set*.

Case IV. When little or no gypsum has been added to a ground portland cement clinker, the hydration of C_3A is rapid and the hexagonal-plate calcium aluminate hydrates start forming in large amounts soon after the addition of water, causing almost an instantaneous set. This phenomenon, known as *flash set*, is associated with large heat evolution and poor ultimate strength.

Case V. When the C_3A in cement is of low reactivity, as is the case in partially hydrated or carbonated cements which have been improperly stored, and at the same time a large amount of calcium sulfate hemihydrate is present in the cement, the solution phase will contain a low concentration of aluminate ions but will quickly become supersaturated with respect to calcium and sulfate ions. This situation will lead to the rapid formation of large crystals of gypsum with a corresponding loss of consistency. The phenomenon, called *false set*, is not associated with large heat evolution and can be remedied by vigorous mixing of the cement paste with or without additional water.

Although gypsum is added to cement to serve as a retarder, what is known as the *optimum gypsum content of cement* is generally determined from standard tests which show maximum cement strength and minimum shrinkage at given ages of hydration. As discussed next, the sulfate ions contributed to the solution phase by gypsum have a retarding effect on the hydration of the aluminate compounds but an accelerating effect on the hydration of the silicates that are the principal compounds in portland cement. Therefore, depending on the composition of a cement, a specific gypsum content is indicated for optimum performance of the cement.

Research has shown that, unlike the suppression of solubility of the aluminate compounds, the solubility of the calcium silicate compounds is actually increased in the presence of sulfate ions in the solution phase. Typical data on the effect of gypsum addition on the strength development rate of pure alite cements are shown in Table 6-3. In conclusion, although the primary purpose of gypsum in portland cement is to retard the hydration of aluminates, a side effect is the acceleration of alite hydration, without which the industrial cements would harden at a slower rate.

6.3.4 Hydration of the Silicates

The hydration of C_3S and β–C_2S in portland cement produces a family of calcium silicate hydrates which are structurally similar but vary widely in calcium/silica ratio and the content of chemically combined water. Since the structure determines the properties, the compositional differences among the calcium silicate hydrates have some effect on their physical characteristics.

The microstructure and properties of calcium silicate hydrates formed in portland cement pastes were described in Chap. 2. In general, the material is poorly crystalline and forms a porous solid which exhibits characteristics of a rigid gel. In the literature, this gel has sometimes been referred to as *tobermorite gel*, after a naturally occurring mineral of seemingly similar structure. Since the chemical composition of the calcium silicate hydrates in hydrating portland cement pastes varies with the water-cement ratio, temperature, and age of hydration, it has become rather customary to refer to these hydrates simply as C-S-H, a notation that does not imply a fixed chemical composition. On complete hydration the approximate composition of the material may be assumed as $C_3S_2H_3$; this composition is therefore used for stoichiometric calculations.

	Type I/II* Portland Cement	Alite Cement*	
		No Gypsum	3% gypsum
Setting time† (h)			
Initial	3.0	8.5	4.5
Final	6.0	11.5	7.5
Heat of hydration† (cal/g)			
3 days	61	59	63
7 days	75	61	66
28 days	83	85	81
Compressive strength† [psi (MPa)]			
3 days	1940	1250	1610
	(13.4)	(8.62)	(11.0)
7 days	3100	2060	2440
	(21.4)	(14.2)	(16.8)
28 days	5070	3650	4010
	(34.9)	(25.2)	(27.6)
90 days	5740	5360	5375
	(36.9)	(36.9)	(37.0)

*Alite cement made by grinding a laboratory preparation of high-purity monoclinic alite to fineness 330 m^2/kg Blaine. An industrial portland cement meeting the requirements of both ASTM Types I and II, and a fineness of 330 m^2/g Blaine, was included for reference purposes.

†ASTM Methods C 266, C 186, and C 109 were used for determination of setting time, heat of hydration, and compressive strength, respectively.

source: Data from Mehta, P.K., D. Pirtz, and M. Polivka, *Cem. Concr. Res.*, Vol. 9, pp. 439–450, 1979.

Table 6-3 Accelerating Effect of Gypsum on Setting Time, Heat of Hydration, and Strength of Alite

The stoichiometric reactions for fully hydrated C_3S and C_2S pastes can therefore be expressed as

$$2C_3S + 6H \rightarrow C_3S_2H_3 + 3CH \tag{6-8}$$

$$2C_2S + 4H \rightarrow C_3S_2H_3 + CH \tag{6-9}$$

In addition to the fact that similar reaction products are formed on hydration of both the tricalcium silicate and the dicalcium silicate present in portland cement, the following points should be noted.

First, stoichiometric calculations show that the C_3S hydration would produce 61 percent $C_3S_2H_3$ and 39 percent calcium hydroxide, whereas the C_2S hydration would produce 82 percent $C_3S_2H_3$ and 18 percent calcium hydroxide. As the high specific surface area and, consequently, the adhesive property of the hydrated cement paste are essentially due to the formation of the calcium silicate hydrate, it is to be expected that the ultimate strength of a high-C_3S portland cement would be lower than that of a high-C_2S cement. This, indeed, is confirmed by data from several research investigations.

Second, if the durability of a hardened cement paste to acidic and sulfate waters is reduced due to the presence of calcium hydroxide, it may be expected that a cement containing a high proportion of C_2S will be more durable to acidic and sulfate environments compared to a cement containing a high proportion of C_3S. This conclusion is also generally confirmed by laboratory and field experience. From the standpoint of resistance to chemical attack, some cement standards attempt to limit the maximum permissible C_3S in cement, others recommend the use of pozzolans in order to remove the excess calcium hydroxide from the hydrated cement paste. Third, it can be calculated from the above stoichiometric equations that C_3S and C_2S require 24 and 21 percent water, respectively, for complete hydration.

The stoichiometric equations of C_3S and C_2S hydration do not tell anything about the reaction rates. From the standpoint of instability of the crystal structure described earlier and the heat of hydration data given below, it is apparent that C_3S will hydrate at a faster rate than C_2S. In the presence of gypsum, C_3S in fine particles of cement begins to hydrate within an hour of the addition of water to cement, and thus contributes to the final time of set and the early strength of a hydrated cement paste. The relatively quicker rate of C_3S hydration in finely ground portland cements has become an important factor in the production of high-early-strength portland cements, as will be discussed later.

6.3.5 Models for the Structure of C-S-H

Since the start of the 20th century, after Le Chatelier in 1919 argued that the principal binding phase of cement is C-S-H, there have been several investigations into the C-S-H structure; characterization is often done for C-S-H formed in dilute suspension. These were referred to as C-S-H (I) and (II), being structurally imperfect forms of two naturally occurring minerals, 1.4 nm tobermorite and jennite, with low and high Ca-Si ratio, respectively. The average Ca-Si ratio of C-S-H gel in a hardened cement paste is around 1.7; however, the local Ca-Si ratio varies. The variation is due to several causes, such as starting materials, reaction time, temperature, water-solid ratio, particle size distribution, and so on.

It is difficult to have one model to precisely represent all the observed properties and structure of C-S-H in cement paste. However, it is now commonly accepted that the C-S-H is a layered structure, which is built up by the three distinct modules (see Fig. 6-10):

- Silicate anion chain structure made of the repetition of a pattern of three tetrahedra ("dreierketten")
- Ribbons of edge-sharing calcium
- Interlayer water and/or calcium ions

From these basic assumptions and with emphasis given to determination using Ca-Si ratio, there are to date many models for C-S-H; however, they can be split into two main categories, depending on the assumptions made on how the Ca-Si ratio of C-S-H is raised to 1.7–1.8 as observed experimentally in portland cement pastes. The two categories are tobermorite/jennite (T/J) type and tobermorite/calcium hydroxide (T/CH) type structure.

Some of the first models for C-S-H were those of Powers and Brownyard,[7] accounting for unreacted cement, free water, the hydration product and chemical shrinkage, and Feldman-Sereda (F-S model), explaining humidity dependence of the mechanical properties of hardened cement paste, such as strength and modulus of elasticity.[8] This was extended upon by Taylor[9,10] to produce one of the most widely accepted and

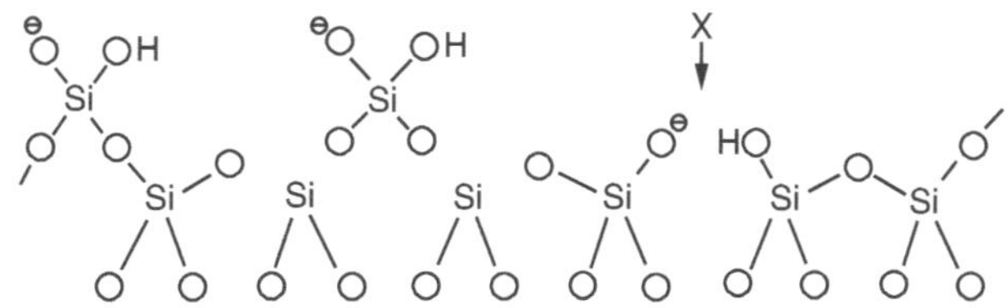

Figure 6-10 The silicate anion structure (dreierkette) consists of paired (lower row) and bridging (upper row) tetrahedral. A bridging tetrahedron is missing at X. (After Taylor, H.F.W., Proposed Structure for Calcium Silicate Hydrate Gel, *J. Am. Ceram.* Soc., Vol. 69, 464–467, 1986.)

referred to models for C-S-H based on the T/J structure. The model suggests that the majority of the disordered layers are that of structurally imperfect jennite with a small proportion being 1.4 nm tobermorite altered by the omission of bridging tetrahedra, especially in the tobermorite (Fig. 6-10). Splitting the theoretically infinite chain lengths of tobermorite and jennite, forming dimers, pentamers, etc. The omission of bridging tetrahedra could account for the difference of Ca-Si ratio in C-S-H.

These fundamental models have since been edited and enhanced to produce several more possibilities.[11–15] Jennings[16] proposed a colloid model that describes C-S-H as an aggregation of precipitated colloidal-size particles, of which there are considered to be two types: high and low density C-S-H, with low density forming first. Richardson and Groves[17,18] developed a "general model" that assumes C-S-H is a highly disordered dreierkette layered structure, with the layers containing variable amounts of CH and finite silicate chains. After further development of the general model, Richardson and Groves allow for the addition of different trace elements into the structural model to allow for local composition variation.

6.4 Heat of Hydration

The compounds of portland cement are nonequilibrium products of high-temperature reactions and are therefore in a high-energy state. When a cement is hydrated, the compounds react with water to acquire stable, low-energy states and the process is accompanied by the release of energy in the form of heat. In other words, the hydration reactions of portland cement compounds are *exothermic*.

The significance of heat of cement hydration in concrete technology is manifold. The heat of hydration can sometimes be a hindrance (e.g., mass concrete structures), and at other times a help (e.g., winter concreting when ambient temperatures may be too low to provide the activation energy for hydration reactions). The total amount of heat liberated and the rates of heat liberation from hydration of the individual compounds can be used as indices of their reactivity. As discussed below, the data from heat of hydration studies can be used for characterizing the setting and hardening behavior of cements, and for predicting the temperature rise.

By using a conduction calorimeter, Lerch[19] recorded the rate of heat evolution from cement pastes during the setting and early hardening period. A typical plot of the data is shown in Fig. 6-11. In general, on mixing cement with water, a rapid heat evolution (ascending portion of peak A) lasting a few minutes occurs. This probably represents the alite hydration and the heat of solution of aluminates and sulfates. This initial heat

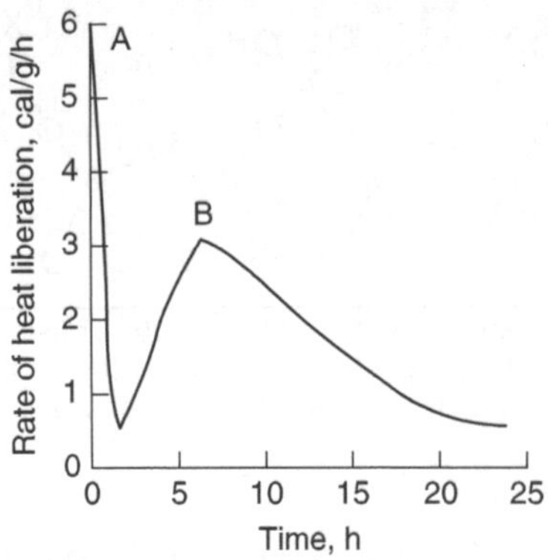

Figure 6-11 Heat liberation rate of a portland cement paste during the setting and early hardening period.

evolution ceases quickly (descending portion of peak A) when the solubility of aluminates is depressed in the presence of sulfate in the solution. The next heat evolution cycle, culminating in the second peak after about 4 to 8 h of hydration for most portland cements, represents the heat of formation of ettringite (ascending portion of peak B). The heat evolution period includes some heat of solution due to C_3S and heat of formation of C-S-H. The paste of a properly retarded cement will retain much of its plasticity before the commencement of this heat cycle and will stiffen and show the *initial set* (beginning of solidification) before reaching the apex at B, which corresponds to the *final set* (complete solidification and beginning of hardening).

From analysis of the heat of hydration data on a large number of cements, Verbeck and Foster[20] computed the relative rates of heat evolution due to the four principal compounds of portland cement (Table 6-4). Since the heat of hydration of cement is an additive property, it can be predicted from an expression of the type:

$$H = aA + bB + cC + dD \tag{6-10}$$

where H represents the heat of hydration at a given age and under given conditions; A, B, C, and D are the percentage contents of C_3S, C_2S, C_3A, and C_4AF present in the cement; and a, b, c, and d are coefficients representing the contribution of 1 percent of the corresponding compound to the heat of hydration. The values of the coefficients will be different for the various ages of hydration.

	Heats of hydration at the given age (cal/g)		
Compound	**3 Days**	**90 Days**	**13 Years**
C_3S	58	104	122
C_2S	12	42	59
C_3A	212	311	324
C_4AF	69	98	102

Table 6-4 Heat of Hydration of Portland Cement Compounds

For a typical portland cement, it appears that approximately 50 percent of the potential heat is liberated within the first 3 days, and 70 percent within the first 7 days of hydration. Type I portland cements meeting ASTM C 150 Standard Specification generally produce 330 to 370 kJ/kg (80 to 90 cal/g) in 7 days.

ASTM C 150 contains standard specifications for a low-heat portland cement (Type IV) and a moderate-heat portland cement (modified Type II) with 7-day heat of hydration limits of 250 kJ/kg (60 cal/g), and 290 kJ/kg (70 cal/g), respectively. Such portland cements are no longer readily available. Hydraulic cements meeting similar heat of hydration requirements, that is, a Type LH (low-heat) cement and a Type MH (moderate-heat) cement can now be marketed under a performance specification (ASTM C 1157) that does not restrict the composition or the constituents of the cement. Blended portland cements containing large proportions of pozzolans or slags can readily satisfy the performance limits specified by ASTM C 1157.

6.5 Physical Aspects of the Setting and Hardening Process

The chemical aspects of the hydration reactions of portland cement compounds have already been discussed. For application to the concrete construction practice it is desirable to review the physical aspects, such as stiffening, setting, and hardening, which are different manifestations of the ongoing hydration process.

Stiffening is the loss of consistency by the plastic cement paste, and is associated with the slump loss phenomenon in concrete. It is the free water in a cement paste that is responsible for its plasticity. The gradual loss of free water from the system as a result of the formation of hydration products, surface adsorption by poorly crystalline products such as ettringite and C-S-H, causes the paste to stiffen and, finally, to set and harden.

The term *setting* refers to the solidification of the plastic cement paste. The beginning of solidification, called *the initial set*, marks the point in time when the paste has become unworkable. Accordingly, concrete placement, compaction, and finishing operations are difficult beyond this stage. The paste does not solidify suddenly; it requires considerable time to become fully rigid. The time taken to solidify completely marks *the final set*, which should not be too long in order to avoid delays in the construction process. Almost universally, the initial and the final setting times are determined by the Vicat apparatus, which measures the resistance of a cement paste of a standard consistency to the penetration of a needle under a total load of 300 g. The initial set is an arbitrary time in the setting process which is said to have been reached when the needle is no longer able to pierce a 40-mm-deep cement paste pat to approximately 5 to 7 mm from the bottom. The final set is said to be reached when the needle makes an impression on the surface of the paste but does not penetrate. ASTM C 150, *Standard Specification for Portland Cement*, requires the initial setting time to be not less than 45 min, and the final setting time to be not more than 375 min, as determined by the Vicat Needle (ASTM C 191).

At the final set, portland cement paste has little or no strength because it represents only the beginning of the hydration of C_3S, the principal compound present. Once the C_3S hydration starts, the reaction continues rapidly for several weeks. The progressive filling of void spaces in the paste with reaction products results in a decrease of porosity and permeability, and an increase in strength. In concrete technology, the phenomenon of strength gain with time is called *hardening*. Figure 6-12 shows a graphical presentation of the relationship between the chemistry of the hydration process of a normal

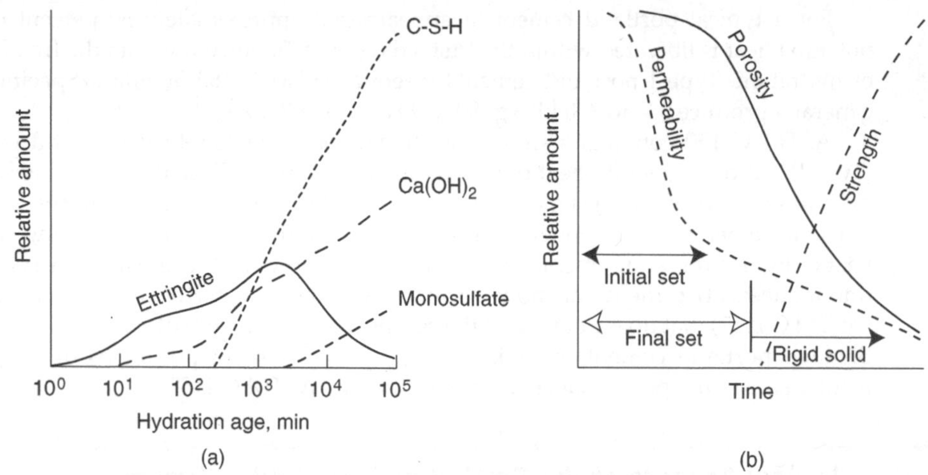

FIGURE 6-12 (*a*) Typical rates of formation of hydration products in an ordinary portland cement paste; (*b*) influence of formation of hydration products on setting time, porosity, permeability, and strength of cement paste. [(*a*) Adapted from Soroka, J., *Portland Cement Paste and Concrete*, Macmillan, London, p. 35, 1979.]

portland cement paste, and the physical phenomena of gradual stiffening, setting, and hardening with a corresponding decrease of porosity and permeability.

6.6 Effect of Cement Characteristics on Strength and Heat of Hydration

Since the rates of reactivity of individual portland cement compounds with water vary considerably, it is possible to change the strength development characteristics of cements simply by altering the compound composition. For instance, the early strengths at 3, 7, and 28 days would be high if the cement contains relatively large amounts of C_3S and C_3A; and the early strength would be low if the cement contains a larger proportion of C_2S. Also, from theoretical considerations [Eqs. (6-8) and (6-9)], the ultimate strength of a high-C_2S cement should be greater than that of a low-C_2S cement. Laboratory studies confirm this (Fig. 6-13*a*). Also, as the compound composition of the cement affects the heat of hydration, it is to be expected that cements containing high C_2S will not only exhibit slow hardening but also less heat production (Fig. 6-13*b*). Modern cements tend to be higher in C_3S, and finer, than those made in the past, because the construction industry today is focused on early strength to meet construction schedules.

The rates of strength development and heat evolution can also be controlled by controlling the fineness of cement. For instance, with a given compound composition, by making a change in the surface area of the cement from 300 to 500 m^2/kg Blaine, it was possible to increase the 1-, 3-, and 7-day compressive strengths of the cement mortar by about 50 to 100 percent, 30 to 60 percent, and 15 to 40 percent , respectively. Typical data on the influence of fineness on strength are shown in Fig. 6-13*c*. Additional data on the influence of compound composition, fineness, and hydration temperature on heat development are shown in Fig. 6-14.

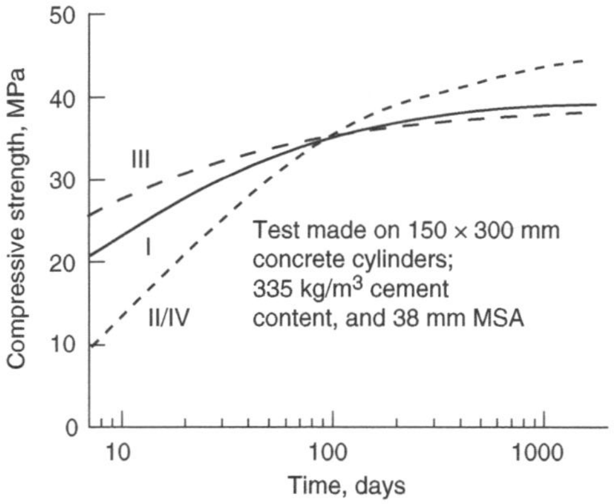

(a)

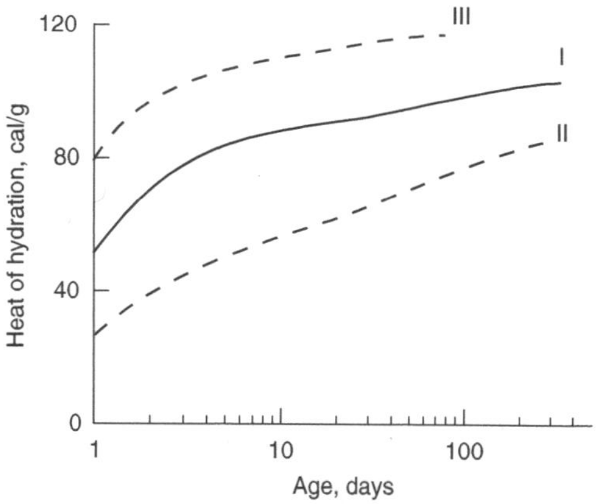

Cement No.	Compound composition			
	C_3S	C_2S	C_3A	C_4AF
I	49	25	12	8
II*	30	46	5	13
III	56	15	12	8

* Cement II corresponds to the ASTM type IV portland cement.

(b)

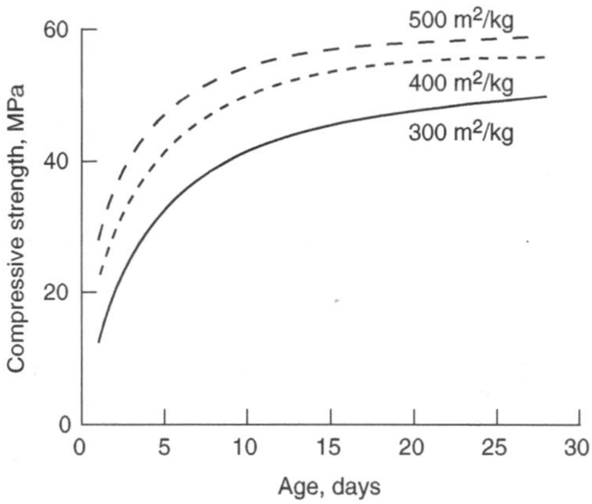

(c)

FIGURE 6-13 (*a*) Influence of cement composition on strength; (*b*) influence of cement composition on heat of hydration; (*c*) influence of cement fineness on strength. [Data for (*a*) and (*b*) are adapted from *Concrete Manual*, U.S. Bureau of Reclamation, 1975, pp. 45–46; data for (*c*) is taken from *Beton-Bogen*, Aalborg Cement Company, Aalborg, Denmark, 1979.]

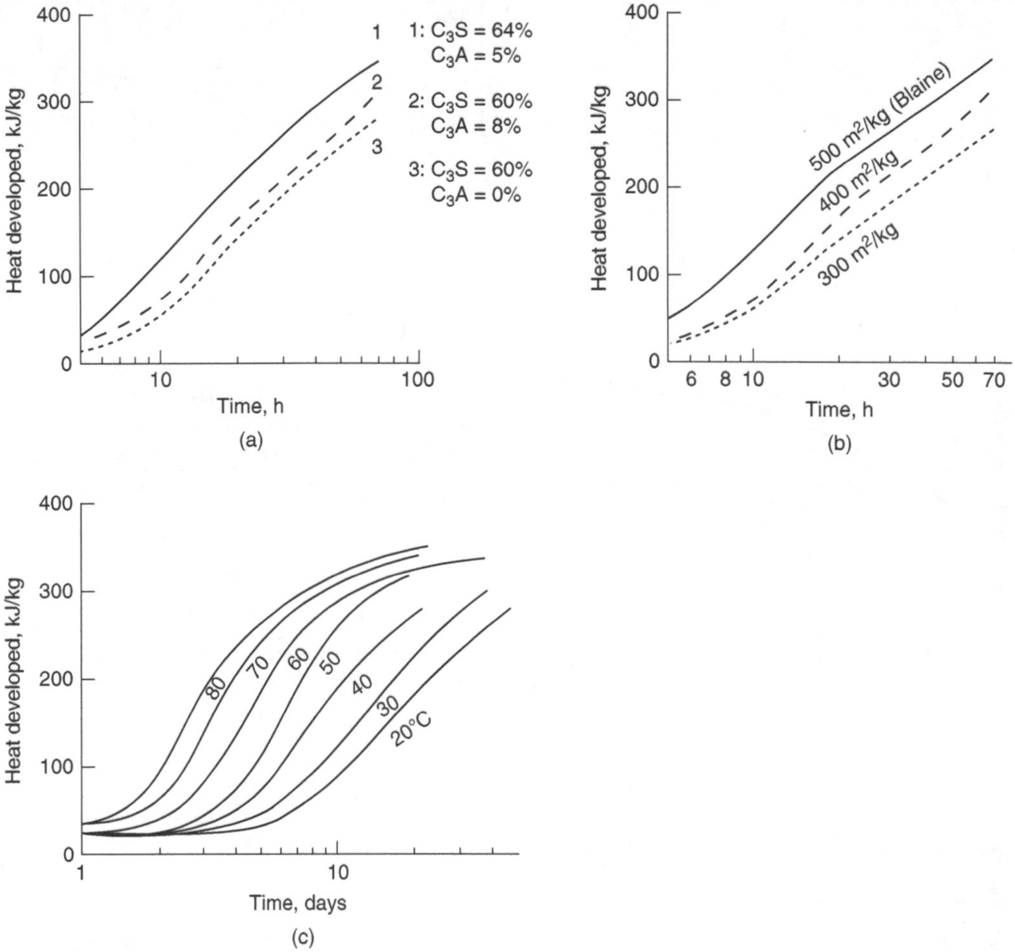

FIGURE 6-14 Influence of compound composition, fineness, and hydration temperature on heat development in cement pastes (0.4 water-cement ratio). (From *Beton-Bogen*, Aalborg Cement Company, Aalborg, Denmark.)

Both the rate and the total amount of heat developed during the hydration of cement are influenced by the compound composition, fineness, and temperature of hydration.

6.7 Types of Portland Cement

A summary of the main characteristics of the principal compounds in portland cement is shown in Table 6-5. From the knowledge of relative rates of reactivity and products of hydration of the individual compounds it is possible to design cements with special characteristics such as high early strength, low or moderate heat of hydration, and high or moderate sulfate resistance. Accordingly, ASTM C 150/C 150M-12, *Standard Specification for Portland Cement*, covers the following 10 types of portland cement:

Approximate Composition	$3CaO \cdot SiO_2$	β-$2CaO \cdot SiO_2$	$3CaO \cdot Al_2O_3$	$4CaO \cdot Al_2O_3 \cdot Fe_2O_3$
Abbreviated formula	C_3S	β-C_2S	C_3A	C_4AF
Common name	Alite	Belite	—	Ferrite phase, F_{ss}
Principal impurities	MgO, Al_2O_3, Fe_2O_3	MgO, Al_2O_3, Fe_2O_3	SiO_2, MgO, alkalies	SiO_2, MgO
Common crystalline form	Monoclinic	Monoclinic	Cubic, orthorhombic	Orthorhombic
Proportion of compounds present (%):				
Range	35–65	10–40	0–15	5–15
Average in ordinary cement	55	20	8	8
Rate of reaction with water	Fast	Slow	Fast	Moderate
Contribution to strength:				
Early age	Good	Poor	Good	Good
Ultimate	Good	Excellent	Medium	Medium
Heat of hydration	Medium	Low	High	Medium
Typical (cal/g)	120	60	320	100

TABLE 6-5 Principal Compounds of Portland Cement and their Characteristics

Type I. For use when the special properties specified for any other type are not required. No limits are imposed on any of the four principal compounds.

Type IA. Air-entraining Type I cement, where air entrainment is desired (e.g., for making frost-resisting concrete).

Type II. For general use, especially when moderate sulfate resistance is desired. The C_3A content of the cement is limited to a maximum of 8 percent. Also, an additional maximum limit of 58 percent of the sum of C_3S and C_3A applies when a moderate heat of hydration is desired and test data for the heat of hydration are not available.

Type IIA. Air-entraining Type II cement, where air entrainment is desired.

Type II (MH). For general use, especially when moderate heat of hydration and moderate sulfate resistance are desired.

Type II (MH) A. Air entraining cement for the same use as Type II (MH).

Type III. For use when high early strength is desired. To ensure that the high strength is not due mainly to the hydration products of C_3A, the Specification limits the C_3A content of the cement to a maximum of 15 percent. It may be noted from Fig. 6-13 that generally the high early strength of the Type III portland cement is partly due to the higher specific surface that is approximately 500 m^2/kg Blaine, instead of 330 to 400 m^2/kg typical for Type I portland cement.

Type IIIA. Air-entraining Type III cement.

Type IV. For use when a low heat of hydration is desired. As C_3S and C_3A produce high heats of hydration, and C_2S produces much less heat, the specification calls for maximum limits of 35 and 7 percent on C_3S and C_3A, respectively, and requires a minimum of 40 percent C_2S.

Type V. For use when high sulfate resistance is desired. The specification calls for a maximum limit of 5 percent C_3A to be applied when the sulfate expansion test is not required.

Some cements are designated with a combined classification such as Type I/II, indicating that the cement meets the requirements of both types. More than 90 percent of the portland cement produce in the United States meets ASTM Types I and II requirements. It should be noted that the hydration product of cements with more than 5 percent potential C_3A, as calculated by Bogue equations, contains monosulfate hydrate which is unstable when exposed to a sulfate solution. Ettringite is the stable product in a sulfate environment, and conversion of the monosulfate to ettringite is generally associated with expansion and cracking.

Although ASTM C 150/C 150M covers the production and use of air-entraining portland cements, concrete producers prefer cements without entrained air because the application of air-entraining admixtures during concrete manufacture offers better control for obtaining the desired amount and distribution of air in the product. Consequently, there is little demand for the air-entrained cements. Similarly, low heat portland cement is no longer made in the United States because the use of mineral admixtures in concrete offers, in general, a less expensive way to control the temperature rise. The setting and hardening characteristics, fineness, and compound composition of ASTM Type I and Type II cements are very similar except that the latter has a maximum 8 percent limit on the C_3A content. ASTM Type III is similar in compound composition to ASTM Type I except that the former is more finely ground in order to achieve higher strength at early ages.

Typical compound composition of the commonly available portland cements in the United States is as follows: 50 to 60 percent C_3S, 10 to 25 percent C_2S, 5 to 12 percent C_3A (except Type V cement must have less than 5 percent C_3A), and 6 to 15 percent C_4AF. Important features of the *physical requirements according to ASTM C 150/150 M*, with a reference to methods of testing, are summarized in Table 6-6. The test methods and specifications are useful for the purpose of quality control of cement; they must not be used to predict properties of concrete which, among other factors, are greatly influenced by the water-cement ratio, curing temperature, and cement-admixture interactions. For instance, compared to the standard test conditions, the time of set of a cement will increase with increasing water-cement ratio, and will decrease with increase in the curing temperature.

Many *cement standards of the world* are similar to ASTM C 150/C 150M but vary in minor details. Some cement standards do not differentiate between the ASTM Types I and II portland cements. Also, most cement standards do not follow the ASTM C 151 autoclave expansion test and the C 150 soundness specification, because the hydration behavior of portland cement is considerably distorted under the autoclaving conditions in the test and a correlation has never been demonstrated between the maximum permissible expansion according to the test and the soundness of cement in service.[21] Instead, the Le Châtelier's test, which involves exposure of a cement paste to boiling water (not autoclaving conditions), is preferred.

Requirement Specified by ASTM C 150/C 150M	Type I	Type II	Type II (MH)	Type III	Type V	Method of Test
Fineness: minimum (m^2/kg)	260	260	260	None	260	ASTM Method C 204 covers determination of fineness of cements using Blaine Air Permeability Apparatus. Fineness is expressed in terms of specific surface of the cement.
Soundness: maximum, autoclave expansion (%)	0.8	0.8	0.8	0.8	0.8	ASTM Method C 151 covers determination of soundness of cements by measuring expansion of neat cement paste prisms cured normally for 24 h and subsequently at 2 MPa (295 psi) steam pressure in an autoclave for 3 h.
Time of setting* Time of setting not less than (min)	45	45	45	45	45	ASTM Method C 191 covers determination of setting time of cement pastes by Vicat apparatus. Initial setting time is obtained when the 1-mm needle is able to penetrate the 35-mm depth of a 40-mm-thick pat of the cement paste. Final setting time is obtained when a hollowed-out 5-mm needle does not sink visibly into the paste.
Time of setting not more than (min)	375	375	375	375	375	
Compressive strength: minimum [MPa (psi)]						ASTM Method C 109 covers determination of compressive strength of mortar cubes, composed of 1 part cement, 0.485 part water, and 2.75 parts graded standard sand by weight; moist cured for 1 day in molds, and in water after demolding.
1 day	—	—	—	12.0 1740	—	
3 days	12.0 1740	10.0* 1450	10.0 (1450)	24.0 3480	8.0 1160	
7 days	19.0 2760	17.0* 2470	17.0 (2470)	—	15.0 2180	
28 days	—	—	—	17.0 (2470)	21.0 3050	

*The time of setting is that described as the initial setting time in test method C191.

Table 6-6 Principal Physical Requirements and Essential Features of ASTM Test Methods for Portland Cements

6.8 Special Hydraulic Cements

6.8.1 Classification and Nomenclature

Portland cements do not satisfy all the needs of the concrete industry; therefore, special cements have been developed to meet certain needs. Compared to portland cement, their volume is small in the United States, however structural engineers should be familiar with special cements and their unique characteristics.

Except for calcium aluminate cement, the special hydraulic cements may be considered as modified portland cements in the sense that they are made either by altering the compound composition of portland cement clinker, or by blending certain additives with portland cement, or by doing both. A clear classification of the special cements is difficult; however, in the American practice the use of the term *blended portland cement* is confined to blends of portland cements with either rapidly cooled blast-furnace slag or pozzolanic material such as fly ash. Other special cements are generally classified under the term *modified portland cement* because they are made by modification of the compound composition of portland cement clinker. The hydraulic calcium silicates, C_3S and β-C_2S, continue to be the primary cementitious constituents; only the aluminate and the ferrite phases are suitably altered to obtain the desired properties.

The main exception to the blended and modified portland cements is calcium aluminate cement, which does not derive its cementing property from the presence of hydraulic calcium silicates. Noteworthy special cements, their compositions, and major applications are summarized in Table 6-7. Important features, hydration characteristics and properties of the cements listed in the table, are discussed next.

6.8.2 Blended Portland Cements

Cost saving was probably the original reason for the development of blended portland cements. However, the impetus to rapid growth in the production of blended cements in many countries in Europe and Asia came as a result of their energy-saving and clinker saving potential. Also, in certain respects, the blended cements perform better than portland cement, and also improvement in several important durability properties. Presently, the production of blended portland-limestone cements, slag cements, and pozzolan cements in Europe and Asia surpassed the production of pure portland cements. In the United States the production of blended cements is still in infancy; however, there is a growing interest to use pozzolanic (e.g., fly ash) and cementitious materials (e.g., ground blast-furnace slag) as mineral admixtures in concrete. The composition and properties of pozzolanic and cementitious materials are described in Chap. 8.

ASTM C 595, Standard Specification for Blended Hydraulic Cements, covers four main classes of cements: portland blast-furnace slag cement (Type IS), portland pozzolan cement (Type IP), portland-limestone cement (Type IL), ternary blended cement (Type IT). According to the Specification, *Type IS cement* consists of an intimate and uniform blend of portland cement and fine granulated blast-furnace slag in which the slag constituent is between 25 and 70 percent of the mass of portland blast-furnace slag cement. *Blast-furnace slag* is a nonmetallic product consisting essentially of silicates and aluminosilicates of calcium and other bases, formed as a by-product of iron extraction from oxide ores; *granulated slag* is the glassy or noncrystalline product which is formed when molten blast-furnace slag is rapidly chilled, as by immersion in water. *Type IP*

Classification and Types	Composition	Major Uses
Blended portland cements Portland blast-furnace slag cement (ASTM Type IS) Portland pozzolan cement (ASTM Type IP)	Uniform blend of granulated blast-furnace slag or a pozzolan or both with portland cement. Sometimes it may contain 5–20% pulverized limestone. Industrial Type IS cements contain typically 30–40 percent slag, while Type IP cements contain 20–25 percent pozzolan. Compared to portland cement, both types are ground to finer particle size to partly compensate for the loss of early strength.	1. Low heat of hydration 2. Excellent durability when properly designed and cured 3. Energy-saving and resource-conserving, and generally less expensive than portland cement
Expansive cements Type K Type M Type S Type O	Consist essentially of portland cement with an expansive additive. Types K, M, and S cements, covered by ASTM C 845, derive their expansion by ettringite formation from $C_4A_3\bar{S}$, CA, and C_3A, respectively. Hard-burnt CaO is the expansive agent in Type O cements.	1. Production of crack-resistant concrete by offsetting the tensile stress due to drying shrinkage 2. Production of chemically prestressed concrete elements 3. Demolition of old concrete without shattering
Rapid setting and hardening cements Regulated set cement (RSC) (or jet cement) Very high early strength cement (VHE) Rapid setting and hardending, High-iron cement (HIC) Ultra high early strength cement (UHE)	Derive rapid setting and hardening properties from compounds capable of forming a large amount of ettringite rapidly and C-S-H subsequently. For ettringite formation, the main source of aluminate ions is a calcium fluoroaluminate in RSC, while it is $C_4A_3\bar{S}$ in VHE and HIC. UHE is a high-C_3S portland cement containing ultra fine particles.	1. Emergency repairs, shotcreting 2. Fabrication of precast-prestressed concrete products without steam curing 3. Agglomeration of particulate matter in mining and metallurgical industries
Oil-well cements	Consist of portland cements with little or no C_3A; relatively coarser particles, and with or without a retarder present.	To allow time for placement of cement slurry, the thickening time at service temperature is retarded:
API Class A-C	Low-C_3A cements without any retarder; Class C is sulfate resistant	For well depths up to 6000 ft or 1830 m (80–170°F or 27–77°C)
API Class F	Low-C_3A cement with retarder	For well depths 10,000–16,000 ft or 3048–4877 m (230–320°F or 110–160°C)
API Class G, H	Essentially coarse-ground ASTM Types II and V portland cement, without retarder	For well temperatures (80–200°F or 27–93°C)
API Class J	Essentially β-C_2S and pulverized silica sand	For well depths below 20,000 ft or 6100 m (>350°F or 177°C)
White and colored cements	Consist of portland cements with little or no iron present (F_{ss} <1 percent). Colored cements are produced by adding suitable pigments to white cement.	Production of architectural concrete
Calcium aluminate cements	Consist essentially of pulverized clinker containing hydraulic calcium aluminates, such as $C_{12}A_7$, CA, and CA_2.	1. High-temperature concrete 2. Emergency repairs, especially in cold weather

TABLE 6-7 Special Hydraulic Cements: Their Composition and Uses

cement consists of an intimate and uniform blend of portland cement (or portland blast-furnace slag cement), and fine pozzolan in which the pozzolan content is between 15 and 40 percent of the mass of the total cement. A *pozzolan* is defined by ASTM C125 as a siliceous or siliceous and aluminous material, which in itself possesses little or no cementing property but will, in a finely divided form and in the presence of water chemically react with calcium hydroxide *at ordinary temperatures* to form compounds possessing cementitious properties.

Compared to pozzolans, finely ground granulated blast-furnace slag is self-cementing; that is, it does not require calcium hydroxide to form cementitious products such as C-S-H. However, when granulated blast-furnace slag hydrates by itself, the amount of cementitious products formed and the rates of formation are insufficient for the use of the material by itself for structural applications. When used in combination with portland cement, the hydration of slag is accelerated due to the presence of calcium hydroxide and gypsum. During the hydration of Type IS cement, some calcium hydroxide produced by the portland cement component is consumed by the slag component of the cement. Due to general similarity in the hydration reaction and microstructure of the hardened Type IS and Type IP cement pastes, it is desirable to discuss the hydration characteristics and properties of the two types of cement together.

The Pozzolanic Reaction and Its Significance

With respect to the main C-S-H-forming reaction, a comparison between portland cement and portland-pozzolan cement is useful for the purpose of understanding the reasons for differences in their behavior:

Portland cement

$$C_3S + H \xrightarrow{\text{fast}} \text{C-S-H} + CH$$

Portland-pozzolan cement

$$\text{Pozzolan} + CH + H \xrightarrow{\text{slow}} \text{C-S-H}$$

The reaction between a pozzolan and calcium hydroxide is called the *pozzolanic reaction*. The technical consequences of using pozzolan cements and slag cements is derived mainly from three features of the pozzolanic reaction. First, the reaction is slow; therefore, the rates of heat liberation and strength development will be accordingly slow. Second, the reaction is lime-consuming instead of lime-producing, which has an important bearing on the durability of the hydrated paste in acidic environments. Third, pore size distribution studies of hydrated IP and IS cements have shown that the reaction products are very efficient in filling up capillary spaces, thus improving the strength and impermeability of the system. Pore size distribution data of portland pozzolan cements containing a Greek pozzolan (Santorini earth) are shown in Fig. 6-15, and a diagrammatic representation of the *pore refinement process* associated with the pozzolanic reaction is shown in Fig. 6-16.

In addition to contributing reactive silica, slags and most pozzolans also contribute reactive alumina, which, in the presence of calcium hydroxide and sulfate ions in the system, also forms cementitious products such as C-(A)-S-H, C_4AH_{13}, AF_t, and AF_m.

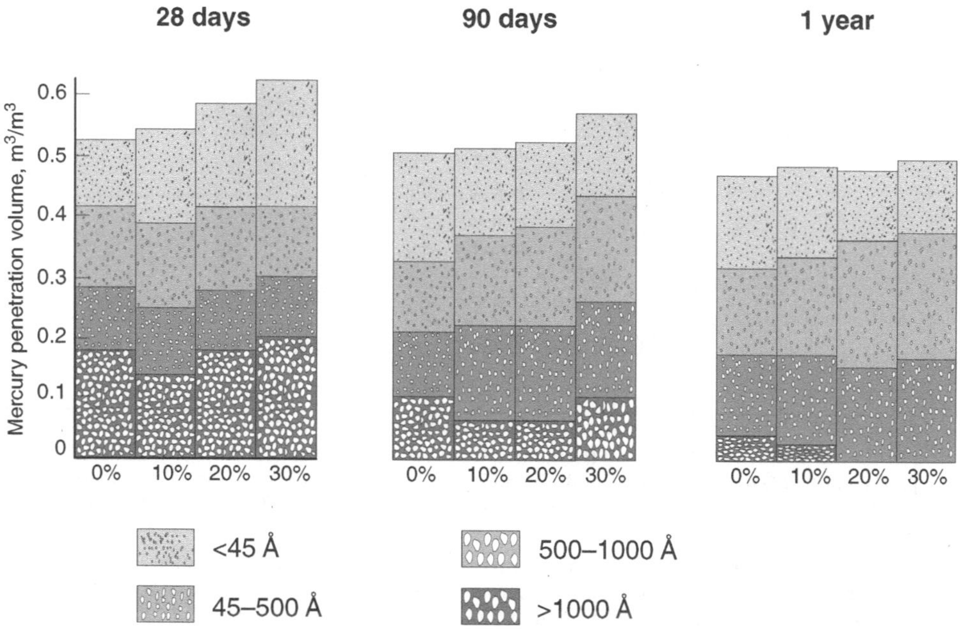

FIGURE 6-15 Changes in pore size distribution of cement pastes with varying pozzolan content. (Reprinted with permission from Mehta, P. K., *Cem. Concr. Res.*, Vol. 11, No. 4, Pergamon Press, New York.)

In a laboratory investigation portland pozzolan cements containing 10, 20, or 30 weight percent of a Greek natural mineral pozzolan were hydrated at a given water-cement ratio, and the pore size distributions were determined at 28, 90, and 365 days by mercury penetration porosimetry. With 20 or 30 percent pozzolan content, no large pores (>0.1 μm) were found in the pastes cured for 1 year. Water permeability tests showed that these cement pastes were much more impermeable than the reference portland cement paste.

Properties of Types IP and IS vary widely depending on the curing conditions and the proportions as well as physical-chemical characteristics of the constituent materials present. The properties described below may therefore be considered as indicative of a general trend.

Heat of Hydration

Figure 6-17 shows the effect of increasing amounts of pozzolan on the heat of hydration of portland-pozzolan cements. Type IS cements containing more than 50 percent slag show approximately 60 cal/g heat of hydration at 7 days, which is comparable to cements with 30 percent pozzolan.

Strength Development

Figure 6-18*a* shows strength development rates up to 1 year in cements containing 10, 20, or 30 percent pozzolan, and Fig. 6-17*b* shows similar data for cements containing 40, 50, or 60 percent granulated slag. In general, pozzolan cements are somewhat slower than slag cements in developing strength; whereas the slag in Type IS cements usually makes a significant contribution to the 7-day strength, a Type IP cement containing an ordinary pozzolan shows strength gain from the pozzolanic constituent only after 7 days of hydration. When adequately reactive materials are used in moderate proportion

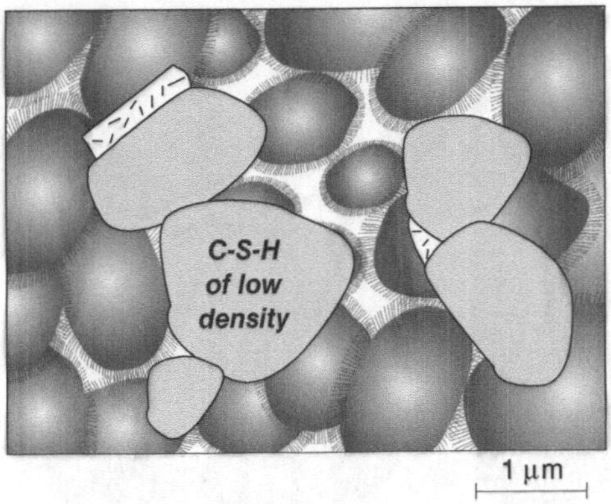

FIGURE 6-16 Diagrammatic representation of well-hydrated cement pastes made with a portland pozzolan cement. Compared to a portland cement paste (see Fig. 2-6 for identification of the phases present) it is shown here that, as a result of the pozzolanic reaction, the capillary voids are either eliminated or reduced in size, and crystals of calcium hydroxide are replaced with additional C-S-H of a lower density.

On the basis of scanning electron microscopic and pore-size distribution studies of hydrated cement pastes both with and without a pozzolan, it is possible to conclude that there are two physical effects of the chemical reaction between the pozzolanic particles and calcium hydroxide: (i) pore-size refinement and (ii) grain-size refinement. The formation of secondary hydration products (mainly calcium silicate hydrates) around the pozzolan particles tends to fill the large capillary voids with a microporous, low-density material. The process of transformation of a system containing large capillary voids into a microporous product containing numerous fine pores is referred to as "pore-size refinement." Also, nucleation of calcium hydroxide around the fine and well distributed particles of pozzolan will have the effect of replacing the large and oriented crystals of calcium hydroxide with numerous, small, and less oriented crystals plus poorly crystalline reaction products. The process of transformation of a system containing large grains of a component into a product containing smaller grains is referred to as "grain-size refinement." Both the pore size and the grain-size refinement processes strengthen the cement paste.

From the standpoint of impermeability and durability, the effects of the pozzolanic reaction are probably more important in concrete than in the hydrated cement paste. As discussed in Chap. 5, the permeability of concrete is generally much higher than the permeability of cement paste because of microcracks in the cement paste-aggregate interfacial transition zone. It is suggested that the process of pore-size and grain-size refinement strengthens the cement paste in the transition zone, thus reducing the microcracks and increasing the impermeability of concrete.

(e.g., 15 to 30 percent pozzolan or 25 to 50 percent slag), and moist curing is available for long periods, the ultimate strengths of Types IP and IS cements are higher than the strength of the reference portland cement without the blending materials. This is because of the pore refinement associated with the pozzolanic reaction and the increase in C-S-H and other hydration products at the expense of calcium hydroxide.

Durability

Compared to portland cement, the superior durability of Type IP cement to sulfate and acidic environments is due to the combined effect of higher impermeability, and lower calcium hydroxide content of the hydrated cement paste (Fig. 6-19*a*). In one investigation it was found that, compared to portland cement, the depth of penetration of water was reduced by about 50 percent in 1-year-old pastes of cements containing 30 mass

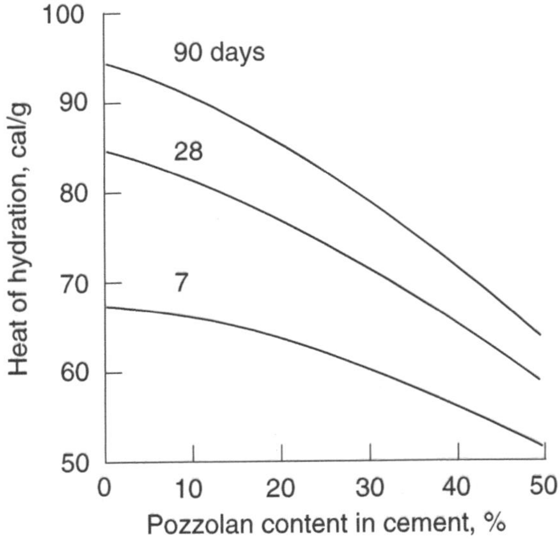

FIGURE 6-17 Effect of substituting an Italian natural pozzolan on the heat of hydration of portland cement. (From Massazza, F., and U. Costa, *Il Cemento*, Vol. 76, p. 13, 1979.)

percent of a Greek volcanic ash. Also, a 1-year-old paste of the reference portland cement contained 20 percent calcium hydroxide, whereas there was only 8.4 percent calcium hydroxide in a similarly hydrated paste of the cement containing 30 mass percent of the Greek pozzolan.

Type IS cements behave in a similar manner. Figure 6-19*b* shows the effect of increasing the slag content on the amount of calcium hydroxide in portland blast-furnace slag cements at 3 and 28 day after hydration. At about 60 percent slag content, the amount of calcium hydroxide becomes so low that even slags containing large amounts of reactive alumina can be safely used to make sulfate-resisting cements. It may be recalled (see Chap. 5) that the rate of sulfate attack depends on the permeability, and the amount of calcium hydroxide and reactive alumina phases present. Some high-alumina slags and fly ashes tend to increase the content of calcium aluminate hydrates and monosulfate (which are vulnerable to sulfate attack) in the hydrated cement paste. Because the presence of significant amounts of calcium hydroxide in the system is necessary for the ettringite-related expansion to occur, both laboratory and field experience show that IS cements containing 60 to 70 percent or more slag are highly resistant to sulfate attack, irrespective of the C_3A content of portland cement and the reactive alumina content of the slag.

In regard to the deleterious expansion associated with the alkali-aggregate reaction, combinations of high-alkali portland cement with pozzolan or slag are generally known to produce durable products (Fig. 6-20). Sometimes the alkali content of pozzolans and slags can be high, but if the alkali-containing mineral is not soluble in the high-pH environment of portland-cement concrete, the high-alkali content of the blended cement generally does not cause any problem. Also the C-S-H gel with a low C/S, produced by the pozzolanic reaction, can adsorb some of the alkalis.

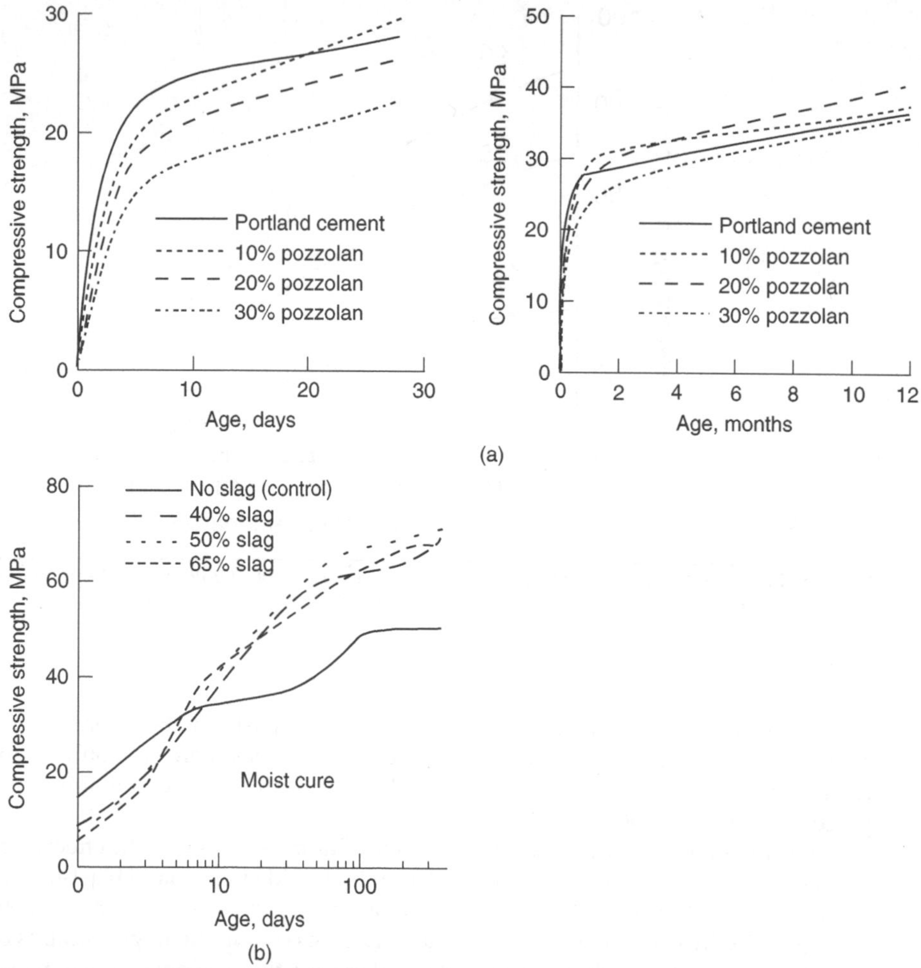

FIGURE 6-18 Strength of blended cement containing a pozzolan or a blast-furnace slag. [(*a*) Reprinted with permission from Mehta, P.K., *Cem. Concr. Res.*, Vol. 11, No. 4, Pergamon Press; (*b*) reprinted with permission from Hogan, F.J. and J.W. Meusel, *Cem. Concr. Aggregates*, Vol. 3, No. 1, 1981, ASTM, Philadelphia, PA.]

The upper figures show the compressive strength of portland cements (<400 m^2/kg Blaine) made with a Greek natural mineral pozzolan. The lower figure shows the compressive strengths of portland blast-furnace slag cements (>500 m^2/kg) made with an American granulated blast-furnace slag.

6.8.3 Expansive Cements

Expansive cements are hydraulic cements which, unlike portland cement, expand during the early hydration period after setting. Large expansion occurring in an unrestrained cement paste can cause cracking; however, if the expansion is properly restrained, its magnitude will be reduced and a prestress will develop. When the magnitude of expansion is small such that the prestress developed in concrete is on the order of 15 to 100 psi

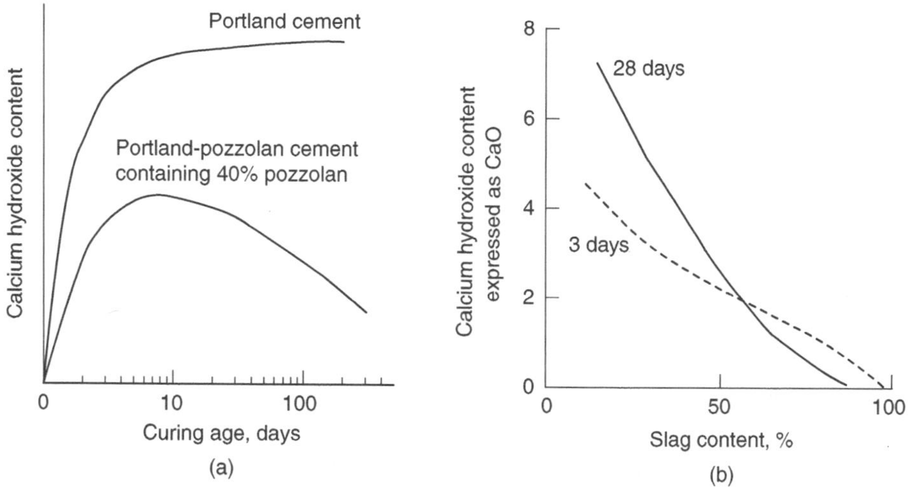

FIGURE 6-19 (*a*) Effect of curing age on the calcium hydroxide content of a cement-sand mortar made with a portland-pozzolan cement. (*b*) Effect of curing age and proportion of slag on the lime content of the portland-slag cement paste. [Based on Lea, F.M., *The Chemistry of Cement and Concrete*, Chemical Publishing Company, New York, pp. 442, 481, 1971, by permission of Edward Arnold (Publishers)].

In the case of portland-pozzolan and portland-blast-furnace slag cements the reduction of calcium hydroxide in the hydrated cement paste, which is due to both the dilution effect and the pozzolanic reaction, is one reason that concrete made from such cements tends to show superior resistance to sulfate and acidic environments. Initially, with curing the calcium hydroxide content of the cement increases due to hydration of the portland cement present; however, later it begins to drop with the progress of the pozzolanic reaction. Depending on curing conditions, portland-blast-furnace slag cements with 60 percent or more slag may contain as little as 2 to 3 percent calcium hydroxide; portland-pozzolan cement products contain higher calcium hydroxide because the reactive pozzolan content may range between 20 to 30 percent in a cement containing 40 percent pozzolan.

15 to 100 psi, which is usually adequate to offset the tensile stress from restrained drying shrinkage, the cement is known as *shrinkage compensating*. Cements of this type have proved very useful for making crack-free pavements and slabs. When the magnitude of expansion is large enough to produce prestress levels on the order of 6.9 MPa (1000 psi), the cement is called *self-stressing* and can be used for the production of chemically prestressed concrete elements.

Formation of ettringite and hydration of hard-burnt CaO are two phenomena known to cement chemists that can cause disruptive expansion in concrete (Chap. 5). Both phenomena have been harnessed to produce expansive cements. The cement produced by grinding a sulfoaluminate-type clinker is called *Type K expansive cement*. Developed originally by Alexander Klein of the University of California at Berkeley in the 1960s, the sulfoaluminate-type clinker is a modified portland cement clinker containing significant amounts of $C_4A_3\bar{S}$ and $C\bar{S}$, in addition to the principal cementitious compounds such as C_3S and C_2S. To achieve better control of expansion in industrial expansive cements, it is customary to blend a suitable proportion of the sulfoaluminate clinker with normal portland cement clinker.

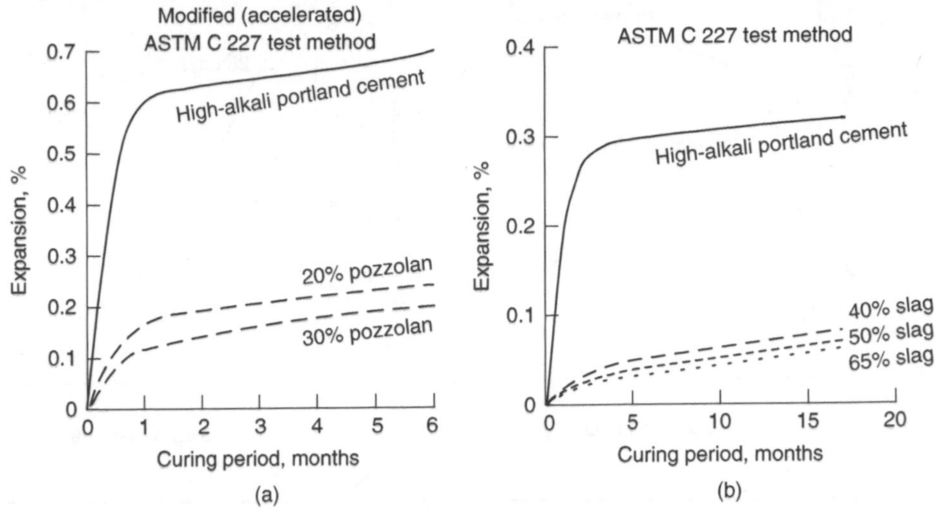

FIGURE 6-20 Influence of pozzolan or slag addition on alkali-aggregate expansion. [(*a*) From Mehta, P.K., Cem. Concr. Res., Vol. 11, No. 4, Copyright 1981, Pergamon Press, New York; (*b*), reprinted with permission from Hogan, F.J. and J.M. Meusel, *Cem. Concr. Aggregates*, Vol. 3, No. 1, 1981, ASTM, Philadelphia, PA.]

Pozzolans and slags are generally very effective in reducing the expansion due to the alkali-aggregate reaction. Santorin Earth from Greece was used for the test data shown in part (a); a granulated blast-furnace slag from the United States was used for the test data shown in part (b). As different test methods were used, the data in the two figures are not directly comparable; however, the trend is similar in both cases.

Type K expansive cement used in the U.S. construction practice is covered by ASTM Standard C 845. ASTM C 845 covers two other expansive hydraulic cements which also derive their expansion characteristic from ettringite but are not produced in the United States. The cements differ from the Type K cement and from each other with respect to the source of aluminate ions for ettringite formation. *Type M expansive cement* is a mixture of portland cement, calcium aluminate cement (with CA as the principal compound), and calcium sulfate. *Type S expansive cement* is composed of a very high C_3A portland cement (approximately 20 percent C_3A) and large amounts of calcium sulfate. The stoichiometry of the expansive reactions in the three cements can be expressed as follows:

Type K $$C_4A_3\bar{S} + 8C\bar{S} + 6CH + 90H \rightarrow 3C_6A\bar{S}_3H_{32} \quad (6\text{-}11)$$

Type M $$CA + 3C\bar{S} + 2CH + 30H \rightarrow C_6A\bar{S}_3H_{32} \quad (6\text{-}12)$$

Type S $$C_3A + 3C\bar{S} + 2CH + 30H \rightarrow C_6A\bar{S}_3H_{32} \quad (6\text{-}13)$$

The CH in the above reactions is provided by the portland cement hydration although Type K clinkers generally contain some free CaO. Initially developed by the

Onoda Cement Company of Japan, the expansive portland cement deriving its expansion from hard-burnt CaO is called *Type O expansive cement*.

Compared to portland cements, the ettringite-forming expansive cements are quick setting and prone to suffer rapid slump loss. However, they show excellent workability. These properties can be anticipated from the large amounts of ettringite formed and the water-imbibing characteristic of the ettringite. Other properties of expansive cement concretes are similar to portland cement concrete except durability to sulfate attack. Type K shrinkage-compensating cements made by blending ASTM Types II or Type V portland cement show excellent durability to sulfate attack because they contain little reactive alumina or monosulfate after hydration. Types M and S cement products usually contain significant amounts of compounds that are vulnerable to sulfate attack and therefore are not recommended for use in sulfate environments. A review of the properties and applications of expansive cement concrete is included in Chap. 12.

6.8.4 Rapid Setting and Hardening Cements

It may be noted that ASTM Type III cement is *rapid hardening* (i.e., high early strength) but not *rapid setting* because the initial and final setting times of the cement are generally similar to Type I portland cement. For applications such as emergency repair of leaking joints and shotcreting, hydraulic cements are needed that not only are rapid hardening but also rapid setting. Setting times as low as 10 minutes can be achieved by using mixtures of either portland cement and plaster of paris ($CaSO_4 \cdot ½H_2O$), or portland cement and calcium aluminate cement. The durability and ultimate strength of the hardened product are generally low.

During the 1970s, a new generation of cements were developed which derive rapid setting and hardening characteristics from ettringite formation. After the initial rapid hardening period, these cements continue to harden subsequently at a normal rate due to the formation of C-S-H from hydraulic calcium silicates.

Regulated-set cement, also called *Jet cement* in Japan, is manufactured under patents issued to the U.S. Portland Cement Association. Using a modified portland cement clinker containing mainly alite and a calcium fluoroaluminate ($11CaO \cdot 7Al_2O_3 \cdot CaF_2$), a suitable proportion of the clinker is blended with normal portland cement clinker and calcium sulfate so that the final cement contains 20 to 25 percent of the fluoroaluminate compound and about 10 to 15 percent calcium sulfate. The cement is generally very fast setting (2 to 5 min setting time), but the setting can be retarded by using citric acid, sodium sulfate, calcium hydroxide, and other retarders.

The high reactivity of the cement is confirmed by the high heat of hydration (100 to 110 cal/g at 3 days), and over 6.9 MPa (1000 psi) compressive strength (ASTM C 109 mortar) at 1 h after hydration. The ultimate strength and other physical properties of the cement are comparable to those of portland cement except that due to the high content of the reactive aluminate, the sulfate resistance is poor. Studies at the concrete laboratory of the U.S. Army Engineer Waterways Experiment Station[22] have shown that the high heat of hydration of the regulated-set cement can help produce concrete with adequate strength even when the concrete is placed and cured at temperatures as low as –9.5°C (15°F).

In addition to regulated set cements, two other modified portland cements derive their rapid setting and hardening characteristics from the formation of large amounts of ettringite during the early hydration period. With the very high-early-strength (VHE) cement, $C_4A_3\bar{S}$ is the main source of aluminate for the ettringite formation whereas with high-iron cement (HIC) both $C_4A_3\bar{S}$ and a reactive C_4AF provide the aluminate ions. Although there are certain basic differences in their composition, both cement types exhibit setting time and strength development rates that are suitable for emergency repair jobs and for application to precast and prestressed concrete products. In the precast concrete industry, quick turnover of forms is an economic necessity. Rapid setting and hardening cements should have a considerable appeal to the construction industry because under normal curing temperatures (i.e., without steam curing) they are capable of developing compressive strengths of 15 and 25 MPa within 8 and 24 h, respectively, with about 50 MPa ultimate strength. A belite-ferrite-sulfoaluminate type cement with a potential compound composition of 50 percent C_2S, 30 percent $C_4A_3\bar{S}$ and C_4AF, and 20 percent $C\bar{S}$ gave 15.6, 28.3, 35.7, and 54 MPa compressive strength at 8-h, 1-d, 7-d, and 120-day, respectively. This is truly an energy-saving cement because the clinker formation temperature was 250°C lower than the temperatures normally used for portland-cement clinker manufacture.[23]

6.8.5 Oil-Well Cements

As discussed below, oil-well cements are not used for making structural concrete. However, approximately 5 percent of the total portland cement produced in the United States is consumed by the petroleum industry, therefore it may be desirable to have an idea of their composition and properties.

Once an oil well (or gas well) has been drilled to the desired depth, cementing a steel casing to the rock formation offers the most economic way to achieve the following purposes:

- To prevent unwanted migration of fluids from one formation to another
- To prevent pollution of the valuable oil zone
- To protect the casing from external pressures that may be able to collapse it
- To protect the casing from possible damage due to corrosive gases and water

For the purposes of cementing a casing, a high water-cement ratio mortar or cement slurry is pumped to depths which, in some instances, may be below 6100 m and where the slurry may be exposed to temperatures above 204°C and pressures above 140 MPa. In the U.S. Gulf coast region, the static bottom hole temperature increases by 0.8°C for every 30 m of well depth. It is desired that the slurry must remain sufficiently fluid under service conditions for the several hours needed to pump it into position, and then harden quickly. Oil-well cements are modified portland cements that are designed to serve this need.

Nine classes of oil-well cements (Classes A to J in Table 6-7) that are applicable for use at different well depths are covered by the American Petroleum Institute (API) Standard 10A. The discovery that the *thickening time of cement slurries at high temperatures* can be increased by reducing the C_3A content and fineness of ordinary portland cement (i.e., by using coarsely ground cement) led to the development of initial oil-well cements. Later, it was found that for applications above 82°C, the cement must be further retarded by addition of lignosulfonates, cellulose products, or salts of acids containing one or more hydroxyl groups (Chap. 10). Subsequently, it was also discovered that with oil-well temperatures above 110°C, the CaO/SiO_2 ratio of the cement hydration product must be lowered to below 1.3 by the addition of silica flour in order to achieve high strength after hardening. These findings became the basis for the development of numerous cement additives for application to the oil-well cement industry.

The petroleum industry generally prefers the basic low-C_3A, coarse-ground portland cements (API Classes G and H), to which one or more admixtures of the types listed below are added at the site:

1. *Cement retarders.* To increase the setting time of cement and allow time for placement of the slurry
2. *Cement accelerators.* To reduce the setting time of cement for early strength development when needed (i.e., in permafrost zone)
3. *Lightweight or heavyweight additives.* To reduce or increase the weight of the column of cement slurry as needed
4. *Friction reducers.* To allow placement of slurry with less frictional pressure (2 to 3 percent bentonite clay is commonly used for this purpose)
5. *Low water-loss additives.* To retain water in the slurry when passing permeable zones downhole (i.e., latex additives)
6. *Strength-retrogression reducers.* To reduce the CaO/SiO_2 ratio of the hydration product at temperatures above 110°C (i.e., silica flour or pozzolans)

Since organic retarders are unstable at high temperatures, API Class J cement represents a relatively recent development in the field of modified portland cements that can be used for case-cementing *at temperatures above 150°C without the addition of a retarder*. The cement, composed mainly of a β-C_2S clinker, is ground to about 200 m^2/kg Blaine, with 40 mass percent silica flour. It may be noted that slurry thickening times and strength values for oil-well cements are determined according to special procedures set out in API RP-10B, *Recommended Practice for Testing Oil-Well Cements and Cement Additives*.

6.8.6 White and Colored Cements

The universally gray color of portland cement products limits an architect's opportunity for creating surfaces with aesthetic appeal. A white cement, with exposed-aggregate finish, can be useful in creating desired aesthetic effects. Furthermore, by adding appropriate pigments, white cement can be used as a base for producing cements with varying colors.

White cement is produced by pulverizing a white portland-cement clinker. The gray color of ordinary portland-cement clinker is generally due to the presence of iron. Thus by lowering the iron content of clinker, light-colored cements can be produced. When the total iron in clinker corresponds to less than 0.5 percent Fe_2O_3, and the iron is held in the reduced Fe^{2+} state, the clinker is usually white (see the story below). These conditions are achieved in cement manufacturing by using iron-free clay and carbonate rock as raw materials, special ball mills with ceramic liners and balls for grinding the raw mix, and clean fuels such as oil or gas for production of clinker under a reducing environment in the high-temperature zone of the cement rotary kiln. Consequently, white cements are approximately three times as expensive as conventional portland cement.

The Princess and the Fool

The importance of the reducing environment in making white-cement clinker is underscored by an experience that the author (PKM) had during a consultation visit with a South American cement plant. The raw-mix contained more iron than normally acceptable, and the clinker from the kiln was persistently off-white. In order to prolong the reducing environment around the clinker particles by increasing the amount of oil sprayed on hot clinker before leaving the burning zone, I requested a heat-resisting steel pipe of a larger diameter. Since there was none in stock and the cement plant was located far away from any city, I was getting nowhere while the low-iron raw-mix specially made for this experiment was running out fast.

The language problem added to the difficulty. I could not speak Spanish and the plant foreman did not understand English. To emphasize my need for one pipe with a larger diameter I raised one finger. In response, the foreman waved two fingers into my face. I stopped arguing because his action brought to my mind an old story from the Sanskrit literature. A king in ancient India had a very beautiful daughter, named Tilotama, who refused to marry until she could find someone wiser than herself. When many scholarly princes failed to win her in debates on philosophical and religious issues, they decided to play a practical joke. A dumb and stupid man was dressed in scholarly robes and presented to her for a debate. The princess raised one finger and the fool, assuming that the princess was threatening to poke one of his eyes, raised two fingers. The judges, interpreting one finger to mean that God is the only important thing in the universe and two fingers to mean that nature is equally important as it reveals the glory of God, awarded victory to the fool. What the foreman really meant was that he would like to install two pipes of the smaller diameter because he did not have a pipe with a larger diameter. When the thought of Tilotama's fool trying to blind him in both eyes came to my mind, I yielded without further argument. The foreman installed the two small pipes for spraying oil on hot clinker. Subsequently, the whitest clinker I have ever seen came out of the kiln.

Colored cements fall into two groups; most are derived from the addition of a pigment to white cement, but some are produced from clinkers having the corresponding colors. A buff-colored cement marketed in the United States under the name *warm tone cement* is produced from the clinker made from a portland cement raw mix containing a higher iron content (approximately 5 percent Fe_2O_3) than normal, and processed under reducing conditions.

For producing colored cements with pigment, it should be noted that not all the pigments that are used in the paint industry are suitable for making colored cements. To be suitable, a pigment should not be detrimental to the setting, hardening, and durability characteristics of portland cement, and should produce durable color when exposed to light and weather. Red, yellow, brown, or black cements can be produced by intergrinding 5 to 10 weight percent iron-oxide pigments of the corresponding color with a white clinker. Green and blue-colored cements can be made by using chromium oxide and cobalt blue, respectively.

6.8.7 Calcium Aluminate Cement

Compared to portland cement, calcium aluminate cement (CAC) possesses many *unique properties*, such as high early strength, ability to harden even under low-temperature conditions, and superior durability to sulfate attack. However, several structural failures due to gradual strength loss of concrete containing CAC have led to limitations on the use of this cement for structural applications. In most countries, CAC is now used mainly for making castable refractory lining for high-temperature furnaces.

According to ASTM C 219 definitions, *calcium aluminate cement* is the product obtained by pulverizing calcium aluminate cement clinker; the clinker is a partially fused or a completely fused product consisting of hydraulic calcium aluminates. Thus unlike portland and modified portland cements, in which C_3S and C_2S are the principal cementing compounds, CAC contains monocalcium aluminate (CA) as the principal cementing compound with $C_{12}A_7$, CA, C_2AS, β-C_2S, and F_{ss} as minor compounds. Typically, the chemical analysis of ordinary CAC shows approximately 40 percent Al_2O_3, and some cements contain even higher alumina content (50 to 80 percent); therefore, this cement is also called *high-alumina cement* (HAC).

Bauxite, a hydrated alumina mineral, is the commonly used source of alumina in raw materials for the manufacture of CAC. Most bauxite ores contain a considerable amount of iron as an impurity, and this accounts for the 10 to 17 percent iron (expressed as Fe_2O_3) usually present in ordinary CAC. This is why, unlike portland cement clinker, the CAC clinker is in the form of a completely fused melt that requires a specially designed furnace. This is also the reason why in France and Germany the cement is called *ciment fondu* and *tonerdeschmelz zement*, respectively. CAC cements meant for making very high-temperature furnace linings, contain very low iron and silica, and can be made by sintering in a rotary kiln.

Like portland cement, the properties of CAC are dependent on the hydration characteristics of the cement and the microstructure of the hydrated cement paste. The principal cement compound is CA which usually ranges between 50 and 60 percent. Although CAC products have setting times comparable to ordinary portland cement, the rate of strength development at early ages is quite high due to the high reactivity of CA. Within 24 h of hydration, the strength of a normally cured CAC concrete can attain values equal to or exceeding the 7-day strength of ordinary portland cement (Fig. 6-21*a*). Also, the strength gain characteristic under subzero curing conditions (Fig. 6-21*b*) is much better than with portland cement; hence the material is quite attractive for cold

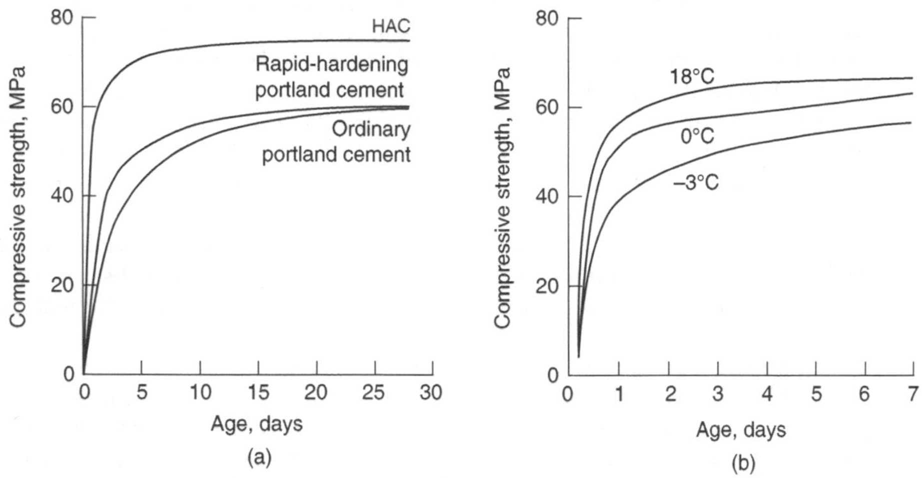

FIGURE 6-21 (*a*) Strength development rates for various cements at normal temperature; (*b*) effect of low curing-temperatures on the strength of high-alumina cement concrete. [From Neville, A.M., in Progress in Concrete Technology, Malhotra, V.M., ed., CANMET, Ottawa, Canada, pp. 293–331, 1980.]

Calcium aluminate or high-alumina cements are able to develop very high strengths in relatively short periods of time. Unlike portland cements, they can develop high strengths even at lower than normal temperatures.

weather applications. It may be noted that the rate of heat liberation from a freshly hydrated CAC can be as high as 9 cal/g per hour, which is about three times the rate of high-early strength portland cement.

The composition of the hydration products shows a time-temperature dependency; the low-temperature hydration product (CAH_{10}) is thermodynamically unstable, especially in warm and humid storage conditions, under which a more stable compound, C_3AH_6, is formed (as illustrated by the equations below). Laboratory and field experience with CAC concrete show that on prolonged storage the hexagonal CAH_{10} and C_2AH_8 phases do in fact tend to convert to the cubic C_3AH_6. As a *consequence of the CAH_{10}–C_3AH_6 conversion*, a hardened CAC paste can show more than 50 percent reduction in the volume of solids, which causes an increase in porosity (Fig. 6-22*a*) and a loss in strength associated with this phenomenon (Fig. 6-22*b*).

$$CA + H \begin{cases} \xrightarrow{<10^\circ C} & CAH_{10} \\ \xrightarrow{<10\text{–}20^\circ C} & C_2AH_8 + AH_3 \\ \xrightarrow{<30^\circ C} & C_3AH_6 + 2AH_3 \end{cases}$$

$$3CAH_{10} \rightarrow C_3AH_6 + 2AH_3 + 18H \uparrow$$

	$3CAH_{10}$	C_3AH_6	$2AH_3$
Mol. wt., g	1014	378	312
g/cm^3	1.72	2.52	2.4
Mol. vol., cm	590	150	136

Formerly, it was assumed that the strength-loss problem in concrete could be ignored when low water-cement ratios were used, and the height of casting was limited to reduce the temperature rise due to heat of hydration. The data in Fig. 6-22*b* show that this may not be the case. The real problem is not that the residual strength is inadequate for structural purposes but that, as a result of the increase in porosity, the resistance to atmospheric carbonation and to corrosion of the embedded steel in concrete is reduced.

From hydration reaction of CAC, it may be noted that there is no calcium hydroxide in the hydration product; this feature also distinguishes CAC from portland cement and is the reason why CAC concrete shows *excellent resistance to acidic environments* (pH 4 to 6), seawater, and sulfate waters. As discussed below, the absence of calcium hydroxide in hydrated CAC is also beneficial for the use of the material in making high-temperature concrete.

In practice, the use of portland cement for concrete exposed to high temperature is limited to about 500°C, because at higher temperatures the free CaO formed on decomposition of calcium hydroxide causes the concrete to become unsound on exposure to moist air or water. Not only does CAC not produce any calcium hydroxide on hydration but also, at temperatures above 1000°C, CAC is capable of developing a *ceramic bond*, which is as strong as the original hydraulic bond. The green or the unfired strength of the CAC concrete drops considerably during the first-heating cycle due to the CAH_{10}-to-C_3AH_6 conversion phenomenon. With a high cement content in the concrete, however, the green strength may be adequate to prevent damage until the strength increases again due to the development of the ceramic bond (Fig. 6-23).

(a)

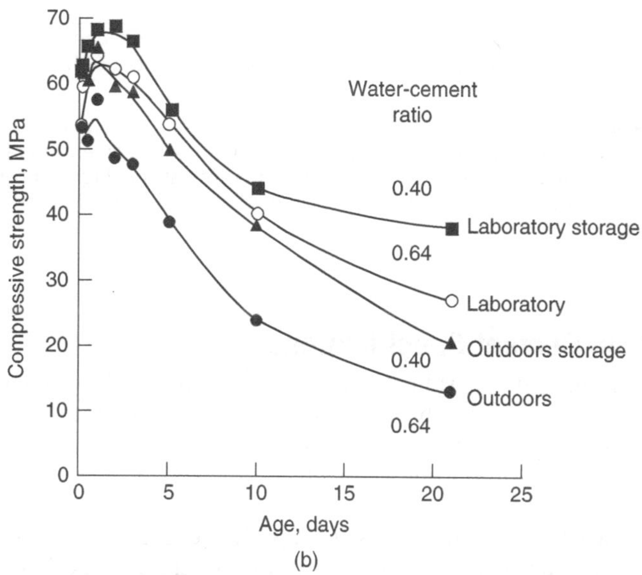

(b)

Figure 6-22 (*a*) Scanning electron micrograph of a partially converted calcium aluminate cement system; (*b*) influence of water-cement ratio on the long time strength of calcium aluminate cement concretes. [(*a*) From Mehta, P.K., and G. Lesnikoff, *J. Am. Ceram. Soc.*, Vol. 54, No. 4, pp. 210–212, 1971, reprinted with permission of American Ceramics Society; (*b*) From Neville, A., *High Alumina Cement Concrete*, Halstead Press, New York, p. 58, 1975, reprinted with permission from Construction Press (Longman Group Ltd.).]

Calcium aluminate cement concretes are generally not recommended for structural use. This is because the principal hydration product, CAH_{10}, is unstable under ordinary conditions. It gradually transforms into a stable phase, C_3AH_6, which has a cubic structure and is denser. The CAH_{10}-to-C_3AH_6 conversion is associated with a large increase in porosity and therefore a corresponding decrease in strength.

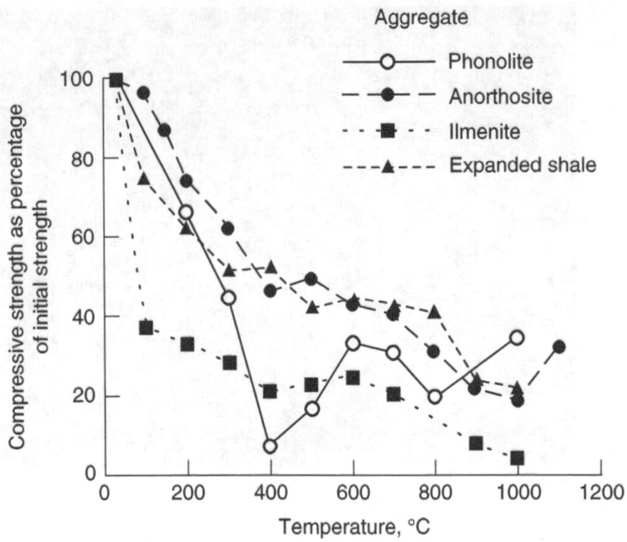

FIGURE 6-23 Effect of temperature rise on strength of calcium aluminate cement concretes. [From Neville, A.M., in *Progress in Concrete Technology*, Malhotra, V.M., ed., CANMET, Ottawa, Canada, pp. 293–331, 1980.]

Calcium aluminate cement concretes mostly find application in monolithic refractory linings for high-temperature furnaces. With increasing temperatures, the cement hydration products decompose and this causes a loss in strength. However, at very high temperatures, the strength increases due to the formation of a stable sintered material (ceramic bond).

6.9 Trends in Cement Specifications

Most countries in the world produce a variety of hydraulic cements according to their national standards. Usually, there are separate specifications governing portland cements and different types of blended portland cements that prescribe their blending constituents, their proportions and physical characteristics. Although national standards are constantly under review, it seems that the customary minor revisions are no longer sufficient to meet the needs of a rapidly changing world. As a result, worldwide, the standards specifications for cement are undergoing fundamental change that is reflected by recent developments in Europe and North America, as described below.

In 1992, all member states of the European Union decided to establish a *single market* for their products. As cement is one of their most important construction products, the harmonization of the national standards was a formidable task which was accomplished in April 2002 with the release of *EN 197—a single standard to replace earlier standards and certification codes for all types of portland and blended-portland cements throughout Europe* (see Table 6-8). EN 197-1 outlines the specification requirements for 27 different cements that are classified into five main cement types, described as follows:

1. CEM I covers traditional portland cements comprising at least 95 percent portland-cement clinker and up to 5 percent additional constituents (such as gypsum).

<table>
<tr><th>Type</th><th>Name</th><th>Designation</th><th>Clinker K</th><th>GGBF Slag S</th><th>Silica Fume D</th><th colspan="2">Pozzolan</th><th colspan="2">Fly Ash</th><th>Calcined Shade T</th><th>Lime-stone L*</th><th>Secondary Ingredients</th></tr>
<tr><th></th><th></th><th></th><th></th><th></th><th></th><th>Natural P</th><th>Industrial Q</th><th>Siliceous V</th><th>Calcitic W</th><th></th><th></th><th></th></tr>
<tr><td>I</td><td>Portland Cement</td><td>I</td><td>95-100</td><td>—</td><td>—</td><td>—</td><td>—</td><td>—</td><td>—</td><td>—</td><td>—</td><td>0-5</td></tr>
<tr><td>II</td><td>Slag</td><td>II-A/S</td><td>80-94</td><td>6-20</td><td>—</td><td>—</td><td>—</td><td>—</td><td>—</td><td>—</td><td>—</td><td>0-5</td></tr>
<tr><td></td><td>Portland Cement</td><td>II-B/S</td><td>65-79</td><td>21-35</td><td>—</td><td>—</td><td>—</td><td>—</td><td>—</td><td>—</td><td>—</td><td>0-5</td></tr>
<tr><td></td><td>Microsilica Portland Cement</td><td>II-A/D</td><td>90-94</td><td>—</td><td>6-10</td><td>—</td><td>—</td><td>—</td><td>—</td><td>—</td><td>—</td><td>0-5</td></tr>
<tr><td></td><td>Pozzolan Portland Cement</td><td>II-A/P</td><td>80-94</td><td>—</td><td>—</td><td>6-20</td><td>—</td><td>—</td><td>—</td><td>—</td><td>—</td><td>0-5</td></tr>
<tr><td></td><td></td><td>II-B/P</td><td>65-79</td><td>—</td><td>—</td><td>21-35</td><td>—</td><td>—</td><td>—</td><td>—</td><td>—</td><td>0-5</td></tr>
<tr><td></td><td></td><td>II-A/Q</td><td>80-94</td><td>—</td><td>—</td><td>—</td><td>6-20</td><td>—</td><td>—</td><td>—</td><td>—</td><td>0-5</td></tr>
<tr><td></td><td></td><td>II-B/Q</td><td>65-79</td><td>—</td><td>—</td><td>—</td><td>21-35</td><td>—</td><td>—</td><td>—</td><td>—</td><td>0-5</td></tr>
<tr><td></td><td>Fly Ash Portland Cement</td><td>II-A/V</td><td>80-94</td><td>—</td><td>—</td><td>—</td><td>—</td><td>6-20</td><td>—</td><td>—</td><td>—</td><td>0-5</td></tr>
<tr><td></td><td></td><td>II-B/V</td><td>65-79</td><td>—</td><td>—</td><td>—</td><td>—</td><td>21-35</td><td>—</td><td>—</td><td>—</td><td>0-5</td></tr>
<tr><td></td><td></td><td>II-A/W</td><td>80-94</td><td>—</td><td>—</td><td>—</td><td>—</td><td>—</td><td>6-20</td><td>—</td><td>—</td><td>0-5</td></tr>
<tr><td></td><td></td><td>II-B/W</td><td>65-79</td><td>—</td><td>—</td><td>—</td><td>—</td><td>—</td><td>21-35</td><td>—</td><td>—</td><td>0-5</td></tr>
<tr><td></td><td>Calcined Portland Cement</td><td>II-A/T</td><td>80-94</td><td>—</td><td>—</td><td>—</td><td>—</td><td>—</td><td>—</td><td>6-20</td><td>—</td><td>0-5</td></tr>
<tr><td></td><td></td><td>II-B/T</td><td>65-79</td><td>—</td><td>—</td><td>—</td><td>—</td><td>—</td><td>—</td><td>21-35</td><td>—</td><td>0-5</td></tr>
<tr><td></td><td>Limestone Portland Cement</td><td>II-A/L</td><td>80-94</td><td>—</td><td>—</td><td>—</td><td>—</td><td>—</td><td>—</td><td>—</td><td>6-20</td><td>0-5</td></tr>
<tr><td></td><td></td><td>II-B/L</td><td>65-79</td><td>—</td><td>—</td><td>—</td><td>—</td><td>—</td><td>—</td><td>—</td><td>21-35</td><td>0-5</td></tr>
<tr><td></td><td>Composed Portland Cement</td><td>II-A/M</td><td>80-94</td><td colspan="9">← 6-20 →</td></tr>
<tr><td></td><td></td><td>II-B/M</td><td>65-79</td><td colspan="9">← 21-35 →</td></tr>
<tr><td>III</td><td>Blastfurnace Slag Cement</td><td>III-A</td><td>35-64</td><td>36-65</td><td>—</td><td>—</td><td>—</td><td>—</td><td>—</td><td>—</td><td>—</td><td>0-5</td></tr>
<tr><td></td><td></td><td>III-B</td><td>20-34</td><td>66-80</td><td>—</td><td>—</td><td>—</td><td>—</td><td>—</td><td>—</td><td>—</td><td>0-5</td></tr>
<tr><td></td><td></td><td>III-C</td><td>5-19</td><td>81-95</td><td>—</td><td>—</td><td>—</td><td>—</td><td>—</td><td>—</td><td>—</td><td>0-5</td></tr>
<tr><td>IV</td><td>Pozzolanic Cement</td><td>IV-A</td><td>65-89</td><td>—</td><td colspan="4">← 11-35 →</td><td>—</td><td>—</td><td>—</td><td>0-5</td></tr>
<tr><td></td><td></td><td>IV-B</td><td>45-64</td><td>—</td><td colspan="4">← 36-55 →</td><td>—</td><td>—</td><td>—</td><td>0-5</td></tr>
<tr><td>V</td><td>Composite Cement</td><td>V-A</td><td>40-64</td><td>18-30</td><td>—</td><td colspan="3">← 18-30 →</td><td>—</td><td>—</td><td>—</td><td>0-5</td></tr>
<tr><td></td><td></td><td>V-B</td><td>20-39</td><td>31-50</td><td>—</td><td colspan="3">← 31-50 →</td><td>—</td><td>—</td><td>—</td><td>0-5</td></tr>
</table>

*If the local organic content (TOC) is lower than 0.2% the symbol used is LL instead of L (TOC < 0.50%)

TABLE 6-8 Cements According to EN 197-1: Types and Composition

2. CEM II covers 19 varieties of blended portland cements containing at least 65 percent portland-cement clinker. A letter notation identifies the blending constituents that include blast furnace slag, siliceous fly ash, calcareous fly ash, natural uncalcined pozzolan, natural calcined pozzolan, burnt shale, limestone, and silica fume. Each cement type is available either with 6 to 20 percent or 21 to 35 percent of the blending constituent by mass except the silica-fume cements, which contain 6 to 10 percent silica fume.
3. CEM III covers three varieties of blended portland-slag cements containing more than 35 percent granulated blastfurnace slag, namely 36 to 65 percent, 66 to 80 percent, and 81 to 95 percent slag.
4. CEM IV covers two portland-pozzolan cements containing 11 to 35 percent or 36 to 55 percent pozzolan.
5. CEM V covers two composite portland cements containing either 36 to 60 percent or 61 to 80 percent of a mixture of blending components, namely, blastfurnace slag, fly ash, and other pozzolans.

EN 197-1 also provides six strength grades according to which the cements may be manufactured for marketing. Besides the three customary strength grades, 32.5, 42.5, and 52.5 (minimum 28-day compressive strength, MPa), a cement may also be classified as a rapid-hardening or normal-hardening on the basis of its early strength characteristics.

ASTM International provides for 10 types of portland cements (covered by ASTM Standard C 150/C 150M) and eight types of blended portland cements (covered by ASTM Standard C 595), which contain restrictions on chemical composition, physical properties, and characteristics as well as proportions of blending materials. As with the CEM II category of EN197, ASTM cement specifications are being amended to permit the use of limestone as a blending material in blended portland cements. Due to the cumbersome, prescriptive requirements, hardly any blended cements meeting the ASTM C 595 requirements are being manufactured in the United States. Instead, blending materials are added at the ready-mixed concrete batching plant to produce concrete mixtures meeting certain performance standards. With regard to the ASTM Standard C 150 covering portland cements, a 1998 survey of the U.S. cement manufacturers conducted by the Portland Cement Association,[24] shows that, in general, the cement industry is making only one type of cement-clinker which meets the requirements of the Types I, II, and III cements, except that in the case of Type II cement, the C_3A content is somewhat lower than 8 percent. With all three cement types, the mean compound composition was 56 percent C_3S and 17 percent C_2S. The mean Blaine fineness for Types I and II cements was 380 m^2/kg, whereas it was 547 m^2/kg for the Type III cement.

In 1992, a performance-based standard for blended hydraulic cements, C 1157, was issued by ASTM. Unlike, ASTM C 595, this specification contains no restrictions whatsoever on the composition of blended cements or the proportion of their constituents. Also, there are no requirements on the physical-chemical properties of the constituents. In 1998, ASTM C 1157 was amended to include portland cements. Thus, this is a performance-based cement standard that covers all hydraulic cements. ASTM C 1157 classifies cements by type based on specific performance requirements such as general use, high-early strength, resistance to attack by sulfates, and heat of hydration.

The six cement types conforming to this specification, along with some of the key requirements are as follows:

1. *Type GU—General Use Hydraulic Cement.* Minimum compressive strength 13, 20 and 28 MPa at age 3, 7, and 28 days, respectively.
2. *Type HE—High Early Strength.* Minimum compressive strength 12 and 24 MPa at age 1 and 3 days, respectively.
3. *Type MS—Moderate Sulfate Resistance.* 0.1 percent maximum expansion in 6 months with mortar bars immersed in a standard sulfate solution (ASTM C 1012).
4. *Type HS—High Sulfate Resistance.* 0.05 percent maximum expansion in 6 months with mortar bars immersed in a standard sulfate solution (ASTM C 1012).
5. *Type MH—Moderate Heat of Hydration.* 290 kJ/kg (70 cal/g) max., heat of hydration in 7 days.
6. *Type LH—Low Heat of Hydration.* 250 kJ/kg (60 cal/g) max., heat of hydration in 7 days.

The performance-based cement standards, such as ASTM C 1157, are expected to play a major part in the future development of multi-component hydraulic cements containing large amounts of industrial by-products and a correspondingly smaller proportion of portland cement clinker. The manufacturing process for portland cement clinker is not only energy-intensive and expensive, but also produces large amounts of CO_2, which is a primary greenhouse gas. Therefore, in the future it is expected that the use of pure portland cement would be limited to special applications, whereas performance-based blended portland cements with low portland clinker content will find increasing use for all types of concrete construction.

Test Your Knowledge

6.1 When producing a certain type of portland cement it is important that the oxide composition remains uniform. Why?

6.2 In regard to sulfate resistance and rate of strength development, evaluate the properties of a portland cement which has the following chemical analysis: SiO_2 = 20.9 percent; Al_2O_3 = 5.4 percent; Fe_2O_3 = 3.6 percent; CaO = 65.1 percent; MgO = 1.8 percent; and SO_2 = 2.1 percent.

6.3 What do you understand by the following terms: alite, belite, periclase, langbeinite, plaster of paris, tobermorite gel?

6.4 Why is C_3S more reactive, and γ-C_2S nonreactive with water at normal temperatures? MgO and CaO have similar crystal structures, but their reactivities are very different from each other. Explain why.

6.5 What is the significance of fineness in cement? How is it determined? Can you give some idea of the fineness range in industrial portland cements?

6.6 Why is gypsum added to the cement clinker? Typically, how much is the amount of added gypsum?

6.7 The presence of high free-lime in portland cement can lead to unsoundness. What is meant by the term, "unsoundness"? Which other compound can cause unsoundness in portland cement products?

6.8 Approximately, what is the combined percentage of calcium silicates in portland cement? What are the typical amounts of C_3A and C_4AF in ordinary (ASTM Type I) portland cement?

6.9 Which one of the four major compounds of portland cement contributes most to the strength development during the first few weeks of hydration? Which compound or compounds are responsible for rapid stiffening and early setting problems of the cement paste?

6.10 Discuss the major differences in the physical and chemical composition between an ordinary (ASTM Type I) and a high early strength (ASTM Type III) portland cement.

6.11 Why do the ASTM Specifications for Type IV cement limit the minimum C_2S content to 40 percent and the maximum C_3A content to 7 percent?

6.12 Explain which ASTM type cement would your use for:
(*a*) Cold-weather construction
(*b*) Construction of a dam
(*c*) Making reinforced concrete sewer pipes

6.13 The aluminate-sulfate balance in solution is at the heart of several abnormal setting problems in concrete technology. Justify this statement by discussing how the phenomena of quick-set, flash set, and false set occur in freshly hydrated portland cements.

6.14 Assuming that the chemical composition of the calcium silicate hydrate formed on hydration of C_3S or C_2S corresponds to $C_3S_2H_3$, make calculations to show the proportion of calcium hydroxide in the final products and the amount of water needed for full hydration.

6.15 Define the terms *initial set* and *final set*. For a normal portland cement draw a typical heat evolution curve for the setting and early hardening period, label the ascending and descending portions of the curve with the underlying chemical processes at work, and show the points where the initial set and final set are likely to take place.

6.16 Discuss the two methods that the cement industry employs to produce cements having different rates of strength development or heat of hydration. Explain the principle behind the maximum limit on the C_3A content in the ASTM C 150 Standard Specification for Type V portland cements.

6.17 With the help of the "pozzolanic reaction," explain why under given conditions, compared to portland cement, portland pozzolan and portland blast-furnace slag cements are likely to produce concrete with higher ultimate strengths and superior durability to sulfate attack.

6.18 What is the distinction between shrinkage-compensating and self-stressing cements? What are Types K, M, S, and O expansive cements? Explain how the expansive cements function to make concrete crack-free.

6.19 Write short notes on the compositions and special characteristics of the following cements: regulated-set cement, very high early strength cement, API Class J cement, white cement, and calcium aluminate cement.

6.20 Discuss the physical-chemical factors involved in explaining the development of strength in products containing the following cementitious materials, and explain why portland cement has come to stay as the most commonly used cements for structural purposes:

(*a*) lime
(*b*) plaster of Paris
(*c*) calcium aluminate cement

References

1. Rietveld, H.M. A profile refinement method for nuclear and magnetic structures, *Journal of Applied Crystallography*, Vol. 2, pp. 65–71, 1969.
2. Le Saoût, G., V. Kocaba, and K. Scrivener, Application of the Rietveld method to the analysis of anhydrous cement, *Cement and Concrete Research*, Vol. 41, pp. 133–148, 2011.
3. K. Scrivener, Quantitative study of Portland cement hydration by X-ray diffraction/Rietveld analysis and independent methods, *Cement and Concrete Research*, Vol. 34, pp. 1541–1547, 2004.
4. Lea, F.M. *The Chemistry of Cement and Concrete*, Chemical Publishing Company, New York, pp. 317–337, 1971.
5. Brunauer, S., and L.E. Copeland, *The Chemistry of Concrete*, Sci. Am., April 1964.
6. Richardson, I.G., Tobermorite/jennite- and tobermorite/calcium hydroxide-based models for the structure of C-S-H: applicability to hardened pastes of tricalcium silicate, [beta]-dicalcium silicate, Portland cement, and blends of Portland cement with blast-furnace slag, metakaolin, or silica fume, *Cement and Concrete Research*, Vol. 34, pp. 1733–1777, 2004.
7. Powers, T.C., and T.L. Brownyard, Studies of the physical properties of hardened Portland cement paste, Skokie, IL, USA, 1947.
8. Feldman R.F., and P.J. Sereda, A new model of hydrated cement and its practical implications, *Engineering Journal*, Vol. 53, p. 7, 1970.
9. Taylor, H.F.W., Proposed Structure for Calcium Silicate Hydrate Gel, *Journal of the American Ceramic Society*, pp. 464–467, 1986.
10. Taylor, H.F.W., Nanostructure of C-S-H: Current status, *Advanced Cement Based Materials*, Vol. 1, pp. 38–46, 1993.
11. Fujii, K., and W. Kondo, Estimation of Thermochemical Data for Calcium Silicate Hydrate (C-S-H), *Communications of the American Ceramic Society*, pp. C-220–C-221, 1983.
12. Cong, X., and R.J. Kirkpatrick, 29Si MAS NMR study of the structure of calcium silicate hydrate, *Advanced Cement Based Materials*, Vol. 3, pp. 144–156, 1996.
13. Cong, X., and R.J. Kirkpatrick, 29Si and 17O NMR investigation of the structure of some crystalline calcium silicate hydrates, *Advanced Cement Based Materials*, Vol. 3, pp. 133–143, 1996.
14. Grutzeck, M.W., A new model for the formation of calcium silicate hydrate (C-S-H), *Materials Research Innovations*, Vol. 3, pp. 160–170, 1999.
15. Grutzeck, M.W., S. Kwan, B.J.L., A. Benesi, A sorosilicate model for calcium silicate hydrate, *Journal of Materials Science*, Vol. 18, pp. 217–220, 1999.
16. Jennings, H.M., Colloid model of C-S-H and implications to the problem of creep and shrinkage, *Materials and Structures*, Vol. 37, pp. 59–70, 2004.
17. Richardson, I.G., and G.W. Groves, The incorporation of minor and trace elements into calcium silicate hydrate (C-S-H) gel in hardened cement pastes, *Cement and Concrete Research*, Vol. 23, pp. 131–138, 1993.

18. Richardson, I.G., and G.W. Groves, Models for the composition and structure of calcium silicate hydrate (C-S-H) gel in hardened tricalcium silicate pastes, *Cement and Concrete Research*, Vol. 22, pp. 1001–1010, 1992.
19. Lerch, W., *Proceedings of the American Society for Testing and Materials*, Vol. 46, p. 1252, 1946.
20. Verbeck, G.J., and C.W. Foster, *Proceedings of the American Society for Testing and Materials*, Vol. 50, p. 1235, 1950.
21. Mehta, P.K., *ASTM STP 663*, pp. 35–60, 1978.
22. Hoff, G.C., B.J. Houston, and F.H. Sayler, *U.S. Army Engineer Waterway Experiment Station*, Vicksburg, MS, Miscellaneous Paper C-75-5, 1975.
23. Mehta, P.K., World Cement Technology, pp. 166–177, May 1980.
24. Tennis, P.D., *Concr. Tech. Today*, Vol. 20, No. 2, Portland Cement Association, August 1999.

Suggestions for Further Study

Hewlett P., C., ed., *Lea's Chemistry of Cement and Concrete*, 4th ed., Arnold, London, p. 1053, 1998.

Juenger, M.C.G., F. Winnefeld, J.L. Provis, and Ideker, *J. Advances in alternative cementitious binders*. Cement and Concrete Research, Vol. 41, no. 12: pp. 1232-1243, 2011.

Locher F.W., Cement: Principles of Production and use, Verlag Bau+Technik GmbH, Dusseldorf, 2006.

Malhotra, V.M., ed., *Progress in Concrete Technology*, CANMET, Ottawa, 1994.

Mangabhai R.J., and Glasser F.P. eds, *Calcium Aluminates 2001*, The University Press, Cambridge, 2001. Neville, A.M., *History of high-alumina cement. Part 1: Problems and the Stone report. Proceedings of the Institute of Civil Engineers - Engineering History and Heritage 162(EH2)*, Vol. 81-91, 2009.

Neville, A.M., *History of high-alumina cement. Part 1: Problems and the Stone report. Proceedings of the Institute of Civil Engineers - Engineering History and Heritage 162(EH2)*, Vol. 81-91, 2009.

Newman, J., and B.S., Choo, eds., *Advanced Concrete Technology: Constituent Materials*, Butterworth-Heinemann, Oxford, 2003.

Skalny, J.P., ed., *Material Science of Concrete*, The American Ceramic Society, 1989; Cement Production and Cement Quality by V. Johansen; Hydration Mechanisms by E.M. Gartner and J.M. Gaidis; The Microtextures of Concrete by K.L. Scrivener.

Taylor, H.W.F., *Cement Chemistry*, 2d ed., T. Telford, 459 p., 1997.

CHAPTER 7
Aggregates

Preview

Aggregate is relatively inexpensive and does not enter into complex chemical reactions with water; it has been customary, therefore, to treat it as an inert filler in concrete. However, due to increasing awareness of the role played by aggregates in determining many important properties of concrete, the traditional view of the aggregate as an inert filler is being seriously questioned.

Aggregate characteristics that are significant for making concrete include porosity, grading or size distribution, moisture absorption, shape and surface texture, crushing strength, elastic modulus, and the type of deleterious substances present. These characteristics are derived from mineralogical composition of the parent rock (which is affected by geological rock-formation processes), exposure conditions to which the rock has been subjected to before mining, and the type of equipment used for producing the aggregate. Therefore, fundamentals of rock formation, classification and description of rocks and minerals, and industrial processing factors that influence aggregate characteristics are briefly described in this chapter.

Natural mineral aggregates, which comprise over 90 percent of the total aggregates used for making concrete, are described in more detail. Due to their greater potential use, the aggregates from industrial by-products such as blast-furnace slag, fly ash, municipal waste, and recycled concrete are also described. Finally, the principal aggregate characteristics that are important for concrete making are covered in detail.

7.1 Significance

From Chap. 6 we know that cements consist of chemical compounds that enter into chemical reactions with water to produce complex hydration products with adhesive property. Unlike cement, although the aggregate in concrete occupies 60 to 80 percent of the volume, it is frequently looked upon as an inert filler and therefore not much attention is given to its possible effect on properties of concrete. The considerable influence that the aggregate component can exercise on the strength, dimensional stability, and durability of concrete has been discussed in Chaps. 3, 4, and 5, respectively. In addition to these important properties of hardened concrete, the aggregate also plays a major role in determining the cost and workability of concrete mixtures (Chap. 9); therefore, *it is inappropriate to treat the aggregate with any less respect than cement.*

7.2 Classification and Nomenclature

Classification of aggregates according to particle size, bulk density, or source have given rise to a special nomenclature, which should be clearly understood. For instance, the term *coarse aggregate* is used to describe particles larger than 4.75 mm (retained on No. 4 sieve), and the term *fine aggregate* is used for particles smaller than 4.75 mm. Typically fine aggregates contain particles in the size range 75 μm (No. 200 sieve) to 4.75 mm, and coarse aggregates from 4.75 to about 50 mm, except for mass concrete that may contain particles up to 150 mm.

Most natural mineral aggregates, such as sand and gravel, have a bulk density of 1520 to 1680 kg/m^3 (95 to 100 lb/ft^3) and produce *normal-weight* concrete with approximately 2400 kg/m^3 (150 lb/ft^3) unit weight. For special needs, aggregates with lighter or heavier density can be used to make correspondingly lightweight and heavyweight concretes. Generally, the aggregates with bulk densities less than 1120 kg/m^3 (70 lb/ft^3) are called *lightweight* and those weighing more than 2080 kg/m^3 (130 lb/ft^3) are called *heavyweight*.

For the most part, concrete aggregates are comprised of sand, gravel, and crushed rock derived from natural sources. These are referred to as *natural mineral aggregates*. On the other hand, thermally processed materials such as expanded clay and shale, which are used for making lightweight concrete, are called *synthetic aggregates*. Aggregates made from industrial by-products (e.g., blast-furnace slag and fly ash) also belong to this category. Municipal wastes and recycled concrete from demolished buildings and pavements are also being investigated for use as aggregate for concrete.

7.3 Natural Mineral Aggregates

Natural mineral aggregates form the most important class of aggregates for making portland cement concrete. Approximately half of the total coarse aggregate consumed by the concrete industry in the United States consists of gravel; most of the remainder is crushed rock. Carbonate rocks comprise about two-thirds of the crushed aggregate; sandstone, granite, diorite, gabbro, and basalt make up the rest. Natural silica sand is predominantly used as fine aggregate, even with most lightweight concrete. Natural mineral aggregates are derived from rocks of several types and most rocks are themselves composed of several minerals. A *mineral* is defined as a naturally occurring inorganic substance of more or less definite chemical composition and usually of a specific crystalline structure. An elementary review of aspects of rock formation and the classification of rocks and minerals is essential for understanding why some materials are more abundantly used as aggregates than others, and also understanding the microstructure-property relations in different aggregate types.

7.3.1 Description of Rocks

According to their origin, rocks are classified into three major groups: igneous, sedimentary, and metamorphic; these groups are further subdivided according to mineralogical and chemical composition, texture or grain size, and crystal structure.

Igneous rocks are formed by cooling of the magma (molten rock matter) either above, or below, or near the earth's surface. The degree of crystallinity and the grain size of igneous rocks, therefore, vary with the rate at which magma was cooled at the time of rock formation. It may be noted that grain size has a significant effect on the rock characteristics; rocks having the same chemical composition but different grain size may behave differently under the same condition of exposure.

Magma intruded at great depths cools at a slow rate and forms completely crystalline minerals with coarse grains (>5 mm grain size); rocks of this type are called *intrusive* or *plutonic*. Due to quicker cooling, the rocks formed near the surface of the earth contain minerals with smaller crystals. These fine-grained rocks (1 to 5 mm grain size) may also contain some glass and are called *shallow-intrusive* or *hypabyssal*. Rapidly cooled magma, as in the case of rocks formed by volcanic eruptions, contains mostly noncrystalline or glassy matter; the glass may be dense (quenched lava) or cellular (pumice), and the rock type is called *extrusive* or *volcanic*.

Also, a magma may be supersaturated, saturated, or undersaturated with respect to the amount of silica present for mineral formation. From a supersaturated magma, the free or uncombined silica crystallizes out as quartz after the formation of minerals such as feldspars, mica, and hornblende. In saturated or unsaturated magma, the silica content is insufficient to form quartz. This leads to a classification of igneous rocks based on the total SiO_2 present; rocks containing more than 65 percent SiO_2, 55 to 65 percent SiO_2, and less than 55 percent SiO_2 are called *acid, intermediate*, and *basic*, respectively. Again, the classifications of igneous rocks on the basis of crystal structure and silica content are useful because it is the combination of the acidic character and grain size of the rock that seems to determine whether an aggregate would be vulnerable to alkali attack in portland-cement concrete.

Sedimentary rocks are stratified rocks that are usually laid down under water but are, at times, accumulated by wind and glacial action. The siliceous sedimentary rocks are derived from existing igneous rocks. Depending on their method of deposition and consolidation, it is convenient to subdivide them into three groups: (1) mechanically deposited either in an unconsolidated or physically consolidated state, (2) mechanically deposited and consolidated usually with chemical cements, and (3) chemically deposited and consolidated.

Gravel, sand, silt, and *clay* are the important members of the group of unconsolidated sediments. Although the distinction between these four members is made on the basis of particle size, there is a general trend in the mineral composition. Gravel and coarse sands usually consist of rock fragments; fine sands and silt consist predominately of mineral grains, and clays consist exclusively of mineral grains.

Sandstone, quartzite, and *graywacke* belong to the second category. The main difference between sandstone and quartzite is that the first is a sedimentary stone and the second is the result of its metamorphic transformation. The action of pressure and temperature during the metamorphism leads to a more "consolidated" structure. The cementing or interstitial materials of sandstone may be opal (silica gel), calcite, dolomite, clay, or iron hydroxide. *Graywackes* belong to a special class of sandstone, which contains angular and sand-size rock fragments in an abundant matrix of clay, shale, or slate.

Chert and *flint* belong to the third group of siliceous sedimentary rocks. Chert is usually fine-grained and can vary from porous to dense. Dense black or gray cherts, which are quite hard, are called flint. In regard to mineral composition, chert consists of poorly crystalline quartz, chalcedony, and opal; often all three are present.

Limestones are the most widespread of carbonate rocks. Limestone normally contains both calcite and dolomite. To express the predominance of one or other mineral, normally the rock is classified as calcitic limestone or dolomitic limestone. Usually, they contain both the calcium and magnesium carbonate minerals in various proportions, and significant amounts of noncarbonate impurities, such as clay and sand.

It should be noted that compared to igneous rocks, the aggregates produced from stratified sediments can vary widely in characteristics, such as the shape, texture, porosity, strength, and soundness. This is because the conditions under which they are consolidated vary widely. The rocks tend to be porous and weak when formed under a relatively low pressure. They are dense and strong if formed under a high pressure. Some limestones and sandstones may have less than a 100-MPa crushing strength that makes them unsuitable for use in high-strength concrete. Also, compared to igneous rocks, sedimentary rocks frequently contain impurities, which at times, jeopardize their use as aggregate. For instance, limestone, dolomite, and sandstone may contain opal or clay minerals which adversely affect the behavior of aggregate under certain conditions of exposure.

Metamorphic rocks are igneous or sedimentary rocks that have changed their original texture, crystal structure, or mineralogical composition in response to physical and chemical conditions below the earth's surface. Common rock types belonging to this group are marble, schist, phyllites, and gneiss. The rocks are dense but frequently foliated. Some phyllites, gneiss, and schists are reactive with the alkalies present in portland cement paste.

Earth's crust consists of 95 percent igneous and metamorphic and of 5 percent sedimentary rocks. Approximately, sedimentary rocks are composed of 4 percent shale, 0.75 percent sandstone, and 0.25 percent limestone. As sedimentary rocks cover 75 percent of the earth's landed area, a large amount of the natural mineral aggregates used in concrete namely sand, gravel, and crushed rocks are derived from sedimentary rocks. Although some sedimentary deposits are up to 13 km thick, over the continental areas the average is about 2300 m.

7.3.2 Description of Minerals

ASTM Standard C 294 contains the descriptive nomenclature that is useful for understanding the terms used to designate aggregate constituents. Based on this standard, a brief description of the constituent minerals that commonly occur in natural rocks is given below.

Silica Minerals

Quartz is a very common hard mineral composed of SiO_2. The hardness of quartz as well as that of feldspar is due to the framework Si-O structure, which is very strong. Quartz is present in acidic-type igneous rocks (>65 percent SiO_2), such as granite and rhyolites. Due to its resistance to weathering, it is an important constituent of many sand and gravel deposits, and sandstones. *Tridymite* and *cristobalite* are also crystalline silica minerals but are metastable at ordinary temperature and pressure, and are rarely found in nature except in volcanic rocks. Noncrystalline materials are referred to as *glass*.

Opal is a hydrous silica mineral (3 or 9 percent water) that appears noncrystalline by optical microcopy but may show short-order crystalline arrangement by x-ray diffraction analysis. It is usually found in sedimentary rocks, especially chert, and is the principal constituent of diatomite. *Chalcedony* is a porous silica mineral, generally containing microscopic fibers of quartz. The properties of chalcedony are intermediate between those of opal and quartz.

Silicate Minerals

Feldspars, amphibole, pyroxene and olivine (ferromagnesium silicates), mica, and clay minerals belong to this category. The minerals of the *feldspar group* are the most abundant rock-forming minerals in the earth's crust and are important constituents of igneous, sedimentary, and metamorphic rocks. They are almost as hard as quartz, and various members of the group are differentiated by chemical composition and crystallographic properties.

Orthoclase, sanidine, and microcline are potassium aluminum silicates, which are frequently referred to as the *potash feldspars*. The *plagioclase* or soda-lime feldspars include sodium aluminum silicates (albite), calcium aluminum silicates (anorthite), or both. The alkali feldspars containing potassium or sodium occur typically in igneous rocks of high silica content, such as granites and rhyolites, whereas those of higher calcium content are found in igneous rocks of lower silica content such as diorite, gabbro, and basalt.

Ferromagnesium minerals, which occur in many igneous and metamorphic rocks, consist of silicates of iron or magnesium or both. Hornblende and augite are members of the amphibole and pyroxene groups, respectively. Olivine is a common mineral of this class, which occurs in igneous rocks of relatively low silica content.

Muscovite, biotite, chlorite, and vermiculite, which form the group of *micaceous minerals*, also consist of silicates of iron and magnesium, but their internal sheet structure arrangement is responsible for the tendency to split into thin flakes. The micas are abundant and occur in all three major rock groups.

The *clay mineral* group covers sheet-structure silicates less than 2 μm (0.002 mm) in grain size. The clay minerals, which consist mainly of hydrous aluminum, magnesium, and iron silicates, are major constituents of clays and shales. Some clays undergo large expansions on wetting. Clays and shales are therefore not directly used as concrete aggregates. However, clay minerals may be present as contaminants in a natural mineral aggregate.

Carbonate Minerals

The most common carbonate mineral is *calcite* or calcium carbonate, $CaCO_3$. The other common mineral, *dolomite*, consists of equimolecular proportions of calcium carbonate and magnesium carbonate (corresponding to 54.27 and 45.73 percent by mass $CaCO_3$ and $MgCO_3$, respectively). Both carbonate minerals are softer than quartz and feldspars.

Sulfide and Sulfate Minerals

The sulfides of iron (e.g., *pyrite*, *marcasite*, and *pyrrohotite*) are frequently present in natural aggregates. Marcasite, which is found mainly in sedimentary rocks, readily oxidizes to form sulfuric acid and iron hydroxides. The formation of acid is undesirable, due to potential for corrosion of steel in prestressed and reinforced concrete structures. Marcasite and some forms of pyrite and pyrrohotite are suspected of being responsible for expansive reactions in concrete, causing cracks and pop-outs.

Gypsum (hydrous calcium sulfate) and *anhydrite* (anhydrous calcium sulfate) are the most abundant sulfate minerals that may be present as impurities in carbonate rocks and shales. Sometimes found as coatings on sand and gravel, gypsum and anhydrite increase the chances of internal sulfate attack in concrete.

As large amounts of concrete aggregate are derived from the sedimentary and igneous rocks, a description of the rock types in each class, principal minerals present, and characteristics of the aggregates are summarized in Tables 7-1 and 7-2, respectively.

Rock Type	Common Name	Principal Minerals Present	Aggregate Characteristics
Siliceous Rocks Mechanically deposited either in an unconsolidated or physically consolidated state.	Cobbles (>75 mm) Gravel (4.75–75 mm) Sand (0.075–4.75 mm) Silt (0.002–0.075 mm) Clay (<0.002 mm) Shale (consolidated clay)	All types of rocks and minerals may be present in cobbles, gravel, and sand. Silt consists predominately of grains of silica and silicate minerals. Clays are composed largely of a group of clay minerals.	Since natural cobbles, gravel, and sand are derived from geological weathering processes, they consist of hard rocks and minerals that often have a rounded shape and a smooth surface. When uncontaminated with clay and silt, they make strong and durable aggregates for concrete. Shales may appear to be hard, but they produce platy fragments.
Mechanically deposited and consolidated usually with chemical cements	Sandstone	Sand-size fragments consisting mainly of quartz and feldspar, usually cemented with opal, calcite, dolomite, clay, or iron hydroxide.	Generally, sandstones produce aggregates of satisfactory quality. Like carbonate rocks, the porosity, moisture absorption capacity, strength, and durability of sandstones can vary widely and will therefore affect properties of the aggregate.
	Graywacke	Graywackes are gray sandstones containing angular and sand-sized rock fragments in an abundant matrix of clay, shale, or slate.	
Chemically deposited and consolidated	Chert, flint	Chert consist of poorly crystalline quartz, chalcedony, or opal; often, all three are present. Flint is the name given to dense varieties of chert.	Dense cherts make good aggregates. Predominately opaline or chalcedonic cherts are capable of reacting with the alkalies in portland cement paste.
Carbonate Rocks	Calcitic limestone Dolomitic limestone Dolomitic calcite Calcitic dolomite Arenaceous limestone or dolomite Argillaceous limestone or dolomite	Predominately calcite Predominately dolomite 50–90% calcite; rest is dolomite 50–90% dolomite; rest is calcite Carbonate rocks containing 10–50% sand Carbonate rocks containing 10–50% clay	Carbonate rocks are softer than siliceous sedimentary rocks. However, they generally produce aggregates of satisfactory quality. Like sandstone, the porosity, moisture absorption capacity, strength, and durability of carbonate rocks can vary widely and will therefore affect the properties of the aggregate. Being stratified rocks, they tend to produce flat or elongated fragments.

Table 7-1 Characteristics of Aggregates from Sedimentary Rocks

Rock Type	Common Name	Principal Minerals Present	Aggregate Characteristics
Intrusive or plutonic	Granite	Quartz, feldspar (O, P),* mica	The rocks of this group generally make excellent aggregate because: (1) They are medium to coarse-grained, strong, and produce equidimensional fragments on crushing. (2) They have very low porosity and moisture absorption.
	Syenite	Feldspar (O, P), hornblende, biotite	
	Diorite	Feldspar (P), hornblende, biotite	
	Gabbro	Hornblende, augite, feldspar (P)	
	Diabase or dolerite	Same minerals as in gabbro, but medium-to-fine grained	
	Trap rock	Gabbro, diabase, and basalt	
Shallow-intrusive or hypabyssal	Felsite group: rhyolite, trachyte, andesite	The mineral composition of the rocks of the felsite group—rhyolite, trachyte, and andesite—is equivalent of granite, syenite, and diorite, respectively.	The rocks of this group are fine-grained and hard and make good aggregate except that felsite rocks, when microcrystalline or containing natural glass, are reactive with the alkalies in portland cement concrete.
	Basalt	With respect to mineral composition, basalt is the shallow-intrusive or extrusive equivalent of gabbro and diabase.	
Extrusive or volcanic	Obsidian	A dense, dark, natural glass of high silica content	Obsidian and pitchstone are dense and hard, but do not occur commonly.
	Pitchstone	Natural glass containing up to 10% water	
	Perlite	High-silica glass with onionlike texture and pearly luster; contains 2 to 5% water	Perlite is generally used for making insulating concretes after its structure is altered to a pumicelike vesicular structure by heat treatment.
	Pumice	Porous glass with elongated voids	Pumice, scoria, and tuffs are porous and weak, and are useful for making lightweight and insulating concretes.
	Scori	Porous glass with spherical voids	
	Tuff	Porous glass formed by consolidation of volcanic ash	

*The abbreviations O and P stand for orthoclase and plagioclase feldspar, respectively.

TABLE 7-2 Characteristics of Aggregates from Igneous Rocks

7.4 Lightweight Aggregate

Aggregates that weigh less than 1120 kg/m^3 (70 lb/ft^3) are generally considered lightweight, and find application in the production of various types of lightweight concretes. The light weight of the aggregate is due to the cellular or highly porous microstructure. It may be noted that cellular organic materials such as wood chips should not be used as aggregate because they would not be durable in the moist alkaline environment within portland-cement concrete.

Natural lightweight aggregates are made by crushing igneous volcanic rocks such as pumice, scoria, or tuff. Synthetic lightweight aggregates are manufactured by thermal treatment of a variety of materials, for instance, clays, shale, slate, diatomite, pearlite, vermiculite, blast-furnace slag, and fly ash.

Actually, there is a whole spectrum of lightweight aggregates (Fig. 7-1) weighing from 80 to 900 kg/m^3 (5 to 55 lb/ft^3). Very porous aggregates, at the lighter end of the spectrum, are generally weak and are therefore more suitable for making nonstructural insulating concretes. At the other end of the spectrum are those lightweight aggregates that are relatively less porous. When the pore structure consists of uniformly distributed fine pores, the aggregate particles are strong and therefore capable of producing structural concrete. ASTM has separate specifications covering lightweight aggregates for use in structural concrete (ASTM C 330), insulating concrete (ASTM C 332),

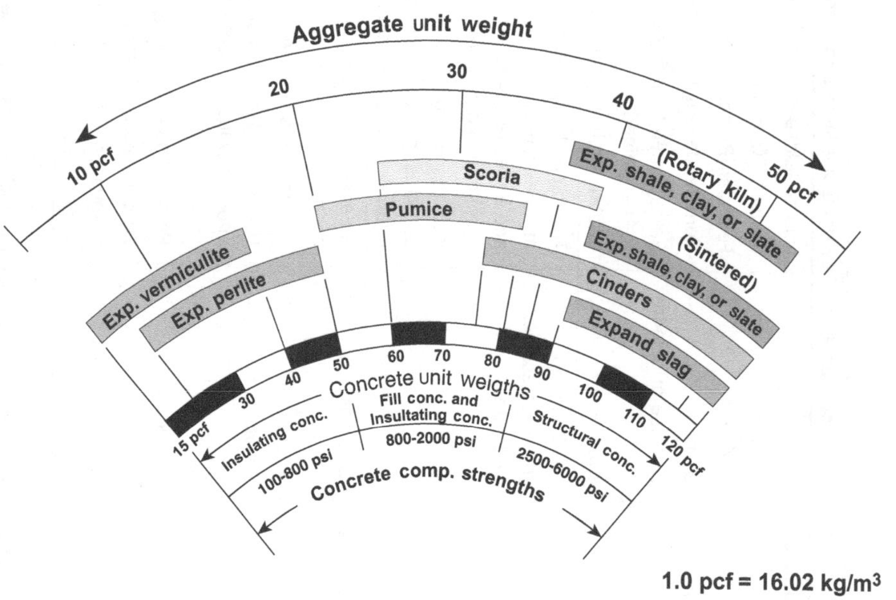

FIGURE 7-1 Lightweight aggregate spectrum. (Adapted from Litvin, A., and A.E. Fiorato, *Concr. Int.*, Vol. 3, No. 3, p. 49, 1981.)

and concrete for production of masonry units (ASTM C 331). These specifications contain requirements for grading, undesirable substances, and unit weight of aggregate. They also contain requirements for unit weight, strength, and drying shrinkage of concrete containing the aggregate. Properties of lightweight-aggregate concrete are described in Chap. 12.

7.5 Heavyweight Aggregate

Compared to normal-weight aggregate concrete with a typical unit weight of 2400 kg/m^3 (150 lb/ft^3), heavy-weight concretes weigh from 2900 to 6100 kg/m^3 (180 to 380 lb/ft^3), and are primarily used for making nuclear radiation shields. Heavy-weight aggregates (i.e., those that have a substantially higher density than normal-weight aggregate) are used for the production of heavy-weight concrete. Natural rocks suitable for heavy-weight aggregate consist predominately of two barium minerals, several iron ores, and a titanium ore (Table 7-3).

A synthetic product called ferro-phosphorus slag can also be used as heavy-weight aggregate. ASTM C 637 and C 638, which cover Standard Specifications and Descriptive Nomenclature, respectively, of aggregates for radiation-shielding concrete warn that ferro-phosphorus aggregate, when used for making portland cement concrete, will generate flammable and possibly toxic gases that can develop high pressures. Hydrous iron ores and boron minerals and frits are at times included in the aggregates for making heavyweight concretes because boron and hydrogen are very effective in neutron attenuation (capture). Steel punchings, sheared iron bars, and iron shots have also been investigated as heavyweight aggregates; however, the tendency of aggregate to segregate in a concrete mixture increases with the density of the aggregate.

Type of Aggregate	Chemical Composition of Principal Mineral	Specific Gravity of Pure Mineral	Typical Bulk Density [kg/m^3 (lb/ft^3)]
Witherite	$BaCO_3$	4.29	2320 (145)
Barite	$BaSO_4$	4.50	2560 (160)
Magnetite	Fe_3O_4	5.17	2720 (170)
Hematite	Fe_2O_3	4.9–5.3	3040 (190)
Lepidocrocite Geothite Limonite	Hydrous iron ores containing 8–12% water	3.4–4.0	2240 (140)
Ilmenite	$FeTiO_3$	4.72	2560 (160)
Ferrophosphorus	Fe_3P, Fe_2P, FeP	5.7–6.5	3680 (230)
Steel aggregate	Fe	7.8	4480 (280)

Table 7-3 Composition and Density of Heavyweight Aggregates

7.6 Blast-Furnace Slag Aggregate

Slow cooling of blast-furnace slag in ladles, pits, or iron molds yields a product that can be crushed and graded to obtain dense and strong particles suitable for use as concrete aggregate. The properties of the aggregate vary with the composition and rate of cooling of slag. Acid slags generally produce a denser aggregate, and basic slags tend to produce a vesicular or honeycombed structure with a lower apparent specific gravity (2 to 2.8). On the whole, the bulk density of slowly cooled slags, which typically ranges from 1120 to 1360 kg/m^3, is somewhere between normal-weight natural aggregate and structural lightweight aggregate. These aggregates are widely used for making precast concrete products such as masonry blocks, channels, and fence posts.

The presence of excessive iron sulfide in the slag may cause color and durability problems in concrete products. Under certain conditions, sulfide can be converted to sulfate which is undesirable from standpoint of potential sulfate attack on concrete. British specifications limit the content of acid-soluble SO_3, and total sulfide sulfur in slag to 0.7 and 2 percent, respectively. It should be noted that blast-furnace slags have also been used for the production of lightweight aggregate meeting ASTM C 330 or C 331 requirements. For this purpose, molten slag is treated with limited amounts of water or steam, and the product is called *expanded* or *foamed slag*.

7.7 Aggregate from Fly Ash

Fly ash consists essentially of small spherical particles of aluminosilicate glass, which is produced by combustion of pulverized coal in thermal power plants. As large quantities of the ash remain unutilized in most countries of the world, attempts have been made to use the ash for making lightweight aggregate. In a typical manufacturing process, fly ash is pelletized and then sintered in a rotary kiln, shaft kiln, or a traveling grate at temperatures in the range 1000 to 1200°C. Variations in the fineness and carbon content of fly ash are a major problem in controlling the quality of sintered fly-ash aggregate.

7.8 Aggregates from Recycled Concrete

The generation of construction and demolition waste (CDW) is increasing all over the world. In 2009 the European Union generated nearly 850 million tons of CDW, and in 2003 the United States generated almost 170 million tons of CDW. Although CDW is mainly composed of concrete, mortar, ceramic material, and stones, it may also be contaminated with gypsum, asphalt, metal, wood, paper, and plastic. This contamination needs to be reduced or eliminated during the aggregate recycling process. It is becoming more common to separate the residue during its generation; therefore, the quality of recycled aggregate is improving. The recycled aggregate is obtained by crushing and screening the mineral part of CDW using the same equipment used to process natural aggregate. Since CDW is often composed of a number of materials, for structural concrete the recycled aggregate often comes from concrete waste. Even so, recycled concrete aggregate (RCA) shows some differences when compared to natural aggregate, which is mostly due to the mortar and cement paste attached to the original aggregate

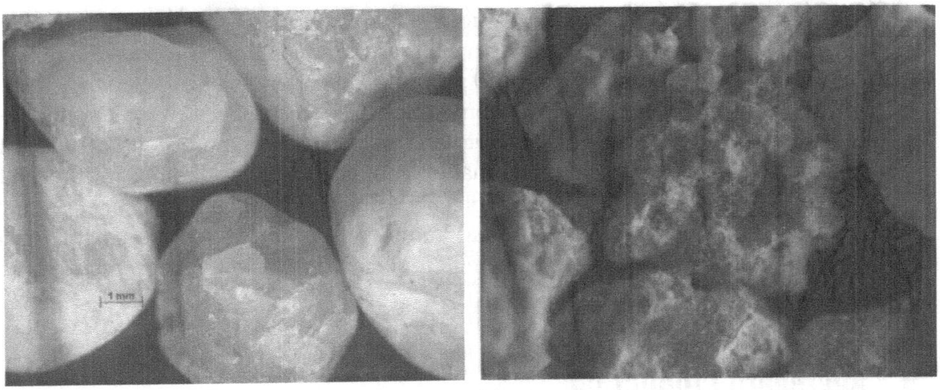

(a) (b)

FIGURE 7-2 (*a*) Natural aggregate, (*b*) recycled aggregate. (*Photograph courtesy of Professor Monica Leite.*)

Figures 7-2a and b illustrate the shape and the surface texture of natural and recycled aggregates, respectively. Because recycled aggregate is formed mainly from a conglomerate of the natural aggregate particles and the cement paste, its surface is rougher compared to natural aggregate.

(see Fig. 7-2). This residual mortar content varies from 23 to 55 percent, depending on the size of the RCA.[1] Smaller grain sizes show higher residual mortar content. Recycled concrete aggregate has lower density and higher porosity, higher water absorption rate, higher loss percentage when subjected to the Los Angeles abrasion test, and a rougher surface compared to natural coarse aggregate. These characteristics of RCA affect directly the properties of the recycled concrete. Table 7-4 summarizes the differences in properties between conventional and recycled concrete.

Property	Difference
Compressive strength	Decrease up to 25%
Splitting and flexural tensile strength	Decrease up to 10%
Modulus of elasticity	Decrease up to 45%
Drying shrinkage	Increase up to 70%
Creep	Increase up to 50%
Water absorption	Increase up to 50%
Depth of carbonation	Similar
Freezing and thawing resistance	Decreased
Chloride penetration	Same or slightly increased
Slump	Same or slightly decreased

Marinkovi, S.B.[2] et al., *Innovative Materials and Techniques in Concrete Construction*, M.N. Fardis, ed., Springer, London, p. 379, 2012.

TABLE 7-4 Differences in Recycled Concrete Properties Compared to Concrete with Natural Aggregate, Using the Same Water-Cement Ratio and 100 Percent of Aggregate Replacement

Based on the analysis of recent studies, Vasquez and Goncalvez[3] reported that when using 20 percent or less of RCA, the mechanical properties of the concrete remains the same; DIN[4] estimates that for up to 40 percent the properties of recycled concrete shows no significant difference from the conventional concrete. Because of the need to find new sources of aggregate for concrete and to recycle large amounts of waste that is generated each year, some countries have established new regulations for production of structural recycled concrete (see Table 7-5). Spain, Portugal, and Germany are now using design codes for conventional reinforced concrete structures for structures wherein the natural coarse aggregate is replaced with a low replacement percentage of RCA.

7.9 Aggregate Production

Deposits of coarse-grained soil are a good source of *natural sand and gravel*. Since soil deposits usually contain varying quantities of silt and clay, which adversely affect the properties of both fresh and hardened concrete, these contaminants must be removed by washing or dry screening. The choice of the process between washing and dry screening of silt and clay will clearly influence the amount of deleterious substances in the aggregate; for instance, clay coatings may not be removed as efficiently by dry screening as by washing.

Generally, crushing equipment is a part of the aggregate production facilities because oversize-gravel may be crushed and blended suitably with the uncrushed material of similar size. Again, the choice of crushing equipment may determine the shape of particles. With laminated sedimentary rocks, jaw-and-impact-type crushers tend to produce flat particles. The importance of proper aggregate grading is so well recognized now that modern aggregate plants, whether they are producing sand and gravel or crushed rock, have the necessary equipment to carry out operations involving crushing, cleaning, size separation, and combining two or more fractions to meet the customer specifications. A photograph of a modern aggregate processing plant is shown in Fig. 7-3.

Synthetic lightweight aggregates such as expanded clays, shales, and slate are produced by heat treatment of suitable materials. The raw material is either crushed and sized, or ground and pelletized, and then exposed to temperatures generally of the order of 1000 to 1100°C such that a portion of the material fuses to provide a viscous melt. As a result of chemical decomposition of some of the constituents present in raw materials, the evolved gases are entrapped by the viscous melt, thus expanding the sintered mass. Generally, either carbonaceous matter or a carbonate mineral is the source of these gases; alkalies and other impurities in clay or shale are responsible for the melt formation at low temperatures. The heat treatment is carried out in a gas or an oil-fired rotary kiln similar to those used for the manufacture of portland cement. Many plants vacuum-saturate the product with moisture before delivery to the customer in order to facilitate better control on the consistency of fresh concrete.

The properties of aggregate are greatly influenced by the external coating on the aggregate particle and the internal void distribution. Modern lightweight aggregate plants crush, grind, blend, and pelletize materials to obtain a uniform distribution of fine pores, which is necessary for producing concrete products possessing high crushing strength. Stable and impervious glassy coatings tend to reduce the moisture absorption

Standard	Classification	Composition	Maximum Water Absorption Rate (%)	Oven Dry Density (kg/m³)	Maximum Strength Class (MPa)	Replacement Percentage of Coarse Recycled Aggregate		Use of Fine Recycled Aggregate
RILEM TC 121-DRG (1994)[5]	Type I	Aggregates from masonry rubble	20	≥1500	C16/20	Size > 4 mm Up to 100%		Size < 4 mm—not recommended
	Type II	Aggregates from concrete	10	≥2000	C50/60	Size > 4 mm Up to 100%		
	Type III	Mixture = 80% Natural + 20% (max. recycled) aggregate	–	≥2400	No upper limit	Size >4 mm Up to 20%		
Portugal E 471 (2006)[6]	ARB 1	≥90% aggregates from concrete	7	≥2200	C40/50	25%		Not recommended
	ARB 2	≥70% aggregates from concrete	7	≥2200	C35/45	20%		
Spain EHE (2008) – Annex 15[7]	RCA	Aggregates from concrete	7	≥2100	≤40	Size >4 mm Up to 20%		Not recommended
GermanyDIN 4226-100[4]: Recycled aggregates	Type I Concrete chippings/ concrete crushed sand	≥90% of concrete and natural aggregates according to DIN 4226-1	10	≥2000	According to DIN EN 206-1 and DIN 1045-2 - C30/37	WO	≤45	Size <2 mm—not recommended
						WF	≤35	
	Type II Construction chippings/ construction crushed sand	≥70% of concrete and natural aggregates according to DIN 4226-1	15	≥2000		WO	≤35	
						WF	≤25	

WO – Normal cured concrete that remains mostly dry during use. WF – Concrete that in use is frequently wet or wet over extended periods of time.

TABLE 7-5 Requirements for Recycled Aggregate and Structural Recycled Concrete

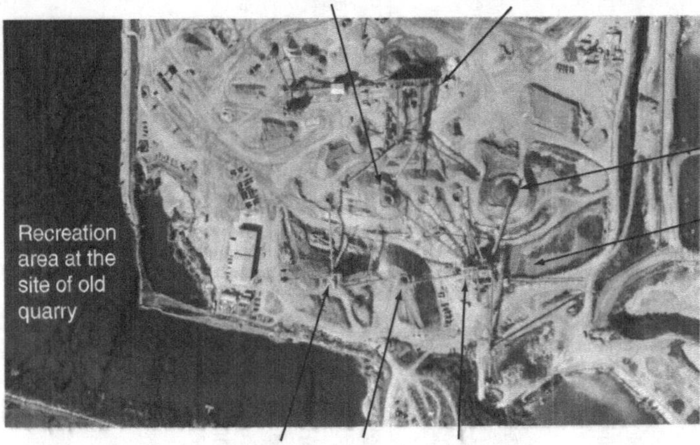

FIGURE 7-3 Aerial view of sand and gravel processing plant. (Photograph courtesy of RMC Pacific Materials, Pleasanton, CA.)

capacity of the aggregate which favorably affects the water demand of fresh concrete and dimensional stability of hardened concrete.

7.10 Aggregate Characteristics and Their Significance

A knowledge of certain aggregate characteristics (i.e., density, grading, and moisture state) is required for proportioning concrete mixtures (Chap. 9). Porosity or density, grading, shape, and surface texture determine the properties of plastic concrete mixtures. In addition to porosity, the mineralogical composition of aggregate affects its crushing strength, hardness, elastic modulus, and soundness, which in turn, influence various properties of hardened concrete containing the aggregate. From a diagram illustrating the various interrelations (Fig. 7-4) it is evident that the aggregate characteristics that are significant to concrete properties are derived from the microstructure of the parent rock, prior exposure conditions, and processing factors.

Generally, aggregate properties affect not only the concrete mixture proportions but also the behavior of fresh and hardened concrete. Due to considerable overlap between the two, it is more appropriate to divide the study of aggregate properties into three categories that are based on microstructural and processing factors:

1. *Characteristics dependent on porosity*: density, moisture absorption, strength, hardness, elastic modulus, and soundness
2. *Characteristics dependent on prior exposure and processing factors*: particle size, shape, and texture
3. *Characteristics dependent on chemical and mineralogical composition*: strength, hardness, elastic modulus, and deleterious substances present

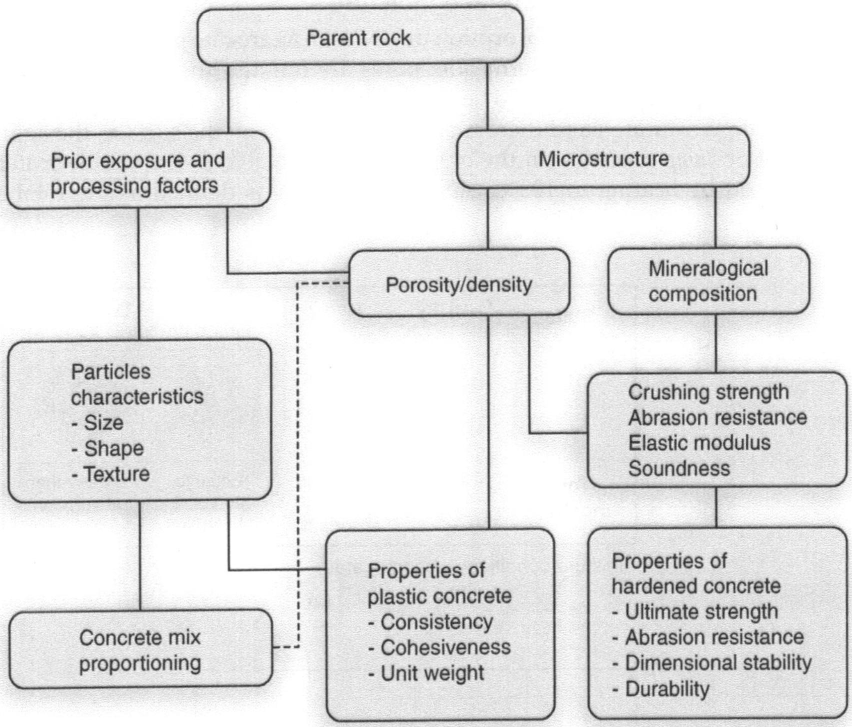

FIGURE 7-4 Diagram illustrating how microstructure, prior exposure, and processing factors determine aggregate characteristics that affect mix proportions and properties of fresh as well as hardened concrete.

7.10.1 Density and Apparent Specific Gravity

For the purpose of proportioning concrete mixtures it is not necessary to determine the true specific gravity of an aggregate. Natural aggregates are porous; porosity values up to 2 percent are common for intrusive igneous rocks, up to 5 percent for dense sedimentary rocks, and 10 to 40 percent for very porous sandstones and limestone. For the purpose of mix proportioning, it is desired to know the space occupied by the aggregate particles, inclusive of the pores existing within the particles. Therefore, determination of the *apparent specific gravity*, which is defined as the density of the material including the internal pores, is sufficient. The apparent specific gravity for many commonly used rocks ranges between 2.6 and 2.7; typical values for granite, sandstone, and dense limestone are 2.69, 2.65, and 2.60, respectively.

For the purpose of concrete mixture proportioning, in addition to apparent specific gravity, data are usually needed on *bulk density* which is defined as the mass of the aggregate fragments that would fill a unit volume. The phenomenon of bulk density arises because it is not possible to pack aggregate fragments together such that there is no void space. The term *bulk* is used since the volume is occupied by both *aggregates* and *voids*. The approximate bulk density of aggregates commonly used in normal-weight concrete ranges from 1300 to 1750 kg/m^3 (80 to 110 lb/ft^3).

7.10.2 Absorption and Surface Moisture

Various states of moisture absorption in which an aggregate particle can exist are shown in Fig. 7-5*a*. When all the permeable pores are full and there is no water film on the surface, the aggregate is said to be in the *saturated-surface dry condition* (SSD); when the aggregate is saturated and there is also free moisture on the surface, the aggregate is in the *wet* or *damp condition*. In the *oven-dry condition*, all the evaporable water has been driven off by heating to 100°C. *Absorption capacity* is defined as the total amount of

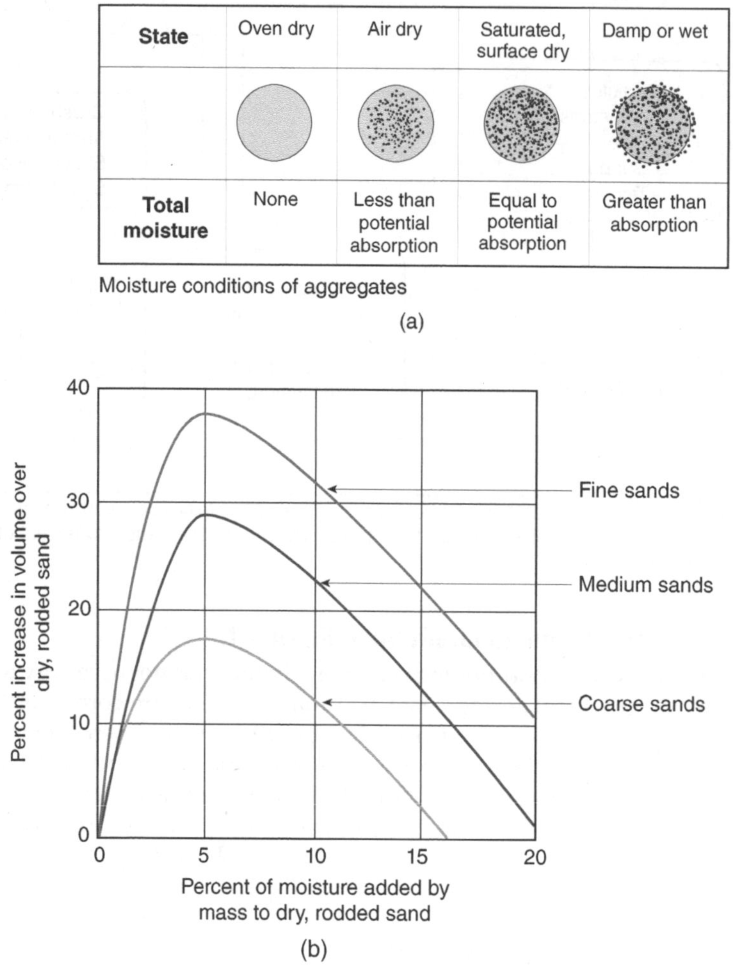

FIGURE 7-5 (*a*) Aggregate in various moisture states; (*b*) bulking due to moisture in fine aggregate. (From *Design and Control of Concrete Mixtures, 13th ed.*, Portland Cement Association, Skokie, IL., pp. 36–37,1988.)

The figure at the top illustrates how the concept of saturated-surface dry (SSD) condition is useful in determining potential absorption or free moisture in aggregate. The figure at the bottom shows that surface moisture on fine aggregate can cause considerable bulking that varies with the amount of moisture present and the aggregate grading.

moisture required to bring an aggregate from the oven-dry to the SSD condition. *Effective absorption* is defined as the amount of moisture required to bring an aggregate from the air-dry to the SSD condition. The amount of water in excess of the water required for the SSD condition is referred to as the *surface moisture*. The absorption capacity, effective absorption, and surface moisture data are invariably needed for correcting the batch water and aggregate proportions in concrete mixtures made from stock materials (Chap. 9). As a first approximation, the absorption capacity of an aggregate, which is easily determined, can be used as a rough measure of porosity and strength.

Normally, the moisture correction values for intrusive igneous rocks and for dense sedimentary rocks are very low, but they are high in the case of porous sedimentary rocks, lightweight aggregates, and wet sands. For example, typically, the effective absorption values of trap rock, porous sandstone, and expanded shale aggregates are ½, 5, and 10 percent, respectively.

Damp sands may suffer from a phenomenon known as *bulking*. Depending on the amount of moisture and aggregate grading, a considerable increase in the bulk volume of the sand can occur (Fig. 7-5*b*). Because the surface tension of water keeps the particles apart, fine sands show more bulking. Since most sands are delivered at the job site in a damp condition, wide variations can occur in the batch quantities if batching is done by volume. For this reason, proportioning of concrete mixture by mass has become the standard practice in most countries.

7.10.3 Crushing Strength, Abrasion Resistance, and Elastic Modulus

The crushing strength, abrasion resistance, and elastic modulus of aggregate are interrelated properties that are greatly influenced by porosity. Aggregates from natural sources that are commonly used for making normal-weight concrete, are generally dense and strong; therefore, they are seldom a limiting factor to strength and elastic properties of concrete. Typical values of the crushing strength and dynamic elastic modulus for most granite, basalt, trap rock, flint, quartzitic sandstone, and dense limestone aggregates are in the range 210 to 310 MPa (30 to 45 $\times 10^3$ psi) and 70 to 90 GPa (10 to 13 $\times 10^6$ psi), respectively. With regard to sedimentary rocks, the porosity varies over a wide range, as will the crushing strength and related characteristics. In one investigation involving 241 limestones and 79 sandstones, while the maximum crushing strengths for each rock type were of the order of 240 MPa (35,000 psi), some limestones and sandstones showed as low as 96 MPa (14,000 psi) and 48 MPa (7000 psi) crushing strength, respectively.

7.10.4 Soundness

An aggregate is considered *unsound* when the volume changes in aggregate induced by weather (e.g., alternate cycles of wetting and drying, or freezing and thawing), result in the deterioration of concrete. Unsoundness is shown generally by rocks having a certain characteristic pore structure. Concretes containing some cherts, shales, limestones, and sandstones have been found susceptible to damage by frost action or by salt crystallization within the aggregate particle. Although high moisture absorption is often used as an index for unsoundness, many aggregates such as pumice and expanded clays can absorb large amounts of water but remain sound. Unsoundness is therefore related to pore size distribution rather than to the total porosity of aggregate. A pore size distribution that allows the aggregate particles to get saturated on wetting (or thawing in the case of frost attack), but prevent easy drainage on drying (or freezing) is capable of causing high hydraulic pressure within the aggregate particles. Soundness of aggregate to weathering action is determined by ASTM Method C 88, which describes a standard procedure for directly determining the resistance of aggregate to disintegration on exposure to five wet-dry cycles; a saturated sodium or magnesium sulfate solution is used for the wetting cycle.

In the case of frost attack, in addition to pore size distribution and degree of saturation, as described in Chap. 5, there is a critical aggregate size below which high internal stresses capable of cracking the particle will not occur. For most aggregate, this critical aggregate size is greater than the normal size of coarse aggregates used in the construction practice; however, for some poorly consolidated rocks (sandstones, limestones, cherts, shales), this size may be in the 12 to 25 mm range.

7.10.5 Size and Grading

Grading is the distribution of particles of a granular material among various size ranges, usually expressed in terms of cumulative percentage larger or smaller than each of a series of sizes of sieve openings, or the percentage between certain range of sieve openings. ASTM C 33 grading requirements for coarse and fine aggregates are shown in Tables 7-6 and 7-7, respectively.

There are several *reasons for specifying grading limits and maximum aggregate size*; the most important is their influence on workability and cost. For example, very coarse sands produce harsh and unworkable concrete mixtures, and very fine sands increase the water requirement (therefore, the cement requirement for a given water-cement ratio) and are uneconomical. Aggregates that do not have a large deficiency or excess of any particular size produce the most workable and economical concrete mixtures.

The maximum size of aggregate is conventionally designated by the sieve size on which 15 percent or more particles are retained. In general, the larger the maximum aggregate size, the smaller will be the surface area per unit volume which has to be covered by the cement paste of a given water-cement ratio. Since the price of cement may be 10 to 15 times as much as the price of aggregate, any action that saves cement without reducing the strength and workability of concrete can result in significant economic benefit. In addition to cost economy, there are other factors that govern the choice of maximum aggregate size for a concrete mixture. According to one rule of thumb used in the construction industry, the maximum aggregate size should not be larger than one-fifth of the narrowest dimension of the form in which the concrete is to be placed; also, it should not be larger than three-fourths of the maximum clear distance between the reinforcing bars. As large particles tend to produce more microcracks in the interfacial transition zone between the coarse aggregate and cement paste, with high-strength concrete mixtures the maximum aggregate size is generally limited to 19 mm.

Similarly, aggregate grading has also considerable effect on the cement paste requirement of a concrete mixture. The effect of aggregate packing with particles of varying sizes is demonstrated in Fig. 7-6*a*. One beaker is filled with 25-mm particles of relatively uniform size and shape; a second beaker is filled with a mixture of 25- and 9-mm particles. Below each beaker is a graduated cylinder holding the amount of water required to fill the voids in beaker above. It is evident that by combining two different aggregate sizes, the void content is decreased. If particles of several more sizes smaller

Size Number	Nominal Size (sieves with square openings)	Amounts Finer Than Laboratory Sieve (square openings) (wt %)												
		100 mm (4 in.)	90 mm (3½ in.)	75 mm (3 in.)	63 mm (2½ in.)	50 mm (2 in.)	37.5 mm (1½ in.)	25.0 mm (1 in.)	19.0 mm (¾ in.)	12.5 mm (½ in.)	9.5 mm (3/8 in.)	4.75 mm (No. 4)	2.36 mm (No. 8)	1.18 mm (No. 16)
1	90–37.5 mm (3½–1½ in.)	100	90–100	—	25–60	—	0–15	—	0–5	—	—	—	—	—
2	63–37.5 mm (2½–1½ in.)	—	—	100	90–100	35–70	0–15	—	0–5	—	—	—	—	—
3	50–25.0 mm (2–1 in.)	—	—	—	100	90–100	35–70	0–15	—	0–5	—	—	—	—
357	50–4.75 mm (2 in.–No. 4)	—	—	—	100	95–100	—	35–70	—	10–30	—	0–5	—	—
4	37.5–19.0 mm (1½–¾ in.)	—	—	—	—	100	90–100	20–25	0–15	—	0–5	—	—	—
467	37.5–4.75 mm (1½ in.–No. 4)	—	—	—	—	100	95–100	—	35–70	—	10–30	0–5	—	—
5	25.0–12.5 mm (1–½ in.)	—	—	—	—	—	100	90–100	20–55	0–10	0–5	—	—	—
56	25.0–9.5 mm (1–3/8 in.)	—	—	—	—	—	100	90–100	40–85	10–40	0–15	0–5	—	—
57	25.0–4.75 mm (1 in.–No. 4)	—	—	—	—	—	100	95–100	—	25–60	—	0–10	0–5	—
6	19.0–9.5 mm (¾–3/8 in.)	—	—	—	—	—	—	100	90–100	20–25	0–15	0–5	—	—
67	19.0–4.75 mm (¾ in.–No. 4)	—	—	—	—	—	—	100	90–100	—	20–55	0–10	0–5	—
7	12.5–4.75 mm (½ in.–No. 4)	—	—	—	—	—	—	—	100	90–100	40–70	0–15	0–5	—
8	9.5–2.36 mm (1-3/8 in.–No. 8)	—	—	—	—	—	—	—	—	100	85–100	10–30	0–10	0–5

Source: Reprinted, with permission, from the 1991 *Annual Book of ASTM Standards*, Section 4, Vol. 04.02. Copyright, ASTM 1916 Race Street , Philadelphia, PA.

TABLE 7-6 Grading Requirements for Coarse Aggregates

Sieve (Specification E 11)	Percent Passing
9.5 mm (3/8 in.)	100
4.75-mm (No. 4)	95–100
2.36-mm (No. 8)	80–100
1.18-mm (No. 16)	50–85
600-μm (No. 30)	25–60
300-μm (No. 50)	10–30
150-μm (No. 100)	2–10

Source: Reprinted, with permission, from *Annual Book of ASTM Standards*, Section 4, Vol. 04.02, ASTM, Philadelphia, PA, 1991.

TABLE 7-7 Grading Requirements for Fine Aggregates

than 9 mm are added to the 25-mm and 9-mm aggregate mixture, a further reduction in the void content will result. In practice, low void contents are achieved by using smoothly graded coarse aggregates with suitable proportions of graded sand (Fig. 7-6*b*). The data show test results in which as low as 21 percent void content was obtained when 40 percent sand was mixed with a well-graded, 4.75 to 37 mm (No. 4 to 1½) gravel. From standpoint of obtaining high workability of concrete mixtures, it is well known that the smallest percentage of voids corresponding to the greatest dry-rodded density with given aggregates is not the most satisfactory; the volume of cement paste to fill the voids should be somewhat more.

In practice, an empirical factor called the fineness modulus is often used as an index of the fineness of aggregate. The *fineness modulus* is computed from screen analysis data by adding the cumulative percentages of aggregate retained on each of a specified series of sieves, and dividing the sum by 100. The sieves used for determining the fineness modulus are: No. 100 (150 μm), No. 50 (300 μm), No. 30 (600 μm), No. 16 (1.18 mm), No. 8 (2.36 mm), No. 4 (4.75 mm), 3/8 in. (9.5 mm), ¾ in. (19 mm), 1½ in. (37.5 mm), and larger, increasing in the ratio of 2 to 1. Examples of the method for determining the fineness modulus of fine aggregates from three different sources are shown by the tabulated data in Fig. 7-7, together with a typical grading curve. It may be noted that the higher the fineness modulus, the coarser the aggregate.

7.10.6 Shape and Surface Texture

The shape and surface texture of aggregate particles influence the properties of fresh concrete more than hardened concrete. Compared to smooth and rounded particles, rough-textured, angular, and elongated particles require more cement paste to produce workable concrete mixtures, thus increasing the cost.

Shape refers to geometrical characteristics such as round, angular, elongated, or flaky. Particles shaped by attrition tend to become *rounded* by losing edges and corners. Wind-blown sands, as well as sand and gravel from seashore or riverbeds generally have a well-rounded shape. Crushed intrusive rocks possess well-defined edges and corners and are called *angular*. They generally produce equidimensional particles. Laminated limestones, sandstones, and shale tend to produce elongated and flaky fragments,

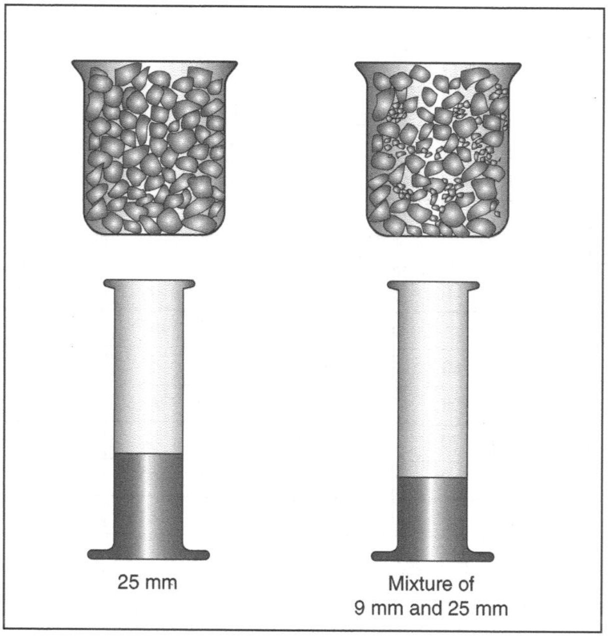

(a)

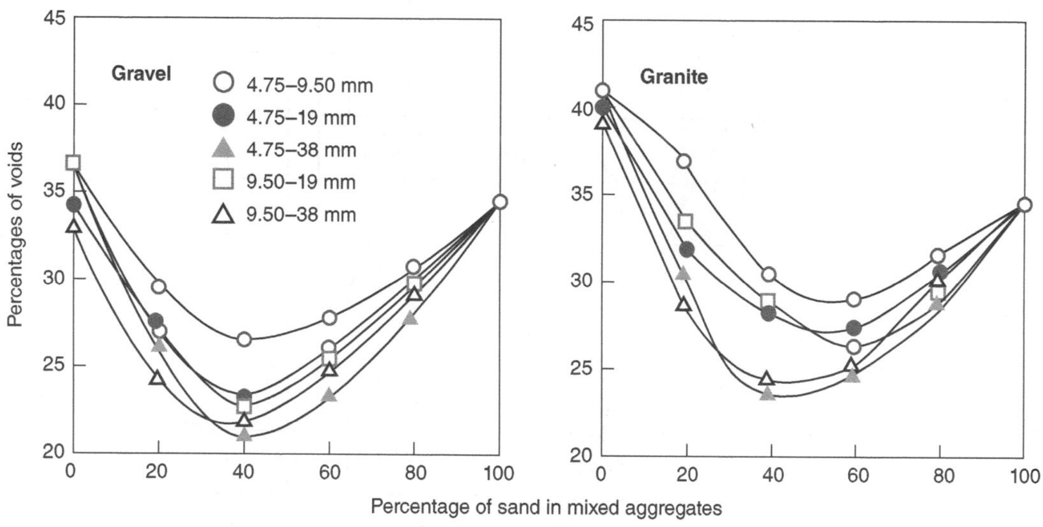

(b)

FIGURE 7-6 Reduction in the volume of voids on mixing fine and coarse aggregates. [(*a*) From *Design and Control of Concrete Mixtures, 13th ed.*, Portland Cement Association, Skokie, IL., p. 33, 1988; (*b*) from Walker, S., Circular 8, National Sand and Gravel Association, 1930.]

Date	September 7, 2013			September 7, 2013			September 7, 2013		
Source	A (fine sand for blending)			B (Concrete sand)			C (Concrete sand)		
Sample wt.	455 g			450 g			456 g		
Sieve size	Weight retained	Retained, %		Weight retained	Retained, %		Weight retained	Retained, %	
		Individual	Cumulative		Individual	Cumulative		Individual	Cumulative
No. 4	0	0	0	0	0	0	0	0	0
8	0	0	0	40.5	9.1	9	42.1	9.2	9
16	2.8	0.6	1	86.0	19.1	28	137.0	30.2	39
30	10.1	2.2	3	94.5	21.0	49	112.1	24.7	64
50	259.2	56.9	60	135.9	30.2	79	84.9	18.7	83
100	173.1	38.0	98	77.0	17.1	96	48.8	10.8	94
200	5.6	1.2	99	13.5	3.0	99	29.1	6.4	100
Pan	3.3	0.7	100	2.1	0.5	100	1.0	0.2	100
Total	454.1	F.M.	1.62	449.5	F.M.	2.61	455.0	F.M.	2.89

(a)

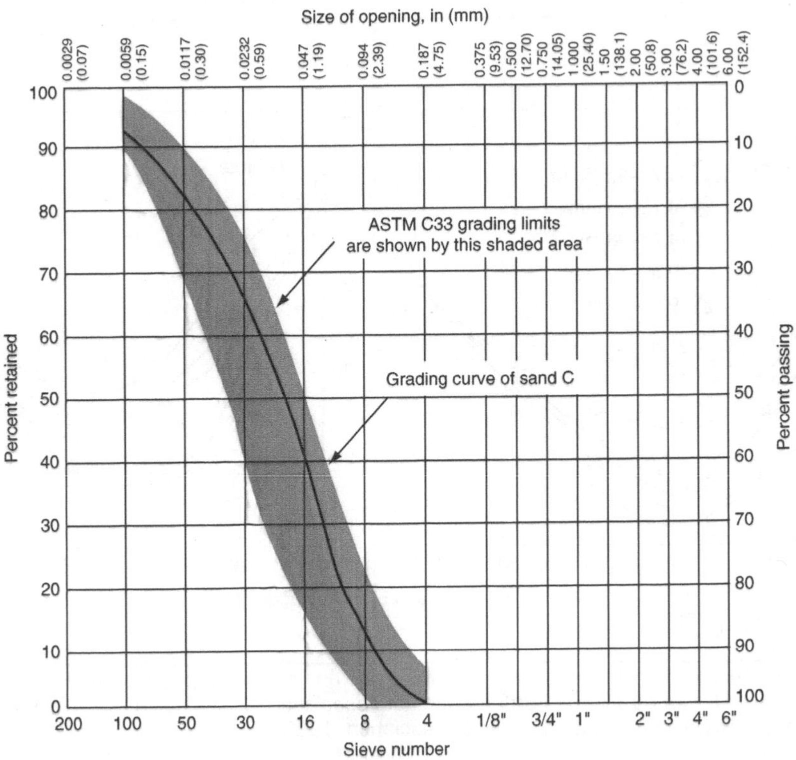

(b)

Figure 7-7 (*a*) Determination of the fineness modulus from sieve analysis data; (*b*) typical grading curve for sand with ASTM C 33 grading limits.

especially when jaw crushers are used for crushing. Those particles in which thickness is small relative to two other dimensions are referred to as *flat* or *flaky*, while those which are considerably bigger in length than the other two dimensions are called *elongated*. Another term sometimes used to describe the shape of coarse aggregate is *sphericity* which is defined as a ratio of surface area to volume. Spherical or well-rounded particles have low sphericity, but elongated and flaky particles possess high sphericity.

Photographs of particles of various shapes are shown in Fig. 2-3. Aggregates should be relatively free of flat and elongated particles. Elongated, blade-shaped aggregate particles should be avoided or limited to a maximum of 15 percent by mass of the total aggregate. This requirement is important not only for coarse aggregate but also for manufactured sands (made by crushing stone), containing elongated grains, which produce very harsh concrete.

The classification of *surface texture*, defined as the degree to which the aggregate surface is smooth or rough, is based on visual judgment. Surface texture of aggregate depends on the hardness, grain size, and porosity of the parent rock and its subsequent exposure to forces of attrition. Obsidian, flint, and dense slags show a smooth, glassy texture. Sand, gravels, and chert are smooth in their natural state. Crushed rocks such as granite, basalt, and limestone show a rough texture. Pumice, expanded slag, and sintered fly ash show a honeycombed texture with visible pores.

There is some evidence that, during early age, the strength of concrete, particularly the flexural strength, can be affected by the aggregate texture; a rougher texture seems to help the formation of a stronger physical bond between the cement paste and aggregate. At a later age, with a stronger chemical bond between the paste and the aggregate, this effect may not be so important.

7.10.7 Deleterious Substances

Deleterious substances are those that are present as minor constituents of either fine or coarse aggregate but are capable of adversely affecting the workability, setting and hardening, and durability characteristics of concrete. A list of harmful substances, their possible effects on concrete, and ASTM C 33 Specification limits on the maximum permissible amounts are shown in Table 7-8.

In addition to the materials listed in Table 7-8, there are other substances that may have deleterious effect involving chemical reactions in concrete. For both fine and coarse aggregates, ASTM C 33 requires that "aggregate for use in concrete that will be subject to wetting, extended exposure to humid atmosphere, or contact with moist ground shall not contain any materials that are deleteriously reactive with the alkalies in the cement in an amount sufficient to cause excessive expansion except that if such materials are present in injurious amounts, the aggregate may be used with a cement containing less than 0.6 percent alkalies or with the addition of a material that has been shown to prevent harmful expansion due to the alkali-aggregate reaction." A description of the alkali-aggregate reaction, including a list of the reactive aggregates, is given in Chap. 5.

Iron sulfides, especially marcasite, are present as inclusions in certain aggregates and have been found to cause an expansive reaction. In the lime-saturated environment of portland cement concrete, reactive iron sulfides can oxidize to form ferrous sulfate, which causes sulfate attack on concrete and corrosion of the embedded steel. Aggregates contaminated with gypsum or other soluble sulfates, such as magnesium, sodium, or potassium sulfate, also promote sulfate attack.

Substance	Possible Harmful Effects on Concrete	Maximum Permitted (wt %) Fine Aggregate	Coarse Aggregate*
Material finer than 75-μm (No. 200) sieve	Affect workability; increase water requirement		
Concrete subject to abrasion		3†	1
All other concrete		5†	
Clay lumps and friable particles	Affect workability and abrasion resistance	3	5
Coal and lignite	Affect durability; cause staining		
Where surface appearance of concrete is important		0.5	0.5
All other concrete		1.0	
Chert (less than 2.4 specific gravity)	Affect durability	—	5

*ASTM C33 limits for deleterious substances in coarse aggregate vary with the conditions of exposure and type of concrete structure. The values shown here are for outdoor structures exposed to moderate weather conditions.

†In the case of manufactured sand, if the material finer than 75-μm size consists of dust of fracture, essentially free of clay or shale, these limits may be increased to 5 and 7 percent, respectively.

Source: Reprint, with permission, from the 1991 *Annual Book of ASTM Standards*, Section 4, Vol. 04.02. Copyright ASTM, 1916 Race Street, Philadelphia, PA.

TABLE 7-8 Limits for Deleterious Substances in Concrete Aggregates

Cases of failure of concrete to set were reported[8] from two block-making plants in southern Ireland. The problem was attributed to the presence of significant amounts of lead and zinc (mostly in the form of sulfides) in the calcite aggregate. Those blocks had failed to set which contained either 0.11 percent or more lead compound or 0.15 percent or more zinc compound by mass of concrete. Soluble lead or zinc salts are such powerful retarders of portland cement hydration that experimental concretes made with samples of contaminated aggregate failed to develop any strength after 3 days of curing. It should be noted that concrete setting and hardening problems can also be caused by organic impurities in aggregate, such as decayed vegetable matter that may be present in the form of organic loam or humus.

Test Your Knowledge

7.1 The aggregate in concrete is looked down upon as an "inert filler." Explain why this viewpoint is erroneous.

7.2 What is the distinction between the terms *rocks* and *minerals*? Write a short report on the influence of rock-forming process on characteristics of aggregates derived from the rock. In particular, explain why:

(*a*) Basalts, which are generally fine-grained or glassy, are not alkali-reactive.
(*b*) Limestones tend to form flat aggregate particles.
(*c*) Pumice is useful for the production of lightweight aggregate.

7.3 What do you know about the following rocks and minerals: dolomite, graywacke, flint, opal, plagioclase, smectites, and marcasite?

7.4 Give typical ranges of aggregate unit weights for making structural lightweight, normal-weight, or heavyweight concretes. What types of natural and synthetic aggregates are used for making lightweight masonry blocks and insulating concrete; what types of natural minerals are useful for producing heavyweight aggregates?

7.5 You are the civil engineer in charge of rehabilitating some old concrete pavements in your area. In a brief note to your superiors, discuss the equipment needed, deleterious constituents to be avoided, and the cost economy of using the crushed concrete from old pavements as a source of aggregates for the construction of new pavements.

7.6 How are aggregates made from expandable clays, fly ash, and blast-furnace slag? What are some of the interesting characteristics of the products?

7.7 In concrete technology, what distinction is made between the terms *specific gravity* and *bulk density*? With the help of suitable sketches, explain the following terms and discuss their significance: absorption capacity, saturated-surface-dry condition, damp condition.

7.8 What is the cause of the *bulking* phenomenon and what role does it play in concrete manufacturing practice?

7.9 List any three characteristics of concrete aggregate and discuss their influence on both the properties of fresh concrete and hardened concrete.

7.10 Briefly discuss the following propositions:

(*a*) Specific gravity and absorption capacity are good indicators of aggregate quality.

(*b*) The crushing strength of an aggregate can have a large influence on the compressive strength of concrete.

(*c*) Using natural mineral aggregates, the unit weight of structural quality concrete can be varied between 1600 and 3200 kg/m^3 (2700 – 5400 lb/yd^3).

(*d*) Pores in aggregate that are smaller than 4 μm can become *critically saturated.*

7.11 (*a*) If the dry-rodded unit weight of an aggregate is 1680 kg/m^3 (105 lb/ft^3) and its specific gravity is 2.65, determine the void content. (*b*) A sample of sand weighs 500 g and 480 g in "as received" and oven-dried condition, respectively. If the absorption capacity of the sand is 2 percent, calculate the percentage of free moisture.

7.12 Define the terms *grading* and *maximum aggregate size*, as used in concrete technology. What considerations control the choice of the maximum aggregate size of aggregate in a concrete? Discuss the reason why grading limits are specified.

7.13 Assuming that the workability of concrete is of no consequence, would sand be necessary in concrete mixtures? What is the significance of *fineness modulus*? Calculate the fineness modulus of sand with the following sieve analysis:

Weight retained on No. 8 sieve, g = 30
Weight retained on No. 16 sieve, g = 70
Weight retained on No. 30 sieve, g = 125
Weight retained on No. 50 sieve, g = 135
Weight retained on No. 100 sieve, g = 120
Weight retained on No. 200 sieve, g = 20
Is this sand suitable for making concrete?

7.14 With the help of appropriate sketches, explain the terms *sphericity*, *flat particles*, and *angular particles*. Explain how the surface texture of fine aggregate may influence the properties of concrete.

7.15 When present as deleterious substances in concrete aggregates, how can the following materials affect properties of concrete: clay lumps, silt, zinc sulfide, gypsum, humus?

References

1. Sanchez de Juan, M., and Gutierrez, P.A. Influence of recycled aggregate quality on concrete properties. In: Vázquez, E. et al. (eds.). *The Use of Recycled Materials in Building and Structures*, Barcelona, Nov. 8–11, 2004.
2. Marinkovi, S.B. et al., *Innovative Materials and Techniques in Concrete Construction*, M.N. Fardis, ed., Springer, London, p. 379, 2012.
3. Vasquez, E., and Goncalvez, A. Recycled aggregate in concrete. In: Hendriks, C.F., Janssen, G.M.T., Vasquéz, E. *Use of Recycled Materials*. RILEM TC 198-URM: Final Report, p. 74, 2005.
4. DIN. Aggregates for mortar and concrete; Part 100: Recycled aggregates. DIN4226–100, 2002.
5. RILEM TC 121-DRG. RILEM recommendation: specifications for concrete with recycled aggregates. *Mater Struct.* 27, 557–559, 1994.
6. LNEC E 471. Guiaparaautilização de agregadosrecicladosgrossosembetões de liganteshidráulicos. LaboratórioNacional de Engenharia Civil, Portugal, 2006.
7. Instruccion de hormigon structural (EHE-08), Spain, 2008.
8. Concrete Construction Publications, *Concrete Construction*, Vol. 22, No. 4, 1977.

Suggestions for Further Study

ACI Committee 221, Report 221R Guide for Use of Normal-weight Aggregates in Concrete, 221.1R: Report on Alkali-Aggregate Reactivity *ACI Manual of Concrete Practice, Part 1*, 2012.

Alexander, M., and Mindess, S., *Aggregates in Concrete*, Taylor & Francis, New York, 2005.

ASTM Volume 04.02 Concrete and Aggregates, ASTM International, 2013.

Brito, J., and N. Saikia, *Recycled Aggregate in Concrete: Use of Industrial, Construction and Demolition Waste*, Springer-Verlag, London, 2013.

Dolar-Mantuani, L., *Handbook of Concrete Aggregates*, Noyes Publications, Park Ridge, NJ, 1983.

Lamond, J., and J. Peilert, *Significance of Tests and Properties of Concrete and Concrete-Making Materials*, ASTM International, 2006.

Newman, J., and Choo, B.S., eds., *Advanced Concrete Technology: Constituent Materials*, Elsevier, 2003.

Orchard, D.F., *Properties and Testing of Aggregates, Concrete Technology*, Vol. 3, Wiley (Halstead Press), New York, 1976.

Popovics, S., *Concrete Materials: Properties, Specifications, and Testing*, Noyes Publications, Park Ridge, NJ, 1992.

CHAPTER 8
Admixtures

Preview

The recognition that properties of concrete, in both the fresh and hardened states, can be modified by adding certain materials to concrete mixtures has been responsible for the large growth of the concrete admixtures industry during the last 60 years. Hundreds of products are being marketed today, and in developed countries 80 to 90 percent of all concrete produced contains one or more admixtures. Therefore, construction engineers should have some knowledge of the advantages and limitations of commonly used admixtures.

Admixtures vary widely in composition, from surfactants and soluble salts to polymers and insoluble minerals. Generally, they are used in concrete to improve workability, accelerate or retard setting time, control strength development, and enhance the durability to frost action, thermal cracking, alkali-aggregate expansion, sulfate attack, and resistance to corrosion of the reinforcement. Important classes of concrete admixtures, their physical-chemical characteristics, mechanisms of action, applications, and side effects are described in this chapter.

8.1 Significance

ASTM C 125 defines an *admixture* as a material other than water, aggregates, hydraulic cements, and fiber reinforcement that is used as an ingredient of concrete or mortar and added to the batch immediately before or during mixing. ACI Committee 212[1] lists 20 important purposes for which admixtures are used, for example, to increase the plasticity of concrete without increasing the water content, to reduce bleeding and segregation, to retard or accelerate the time of set, to accelerate the rates of strength development at early ages, to reduce the rate of heat evolution, and to increase the durability of concrete to specific exposure conditions. The realization that important properties of concrete, can be modified by application of admixtures, gave such an impetus to the admixture industry that within 20 years after the beginning of development of the industry in the 1940s, nearly 275 different products were marketed in England and 340 in Germany.[2]

8.2 Nomenclature, Specifications, and Classifications

Admixtures vary widely in chemical composition. Many perform more than one function, therefore, it is difficult to classify them simply according to their functions. *Chemical admixtures* can be divided into two types. Some chemicals begin to act on the cement-water system instantaneously by influencing the surface tension of water and by adsorbing on the surface of cement particles; others break up into their ionic constituents and affect the chemical reactions between cement compounds and water, from several minutes to several hours after the addition. Finely ground materials which are used to supplement or replace part of the portland cement either from natural sources or generated as by-products of some industries, are called *mineral admixtures*. The physical effect of the presence of these mineral admixtures on the rheological behavior of fresh concrete becomes immediately apparent, but it can take several days to several months for the chemical effects on hydration to manifest.

The soluble salts and polymers, both surface-active agents and others, are added to concrete in very small amounts (often less than 2 percent by weight of cementitious material) for purposes such as entrainment of air, plasticization of fresh concrete mixtures, or control of setting time. By plasticizing a concrete mixture, it is possible either to increase the consistency without increasing the water content, or to reduce the water content while maintaining a given consistency. Therefore, in the United States, the *plasticizing chemicals* are called *water-reducing admixtures*.

The ASTM has separate specifications for air-entraining chemicals and for water-reducing and/or set-controlling chemicals. ASTM C 260, *Standard Specification for Air-Entraining Admixtures for Concrete*, sets limits on the effect that a given admixture under test may exert on bleeding, time of set, compressive and flexural strengths, drying shrinkage, and freeze-thaw resistance of concrete when compared to a reference air-entraining admixture. ASTM C 494, *Standard Specification for Chemical Admixtures for Concrete*, divides the water-reducing and/or set-controlling chemicals into the following eight types: *Type A*, water-reducing; *Type B*, retarding; *Type C*, accelerating; *Type D*, water-reducing and retarding; *Type E*, water-reducing and accelerating; *Type F*, high-range water-reducing; *Type G*, high-range water-reducing and retarding; and *Type S*, specific performance admixtures. The distinction between the normal (Types, A, D, and E) and the high-range water-reducing agents, also known as superplasticizers, is that, compared to the reference concrete mixture of a given consistency, the former should be able to reduce at least 5 percent of mixing water and the latter at least 12 percent. The specification also sets limits on the time of set, compressive and flexural strengths, and drying shrinkage. *Specific performance admixture (Type S)* provides a desired performance characteristic other than reducing water content, or changing the time of setting of concrete. Shrinkage reduction, mitigation of alkali-silica reaction, and viscosity modification are examples of specific performance characteristics.

Mineral admixtures are usually added to concrete in larger amounts than is the case for chemical admixtures. Besides cost reduction, environmental benefit of recycling waste products, and enhancement of workability of fresh concrete, they can improve the resistance of concrete to thermal cracking, alkali-aggregate expansion, and chloride or sulfate attack. Natural pozzolanic materials and industrial by-products, such as fly ash and slag, are commonly used mineral admixtures. Due to numerous benefits associated with their use, they are also known as *supplementary cementing materials*. In fact, with more than 50 percent fly ash or iron blast-furnace slag present as a cementitious component in some concretes, it may be more appropriate to use the term *complementary cementing materials*.

ASTM has separate classifications covering natural pozzolans, fly ashes, silica fume, and iron blast-furnace slag. ASTM C 618, *Standard Specification for Fly Ash and Raw or Calcined Natural Pozzolan for Use as a Mineral Admixture in Portland Cement Concrete*, covers the following three classes of mineral admixtures: *Class N*, raw or calcined pozzolans such as diatomaceous earths, opaline cherts and shales, tuffs, and volcanic ashes or pumicite, and calcined materials such as clays and shales; *Class F*, fly ash normally produced from the burning of anthracite or bituminous coal; *Class C*, fly ash which is normally produced from the burning of lignite or subbituminous coal (in addition to being pozzolanic, this fly ash is also cementitious due to the relatively high Ca content in relation to Class F). The most important difference between *Class C* and *Class F* fly ashes lies in the calcium content, which is generally much higher in Class C than Class F fly ashes. The ASTM C 618 specification sets limits on fineness, water requirement, pozzolanic activity, soundness, and chemical constituents, which are important factors in determining the value of a fly ash in concrete. Among others, a serious objection to ASTM C 618 are the arbitrary chemical requirements that have no proven relationship with performance of the mineral admixture in concrete. Silica fume, covered by ASTM 1240-12, is a very fine pozzolanic material composed mostly of amorphous silica. This is a by-product in the production of elemental silicon or ferrosilicon alloys.

ASTM C 989, *Standard Specification for Ground Iron Blast-Furnace Slag for Use in Concrete and Mortars*, is refreshingly free from cumbersome chemical and physical requirements. Unlike ASTM C 618, this standard takes the approach that when a material meets the performance specification, there should be no further need for prescriptive specifications. At the heart of ASTM C 989 is a slag activity test with portland cement that is intended to grade slags based on their contribution to the strength of a slag-portland cement mixture. According to the test method, slag activity is evaluated by determining the compressive strength of both portland cement mortars and corresponding mortars made with the same amount of a 50-50 weight combination of slag and portland cement. A strength index is calculated as the percentage strength of the slag-portland cement mortar relative to the strength of the reference portland cement mortar. The specification covers three strength grades of finely ground granulated iron blast-furnace slag for use as cementitious material: Grades 80, 100, and 120 correspond to a minimum strength index of 75, 95, and 115 percent in the 28-day cured mortars, respectively. Grades 100 and 120 also have a 7-day strength index requirement.

As no single classification can cover all concrete admixtures, to discuss their composition, mechanism of action, and applications, it is convenient to group these materials into the following three categories: (1) surface-active chemicals, (2) set-controlling chemicals, and (3) mineral admixtures.

8.3 Surface-Active Chemicals

8.3.1 Nomenclature and Chemical Composition

Surface-active chemicals, also known as *surfactants*, cover admixtures generally used for air entrainment or reduction of water in concrete mixtures. An *air-entraining admixture* is defined as a material that is used as an ingredient of concrete for the purpose of entraining air. A *water-reducing admixture* is an admixture that reduces the quantity of mixing water required to produce concrete of a given consistency.

Surface-active admixtures consist essentially of long-chain organic molecules, one end of which is *hydrophilic* (water-attracting) and the other *hydrophobic* (water-repelling). The hydrophilic end contains one or more polar groups, such as $-COO^-$, $-SO_3^-$, or $-NH_3^+$. In concrete technology, mostly anionic admixtures are used, either with a nonpolar chain or with a chain containing some polar groups. The former serves as air-entraining, and the latter as water-reducing, admixtures. As explained below, the surfactants become adsorbed at the air-water and cement-water interfaces with a molecular orientation that determines whether the predominant effect is entrainment of air or plasticization of the cement-water system.

Surfactants used as air-entraining admixtures generally consist of salts of wood resins, proteinaceous materials and petroleum acids, and some synthetic detergents. Surfactants used as plasticizing admixtures usually are salts and modifications, and derivatives of lignosulfonic acid, hydroxylated carboxylic acids, and polysaccharides or any combinations of the foregoing three, with or without other subsidiary constituents. The first generation of superplasticizers or high range water-reducing admixtures, which are also discussed below, consist of sulfonated salts of melamine or naphthalene formaldehyde condensates. However, the now more popular "new generation" superplasticizers, polycarboxylic ether-based superplasticizers have gained in popularity due to certain advantages.

8.3.2 Mechanism of Action

Air-Entraining Surfactants

The chemical formula of typical air-entraining surfactant, which consists of a nonpolar hydrocarbon chain with an anionic polar group, is shown in Fig. 8-1*a*; the mechanism of action is illustrated in Fig. 8-1*b*. Lea[3] describes the mechanisms by which air voids are

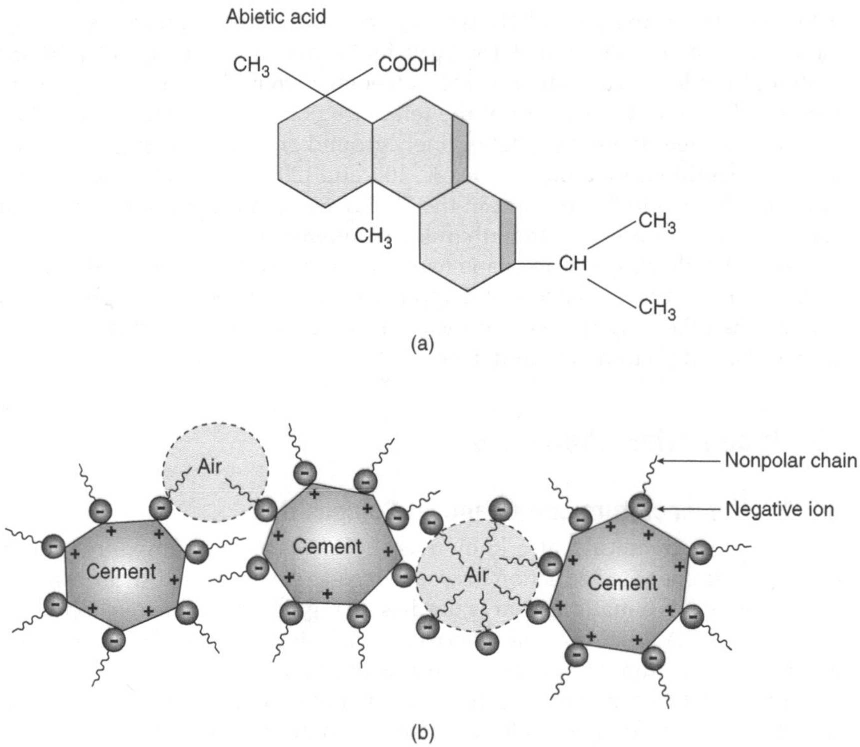

FIGURE 8-1 (*a*) Formula of a typical air-entraining surfactant derived from pine oil or tall oil processing; (*b*) mechanism of air entrainment when an anionic surfactant with a nonpolar hydrocarbon chain is added to the cement paste. (Adapted from Kreijger, P.C., *Admixtures*, Construction Press, Lancaster, London, 1980.)

entrained and stabilized when a surfactant is added to the cement-water system: *At the air-water interface the polar groups are oriented toward the water phase lowering the surface tension, promoting bubble formation, and counteracting the tendency for the dispersed bubbles to coalesce. At the solid-water interface where directive forces exist in the cement surface, the polar groups become bound to the solid with the nonpolar groups oriented toward the water, making the cement surface hydrophobic so that air can displace water and remain attached to the solid particles as bubbles.*

Water-Reducing Surfactants

The formulas of three typical plasticizing surfactants are shown in Fig. 8-2*a*. Note, in contrast to air-entraining surfactants, in the case of plasticizers the anionic polar group

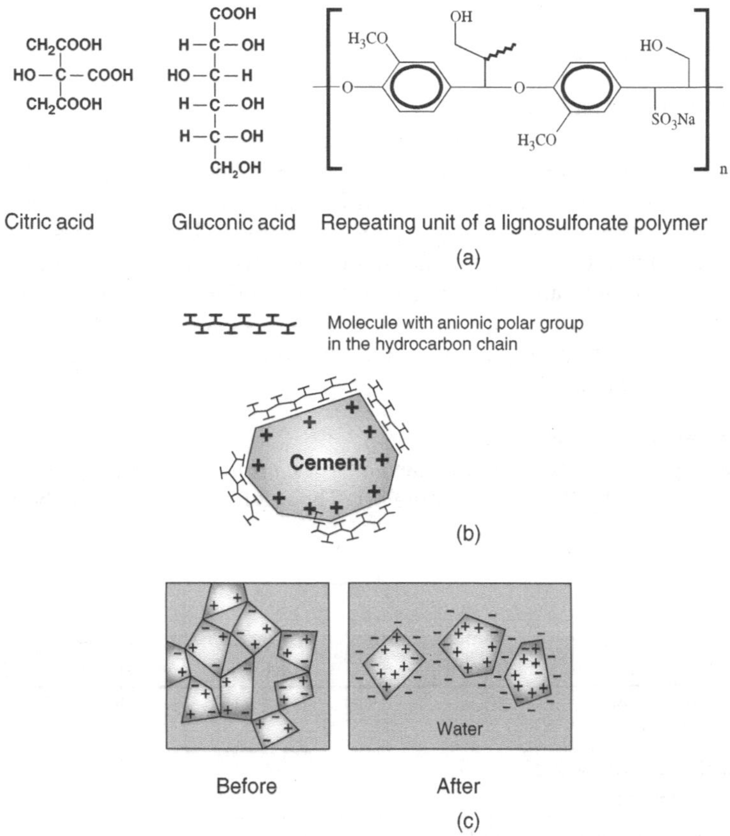

FIGURE 8-2 (*a*) Formulas of typical surfactants used as water-reducing admixtures. These are hydrocarbons containing anionic polar groups; (*b*) when a surfactant with several anionic polar groups in the hydrocarbon chain is added to the cement-water system, the polar chain gets absorbed on the surface of cement particle. Thus, not only the surface tension of water is lowered but also the cement particles are made hydrophilic; (*c*) diagrammatic representation of floc formation by cement particles before the addition of a water-reducing surfactant, and dispersion of flocs after the addition. (Adapted from Kreijger, P.C., in *Admixtures*, Construction Press, Lancaster, London, 1980.)

is joined to a hydrocarbon chain that itself is polar or hydrophilic (i.e., several OH groups are present in the chain). When a small quantity of water is added to cement without the presence of a surfactant a well-dispersed system is not attained because, first, the water possesses high surface tension (hydrogen-bonded molecular structure), and second, the cement particles tend to cluster together or form flocs (attractive force exists between positively and negatively charged edges, corners, and surfaces when crystalline minerals or compounds are finely ground). A diagram representing such a flocculated system is shown in Fig. 8-2*c*.

When a surfactant with a hydrophilic chain is added to the cement-water system, the polar chain is adsorbed alongside the cement particle; in this case the surfactant directs a polar end, instead of a nonpolar end toward the water, thus lowering the surface tension of water and making the cement particle hydrophilic (Fig. 8-2*b*). As a result of layers of water dipoles surrounding the hydrophilic cement particles, their flocculation is prevented and a well-dispersed system is obtained (Fig. 8-2*c*).

8.3.3 Applications

Air-Entraining Admixtures

The most important application of air-entraining admixtures is for concrete mixtures designed to resist freezing and thawing cycles (Chap. 5). A side effect of entrained air is the improved workability of concrete mixtures, particularly those containing less cement and water, rough-textured aggregates, or lightweight aggregates. Air entrainment is, therefore, commonly used in making mass concrete and lightweight concrete mixtures. Note that, because air-entraining surfactants render the cement particles hydrophobic, any overdose of the admixture would cause an excessive delay in cement hydration. Also, depending on the amount of the air in the concrete mixture, there would be a corresponding strength loss.

Water-Reducing Admixtures

A variety of aims that can be achieved by the application of water-reducing admixtures are illustrated by the data in Table 8-1. The reference concrete without any admixture,

Test Series	Cement Content (kg/m^3)	Water-Cement Ratio	Slump (mm)	Compressive Strength (MPa)	
				7 Days	28 Days
A—Reference concrete (no admixture)	300	0.62	50	25	37
Water-reducing admixture is added for the purpose of:					
B—Consistency increase	300	0.62	100	26	38
C—Strength increase	300	0.56	50	34	46
D—Cement saving	270	0.62	50	25.5	37.5

Source: Hewlett, P.C., M.R. Rixom, ed., in *Concrete Admixtures: Use and Applications*, Construction Press, Lancaster, London, p. 16, 1978. By permission of Longman.

TABLE 8-1 Benefits Achieved by Using Water-Reducing Admixtures

Series A, had a 300-kg/m^3 cement content and a 0.62 water-cement ratio; the fresh concrete showed 50-mm slump, and the hardened concrete gave 25 and 37 MPa compressive strength at 7 and 28 days, respectively.

With Test Series B, the purpose was to *increase the consistency* of the reference concrete mixture without adding more cement or water. This was easily achieved by incorporating a small dosage of the water-reducing admixture. Such an approach is useful when concrete is to be placed in heavily reinforced sections by pumping.

With Test Series C, the object was to *achieve higher compressive strengths* without increasing the cement content or reducing the consistency of the reference concrete mixture. Incorporating the same amount of the water-reducing admixture as used with Series B made it possible to reduce the water content by 10 percent (from 186 to 168 kg/m^3) while maintaining the 50-mm slump. As a result of reduction in the water-cement ratio, the 7-day compressive strength increased from 25 to 34 MPa and the 28-day strength from 37 to 46 MPa. This approach may be needed when job specifications limit the maximum water-cement ratio but require high early strength to develop.

Test Series D demonstrates how the addition of the water-reducing admixture made it possible to affect a 10 percent *cement saving* without compromising either the consistency or the strength of the reference concrete. Besides cost economy, such cement savings may be important when reduction of the temperature rise in mass concrete is the primary goal.

It should be noted, however, from the foregoing description of benefits resulting from the application of water-reducing admixtures, that *all of the three benefits are not available at the same time.*

The *period of effectiveness of surfactants* is rather limited because soon after the onset of hydration reactions between portland cement compounds and water, large amounts of products, such as ettringite, begin to form. The cement hydration products engulf the small quantity of the surfactant present in the system. Thus the ambient temperature and cement fineness and composition, especially the C_3A, SO_3, and alkali contents, which control the rate of ettringite formation, may have an effect on cement-admixture interactions. Obviously, the *amount or concentration of the admixture* in the system will also determine the effect. Larger amounts of an admixture than normally needed for the plasticizing or water-reducing effect may retard the time of set by preventing the hydration products from forming bonds. Thus, depending on the dosage, most surfactants can serve simultaneously as water reducers and set retarders. Commercial water reducers may contain accelerating agents to offset the retarding tendency when this is unwanted. Except for possible retardation of the setting time, other mechanical properties of concrete are not significantly affected by the presence of water-reducing agents; in fact, early strength development may be somewhat accelerated due to better dispersion of the cement particles in water. Durability is also increased as large voids are reduced due to improved dispersion.

Also, some water-reducing admixtures, especially those derived from lignin products, are known to entrain considerable air; to overcome this problem, commercial lignin-based admixtures usually contain air-detraining agents. As commercial products may contain many unknown ingredients, it is always desirable to make a laboratory investigation of the cement-admixture compatibility before using a new admixture or a combination of two or more admixtures.

8.3.4 Superplasticizers

Superplasticizers, also called high range water-reducing admixtures because of their ability to reduce the mixing water in a given concrete mixture three to four times more effectively compared to normal water-reducing admixtures, were developed in the 1970s and have found wide acceptance in the concrete construction industry. They consist of long-chain, high-molecular-weight (20,000 to 30,000) anionic surfactants with a large number of polar groups in the hydrocarbon chain. When adsorbed on cement particles, the surfactant imparts a strong negative charge, which helps to lower the surface tension of the surrounding water considerably and greatly enhances the fluidity of the system.

Compared to normal water-reducing admixtures, relatively large amounts of superplasticizers, up to 1 percent by weight of cement, can be incorporated into concrete mixtures without causing excessive bleeding and set retardation, in spite of a consistency on the order of 200- to 250-mm slump. It is probably the colloidal size of the long-chain particles of the admixture that obstructs the bleed-water flow channels in concrete, so that segregation is generally not encountered in superplasticized concretes. Excellent dispersion of the cement particles in water (Fig. 8-3) accelerates the rate of hydration; therefore, retardation is rarely observed unlike with plasticizers; instead, acceleration of setting and hardening is a common occurrence. In fact, the first generation of superplasticizers acquired a bad reputation for rapid loss of consistency or slump. Currently available products often contain lignosulfonate or other retarding materials to offset the rapid loss of high consistency achieved immediately after the addition of the admixture.

Compared to 5 to 10 percent reduction in the mixing water that is possible by the application of ordinary plasticizing admixtures, a water reduction in the 20 to 30 percent range can generally be achieved while maintaining high consistency in the reference concrete. The increase in mechanical properties (i.e., compressive and flexural strength) is generally commensurate with reduction in the water-cement ratio. Frequently, due to a greater rate of cement hydration in the well-dispersed system, concrete mixtures containing superplasticizers show even higher compressive strengths at 1, 3, and 7 days

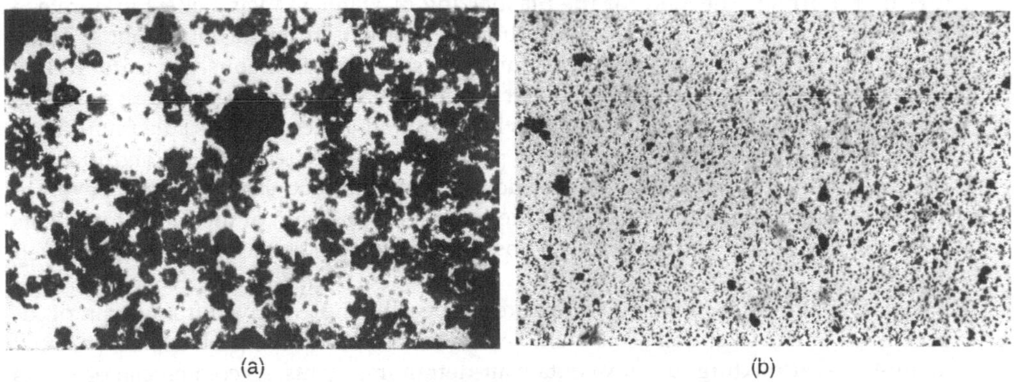

FIGURE 8-3 (*a*) Photomicrograph of flocculated cement particles in a portland cement-water suspension with no admixture present; (*b*) photomicrograph of the system after it is dispersed with the addition of a superplasticizing admixture. (*Photographs courtesy of M. Collepardi, Enco, Italy.*)

Test	Cement Content (kg/m³)	Water-Cement Ratio	Slump (mm)	Compressive Strength (MPa)			
				1 Day	3 Days	7 Days	28 Days
Reference concrete (no admixture)	360	0.60	225	10	21	32	45
Concrete of same consistency as above but containing less water and 2 percent superplasticizer by weight of cement	360	0.45	225	20	35	43	55
Concrete of same water-cement ratio but containing no superplasticizer, and having a lower slump than the preceding mixture	360	0.45	30	16	28	37	52

TABLE 8-2 Examples of High Early Strengths Achieved by the Use of Superplasticizing Admixtures

than the reference concrete having the same water-cement ratio (Table 8-2). This is of special importance in the precast concrete industry where high early strengths are required for fast turnover of the formwork. By using high cement content and water-cement ratios much lower than 0.4, it is possible to achieve even more rapid strength development rates while keeping the later age strength. Currently the most commonly used superplasticizers are polymelamine sulfonate (PMS), polynaphthalene sulfonate (PNS), and the new generation products based on polycarboxylates (PC) (see Fig. 8-4) and polymethacrylates.

In the 1990s, a new generation of superplasticizers was developed[4] that included polyacrylates, polycarboxylates, and polyethylene-based copolymers, which have a comb-like molecular structure. Instead of electrostatic repulsion as the dominant cement dispersion mechanism, inhibition of reactive sites through dispersion is the dominating mechanism by which these new generation of superplasticizers work. In steric repulsion, short-range physical barriers are created between the cement particles.

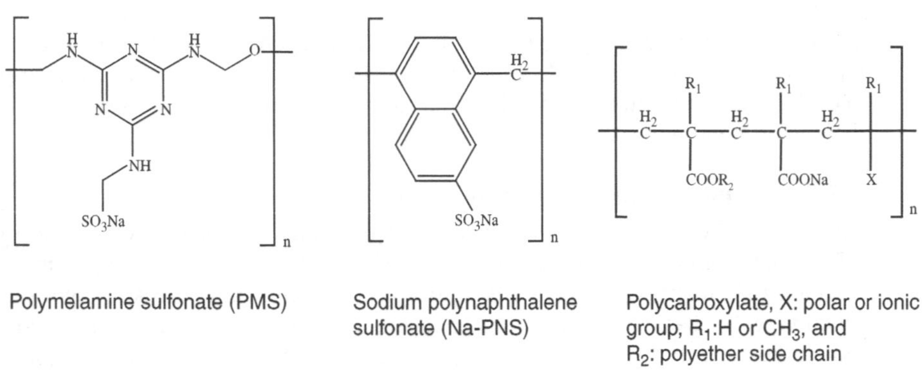

FIGURE 8-4 Different types of superplasticizers showing the repetitive molecular unit of Na-PMS; Na-PNS and polycarboxylate.

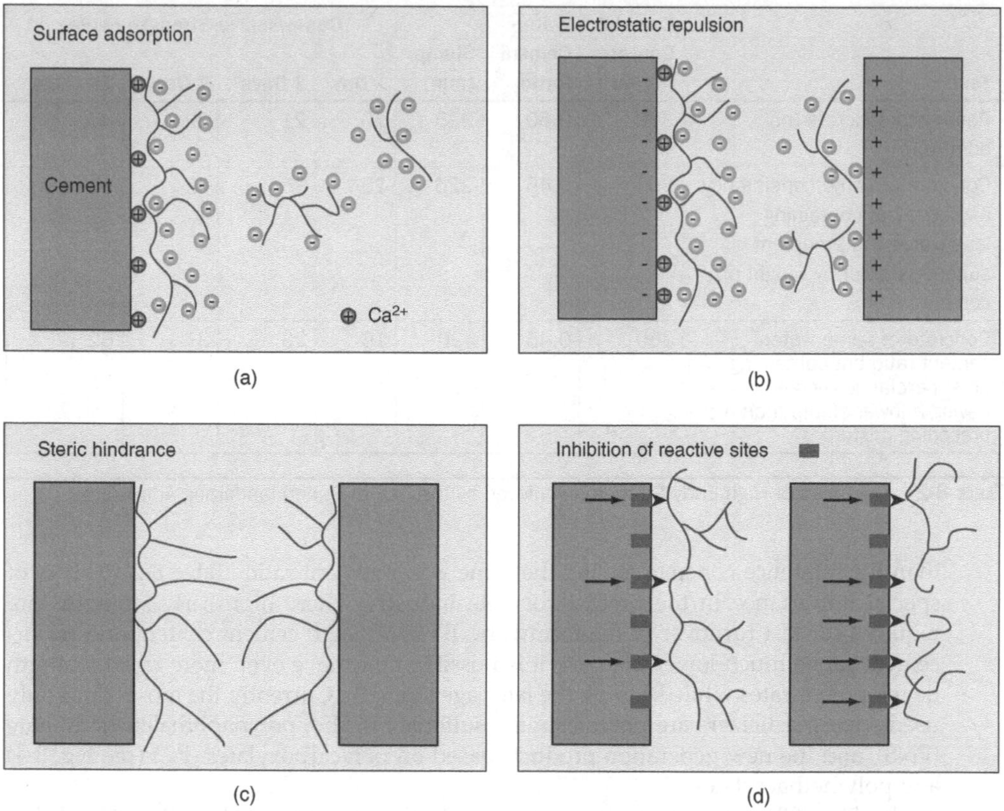

FIGURE 8-5 Diagrammatic illustration of mechanisms by which superplasticizers can disperse cement particles in the cement-water system. (*Courtesy of C. Jolicoeur.*)

One side of the polymer chain gets adsorbed on the surface of the cement grain, while the long unadsorbed side creates the steric repulsion. The grafted side chains of comb superplasticizers protrude and extend from the adsorbed-cement-surface site to hinder neighboring cement particles reaching the range within which van der Waal's force of attraction are effective. Figure 8-5 illustrates these mechanisms.[5] As the effect of steric repulsion lasts larger than electrostatic repulsion, the former has a greater influence over the slump retention, even with dosages that are considerably lower than the naphthalene or melamine-sulfonate type superplasticizers.

8.3.5 Shrinkage Reducing Admixtures[1]

Since their introduction to the construction community, shrinkage reducing admixtures (SRAs) have been extensively advocated as a methodology for reducing shrinkage and cracking in cement-based materials.[6] Introduced roughly three decades ago, the SRAs function on the basic premise that by reducing the surface tension of the pore fluid, they reduce the magnitude of capillary stresses developed, and thus mitigate drying shrinkage.

[1]The authors thank Professor Gaurav Sant for his contribution to this section.

Commercial admixtures are commonly reported as glycol ether blends, propylene glycol ether mixtures, or as polyoxyalkylene ethers. The polar head is attracted by polar solvents, such as water. The hydrophobic tail is nonpolar, which is composed of a hydrocarbon chain that is attracted by nonpolar solvents (oil) but is repelled by other polar substances (e.g., water). Depending on their molecular structures, SRAs are simultaneously soluble in either polar and/or nonpolar solvents.

As a function of their molecular chemistry—having a polar head and nonpolar tail—when dissolved in water, SRAs are expelled to mismatched interfaces (e.g., a water-air, or water-oil interface) such that the polar head resides proximal (closer) to the polar component (i.e., water) while the nonpolar tail resides proximal to the nonpolar component (i.e., air or oil). By adsorbing at such mismatched interfaces, surfactants reduce the *degree of dissimilarity* (e.g., between water and air) at the interface, which results in a reduction in the interfacial free energy. Thus the addition of SRAs reduces the surface tension (i.e., interfacial energy) of the water-air (i.e., liquid-vapor) interface. That said, the extent of surface tension reduction is obviously finite and is controlled by the number of SRA molecules that can be accommodated at the interface. Therefore, SRAs produce large reductions in surface tension at small dosages, and as the dosage is elevated, the surface tension of the solution achieves a plateau. Once the liquid-vapor (i.e., water-air) interface is saturated, any excess SRA molecules that cannot be accommodated remain mobile in the liquid phase. At low concentrations, these molecules remain discrete as monomers in solution; however, at higher concentrations, the monomers aggregate to form micelles as a means of reducing undesirable interactions between the nonpolar tail and water molecules (see Fig. 8-6). The aggregation sequence is correlated with a surfactant concentration known as the critical micelle concentration (CMC) such that (a) at dosages lower than the CMC, surfactant (SRA) molecules are

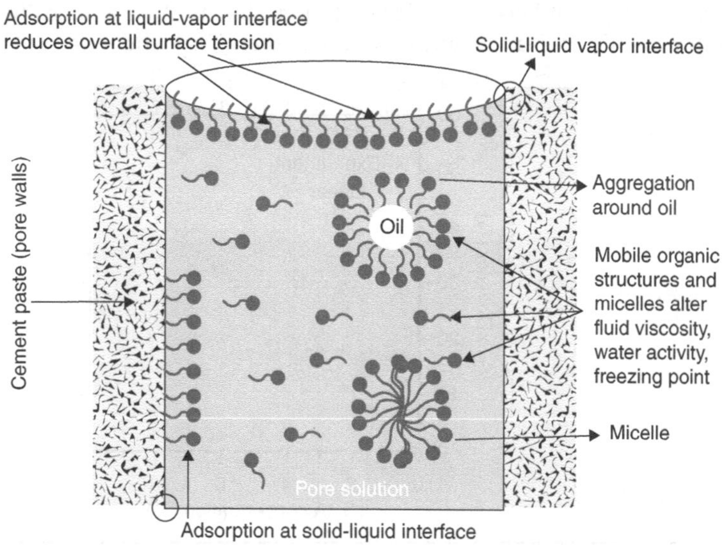

Figure 8-6 The interactions of surfactants in cementing materials. (Adapted from Zana, R., "Introduction to surfactants and surfactant self-assemblies," in Zana, R., ed., *Dynamics of Surfactant Self-Assemblies*, Taylor & Francis, Boca Raton, Florida, pp. 1–35, 2005.)

adsorbed at the liquid-vapor interface and the surface tension progressively reduces while (b) at concentrations higher than the CMC, additional surfactant molecules in solution aggregate to form micelles and do not contribute to surface tension reductions. In general and as a function of their dosage level, that is, adsorption at liquid-vapor interfaces, micelle formation, or both, surfactants (i.e., SRAs) can reduce the surface tension at the liquid-vapor interface, and/or increase the viscosity and depress the water activity of water-SRA solutions.

The most common use of SRA in concrete is through integral mixing, wherein the admixture is first dissolved into the mixing water prior to contact with the solids (i.e., cement, aggregates, etc.). When SRAs are added to concrete mixtures, their dosage is reported as a mass percentage of the cement or as a mass percentage of the total liquid (i.e., water + SRA) content of the system. The latter is a more appropriate means of dosing SRAs, since this reflects the surface tension reduction that may be produced. In commercial applications, dosages between 1 and 5 percent (i.e., of the liquid phase, excluding other admixtures) are common.

Incorporating SRAs often delays slightly the time of initial set and final set in cementing systems. These effects have been related to the tendency of SRAs to slightly retard hydration reactions. The influences of SRA on the mechanical properties have also been evaluated. Most studies report a slight reduction in strength and stiffness for water-cured samples. For example, Brooks and Jiang[7] reported lower compressive, tensile, and flexural strengths, in addition to a decrease in the elastic moduli in cementing materials containing SRAs. Similar reductions were reported by Cope and Ramey, as well as Folliard and Berke.[8,9] In addition, the rate of strength gain is also slightly reduced in SRA containing materials at early ages due to the retardation induced in early age reactions.

The introduction of SRAs to concrete mixtures significantly influences the rate of fluid (liquid) penetration. Cementitious materials containing SRAs show a dramatically slower rate of fluid penetration, wherein the rate of penetration scales are a function of the surface tension and viscosity of the solution. Since SRAs dramatically elevate solution viscosities (see Fig. 8-7) and depress surface tension, they tend to reduce capillary suction

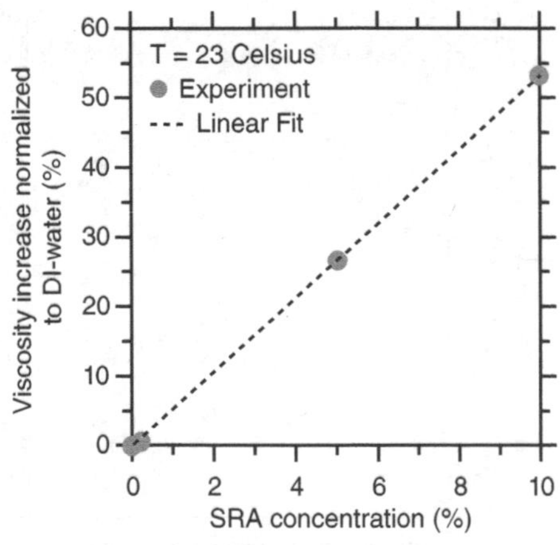

Figure 8-7 Normalized increase in solution viscosity of SRA-water solutions as a function of SRA concentration. [Sant, G., A. Eberhardt, D. Bentz, and W.J. Weiss, The influence of shrinkage reducing admixtures (SRAs) on moisture movement in cementitious materials at early ages, ASCE, *J. Mater. Civil Eng.*, Vol. 22, No. 3, pp. 277–286, 2010.]

effects that result in lower sorptivities, moisture diffusivities, and retarded fluid penetration. Recent work by Sant et al.[10] and Bentz et al.[11] has clearly demonstrated the benefits of these effects. By means of this viscosity amplification, specimens containing SRAs have been shown to be much more resistant than traditional materials to the penetration of chloride ions and sulfate attack. This is especially true when used in conjunction with light weight porous aggregates that are provided for internal curing (see Chap. 12).

In addition, SRAs significantly affect drying profiles and drying rates of fresh concretes. Bentz et al.[12] observed reduced evaporation from mixtures containing SRAs and deduced that the initial drying results in the formation of an SRA-rich (solution) layer, which then hinders onward evaporation. This outer SRA-rich layer dries preferentially due to its lower surface tension protecting the inner layers of the system from drying and plastic shrinkage damage. Note that by increasing the viscosity of a solution, SRAs reduce the flow rate of water to the surface of the material, even at equivalent capillary pressures. The influence of SRA on plastic shrinkage was also studied by Lura et al.,[13] who noted that SRAs do not substantially influence the rate of evaporation of a liquid. In mortars, containing SRAs, the reduced surface tension of the pore solution (mixing water) results in less fluid being forced out of the porous network once the layer of bleed water at the surface is consumed by evaporation. This results in reduced total moisture evaporation, settlement, and capillary and crack-inducing stresses at the topmost layer of the mortar, thereby reducing the risk of plastic cracking.

Autogenous shrinkage is the external (bulk) shrinkage experienced by a cementitious paste, mortar, or concrete under sealed, isothermal conditions. Autogenous shrinkage is caused by the consumption of water during hydration (i.e., self-desiccation) and chemical shrinkage, which results in the formation of curved menisci and vapor spaces; it is most relevant to cementitious materials having a low w/c, especially for w/c <0.42 (see Chap. 12). Thus, autogenous shrinkage is driven by internal drying that induces a relative humidity (RH) decrease and the development of capillary stresses that cause bulk shrinkage in a hydrating cementitious material. Sant et al.[14] showed that autogenous shrinkage measurements are very sensitive to the start-time of the measurements; however, when measured properly, SRAs substantially reduce autogenous shrinkage (see Fig. 8-8). Bentz et al.[15] measured the autogenous shrinkage

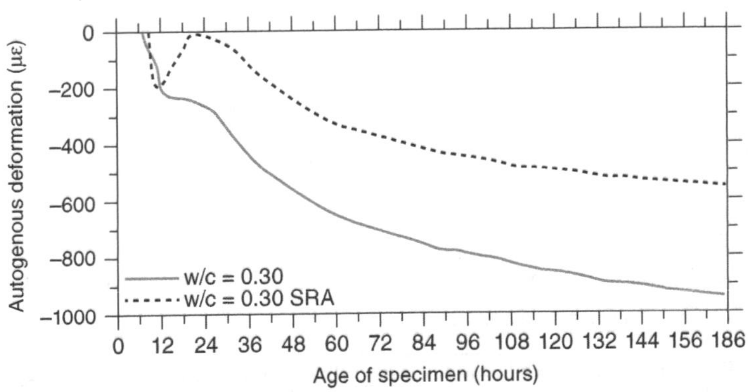

Figure 8-8 The influence of shrinkage reducing admixtures on autogenous deformations. (After Sant, G., B. Lothenbach, P. Juilland, G. Lesaout, K. Scrivener, and W.J. Weiss, The origin of early-age expansions in cementitious materials containing shrinkage reducing admixtures, *Cem. Concr. Res.*, Vol. 41, No. 3, pp. 218–229, 2011.)

of low w/c mortars containing silica fume and noted that SRAs reduced autogenous shrinkage by more than a factor of two. Sant et al.[16] demonstrated that pastes containing SRAs show an expansion shortly after set. This expansion was correlated with the enhanced crystallization of specific phases, an effect which is sensitive to cement chemistry. Thus, by a combination of two effects, inducing expansions and reducing self-desiccation, SRAs reduce autogenous shrinkage in hydrating cementitious systems.

In addition to the autogenous phenomena, cementitious materials also shrink due to water loss to the external environment; this is commonly referred to as "drying shrinkage." This loss of water is contingent upon the external RH being lower than the internal RH in the concrete. [The reverse is also true: the concrete gains water from its environment and expands (slightly).] This loss of water occurs as the concrete attempts to achieve a water potential equivalent to its ambient (i.e., external) environment. Thus upon drying as water is lost from the pores, the interfacial liquid-vapor meniscus becomes increasingly curved as it recedes into smaller and smaller sized pores. Therefore, the lower the external RH, the greater the quantity of water lost from the concrete and greater the range of pore sizes being emptied. While large pores empty first, small pores empty later in relation to their size and the RH as governed by the Kelvin-Laplace relations (see Chap. 12). In general, the drying of large pores generates small capillary stresses, while the emptying of smaller pores generates progressively larger stresses in relation to the pore size. Thus, small pores drive shrinkage in cementing materials.

The majority of research has indicated that the use of SRAs results in a reduction in drying shrinkage. While this reduction in shrinkage depends on the concentration of SRA used, the extent of reduction is reasonably independent of the w/c. The research of Weiss et al.[17] suggests that the addition of a SRA does not alter the total amount of water that is lost due to drying, except between specific RH ranges: 65 and 80 percent RH. However, the addition of a SRA does alter the rate of moisture loss (see Fig. 8-9).

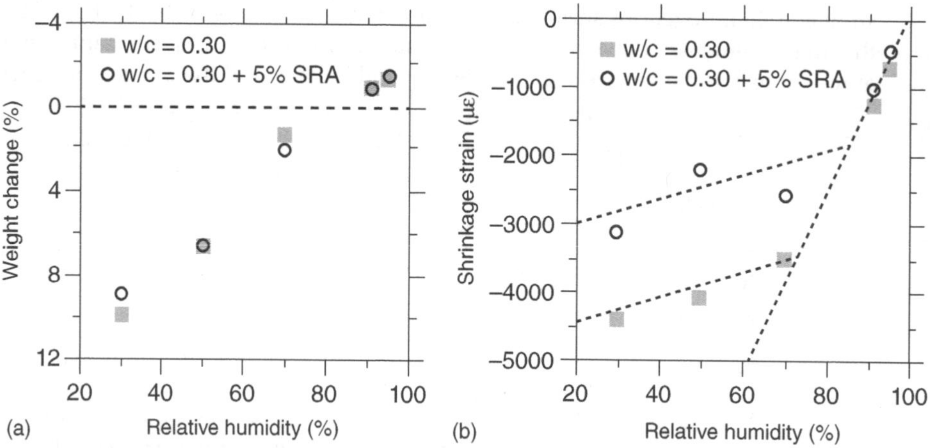

FIGURE 8-9 (*a*) Weight change and (*b*) drying shrinkage measured in cement pastes with and without SRA exposed to drying after 7 days of sealed curing. (From Sant, G. Examining volume changes, stress development and cracking in cement-based materials at early-ages, MSCE thesis, Purdue University, West Lafayette, IN, 2007; Weiss, W.J., P. Lura, F. Rajabipour, and G. Sant, Performance of shrinkage reducing admixtures at different humidities and at early ages, *ACI Mater. J.*, Vol. 105, No. 5, pp. 478–486, 2008.)

Interestingly, despite similar levels of equilibrium water loss, SRA-containing materials show substantially reduced shrinkage at lower RHs—70, 50, and 30 percent RH—as compared to systems without SRAs; note that shrinkage percentages measured at RHs in excess of around 80 percent RH are quite similar. It is assumed that SRAs alter the range of RHs, where capillary stresses control the shrinkage response of concrete. In general, by reducing shrinkage SRAs reduce and delay the potential for shrinkage cracking, the width of cracks formed, and the spacing between cracks formed in restrained concretes. All of these will factors improve the overall durability of concrete.

8.4 Set-Controlling Chemicals

8.4.1 Nomenclature and Composition

Besides certain types of surfactants already described, there are a large number of chemicals that can be used as retarding admixtures. In addition, there are chemicals that can accelerate the setting time and the rate of strength development at early ages. Interestingly, some chemicals act as retarders when used in small amounts (e.g., 0.3 percent by weight of cement), but in large dosage (e.g., 1 percent by weight of cement) they behave as accelerators.

Forsen[18] was the first to present a comprehensive analysis of the action of chemical admixtures on the setting of portland cement. He divided retarders into several groups according to the type of curve obtained when initial setting time was plotted against the concentration of the retarder in the cement-water system. A modified version of Forsen's classification covering both retarders and accelerators is shown in Fig. 8-10. The composition of the commonly used chemicals under each class is also shown in the figure.

8.4.2 Mechanism of Action

It is generally accepted now that the early reactions of portland cement compounds with water occur through solution; that is, the anhydrous compounds first ionize and then the hydration products form in the solution. Due to their limited solubility, the

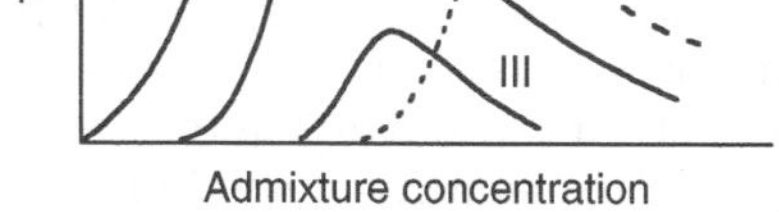

Figure 8-10 Classification and composition of set-controlling chemicals: Class I, $CaSO_4 \cdot 2H_2O$; Class II, $CaCl_2$, $Ca(NO_3)_2$; Class III, K_2CO_3, Na_2CO_3, Na_2SiO_3; Class IV, (1) surfactants with polar groups in the hydrocarbon chain (i.e., gluconates, lignosulfates, and sugars), (2) sodium salts of phosphoric, boric, oxalic, or hydrofluoric acid, and (3) zinc or lead salts; Class V, salts of formic acid and triethanolamine. (From Lea, F.M. *The Chemistry of Cement and Concrete*, Chemical Publishing Company, N.Y., 1974.)

hydration products crystallize out and the stiffening, setting, and hardening phenomena with portland cement pastes are directly related to different stages of the progressive crystallization process. It is therefore reasonable to assume that, by adding certain soluble chemicals to the portland cement-water system, one can influence either the rate of ionization of cement compounds or the rate of formation of the hydration products. Consequently this will affect the setting and hardening characteristics of the cement paste. According to Joisel[19] the action of set-controlling chemicals on portland cement can be attributed mainly to dissolution of the anhydrous constituents rather than to the crystallization of the hydrates.

To understand the mechanism of acceleration or retardation, consider a hydrating portland cement paste as being composed of calcium cations and silicate and aluminate anions, the solubility of each being dependent on the type and concentration of the acid and base ions present in the solution. Because most chemical admixtures readily ionize in water, it is possible to alter the type and concentration of the ionic constituents in the solution phase by adding these admixtures to the cement-water system, thus influencing the dissolution of the cement compounds according to the following guidelines proposed by Joisel:

1. An accelerating admixture must promote the dissolution of the cations or anions from the cement. As there are two predominant anions to dissolve, the accelerator should promote the dissolution of that constituent which has the lowest dissolution rate during the early hydration period (i.e., the silicate).
2. A retarding admixture must impede the dissolution of the cement cations or anions, preferably that anion which has the highest dissolution rate during the early hydration period (i.e., the aluminate).
3. The presence of monovalent cations in solution (i.e., K^+ or Na^+) reduces the solubility of Ca^{2+} ions but tends to promote the solubility of silicate and aluminate ions. In small concentrations, the former effect is dominant; in large concentrations, the latter effect becomes dominant.
4. The presence of certain monovalent anions in solution (i.e., Cl^-, NO_3^-) or SO_4^{2-} reduces the solubility of silicates and aluminates but tends to promote the solubility of calcium ions. In small concentrations, the former effect is dominant; in large concentrations, the latter effect becomes dominant.

From the foregoing principles it can be concluded that the overall outcome when a chemical admixture is added to a portland cement-water system will be determined by a number of both complementary and opposing effects, which are dependent on the type and concentration of ions contributed to the solution by the admixture. When used in small concentration (e.g., 0.1 to 0.3 percent by mass of cement), the salts of weak bases and strong acids (e.g., $CaCl_2$) or strong bases and weak acids (e.g., K_2CO_3), the retardation effect on the solubility of calcium and aluminate ions from the cement is the more dominant effect than the acceleration of solubility of silicate ions; hence the overall result is retardation. With larger concentrations (e.g., 1 percent or more) of these salts, the accelerating effect of the ions in solution on the solubility of the silicate and the aluminate ions of cement become more dominant than the retarding effects; thus, it is possible for the same salt to change its role and become an accelerator instead of a retarder (Fig. 8-5). It should be noted that $CaCl_2$, 1 to 3 percent by mass of cement, is the most commonly used accelerator for unreinforced concrete.

Gypsum ($CaSO_4 \cdot 2H_2O$) is the salt of a weak base and a strong acid, but it does not show the retarder-to-accelerator, role-reversal phenomenon on the setting time of cement when the amount of gypsum added to a portland cement paste is gradually increased. This is because the solubility of gypsum in water is very low (2 g/L at 20°C). Until the sulfate ions from gypsum go into solution, they will not be able to accelerate the solubility of calcium from the cement compounds. With the gradual removal of sulfate ions from the solution due to crystallization of calcium sulfoaluminate hydrates (mostly ettringite), more gypsum goes into solution; this has a beneficial effect on the C_3S hydration and therefore on the rate of strength development. However, if instead of gypsum, a large amount of sulfate is introduced in a highly soluble form (e.g., plaster of paris or hemihydrate), both the setting time and early strengths will be accelerated.

It is expected that low molecular-weight organic acids and their soluble salts, which are weakly acidic, would serve as accelerators because of their ability to promote the solution of calcium ions from the cement compounds. Actually, calcium formate, $Ca(HCOO)_2$, and formic acid, HCOOH, are accelerators, but other acids with a long hydrocarbon chain generally act as retarders by inhibiting bond formation among the hydration products. Triethanolamine, $N(CH_2–CH_2OH)_3$, is another organic chemical that, in small amounts (0.1 to 0.5 percent) is used as an accelerating ingredient in some water-reducing admixtures because of its ability to accelerate the hydration of C_3A (and formation of ettringite). However, triethanolamine tends to retard C_3S hydration and therefore reduces the rate of strength development. Both organic accelerators play an important part in prestressed and reinforced concrete applications where the use of accelerating admixtures containing chloride is considered undesirable.

Chemical admixtures listed under Class IV (Fig. 8-10) act as powerful retarders by mechanisms other than those discussed above. Surfactants, such as gluconates and lignosulfonates, act as retarders by delaying bond formation among the hydration products; others reduce the solubility of the anhydrous constituents from cement by forming insoluble and impermeable products around the particles. Sodium salts of phosphoric, boric, oxalic, and hydrofluoric acid are soluble, but the calcium salts are highly insoluble and therefore readily form in the vicinity of hydrating cement particles. Once insoluble and dense coatings are formed around the cement grains, further hydration slows down considerably. Phosphates are commonly used as an ingredient of commercial set-retarding admixtures.

8.4.3 Applications

Accelerating Admixtures

According to the report by ACI Committee 212: Accelerating admixtures are useful for modifying the properties of portland cement concrete, particularly in cold weather, to: (a) expedite the start of finishing operations and, when necessary, the application of insulation for protection; (b) reduce the time required for proper curing and protection; (c) increase the rate of early strength development so as to permit earlier removal of forms and earlier opening of the construction for service; and (d) permit more efficient plugging of leaks against hydraulic pressures.[1]

Because calcium chloride is by far the best known and most widely used accelerator, the effects of $CaCl_2 \cdot 2H_2O$ additions in amounts from 0.5 to 2.0 percent by weight of cement on setting times and relative compressive strengths are shown in Fig. 8-11.

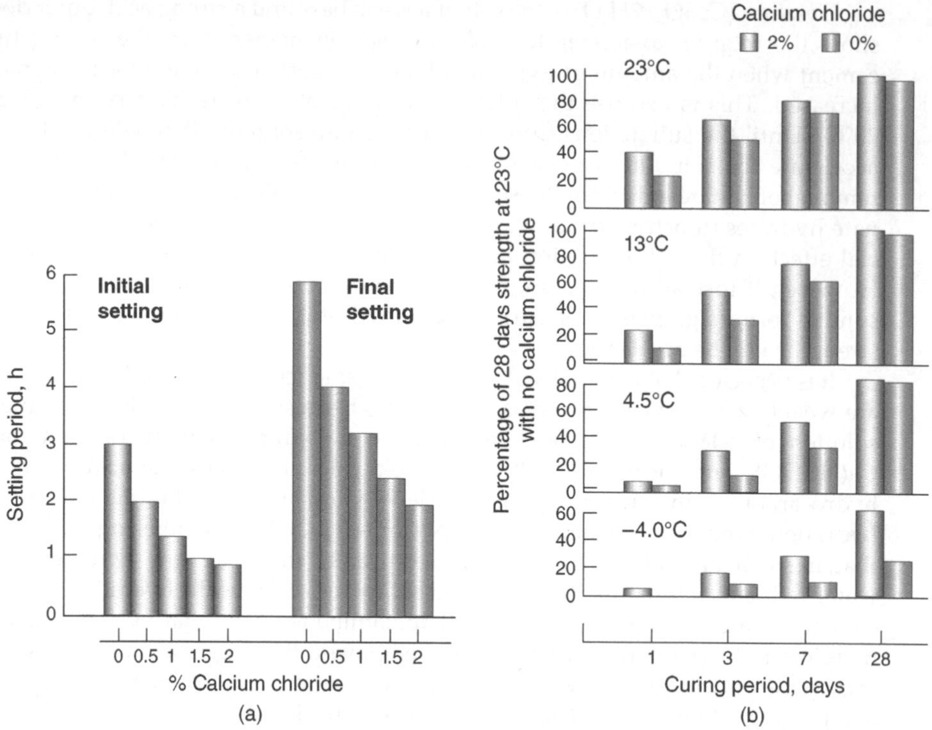

FIGURE 8-11 Effect of calcium chloride addition on (*a*) setting time of portland cement (*b*) effect of calcium chloride addition on strength at various curing temperatures. (From Ramachandran, V.S., *Progress in Concrete Technology*, Malhotra, V.M., ed., CANMET, Ottawa, pp. 421–450, 1980.)

Properties of concrete as affected by the use of calcium chloride are summarized in Table 8-3.

Retarding Admixtures

According to ACI Committee 212, the following applications of retardation of setting are particularly important in the construction practice:

1. Adverse ambient temperature conditions such as hot weather concreting. Extensive use is made of retarding admixtures to permit proper placement and finishing, as well as to overcome the damaging and accelerating effects of high temperature.
2. Control of setting in large structural units to keep concrete workable throughout the entire placing period. This is particularly important for the elimination of cold joints and discontinuities in large structural units. Also, control of setting may prevent cracking of concrete beams, bridge decks, and composite construction due to form deflection associated with the placement of adjacent units. Adjustments of the dosage as placement proceeds can permit various portions of a unit, a large post-tensioned beam for example, to attain a given level of early strength at approximately the same time.

Property	General Effect	Remarks
Setting	Reduces both initial and final setting	ASTM Standard requires that initial and final setting times should occur at least 1 h earlier with respect to reference concrete.
Compressive strength	Significantly increases the compressive strength in the first 3 days of curing (gain may be about 30–100%)	ASTM Standard requires an increase of at least 125% over control concrete at 3 days. At 6–12 months, requirement is only 90% of control specimen
Tensile strength	A slight decrease at 28 days	
Flexural strength	A decrease of about 10% at 7 days	This figure may vary depending on the starting materials and method of curing. The decrease may be more at 28 days.
Heat of hydration	An increase of about 30% in 24 h	Total amount of heat at longer times is almost the same as that evolved by reference concrete.
Resistance to sulfate attack	Reduced	Can be overcome by use of Type V cement with adequate air entrainment.
Alkali-aggregate reaction	Aggravated	Can be controlled by use of low-alkali cement or pozzolan.
Corrosion	Causes no problems in normal reinforced concrete if adequate precautions taken. Dosage should not exceed 1.5% $CaCl_2$ and adequate cover to be given. Should not be used in concrete containing a combination of dissimilar metals or where there is a possibility of stray currents	Calcium chloride admixture should not be used in prestressed concrete or in concrete containing a combination of dissimilar metals. Some specifications do not allow the use of $CaCl_2$ in reinforced concretes.
Shrinkage and creep	Increased	
Volume change	Increase of 0–15% reported	
Resistance to damage by freezing and thawing	Early resistance improved	At later ages may be less resistant to frost attack.
Watertightness	Improved at early ages	
Modulus of elasticity	Increased at early ages	At longer periods almost the same as reference concrete.
Bleeding	Reduced	

Source: Ramachandran, V.S., in Progress in *Concrete Technology*, Malhotra, V.M., ed., CANMET, Ottawa, pp. 421–450, 1980.

Table 8-3 Some of the Properties Influenced by the Use of Calcium Chloride Admixture in Concrete

Admixture Dosage by Weight of Cement (L/kg)	Setting Time (h) (ASTM C 403)		Water-Cement Ratio	Compressive Strength (MPa)		
	Initial	Final		3 Day	7 Day	28 Day
0	4.5	9	0.68	20.3	28.0	37.0
0.14	8	13	0.61	28.0	36.5	46.8
0.21	11.5	16	0.58	29.6	40.1	49.7

*According to ASTM C 494, Type D admixtures are both retarding and water reducing.
Source: Based on Hewlett, P.C., in *Concrete Admixtures: Use and Applications*, Rixom, M.R., ed., Construction Press, Lancaster, London, p. 18, 1978. By permission of Longman.

Table 8-4 Effects of an ASTM Type D Admixture* on Setting Time and Strength

Test	Ambient Temperature (°C)	Concrete Slump (mm)					
		0	1 h	2 h	3 h	4 h	5 h
Control concrete (no admixture)	20	127	89	76	57	38	32
Concrete with admixture	20	127	127	114	102	70	57
Control concrete (no admixture)	43	114	57	7	0	0	0
Concrete with admixture	43	127	70	25	19	13	0

Source: Based on Hewlett, P.C., *Concrete Admixtures: Use and Applications*, Rixom, M.R., ed., Construction Press, Lancaster, London, p. 18, 1978. By permission of Longman.

Table 8-5 Effect of an ASTM Type D Admixture on Consistency at a High Ambient Temperature

Effects of addition of a retarding and water-reducing admixture (Type D, ASTM C 494) on the setting time and strength of concrete, at normal temperature, are shown by the data in Table 8-4. The effectiveness of the admixture in maintaining concrete in a workable condition for longer periods, even at a higher than normal ambient temperature, is illustrated by the data in Table 8-5.

8.5 Mineral Admixtures

8.5.1 Significance

Mineral admixtures are often finely divided siliceous materials that are added to concrete in relatively large amounts, generally, in the range 20 to 70 percent by mass of the total cementitious material. Although natural pozzolans in the raw state or after thermal activation are being used in some parts of the world, due to economic and environmental considerations many industrial by-products have become the primary source of mineral admixtures in concrete.

Power plants* using coal as fuel, and metallurgical furnaces producing cast iron, silicon metal, and ferrosilicon alloys are the major sources of by-products being produced

*Small power plants using rice husks as fuel are being used increasingly in many countries; under controlled combustion conditions they produce a highly pozzolanic ash.

at the rate of millions of tons every year in many countries. Dumping of these by-products into landfills and streams amounts to a waste of the material and causes serious environmental pollution. Disposal as concrete aggregate or for roadbase construction is a low-value use that does not utilize the pozzolanic and cementitious potential of these materials. With proper quality control, large amounts of many industrial by-products can be incorporated into concrete, either in the form of blended portland cement or as mineral admixtures. Whenever a pozzolanic and/or cementitious by-product can be used as a partial replacement for portland cement in concrete, it represents significant energy and cost savings.

The mechanism by which the pozzolanic reaction exercises a beneficial effect on the properties of concrete is the same irrespective of whether a pozzolanic material is added to concrete in the form of a mineral admixture or as component of a blended portland cement. From a description of the pozzolanic reaction and properties of blended cements (Chap. 6) it is clear that the engineering benefits likely to be derived from the use of mineral admixtures in concrete include improved resistance to thermal cracking due to low heat of hydration, enhancement of ultimate strength and impermeability due to pore refinement, strong interfacial transition zone, and very high durability to sulfate attack and alkali-aggregate expansion.

8.5.2 Classification

Some mineral admixtures are pozzolanic (e.g., low-calcium, Class F fly ash), some are cementitious (e.g., granulated iron blast-furnace slag), whereas others are both cementitious and pozzolanic (e.g., high-calcium, Class C fly ash). A classification of mineral admixtures according to their pozzolanic and/or cementitious characteristics according to Mehta[20] is shown in Table 8-6. The table also contains a description of mineralogical composition and particle characteristics as these two properties, rather than simply the chemical composition or the source of the material, determine the effect of a mineral admixture on the behavior of concrete containing the admixture.

For the purposes of a detailed description of the important mineral admixtures given below, the materials are divided into two groups:

1. *Natural materials.* Those materials that have been processed for the sole purpose of producing a pozzolan. Processing usually involves crushing, grinding, and size separation; in some cases it may also involve thermal activation.
2. *By-product materials.* Those materials that are not the primary products of the industry producing them. Industrial by-products may or may not require any processing (e.g., drying and pulverization) before use as mineral admixtures.

8.5.3 Natural Pozzolanic Materials

Most natural pozzolans are derived from volcanic rocks and minerals. During explosive volcanic eruption, quick cooling of the magma that is composed mainly of aluminosilicates results in the formation of glassy or vitreous phases with a disordered structure. Due to the simultaneous evolution of dissolved gases when magma is solidifying, the solidified matter often acquires a porous texture and high surface area which enhances its chemical reactivity. Aluminosilicates with disordered structure are not stable on exposure to alkaline solutions. This is why volcanic glasses exhibit pozzolanic activity with lime or portland cement in an aqueous environment.

Classification	Chemical and Mineralogical Composition	Particle Characteristics
Cementitious and Pozzolanic		
Granulated blast-furnace slag (cementitious)	Mostly silicate glass containing mainly calcium, magnesium, aluminium, and silica. Crystalline compounds of melilite group may be present in small quantity.	Unprocessed material is of sand size and contains 10–15% moisture. Before use it is dried and ground to particles less than 45 μm (usually about 500 m^2/kg Blaine). Particles have rough texture.
High-calcium fly ash (cementitious and pozzolanic)	Mostly silicate glass containing calcium, magnesium, aluminum, and alkalies. The small quantity of crystalline matter present generally consists of quartz and C_3A; free lime and periclase may be present; $C\bar{S}$ and $C_4A_3\bar{S}$ may be present in the case of high-sulfur coals. Unburnt carbon is usually less than 2%.	Powder corresponding to 10–15% particles larger than 45 μm (usually 300–400 m^2/kg Blaine). Most particles are solid spheres less than 20 μm in diameter. Particle surface is generally smooth but not as clean as low-calcium fly ashes.
Highly Active Pozzolans		
Silica fume	Consist essentially of pure silica in noncrystalline form.	Extremely fine powder consisting of solid spheres of 0.1 μm average diameter (about 20 m^2/g surface area by nitrogen adsorption).
Rice husk ash	Consist essentially of pure silica in noncrystalline form.	Particles are generally less than 45 μm but they are highly cellular (40–60 m^2/g surface area by nitrogen adsorption).
Normal Pozzolans		
Low-calcium fly ash	Mostly silicate glass containing aluminum, iron, and alkalies. The small quantity of crystalline matter present generally consists of quartz, mullite, sillimanite, hematite, and magnetite.	Powder corresponding to 15–30% particles larger than 45 μm (usually 200–300 m^2/kg Blaine). Most particles are solid spheres with average diameter 20 μm. Cenospheres and plerospheres may be present.
Natural materials	Usually volcanic soils. Besides aluminosilicate glass, natural pozzolans contain quartz, feldspar zeolites, and/or mica.	Particles are ground to mostly under 45 μm and have rough texture.
Metakaolin	Kaolinitic clay calcined at 650–800°C to enhance pozzolanic activity.	It may be noted that HRM (highly reactive metakaolin) for commercial use is available in the United States and this material is made by calcination of a highly pure kaolinite clay and by grinding the product to a very fine particle size (surface area 12 m^2/g by nitrogen adsorption).
Weak Pozzolans		
Slowly cooled blast-furnace slag, bottom ash, boiler slag, field burnt rice husk ash	Consist essentially of crystalline silicate materials, and only a small amount of noncrystalline matter.	The materials must be pulverized to very fine particle size in order to develop some pozzolanic activity. Ground particles are rough in texture.

TABLE 8-6 Classification, Composition, and Particle Characteristics of Mineral Admixtures for Concrete

Alteration of volcanic glass under hydrothermal conditions leads to the formation of zeolite minerals which are compounds of the type $(Na_2,K_2,Ca)O{\cdot}Al_2O_3{\cdot}(2\text{–}10)SiO_2{\cdot}xH_2O$. This product, called *volcanic tuff*, is characterized by a compact texture. Zeolite minerals in finely ground tuffs are able to react with lime by a base-exchange reaction.

In nature, progressive alteration of the aluminosilicates in volcanic glass is believed to be responsible for the formation of clay minerals which are not pozzolanic unless the crystalline structure of the aluminosilicate hydrates in clay is converted into an amorphous or disordered structure by heat treatment.

Diatomaceous earths consist of opaline or amorphous hydrated silica derived from the skeletons of diatoms, which are tiny water plants with cell walls composed of silica shells, and where these skeletons have been deposited in geological layers over many millions of years. The material is pozzolanic when pure, but is generally found contaminated with clay minerals and, therefore, has to be thermally activated to enhance its pozzolanic reactivity.

It is difficult to classify natural pozzolans because the materials seldom contain only one reactive constituent. However, based on the principal reactive constituent present, a classification can be made into volcanic glasses, volcanic tuffs, calcined clays or shales, and diatomaceous earths. A description of their formation processes and relevant characteristics is given below.

Volcanic Glasses

Santorini Earth of Greece, Bacoli Pozzolan of Italy, and Shirasu Pozzolan of Japan are examples of pozzolanic materials that derive their lime-reactivity characteristic mainly from the unaltered aluminosilicate glass. A photograph of the pozzolan quarry on Santorini Island is shown in Fig. 8-12*a*, and the pumiceous or porous texture that accounts for the high surface area and reactivity is evident from the scanning electron

(a) (b)

Figure 8-12 (*a*) Quarry for mining the pozzolan on Santorini Island, Greece; (*b*) scanning electron micrograph of the porous structure of the pozzolan.

micrograph shown in Fig. 8-12*b*. Generally, small amounts of nonreactive crystalline minerals, such as quartz, feldspar, and mica, are found embedded in the glassy matrix.

Volcanic Tuffs

The Roman pozzolan in Italy (Fig. 8.13), which was used in the construction of the concrete dome and the walls of the Pantheon, and the trass of Rheinland and Bavaria (Germany) are typical volcanic tuffs that are a product of hydrothermal alteration of volcanic glass. The zeolite tuffs with their compact texture are fairly strong, possessing compressive strengths on the order of 10 to 30 MPa. The principal zeolite minerals are phillipsite and herschelite. After the compact mass is ground to a fine particle size, the zeolite minerals show considerable reactivity with lime, and develop cementitious characteristics similar to pozzolans containing volcanic glass.

Calcined Clays or Shales

Volcanic glasses and tuffs do not require heat treatment to enhance their pozzolanic property. However, clay and shales will not show appreciable reactivity with lime unless the crystal structures of the clay minerals present are destroyed by heat treatment. Temperatures on the order of 600 to 900°C, in oil-, gas-, or coal-fired rotary kilns may be required. The pozzolanic activity of the product is due mainly to the amorphous or disordered aluminosilicate structure of clay as a result of the thermal treatment. Metakaolin is a pozzolan of enhanced reactivity by calcination and by finely grinding the calcined product. *Surkhi*, a pozzolan made in India by pulverization of fired-clay bricks, belongs to this category. It should be obvious why heat treatment of clays and shales that contain large amounts of quartz and feldspar would not produce a good pozzolan. In other words, pulverization of fired-clay bricks made from any clay may not yield a suitable mineral admixture for concrete.

Diatomaceous Earth

This group of pozzolans is characterized by materials of organogen origin. Diatomite is a hydrated amorphous silica composed of skeletal shells from the cell walls of microscopic aquatic algae. The largest known deposit is in California. Other large deposits are reported in Algeria, Canada, Germany, and Denmark. Diatomites are highly reactive to lime, but their porous microstructure is responsible for high water requirement, which is harmful for the strength and durability of concrete containing this pozzolan. Furthermore, diatomite deposits, such as Moler Earth in Denmark, generally contain large amounts of clay and therefore must be thermally activated before use in order to enhance the pozzolanic activity.

8.5.4 By-Product Materials

Ashes from the combustion of coal and some crop residues such as rice hull and rice straw, silica fume from certain metallurgical operations, and granulated slag from both ferrous and nonferrous metal industries are among the industrial by-products that are suitable for use as mineral admixtures in portland cement concrete. Countries such as China, India, the United States, Russia, Germany, South Africa, and the United Kingdom are among the biggest producers of coal fly ash which at the current rate of production, some 900 million tons a year, constitutes the largest volume industrial waste product in the world. The worldwide production of fly ash is predicted to reach 2000 million tons a year by 2020. Norway is the principal producer of silica fume, while granulated blast-furnace slag is available in many countries. In addition to these materials, China,

FIGURE 8-13 Ancient Roman pozzolanic concretes: (*a*) Great Hall of the Markets of Trajan in Rome, constructed about 105–112 AD (*photograph courtesy of M. Jackson*); (*b*) 20 cm diameter drill core of the 1-m-thick wall of the Great Hall showing volcanic tuff and brick coarse aggregate and the pozzolanic mortar (*photograph courtesy of C. Hagen*); (*c*) submarine concrete block at Caesarea harbor, constructed on the coast of Israel in late first century AD (*photograph courtesy of J.P. Oleson*); (*d*) 9 cm diameter drill core of maritime concrete from the Bay of Pozzuoli near Naples, first century BC, showing the volcanic tuff coarse aggregate and the pumiceous pozzolanic mortar with relict clasts of lime now composed of aluminous tobermorite (*photograph courtesy of C. Hagen*).

The pozzolanic concretes of the monuments of ancient Rome and the seawater harbor constructions that Romans built in the Mediterranean sea are some of the most durable cementitious materials on the planet. They remain resistant to decay, even after 2000 years exposure to groundwater saturation, strong variations in relative humidity, and submersion in seawater. The concretes are composed of large, decimeter-sized chunks of volcanic tuff, a glassy pyroclastic rock with zeolite cements, and a sophisticated mortar produced from lime (CaO), calcined from limestone in kilns, and alkali- and alumina-rich volcanic ash pozzolans (Figs. 8.13a and b). Romans developed the mortar formulations for the monuments over about 150 years. They eventually selected a specific scoriaceous ash erupted from nearby Alban Hills volcano[21,22] shown in the drill core of the 1-m-thick wall that supports the Great Hall of the Markets of Trajan in Rome. Submarine concrete harbor structures also have pyroclastic rock aggregates. The mortar contains the pulvis volcanic ash described by Vitruvius and Pliny the Elder, which was "quarried in the vicinity of Baiae and territories surrounding the municipalities of Mount Vesuvius" (de Architectura 2.6.1, Naturalis Historia 35.166). Ancient Puteoli was one of these municipalities, and its modern name, Pozzuoli, is the root of the word pozzolan. The principal reactive components are volcanic glass and surface coatings of opal, clay, and zeolite. The white inclusions are relict clasts of lime now composed of aluminous tobermorite, an important, but rare, mineral.[23]

India, and other Asian and South American countries have the potential for producing large amounts of rice husk ash. The production and properties of important by-product materials are described below.

Fly Ash

During the combustion of powdered coal in modern thermal power plants, as coal passes through high-temperature zone in the furnace the volatile matter and carbon are burned off while most of the mineral impurities such as clays, quartz, and feldspar melt at the high temperature. The fused matter is quickly transported to low-temperature zones where it solidifies as spherical particles of glass. Some of the minerals agglomerate forming the bottom ash, but most of the fine particles fly out with the flue gas stream and are called *fly ash* (pulverized fuel ash in the U.K.). This ash is subsequently removed from the gas by cyclone separation, electrostatic precipitation, and bag-house filtration.

From the standpoint of significant differences in mineralogical composition and properties, fly ashes can be divided into two categories, differing from each other mainly in calcium content. Experience shows that the ash in the first category, containing approximately 10 percent or less than 15 percent analytical CaO, is generally a product of the combustion of anthracite and bituminous coals. The ash in the second category, typically containing 15 to 40 percent analytical CaO, is generally a product of combustion of lignite and subbituminous coals.

The low-calcium fly ashes, Class F, due to the high proportions of silica and alumina present, consist principally of aluminosilicate glass. In the furnace, when large spheres of molten glass do not get cooled rapidly and uniformly, sillimanite ($Al_2O_3 \cdot SiO_2$) or mullite ($3Al_2O_3 \cdot 2SiO_2$) may crystallize as slender needles in the interior of glassy spheres. This kind of partial devitrification of glass in low-lime fly ashes accounts for the presence of crystalline aluminosilicate minerals. Also, depending on the fineness to which a coal has been ground before combustion, remnants of α-quartz are likely to be present in all fly ashes. In fact, X-ray diffraction analyses have confirmed that the principal crystalline minerals in low-calcium fly ashes are quartz, mullite, and hematite and/or magnetite. Since these crystalline minerals are nonreactive at ordinary temperature, their presence in large proportions, at the cost of the noncrystalline component or glass, tends to reduce the pozzolanic activity of the fly ash.

Compared to low-calcium fly ashes, the high-calcium variety is, in general, more reactive because it contains more of the calcium in the form of reactive crystalline compounds, such as C_3A, $C\overline{S}$, and $C_4A_3\overline{S}$. In addition, the principal constituent (i.e., the noncrystalline phase) contains enough calcium ions to enhance the reactivity of the aluminosilicate glass. Most fly ashes from modern furnaces, whether of the low-calcium or the high-calcium variety, contain approximately 60 to 85 percent glass, 10 to 30 percent crystalline compounds, and up to about 5 percent unburnt carbon. Carbon is generally present in the form of cellular particles larger than 45 μm. More than 5 percent carbon in a fly ash meant for use as a mineral admixture in concrete is considered undesirable because the cellular particles of carbon tend to increase both the water requirement for a given consistency and the admixture requirement for air entrainment.

Micrographic evidence presented in Fig. 8-14 shows that most of the particles in fly ash occur as *solid spheres of glass*. Sometimes a small number of hollow spheres, called *cenospheres* (completely empty) and *plerospheres* (packed with numerous small spheres), may also be present. Typically, the spherical particles in low-calcium fly ash look cleaner than those in high-calcium fly ash. As alkalies and sulfate tend to occur in a relatively

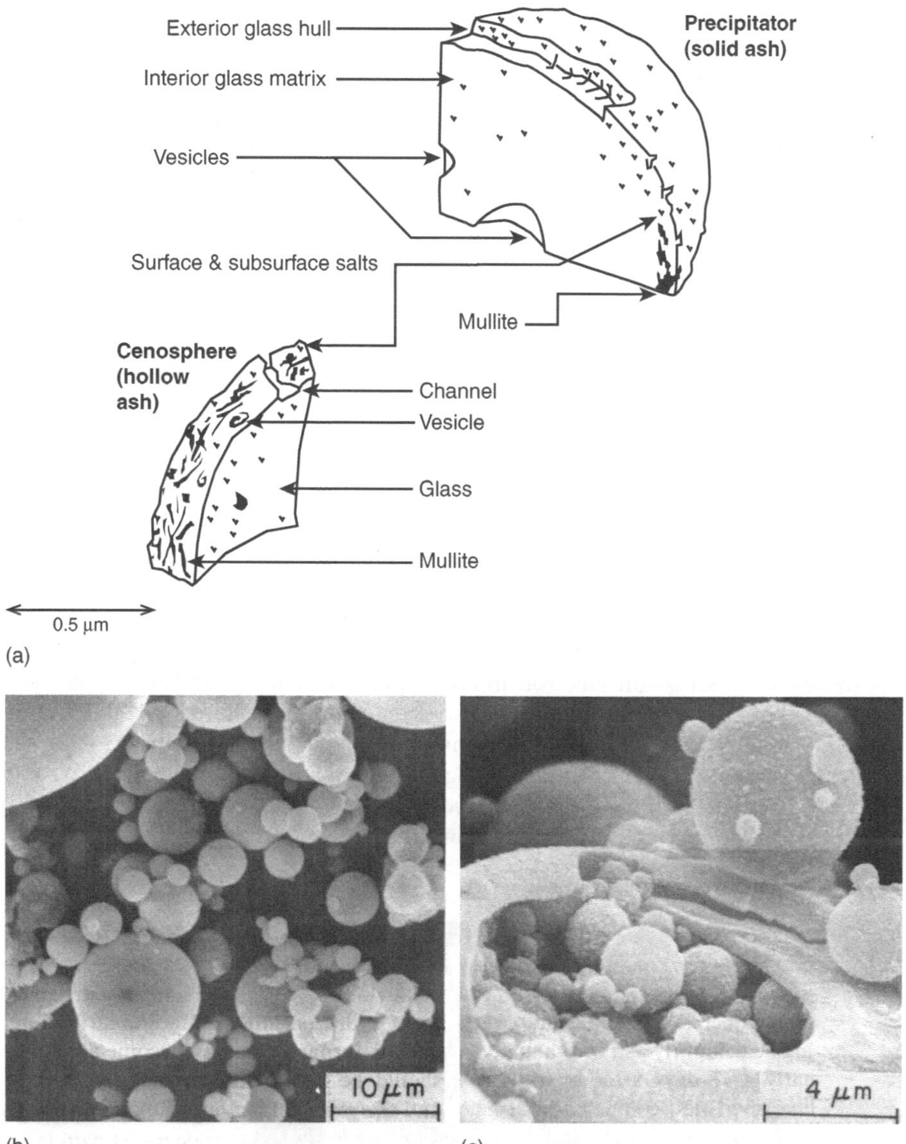

FIGURE 8-14 (*a*) Representation of the submicroscopic characteristics of fly ash particles (after Dudas, M.J., and C.J. Warren, *Geoderma*, Vol. 40, pp. 101–114, 1987). Scanning electrons micrographs of a typical Class F fly ash: (*b*) spherical glassy particles; (*c*) a plerosphere.

larger proportion in high-calcium fly ashes, the deposition of alkali sulfates on the surface of spherical particles accounts for their dirty appearance. Particle size distribution studies show that the particles in a typical fly ash vary from <1 μm to nearly 100 μm in diameter, with more than 50 percent by mass less than 20 μm (Fig. 8-15). The particle size distribution, morphology, and surface characteristics of the fly ash selected for use

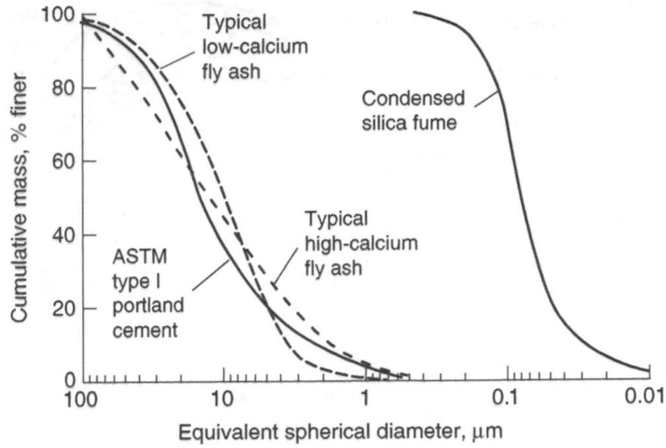

FIGURE 8-15 Comparison of particle size distribution of portland cement, fly ash, and condensed silica fume.

as a mineral admixture exercise a considerable influence on the water requirement and workability of fresh concrete, and rate of strength development in hardened concrete.

The hydration reaction of cement paste containing fly ash is complex. Figure 8-16 shows a transmission electron micrograph of a white portland cement paste with 50 percent fly ash replacement that had been cured for 9 years. Note that in this sample there are large areas of unreacted fly ash, where the glassy structure of the fly ash grain is still present. This micrograph is a good example of the uneven reactivity of the fly ash.[24] The reaction product intermixed with glassy unreacted fly ash is inner product (Ip), as it is formed in the original grain-filled space. In this case the Ip present is not as

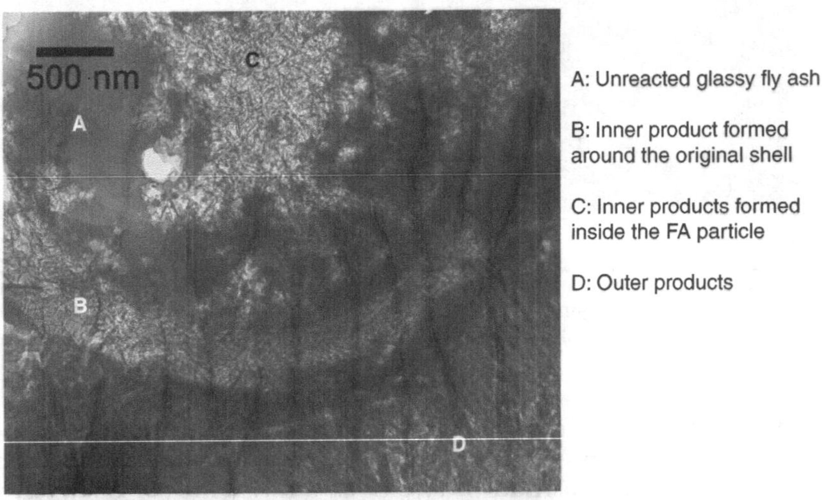

FIGURE 8-16 Nine-year-old white portland cement paste with 50 percent fly ash replacement. (*Micrograph courtesy of Rae Taylor—unpublished work.*)

dense as would be expected at this age of hydration, but is not as featureless as Ip normally is at earlier ages. A very small amount of outer-product (Op) C-S-H can be seen in this micrograph in the bottom right corner. This is identifiable as Op as it forms in the original water-filled space of the sample.

Iron Blast-Furnace Slag

In the production of cast iron (also called pig iron) when the slag is cooled slowly in air, the mineral components are usually present as crystalline melilites (C_2AS-C_2MS_2 solid solution) that do not react with water at ordinary temperature. When ground to very fine particles, the material will be weakly cementitious and pozzolanic. However, if the liquid slag is rapidly quenched from a high temperature by either water or a combination of air and water, most of the lime, magnesia, silica, and alumina are held in a noncrystalline or glassy state. The water-quenched product is called *granulated slag* due to the sand-size particles; the material quenched by air and a limited amount of water is in the form of pellets and is called *pelletized slag*. Normally, the former contains more glass; however, when ground to 400 to 500 m^2/kg Blaine, both products develop satisfactory cementitious properties.

Although high-calcium fly ashes are of relatively recent origin and the production and use of granulated blast-furnace slag in cement products is more than 100 years old, there are similarities in the mineralogical character and reactivity of the two materials. Both are essentially noncrystalline, and the reactivity of the high-calcium glass in both cases appears to be similar. Compared to low-calcium fly ash, which usually does not make much contribution to the strength of portland-cement products for two weeks after hydration, significant strength contribution by high-calcium fly ash or by granulated iron blast-furnace slag can generally occur as early as 7 days after hydration. When used in large amounts, some high-calcium fly ashes tend to retard the setting time of blended portland cement-fly ash mixtures. Although particle size characteristics, composition of glass, and the glass content are primary factors determining the reactivity of fly ash and slags, it may be noted that the reactivity of the glass itself varies with the thermal history of the material. A glass chilled from a higher temperature and at a faster rate will have a more disordered structure and will therefore be more reactive.

Generally, both with fly ash and slag, particles of less than 10 μm contribute to early strength of concrete up to 28 days; particles of 10 to 45 μm contribute to later strength, and particles coarser than 45 μm are difficult to hydrate. As the slag obtained after granulation is very coarse and also moist, it is dried and pulverized to particles mostly under 45 μm, corresponding to approximately 400 to 500 m^2/kg Blaine surface area.

The use of ground granulated blast furnace slag can alter significantly the morphology of cement paste. Figure 8-17 shows two types of C-S-H morphology: the hydration of portland cement produces C-S-H with fibril morphology, while the slag grains produce hydration products with a foil-like morphology.

Silica Fume

Silica fume is a by-product of the induction arc furnaces in the silicon metal and ferrosilicon alloy industries. Reduction of quartz to silicon at temperatures up to 2000°C produces SiO vapors, which oxidize and condense in the low-temperature zone of the furnace to tiny spherical particles (Fig. 8-18) consisting of noncrystalline silica. The material removed by filtering the outgoing gases in bag filters possesses an average

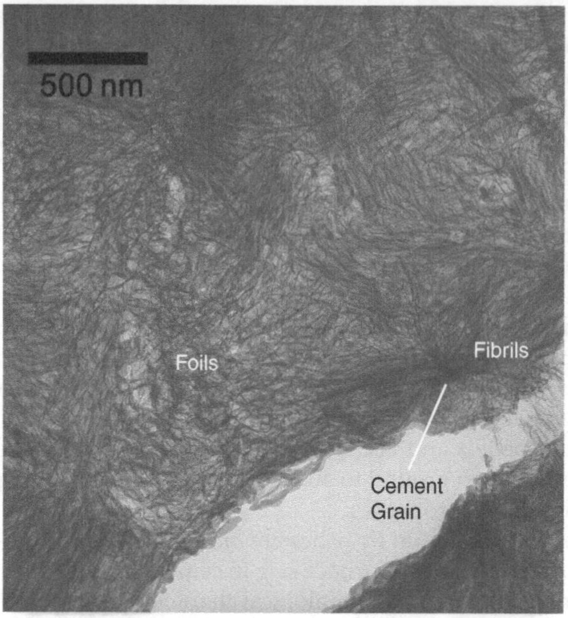

FIGURE 8-17 Paste composed of 10 percent portland cement and 90 percent ground granulated blast furnace slag cured for 3 months with both foil-like (crumped zones) and fibril morphology scale bar is 200 nm. (*Micrograph courtesy of Rae Taylor—unpublished work.*)

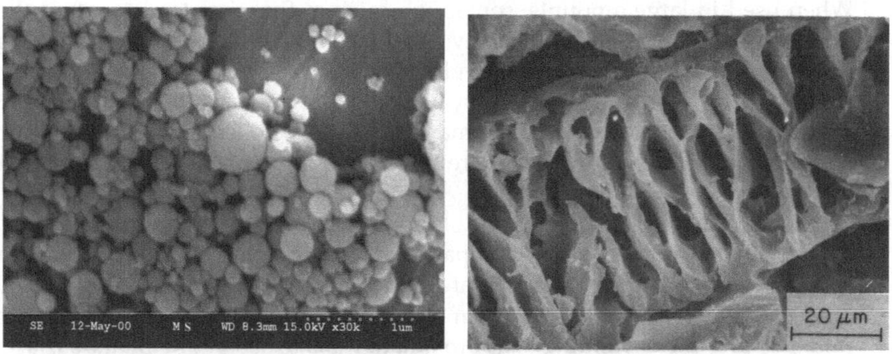

(a) (b)

FIGURE 8-18 Scanning electron micrographs of silica fume (left); and rice husk ash (right). [Silica fume micrograph courtesy of Elken company and rice hush ash micrograph courtesy of P. C. Aitcin, University of Sherbrooke.]

diameter on the order of 0.1 μm and surface areas in the range 15 to 25 m^2/g. Compared to normal portland cement and typical fly ashes, silica fume samples show particle size distributions that are two orders of magnitude finer (Fig. 8-15). The material is highly pozzolanic, but it is hard to handle and it increases the water requirement in concrete appreciably unless high range water-reducing admixtures are used. The by-products from the silicon metal and the ferrosilicon alloy industries, producing alloys with 75 percent or higher silicon content, contain 85 to 95 percent noncrystalline silica. The by-product from the production of ferro-silicon alloy with 50 percent silicon contains a much lower silica content and is unsuitable for use as a pozzolanic material.

Rice Husk Ash

Rice husks, also called rice hulls, are the shells produced during the dehusking operation of paddy rice. As they are bulky, the husks present an enormous disposal problem for centralized rice mills. Each ton of paddy rice produces about 200 kg of husk which, on combustion, yield approximately 40 kg ash. The ash formed during open-field burning or uncontrolled combustion in furnaces generally contains a large proportion of less crystalline reactive silica minerals such as cristobalite and tridymite, and must be ground to a very fine particle size in order to develop some pozzolanic activity. On the other hand, a highly pozzolanic ash can be produced by controlled combustion, where silica is retained in a noncrystalline form and in a cellular microstructure (Fig. 8-18). Commercially produced samples of this type of rice husk ash show 50 to 60 m^2/g surface area by nitrogen adsorption. Mehta[25] reported the effect of processing conditions on the characteristics of rice husk ash and the beneficial effects of the amorphous ash on concrete properties. The durability aspects of blended portland cement-rice husk ash mixtures are examined by Mehta and Folliard.[26] Zhang and Malhotra[27] have also confirmed that with regard to pozzolanic activity the reactive rice husk ash is similar to silica fume.

8.5.5 Applications

The mechanisms by which properties of concrete are improved by the use of combinations of portland cement and pozzolanic materials are described in Chap. 6. Selected applications of mineral admixtures are described next to illustrate the principles already discussed.

Workability Improvement

With fresh concrete mixtures that show a tendency to bleed or segregate, it is well known that incorporation of finely divided particles generally improves the workability by reducing the size and volume of voids. The finer a mineral admixture, the lesser the amount needed for enhancement of the cohesiveness and workability of freshly-mixed concrete.

The small size and the glassy texture of fly ash and slag makes it possible to reduce the amount of water required for a given consistency. Many investigations including the one reported by Berry and Malhotra[28] showed that by substituting 30 percent of the cement with a fly ash, 7 percent less water was required than the control concrete mixture of equal slump. With high-volume fly ash mixtures containing 50 to 60 percent Class F or C fly ash by mass of the cementitious material typical water reductions of the order of 15 to 20 percent have been reported.[29] Reduction of segregation and bleeding by the use of mineral admixtures is of considerable importance when concrete is placed by pumping. The improvements in cohesiveness and finishability are particularly valuable in lean concrete mixtures or those made with aggregates that are deficient in fine particles.

Note that although all mineral admixtures tend to improve the cohesiveness and workability of fresh concrete, many do not possess the water-reducing capability of fly ash and slag. For a given consistency of concrete, the use of very high surface area materials such as pumicite, rich husk ash, and silica fume increases the water requirement.

Durability to Thermal Cracking

Assuming that due to heat of hydration the maximum temperature in a massive structure is reached within about 3 days of the concrete placement, the use of ordinary mineral admixtures (i.e., natural pozzolan, fly ash, or slag) offers the possibility of reducing the temperature rise almost in direct proportion to the amount of portland cement replaced by the admixture. This is because, under normal conditions, these admixtures do not react to a significant degree for several days. As a rule of thumb, the total heat of hydration produced by the pozzolanic reactions involving mineral admixtures is considered to be half as much as the average heat produced by the hydration of portland cement.

Portland cement replacement by fly ash has been practiced in the United States since the 1930s. In mass concrete construction, where low cement content and fly ash proportions as high as 60 to 100 percent by weight of portland cement are now commonly employed, the first successful application was in 1948 for building the Hungry Horse Dam, Montana. More than 3 million cubic yards (2.3 million m^3) of concrete were used; some of the concrete contained 32 percent portland cement replaced by fly ash that was shipped from the Chicago area. More recently, fly ash was used in concrete for the Dworshak Dam, Idaho, which is a 7-million cubic yard (5.4 million m^3) concrete structure.

Furthermore, construction engineers should be aware of another advantage from the use of mineral admixtures when concrete is to be exposed to considerably higher than normal temperatures, either because of the heat of cement hydration or any other reason. In contrast to specimens cured in the laboratory, the cores from field concrete mixtures that do not contain any mineral admixture are likely to undergo a strength loss due to thermal microcracking. Compared to laboratory-cured cylinders, Manmohan and Mehta[30] found that the 7-day compressive strength of cores taken from a high-volume fly ash concrete were considerably greater. While high-temperature exposures can be harmful to portland cement concrete, the concrete containing a high proportion of a mineral admixture benefits from thermal activation (acceleration of the pozzolanic reaction). For the intake pressure tunnel of Kurobegowa Power Station in Japan, where the concrete is located in hot base rock (100 to 160°C), the use of 25 percent fly ash as a cement replacement in the concrete mixture showed a favorable effect on the strength.

Durability to Chemical Attack

The permeability of concrete plays a fundamental role in determining the rate of deterioration due to destructive chemical actions, such as the alkali-aggregate expansion and attack by acidic or sulfate solution (Chap. 5). As the pozzolanic reaction involving mineral admixtures causes pore refinement that reduces the permeability of concrete, both field and laboratory studies have shown considerable improvement in the chemical durability of concrete containing mineral admixtures.

Investigations in the 1950s by R. E. Davis at the University of California, Berkeley, on the permeability of concrete pipes both with and without fly ash showed that the permeability of concrete containing 30 percent* of a low-calcium fly ash was higher at

*As mineral admixtures are used in large amounts, it is customary to show their proportion in concrete as a mass percent of the total cementitious material (i.e., cement + mineral admixture).

28 days after casting, but was considerably lower at 6 months. Research[31] has confirmed that, with cement pastes containing 10 to 30 percent of a low-calcium fly ash, significant pore refinement occurred during the 28- to 90-day curing period, which resulted in a drastic reduction of the coefficient of permeability from 11 to 13×10^{-11} to 1×10^{-11} cm/s. In the case of cement pastes containing 10 to 30 percent of either rice husk ash or silica fume, or 70 percent of granulated blast-furnace slag, even at the early age of 28 days after the hydration, the system was found to be almost *impermeable*. These admixtures are therefore excellent for improving the durability to chemical attack.

Depending on the individual characteristics of the mineral admixture used, usually combinations of a high-alkali portland cement with either 40 to 65 percent granulated blast-furnace slag, 30 to 40 percent low-calcium fly ash, or 20 to 30 percent natural pozzolan, have been found to be effective in limiting the alkali-aggregate expansion to very low levels. In California, as aggregate deposits in many parts of the state contain alkali-reactive minerals, the Department of Water Resources has made it a standard practice to include pozzolanic admixtures in all concrete used for hydraulic structures. The use of large amounts of mineral admixtures to reduce the alkali-aggregate expansion is sometimes objectionable when high early strength is needed. Highly active pozzolans, such as rice husk ash and silica fume, when used in amounts as low as 10 percent are found to be very effective in inhibiting the alkali-aggregate expansion without reducing the strength at early ages. In Iceland, where only high-alkali portland cement is available and the aggregates are generally reactive, it is customary to blend 7 percent silica fume with all portland cement.

The published literature contains sufficient evidence that, in general, the incorporation of mineral admixtures improves the resistance of the concrete to acidic water, sulfate water, and seawater. This is due mainly to the pozzolanic reaction, which is accompanied by a reduction in permeability as well as a reduction in the calcium hydroxide content of the hydrated product. However, as discussed below, not all portland cement-slag or fly-ash combinations are found satisfactory in combating the sulfate attack in concrete.

In the 1960s and 1970s, extensive studies at the U.S. Bureau of Reclamation on concretes containing 30 percent low-calcium fly ashes showed greatly improved sulfate resistance to a standard sodium sulfate solution; however, the use of high-calcium fly ashes generally reduced the sulfate resistance. The explanation for this behavior probably lies in the mineralogical composition of the fly ash. In addition to the permeability and the calcium hydroxide content of the cement paste, the sulfate resistance is governed by the amount of reactive alumina present. Fly ashes containing a large proportion of reactive alumina in glass or in crystalline constituents would not be expected to reduce the sulfate resistance of concrete. In general, high-calcium fly ashes containing highly reactive alumina in the form of C_3A or $C_4A_3\bar{S}$ have been found to be less suitable than low-calcium fly ashes for improving the sulfate resistance of concrete.

The sulfate resistance of portland cement concrete containing granulated blast-furnace slag depends on the amount of slag used and the alumina content of the slag. The following excerpt from an appendix to the ASTM C 989 (*Standard Specification for Ground Iron Blast-Furnace Slag for Use in Concrete and Mortars*) explains why:

> *Effect of Ground Slag on Sulfate Resistance*. The use of ground slag will decrease the C_3A content of the cementing materials and decrease the permeability and calcium hydroxide content of the mortar or concrete. Tests have shown that the alumina content of the slag also influences the sulfate resistance at low slag-replacement percentages. The data from these studies of laboratory

exposures of mortars to sodium and magnesium sulfate solutions provide the following general conclusions.

The combinations of ground slag and portland cement, in which the slag content was greater than 60 to 65 percent, had high sulfate resistance, always better than the portland cement alone, irrespective of the alumina content of the slag. The low-alumina (11 percent) slag tested, increased the sulfate resistance independently of the C_3A content of the cement. To obtain adequate sulfate resistance, higher slag percentages were necessary with the higher C_3A cements. The high-alumina (18 percent) slag tested, adversely affected the sulfate resistance of portland cements when blended in low percentages (50 percent or less). Some tests indicated rapid decreases in resistance for cements in the 8 and 11 percent C_3A ranges with slag percentages as low as 20 percent or less in the blends.

Production of High-Strength and High-Performance Concrete

Since the 1980s mineral admixtures have played a very important part in the production of *high-performance concrete* and its precursor *high-strength concrete*. Now, they are finding application in the development of *self-consolidating concrete mixtures*. These three concrete types will be described in detail in Chap. 12. Briefly discussed here are some of the underlying concepts to help understand the potential contribution of mineral admixtures to the present-day concrete construction practice.

The published literature tends to describe high-performance concrete as any concrete that has at least one unusual property compared to ordinary concrete. Ideally, *a high-performance concrete mixture should meet all the following characteristics*:

- High constructability (high consistency without segregation)
- High resource productivity through high durability (low permeability and high resistance to cracking)
- High ultimate strength and moderate early strength (note that very high early strength concrete mixtures tend to crack)
- High environmental friendliness (through the maximum possible use of industrial by-products)
- Cost-effectiveness (the concrete should be a product of simple technology and conventional materials)

High-strength and high-performance concrete mixtures required for tall buildings and concrete bridge decks could not be produced using conventional portland cement alone. Concrete mixtures containing very high cement content are prone to cracking from thermal and drying shrinkage; those with too much mixing water tend to bleed and segregate. Therefore, only the use of mineral admixtures like fly ash and water-reducing admixtures in combination with portland cement provides a satisfactory solution to the problem. In 1984, Malhotra[32] concluded that *the development of high-strength concrete for use in tall buildings in Chicago area has shown that the use of fly ash is almost mandatory to achieve strengths greater than 60 MPa*. Concrete mixtures with 500 kg/m^3 portland cement, 60 kg/m^3 Class F fly ash, and 0.33 w/cm gave 72 MPa compressive strength in 56 days. Cook[33] reported similar strength results with structural elements used in the construction of all buildings in Houston, Texas; the concrete mixture contained 400 kg/m^3 portland cement and 100 kg/m^3 Class C fly ash. Using silica fume, a superplasticizer and aggregate mixtures with high packing density, ultrahigh strength concrete mixtures (>200 MPa compressive strength) are being commercially used in many countries.

8.6 Concluding Remarks

For ready reference purposes, a summary of the commonly used concrete admixtures, their primary function, principal active ingredients, applicable ASTM Standard Specification, and possible side effects are presented in Table 8-7.

In the 1940s and 1950s, efforts to promote the introduction of admixtures in concrete on a large scale met with considerable resistance because there was little understanding of their mode of action, leading to many unsatisfactory experiences. Today, the situation is different. Admixtures have become such an integral part of concrete that in the near future the definition of concrete should be revised to include admixture as a primary component of concrete mixtures.

Problems associated with the misuse of admixtures, however, continue to arise. The genesis of most of the problems appears to lie in the incompatibility between a particular admixture and a cement composition, or between two or more admixtures that may be present simultaneously. Surfactants such as air-entraining chemicals, lignosulfonates, and superplasticizers are especially sensitive to interaction effects among the aluminate, sulfate, and alkali ions in the solution phase at the beginning of cement hydration. Loss of air or proper air-void spacing in concrete containing a superplasticizer or an exceedingly fine mineral admixture is a matter of serious concern to the concrete industry. Therefore, it is highly recommended to carry out laboratory tests involving field materials and conditions before the actual use of admixtures in concrete construction, particularly when large projects are undertaken or when the concrete-making materials are subject to significant variations in quality.

Finally, admixtures can certainly enhance the properties of a concrete but should not be expected to compensate for the poor quality of concrete ingredients or poor mixture proportioning.

Test Your Knowledge

8.1 Why are plasticizing admixtures called *water reducing*? What is the distinction between normal and high-range water-reducing admixtures according to the ASTM Standard Specification?

8.2 Can you list and define the eight *types* of chemical admixtures, four *classes* of mineral admixtures, and three *grades* of iron blast-furnace slag that are used as admixtures for concrete?

8.3 After reviewing the ASTM C 618 and C 989 Standard Specifications and other published literature, write a critical note comparing the two standards.

8.4 What are the essential differences in composition and mode of action between the surfactants used for air entrainment and those used for water reduction?

8.5 Some manufacturers claim that application of water-reducing admixtures can lower the cement content and increase the consistency and strength of a reference concrete mixture. Explain why all three benefits may not be available at the same time.

8.6 Commercial lignin-based admixtures when used as water-reducing agents may exhibit certain side effects. Discuss the possible side effects and explain how they are corrected.

8.7 In their composition and mechanism of action, how do the superplasticizers differ from the normal water-reducing admixtures? Addition of 1 to 2 percent of a normal water-reducing agent

Primary Function	Principal Active Ingredient/ASTM Specification	Side Effects
Water-Reducing		
Normal	Salts, modifications and derivatives of lignosulfonic acid, hydroxylated carboxylic acids, and polyhydroxy compounds. ASTM C 494 (Type A).	Lignosulfonates may cause air entrainment and strength loss; Type A admixtures tend to be set retarding when used in high dosage.
High range	Sulfonated naphthalene or melamine formaldehyde condensates. ASTM C 494 (Type F).	Early slump loss; difficulty in controlling void spacing when air entrainment is also required.
Set-Controlling		
Accelerating	Calcium chloride, calcium formate, and triethanolamine. ASTM C 494 (Type C).	Accelerators containing chloride increase the risk of corrosion of the embedded metals.
Retarding	Same as in ASTM Type A; compounds such as phosphates may be present. ASTM C 494 (Type B).	
Water-Reducing and Set-Controlling		
Water-reducing and retarding	Same as used for normal water reduction. ASTM C 494 (Type D).	See Type A above.
Water-reducing and accelerating	Mixtures of Types A and C. ASTM C 494 (Type E).	See Type C above.
High-range water-reducing and retarding	Same as used for Type F with lignosulfonates added. ASTM C 494 (Type G).	See Type F above.
Workability-Improving		
Increase in consistency	Water-reducing agents [e.g., ASTM C 494 (Type A)].	See Type A above.
Reduce segregation	(a) Finely divided minerals (e.g., ASTM C 618) (b) Air-entraining surfactants (ASTM C 260).	Loss of early strength when used as cement replacement. Loss of strength. Loss of strength.
Strength-Increasing		
With water-reducing admixtures	Same as listed under ASTM C 494 (Types A, D, F, and G).	See Types A and F above.
With pozzolanic and cementitious admixtures	Same as listed under ASTM C 618 and C 989.	Workability and durability may be improved
Durability-Improving		
Frost action	Wood resins, proteinaceous materials, and synthetic detergents (ASTM C 260).	Strength loss.
Thermal cracking Alkali-aggregate expansion Sulfate and acidic solutions	Fly ashes, and raw or calcined natural pozzolans (ASTM C 618); granulated and ground iron blast-furnace slag (ASTM C 989); fly ash, condensed silica fume; rice husk ash.	Loss of strength at early ages, except when highly pozzolanic admixtures are used in conjunction with a superplasticizing agent.

TABLE 8-7 Commonly Used Concrete Admixtures

to a concrete mixture may cause segregation and severe retardation. These effects do not take place in the superplasticized concrete. Explain why.

8.8 When added to portland cement paste in very small amounts, calcium chloride acts as a retarder, but in large amounts it behaves as an accelerator. Can you explain the phenomenon?

8.9 Why doesn't calcium sulfate behave like an accelerator for portland cement similar to calcium chloride?

8.10 As an accelerator, why isn't sodium chloride as effective as calcium chloride?

8.11 Mineral acids are accelerators for portland cement, but organic acids do not show a consistent behavior. Explain why.

8.12 Formic acid is an accelerator, while gluconic acid is a retarder. Explain why.

8.13 What type of admixtures would you recommend for concreting in: (i) hot weather, (ii) cold weather.

8.14 When used as an accelerator, what effect would calcium chloride have on the mechanical properties, dimensional stability, and durability of concrete?

8.15 State several important reasons why it is desirable to use pozzolanic admixtures in concrete.

8.16 Why are clays and shales heat treated to make them suitable for use as a pozzolan?

8.17 Name some of the commonly available industrial by-products that show pozzolanic or cementitious properties when used in combination with portland cement.

8.18 What do you know about the origin and characteristics of the following mineral admixtures: pumice, zeolitic tuff, rice husk ash, and silica fume?

8.19 Compare and contrast industrial fly ashes and ground iron blast-furnace slag with respect to mineralogical composition and particle characteristics.

8.20 Explain the mechanism by which mineral admixtures are able to improve the pumpability and finishability of concrete mixtures. In the amounts normally used, some mineral admixtures are water reducing whereas others are not. Discuss the subject with the help of examples.

8.21 Discuss the mechanisms by which mineral admixtures improve the durability of concrete to acidic waters. Why is that not all fly ash-portland cement or slag-portland cement combinations may turn out to be sulfate-resisting?

8.22 What maximum strength levels have been attained in recently developed high-strength concrete mixtures? Explain the role played by admixtures in the development of these concretes.

8.23 What is high-performance concrete?

References

1. ACI Committee 212, Admixtures for Concrete, *ACI Manual of Concrete Practice*, American Concrete Institute, Farmington Hills, MI, 2005.
2. Mielez, R.C., *Concr. Int.*, Vol. 6, No. 4, pp. 40–53, 1984.
3. Lea, F.M., *The Chemistry of Cement and Concrete*, Chemical Publishing Company, New York, p. 596, 1971.

4. Malhotra, V.M., ed., *Proceedings of the International Conference on Superplasticizers and Other Chemical Admixtures in Concrete*, SP-195, American Concrete Institute, Farmington Hills, MI, 2000.
5. Spiratos, N., and C. Jolicoeur, ACI, SP-195, American Concrete Institute, Farmington Hills, MI, 2000.
6. ACI Committee 231-R10, "Report on early-age cracking: causes, measurement, and mitigation," American Concrete Institute, p. 46, 2010.
7. Brooks, J.J., and X. Jiang, The influence of chemical admixtures on restrained drying shrinkage of concrete, *ACI SP-173: Chemical Admixtures*, pp. 249–265, 1994.
8. Cope, B.L., and G.E. Ramey, Reducing drying shrinkage of bridge deck concrete, *Concrete International*, Vol. 23, No. 8, pp. 76–82, 2001.
9. Folliard, K.J., and N.S. Berke, Properties of high-performance concrete containing shrinkage reducing admixtures, *Cem. Concr. Res.*, Vol. 27, No. 9, pp. 1357–1364, 1997.
10. Sant, G., A. Eberhardt, D. Bentz, and W.J. Weiss, Influence of shrinkage-reducing admixtures on moisture absorption in cementitious materials at early ages, ASCE, *J. Mater. Civil Eng.*, Vol. 22, No. 3, pp. 277–286, 2010.
11. Bentz, D.P., K.A. Snyder, L.C. Cass, and M.A. Peltz, Doubling the service life of concrete I: reducing ion mobility using nanoscale viscosity modifiers, *Cem. Concr. Composites*, Vol. 30, No. 8, pp. 674–678, 2008.
12. Bentz, D.P., K.H. Hansen, and M.R. Geiker, Shrinkage reducing admixtures and early age desiccation in cement pastes and mortars, *Cem. Concr. Res.*, Vol. 31, No. 7, pp. 1075–1085, 2001.
13. Lura, P., B. Pease, G.B. Mazzotta, F. Rajabipour, and W.J. Weiss, Influence of shrinkage reducing admixtures on development of plastic shrinkage cracks, *ACI Mater. J.*, Vol. 104, No. 2, pp. 187–194, 2007.
14. Sant, G., P. Lura, and W.J. Weiss, Measurement of volume changes in cementitious materials at early ages: review of testing protocols and interpretation of results, *Trans. Res. Rec.*, Vol. 1979, pp. 21–29, 2006.
15. Bentz, D.P., K.H. Hansen, and M.R. Geiker, Shrinkage reducing admixtures and early age desiccation in cement pastes and mortars, *Cem. Concr. Res.*, Vol. 31, No. 7, pp. 1075–1085, 2001.
16. Sant, G., B. Lothenbach, P. Juilland, G. Lesaout, K. Scrivener, and Weiss, W. J., The origin of early-age expansions in cementitious materials containing shrinkage reducing admixtures, *Cem. Concr. Res.*, Vol. 41, No. 3, pp. 218–229, 2011.
17. Weiss, W.J., P. Lura, F. Rajabipour, and G. Sant, Performance of shrinkage reducing admixtures at different humidities and at early ages, *ACI Mater*. J., Vol. 105, No. 5, pp. 478–486, 2008.
18. Forsen, L., *Proceedings of the International Symposium on Chemistry of Cements*, Stockholm, p. 298, 1983.
19. Joisel, A., *Admixtures for Cement*, published by the author, Soisy, France, 1973.
20. Mehta, P.K., ACI, SP-79, American Concrete Institute, Detroit, pp. 1–35, 1983.
21. Jackson, M.D., J.M. Logan, B.E. Scheetz, D.M. Deocampo, C.G. Cawood, F. Marra, M. Vitti, and L. Ungaro, Assessment of material characteristics of ancient concretes, Grande Aula, Markets of Trajan, Rome: *J. Archaeol. Sci.*, Vol. 36, pp. 2481–2492, 2009.
22. Jackson, M.D., D. Deocampo, F. Marra, and B.E. Scheetz, Mid-Pleistocene volcanic ash in ancient Roman concretes, *Geoarchaeology*, Vol. 25, pp. 36–74, 2010.

23. Jackson, M., G. Vola, D. Vsiansky, J. Oleson, B. Scheetz, C. Brandon, and R. Hohlfelder, Cement microstructures and durability in ancient Roman seawater concretes, *Historic Mortars, Characteristics and Tests*, J. Valek, C. Groot, and J. Hughes, eds. Springer – RILEM, pp. 49–76, 2012.
24. Fernández-Jiménez, A., A. Palomo, and M. Criado, Microstructure development of alkali-activated fly ash cement: a descriptive model, *Cem. Concr. Res.*, Vol. 35, pp. 1204–1209, 2005.
25. Mehta, P.K., U.S. Patent No. 4105459 (Aug. 1978), and 5346548 (Sep. 1994); also in Malhotra, V.M., ed., *Advances in Concrete Technology*, CANMET, Ottawa, Canada, 1994.
26. Mehta, P.K., and K.J. Folliard, ACI, SP-154, American Concrete Institute, Detroit, 1995.
27. Zhang, M.H., and V.M. Malhotra, ibid.
28. Berry, E.E., and V.M. Malhotra, *J. ACI, Proc.*, Vol. 77, No. 2, pp. 59–73, 1980.
29. Malhotra, V.M., and P.K. Mehta, *High-Performance High-Volume Fly Ash Concrete*, 2d ed., Supplementary Cementitious Materials for Sustainable Development, Ottawa, Canada, 2005.
30. Manmohan, D., and P.K. Mehta, *Concr. Int.*, Vol. 24, No. 8, pp. 64–70, 2002.
31. Manmohan, D., and P.K. Mehta, *Cem. Concr. Aggregates*, Vol. 3, No. 1, pp. 63–67, 1981.
32. Malhotra, V.M., *Concr. Int.*, Vol. 6, No. 4, p. 21, 1984.
33. Cook, J.E., *Concr. Int.*, Vol. 4, No. 7, p. 72, 1982.

Suggestions for Further Study

ACI Committee 212, Admixtures for Concrete, *ACI Manual of Concrete Practice*, American Concrete Institute, Farmington Hills, MI, 2005.

Detweiler, R.J., J. Bhatty, and S. Bhattacharja, *Supplementary Cementing Materials for Use in Blended Cements*, Portland Cement Association, Bulletin RD112R, p. 96, 1996.

Helmuth, R., *Fly Ash in Cement and Concrete*, Portland Cement Association, Skokie, IL, 1987.

Lea, F.M., *The Chemistry of Cement and Concrete*, Chemical Publishing Company, New York, pp. 302–310, 414–489, 1971.

Malhotra, V.M., *Proceedings of International Conference on Superplasticizers and Other Chemical Admixtures*, ACI SP-148 (1994), SP-173 (1997), SP-195 (2000), SP-217 (2003), SP-239 (2006), SP-262 (2009), American Concrete Institute, Farmington Hills, MI.

Malhotra, V.M., ed., *Use of Fly Ash, Silica Fume, Slag, and Other Mineral By-products in Concrete, Proc. Symp.*, ACI, SP 79 (1983), SP 91 (1986), SP 114 (1989), SP 132 (1992), SP153 (1995), SP 178 (1998), SP 199 (2001), SP221 (2004), SP242 (2007), American Concrete Institute, Farmington Hills, MI.

Malhotra, V.M., and P.K. Mehta, *High-Performance, High-Volume Fly Ash Concrete for Building Durable and Sustainable Structures*, 4th ed., Supplementary Cementing Materials for Sustainable Development Inc., Ottawa, Canada, 2012.

Malhotra, V.M., and P.K. Mehta, *Pozzolanic and Cementitious Materials*, Gordon and Breech Publishers, Amsterdam, The Netherlands, p. 191, 1996.

Paillere, A.M., ed., *Applications of Admixtures in Concrete*, E & FN Spon, London,1995.

Rixom, R., and N. Mailvaganam, *Chemical Admixtures for Concrete*, E & FN Spon, London, 1999.

Ramachandran, V.S., ed., *Concrete Admixtures Handbook*, Noyes Publications, Park Range, NJ, 1995.

Sabir, B.B., S. Wild, and J. Bai, Metakaolin and calcined clays as pozzolan for concrete: a review, *Cem. Concr. Composites*, Vol. 23, pp. 441–454, 2001.

Snellings, R., G. Mertens, and J. Elsen, Supplementary cementitious materials, *Reviews in Mineralogy & Geochemistry*, Vol. 74, pp. 211–278, 2012.

CHAPTER 9

Proportioning Concrete Mixtures

Preview

To obtain concrete with certain desired performance characteristics, the selection of component materials is the first step. The next step is a process called *mixture proportioning*, which means achieving the right combination of components. Although there are sound technical principles governing mixture-proportioning procedures, for several reasons the process is not entirely in the realm of science. Nevertheless, because concrete composition greatly influences the cost and the properties of the product, it is important that engineers responsible for developing or approving mixture proportions should be familiar with the underlying principles and the commonly used procedures.

This chapter describes the significance and objectives of concrete mix proportioning. General considerations governing cost, workability, strength, and durability are discussed, and the ACI 211.1 *Standard Practice for Selecting Proportions for Normal, Heavy Weight, and Mass Concrete* is described, with a sample computation to illustrate the procedures.

9.1 Significance and Objectives

The *proportioning of concrete mixtures* is the *process* of arriving at the right combination of cement, aggregates, water, and admixtures for making concrete according to given specifications. For reasons described below, *this process is considered an art rather than a science*. Although many engineers do not feel comfortable with matters that cannot be reduced to an exact set of numbers, with an understanding of the underlying principles and, with some practice, the art of proportioning concrete mixtures can be mastered. Given an opportunity, the exercise of this art is very rewarding because the effect of mix proportioning on the cost of concrete and several important properties of both fresh and hardened concrete can be clearly seen.

One purpose of mix proportioning is to obtain a product that will perform according to certain predetermined requirements. Conventionally, the two most essential requirements are *the workability of fresh concrete and the strength of hardened concrete at a specified age*. Workability, which is discussed in more detail in Chap. 10, is the property that determines the ease with which a concrete mixture can be placed, compacted, and finished. Durability is another important property, but it is generally assumed that under normal exposure conditions durability will be satisfactory if the concrete mixture develops the necessary strength. Of course, under severe conditions, such as freeze-thaw cycles or exposure to sulfate water, the proportioning of concrete mixture will require special attention.

Another purpose of mix proportioning is to obtain a concrete mixture satisfying the performance requirements *at the lowest possible cost*. This involves decisions regarding the selection of ingredients that are not only suitable but also available at reasonable prices. The *overall objective* of proportioning concrete mixtures can therefore be summarized as selecting the suitable ingredients among the available materials and determining the most economical combination that will produce concrete with certain minimum performance characteristics.

The tools available to the engineer to achieve this objective are limited. An obvious constraint in concrete mixture proportioning is that within a fixed volume you cannot alter one component independent of others. For example, in a cubic meter of concrete, if the aggregate component is increased, the cement paste component decreases. With concrete-making materials of given characteristics and with given job conditions (i.e., structural design, and equipment for handling concrete), the variables generally under the control of a mix designer are as follows: the cement paste-aggregate ratio in the mixture, the water-cement ratio in the cement paste, the sand-coarse aggregate ratio in the aggregates, and the use of admixtures.

The task of mixture proportioning is complicated by the fact that certain desired properties of concrete may be oppositely affected by changing a specific variable. For example, the addition of water to a stiff concrete mixture with a given cement content will improve the flowability of fresh concrete but at the same time will reduce the strength. In fact, workability itself is composed of two main components [i.e., consistency (ease of flow) and cohesiveness (resistance to segregation)], and both tend to be affected in an opposite manner when water is added to a given concrete mixture. The process of mixture proportioning boils down to the art of balancing various conflicting requirements.

9.2 General Considerations

Before discussing the specific principles underlying the procedures commonly used for mixture proportioning, let us examine some of the general considerations such as cost, workability, strength, and durability of concrete.

9.2.1 Cost

The most obvious consideration when choosing concrete-making materials is that they are technically acceptable and, at the same time, economically attractive. In other words, when a material is available from two or more sources and a significant price differential exists, the least expensive source of supply is usually selected unless there are demonstrable technical reasons that the material will not be suitable for the job at hand.

In spite of the usually small differences in the price of aggregates from various local sources, the overall savings for a large project are worthy of consideration. Assume that a concrete mixture composed of 1800 kg/m^3 of total aggregate is required for a 6 million cubic meter concrete job, and that the two sources capable of furnishing suitable aggregates have a 10-cent/tonne price difference between them. A simple computation will show that a cost saving of over $1 million is possible if the less expensive aggregate is selected.

At times, for traditional or other reasons which may no longer be valid, some specifying agencies continue to require materials for concrete that are more expensive and perhaps unnecessary. For example, requiring the use of a low-alkali portland cement when the locally available cements are of high-alkali type and the aggregates are essentially free from alkali-reactive minerals will increase the cost of concrete due to the extra haulage expense for low-alkali cement. Even when the aggregate under consideration contains alkali-reactive minerals, the use of pozzolanic admixtures in combination with a high-alkali cement may turn out to be the more cost-effective alternative.

A *key consideration* governing many of the principles behind the procedures for proportioning concrete mixtures is the recognition that *cement costs much more than aggregates*; therefore, all possible steps should be taken to reduce the cement content of a concrete mixture without sacrificing the desired performance characteristics of concrete, such as strength and durability.

For the purpose of illustration, let us refer to the data in Fig. 3-6 (Mixtures Nos. 1 and 3). A reduction in the cement content from 530 to 460 lb per cubic yard of concrete at a given water-cement ratio (i.e., without compromising the strength of concrete) made it possible to reduce the cost by $1.55 per cubic yard, because a lower consistency was acceptable for the job. This may well be the case with lightly reinforced or unreinforced concrete structures. The economic implication of reduction in the cement content can be enormous in the projects requiring large amounts of concrete.

Further cost reduction is possible, without compromising the essential performance characteristics of a concrete mixture, *if cheaper and suitable materials are found to replace a percentage of portland cement*. For instance, under most conditions, substitution of pozzolanic or cementitious by-products (such as fly ash or ground granulated iron blast-furnace slag) for portland cement is likely to produce direct savings in the cost of materials. Furthermore, at some point in the future every nation will have to consider the *indirect cost savings* resulting from resource preservation and reduced pollution when these industrial by-products are utilized properly, instead of being dumped into the environment (see Chap. 14).

9.2.2 Workability

Workability of fresh concrete has a direct effect on the pumpability and constructibility because it determines the ease with which a concrete mixture can be handled without harmful segregation. In all likelihood, a concrete mixture that is difficult to place and consolidate will not only increase the cost of handling but will also have poor strength, durability, and appearance. Similarly, mixtures prone to segregate and bleed are more expensive to finish and will yield less durable concrete. Thus, workability can affect both the cost and the quality of concrete mixtures.

However, there is a problem. The term *workability* represents many diverse characteristics of fresh concrete that are difficult to measure quantitatively. This is another reason why the proportioning of concrete mixtures for a desirable but not fully definable measure of workability remains an art as well as a science. Clearly, mere knowledge of mixture design procedures is not sufficient without an understanding of the basic principles involved.

General considerations guiding the workability of concrete mixtures are as follows:

- The consistency of concrete should be no more than necessary for the ease of placing, compaction, and finishing.
- The water requirement for a given consistency increases with both sand/coarse aggregate ratio and the amount of fines in the sand. Whenever possible,

the cohesiveness and finishability of concrete should be improved by increasing the sand/coarse aggregate ratio alone rather than by increasing the proportion of fine particles in the sand.

- For concrete mixtures requiring high consistency at the time of placement, the use of water-reducing and set-retarding admixtures should be considered rather than the addition of extra water at the job site; water that has not been accounted for in the mixture proportioning is frequently responsible for the failure of concrete to perform according to design specifications.

9.2.3 Strength and Durability

In Chap. 2 it was shown that strength and impermeability of hydrated cement pastes are mutually related through capillary porosity, which is controlled by the water-cement ratio and the degree of hydration (Fig. 2-11). With the exception of frost resistance, the durability of concrete is generally controlled by permeability. Consequently, *in routine mix designing operations only the workability and strength of concrete are specified; consideration of durability is ignored unless special environmental exposures require it.*

With normally available cements and aggregates, structural concretes of consistency and strength adequate for most purposes, that is, 100- to 150-mm slump and 20 to 40 MPa 28-day compressive strength, can be produced without any difficulty. When strength or durability considerations require a lower water-cement ratio, this is generally achieved by lowering the water demand at a given cement content through control of the aggregate grading and the use of water-reducing admixtures. This approach not only is more economical but also would reduce the chances of cracking due to high thermal shrinkage and high drying shrinkage when the water-cement ratio is lowered by using a high cement content.

9.2.4 Ideal Aggregate Grading

Considerations of cost, workability, strength, and durability may lead to the assumption that the most dense aggregate packing with a minimum content of voids will be the most economical because it requires the least amount of cement paste. This assumption has led to a number of theoretical studies on the *packing density* of granular materials, which is defined as the solid volume in a unit total volume. The objective of such studies has been to obtain mathematical expressions or ideal grading curves that help determine the ideal combination of different size fractions of aggregate particles to produce the minimum void space. De Larrard[1] provides an excellent review of models to predict the packing density of granular mixtures.

Besides being uneconomic, the use of ideal aggregate grading is not prevalent in concrete field practice because often it does not produce the best workability. In the United States, the grading limits specified by ASTM C 33 are usually followed. Not only they are broad and therefore economically attractive, but also are based on practical experience with a large number of concrete mixtures. Using aggregates outside the limits of ASTM C 33 have caused workability problems and produced large voids in concrete. However, using aggregates that meet the requirements of ASTM C 33 may not necessarily produce satisfactory concrete mixtures because the grading limits happen to be too broad to guarantee optimum packing density. Shilstone[2] reported that combined mixture containing the coarse and the fine aggregates is often deficient of particles in the size range 4.75 to 9.5 mm. This can be remedied by substituting a portion (e.g., 15 to 30 percent by mass) of the coarse aggregate with pea-size (4.75 to 9.5 mm) gravel or crushed rock.

9.3 Specific Principles

When reviewing the following specific principles for selecting concrete mixture proportions, it will be helpful to remember again that the underlying goal is to strike a reasonable balance between the workability, strength, durability, and cost of concrete.

9.3.1 Workability

As already stated, workability embodies certain characteristics of fresh concrete, such as consistency and cohesiveness. *Consistency*, broadly speaking, is a measure of the wetness of the concrete mixture, which is commonly evaluated in terms of slump (i.e., the wetter the mixture, the higher the slump). If the water content is a key factor affecting the cost economy, it should be noted that there is almost a direct proportionality between the slump and the water content, with a given set of materials. To obtain the specified slump, the mixture water requirement generally decreases as: (1) the maximum size of a well-graded aggregate is increased; (2) the content of angular and rough-textured particles in the aggregate is reduced; (3) the amount of entrained air in the concrete mixture is increased; and (4) coal fly ash is used as a partial replacement for a cement.

Cohesiveness is a measure of compactibility and finishability, which is generally evaluated by trowelability and visual judgment of resistance to segregation. In trial mixtures when cohesiveness is judged as poor, it can usually be improved by taking one or more of the following steps: increase the sand/coarse aggregate ratio, partially replace the cement or sand with coal fly ash, and increase the cement paste/aggregate ratio. Obviously, due to its lower density, fly ash has the ability to increase the cement mortar/aggregate ratio by volume without an increase in the cement, water, or sand content of the mixture.

As the slump of fresh concrete is a measure of the ease with which the concrete mixture flows during the placement, and as the test for slump is simple and quantitative, most *mix-design procedures rely on slump as a crude index of workability*; it is assumed that mixtures containing adequate cement content (with or without mineral admixtures) and well-graded aggregate will have a satisfactory degree of cohesiveness. It should be noted that several laboratory trial mixtures are usually necessary before arriving at a qualitative notion of workability judged as satisfactory for a given job. Due to differences in equipment, further adjustment in the mixture proportions may be needed after a field trial or after some experience with full-size batch leads. This is yet another reason why past experience is recognized as so important in concrete mix-proportioning.

It is worth mentioning here that there are no standard requirements for workability because they may vary from one job to another, depending on the type of construction and the equipment used to transport and consolidate concrete. For example, the workability of concrete desired for a slip-formed unreinforced pavement will not be the same as for a congested reinforced column, and the workability desired for pumped concrete in a high-rise structure will not be the same as for mass concrete placed by crane or belt conveyor.

9.3.2 Strength

From the standpoint of structural safety, the strength of concrete specified by the designer is treated as the minimum required strength. Therefore, to account for variations in materials; methods of mixing, transportation, and placement of concrete; and curing and testing of concrete specimens, ACI Building Code 318 requires a certain degree of strength overdesign, which is based on statistical considerations. In other words, depending on the variability of test results, the mixture proportions selected must yield a mean or average strength higher than the minimum or the specified strength. The procedure for determining the average strength from a specified strength value is given in the Appendix at the end of this chapter. It should be noted that the average strength, not the specified strength, is used in mixture design calculations.

Although other factors also influence strength, the tables and charts used for the purposes of mixture proportioning assume that *strength is solely dependent on the water-cement ratio and the content of entrained air in concrete.* A more accurate relationship between the strength and water-cement ratio for a given set of materials and conditions may be available from past experience or should be developed from trial mixtures. Depending on the moisture state of the aggregate, corrections in the amounts of mixing water, sand, and coarse aggregate are necessary to make sure that the water-cement ratio in the concrete mixture is correct.

9.3.3 Durability

As stated earlier, when concrete is subject to normal conditions of exposure, *the mix-proportioning procedures ignore durability* because strength is considered to be an index of general durability. However, under conditions that may tend to shorten the service life of concrete, its durability may be enhanced by special considerations in mixture proportioning. For example, entrained air is required with all exposed concrete in climates where freezing and thawing cycles occur. Concrete exposed to chemical attack by deicing salts or acidic or sulfate waters may require the use of water-reducing and mineral admixtures. In such a situation, although a higher water-cement ratio would have satisfied the strength requirement, a lower water-cement ratio is usually specified considering the exposure conditions.

9.4 Procedures

Numerous procedures for computing the concrete mixture proportions are available in most countries of the world. Mathematical approaches to determine the correct proportion of component materials of a concrete mixture meeting a given set of specifications generally do not work because the materials vary widely in their characteristics. This explains why there is a large number of empirical methods based on extensive test data developed from local materials. A comprehensive review of the British and French procedures is contained in *Concrete Mixture Proportioning*.[1] The method recommended by ACI Committee 211,[3] is popular in the United States and many other countries in the world. The general principles underlying this method are described below.

The *weight method* is considered less exact but does not require the information on the specific gravity of the concrete-making materials. The *absolute volume method* is considered more exact. Both procedures involve a sequence of nine steps given below, the first six steps being common. To the extent possible, the following background data should be gathered before starting the calculations:

- Sieve analysis of fine and coarse aggregate; fineness modulus
- Dry-rodded unit weight of coarse aggregate
- Bulk specific gravity of materials

- Absorption capacity or free moisture in the aggregate
- Variations in the approximate mixing water requirement with slump, air content, and grading of the available aggregates
- Relationship between strength and water-cement ratio for available combinations of cement and aggregate
- Job specifications if any [e.g., maximum water-cement ratio, minimum air content, minimum slump, maximum size of aggregate, and strength at early ages (normally, 28-day strength is specified)].

Regardless of whether the concrete characteristics are prescribed by specifications or left to the mixture designer, the batch weights can be computed using the following sequence of steps: The data in Tables 9-1 to 9-6 are in the U.S. customary units. The same data are presented in metric units in Tables 9-1A through 9-3A, 9-5A, and 9-6A (Sec. 9.6).

Step 1: Choice of Slump If the slump is not specified, a value appropriate for the job can be selected from Table 9-1. Mixtures with the stiffest possible consistency that can be easily placed and compacted without segregation should be used. Concrete mixtures to be placed by pumping are typically designed for 4 in. (100 mm) to 6 in. (150 mm) slump.

Step 2: Choice of Maximum Size of Aggregate For the same volume of coarse aggregate, using a large maximum size of a well-graded aggregate will produce less void space than a smaller size, thereby reducing the mortar requirement in the concrete mixture. Generally, the maximum size of coarse aggregate should be the largest that is economically available and consistent with the dimensions of the structure. ACI recommends that, in no event, should the maximum aggregate size exceed one-fifth of the narrowest dimension between the sides of the forms, one-third the depth of slabs, or three-fourths of the minimum clear spacing between reinforcing bars.

Step 3: Estimation of the Mixing Water Content and Air Content According to ACI recommendations, the quantity of mixing water per unit volume of concrete required to produce a given slump is dependent mainly on the maximum particle size of the aggregate and whether or not the mixture has entrained air. Accordingly, Table 9-2 has separate

Types of Construction	Slump (in.)	
	Maximum*	**Minimum**
Reinforced foundation walls and footings	3	1
Plain footings, caissons, and substructure walls	3	1
Beams and reinforced walls	4	1
Building columns	4	1
Pavements and slabs	3	1
Mass concrete	2	1

*May be increased by 1 in. for consolidation methods other than vibration.
Source: Reproduced with permission from the American Concrete Institute.

TABLE 9-1 Recommended Slump for Various Types of Construction set of mixing water data for both non-air-entrained concrete and air-entrained concretes. The data in the table also show the approximate amount of entrapped air expected in non-air-entrained concrete and the recommended levels of air content for the mixture requiring purposely entrained air for frost resistance. Guidelines are provided to accommodate mixing water reductions resulting from the use of well-rounded aggregate and water-reducing chemical admixtures.

Step 4: Selection of Water-Cement Ratio Because different aggregates and cement types may produce different strength at the same water-cement ratio, it is desirable to develop the relationship between strength and water-cement ratio for the materials to be used actually. In the absence of such data, values shown for concrete with ASTM Type I portland cement (Table 9-3) can be used for trial mixtures. The water-cement ratio obtained from the table may have to be reduced depending on any durability requirements in addition to strength (Table 9-4). For instance, with structures exposed to frost action and a moist environment, a maximum w/c of 0.50 is permitted (0.45 w/c for thin sections). With structures exposed to seawater or sulfates, the maximum permissible w/c is 0.45 (0.40 for thin sections).

Step 5: Calculation of the Cement Content The cement content can be computed by dividing the mixing water content from Step 3 by the water-cement ratio.

Step 6: Estimation of the Coarse Aggregate Content Economy can be gained by using the maximum possible volume of coarse aggregate on a dry-rodded basis per unit volume of concrete. Data from a large number of tests have shown that with properly graded materials the finer the sand and the larger the size of the coarse aggregate particles, the higher is the volume of the coarse aggregate that can be used to produce a concrete mixture of satisfactory workability. From the data in Table 9-5, the volume of coarse aggregate in a unit volume of concrete can be estimated from its maximum aggregate size and the fineness modulus of fine aggregate. This volume is converted to the dry weight of coarse aggregate by multiplying with the experimentally determined value of the dry-rodded unit weight.

Step 7: Estimation of the Fine Aggregate Content After completing Step 6, all the ingredients of the concrete mixture have been estimated except the fine aggregate. Its quantity is determined by difference, either by the "weight" method or by the "absolute volume."

According to the *weight method*, if the unit weight of fresh concrete is known from previous experience, then the required weight of fine aggregate is simply the difference between the unit weight of concrete and the total weights of water, cement, and coarse aggregate. In the absence of a reliable estimate of the unit weight of concrete, the first estimate for a concrete mixture of moderate strength, medium slump and approximately 2.7 aggregate specific gravity can be obtained from Table 9-6. Experience shows that even a rough estimate of the unit weight is adequate for making trial batches.

In the case of the *absolute volume method*, the total volume displaced by the known ingredients (i.e., water, air, cement, and coarse aggregate) is subtracted from the unit volume of concrete to obtain the required volume of fine aggregate. This, in turn, is converted to weight units by multiplying it by the density of the material.

Step 8: Adjustments For the Aggregate Moisture Generally, the stock aggregates are moist; without moisture correction the actual water-cement ratio of the trial mix will be higher

	Water, lb/yd^3 of Concrete for Indicated Nominal Maximum Sizes of Aggregate							
Slump, in.	**3/8 in.***	**½ in.***	**¾ in.***	**1 in.***	**1½ in.***	**2 in.*,†**	**3 in.†**	**6 in.†**
Non-Air-Entrained Concrete								
1 to 2	350	335	315	300	275	260	220	190
3 to 4	385	365	340	325	300	285	245	210
6 to 7	410	385	360	340	315	300	270	—
More than 7*	—	—	—	—	—	—	—	—
Approximate amount of entrapped air in non-air-entrained concrete, percent	3	2.5	2	1.5	1	0.5	0.3	0.2
Air-Entrained Concrete								
1 to 2	305	295	280	270	250	240	205	180
3 to 4	340	325	305	295	275	265	225	200
6 to 7	365	345	325	310	290	280	260	—
More than 7*	—	—	—	—	—	—	—	—
Recommended averages total air content, percent for level of exposure:								
Mild exposure	4.5	4.0	3.5	3.0	2.5	2.0	1.5‡,§	1.0‡,§
Moderate exposure	6.0	5.5	5.0	4.5	4.5	4.0	3.5‡,§	3.0‡,§
Severe exposure§	7.5	7.0	6.0	6.0	5.5	5.0	4.5‡,§	4.0‡,§

*The quantities of mixing water given for air-entrained concrete are based on typical total air content requirements as shown for "moderate exposure" in the table above.

†The slump values for concrete containing aggregate larger than $1^1/_2$ in. are based on the slump tests made after removal of particles larger than $1^1/_2$ in. by wet-screening.

‡For concrete containing large aggregates that will be wet-screened over the $1^1/_2$ in. sieve prior to testing for air content, the percentage of air expected in the $1^1/_2$ in. minus material should be as tabulated in the column. However, initial proportioning calculations should include the air content as a percent of the whole.

§When using large aggregate in low cement factor concrete, air entrainment need not be detrimental to strength. In most cases mixing water requirement is reduced sufficiently to improve the water-cement ratio and to thus compensate for the strength-reducing effect of air-entrained concrete. Generally, therefore, for these large nominal maximum sizes of aggregate, air contents recommended for extreme exposure should be considered even though there may be little or no exposure to moisture and freezing.

TABLE 9-2 Approximate Mixing Water and Air Content Requirements for Different Slumps and Nominal Maximum Sizes of Aggregates

Water-Cement Ratio, by Weight		
Compressive Strength at 28 Days (psi)*	Non-Air-Entrained Concrete	Air-Entrained Concrete
6000	0.41	—
5000	0.48	0.40
4000	0.57	0.48
3000	0.68	0.59
2000	0.82	0.74

*Values are estimated average strengths for concrete containing not more than percentage of air shown in Table 9-2. For a constant water-cement ratio, the strength of concrete is reduced as the air content is increased. Strength is based on 6 by 12 in. cylinders moist-cured 28 days at 73.4 ± 3°F (23 ± 1.7°C) in accordance with Sec. 9(b) of ASTM C31, for *Making and Curing Concrete Compression and Flexure Test Specimens in the Field.*

Source: Reproduced with permission from the American Concrete Institute.

TABLE 9-3 Relationships between Water-Cement Ratio and Compressive Strength of Concrete

than selected by Step 4, and the saturated-surface dry (SSD) weights of aggregates will be lower than estimated by Steps 6 and 7. The mixture proportions determined by Steps 1 to 7 are assumed to be on an SSD basis. For the trial batch, depending on the amount of free moisture in the aggregates, the mixing water is reduced and the amounts of aggregates correspondingly increased, as shown later by sample computations.

Step 9: Trial Batch Adjustments Because of many assumptions underlying the foregoing theoretical calculations, the mix proportions for the actual materials to be used must be checked and adjusted by means of laboratory trials consisting of small batches

Exposure	Water Soluble Sulfate* (SO_4) in Soil (% by Mass*)	Sulfate* (SO_4) in Water (ppm)	Cementitious Material Requirements	Water-Cementitious Materials Ratio, Maximum†
Class 0	0.00–0.10	0–150	—	—
Class 1	0.10–0.20	150–1500	Type II or equivalent	0.50
Class 2	0.20–2.00	1500–10,000	Type V or equivalent	0.45
Class 3	Over 2.00	Over 10,000	Type V + pozzol and or slag	0.40

*Sulfate expressed as SO_4 is related to sulfate expressed as SO_3 as in reports of chemical analysis of cement as $SO_3 \times 1.2 = SO_4$.

†When chlorides or other depassivating agents are present in addition to sulfate, a lower water-cement ratio may be necessary to reduce corrosion potential of embedded items. Refer to Chap. 5.

Source: ACI Committee 201, Report 201.2R-08: Guide to Durable Concrete, *ACI Manual of Concrete Practice*, Part 1, Concrete Institute, Farmington Hills, MI, 2008.

TABLE 9-4 Recommendations for Normal Weight Concrete Subject to Sulfate Attack

Maximum Size of Aggregate (in.)	Volume of Dry-Rodded Coarse Aggregate* per Unit Volume of Concrete for Different Fineness Moduli of Sand			
	2.40	2.60	2.80	3.00
3/8	0.50	0.48	0.46	0.44
½	0.59	0.57	0.55	0.53
¾	0.66	0.64	0.62	0.60
1	0.71	0.69	0.67	0.65
1½	0.75	0.73	0.71	0.69
2	0.78	0.76	0.74	0.72
3	0.82	0.80	0.78	0.76
6	0.87	0.85	0.83	0.81

*Volumes are based on aggregates in dry-rodded condition as described in ASTM C29, *Unit Weight of Aggregate*. These volumes are selected from empirical relationships to produce concrete with a degree of workability suitable for usual reinforced construction. For less workable concrete such as required for concrete pavement construction they may be increased about 10 percent. For more workable concrete, such as may sometimes be required when placement is to be by pumping, they may be reduced up to 10 percent.

Source: Reproduced with permission from the American Concrete Institute

Table 9-5 Volume of Coarse Aggregate per Unit of Volume of Concrete

Maximum Size of Aggregate (in.)	First Estimate Concrete Weight* (lb/yd³)	
	Non-Air-Entrained Concrete	Air-Entrained Concrete
3/8	3840	3690
½	3890	3760
¾	3960	3840
1	4010	3900
1½	4070	3960
2	4120	4000
3	4160	4040
6	4230	4120

*Values calculated for concrete of medium richness (550 lb of cement per cubic yard) and medium slump with aggregate specific gravity of 2.7. Water requirements based on values for 3 to 4 in. of slump in Table 9-2. If desired, the estimated weight may be refined as follows when necessary information is available: for each 10-lb difference in mixing water from the Table 9-2 values for 3 to 4 in. of slump, correct the weight per cubic yard 15 lb in the opposite direction; for each 100-lb difference in cement content from 550 lb, correct the weight per cubic yard 15 lb in the same direction; for each 0.1 by which aggregate specific gravity deviates from 2.7, correct the concrete weight 100 lb in the same direction.

Source: Reproduced with permission form the American Concrete Institute.

Table 9-6 First Estimate of Weight of Fresh Concrete

(e.g., 0.01 yd^3 of concrete). Fresh concrete should be tested for slump, workability (freedom from segregation), unit weight, and air content; specimens of hardened concrete cured under standard conditions should be tested for strength at the specified age. After several trials, when a mixture satisfying the desired criteria of workability and strength is obtained, the mixture proportions of the laboratory-size trial batch are scaled up for producing full-size field batches.

9.5 Sample Computations

Job Specifications			
Type of construction	Reinforced concrete footing		
Exposure	Mild (below ground, not exposed to freezing or sulfate water)		
Maximum size of aggregate	1½ in.		
Slump	3 to 4 in.		
Specified 28-day compressive strength	3500 psi		
Characteristics of the Materials Selected			
	Cement, ASTM Type I	**Fine Aggregate**	**Coarse Aggregate**
Bulk specific gravity	3.15	2.60	2.70
Bulk density (lb/ft^3)	196	162	168
Dry-rodded unit weight (lb/ft^3)	—	—	100
Fineness modulus	—	2.8	—
Moisture deviation from SSD condition (%)	—	+2.5	+0.5

Steps 1 to 7: Computing Mix Proportions (SSD Basis, lb/yd^3)

Step 1. Slump = 3 to 4 in. (given).

Step 2. Maximum aggregate size = 1 – ½ in. (given).

Step3. Mixing water content (non-air-entrained concrete) = <u>300</u> lb. Approximate amount of entrapped air = 1 percent (Table 9-2).

Step 4. Average strength from equations in the Appendix (assuming 300 psi standard deviation from past experience) = 3500 + 1.34 × 300 = 3900 psi. Water-cement ratio (Table 9-3) = 0.58.

Step 5. Cement content = 300/0.58 = <u>517</u> lb.

Step 6. Volume fraction of gravel on dry-rodded basis (Table 9-5) = 0.71. Dry-rodded volume of gravel = 0.71 × 27 = 19.17 ft^3. Weight of gravel = 19.17 × 100 = <u>1917</u> lb.

Step 7. Using the *weight method*: unit weight of concrete (Table 9-6) = 4070 lb/yd³. Weight of sand = 4070 – (300 + 517 + 1917) = 1336 lb.

Using the *absolute volume method*:

$$
\begin{aligned}
&\text{Volume displaced by water} = 300/62.4 = 4.81\ \text{ft}^3\\
&\text{Volume displaced by cement} = 517/196 = 2.64\ \text{ft}^3\\
&\text{Volume displaced by gravel} = 1917/168 = 11.43\ \text{ft}^3\\
&\text{Volume displaced by air} = 27 \times 0.01 = 0.27\ \text{ft}^3\\
&\text{Total} \quad 19.15\ \text{ft}^3\\
&\text{Volume displaced by sand} = (27 - 19.15) = 7.85\ \text{ft}^3\\
&\text{Weight of sand} = 7.85 \times 162 = \underline{1272}\ \text{lb}
\end{aligned}
$$

Because the absolute volume method is more exact, the proportions determined by this method will be used.

Step 8. Moisture adjustment for the laboratory trial batch

Material	SSD (lb/yd³)	SSD (lb/0.01 yd³)	Moisture Correction (lb)	Mix Proportions for the First Trial Batch (lb)
Cement	517	5.17	—	5.17
Sand	1272	12.72	12.72 × 0.025 = 0.3	13.02
Gravel	1917	19.17	19.17 × 0.005 = 0.1	19.27
Water	300	3.00	3 – (0.3 + 0.1)	2.60
Total	4006	40.06	Must be equal	40.06

Step 9. Making the first laboratory trial and adjusting the proportions

Measured properties of fresh concrete from the first trial batch:

Slump = 4¾ in.
Workability = slight tendency to segregate and bleed
Unit weight = 148 lb/ft³ (3996 lb/yd³)
Air content = 1%

Action taken for the second trial batch: reduce the gravel by 1/4 lb and increase the sand by the same amount.

Batch weights for the second trial batch:

Cement = 5.17 lb
Sand = 13.27 lb
Gravel = 19.02 lb
Water = 2.60 lb
40.06 lb

Measured properties of fresh concrete from the second trial batch:

$$\text{Slump} = 4 \text{ in.}$$
$$\text{Workability} = \text{satisfactory}$$
$$\text{Unit weight} = 148 \text{ lb/ft}^3$$
$$\text{Air content} = 1\%$$

Three 3- by 6-in. cylinders were cast and moist cured at 73.4 ± 3°F.

Average 28-day compressive strength was 4250 psi, with less than 5 percent variation in strength between the individual cylinders.

Recalculated mix proportions for the full-size field batch are as follows:

	Present Stock (lb/yd³)	Moisture Correction (for Conversion to SSD Condition) (lb)	SSD Basis (lb/yd³)
Cement	517		517
Sand	1327	1327 × 0.025 = 33	1294
Gravel	1902	1902 × 0.005 = 10	1892
Water	260	260 + (33 + 10)	303
Total	4006	Must be equal	4006

9.6 ACI Tables in the Metric Units System

Types of Construction	Slump (mm)	
	Maximum*	Minimum
Reinforced foundation walls and footings	75	25
Plain footings, caissons, and substructure walls	75	25
Beams and reinforced walls	100	25
Building columns	100	25
Pavements and slabs	75	25
Mass concrete	50	25

*May be increased by 25 mm for consolidation methods other than vibration.

Source: Reproduced with permission from the American Concrete Institute.

TABLE 9-1A Recommended Slump for Various Types of Construction

9.7 Proportioning of High-Strength and High-Performance Concrete Mixtures

For a variety of reasons, the ACI 211 procedure for concrete mixture proportioning needs updating. This procedure and the other currently available procedures were developed when concrete mixtures were required to meet rather narrow specifications

Water (kg/m³) of Concrete for Indicated Nominal Maximum Sizes of Aggregate								
Slump, mm	**9.5[*]**	**12.5[*]**	**19[*]**	**25[*]**	**37.5[*]**	**50[*,†]**	**75[†]**	**150[†]**
Non-Air-Entrained Concrete								
25 to 50	207	199	190	179	166	154	130	113
75 to 100	228	216	205	193	181	169	145	124
150 to 175	243	228	216	202	190	178	160	—
Approximate amount of entrapped air in non-air-entrained concrete, percent	3	2.5	2	1.5	1	0.5	0.3	0.2
Air-Entrained Concrete								
25 to 50	181	175	168	160	150	142	122	107
75 to 100	202	193	184	175	165	157	133	119
150 to 175	216	205	197	184	174	166	154	—
Recommended averages total air content, percent for level of exposure:								
Mild exposure	4.5	4.0	3.5	3.0	2.5	2.0	1.5[‡,§]	1.0[‡,§]
Moderate exposure	6.0	5.5	5.0	4.5	4.5	4.0	3.5[‡,§]	3.0[‡,§]
Severe exposure	7.5	7.0	6.0	6.0	5.5	5.0	4.5[‡,§]	4.0[‡,§]

*The quantities of mixing water given for air-entrained concrete are based on typical total air content requirements as shown for "moderate exposure" in the table above.

†The slump values for concrete containing aggregate larger than 40 mm are based on the slump tests made after removal of particles larger than 40 mm by wet-screening.

‡For concrete containing large aggregates that will be wet-screened over the 40-mm sieve prior to testing for air content, the percentage of air expected in the 40-mm minus material should be as tabulated in the column. However, initial proportioning calculations should include the air content as a percent of the whole.

§When using large aggregate in low cement factor concrete, air entrainment need not be detrimental to strength. In most cases mixing water requirement is reduced sufficiently to improve the water-cement ratio and to thus compensate for the strength-reducing effect of air-entrained concrete. Generally, therefore, for these large nominal maximum sizes of aggregate, air contents recommended for extreme exposure should be considered even though there may be little or no exposure to moisture and freezing.

Table 9-2A Approximate Mixing Water and Air Content Requirements for Different Slumps and Nominal Maximum Sizes of Aggregates

Compressive Strength at 28 Days (MPa)[*]	**Water-Cement Ratio, by Weight**	
	Non-Air-Entrained Concrete	**Air-Entrained Concrete**
40	0.42	—
35	0.47	0.39
30	0.54	0.45
25	0.61	0.52
20	0.69	0.60
15	0.79	0.70

Table 9-3A Relationships between Water-Cement Ratio and Compressive Strength of Concrete

Maximum Size of Aggregate (mm)	Volume of Dry-Rodded Coarse Aggregate* per Unit Volume of Concrete for Different Fineness Moduli of Sand			
	2.40	2.60	2.80	3.00
9.5	0.50	0.48	0.46	0.44
12.5	0.59	0.57	0.55	0.53
19	0.66	0.64	0.62	0.60
25	0.71	0.69	0.67	0.65
37.5	0.75	0.73	0.71	0.69
50	0.78	0.76	0.74	0.72
75	0.82	0.80	0.78	0.76
150	0.87	0.85	0.83	0.81

*Volumes are based on aggregates in dry-rodded condition as described in ASTM C29, *Unit Weight of Aggregate*. These volumes are selected from empirical relationships to produce concrete with a degree of workability suitable for usual reinforced construction. For less workable concrete such as required for concrete pavement construction they may be increased about 10 percent. For more workable concrete, such as may sometimes be required when placement is to be by pumping, they may be reduced up to 10 percent.

Source: Reproduced with permission from the American Concrete Institute.

TABLE 9-5A Volume of Coarse Aggregate per Unit of Volume of Concrete

Maximum Size of Aggregate (mm)	First Estimate Concrete Weight* (kg/m³)	
	Non-Air-Entrained Concrete	Air-Entrained Concrete
9.5	2280	2200
12.5	2310	2230
19	2345	2275
25	2380	2290
37.5	2410	2350
50	2445	2345
75	2490	2405
150	2530	2435

*Values calculated for concrete mixture of moderate cement content (330 kg of cement per cubic meter) and medium slump, with aggregate specific gravity of 2.7. Water requirements based on values for 75 to 100 mm of slump in Table 9-2A. If desired, the estimated weight may be refined as follows when necessary information is available: for each 5 g difference in mixing water from the Table 9-2 values for 75 to 100 mm of slump, correct the weight per cubic meter 8 kg in the opposite direction; for each 20 kg difference in cement content from 330 kg, correct the weight per cubic meter 3 kg in the same direction; for each 0.1 by which aggregate specific gravity deviates from 2.7, correct the concrete weight 60 kg in the same direction.

Source: Reproduced with permission from the American Concrete Institute.

TABLE 9-6A First Estimate of Weight of Fresh Concrete

for compressive strength at 28 days (15 to 40 MPa, Table 9-3) and consistency (25 to 100 mm slump, Table 9-1). To satisfy today's high-construction speeds with heavily reinforced structural elements, concrete placement by pumping is the common practice now, and this means that concrete mixtures are designed to have at least 125 to 150 mm slump. Also, high-strength and high-performance concrete mixtures are being designed for compressive strength values from 50 to 100 MPa, which is outside the range of w/c-compressive strength relationship given by ACI 211 (Table 9-3A). Furthermore, the use of mineral admixtures and superplasticizers is much more prevalent now, and ACI 211 guidelines do not adequately deal with concrete mixtures containing these components.

For proportioning of high-performance concrete mixtures containing superplasticizers, mineral admixtures, and 28-day compressive strength values between 65 to 120 MPa, Mehta and Aitcin[4] developed a sequential, eight-step procedure. To provide adequate dimensional stability (e.g., high elastic modulus, and low drying shrinkage and creep), the procedure assumes a fixed ratio of 35 to 65 percent by volume between the cement paste and the aggregate. Note that with these mixtures the slump of concrete is no longer dependent on the water content alone; for high slump the use of a superplasticizing admixture is necessary. Based on experience with a wide variety of high-strength concrete mixtures, Table 12-3 in Chap. 12 shows the relationship between 28-day compressive strength and the total mixing water content in concrete. From a given value of the specified strength the first step involves the selection of the water content from this table.

For a 1 m^3 batch of concrete containing 0.35 m^3 cement paste, having known the volume of water and assuming a certain amount of entrapped or entrained air, the total volume of the cementitious material can be computed by difference. Next, the procedure provides options in the choice of the cementitious material, that is, whether to use portland cement alone or to use partial replacement of the cement by one or more mineral admixtures such as fly ash, slag, and silica fume. To complete the computations for the first trial batch, a 2:3 ratio by volume between the fine aggregate and the coarse aggregate is assumed. Experience shows that from the standpoint of workability in the presence of a relatively high content of the cementitious material in the concrete mixture, it is generally sufficient to have no more than 40 percent of the total aggregate in the form of fine aggregate. From the known values of the absolute volume of all the components of a 1 m^3 concrete mixture, the batch weights for the first laboratory trial are calculated. This trial is used to determine the dosage of the superplasticizer for obtaining the desired consistency and for adjustment of a proper ratio between the coarse and the fine aggregate. In general, depending on the type of the superplasticizer and the physical-chemical characteristics of the cementitious material, the superplasticizer dosage may vary from 1 to 3 L/m^3.

Note that a change of emphasis from the w/cm-strength relation to the water content-durability relation will provide the necessary incentive for incorporation of particle packing concepts into the concrete mixture proportioning methods, as suggested by deLarrard[1] and Shillstone.[2] Furthermore, considerable reductions in the mixing water requirements of conventional concrete mixtures can be realized by incorporating chemical admixtures and high volumes of fly ash (see Chap. 12). It should be obvious that such fundamental shifts in the objectives of concrete mixture proportioning methods are urgently needed to move the concrete industry toward the goal of sustainable development in the 21st century.

Appendix: Methods of Determining Average Compressive Strength from the Specified Strength*

ACI 322, *Building Code Requirements for Structural Plain Concrete*, and ACI 318, *Building Code Requirements for Reinforced Concrete*, specify that concrete shall be proportioned to provide an average compression strength f'_{cr}, which is higher than the specified strength f'_c so as to minimize the probability of occurrence of strengths below f'_c.

When a concrete production facility has a suitable record of 30 consecutive tests of similar materials and condition expected, the standard deviation can be calculated in accordance with the expression

$$S = \left[\frac{\Sigma(x_i - \bar{x})^2}{n-1}\right]^{1/2} \tag{9-1}$$

where S = standard deviation (psi)
x_i = strength value from an individual test
$\bar{x}$ = average strength of n tests
n = number of consecutive strength tests

When data for 15 to 25 tests are available, the calculated value of the standard deviation may be modified according to the following data:

Number of Tests	Multiplication Factor
15	1.16
20	1.08
25	1.03
30 or more	1.00

The required average compressive strength f'_{cr}, which is to be used as the basis for calculating concrete mix proportions, shall be the larger of Eq. (9-2) or (9-3):

$$f'_{cr} = f'_c + 1.34S \tag{9-2}$$

$$f'_{cr} = f'_c + 2.33S - 500 \tag{9-3}$$

Equation (9-2) provides a probability of 1 in 100 that averages of three consecutive tests will be below the specified strength f'_c. Equation (9-3) provides a similar probability of individual tests being more than 500 psi below the specified strength.

When adequate data are not available to establish a standard deviation, the required average strength can be determined from the following:

Specified Compressive Strength, f'_c(psi)	Required Average Compressive Strength, f'_c(psi)
Less than 300	$f'_c + 1000$
3000 – 5000	$f'_c + 1200$
Over 5000	$f'_c + 1400$

*Based on ACI Building Code 318.

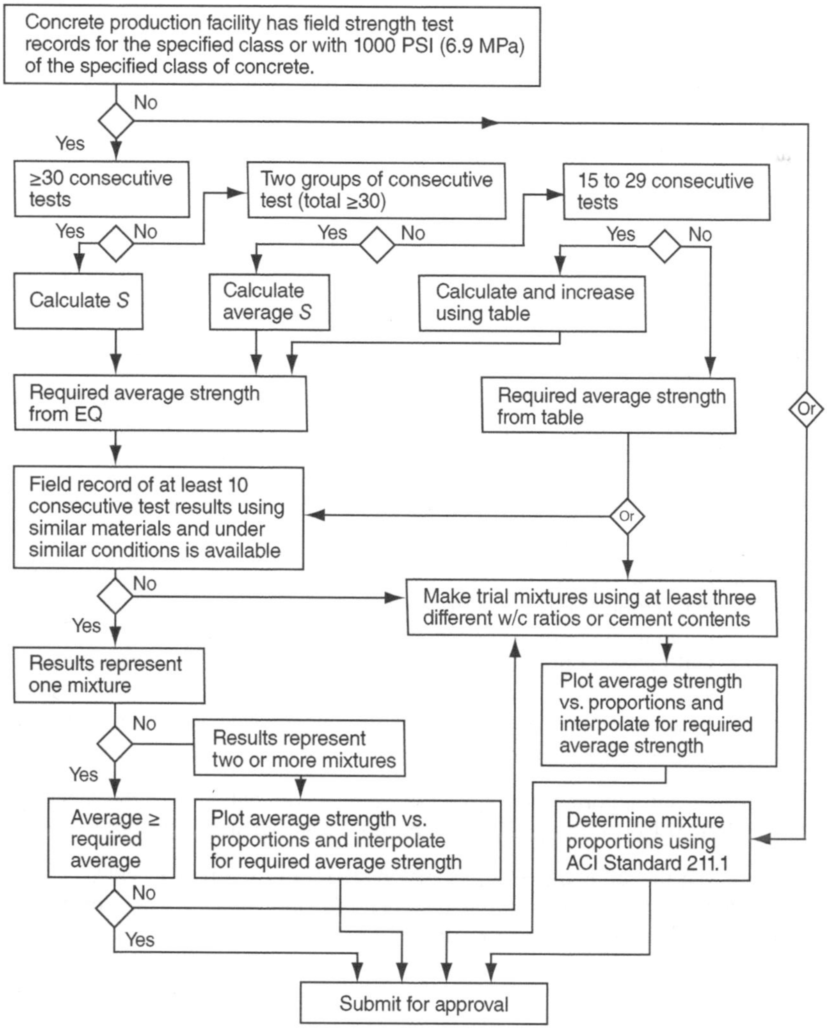

FIGURE 9-1 Flow chart for selection and documentation of concrete proportions. (*Adapted from ACI 318R89. Reproduced by permission.*)

Figure 9-1 gives a flowchart from the ACI Building Code Commentary (318R) outlining the mix selection and documentation procedure based either on field experience or trial mixtures.

Test Your Knowledge

9.1 Explain why the process of proportioning concrete mixtures is still in the realm of art. Have you any ideas on how to make the currently used practice in the United States more scientific?

9.2 You find yourself the project manager for a concrete structure involving several million cubic yards of concrete. Briefly, what tips would you like to pass on to the engineer in charge of mix proportioning on the subject of materials cost reduction? In your answer, emphasize the key ingredient in concrete from the standpoint of cost.

9.3 Why is it not necessary to take into account durability considerations in concrete mix proportioning when the concrete is subject to normal exposure conditions? Give examples of circumstances when durability must be considered in mix designing.

9.4 Theoretically derived ideal gradings of aggregates for maximum density should be the most economical, yet the practice is not followed. Can you explain why?

9.5 In mix designing, why is it desirable to use a minimum amount of water? For a given slump, how can you reduce the amount of water?

9.6 Describe the significance of workability of concrete and the factors affecting the property.

9.7 According to the ACI Building Code 318, selection of mix proportions should be based on the average strength, not the specified strength. Is this justified? Given a specified strength value, what procedures are used to determine the average strength?

9.8 With respect to the ACI 211.1, *Standard Practice for Selecting Proportions for Normal Heavy-Height and Mass Concrete*, explain the principles underlying the following:

(*a*) Estimation of water content.
(*b*) Estimation of coarse aggregate content.
(*c*) Estimation of fine aggregate content by the *weight method.*
(*d*) Estimation of fine aggregate content by the *absolute volume method.*

9.9 Briefly state the influence of maximum aggregate size (i.e., 19 mm vs. 38 mm) on the mixing water content and the cement content of a concrete mixture with a given water-cement ratio of 0.5.

9.10 Why is it important to control the aggregate gradation once the concrete mix design has been selected? How is this gradation control expressed in a specification?

9.11 Given the following SSD proportions (kg/m^3) for a non-air entrained concrete mixture, compute the batch weights for the job when the sand contains 4 percent free moisture and the gravel has 1 percent effective absorption:

cement = 330
water = 180
sand = 780
gravel = 1120

9.12 The proportions by mass for a concrete mixture are given as follows:

cement = 1
water = 0.53
sand = 2.50
gravel = 3.50

If the unit weight is 2400 kg/m^3, compute the cement content.

9.13 Determine the SSD mix proportions of concrete required for an outdoor pavement subject to frequent freeze-thaw cycles. The following data are given:

Specified 28-day compressive strength: 20 MPa
Slump: 75 mm

Coarse aggregate: 25 mm max. size; dry-rodded weight vol. per unit volume of concrete = 0.71
Fine aggregate: 2.8 fineness modulus
Specific gravities of cement, coarse aggregate, and fine aggregate: 3.15, 2.72, and 2.70, respectively.

References

1. Larrard, F. de, *Concrete Mixture Proportioning*, E & FN Spon, London, p. 421, 1999.
2. Shilstone, J.M., *Concr. Int.*, Vol. 12, No. 6, pp. 33–39, 1990.
3. Standard Practice for Selecting Proportions for Normal, Heavy-Weight, and Mass Concrete, ACI 211.1 Report, *ACI Manual of Concrete Practice*, Part 1, 1997.
4. Mehta, P.K., and P.C. Aitcin, *Cem. Concr. Aggregates*, Vol. 12, No. 2, pp. 70–78, 1990.

Suggestions for Further Study

ACI Committee 318, Building Code Requirements for Reinforced Concrete, *Building Code Commentary*, ACI 318-11, Concrete Institute, Farmington Hills, MI, 2011.

ACI Committee Report 211-1: Standard Practice for Selecting Proportions for Normal, Heavyweight, and Mass Concrete, Report 211.2: Standard Practice for Selecting Proportions for Structural Lightweight Concrete, Report 211.3R: Guide for Selecting Proportions for No-Slump Concrete, *ACI Manual of Concrete Practice*, Part 1, Concrete Institute, Farmington Hills, MI, 2012.

Bittencourt, R.M., J.T.F. Fontoura, W.P. de Andrade, and P.J.M. Monteiro, Mass Concrete Mixtures based on Fineness Modulus and Geometrical Gradation, *J. Mat. Civil Eng.*, Vol.13, pp. 33–40, Jan-Feb, 2001.

Concrete Optimization Tool, http://ciks.cbt.nist.gov/cost/

Day, K.W., *Concrete Mix Design, Quality Control, and Specifications*, 3rd edition, Taylor & Francis, New York, 2006.

Kett, I., *Engineered Concrete: Mix Design and Test Methods*, 2d ed., CRC Press, Boca Raton, 2010.

Monteiro, P.J.M., P.R. L. Helene, and S.H. Kang, Designing Concrete Mixtures for Strength, Elastic Modulus and Fracture Energy, *Mat. Struc.*, Vol. 26, pp. 443–452, Oct, 1993.

Neville, A.M., *Properties of Concrete*, 5th ed., Pearson Education Limited, Harlow, England, 2011.

Wilson, M.L., and S.H. Kosmatka, *Design and Control of Concrete Mixtures*, 15th ed., Portland Cement Association, 2011.

CHAPTER 10
Concrete at Early Age

Preview

Selection of proper materials and mixture proportions are important steps in producing a concrete meeting the requirements of strength and durability in a structural member. However, this goal will remain elusive if adequate attention is not paid to the processing operations to which concrete is subjected at early age. The term early age covers only an insignificant amount of time (e.g., first 7 days after production) in the total life of concrete but during this period numerous operations are performed such as mixing, transport to the job site, placement in the forms, consolidation, finishing, curing, and removal of formwork. These operations are affected by the characteristics of fresh concrete, like workability and setting time. Obviously, the control of both early-age operations and properties of fresh concrete is essential to ensure that the finished element is structurally adequate for the purpose for which it was designed.

A detailed description of the operations and equipment used for batching, mixing, conveying, placing, consolidation, and finishing operations for fresh concrete is beyond the scope of this book. Only the basic methods and their significance are described in this chapter. The significance and control of properties of fresh concrete, such as workability, slump loss, segregation and bleeding, plastic shrinkage, setting time, and temperature of fresh concrete are discussed. Finally, as effective and economical tools of modern quality assurance programs, the accelerated strength testing procedures and statistical quality control charts are briefly discussed.

10.1 Definitions and Significance

Deficiencies in freshly made concrete such as loss of workability at or before the placement, segregation and bleeding during the consolidation, or an unusually slow rate of maturity (strength gain) can impair the end product and reduce its service life. In this respect, concrete resembles a human child. To develop into a healthy person, a newborn baby needs special attention during the early period of growth. However, in both cases there is no clear definition of *how early is the early age*. Addressing this question, S.G. Bergstrom of the Swedish Cement and Concrete Research Institute said:

> Time is not a very good measure of "early." The time when the concrete has reached a certain maturity, is dependent on so many factors: cement type, reactivity of the cement, temperature, admixtures, etc. The time factor is not significant in the general case if you are not specifying

> the case very carefully. Then of course the degree of hydration gives a much better indication, which however is not always available if we deal with the practical side. You can also use another more practical definition perhaps, giving the time the property you are interested in has reached the level you need. All times earlier than that level are evidently early ages; which means that the definition depends on the way you will use concrete. The form stripper would say that he needs about 15 MPa, whereas a slipformer does not need as much as that. These two have quite different concepts about early age. The answer is that there is no universal answer. When we try to define the area where we are going to work, we will, as a rule of thumb and for the normal concrete, in normal situations, say about 24 hours, some say about 48 hours, but that is just to indicate the order of magnitude . . .[1]

A normal concrete mixture (i.e., concrete made with ordinary portland cement and subjected to normal curing) generally takes 6 to 10 h for setting and 1 to 2 days for achieving a strength level when the formwork can be removed. The early age period therefore begins with the freshly mixed concrete of plastic consistency and ends with 1- to 2-day-old concrete that is strong enough to be left unattended (and will become stronger with age).

The early-age period in the life of concrete is insignificantly small compared to the total life expectancy, but during this period it is subjected to many operations that not only are affected by properties of the material but also influence them. For instance, a mixture with poor workability will be hard to mix; on the other hand, too much mixing will reduce the workability. It is beyond the scope of this book to describe in detail the operations and the equipment used, but engineers should be familiar with the sequence of main operations, their effect on the characteristics of concrete, and some of the terminology that is used in field practice.

In general, the sequence of main operations is as follows: batching, mixing, and conveying the concrete mixture from the point where it is made to job site; placing the plastic concrete at the point where it is needed; compacting and finishing while the mixture is still workable; finally, moist curing to achieve a desired degree of maturity before the formwork is removed. The operations described next are divided into separate categories only for the purpose of understanding their significance and the basic equipment involved; in practice, they may overlap. For example, in the truck-mixing method, the mixing and transporting operations are carried out simultaneously.

Finally, there are aspects of concrete behavior at early age that cannot be considered as intrinsic to the material but are important because of their effect on the long-term performance of a concrete structure. They include workability, rate of slump loss, segregation and bleeding, plastic shrinkage, setting time, and curing temperature. In practice, many of them are interrelated; however, for the purposes of achieving a clear understanding of their significance and control, they will be discussed individually.

10.2 Batching, Mixing, and Transport

Batching is the process of measuring and introducing into the mixer the ingredients for a batch of concrete. Most specifications require that *batching of concrete ingredients* be carried out by mass rather than by volume. This is because *bulking of damp sand* causes inaccuracies in measurement. Water and liquid admixtures can be batched accurately either by volume or weight. As discussed later, in many countries most concrete today is batched and mixed by ready-mixed concrete plants, where the batching is generally automatic or semiautomatic rather than manual.

Improper handling and mixing of fresh concrete mixtures that are not uniform in appearance is attributable to inadequate *mixing*. Therefore, accurately proportioned concrete ingredients must be mixed thoroughly into a homogeneous mass. Depending on the cost economy, type of construction, and amount of concrete required, the mixing operation can either be performed on site or in a central off-site facility (ready-mixed concrete plant). On-site mixers can be either stationary or paving type.

Ready-mixed concrete is defined as concrete that is manufactured for delivery to a purchaser in a plastic and unhardened state. During the last 60 years, the ready-mixed concrete industry has experienced tremendous growth worldwide. For example, in the United States there are some 3700 companies operating 10,000 plants that furnish over two-thirds of the total concrete consumed in the country.[2] Six or seven percent of the companies produce 50 percent of the ready-mixed concrete and each operates more than 100 truck mixers. Concrete is batched and mixed in accordance with ASTM C 94, *Standard Specification for Ready-Mixed Concrete*. Most of the plants are equipped with automatic batching system and controls made possible by the use of microprocessors and computers. Truck mixing and transport rather than central mixing is still the commonly used method of mixing in some countries although, due to better quality control most ready-mixed concrete plants now use trucks for the transportation of centrally mixed concrete (Fig. 10-1). Stationary mixers of sizes up to 9 m^3 (12 yd^3) which can be of the tilting or the nontilting type, open-top revolving blade or paddle type with rear or with front discharge, are commonly used. In the past 25 years, for important jobs there has been a trend to move away from the prescriptive to the performance-based specifications; also, ready-mixed concrete producers are assuming greater responsibility for mixture proportioning and quality control.

FIGURE 10-1 Centrally mixed concrete in a ready-mixed concrete plant. (Photograph courtesy of RMC Industries, San Francisco, CA.)

Transportation of the ready-mixed concrete to a job site should be done as quickly as possible to minimize stiffening to the point that, after the placement, full consolidation and proper finishing become difficult. The causes and control of stiffening or loss of consistency, which is also referred to as *slump loss*, are discussed later. Under normal conditions there is usually a negligible loss of consistency during the first 30 min after the beginning of cement hydration. When concrete is kept in a slow state of agitation or is mixed periodically, it undergoes some slump loss with time but, normally this does not present a serious problem for the placement and consolidation of freshly made concrete within 90 min. However, as discussed next, attention must be paid to possible delays in transporting and placing concrete under hot and dry weather conditions.

A summary of commonly used methods and equipment for mixing and transport of concrete is presented in Table 10-1. According to the Portland Cement Association:

> There have been few, if any, major changes in the principles of conveying concrete in the last 40 years. What has changed is the technology that led to development of better machinery to do the work more efficiently. The wheelbarrow has become the power buggy; the bucket hauled over a pulley has become the hoist; and the horse drawn wagon is now the ready-mixed concrete truck. For some years, concrete was placed in reinforced concrete buildings by means of a tower and long chutes.... As concrete-frame buildings became taller, the need to bring reinforcement and formwork as well as concrete to higher levels led to the development of the tower crane—a familiar sight on the building skyline today... The mobile pump with hydraulic placing boom is probably the single most important innovation in concrete handling equipment.[3]

In choosing the method and equipment for transporting and placing concrete, a primary objective is to ensure that concrete will not segregate. The causes, significance, and control of segregation (i.e., the tendency of the coarse aggregate to separate from the mortar) are discussed later.

10.3 Placing, Compacting, and Finishing

After arrival at the job site, *the ready-mixed concrete should be placed as near as possible to its final position*. Belt conveyers, truck-mounted chutes, and mobile-boom pumps are among the most commonly used today for *concrete placement* (Fig. 10-2). To minimize segregation, concrete should not be moved over too long a distance during the placement into forms. In general, the concrete mixture is deposited in horizontal layers of uniform thickness, and each layer is thoroughly compacted before the next is placed. The rate of placement is kept rapid enough so that the layer immediately below is still plastic when a new layer is deposited. This prevents cold joints, flow lines, and planes of weakness that occur when fresh concrete is placed on hardened concrete.

Consolidation or *compaction* is the process of molding concrete within the forms and around embedded items and reinforcing steel to eliminate pockets of empty space and entrapped air. This operation can be carried out by hand *rodding* and *tamping*. However, now it is carried out by mechanical methods such as power tampers and vibrators that make it possible to place stiff mixtures with low water-cement ratio

Equipment	Type and Range of Work for Which Equipment is Best Suited	Advantages	Points to Watch For:
Truck agitator	Used to transport concrete for all uses in pavements, structures, and buildings. Haul distances must allow discharge of concrete within $1^1/_2$ h, but limit may be waived.	Truck agitators usually operate from central mixing plants where quality concrete is produced under controlled conditions. Discharge from agitators is well controlled. There is uniformity and homogeneity of concrete on discharge.	Timing of deliveries to suit job organization. Concrete crew and equipment must be ready onsite to handle concrete in large batches.
Truck mixer	Used to mix and transport concrete to job site over short and long hauls. Hauls can be any distance.	No central mixing plant needed, only a batching plant since concrete is completely mixed in truck mixer. Discharge is same as for truck agitator.	Control of concrete quality is not as good as with central mixing. Slump tests of concrete consistency are needed on discharge. Careful preparations are needed for receiving the concrete.
Nonagitating truck	Used to transport concrete on short hauls.	Capital cost of nonagitating equipment is lower than that of truck agitators or mixers.	Concrete slump should be limited. Possibility of segregation. Height is needed for high lift of truck body upon discharge.
Mobile continuous mixer	Used for continuous production of concrete at job site.	Combination materials transporter and mobile mixing system for quick, precise proportioning of specified concrete. One-man operation.	Trouble-free operation requires good preventive maintenance program on equipment. Materials must be identical to those in original mix-design proportioning.
Crane	The right tool for work above ground level.	Can handle concrete, reinforcing steel, formwork, and sundry items in high-rise, concrete-framed buildings.	Has only one hook. Careful scheduling between trades and operations are needed to keep it busy.
Buckets	Used on cranes and cableways for construction of buildings and dams. Convey concrete direct from central discharge point to formwork or to secondary discharge point.	Enable full versatility of cranes and cableways to be exploited Clean discharge. Wide range of capacities.	Select bucket capacity to conform with size of the concrete batch and capacity of the placing equipment. Discharge should be controllable.
Barrows and buggies	For short flat hauls on all types of on-site concrete construction especially where accessibility to work area is restricted.	Very versatile and therefore ideal inside and on job sites where placing conditions are constantly changing.	Slow and labor intensive.

TABLE 10-1 Methods and Equipment for Handling Concrete

Equipment	Type and Range of Work for Which Equipment is Best Suited	Advantages	Points to Watch For:
Chutes	For conveying concrete to lower level, usually below-ground level, on all types of concrete construction.	Low cost and easy to maneuver. No power required, gravity does most of the work.	Slopes range between 1 to 2 and 1 to 3 and chutes must be adequately supported in all positions. Arrange for discharge at end (downpipe) to prevent segregation.
Belt conveyors	For conveying concrete horizontally or to a higher level. Usually used between main discharge point. Not suitable for conveying concrete directly into formwork.	Belt conveyors have adjustable reach, traveling diverter, and variable speed both forward and reverse. Can place large volumes of concrete quickly when access is limited.	End-discharge arrangements needed to prevent segregation. Leave no mortar on return belt. In adverse weather (hot, windy) long reaches of belt need cover.
Pneumatic guns	Used where concrete is to be placed in difficult locations and where thin sections and large areas are needed.	Ideal for placing concrete in free-form shapes, for repairing and strengthening buildings, for protective coatings, and thin linings.	Quality of work depends on skill of those using equipment. Only experienced nozzlemen should be employed.
Concrete pumps	Used to covey concrete direct from central discharge point to formwork or to secondary discharge point.	Pipelines take up little space and can be readily extended. Deliver concrete in continuous stream. Mobile-boom pump can move concrete both vertically and horizontally.	Constant supply of fresh, plastic concrete is needed with average consistency and without any tendency to segregate. Care must be taken in operating pipeline to ensure an even flow and to clean out at conclusion of each operation. Pumping vertically, around bends, and through flexible hose will considerably reduce the maximum pumping distance.
Dropchutes	Used for placing concrete in vertical forms of all kinds. Some chutes are in one piece, while others are assembled from a number of loosely connected segments.	Dropchutes direct concrete into form-work and carry it down to bottom of forms without segregation. Their use avoids spillage of grout and concrete on the form sides, which is harmful when off-the-form surfaces are specified. They also will prevent segregation of coarse particles.	Dropchutes should have sufficiently large, splayed-top openings into which concrete can be discharged without spillage. The cross section of dropchute should be chosen to permit inserting into the formwork out interfering with reinforcing.

Table 10-1 Methods and Equipment for Handling Concrete (*Continued*)

Equipment	Type and Range of Work for Which Equipment is Best Suited	Advantages	Points to Watch For:
Tremies	For placing concrete under water.	Can be used to funnel concrete down through the water into the foundation or other part of the structure being cast.	Precautions are needed to ensure the tremie discharge end is always buried in fresh concrete, so that a seal is preserved between water and concrete mass. Diameter should be 10 to 12 in. (200 to 300 mm) unless pressure is available. Concrete pumps can be used. Concrete mixture needs more cement, $6^1/_2$ to 8 bags per cubic yard (363 to 466 kg/m^3), and greater slump, 6 to 9 in. (150 to 230 mm), because concrete must flow and consolidate without any vibration.
Screw spreaders	Used for spreading concrete over flat areas as in pavements.	With a screw spreader a batch of concrete discharged from bucket or truck can be quickly spread over a wide area to a uniform depth. The spread concrete has good uniformity before vibration for final compaction.	Screws are usually used as part of a paving train. They should be used for spreading before vibration is applied.

Source: Reproduced from *Design and Control of Concrete Mixtures*, 12th ed., Portland Cement Association, Skokie, IL, pp. 70–71, 1979.

Table 10-1 Methods and Equipment for Handling Concrete (*Continued*)

or high content of coarse aggregate. High-consistency mixtures should be consolidated with care because they are likely to segregate when intensely worked. Vibrators should only be used to compact concrete and not to move it horizontally, as this would cause segregation.

Vibration, whether internal or external, is the most widely used method for compacting concrete. The internal friction between the coarse aggregate particles is greatly reduced on vibration; consequently, the mixture behaves like a liquid and begins to flow into the empty space. One purpose of using internal vibrators (described below) is to force entrapped air out of the concrete by plunging the vibrator rapidly into the mixture and removing it slowly with an up-and-down motion. The rapid penetration forces the concrete upward and outward, thereby helping the air to escape.

Internal or *immersion-type vibrators*, also called *spud* or *poke*r vibrators, are commonly used for compacting concrete in beams, columns, walls, and slabs. Flexible-shaft vibrators usually consist of a cylindrical vibrating head, 19 to 175 mm in diameter and

(a) Conveyor belt

(b) Truck-mounted chute

(c) Shotcrete

(d) Tremie

(e) Pumping

FIGURE 10-2 Placement of concrete as near as possible to its final position prevents segregation. [Photographs courtesy of Jose Marques Filho (*a*), Larry Totten (*c*), San Yao (*d*).]

connected to a driving motor by a flexible shaft. Inside the head an unbalanced weight rotates at high speed, causing the head to revolve in a circular orbit. Small vibrators have frequencies ranging from 10,000 to 15,000 vibrations per minute and low amplitude, between 0.4 and 0.8 mm (deviation from the point of rest); as the diameter increases, the frequency decreases and the amplitude increases. An idealized representation of the sequence of actions during the consolidation of concrete by a high-frequency immersion-type vibrator is shown in Fig. 10-3.

External or *form vibrators* can be securely clamped to the outside of the forms. They are commonly used for compacting thin or heavily reinforced concrete members. While the concrete mixture is still mobile, vibration of a member congested with reinforcement helps to remove air and water that may be entrapped underneath the reinforcing bars thus improving the bond between the bars and concrete. Precasting plants generally

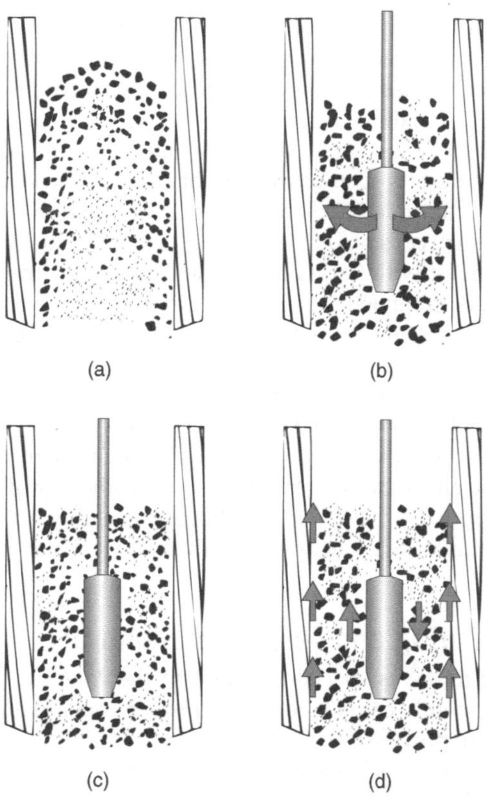

FIGURE 10-3 Idealized representation of the influence of a high-frequency vibrator on concrete consolidation.

(*a*) The mix is introduced into the form. (*b*) The vibrator moves aggregate closer together at the form face and cement-sand mortar begins to move outward; air pockets collect on the faces of forms. (*c*) The mortar continues to move through the coarse aggregate toward the face of the form. (*d*) The movement of mortar toward the face is complete; as the operator moves the vibrator down and up, air bubbles move upward along the form face and out of the concrete. (Illustration courtesy of Concrete Construction, Vol. 17, No. 11, 1972. By permission of Concrete Construction Publications, 426 South Westgate, Addison, IL.)

use vibrating tables equipped with suitable controls so that the frequency and amplitude can be varied according to the size of the member and the consistency of concrete. Surface vibrators such as vibrating screeds are used to consolidate concrete in floors and slabs up to 150 mm thick.

Revibration of concrete an hour or two after the initial consolidation but before setting is sometimes needed in order to weld successive castings together. This helps to remove any cracks, voids, or weak areas created by settlement or bleeding, particularly around the reinforcing steel or other embedded items.

Flatwork such as slabs and pavements require proper *finishing* to produce dense surfaces that will remain maintenance-free. Depending on the intended use, some surfaces require only strike-off and screeding, whereas others may need finishing operations consisting of a *sequence of steps described below, which must be carefully coordinated with the setting and hardening of the concrete mixture.*

Screeding is the process of striking off excess concrete to bring the top surface to the desired grade. With a sawing motion a straight edge is moved across the surface with a surplus of concrete against the front face of the straight edge to fill in low areas. A *Darby* or *bull-float* is used immediately after screeding to firmly embed large aggregate particles and to remove any remaining high and low spots. Bull-floating must be completed before any excess bleed water accumulates on the surface because this is one of the principal causes of surface defects such as dusting or scaling in concrete slabs. When the bleed-water sheen has evaporated and concrete is able to sustain foot pressure with only slight indentation, the surface is ready for floating and final finishing operations. *Floating* is an operation carried out with flat wood or metal blades for the purposes of firmly embedding the aggregate, compacting the surface, and removing any remaining imperfections. Floating tends to bring the cement paste to the surface; therefore, floating too early or for too long can weaken the surface. After floating, the surface may be steel toweled if a very smooth and highly wear resistant surface is desired. Troweling should not be done on a surface that has not been floated. For producing a skid-resistant surface, *brooming* or *scoring* with a rake or a steel-wire broom is done before the concrete has fully hardened (but has become sufficiently hard to retain the scoring). Photographs of various finishing operations are shown in Fig. 10-4. For additional durability and wear resistance, a special surface treatment after the concrete has fully hardened may be considered.

10.4 Concrete Curing and Formwork Removal

Concrete curing deserves special attention in the construction practice because inadequate curing frequently causes the lack of proper strength and durability. The two *objectives of curing* are to prevent the loss of moisture and to control the temperature of concrete for a period sufficient to achieve a desired strength level. When the ambient temperature is sufficiently well above freezing, the curing of pavements and slabs can be accomplished by ponding or immersion; other structures can be cured by spraying or fogging, or moisture-retaining coverings saturated with water, such as burlap or cotton. These methods afford some cooling through evaporation, which is beneficial in hot-weather concreting. Another group of methods are based on prevention of moisture loss from concrete by sealing the surface through the application of waterproof curing paper, polyethylene sheets, or membrane-forming curing compounds. The use of curing compounds is preferred for speedy construction. To achieve satisfactory results the selection of materials and the method of application must be carefully performed.

FIGURE 10-4 Placement and finishing of concrete slabs.

(*a*) Delivery of the concrete; (*b*) screeding of the concrete; (*c*) bull floating for the removal of any high and low spots must be completed before any excess bleed water accumulates on the surface; (*d*) when the bleed-water sheen has evaporated and the concrete sustains foot pressure with only slight indentation, the surface is ready for floating and final finishing operations; (*e*) application of polyethylene sheet on the concrete surface to prevent moisture loss.

When the ambient temperature is low, concrete must be protected from freezing with the help of insulating blankets. In cold weather, the rate of strength gain can be accelerated by curing concrete with live steam, heating coils, or electrically heated forms or pads.

Formwork removal is generally the last operation carried out during the "early-age" period of concrete. The operation has economic implication because, on the one hand, early removal of formwork keeps the construction cost low, while on the other hand, concrete structures have failed when forms were stripped before concrete had attained sufficient strength. Formwork should not be removed until the concrete is strong enough to carry the stresses from both the dead load and the imposed construction load. Also, concrete should be sufficiently hard so that the surface is not injured in any way during the formwork removal or other construction activities. As the strength of a freshly hydrated cement paste depends on the ambient temperature and availability of moisture, it is better to rely on a direct measure of the concrete strength rather than an arbitrarily selected time for the formwork removal. Under normal moist-curing and temperature conditions, conventional concrete mixtures made with ordinary portland cement may gain adequate strength for formwork removal, for example, 6- to 7-MPa compressive strength, in 24 h; mixtures made with a high early strength portland cement in 12 to 15 h, and those containing high volume of slag or fly ash in 48 h. For safety of structures in cold weather, designers often specify a minimum compressive strength before concrete is exposed to freezing. In hot weather, moisture from unprotected concrete may be lost by evaporation, causing interruption in the normal rate of cement hydration and strength gain.

10.5 Workability

10.5.1 Definition and Significance

Workability of concrete is defined in ASTM C 125 as the property determining the effort required to manipulate a freshly mixed quantity of concrete with minimum loss of homogeneity. The term *manipulate* includes the early-age operations of placing, compacting, and finishing. The effort required to place a concrete mixture is determined largely by the overall work needed to initiate and maintain flow, which depends on the rheological property of the lubricant (the cement paste) and the internal friction between the aggregate particles on the one hand, and the external friction between the concrete and the surface of the formwork on the other.

Consistency, measured by the slump-cone test or Vebe apparatus (described below), is used as a simple index for mobility or flowability of fresh concrete. The effort required to compact concrete is governed by the flow characteristics and the ease with which void reduction can be achieved without destroying the stability under pressure.

Stability is an index for both the water-holding capacity (the opposite of bleeding) and the coarse-aggregate-holding capacity (the opposite of segregation) of a plastic concrete mixture. A qualitative measure of these two characteristics is generally covered by the term *cohesiveness*.

It should be apparent by now that *workability is a composite property*, with at least two main components:

- Consistency (describes the ease of flow)
- Cohesiveness (describes the stability or lack of bleeding and segregation characteristics)

Like durability, *workability is not a fundamental property of concrete*; to be meaningful it must be related to the type of construction and the method of placement, compaction, and finishing. A concrete that can readily be placed in a massive foundation without segregation, may be entirely unworkable to form a thin structural member. Concrete judged to be workable when high-frequency vibrators are available for consolidation, would be unworkable if hand tamping is used.

The *significance of workability* in concrete technology is obvious. It is one of the key properties that affect constructibility. Regardless of the sophistication of the mix design procedure used and other considerations, such as cost, a concrete mixture that cannot be placed easily or compacted fully is not likely to yield the expected strength and durability characteristics.

10.5.2 Measurement

The composite nature of workability as a property, and its dependence on the type of construction and methods of placing, compacting, and finishing are the reasons why no single test method can be designed to measure workability. The most universally used test, which measures only the consistency of concrete, is called the *slump test*. For the same purpose, the second test in order of importance is the *Vebe test*, which is more meaningful for mixtures with low consistency. The third test is the *compacting factor test*, which attempts to evaluate the compactibility characteristic of a concrete mixture. The slump test is covered by ASTM C 143, and the other two tests by ACI Standard 211.3. Only brief descriptions of the equipment and procedures are given below.

Slump Test

The equipment for the slump test is indeed very simple. It consists of a tamping rod and a truncated cone, 300 mm height and 100 mm diameter at the top, and 200 mm diameter at the bottom. The cone is filled with concrete and then slowly lifted. The unsupported concrete cone slumps down by its own weight; the decrease in the height of the slumped cone is called the *slump of concrete*. The sequence of steps in the ASTM C 143 test procedure are shown in Fig. 10-5.

The slump test is not suitable for measuring the consistency of a very wet or very dry concrete mixture. Also, it is not a good measure of workability although it is a fairly good measure of the consistency or flow characteristic of plastic concrete. This test is not a satisfactory measure of the rheological behavior of concrete, the main reason why it is popular is that it provides a simple and convenient method for controlling the batch-to-batch uniformity of ready-mixed concrete. For example, a more than normal variation in slump may mean an unexpected change in the mixture proportions, aggregate grading, or moisture in aggregate. The test result enables the ready-mixed concrete plant operator to investigate and remedy the problem.

Vebe Test

The equipment for the test, which was developed by Swedish engineer V. Bährner, is shown in Fig. 10-6*a*. It consists of a vibrating table, a cylindrical pan, a slump cone, and a glass or plastic disk attached to a free-moving rod that serves as a reference end point. The cone is placed in the pan, filled with concrete, and removed. The disk is brought into position on top of the concrete cone, and the vibrating table is set in motion. The time required to remold the concrete, from the conical to the cylindrical shape, is a measure of the consistency and is reported as *Vebe seconds*.

1. Stand on the two-foot pieces of cone to hold in firmly in the place during Steps 1 through 4. Fill cone mold 1/3 full by volume [2-5/8" (67 mm) high] with the concrete sample and rod it with 25 strokes using a round, straight steel rod of 5/8" (16 mm) diameter × 24" (600 mm) long with a hemispherical tip end. Uniformly distribute strokes over the cross section of each layer. For the bottom layer, this will necessitate inclining the rod slightly and making approximately half the strokes near the perimeter (out edge), then progressing with vertical strokes spirally toward the center.

2. Fill cone 2/3 full by volume (half the height) and again rod 25 times with rod just penetrating into, but not through the first layer. Distribute strokes evenly as described in Step 1.

3. Fill cone to overflowing and again rod 25 times with rod just penetrating into, but not through, the second layer. Again distribute strokes evenly.

4. Strikes off excess concrete form top of cone with the steel rod so that the cone is exactly level full. Clean the overflow away from the base of the cone mold.

5. Immediately after completion of Step 4, the operation of raising the mold shall be performed in 5±2 sec by a steady upward lift with no lateral or torsional motion being imparted to the concrete. The entire operation from the start of the filling through removal of the mold shall be carried out without interruption and shall be completed within an elapsed time of 2-1/2 min.

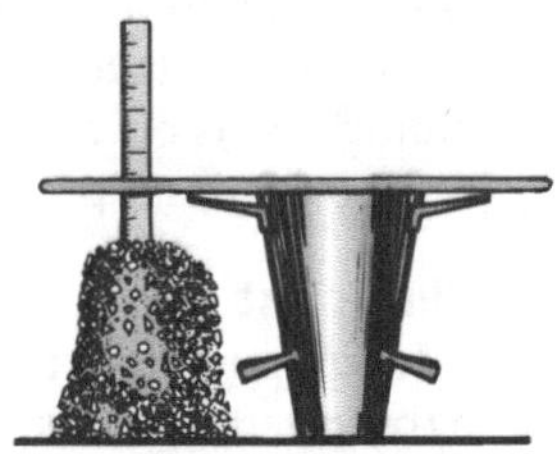

6. Place the steel rod horizontally across the inverted mold so that the rod extends over the slumped concrete. Immediately measure the distance from bottom of the steel rod to the displaced original center of the specimen. This distance, to the nearest 1/4 in (6 mm), is the slump of the concrete. If a decided falling away or shearing off concrete from one side or portion of the mass occurs, disregard the test and make a new test on another portion of the sample.

Figure 10-5 Sequence of steps in the slump test procedure.

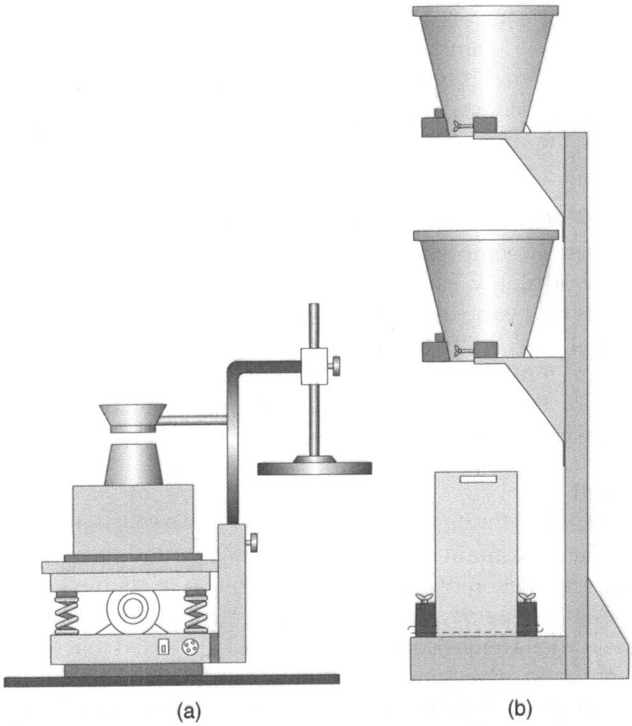

FIGURE 10-6 Equipment for measuring the consistency of concrete: (a) Vebe apparatus; (b) compacting factor apparatus.

Compacting Factor Test

This test, developed in Great Britain, measures the degree of compaction achieved when a concrete mixture is subjected to a standard amount of work. The degree of compaction, called the *compacting factor*, is measured by the density ratio (i.e., the ratio of the density actually achieved in the test to the density of the same concrete when in a fully compacted condition). The apparatus consists essentially of two conical hoppers fitted with doors at the base and placed one above the other (Fig. 10-6*b*), and a 150 by 300 mm cylinder placed below the hoppers. The upper hopper, which is bigger than the lower, is filled with concrete and struck off without compacting. By opening the door at the bottom of the hopper, the concrete is allowed to fall by gravity into the lower hopper that overflows. This ensures that a given amount of concrete is obtained in a standard state of compaction without the influence of human factor. The door of the lower hopper is released and the concrete falls into the cylinder. Excess material is struck off and the net weight of concrete in the known volume of the cylinder is determined, from which the density is easily calculated.

Tattersall Test

Tattersall[4] discussed the principles of measuring the workability of fresh concrete and proposed a two-point test assuming plastic concrete to be a Bingham fluid that follows a close relationship between the plastic viscosity, the rate of shear, and the yield value. The test procedure consists of measuring the power required at three different speeds to operate a mixer under two conditions, namely when empty and when full with a batch of 21 kg of concrete. The values for yield and plastic viscosity are obtained by plotting $(P - P_E)/w$ against w, where w is the speed, P is power under load, and P_E is power when the bowl is empty. Although Tattersall's two-point test gives more information on the rheological characteristics of a fresh concrete mixture, for a field test it does not have the simplicity of the other tests described here.

10.5.3 Factors Affecting the Workability and Their Control

Scanlon[5] presents a comprehensive review of the test procedures and factors influencing the concrete workability. For obvious reasons, instead of workability it is more appropriate to consider how various factors affect consistency and cohesiveness because these two components of workability may be oppositely influenced by changing a particular variable. In general, through their influence on consistency and cohesiveness, the workability of concrete mixtures is affected by water content, cement content, aggregate grading and other physical characteristics, admixtures, and slump loss, as discussed below.

Water Content

ACI 211.1, *Standard Practice for Proportioning Concrete Mixtures* (see Table 9-2), assumes that, for a given maximum size of coarse aggregate, the slump or the consistency of concrete is a direct function of the water content; that is, within limits it is independent of other factors such as aggregate grading and cement content. In predicting the influence of mixture proportions on the consistency, it should be noted that of the three factors, that is, water-cement ratio, aggregate-cement ratio, and water content, only two are independent. For example, when the aggregate-cement ratio is reduced but the water-cement ratio is kept constant, the water content increases and consequently the consistency. On the other hand, when the water content is kept constant but the aggregate-cement ratio is reduced, the water-cement ratio decreases and the consistency is not affected.

Concrete mixtures with very high consistency tend to segregate and bleed, thereby adversely affecting the finishability; mixtures with too low a consistency may be difficult to place and compact, and the coarse aggregate may segregate on placement.

Cement Content

With conventional portland-cement concrete at a given water content, a drastic reduction of the cement content would produce a harsh mixture with poor finishability. Concrete mixtures containing a very high cement content or high proportion of fine particles show excellent cohesiveness but tend to be sticky.

Aggregate Characteristics

The particle size of coarse aggregate influences the water requirement for a given consistency (Table 9-2). Also, very fine sands or angular sands require more water for a given consistency. Alternatively, they will produce harsh and unworkable mixtures at the water content that might have been adequate with a coarse or a well-rounded sand. As a rule of thumb, for similar consistency, concrete needs 2 to 3 percent more sand and 5 to 10 kg/m^3 more mixing water by the absolute volume when crushed sand is used instead of a natural sand.

Admixtures

As already discussed (Table 8-1), when the water content of a concrete mixture is held constant, the addition of a water-reducing admixture increases the consistency.

Entrained air increases the paste volume and improves the consistency of concrete for a given water content (Table 9-2). It also increases cohesiveness by reducing bleeding and segregation. The improvement in consistency and cohesiveness by air entrainment is more pronounced in harsh and unworkable mixtures such as those used in mass concrete, which has a low cement content. Pozzolanic admixtures tend to reduce bleeding and improve the cohesiveness of concrete. Fly ash, when used as a partial replacement for fine aggregate, generally increases the consistency at a given water content.

10.6 Slump Loss

10.6.1 Definitions

Slump loss is defined as the loss of consistency in fresh concrete with elapsed time. This is a *normal phenomenon* with all concrete mixtures because it results from the gradual stiffening and setting of a hydrating portland cement paste, a phenomenon that is associated with the formation of hydration products such as ettringite and calcium silicate hydrates (Chap. 6). Slump loss occurs when the free water from a concrete mixture is removed by hydration reactions resulting in the formation of hydration products and moisture adsorption on their surfaces, and by evaporation.

Under normal conditions, the volume of hydration products during the first 30 min after the addition of water to cement is small and the slump loss is negligible. Thereafter, concrete starts losing slump at a rate determined mainly by elapsed time after hydration, temperature, cement composition, and the admixtures present. Generally, changes in the consistency of concrete up to the time of placement are closely monitored and proper adjustments are made to ensure sufficient consistency for the placement and subsequent operations (e.g., compaction and finishing). Under some conditions, a concrete mixture exhibiting an unusually large loss of slump during the first 30 to 60 min may have the effect of making the mixing, convening, placing, compacting, and finishing operations difficult or, at times, even impossible. In practice, a slump loss prone concrete generally means a product that undergoes a quick and unusually large loss of consistency which is beyond the expected or normal behavior. To overcome the problems with concrete mixtures prone to an unexpected slump loss, certain field practices have evolved, such as starting with a higher initial slump of ready-mixed concrete than is needed at the job site (in order to compensate for the expected slump loss), or adding extra water (within the permissible water-cement ratio) just before the placement and remixing the concrete mixture thoroughly. The latter practice is known as *retempering*.

10.6.2 Significance

The premature stiffening of fresh concrete, depending on when the problem appears, may mean an increase in the mixer drum torque, requirement of extra water in the mixer or at job site, hang-up of concrete within the drum of a truck mixer, difficulty in pumping and placing the concrete, extra labor for handling and finishing operations, and finally loss of production and quality of workmanship, loss of strength, durability, and other properties when the retempering water is excessive or is not mixed thoroughly.

When job site inspection and quality control are lax, construction crews frequently adopt the bad practice of adding extra water to concrete whether it is needed or not. Many failures of concrete to perform satisfactorily have been attributed to the careless addition of the retempering water, which was either poorly mixed or not accounted for in the mixture proportioning calculations. For example,[6] the removal of forms from an unusually large concrete placement revealed areas of severe honey-combing. Construction personnel indicated that quick setting had occurred, primarily during the periods of high ambient temperature. Petrographic analysis of cores revealed that areas of different water-cement ratio were present within a core, indicating that retempering water had been added owing to the slump loss and that incomplete intermixing of the retempering water had occurred. The National Ready Mixed Concrete Association offers this advice: *A wasted load of questionable concrete may represent a tremendous bargain for the company, compared to its possible use and failure to perform.*

10.6.3 Causes and Control

The primary causes of slump-loss problems with concrete are as follows: (1) the use of an abnormal-setting cement; (2) unusually long time for mixing, transporting, placement, compaction, or finishing operation; (3) high temperature of concrete due to excessive heat of hydration and/or the use of concrete-making materials that are stored at a high ambient temperature.

Typical data[7] on the influence of cement composition, elapsed time after hydration, and temperature on the rate of slump loss in normal concrete mixtures are shown in Table 10-2. All concretes contained 307 kg/m^3 (517 lb/yd^3) Type I portland cement, 1040 kg/m^3 (1752 lb/yd^3) coarse aggregate, and 490 kg/m^3 (824 lb/yd^3) fine aggregate. The water content was varied to obtain different initial slumps: approximately 175, 125, and 75 mm (7, 5, or 3 in.). Cement A was a low-alkali cement (0.16 percent equivalent Na_2O)

		Slump (in.)				
		Initial	30 min	60 min	90 min	120 min
Concrete Mix	Cement	Concrete Temperature 70°F (21°C)				
1	A	7½	7	5½	3¾	2¼
2	B	7-1/8	4¾	3¼	2½	1-7/8
3	A	5	4-3/8	3-1/8	2¼	1½
4	B	5¼	3¼	2½	1¾	1¼
5	A	3-5/8	3¼	2-5/8	1-7/8	1-3/8
6	B	3½	2-5/8	2	1½	7/8
		Concrete Temperature 85°F (29°C)				
7	A	7-1/8	5-3/8	4-3/8	2-5/8	1-5/8
8	B	7½	5½	3½	2½	1-3/8
9	A	5½	4½	3-5/8	2-5/8	1-5/8
10	B	5½	4-1/8	2¾	1-7/8	1-1/8
11	A	3½	3½	2½	1-7/8	1-1/8
12	B	3¾	2¼	1-5/8	1-3/8	¾

Source: Based on Previte, R.W., *J. ACI, Proc.*, Vol. 74, No. 8, pp. 361–367, 1977.

Table 10-2 Effect Cement Composition, Elapsed Time, and Temperature on Slump of Concrete Mixtures with Different Initial Slumps with 9 percent C_3A content, whereas Cement B was high-alkali (0.62 percent equivalent Na_2O) with 10.6 percent C_3A content; both had similar SO_3 content and Blaine surface area. The following conclusions were drawn from the investigation:

1. In general, the amount of slump loss was proportional to the initial slump; the higher the initial slump, the higher the slump loss. For example, in the case of Cement A, at the close of the 2-h test at 21°C (70°F), concrete Mix 1 (initial slump 180 mm or 7½ in.) lost 125 mm (5¼ in.) slump, whereas with concrete Mix 3 (initial slump 125 mm or 5 in.) lost 88 mm (3½ in.) slump, and concrete Mix 5 (initial slump 85 mm or 3-5/8 in.) lost 57 mm (2¼ in.) slump. Regardless of the initial slump, the final slump values after 2 h of hydration were of the order of 37 to 50 mm (1¼ to 2 in.). In such a case the method of compensating for the expected slump loss by designing for a higher initial slump is not recommended because the retempering water required at the job site may have the effect of pushing up the water-cement ratio of the concrete mixture to an undesirable level.
2. In general, early slump loss tends to be directly proportional to the temperature of concrete. For example, a comparison of the 180-mm (7-in.) slump concretes made with Cement A at two different temperatures [i.e., 21°C (70°F) (concrete Mix 1) and 29°C (85°F) (concrete Mix 7)] showed that at 30, 60, and 90 min elapsed times, the former lost 13, 28, and 95 mm (½, 1-7/8, and 3¾ in.) slump, while the latter lost 44, 70, and 114 mm (1¾, 2¾, and 4½ in.), respectively.
3. In regard to the effect of cement composition, greater slump loss rates were observed for all test conditions in the case of concretes made with the cement containing higher C_3A and high-alkali content (Cement B). For instance, at 21°C (70°F) and 30, 60, and 90 min elapsed times, concrete Mix 1 lost 13, 28, and 95 mm (½, 1-7/8, and 3¾ in.) slump, while the latter lost 44, 70, and 114 mm (1¾, 2¾, and 4½ in.) compared to 68, 98, and 143 mm (2-2/3, 3-7/8, and 4-5/8 in.), respectively, for concrete Mix 2.

Erlin and Hime[6] reported case histories of unusual slump loss attributable to the cement composition or cement-admixture interaction. In one case, during slip-form construction of a concrete silo, surface irregularities were observed when a light-colored portland cement was used; such irregularities did not occur when a darker cement was used in the initial stages of construction. Workers had noticed higher pumping pressures at the time of placing the concrete containing the light-colored cement. Laboratory analysis revealed that the calcium sulfate in this cement was present in the form of dehydrated gypsum; therefore, the cement showed severe false setting (see Fig. 6-8). This created a condition that caused the concrete surface to *tear* when the forms were slipped.

In another case,[8] during transit the concrete in a ready-mix truck set so severely that it had to be blasted loose. Laboratory tests showed that the concrete contained two or three times the normal dose of an admixture containing triethanolamine which is an accelerator. As soon as the admixture was added, the cement stiffened rapidly and produced considerable heat (i.e., a flash set). From the cement analysis it was found that calcium sulfate was present mostly in the form of natural anhydrite. Thus imbalance in the reactions involving the sulfate and aluminate led to rapid setting (see Fig. 6-9). In yet another incident, because of the presence of a glucoheptanate-type coloring agent in the admixture, retardation of the cement was so severe that no stiffening and setting occurred in 24 h; therefore, the concrete had to be removed the next day. Some water-reducing agents, especially the high-range type or superplasticizers, tend to accelerate slump loss. This is because an efficient dispersion of the cement-water system enhances the rate of formation of the hydration products. Superplasticizers containing excessive sodium sulfate are also known to accelerate the cement hydration and slump loss.

According to Tuthill,[8] problems attributed to slump loss often arise at the very start of a placing operation if mixing is permitted before the formwork is positively ready to receive the concrete, or if the first batches are on the low side of the slump range and are judged too dry to make a safe start without delay, where there is no newly placed concrete into which to work them. Either of these two common problems causes concrete to stay in trucks or buckets, losing slump with time. Delays from the mixing to the placement of concrete can have a serious effect on production rates aside from the direct time loss, especially in operations such as pumping, tunnel lining, slip-formed paving, and tremie concreting, which depend heavily on a uniform consistency of concrete.

Slump-loss problems occur most often in hot weather. The higher the temperature at which a concrete is mixed and placed the more likely it is that slump loss turns out to be the cause of any operating problem. ACI Committee 305 cautions that difficulties may be encountered with concrete at a placing temperature approaching 32°C, and every effort should be made to place it at a lower temperature. In hot and dry weather, it is recommended that aggregate be stored in shaded areas and cooled by sprinkling water. According to Tuthill,[8] the use of chipped ice as a partial or complete replacement for mixing water is the best way to bring down the concrete temperature; each 3 kg of ice will reduce the temperature of 1 m^3 of concrete about 0.7°C.

In conclusion, elimination of every possible delay in concrete handling operations, keeping the temperature of concrete as closed to the 10 to 21°C range as possible, and a laboratory check on the stiffening and setting characteristics of the cement (with or without the admixtures selected for use) are the necessary preventive measures to control slump loss problems.

10.7 Segregation and Bleeding

10.7.1 Definitions and Significance

Segregation is defined as the separation of components of a fresh concrete mixture so that they are no longer uniformly distributed. There are *two kinds of segregation*. The first, which is characteristic of dry concrete mixtures, consists of separation of mortar from the body of concrete. Bleeding, as explained next, is the second form of segregation, which is characteristic of wet concrete mixtures.

Bleeding is defined as a phenomenon whose external manifestation is the appearance of water on the surface after a concrete mixture has been placed and compacted but before it has set (i.e., when sedimentation can no longer take place). Water is the lightest component in a concrete mixture; thus, bleeding is a form of segregation because solids in suspension tend to move downward under the force of gravity. Bleeding results from the inability of the constituent materials to hold all the mixing water in a dispersed state as the relatively heavy solids settle.

It is important to reduce the tendency for segregation in a concrete mixture because full compaction, which is essential for achieving the maximum strength potential, is not possible in a segregated concrete mixture. Furthermore, only some of the bleed-water reaches the surface; a large amount of it gets trapped within concrete. There are some interesting manifestations of this phenomenon. With ordinary reinforced concrete structures, numerous bleed-water pockets, occurring under the coarse aggregate particles and the horizontal reinforcing bars are responsible for weakening these areas. For the same reason, the upper half of a reinforced concrete beam or column may be weaker than the lower half.

Laitance, associated with the external manifestation of bleeding, is caused by the tendency of water rising in the internal channels within concrete, carrying with it very fine particles of cement, sand, and clay (present as a contaminant in aggregate) and depositing them in the form of a scum at the concrete surface. Because the laitance layer contains a very high water-cement ratio, it is porous, soft, and weak. When a floor slab or a pavement suffers from laitance, it may be due to the reason that instead of a hard and durable surface the concrete has a soft surface prone to *dusting*. Hydration products in the porous cement paste of the laitance layer are easily carbonated in air. If laitance occurs at the top of a casting, poor bond to the next casting will result; therefore, laitance on old concrete should always be removed by brushing and washing or by sand blasting before new concrete is placed. The positive role of surface bleed water on plastic shrinkage cracking is discussed later.

10.7.2 Measurement

There are no tests for measuring segregation; visual observation and inspection of cores of hardened concrete are generally adequate to determine whether segregation has occurred. There is, however, an ASTM standard test for the measurement of rate of bleeding and the total bleeding capacity of a concrete mixture. According to ASTM C 232, a sample of concrete is placed and consolidated in a cylindrical container, 250 mm diameter and 280 mm high. The bleed water accumulated on the surface is withdrawn at 10-min intervals during the first 40 min, and thereafter at 30-min intervals. Bleeding is expressed in terms of the amount of accumulated water as a percentage of the net mixing water in the concrete sample.

10.7.3 Causes and Control

A combination of improper consistency, excessive amount of large particles of coarse aggregate with either a too high or a too low density, presence of less fines (due to a low cement content, a low sand content, or a poorly graded sand), and inappropriate placing and compacting methods are among the general causes for segregation and bleeding problems in concrete. Obviously, the problems can be reduced or eliminated by paying attention to the selection of materials, mixture proportioning, and concrete handling and placement methods.

Segregation in dry concrete mixtures can sometimes be reduced by increasing the water content slightly. In most cases, however, proper attention to aggregate grading is required. This may involve a lowering of the maximum size of coarse aggregate and the use of more sand or a finer sand. Increase in the cement content and the use of mineral admixtures and air entrainment are also commonly employed measures in combating the bleeding phenomenon of concrete mixtures. It is interesting to point out that high-C_3A and high alkali cements, which show greater slump loss, tend to reduce bleeding as a result of rapid formation of sulfoaluminate hydrates such as ettringite. When a concrete mixture has to be dropped from considerable height (e.g., in tremie concreting) or discharged against an obstacle, the material should be highly cohesive and extra care is necessary during the placement.

10.8 Early Volume Changes

10.8.1 Definitions and Significance

After fresh concrete has been placed in deep forms, such as the forms for a tall column or a wall, after a few hours the top surface will have subsided. The tendency toward subsidence is also confirmed by the presence of short horizontal cracks. This reduction in volume of fresh concrete is known by terms such as *prehardening, presetting shrinkage,* or *plastic shrinkage,* since the shrinkage occurs while the concrete is still in the plastic state. As a result of prehardening shrinkage, cracks develop over obstructions to uniform settlement, that is, for instance, reinforcing bars and large aggregate particles. In the United States, the term *plastic shrinkage* is usually used with reference to concrete slabs, as discussed below.

With slabs, rapid drying of fresh concrete causes plastic shrinkage when the rate of loss of water from the surface exceeds the rate at which the bleed water is appearing. At the same time, cracks will develop if the concrete near the surface has become too stiff to move but is not strong enough yet to withstand the tensile stress caused by the restrained shrinkage. Typical plastic shrinkage cracks (see Fig. 10-7) are parallel to one another and are 0.3 to 1 m apart and 25 to 50 mm deep.

10.8.2 Causes and Control

A variety of causes contribute to plastic shrinkage in concrete: bleeding or sedimentation, absorption of water by subgrade or forms or aggregate, rapid water loss by evaporation, reduction in the volume of the cement-water system, and bulging or settlement of the formwork. The following conditions, singly or collectively, increase the rate of evaporation of surface moisture and enhance the possibility of plastic-shrinkage cracking: high concrete temperature, low humidity, and high wind velocity. When the rate of evaporation exceeds 1 kg/m^2 per hour (0.2 lb/ft^2 per hour), precautionary measures are necessary to

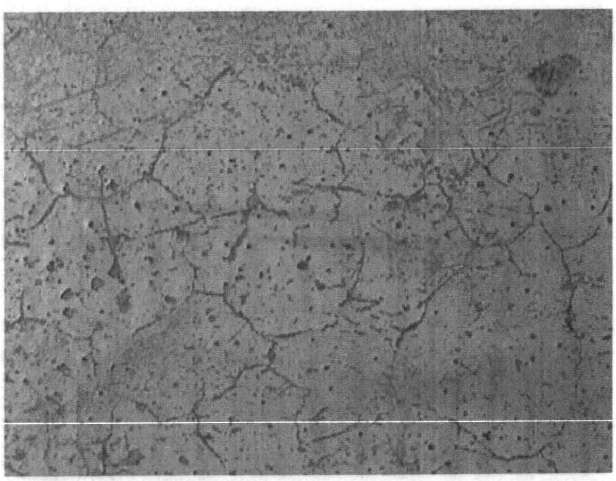

Figure 10-7 Plastic shrinkage cracking in freshly placed concrete. (Photograph courtesy of Carlos Vidella.)

prevent the plastic-shrinkage cracking. The Portland Cement Association[9] has developed a chart (see Fig. 10-8) for determining when precautionary measures should be taken. The measures that should be considered are as follows:

- Moisten the subgrade and forms.
- Moisten aggregates that are dry and absorptive.
- Erect temporary windbreaks to reduce wind velocity over the concrete surface.
- Erect temporary sunshades to reduce concrete surface temperature.
- Keep the fresh concrete temperature low by cooling the aggregate and mixing water.

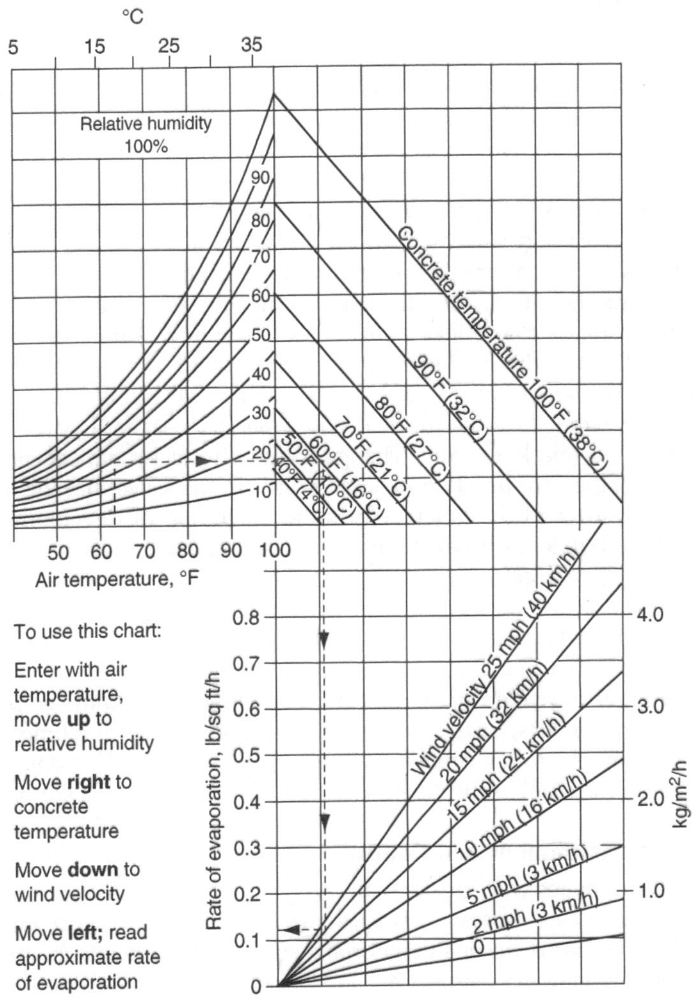

Figure 10-8 Estimating the rate of moisture evaporation from a concrete surface. (From *305R-10: Guide to Hot Weather Concreting, American Concrete Institute, Farmington Hills, MI, 2010.)*

- Protect concrete with temporary coverings such as polyethylene sheeting during any appreciable delay between placing and finishing.
- Reduce the time between placing and start of curing by eliminating delays during construction.
- To minimize evaporation, protect the concrete immediately after finishing by wet burlap, fog spray, or a curing compound.

Settlement cracks in columns and plastic shrinkage cracks in slabs can be eliminated by revibration of concrete when it is still in the plastic state. Revibration also improves the bond between concrete and reinforcing steel, and enhances the concrete strength by relieving the plastic shrinkage stresses around the coarse aggregate particles.

10.9 Setting Time

10.9.1 Definitions and Significance

The reactions between cement and water are the primary cause of the setting of concrete although, for various reasons, discussed later, the setting time of concrete does not coincide with the setting time of the cement with which a concrete mixture has been made. As described in Chap. 6, the phenomena of stiffening, setting, and hardening are the physical manifestations of progressive hydration of cement with time. Also, the initial and the final setting times of cement are the points arbitrarily defined by the method of test. These points indicate the rate of solidification of a freshly mixed cement-water system. Similarly, *setting of concrete* is defined as the onset of solidification in a fresh concrete mixture. Both the initial and the final setting times of concrete are arbitrarily defined by a test method such as the *penetration resistance method* (ASTM C 403), which is described below.

The *initial setting time* and the *final setting time*, as measured by penetration resistance methods, do not mark a specific change in the physical-chemical characteristics of the cement paste; they are purely functional points in the sense that the former defines the limit of handling and the latter defines the beginning of development of mechanical strength. Figure 10-9 illustrates that initial set and final set of concrete measured by ASTM C 403 do not have to coincide exactly with the periods marking the end or the complete loss of workability and the beginning of mechanical strength. Instead, the initial set represents approximately the time at which fresh concrete can no longer be properly mixed, placed, and compacted; the final set represents approximately the time after which strength begins to develop at a significant rate. Obviously, a knowledge of the changes in concrete characteristics, as defined by the initial and final setting times, can be of considerable value in scheduling concrete construction operations. Test data can also be useful in comparing the relative effectiveness of various set-controlling admixtures.

10.9.2 Measurement and Control

For concrete mixtures with greater than zero slump, ASTM C 403, *Test for Time of Setting of Concrete Mixtures by Penetration Resistance*, provides a standard procedure for the measurement of setting time by testing the mortar sieved from a concrete mixture. Briefly, the test consists of removing the mortar fraction from concrete, compacting it in a standard container, and then measuring the force required to cause a needle to penetrate 25 mm into the mortar. The times of set are determined from the rate of solidification curve obtained from a linear plot of data with elapsed time as the abscissa and penetration resistance as the ordinate.

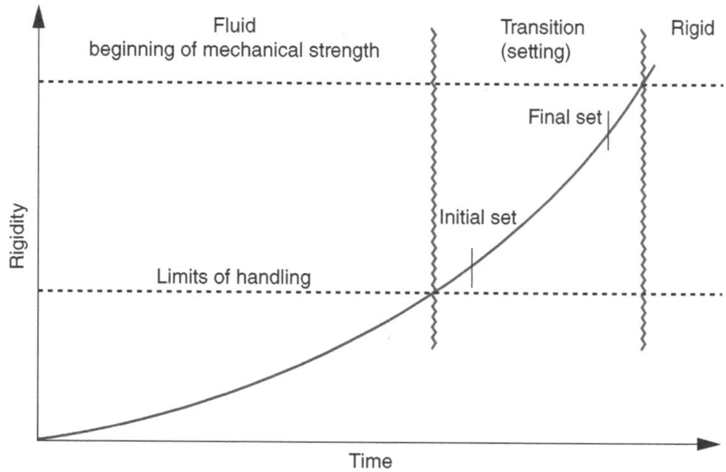

FIGURE 10-9 The progress of setting and hardening in concrete. (From Mindess, S., and J.F. Young, Concrete, p. 401, 1981. Reprinted by permission of Prentice Hall, Englewood Cliffs, NJ.)

Initial and final set are defined as times at which the penetration resistances are 3.5 MPa (500 psi) and 27.6 MPa (4000 psi), respectively. These arbitrarily chosen points do not indicate the strength of concrete; in fact, at 3.5 MPa (500 psi) penetration resistance value the concrete has no compressive strength, while at 27.6 MPa (4000 psi) penetration resistance value the compressive strength may be only about 0.7 MPa (100 psi).

The principal factors controlling the setting time of concrete are cement composition, water-cement ratio, temperature, and admixtures. Cements that are quick setting, false setting, or flash setting will tend to produce concretes with corresponding characteristics. As the setting and hardening phenomena in a hydrating cement paste are influenced by the filling of void space with the products of hydration, the water-cement ratio will obviously affect the initial and the final setting times. However, the setting-time data for a cement paste do not coincide with the setting times of concrete containing the same cement because the water-cement ratios in the two cases are usually different. In general, the higher the water-cement ratio, the longer the time of set.

The effects of cement composition, temperature, and retarding admixtures on typical rates of setting obtained by ASTM C 403 test are shown in Fig. 10-10. When a concrete mixture was made and stored at 10°C instead of 23°C, the initial and the final setting times were retarded approximately by 4 and 7 h, respectively. With cement B and a set-retarding admixture, the retarding effect of the admixture was found to be greater at the higher temperature.

10.10 Temperature of Concrete

10.10.1 Significance

Among other problems, as will be discussed below, in *hot weather*, unprotected concrete is subject to plastic shrinkage cracking. On the other hand, in *cold weather* the low temperature of concrete curing may seriously impede the rate of strength development. Premature removal of formwork (i.e., before the concrete acquires sufficient *maturity* or

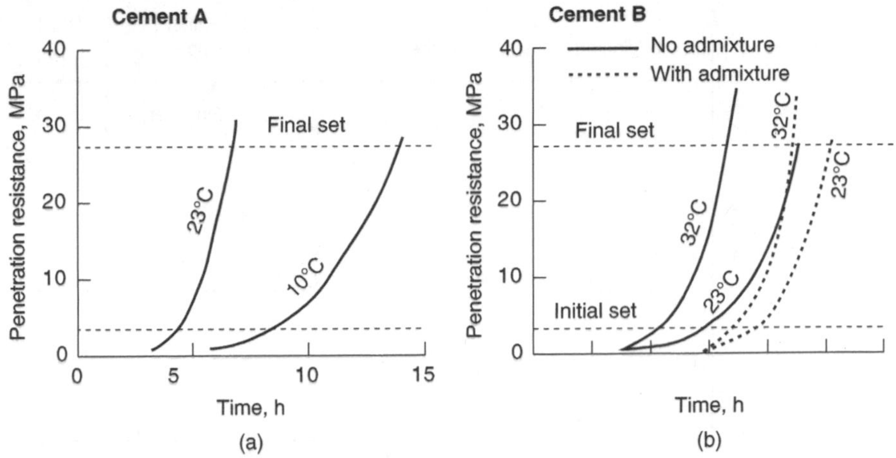

FIGURE 10-10 (*a*) Effect of temperature on initial and final setting times of concrete (ASTM C 403); (*b*) effect of a retarding admixture on setting times of concrete (ASTM C 403). (Reprinted with permission, from Sprouse, J.H., and R.B. Peppler, ASTM STP 169B, pp. 105–121, 1978. ASTM, 1916 Race Street, Philadelphia, PA.)

strength) has led to disastrous consequences in terms of both human and economic losses (see below). The problem usually arises when the construction scheduling decisions are based on laboratory-cured cylinders whereas the actual curing history of the in-place field concrete happens to be very different. Construction engineers should have a general understanding of the possible effects of both lower- and higher-than-normal curing temperatures on properties of concrete at early ages, and the methods of evaluating and controlling them.

> In Kiev, capital of the industrial Ukraine, workers were in a bind to get a building up in the allotted time. The newspaper *Rabochaya Gazeta* said the construction crews fiddled with the architect's plan to cut down the work and then produce a building in record time. When the workers eagerly swung the roof into place, the structure neatly collapsed in a heap. They had left out that part that says "allow the concrete to dry [*cure*]"
>
> *Source*: UPI report
> Published in the San Francisco *Sunday Examiner and Chronicle*,
> January 4, 1976

> On 27 April 1978 a cooling tower under construction at Willow Island in West Virginia, collapsed—killing 51 workers. The contractor was using a slip-formed construction process involving a multilayer scaffold that raises itself up the wall by its own power after anchoring into the hardened concrete of the previous day's work. According to an investigation by the Office of Safety and Health Administration, the accident "could have been prevented if proper engineering practices had been followed." Investigation findings cited that one of the key factors contributing to the collapse was "a failure to make field tests to be sure that the concrete had cured sufficiently before the support forms were removed."
>
> *Source*: Based on a report by Eugene Kennedy
> Published in the San Francisco *Sunday Examiner and Chronicle*,
> December 3, 1978

10.10.2 Cold-Weather Concreting

In the event of little cement hydration, no strength gain occurs when the concrete is frozen and is kept frozen below –10°C. Therefore, fresh concrete must be protected from freezing until adequate strength has been gained.* Disruptive expansion is also prevented when the degree of saturation of concrete has been sufficiently reduced by some progress in the hydration process. Without an external heat source, the heat of cement hydration in large and well-insulated concrete members may be adequate to maintain satisfactory curing temperatures provided that the concrete has been delivered at a proper temperature, and the temperatures of frozen ground, formwork, and reinforcing bars have been taken into consideration.

ACI Committee 306R recommendation for *cold-weather concreting* on placement temperatures for normal-weight concrete is shown in Table 10-3. It may be noted that lower concrete temperatures are permitted for massive sections because with these the heat generated during hydration is dissipated less rapidly than from flatwork. Also, as more heat is lost from the concrete during transport and placement at lower air temperatures, the recommended concrete temperatures are higher for colder weather (see lines 1, 2, and 3 in Table 10-3).

Line	Condition		Sections Less Than 12 in. (300 mm) Thick		Sections 12–36 in. (300 mm–0.9 m) Thick		Sections 36–72 in. (0.9–1.8 m) Thick		Sections over 72 in. (1.8 m) Thick	
			°F	°C	°F	°C	°F	°C	°F	°C
1	Minimum temperature fresh concrete as *mixed* in weather indicated, °F (°C)	Above 30°F (–1°C)	60	16	55	13	50	10	45	7
2		0°F to 30°F (–18°C to –1°C)	65	18	60	16	55	13	50	10
3		Below 0°F (–18°C)	70	21	65	18	60	16	55	13
4	Minimum temperature fresh concrete as *placed* and *maintained*		55	13	50	10	45	7	40	5
5	Maximum allowable *gradual* drop in temperature in first 24 h after end of protection		50	28	40	22	30	17	20	11

*For durability and safe stripping strength of *lightly stressed* members. ACI 306 recommends 1 to 3 day's duration of the temperatures shown in the table, depending on whether the concrete is conventional or the high-early-strength type. For *moderately and fully stressed* members, longer durations are recommended. Also, for the concrete that is *not air-entrained* it is recommended that protection for durability should be at least twice the number of days required for air-entrained concrete.

Source: Adapted from ACI 306–78.

Table 10-3 Recommended Concrete Temperature for Cold-Weather Construction: Air-entrained Concrete*

*A minimum compressive strength of 3.5 MPa (500 psi) prior to freezing is stated in ACI 306R as a criterion for preventing frost damage.

Insufficient curing of concrete can also be detrimental to properties other than strength. Most of the decision making is based on strength because form stripping, prestressing, and other such operations in concrete construction are guided by the strength of concrete on hand. Usually, strength is also the criterion when durability of concrete in early exposure to aggressive waters is of concern. The traditional method for determining safe stripping times is to test laboratory-cured concrete cylinders and strip the forms when the cylinders reach the specified strength. As already stated, this procedure has led to problems when the curing history of the cylinder in the laboratory is considerably different from the curing history of the in-place concrete. In case of weather extremes, test data from field-cured cylinders are preferable. In the report of ACI Committee 306, the *maturity method* is recommended as an alternative to using laboratory or field-cured cylinders.

Control of Concrete Temperature

For cold-weather concreting (Table 10-3), making fresh concrete mixtures at temperatures 21°C (70°F) is not recommended. The higher temperatures do not necessarily offer better protection: first, because at higher temperatures the rate of heat loss is greater, and second, the water requirement for the same consistency is more. Depending on the ambient temperature and transport time form the production site to the job site, the temperature of concrete *as mixed* is maintained at not more than 5.6°C (10°F) above the minimum recommended in Table 10-3. As discussed further, the temperature of fresh concrete is usually controlled by adjusting the temperatures of mixing water and aggregates.

Of all the concrete-making components, mixing water is the easiest to heat. Also, it makes more practical sense to do so because water can store five times as much heat as can the same mass of cement or aggregate. Compared to a specific heat of 1.0 for water, the average specific heat for cement and aggregates is 0.22. At temperatures above freezing, it is rarely necessary to heat the coarse aggregates. At temperatures below freezing, often only the fine aggregate needs to be heated to keep the freshly produced concrete at the required temperature. This is generally accomplished by circulating hot air or steam through pipes embedded in the aggregate stockpile.

Concrete temperature can be measured directly by a mercury thermometer or a bimetallic thermometer. It can also be estimated using the expression

$$T = \frac{0.22(T_a W_a + T_c W_c) + T_w W_w + T_{wa} W_{wa}}{0.22(W_a + W_c) + W_w + W_{wa}} \tag{10-1}$$

where T = temperature of the fresh concrete in °F

T_a, T_c, T_w, and T_{wa} = temperatures of aggregates, cement, mixing water, and free moisture in aggregates, respectively

W_a, W_c, W_w, and W_{wa} = weights (in pounds) of aggregates, cement, mixing water, and free moisture in aggregates, respectively

The formula remains the same in SI units except that °F is changed to °C and pounds to kilograms.

10.10.3 Hot-Weather Concreting

For the purposes of construction problems with structural concrete, ACI Committee 305 defines *hot weather* as any combination of high air temperature, low relative humidity,

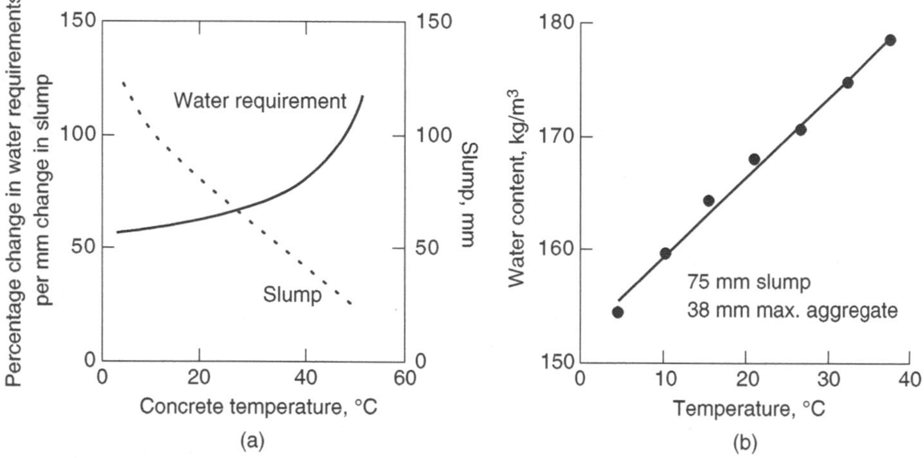

FIGURE 10-11 (*a*) Effect of concrete temperature on the slump and the water requirement to the change slump; (*b*) Effect of ambient temperature on the water requirement of concrete. (305R-10: Guide to Hot Weather Concreting, American Concrete Institute, Farmington Hills, MI, 2010.)

The water requirement of a concrete mixture increases with an increase in the temperature of concrete. As shown in the figure, if the temperature of fresh concrete is increased from 10 to 38°C, the water requirement increases by about 15 kg/m^3 for maintaining 75 mm slump. This increase in the water content can reduce the 28-day compressive strength of concrete by 12 to 15 percent.

and wind velocity tending to impair the quality of fresh or hardened concrete or otherwise resulting in abnormal properties. In addition to the increase in slump loss and plastic-shrinkage cracking, and the decrease of setting time in fresh concrete (already described), hot weather increases the mixing water requirement for a given consistency (Fig. 10-11) and creates difficulty in holding the air in an air-entrained concrete mixture. Retempering of fresh concrete is frequently necessary in hot weather. At times, this causes adverse effects on strength, durability, dimensional stability, and appearance of the hardened concrete. Also, concrete placed and cured at higher than moderate ambient temperatures normally develops high early strength but at 28 days and later ages the strength is usually lower than the same concrete placed and cured at a relatively lower temperature.

Control of Concrete Temperature

As explained earlier, because the mixing water has the greatest effect per unit weight of any of the ingredients on the temperature of concrete, the use of cooled mixing water and/or ice offers the best way of lowering the temperature of concrete. The expression used for determining the temperature of concrete in cold weather by using hot water can be employed for calculating how much cold water will be needed to lower the temperature of a concrete by a given amount. Alternatively, charts such as that shown in Fig. 10-12*a* can be used. The data in Fig. 10-12*a* pertain to a nominal concrete mixture containing 335 kg/m^3 cement, 170 kg/m^3 water, and 1830 kg/m^3 aggregate.

The use of shaved or chipped ice as a substitute for all or part of the required mixing water is the most effective way of reducing the concrete temperature because ice absorbs

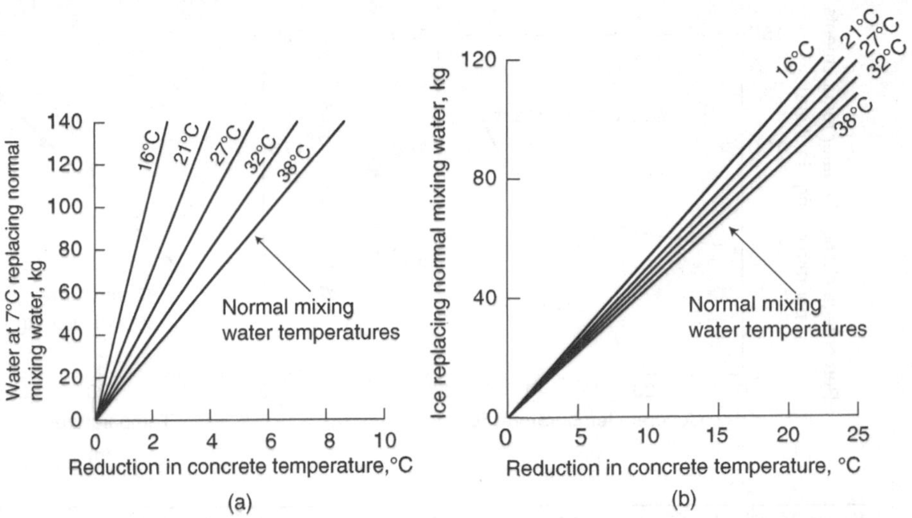

FIGURE 10-12 Determination of reduction in concrete temperature: (*a*) by adding cooled water; (*b*) by adding ice. (305R-10: Guide to Hot Weather Concreting, American Concrete Institute, Farmington Hills, MI, 2010.)

Part (a) shows the effect of cooled (7°C) mixing water, and part (b) shows the effect of ice in mixing water on concrete temperature. Normal mixing water temperatures are shown on the curves. The data are applicable to average mixes made with typical natural aggregates. A comparison of the two figures shows that the use of ice as part of the mixing water is highly effective in reducing the concrete temperature because, on melting, ice absorbs heat at the rate of 80 cal/g).

80 cal/g (144 Btu/lb) on melting. Figure 10-12*b* illustrates the possible reductions in concrete temperature by substitution of varying amounts of ice at 0°C for mixing water at the temperature shown. Figure 10-12 demonstrates that, with normal mixing water at 38°C, there will be a 3.3°C temperature reduction when 54 kg (120 lb) of water at 7°C, replaces the mixing water; the same amount of ice replacing the mixing water would have reduced the temperature of the concrete by 13°C.

10.11 Testing and Control of Concrete Quality

10.11.1 Methods and Their Significance

Engineers representing the owners, designers, and builder of structures are frequently required to develop or approve a *quality assurance* program that, among other things, involves the selection of test methods, statistical analysis of the test results, and follow-up procedures. The objective of such a program is to ensure that a finished concrete element is structurally adequate for the purpose for which it was designed. The size of concrete structures being designed and built today and the speed of modern construction (e.g., over 200 m^3/h placement of concrete in some projects) require that the decision making on acceptance or rejection of concrete quality should not be left to the 28-day compression test, which continues to be the basis for design specifications.

Accelerated strength testing offers one solution to the problem. Increasingly, large projects are using procedures that allow a preliminary assessment 1 or 2 days after placing concrete as to whether the product will reach the required strength level. A low value from an accelerated strength test can warn the contractor of a potential problem and provide an early opportunity for remedial action. In the case where substandard concrete has been placed, it is easier and less expensive to remove it when the concrete is only a few days old rather than when it is 28 days old and probably covered with a superstructure.

A criticism against the testing of concrete samples drawn from the batches before the placement is that the test specimens may not truly represent the quality of concrete in a structure, due possibly to sampling errors and differences in compaction and curing conditions. Also, on large projects the cost of strength testing can be considerable. As an alternative approach to direct strength testing, many in situ/nondestructive test methods have been developed, which provide an excellent means of control of in-place concrete quality. Although in situ/nondestructive tests are not accepted as a complete substitute for direct strength tests, they can reduce the cost of testing for quality control when used in conjunction with core strength tests or standard compression tests.

In large-scale industrial production, an effective and economical system of quality control has to rely on statistical methods of data processing and decision making. A primary statistical tool in concrete quality control programs is the use of control charts which graphically show the results of tests and also contain limit lines indicating the need for action when the plotted data approach the limit lines.

10.11.2 Accelerated Strength Testing

Based on reports by Malhotra[10] and Carino[11] a brief review of the four test procedures covered by ASTM C 684 is discussed below:

Procedure A (Warm-Water Method)

This is the simplest of the four methods and it consists of curing standard cylinders, in a water bath maintained at 35°C for 24 h immediately after molding and while still in molds. A limitation of the method is that strength gain, compared to the 28-day moist-cured concrete at normal temperature, is not high, therefore, job-site testing may be needed. In the mid-1970s, the U.S. Corps of Engineers[12] conducted an extensive study on the evaluation of the warm-water method. It was concluded that accelerated strength testing with this method is indeed a reliable method of routine quality control for concrete.

Procedure B (Boiling-Water Method)

This method consists of normal curing of the concrete cylinders for 24 h, followed by curing in a boiling-water bath at 100°C for $3^1/_2$ h, and then testing 1 h later. The method is the most commonly used of the three procedures because compared to the 24-h warm-water method, the strength gain at $28^1/_2$ h is much higher and concrete cylinders can be transported to a central laboratory for strength testing, thus eliminating the need for an on-site laboratory. In the early 1970s, the method was used successfully to develop concrete mixture proportions in preliminary laboratory studies and to check field concrete in the construction of a large number of dikes, spillways, and a huge underground power station for the Churchill Falls Project in Labrador, Canada.

Procedure C (Autogenous Method)

In this method, immediately after casting the test cylinders are placed in insulated containers made of polyurethane foam and are tested 48 h later. No external heat source is provided, the acceleration of strength gain being achieved by the heat of hydration of cement alone. Again, the strength gain at the end of the curing period is not high, and this method is judged to be the least accurate of the four. It was used as an integral part of the quality control program in the construction of the CN Communication Tower in Toronto, Canada. The project involved placing 30580 m^3 of slip-formed concrete to a height of 475 m. It is believed that the accelerated strength testing played an important role in the quality control of concrete and in the overall structural safety of one of the world's tallest free-standing structure.

Procedure D (High Temperature and Pressure Method)

According to this method, acceleration of strength development is achieved by combination of elevated temperature and pressure using 75 × 150 mm (3 × 6 in.) concrete cylinders. Fresh concrete in the mold is maintained under a compressive stress of 10.3 ± 0.2 MPa (1500 ± 25 psi) and a temperature of 149 ± 3°C (300 ± 35°F) for a period of 3 h. Thereafter, the heater is turned off but the axial stress is maintained and the specimen is allowed to cool for 2 h. At the end of the 5-h test the hardened concrete cylinders are extruded from the molds and tested for compressive strength. According to Carino,[11] the results from this 5-h test correlate well with Procedure B. The test method is especially suitable for concrete mixtures containing pozzolanic admixtures.

10.11.3 Core Tests

In situ/NDT methods provide an effective way of obtaining a considerable amount of preliminary test data at relatively little cost. When these tests indicate internal cracking or zones of weaker concrete, it is necessary to perform direct strength testing on cores obtained from the structure using a rotary diamond drill (ASTM C 42). The core strengths are generally lower than those of standard-cured concrete cylinder, especially in high-strength concrete. In the case of concrete mixtures with high cement content and a correspondingly high heat of hydration, large members of in-place concrete are vulnerable to considerable microcracking in the interfacial transition zone between the coarse aggregate and hydrated cement paste. Consequently, the ratio of core strength to cylinder strength decreases as the strength of concrete increases. The strength of the core will also depend on its position in the structure. Generally, due to the differential bleeding effect, cores taken from near the top of a structural element are weaker than those from the bottom. Chapter 11 presents various nondestructive tests to assess the presence of flaws and delaminations in concrete.

10.11.4 Quality Control Charts

As stated earlier, with high production rates of modern ready-mixed concrete plants or on-site concrete plants for large projects, an effective and economical system of quality control must be based on statistical methods. Statistical procedures are governed by the laws of probability, and for these laws to properly function the first requirement is that the data be gathered by random sampling. The second important statistical concept is that of the frequency distribution follow the bell-shaped normal distribution Gaussian curve

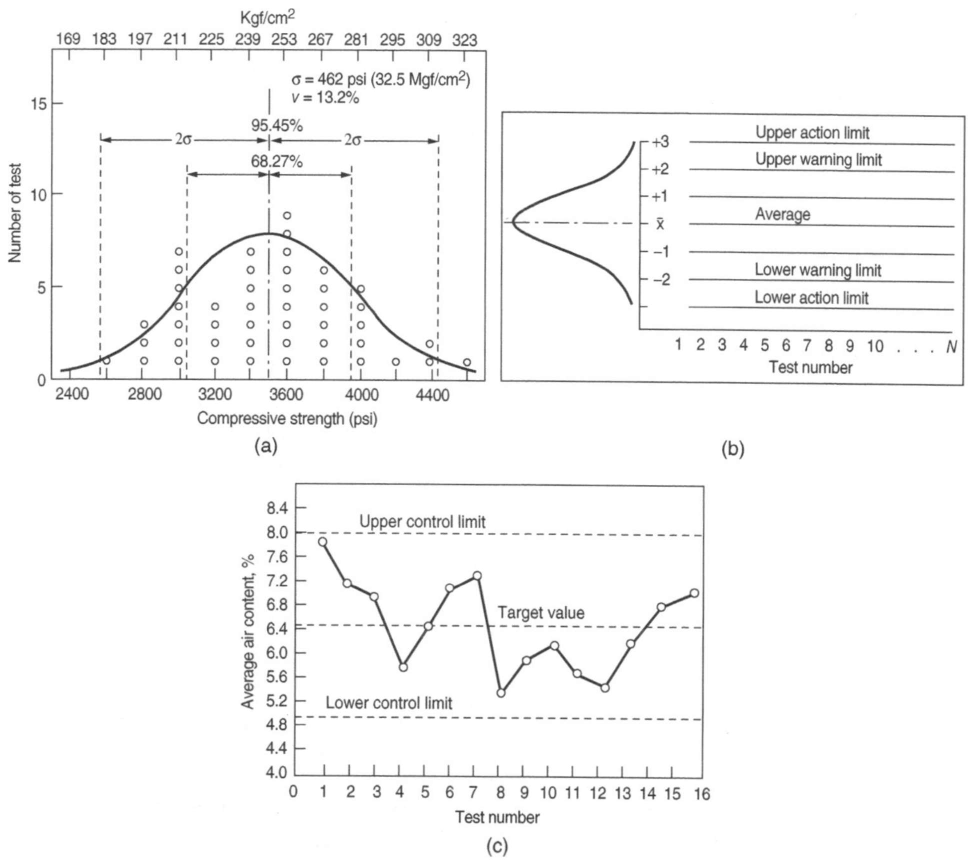

FIGURE 10-13 (*a*) Frequency distribution of strength data and corresponding normal distribution; (*b*) typical statistical control chart; (*c*) X chart for air content. (From ACI Committee 214, Report 214R-77; and Keifer, O. Jr., *Concr. Int.*, Vol. 3, No. 11, pp. 12–16, 1981.)

Statistical quality control charts are based on frequency distribution predicted by the normal-distribution curve. On a typical control chart, the upper and lower control limits may be derived from the normal-distribution curve laid on its side.

(Fig. 10-13*a*). A detailed discussion of the statistical symbols and their definitions is outside the scope of this book. Those interested should refer to any standard textbook on statistics or ASTM Manual on Presentation of Data and Control Chart Analysis (2011).

Statistical quality control utilizes control charts[13] that show graphically the results of a continuous testing program. The charts contain upper- and lower-limit lines that indicate the need for action when the plotted curve approaches or crosses them. The limit lines relate to the normal-distribution curve. In fact, a control chart may be considered as a normal-distribution curve laid on its side (Fig. 10-13*b*). Figure 10-13*c* illustrates the use of control charts in concrete quality control operations.

Based on the report of ACI Committee 214, typical quality control charts for continuous evaluation of strength test data of concrete are shown in Fig. 10-14. Figure 10-14*a* is a plot for *individual strength values;* the line for required average strength, σ_{cr} is obtained

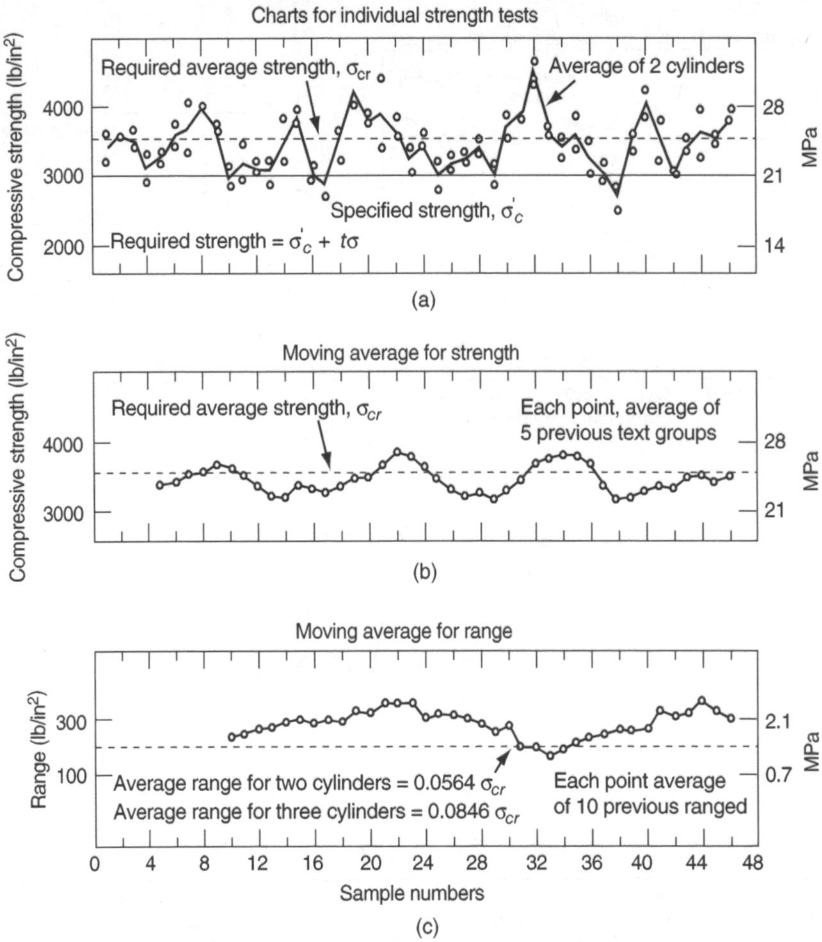

FIGURE 10-14 Typical quality control charts for concrete strength. (From ACI Committee 214, Report 214R-77; and Keifer, O. Jr., Concr. Int., Vol. 3, No. 11, pp. 12–16, 1981.)

from the expression $\sigma_{cr} = \sigma_c + ts$, where σ_c is the specified design strength, t a constant, and s the standard deviation. The chart indicates the range or scatter between individual test values and the number of low values. Unless the trend of individual low values persists, occasional low values may not be significant because they may represent chance variations rather than any problems with materials or testing method. Figure 10-14*b* is a plot of the *moving average for strength*; each point represents the average of the previous five sets of strength tests (each set of strength tests normally represents data from 3 test cylinders). This chart tends to smooth out chance variations and can be used to indicate significant trends due to variations in materials and processes that affect strength. Figure 10-14*c* is a plot of the *moving average for range*, where each point represents the average of the ranges of the 10 previous sets of strength tests. The chart provides a control on the reproducibility of the test procedures. When the range chart indicates poor reproducibility between different sets of data, it is time to check the testing procedures.

10.12 Early Age Cracking in Concrete

In designing reinforced concrete elements it is assumed that concrete will crack due to thermal and humidity cycles; however, by careful design and detailing, cracks can be controlled and crack-widths can be limited. While in principle, thermal shrinkage cracks can be predicted and controlled, extensive *cracking in concrete* can develop due to other causes. It is not easy to distinguish between different crack configurations. Often, a number of laboratory tests and compilation of complete history of the project, including concrete mixture design, placement conditions, curing methods, formwork removal, and loading history is required. Based on a report by the Concrete Society of U.K., crack types are illustrated in Fig. 10-15 and their classification with possible causes and prevention methods are listed in Table 10-4. Most of the causes responsible for nonstructural cracking have been described earlier in this chapter and in Chap. 5. Two other types of nonstructural cracks, namely those due to plastic settlement and crazing which have not been described earlier, are discussed next.

As explained before, plastic settlement cracks occur when bleeding and settlement are high and there is some restraint to the settlement. Methods used to prevent the settlement cracks include: reduction of bleeding, reduction of tendency for settlement

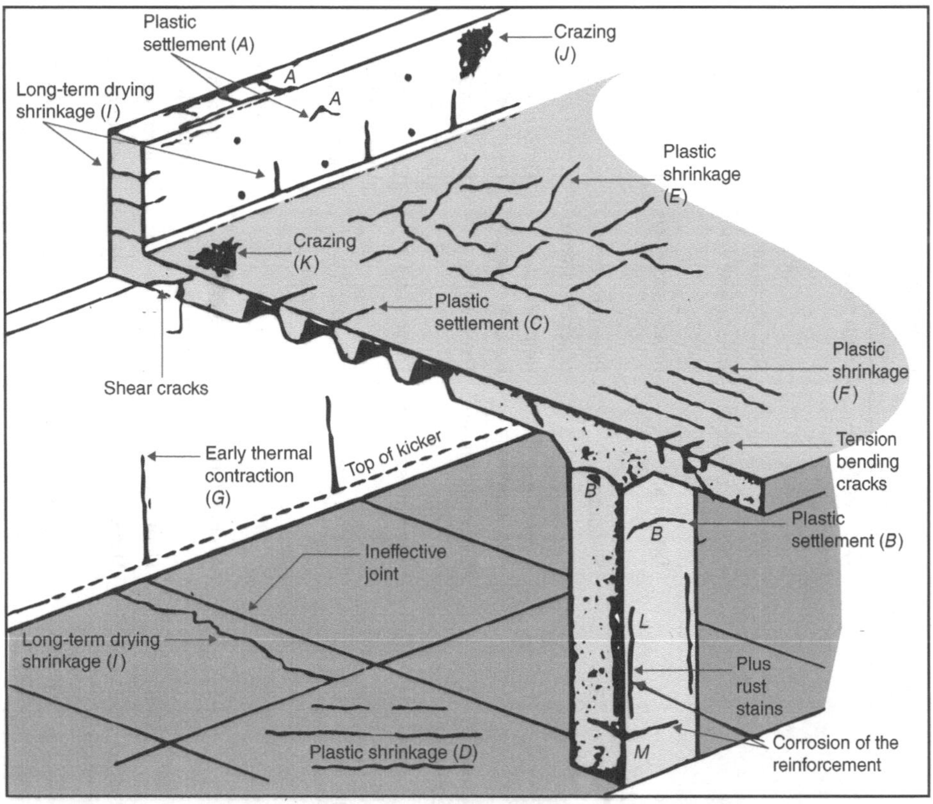

Figure 10-15 Cracks in a hypothetical concrete structure. (Adapted from Concrete Society, *Construction Cracks in Concrete*, The Concrete Society, U.K. Technical Report, No. 22, 1985.)

Type of Cracking	Letter (see Fig. 10-15)	Subdivision	Most Common Location	Primary Cause (Excluding Restraint)	Secondary Causes/ Factors	Remedy (Assuming Basic Redesign Is Impossible) in All Cases Reduce Restraint	Time of Appearance
Plastic settlement	A	Over reinforcement	Deep sections	Excess bleeding	Rapid early drying conditions	Reduce bleeding (air entrainment) or revibrate	10 min to 3 h
	B	Arching	Top of columns				
	C	Change of depth	Trough and waffle slabs				
Plastic shrinkage	D	Diagonal	Roads and slabs	Rapid early drying	Low rate of bleeding	Improve early curing	30 min to 6 h
	E	Random	Reinforced concrete slabs				
	F	Over reinforcement	Reinforced concrete slabs	Ditto plus steel near surface			
Early thermal contraction	G	External restraint	Thick walls	Excess heat generation	Rapid cooling generation	Reduce heat and/or insulate	1 day to 2 or 3 weeks
	H	Internal restraint	Thick slabs	Excess temperature gradients			
Long-term drying shrinkage	I		Thin slabs (and walls)	Inefficient joints	Excess shrinkage Inefficient curing	Reduce water content Improve curing	Several weeks or months
Crazing	J	Against formwork	'Fair faced' concrete	Impermeable formwork	Rich mixes	Improve curing and finishing	1 to 7 days
	K	Floated concrete	Slabs	Overtroweling	Poor curing	much later	sometimes
Corrosion of reinforcement	L	Natural	Columns and beams	Lack of cover	Poor	Eliminate causes listed	More than 2 years
	M	Calcium chloride	Precast concrete	Excess calcium chloride			

Source: Adapted from Concrete Society of U.K., Technical Report No. 22, 1985.

Table 10-4 Classification of Crack Types

by providing adequate restraint, and revibration of concrete. Hairline, discontinuous surface cracking, also called *crazing*, can appear in hardened concrete after several weeks. These cracks are observed particularly during rainy periods when they absorb moisture and pollutants from the atmosphere, giving the disagreeable impression of damage to concrete. In reality, the cracks are quite superficial, perhaps not more than a fraction of a millimeter deep and do not cause structural problems with the exception of opening up later and providing a passage for aggressive agents. Crazing usually occurs as a result of inadequate finishing and curing, particularly in the presence of high humidity gradients between the surface and the bulk of concrete. The use of smooth and impermeable formwork (steel, plastic), or overtrowelling of rich concrete mixtures tends to concentrate the cement paste at the concrete surface that cracks easily due to drying shrinkage, thus producing crazing. Three crack types caused by improper construction practice are shown in Fig. 10-16. A review of structural cracks due to insufficient

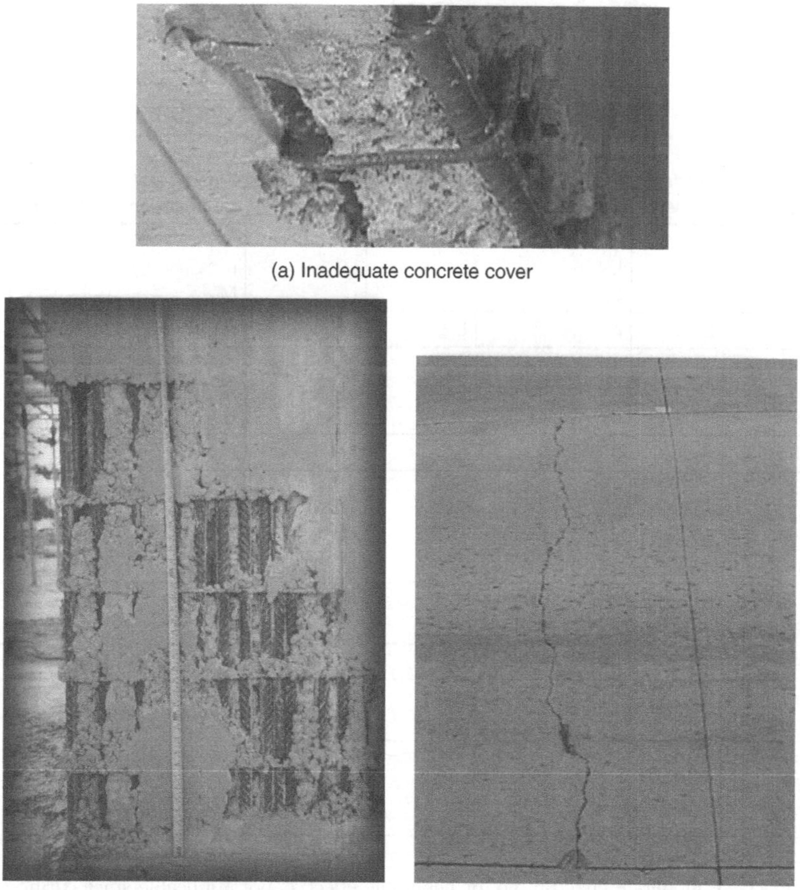

(a) Inadequate concrete cover

(b) Poor consolidation and compaction

(c) Concrete joints too far apart

Figure 10-16 Typical concrete damage caused by improper construction practice. [Photos courtesy from Carlos Videla (a and c) and Paulo Barbosa (b).]

reinforcement or due to application of higher than the designed loads is beyond the scope of this book.

10.13 Concluding Remarks

This chapter demonstrates that various early-age operations, such as placement and compaction, finishing, and curing have an important effect on the properties of concrete. In Chaps. 6 to 9, a similar conclusion was reached regarding the characteristics of cement, aggregate, and admixtures as well as concrete mix proportions that also have an important effect on the properties of concrete. To keep the various factors influencing the properties of concrete in proper perspective, it should be interesting to see, at one glance, their relative significance with respect to some of the major properties of concrete, as shown in Table 10-5.

	Factors						
Properties	**Type of Portland Cement**	**Aggregate Characteristics**	**Type of Admixture**	**Mix Proportions**	**Placing and Compaction**	**Surface Treatment**	**Curing Conditions (Temperature and Humidity)**
Workability							
Consistency	M	L	L	L	n	n	c
Cohesiveness	M	L	L	L	M	n	c
Setting time	L	n	L	M	n	n	c
Strength							
Early	L	n	L	L	L	n	L
Ultimate	n	n	M	L	L	n	L
Permeability	n	L	L	L	L	L	L
Shrinkage							
Plastic	n	n	n	M	M	n	L
Drying	n	L	M	L	n	n	L
Thermal	L	L	L	L	n	M	L
Surface appearance	n	n	n	M	L	L	L
Frost resistance	n	M	L	L	M	M	M
Abrasion resistance	n	L	n	L	L	M	L
Coefficient of thermal expansion	n	L	n	L	n	n	n

L, large effect; M, moderate effect; n, no. or negligible effect; c, not applicable since curing starts after the removal of formwork.

TABLE 10-5 Relative Effects of Material Characteristics, Mix Proportions, and Early-Age Operations on the Properties of Concrete

The information in Table 10-5 is qualitative only, nevertheless it is useful for educational purposes. For instance, it may surprise some engineers to discover that the type of cement influences mainly the setting time, early strength, and heat of hydration (thermal shrinkage of concrete). On the other hand, mixture proportions, placement and compaction, and curing conditions have a far-reaching effect on several important properties of concrete, such as the ultimate strength, permeability, plastic shrinkage, and drying shrinkage.

Test Your Knowledge

10.1 Explain the operations covered by the following terms, and discuss the significance of these operations: retempering, revibration, screeding, bullfloating, and scoring.

10.2 What is the principle behind consolidation of concrete mixtures? Describe the sequence of actions that take place in a fresh concrete mixture when it is exposed to a high-frequency vibrator.

10.3 Explain the two important objects of curing and how they are achieved in (*a*) cold-weather concreting and (*b*) hot-weather concreting.

10.4 How would you define workability? Is workability a fundamental property of fresh concrete? If not, why? What are the principal components of workability and their significance in the concrete construction practice?

10.5 Define the following phenomena, and give their significance and the factors affecting them: slump loss, segregation, and bleeding.

10.6 Suggest at least two methods to reduce "bleeding" of a concrete mixture.

10.7 With the help of a sketch briefly describe the "Vebe test." What is the objective of this test, and when is it more suitable than the slump test for determining the consistency of concrete?

10.8 What are harmful manifestations of plastic shrinkage of concrete in (*a*) reinforced columns and (*b*) slabs? Assuming that the air temperature is 21°C, the concrete temperature is 24°C, and the wind velocity is 30 km/h, determine the rate of evaporation. If this rate is too high from the standpoint of risk of plastic-shrinkage cracking, what precautionary measures would you take? Alternatively, determine the temperature to which concrete must be cooled to reduce the rate of evaporation to a safe limit.

10.9 Why may the setting time of concrete be substantially different from the setting time of the cement from which the concrete is made? Define the initial and the final setting times as measured by the penetration resistance method (ASTM C 403). What is their significance in the concrete construction practice?

10.10 With the help of suitable curves, show how accelerating and retarding admixtures affect the setting time of a concrete mixture.

10.11 Briefly discuss the effect of temperature on the setting time of concrete. What is the most efficient way of reducing the temperature of a fresh concrete mixture? Explain why.

10.12 In the ACI 306R (*Recommended Practice for Cold-Weather Concreting*), explain why higher than ambient concrete temperatures are required placement in cold weather.

10.13 Explain the maturity concept, its application, and its limitations.

10.14 (*a*) For a concrete mixture containing 370 kg of cement, 1830 kg of aggregate (SSD condition), and 190 kg of mixing water, calculate the temperature of concrete, assuming that the cement and the aggregate are at 30°C and the water has been cooled to 5°C. (*b*) For the concrete mixture in part (*a*), calculate the temperature of concrete, assuming that the cement and the aggregate are at 5°C and the water has been heated to 65°C.

10.15 You have recently taken charge of a large project. Write a short note to the attention of the owner on the subject of a concrete quality assurance program, explaining briefly the advantages, disadvantages, and testing costs of the three accelerated testing procedures and the various nondestructive test methods.

10.16 Describe the essential elements of statistical quality control charts. In the case of concrete strength data, explain why moving-average and moving-range charts are more useful than those containing a plot of individual strength values.

References

1. Berstrom, S.G., Conclusion from the Symposium on Concrete at Early Ages, Paris, April 6–8, 1982, *RILEM Bulletin*.
2. Gaynor, R.D., ASTM STP-169C, American Society of Testing and Materials, Philadelphia, PA, pp. 511–521, 1994.
3. *Design and Control of Concrete Mixtures*, 12th ed., Portland Cement Association, Skokie, IL, p. 69, 1979.
4. Tattersall, G.H., *Mag. Concr. Res.*, Vol. 25, No. 84, 1973; and Vol. 28, No. 96, 1976.
5. Scanlon, J.M., *ASTM STP-169C*, American Society of Testing and Materials, Philadelphia, PA, pp. 49–64, 1994.
6. Erlin, B., and W.G. Hime, *Concr. Int.*, Vol. 1, No. 1, pp. 48–51, 1979.
7. Previte, R.W., *J. ACI, Proc.*, Vol. 74, No. 8, pp. 361–367, 1977.
8. Tuthill, L.H., *Concr. Int.*, Vol. 1, No. 1, pp. 30–35, 1970.
9. ACI Committee 306, Cold Weather Concreting, *ACI Manual of Construction Practice*, Concrete Institute, Farmington Hills, MI, 2002.
10. Malhotra, V.M., *Concr. Int.*, Vol. 3, No. 11, pp. 17–21, 1981.
11. Carino, N., Tests and Properties of Concrete, *ASTM STP-169 C*, American Society of Testing and Materials, Philadelphia, PA, 1994.
12. Lamond, J.F., *J. ACI, Proc.*, Vol. 76, No. 4, pp. 399–512, 1979.
13. Neubauer, D.V., ed., *Manual on Presentation of Data and Control Chart Analysis*, 8th ed., ASTM International, West Conshohochen, PA, 2011.

Suggestions for Further Study

ACI Committee 214, 214-02 Report on Evaluation of Strength Test Results of Concrete, *ACI Manual of Construction Practice*, 2012.

ACI Committee 228, 228.1R-03 Report on In-Place Methods for Determination of Strength of Concrete, *ACI Manual of Construction Practice, 2012.*

ACI Committee 305, 305R-99 Report on Hot Weather Concreting, 305R-10: Guide to Hot Weather Concreting, *ACI Manual of Construction Practice*, Part 2, 2012.

ACI Committee 306, 306R-88 Report on Cold Weather Concreting, 306R-10 Guide to Cold Weather Concreting, *ACI Manual of Construction Practice*, Part 2, 2012.

ASTM 169C, American Society for Testing and Materials, Philadelphia, PA, 1994.

ASTM, *Significance of Tests and Properties of Concrete and Concrete-Making Materials*, STP 169B, American Society for Testing and Materials, Philadelphia, PA, Chaps. 7, 9, 13, and 15, 1978.

Design and Control of Concrete Mixtures, 15th ed., Portland Cement Association, Skokie, IL, 2011.

Mindness, S., J.F. Young, and D. Darwin, *Concrete*, 2d ed., Prentice Hall, Englewood Cliffs, NJ, Chaps. 8, 11, and 17, 2002.

Neville, A.M., *Properties of Concrete*, 5th ed., Pearson Education Limited, Essex, England, 2011.

Newman, J., and B.S. Choo, eds., *Advanced Concrete Technology: Processes*, Butterworth-Heinemann, Oxford, 2003.

Powers, T.C., *The Properties of Fresh Concrete*, Wiley, New York, 1968.

Report of RILEM Technical Committee 185-ATC: *Advanced Testing of Cement Based Materials during Setting and Hardening*, H.W. Reinhardt and C.U. Grosse, eds., RILEM Publications, 2005.

Tattersall, G.H., and P.F.G. Banfill, *The Rheology of Fresh Concrete*, Pitman Advanced Publishing Program, Vol. xii, p. 356, 1983.

Tattersall, G.H., *Workability and Quality Control of Concrete*, Chapman and Hall, New York, 1991.

CHAPTER 11

Nondestructive Methods

Preview

Many industrialized nations currently dedicate a considerable portion of the construction budget for restoration, repair, and maintenance of old structures as opposed to new construction. In 1991 the U.S. Department of Transportation reported that $90 billion dollars were required for the rehabilitation and repair of the highway infrastructure system. By 1997, the estimated cost had risen to $212 billion. The anticipated economic impact of an extensive infrastructure repair scheme has produced a renewed interest in improving nondestructive testing methods for assessing concrete structures.

Compared to other structural materials, the progress in the development of advanced nondestructive testing methods for concrete has been slow. Successful techniques for the detection of cracks, flaws, imperfections, and damage in homogeneous materials are of limited value when applied to concrete because of heterogeneities at various length scales that create interferences, such as attenuation, scattering, diffraction, and reflection. Improvements in the computerized data acquisition and manipulation of digital images and in the development of complex theories for heterogeneous media have resulted in new methods that have been successfully tested in the field. These methods are discussed below. The chapter begins with a brief description of the traditional methods used to estimate concrete strength by measuring surface hardness, penetration resistance, and pullout strength of concrete. Maturity methods used to predict compressive strength as function of time and temperature of curing are discussed next.

Owners and designers are beginning to appreciate the importance of building structures with a long service life. For concrete structures this is usually governed by the permeability of the material. Many laboratory procedures are available to assess the permeability and absorption characteristics of concrete; however, there are only a few suitable for use in field conditions. This chapter introduces methods to measure the permeability and surface absorption of concrete in field structures.

Many nondestructive methods use various types of waves to characterize the properties of materials. Therefore, this chapter provides an introduction to the fundamental laws of wave propagation that may aid in understanding future developments in this area. Ultrasonic pulse velocity methods have been used to measure the Young's modulus of elasticity of concrete for a long time. Recent research has used stress wave propagation methods to determine the presence of voids, imperfections, and discontinuities in the concrete mass. Two of the most popular techniques, the impact-echo and the spectral analysis of surface waves, are discussed in this chapter. Also discussed is another powerful nondestructive method that uses acoustic emission techniques to record the waves generated by the creation or propagation of a crack.

Corrosion of reinforced concrete, a complex phenomenon, is described in Chap. 5; however, the fundamentals of electrochemistry of reinforced concrete are presented in this chapter to provide a proper background to electrical and electrochemical test methods of corrosion assessment, namely corrosion potential, polarization resistance, and electrochemical impedance spectroscopy. Next, we describe how electromagnetic waves can detect the presence of reinforcing bars using a cover meter, how to identify delaminations in the concrete using ground-penetrating radar, and how to distinguish heterogeneities in concrete using infrared thermography.

Finally, a brief discussion on tomography of reinforced concrete is included here. While the use of tomography has revolutionized clinical diagnostics in medicine, the use of tomography of concrete in field conditions is still in its infancy, but the method has significant potential to assess the degree of distress existing in concrete structures.

11.1 Surface Hardness Methods

Essentially, the surface hardness method consists of impacting a concrete surface in a standard manner with a given energy of impact and then measuring the size of indentation or rebound. The most commonly used method employs the *Schmidt rebound hammer*, which consists of a spring-controlled hammer that imparts a load on a plunger. At the beginning of the test, the extended plunger is placed in contact with the concrete surface (Fig. 11-1*a*). Next, the outer body of the instrument is pressed against the surface of the concrete making the spring to extend (Fig. 11-1*b*). The latch is released when the spring is fully extended and the hammer moves toward the concrete surface (Fig. 11-1*c*). The hammer impacts upon the plunger, and the spring-controlled mass rebounds taking a rider with it along a guide scale, which is then used to obtain the hammer rebound number (Fig. 11-1*d*). This number depends on the stiffness of the spring and on the selected mass, and therefore a standard procedure has been established and is described in detail in ASTM C 805.

The Schmidt rebound hammer method is simple to use and provides a quick, inexpensive means of checking uniformity of in-place hardened concrete. The results of the test are dependent on the following parameters:

Mix proportions. Although the type of portland cement has little influence on the rebound number, the type and amount of aggregate play a major role on the result. While not a major limitation if the objective is to assess concrete uniformity, it becomes a critical issue if the objective is to obtain a correlation between the rebound number and strength. If this is the case, the aggregate must be identified and a careful calibration curve performed.

Age and type of curing. The relationships established between rebound number and strength are not constant over long periods of time. Also, a special calibration is required when high temperature curing is used.

Surface smoothness. This test requires a smooth and well-compacted surface. Unfortunately, any deviations from these conditions are difficult to determine. As expected, the method is not appropriate for open-textured or exposed aggregate surfaces.

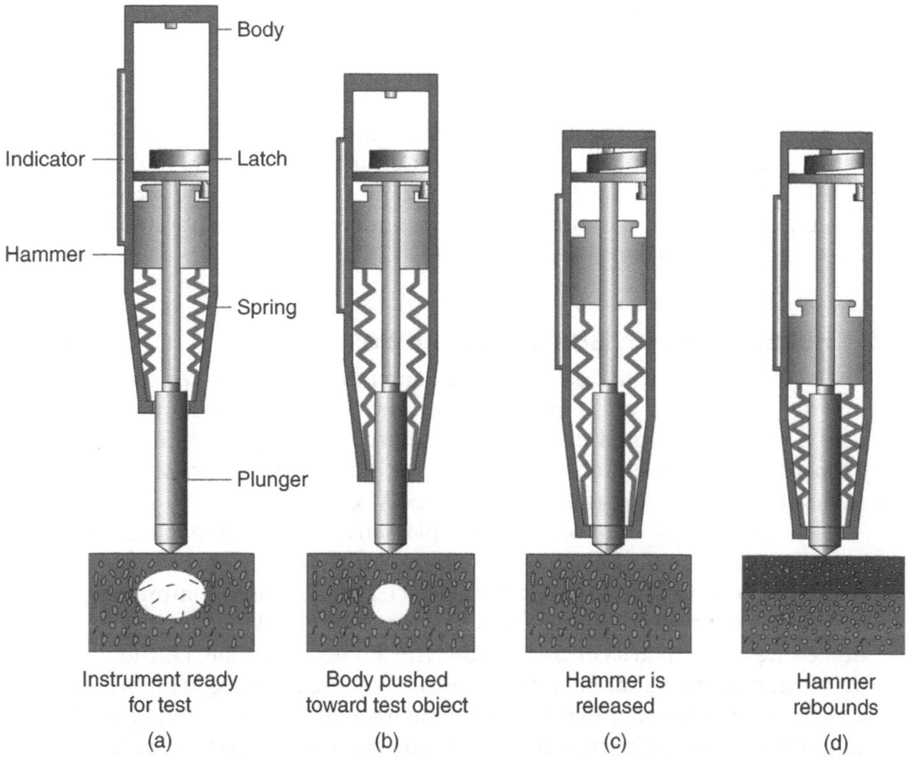

FIGURE 11-1 Schematic diagram illustrating the operation of the rebound hammer. (After ACI 228.1 R-95, In-Place Methods to Estimate Concrete Strength; *ACI Manual of Concrete Practice*, American Concrete Institute, 2002.)

Moisture condition. A wet surface produces a lower rebound number than a dry surface, consequently affecting the strength-rebound number calibration. Bungey[1] has reported that wet surface conditions may underestimate strength up to 20 percent.

Surface carbonation. Calcium carbonate is one of the products of surface carbonation of concrete, which is hard and can increase the rebound number. When testing older concrete structures where carbonation is evident, it is recommended that the carbonation layer be removed in a small area and the results from this area be compared with the results from the rest of the structure.

Stiffness of the member. The stiffness of the concrete specimen in the laboratory or the concrete member in the field should be high enough to prevent vibrations during the impact caused by the hammer. Any vibration will reduce the rebound number, making the strength prediction unreliable.

Location of the plunger. If the plunger is placed over a stiff aggregate, the measurement will yield an unusually high rebound number. Conversely, if the plunger is placed over a large void or a soft aggregate, the test will give a lower rebound number. To address this issue, ASTM C 805 requires that 10 measurements be taken for a test. A reading is discarded if it deviates more than 7 units from the average, and the entire measurement is to be discarded if two readings deviate more than 7 units from the average.

Under ideal field conditions, all these parameters should be accounted for to establish a good correlation between rebound number and strength. In practice, however, it is difficult to know all the variables. According to Malhotra,[2] the accuracy of estimating concrete strength in laboratory specimens with a properly calibrated hammer is ±15 to 20 percent, and in a concrete structure it is ±25 percent.

11.2 Penetration Resistance Techniques

The equipment used to determine the penetration resistance of concrete consists of a powder-activated device. One currently used apparatus, known as the *Windsor probe* uses a powder-activated driver to fire a hardened-alloy probe into the concrete. The exposed length of the probe is a measure of the penetration resistance of concrete. The standard test procedure is described in ASTM C 803.

The type and amount of aggregate play an important role in the penetration resistance, which becomes critical when determining the relationship between penetration resistance and strength. As shown in Fig. 11-2 for the same compressive strength, concrete made with a soft aggregate (i.e., a lower Mohs' scale) will allow a greater penetration of the probe than a concrete made with a hard aggregate. Due to the small volume under testing, the variation in the Windsor probe-test results is higher (as is the case in determining surface hardness) when compared with the variation in standard compressive strength tests on companion specimens. But this method is excellent for measuring

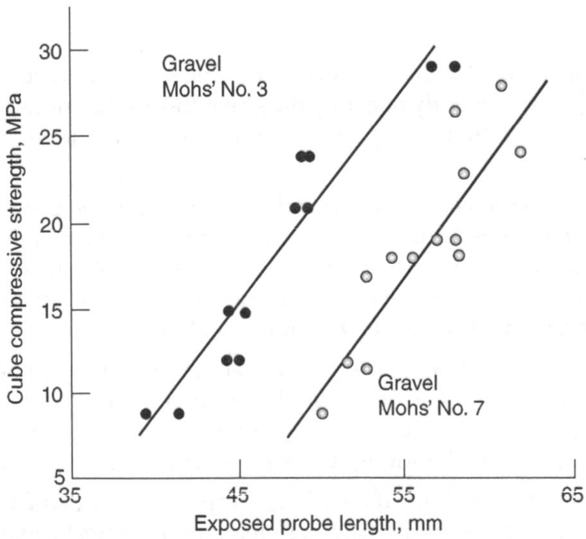

FIGURE 11-2 Compressive strength as a function of exposed probe length. (After ACI 228.1R-95, *In-Place Methods to Estimate Concrete Strength*.)

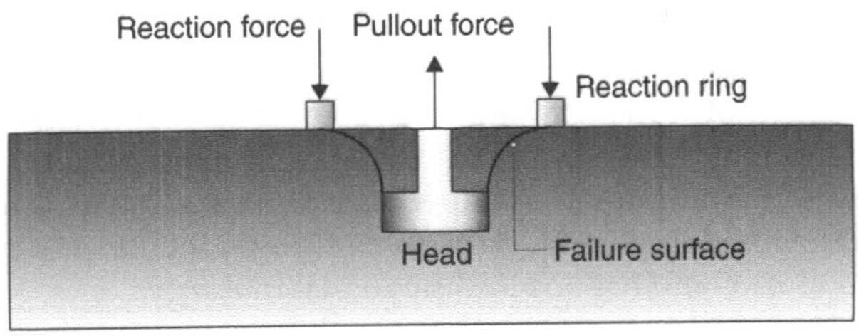

Figure 11-3 Schematic diagram of the pullout test. (After ACI 228.1R-95, *In-Place Methods to Estimate Concrete Strength.*)

the relative rate of strength development of concrete at early ages, especially for determining stripping time for formwork.

11.3 Pullout Tests

A pullout test consists of casting a specially shaped steel insert with an enlarged end into fresh concrete. This steel insert is then pulled out from the concrete and the force required for pullout is measured using a dynamo-meter. A bearing ring is used to confine failure to a well-defined shape (Fig. 11-3). As the steel insert is pulled out, a cone of concrete is also removed, thereby damaging the concrete surface (which must be repaired after the test). If the test is being used to determine the optimum time for safe form-stripping, the pullout assembly need not be torn out of concrete. Instead, the test may be terminated when a predetermined pullout force has been reached on the gage and the forms can be removed safely.

During the pullout test, a complex three-dimensional state of stress develops inside the concrete. Numerical analysis performed before cracking indicates that the principal stresses in the concrete are greatest near the top of the steel insert. It is not clear what mechanism controls the final failure of the pullout test. Proposed failure criteria include: (a) compressive strength of concrete because failure is caused by crushing of the concrete, (b) fracture toughness of concrete, and (c) aggregate interlock across the circumferential crack. Since there is no agreement on what strength the pullout test is measuring, it is recommended to develop a relationship between the pullout test and the compressive strength of concrete. It is important to emphasize that this relationship is only valid for a given test geometry and the concrete mix proportions used in the test. Like the penetration resistance test, the pullout test is an excellent means of determining the strength development of concrete at early ages and safe form-stripping times. Also, the technique is simple and the procedure is quick. The main advantage of pullout tests is that they attempt to measure directly the in situ strength of concrete. The major drawback is that unlike most other in situ tests, the pullout test must be planned in advance. A standard test procedure is described in ASTM C 900. The lok* test originally developed in Denmark is also popular in many countries. A portable hydraulic jack applies the load to the bolt, until failure is reached and the load is quickly released.

*Lok in Danish means punching.

11.4 Maturity Method

Since the degree of cement hydration depends on both time and temperature, the strength of concrete may be evaluated from the concept of maturity, which is expressed as a function of the time and the temperature of curing. It is assumed that batches of the same concrete mixtures of same maturity will attain the same strength regardless of the time-temperature combinations leading to that maturity.

A simple maturity function $M(t)$ can be defined as the product of time and temperature:

$$M(t) = \sum (T_a - T_0)\Delta t \tag{11-1}$$

or in the limit

$$M(t) = \int_0^t (T_a - T_0)\, dt \tag{11-2}$$

where Δt, T_a, and T_0 are time interval, average concrete temperature during the time interval Δt, and the datum temperature, respectively. Traditionally, −10°C or 14°F is assumed to be the datum temperature below which there is no additional gain in strength. ASTM C 1074 recommends a datum temperature of 0°C or 32°F.

The maturity function allows the determination of an equivalent age of curing, t_e, at a reference temperature, T_r:

$$t_e = \frac{\Sigma(T_a - T_0)\Delta t}{(T_r - T_0)} \tag{11-3}$$

Although some researchers have reported good correlation between maturity and compressive strength of concrete, others have questioned the validity of the maturity concept. For instance, the maturity concept does not take into consideration the influences of humidity and temperature of curing at early age. Contrary to the assumption made by the maturity concept, these factors exercise a disproportional effect on strength with time. The effect of curing temperature at early ages on the strength-relationship is shown schematically in Fig. 11-4. Higher curing temperatures cause an acceleration of the hydration reactions, resulting in an increase in the early-strength development. At these early stages, concrete cured with high temperatures will have a higher strength than concrete cured with lower temperatures for the same maturity, computed according to Eq. (11-1). At later stages, the reverse happens. Low-temperature curing produces a more uniform microstructure in the cement paste with low porosity, resulting in concrete with higher ultimate strength.

Subsequent research has been done to reduce the limitations of the maturity function as defined in Eq. (11-1). Instead of the linear relationship between time and temperature, an Arrhenius relationship was determined to be more appropriate. Accordingly, Hansen and Pedersen[3] proposed that the equivalent age as follows:

$$t_e = \sum_0^t e^{\frac{-E}{R}\left[\frac{1}{273+T_a} - \frac{1}{273+T_r}\right]} \Delta t$$

where E is the activation energy and R the universal gas constant. The authors proposed the following values for the activation energy:

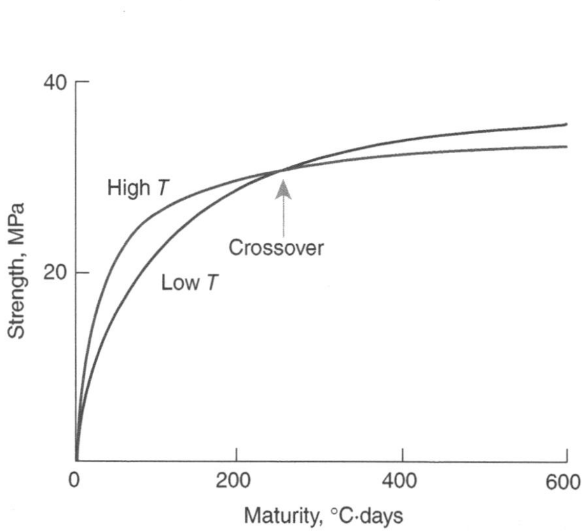

FIGURE 11-4 Influence of curing temperature at early ages on the strength-maturity relationship when Eq. (11-1) is used with $T_0 = -10°C$. This early-age difference can be reduced when better maturity functions are used. (From Carino, N.J., The Maturity Method, in *Handbook on Nondestructive Testing of Concrete*, Malhotra, V.M., and N.J. Carino, Chap. 5, p. 105, CRC Press, Boca Raton, FL, 1991.)

$$\text{for} \quad T_a \geq 20°\text{C}: \quad E = 33{,}500\ \text{J/mol}$$

$$\text{for} \quad T_a < 20°\text{C}: \quad E = 33{,}500 + 1470(20 - T)\text{J/mol}$$

Worth noting is Carino's[4] excellent discussion on the merits and limitations of the other existing formulations.

Before construction begins, a calibration curve is established that plots the relationship between compressive strength and the value M for a series of test cylinders made from particular concrete mixtures but with different combinations of time and curing temperatures. It should be noted that the method must only be used during the curing period and not for existing structures.

11.5 Assessment of Concrete Quality from Absorption and Permeability Tests

The rate of water absorption by capillary suction is a good measure of the quality of a concrete and its potential durability when exposed to aggressive environments. Low values of absorption indicate that aggressive ions will have difficulty penetrating the concrete. Experimental research indicates that the water absorption values are reduced with decrease in the water-cement ratio, increase in the curing time, and increase in the degree of consolidation.[5]

The term permeation is used to describe the mass transport of liquids or gases induced by pressure and concentration gradients or by capillary forces. In concrete, permeation is influenced by the volume and connectivity of the capillary pores in the cement paste matrix. The air permeability of concrete increases when the moisture is eliminated, which in turn increases the connectivity of the pores. The water absorption also increases when the capillary pores are empty.

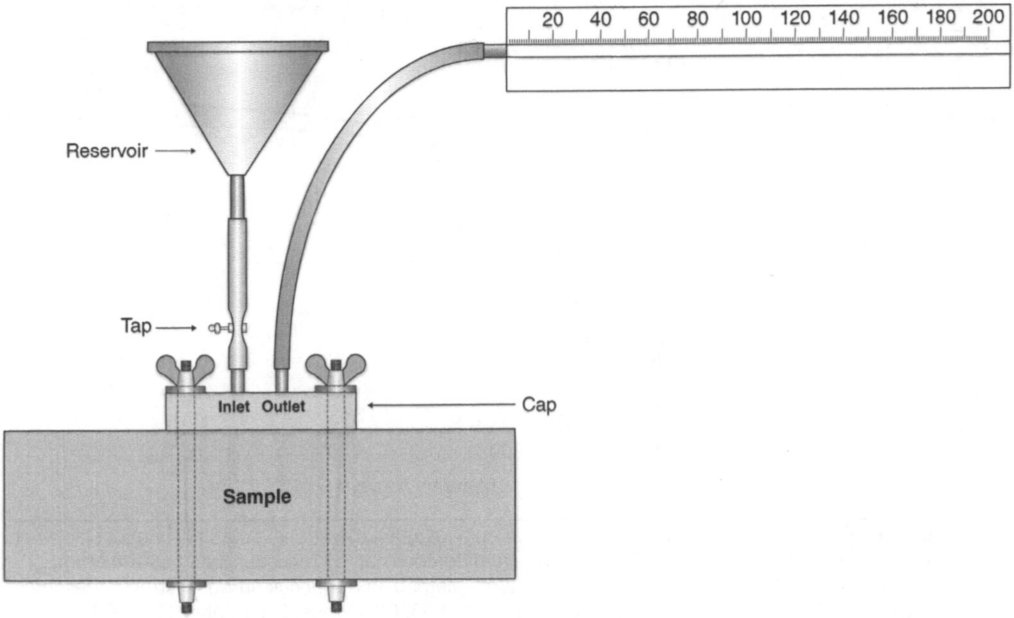

FIGURE 11-5 Configuration of initial surface absorption test. (After ACI 228.2R-98, *Nondestructive Test Methods for Evaluation of Concrete Structures.*)

There are many laboratory tests to measure the capillary suction, resistance to water penetration, and gas permeability. The interested reader should consult the comprehensive review given by Geiker et al.[6] In this section, emphasis will be given to field tests to assess these properties.

There are two basic methods to measure the water absorption under field conditions: the initial surface absorption test (ISAT) and the Figg test. In the ISAT, a constant pressure head is applied to the concrete surface and the resulting rate of water flow into the concrete per unit area is measured.

Figure 11-5 shows the ISAT configuration. A cap seals the concrete surface and connects to a water reservoir and to a calibrated capillary tube. At the beginning of the test, the reservoir tap is opened and water flows and fills the cap. During this stage, the capillary tube should be disconnected from the outlet to guarantee that the air is properly driven out. The water level in the reservoir must be kept constant. Next, the calibrated capillary tube, which is filled with water, is placed at the same horizontal line as the water surface in the reservoir. When inlet tap is closed, the movement of water in the capillary tube is timed, thereby giving the flow.

It is recommended that the concrete be dried before testing. In a field test, it is difficult to guarantee that the concrete is completely dry, and different researchers have proposed the use of in-situ vacuum drying and heaters but no consensus has been reached.[7]

The Figg test[8] consists of drilling a hole perpendicular to the concrete surface (see Fig. 11-6). Different versions of the test use different diameters and depths in the range of 5.5 to 10 mm and 30 to 40 mm, respectively. The hole is carefully cleaned and a disk of foam plugs the free surface of the hole; a final seal of liquid silicone rubber is then applied.

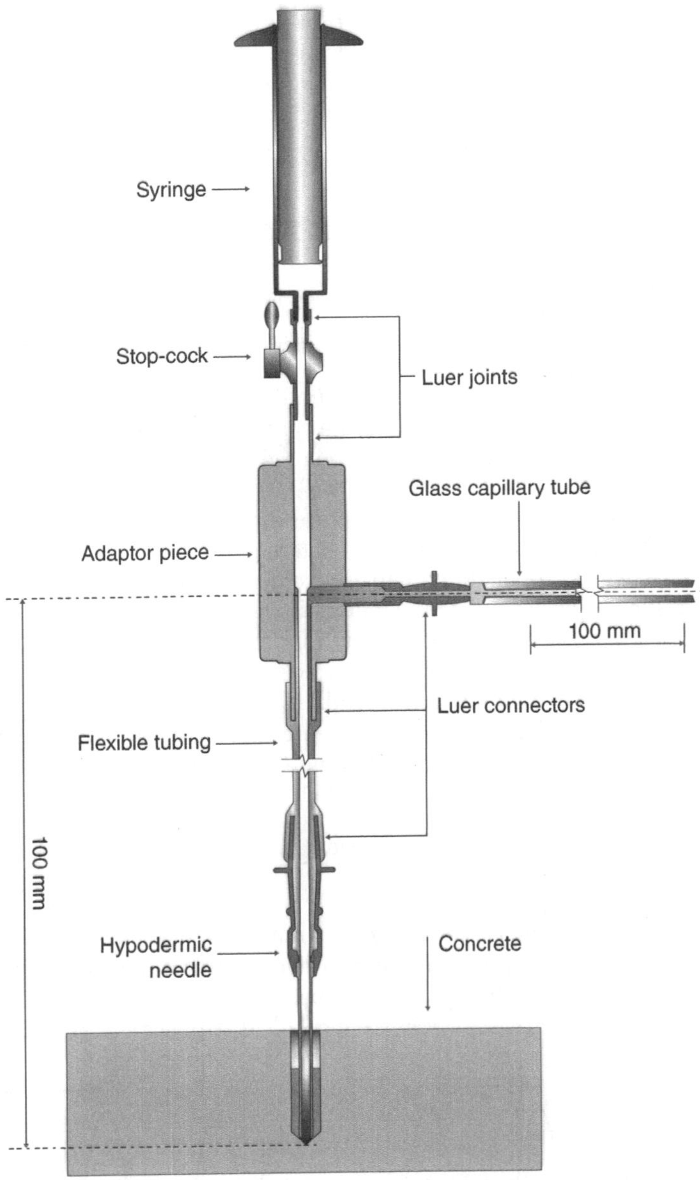

FIGURE 11-6 Configuration of the Figg test. (After ACI 228.2R-98, *Nondestructive Test Methods for Evaluation of Concrete Structures.*)

After the rubber has hardened, a hypodermic needle is inserted through the plug and a water head of normally 100 mm is applied. A calibrated capillary tube determines the volume of absorbed water measured in the same manner as described for the ISAT. The same apparatus can be used to determine the air permeability by replacing the syringe with a vacuum pump and a pressure gauge. The pump reduces the pressure inside the hole, originally at atmospheric pressure, to a specified value (on the order of –55 kPa). At this point, the valve is closed and a measurement is taken of the time it takes for the air flow to reach the hole and increase the pressure to a prescribed value. The time in seconds that it takes to end the test is called the air-permeability index. Less permeable concrete will have a higher air-permeability index.

11.6 Stress Wave Propagation Methods

11.6.1 Theoretical Concepts of Stress Wave Propagation in Solids

To understand the principles of the methodology, it is relevant to review some concepts of stress wave propagation in a continuous medium. For simplicity, consider a periodic sinusoidal wave (Fig. 11-7). The maximum displacement is the amplitude A, the time between two successive wave crests is the period T (Fig. 11-7*a*) and the distance between two successive wave crests is the wavelength λ (Fig. 11-7*b*). The frequency f of the wave is given by $1/T$ and its velocity V by $V = f\lambda$.

The manner whereby a wave reflects and refracts within a solid material can provide vital information about its internal heterogeneity. A particularly useful application of the stress wave propagation method is to determine the existence of layers with different elastic properties inside the material. These layers may be a result of poor compaction or of exposure to aggressive environmental conditions. These methods are also frequently used in geophysics for subsurface mineral exploration and for environmental impact studies of contaminant spread.

When a plane wave hits a horizontal interface with incidence angle θ_1, the reflected angle will be equal to the incidence angle (Fig. 11-8). The relationship between incidence and refracted angles is given by Snell's law:

$$\frac{\sin \theta_1}{\sin \theta_2} = \frac{V_1}{V_2} \tag{11-4}$$

where V_1 and V_2 are the wave velocities in materials 1 and 2, respectively.

In many practical applications, the top layer is often more porous than the core, either because of poor construction practices or because the external surface has been exposed

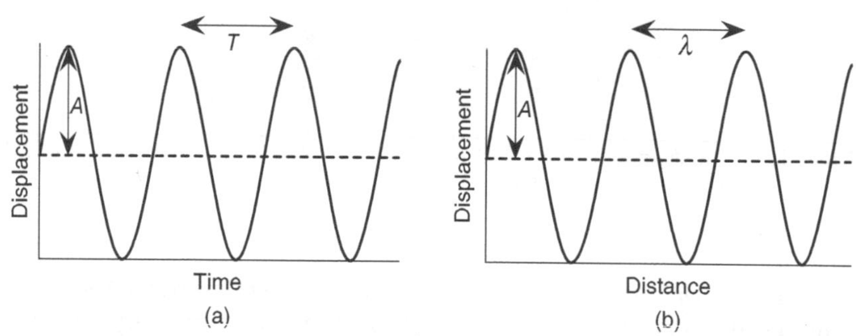

FIGURE 11-7 Periodic sinusoidal waves. The maximum displacement is the amplitude A, the time between two successive wave crests is the period T (*a*), and the distance between two successive wave crests is the wavelength λ (*b*).

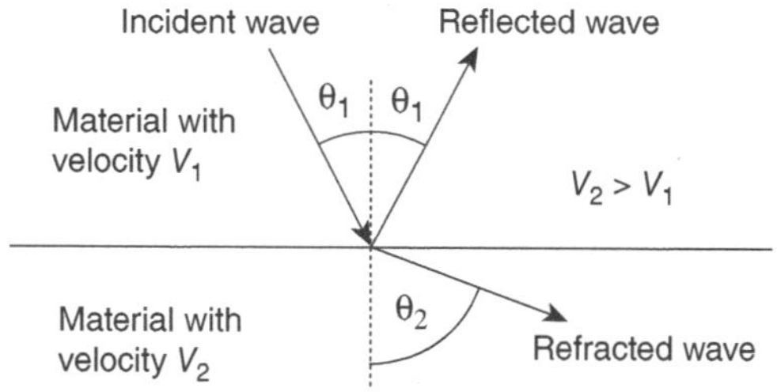

FIGURE 11-8 Reflection and refraction of an incident wave striking an interface between dissimilar materials. The incidence angle is equal to the reflected angle and the relationship between incidence angle θ_1 and refracted angle θ_2 is given by Snell's law. As shown above, when the incident wave penetrates a medium with higher velocity, as shown in the figure, the refracted wave moves away from the normal to the interface ($\theta_2 > \theta_1$).

to aggressive conditions. In this case, V_2 is greater than V_1 and increasing incidence angle θ_1 leads to greater values of angle θ_2 until the refraction angle reaches $\pi/2$. Upon reaching this critical condition, the rays are refracted parallel to the interface. The critical incidence angle θ_{ic} can be computed using Snell's law with the refracted angle $\theta_2 = \pi/2$:

$$\theta_{ic} = \sin^{-1}\left(\frac{V_1}{V_2}\right) \tag{11-5}$$

As discussed in the next section, this equation is important for determining the thickness of horizontal layers. Increasing incidence angles beyond θ_{ic} results in the ray not being refracted but totally reflected.

Simeon Poisson, a French engineer (who also introduced Poisson's ratio), used the equations of the theory of elasticity to demonstrate that only two independent modes of wave propagation are possible in the interior of a homogeneous solid, namely *longitudinal* and *transverse* (or *shear*). In longitudinal waves the particles move back and forth along the direction of wave propagation, similar to sound waves in a fluid, leading to a volume change. In transverse waves the particles move transverse to the direction of wave propagation and cause no volume change.

In 1808, Biot performed the first experiment to determine the velocity of the longitudinal wave in a solid. He used an ingenuous and inexpensive test equipment: a 1000-m iron water pipeline in Paris. Biot rang a bell in one extremity of the pipe and a collaborator measured the time difference between the wave arrival in the pipe and in the air. Because the length of the pipe and the velocity of sound in air were known, it was possible to make a fair estimate of the sound velocity in the metal pipe. Geophysicists were among the pioneers in the experimental study of wave propagation, particularly in regards to measuring waves generated during earthquakes. In an earthquake, longitudinal waves travel faster than the transverse waves, therefore, a seismograph registers the longitudinal waves first. For this reason, longitudinal waves are also called primary or *P waves* and the transverse waves are called secondary or *S waves*.

It is possible to determine the elastic moduli of a homogeneous and isotropic material by measuring the P and S wave velocities:

$$V_p = \sqrt{\frac{K + 4/3G}{\rho}} \tag{11-6}$$

and

$$V_s = \sqrt{\frac{G}{\rho}} \tag{11-7}$$

where ρ = density of the material
K and G = bulk and shear moduli, respectively
V_p and V_s = primary and secondary wave velocities, respectively

Using the relationship between the elastic moduli (see Eq. 13-12), the compression wave velocity can also be expressed in terms of Young's modulus E and Poisson's ratio ν.

$$V_p = \sqrt{\frac{E(1-\nu)}{\rho(1-2\nu)(1+\nu)}} \tag{11-8}$$

and

$$V_s = \sqrt{\frac{E}{2\rho(1+\nu)}} \tag{11-9}$$

As stated before, the longitudinal wave is always faster than the shear wave. This can be easily proven by taking the ratio between the two velocities and noting that the maximum value of Poisson's ratio is 0.5:

$$\frac{V_p}{V_s} = \sqrt{\frac{2(1-\nu)}{1-2\nu}} \tag{11-10}$$

For concrete, 0.2 is a typical value of Poisson's ratio, therefore the velocity ratio for longitudinal and shear waves is 1.63.

The compression and shear waves can change their mode of propagation when they strike an interface between two dissimilar materials. An incident compression (p) wave striking such interface generates reflected compression and shear (s) waves and refracted p and s waves. The angles of incidence, reflected, and transmitted rays are related according to Snell's law:

$$\frac{\sin\theta_1}{V_{p1}} = \frac{\sin\theta_2}{V_{p2}} = \frac{\sin\Phi_1}{V_{s1}} = \frac{\sin\Phi_2}{V_{s2}} \tag{11-11}$$

where V_p and V_s are the compressive and shear wave velocities, respectively, and subscripts 1 and 2 refer to the two dissimilar materials (Fig. 11-9).

Primary and secondary waves travel solid material in all directions. Close to the surface two other types of waves can also be present: Love and Rayleigh. These surface waves are similar to waves produced by throwing a stone into a placid lake. The amplitude of the surface waves decreases exponentially with increasing distance from the surface. That is why a submarine trip, in a stormy weather, becomes more comfortable once the submarine reaches greater depths (around 100 m from the surface waves). Bolt points out that these surface waves are analogous to the sound waves that are trapped near the wall surface in "whispering galleries" such as the dome of St. Paul's Cathedral in London. Only when the ear is placed near the wall can the opposite wall be heard.[9] In the Love wave, the particles move from side to side in a horizontal plane perpendicular

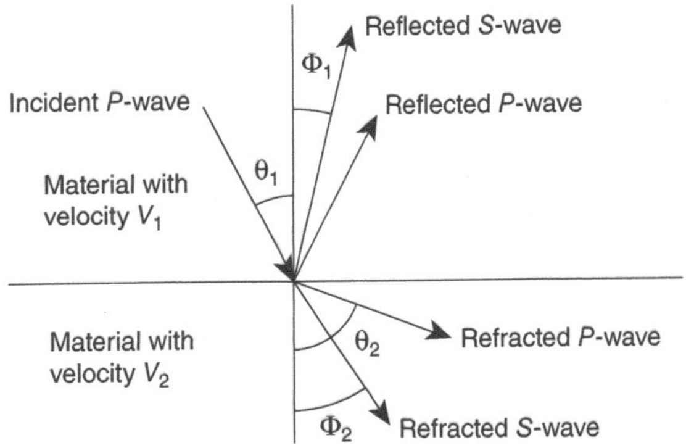

FIGURE 11-9 Conversion of a *P* wave striking an interface between dissimilar materials, always following Snell's law.

to the direction of wave propagation. In the Rayleigh wave the particles vibrate in an elliptical movement. The surface waves can be used to detect imperfections close to the surface of a concrete structure, as it will be described later. Figure 11-10 summarizes the types of waves that may propagate in a structure.

11.6.2 Ultrasonic Pulse Velocity Methods

The ultrasonic pulse velocity method consists of measuring the travel time of a pulse of longitudinal ultrasonic waves passing through the concrete. Longitudinal waves with frequencies in the range of 20 to 150 kHz are normally used. The travel times between the initial onset and reception of the pulse are measured electronically. The path length between transducers divided by the time of travel gives the average velocity of wave propagation. A suitable apparatus and a standard procedure are described in ASTM C 597.

A good acoustic coupling between the surface of the concrete and that of the transducers is critical for the reliable measurements. The transducers can be placed on opposite faces thereby originating a direct transmission, or they can be placed on the same face generating an indirect transmission (Fig. 11-11).

An effective method used to verify the homogeneity of a member is to place a series of receivers along the surface of a thick member of concrete (Fig. 11-12*a*). The transmitter sends the pulse and, according to the Huygen's principle, each point on a wavefront behaves as a point source for generation of secondary spherical waves and creates a series of wavefronts, as indicated in Fig. 11-12*a*. If the material is uniform, a unique straight line is obtained in a time vs. distance plot (Fig. 11-12*b*). If large heterogeneities are present, the plot will deviate from this unique straight line.

Suppose we want to study the presence of horizontal layers that are formed when concrete is exposed to an aggressive environment such as fire. Consider a layer with thickness h and wave velocity V_1, which is lower than the velocity V_2 of the sound concrete (Fig. 11-13*a*). A series of receivers are placed on the surface, as shown in Fig. 11-12*a*. At first, receivers close to the transmitter will only sense the top layer and the time vs. distance plot will be a straight line similar to Fig. 11-12*b* with slope $1/V_1$; but as the distance (or time) increases, the influence from the lower layer is felt. Figure 11-13*a* shows the case where the wave hits the interface at the critical incidence angle θ_{ic} and

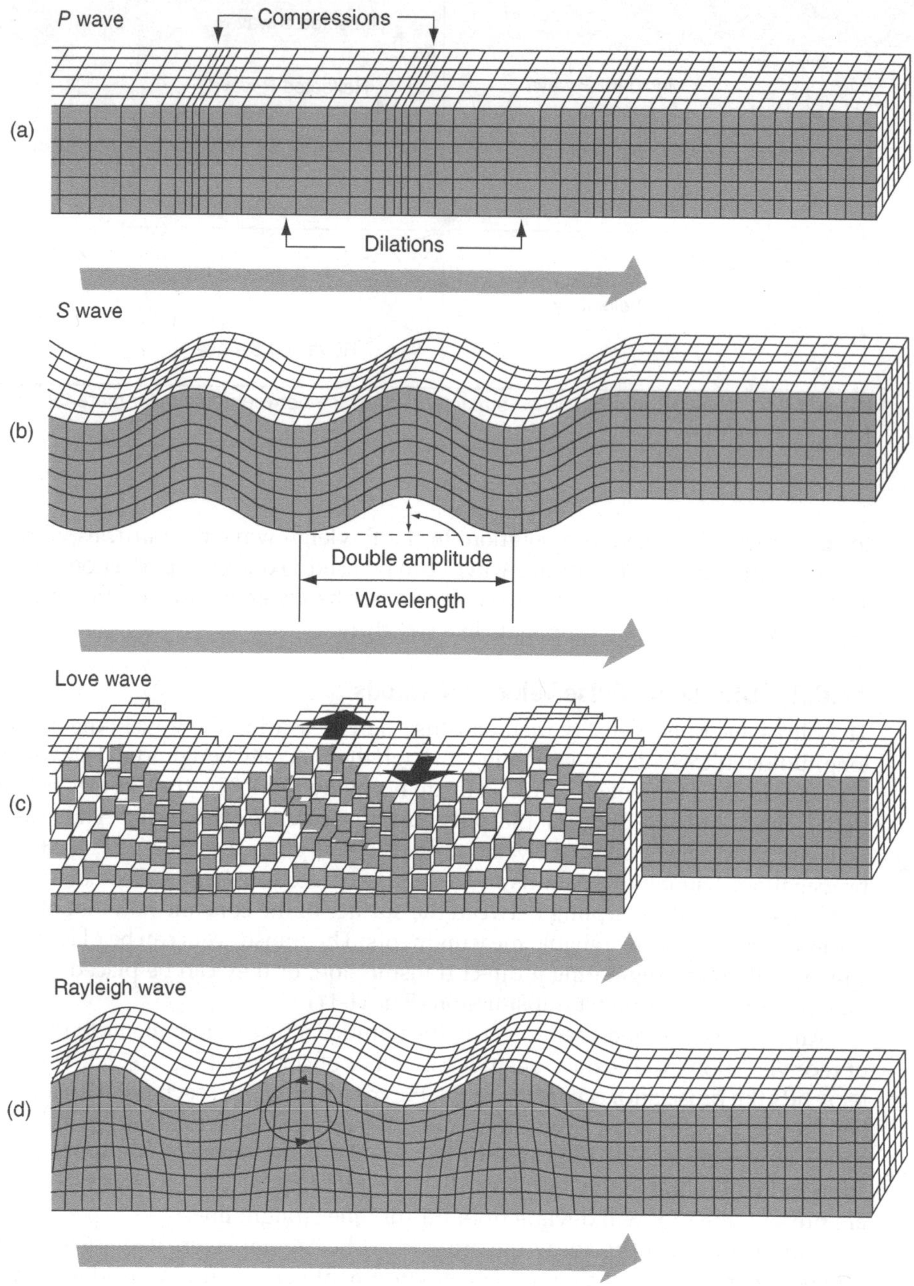

FIGURE 11-10 The two main modes of propagation in the bulk of a material are (*a*) the compression or *P*-wave and (*b*) the shear or *S*-wave. For a *P*-wave, particles move parallel to the direction of wave propagation. For an *S*-wave, particles move perpendicular to the direction of wave propagation. Rayleigh and Love waves can propagate close to a free surface. In a Love surface wave (*c*), the particles have a horizontal transverse movement perpendicular to the direction of wave propagation. The Rayleigh surface wave (*d*) is a combination of *P*- and *S*-waves whereby the particles vibrate in an elliptical movement. (From Bolt, B.A., *Nuclear Explosions and Earthquakes: The Parted Veil*, W.H. Freeman, San Francisco, 1976.)

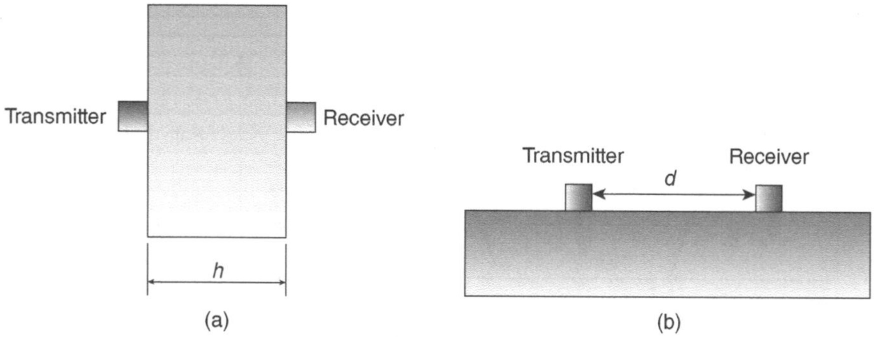

FIGURE 11-11 Configuration of the transmitter and receiver for (*a*) direct and (*b*) indirect transmission.

the refracted angle is parallel to the interface between the two materials. Applying Huygen's principle, the refracted wave will generate secondary waves that will reach the receiver before the direct arrival.

The total time t that the wave takes to travel from A to D is given by

Path 1:

$$t = x/V_1$$

Path 2 (ABCD):

$$t = \frac{2h}{V_1 \cos\theta_{ic}} + \frac{x - 2h\tan\theta_{ic}}{V_2} \tag{11-12}$$

Note that the refracted wave between B and C travels with velocity V_2.

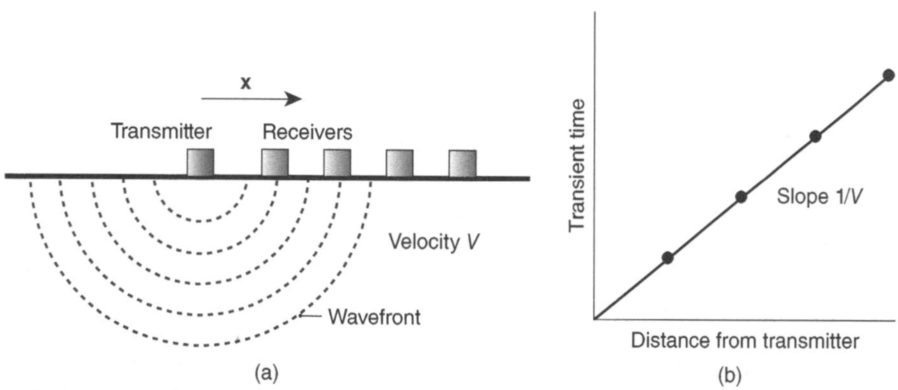

FIGURE 11-12 (*a*) Configuration of many receivers using the indirect transmission method and (*b*) typical plot to determine velocity *V* using the configuration shown in (*a*). The material is assumed to be homogeneous and uniform, compare the wave propagation when the material is not uniform, such as shown in Fig. 11-13, where a low-velocity layer is on top of a high-velocity material.

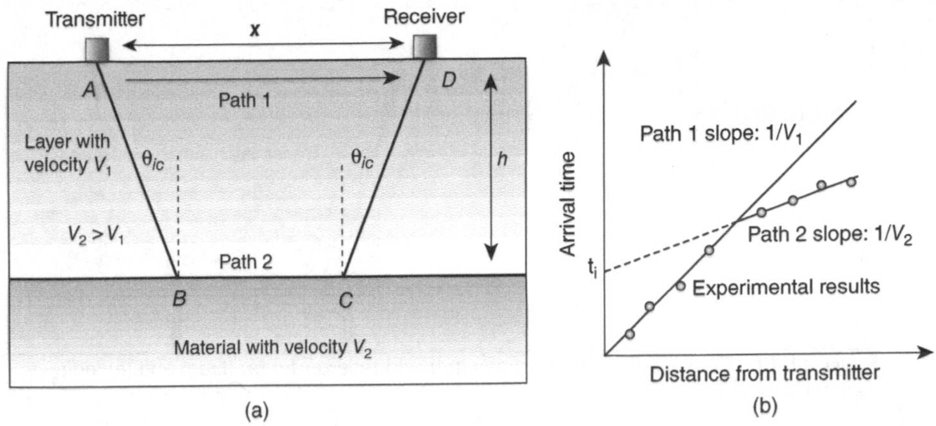

Figure 11-13 Effect of a low-velocity layer on the wave propagation: (*a*) geometric construction for determination of the thickness *h*, (*b*) graphical procedure to determine the value of *ti* and consequently the thickness *h*. See Eq. (11-19).

Using Eq. (11-5)

$$\sin\theta_{ic} = \frac{V_1}{V_2} \tag{11-13}$$

After trigonometric simplifications, Eq. (11-12) can be rearranged as

$$t = \frac{2h\cos\theta_{ic}}{V_1} + \frac{x}{V_2} \tag{11-14}$$

Using Eq. (11-13) and trigonometric relationships, $\cos\theta_{ic}$ can be expressed as a function of the two velocities and the previous equation can be rewritten as

$$t = \frac{2h\sqrt{1-(V_1/V_2)^2}}{V_1} + \frac{x}{V_2} \tag{11-15}$$

or

$$t = \frac{2h\sqrt{V_2^2 - V_1^2}}{V_1 V_2} + \frac{x}{V_2} \tag{11-16}$$

As before, the experimental results are plotted in a time vs. distance plot. The slope of the line is given by the partial derivative of t with respect to x:

$$\frac{\partial t}{\partial x} = \frac{1}{V_2} \tag{11-17}$$

Now it is easy to construct graphical representations of the solution (see Fig. 11-13*b*). By extrapolating the linear curve of slope $1/V_2$ to $x = 0$, the intercept with the vertical axis, of t_i, is obtained. Note that Eq. (11-16) gives for $x = 0$:

$$t_i = \frac{2h\sqrt{V_2^2 - V_1^2}}{V_1 V_2} \tag{11-18}$$

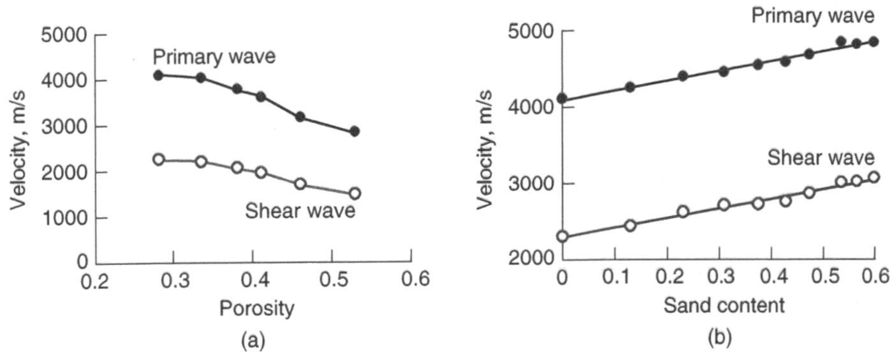

FIGURE 11-14 Effect of porosity and sand concentration on the wave velocities.

and therefore the thickness h is given by

$$h = \frac{t_i V_1 V_2}{2\sqrt{V_2^2 - V_1^2}} \tag{11-19}$$

The method can be extended for multiple and for dipping layers. Burger[10] presents a clear presentation to these approaches.

The wave velocities in concrete are affected by a number of variables. In brief:

Age. As cement hydration continues, the porosity decreases and waves propagate faster in the solid medium (see Fig. 11-14*a*). This property can be used in the laboratory to study the changes in the hydration process as affected by different admixtures, and in the field to monitor the hydration evolution as affected by the existing conditions of temperature and humidity.

Moisture condition. The wave velocities in concrete increase for saturated conditions.

Amount and type of aggregate. Rocks normally used as aggregate in concrete have higher wave velocities than the cement paste, so increasing the amount of aggregate for a given cement paste matrix also increases the average wave velocity of the composite (see Fig. 11-14*b*). The influence of different types of rocks on the effective velocities of the composite can be estimated by using the equations developed in Chap. 13 for the prediction elastic moduli of concrete.

Microcracking. Microcracks form when the concrete member has been exposed to a stress higher than 50 percent of its compressive strength. They can also form if the concrete is exposed to aggressive environmental conditions. Microcracks reduce the elastic moduli of the concrete and, consequently, reduce the wave velocity in its interior. Many analytical expressions are described in Chap. 13.

Presence of reinforcing bar. The presence of reinforcement should be avoided when measuring the wave velocity in concrete. Unfortunately, it is sometimes difficult, if not impossible, to take measurements when no reinforcing bars are close by. The presence of the reinforcement increases the apparent wave velocity of the concrete.

11.6.3 Impact Methods

A simple method of assessing the condition of concrete is to tap the surface with a hammer and listen to the resulting tone. A high-frequency pitch indicates a sound concrete and a low-frequency pitch indicates the presence of flaws. A trained operator can delineate zones of high and low pitch using this method. The disadvantage of the method is that it is dependent on the skill level of the operator and does not provide quantitative information on the amount of damage in the interior of the concrete. To overcome these limitations, different methods were developed (a) to control the duration of the impact force so as to assure the reproducibility of the test and (b) to characterize the surface displacement generated by the impact on concrete.

At the point of impact, spherical compressive and shear waves are generated and travel radially inside the material, while the surface wave travels away from the point of impact. When the compression and shear waves interact with heterogeneity or an external boundary, they are reflected and return to the surface. A transducer placed on the concrete surface can measure the displacements caused by the reflected waves from which the location of the reflecting interface can be determined.

This approach, often called *sonic-echo* or *seismic echo*, has been used successfully to evaluate the integrity of piles and caissons. These long structures permit that the time difference between the impact and the reflection to be large enough to perform reliable analysis. The complexity increases when it is used to detect flaws in relatively thin concrete structures, such as slabs and walls. For such applications, Sansalone[11] developed a method called *impact-echo*. A standard test procedure is described in ASTM C 1383.

The impact-echo test has the following features:

Impact forces generated by steel spheres. One of the critical steps for the success of the impact-echo is to have a reliable source of impact force to strike the concrete surface. There are many sources available when long concrete structures are going to be analyzed, however, for thin members it is required that the contact time be significantly reduced because the duration of the impact must be less than the round-trip travel of the *P*-wave. The use of steel ball bearings is a creative solution to generate low-frequency pulses with short duration but they are still capable of penetrating the concrete member. The analytical theory of spheres hitting a surface is well understood, and it shows that the contact time is proportional to the diameter of the spheres; therefore, it is possible to cover a large spectrum of contact times simply by changing the sizes of the spheres. Sansalone[11] reports that small ball bearings in the range of 4 to 15 mm of diameter generate impacts with a contact time in the range of 15 to 80 μs.

Use of sensitive broadband transducer at the surface. A small conical piezoelectric transducer, originally developed for acoustic emission monitoring of metals, has proven to be successful in measuring small displacements normal to the concrete surface. A thin sheet of lead is used to couple the concrete surface and the transducer.

Analysis of the waveforms in frequency domain. In the previous section, the analysis was performed in the time domain, which is appropriate for ultrasonic testing done at high frequency. The same analysis could be performed here but it would be cumbersome because of the multiple reflections between the surfaces and the flaws. It is more convenient to perform the analysis in the frequency domain using a fast Fourier transform technique. The location of the imperfection becomes rather easy.

In a plate, for instance, the depth of a reflecting interface, h, can be determined as function of the P-wave velocity, V_p, and the peak frequency f:

$$h = \frac{V_p}{2f} \tag{11-20}$$

To validate the method, Sansalone and Carino[12] used laboratory samples containing controlled flaws and carried out many numerical simulations using the finite element method. Figure 11-15 illustrates one of these experimental simulations. In Fig. 11-15*a* the concrete slab contains no defects, so the depth of the reflecting interface h is equal to 0.5 m. For this configuration, the frequency peak is 3.42 kHz. Using Eq. (11-20), the P-wave velocity can be computed:

$$V_p = 2 \times 0.5 \times 3420 = 3420 \text{ m/s} \tag{11-21}$$

For the slab containing a disk-shaped void (Fig. 11-15*b*) the frequency peak was at 7.32 kHz. Using Eq. (11-20), the computed position of the reflecting interface is:

$$h = 3420/(2 \times 7320) = 0.23 \text{ m} \tag{11-22}$$

which is close to the actual depth of 0.25 m.

Spectral Analysis of Surface Waves (SASW)

The methods presented so far have only used P-waves to assess the quality in a concrete structure. Surface waves can also be employed to characterize the interior of a concrete member. It is important to point out that surface waves are not confined to the surface but, rather, are capable of penetrating a finite depth inside the material, sensing its properties. Waves with long wavelengths penetrate deeper than short wavelength waves. It is possible to take advantage of this property to develop nondestructive methods that use surface waves with different frequencies and therefore different wavelengths to probe different depths of the structure.

A convenient way of generating surface waves with a range of frequencies is to hit the surface with a hammer. The high frequency (short wavelength) waves will not penetrate deep and will provide information on the properties of the top layer close to the surface. However, the low frequency (long wavelength) waves will penetrate deeper and therefore their velocity will be influenced by the material properties in the interior. The various frequency components in the R-wave propagate with different velocities in a layered system and they are called phase velocities. These phase velocities can be determined at each frequency by measuring the time it takes to travel between two receivers with a known spacing (see Fig. 11-16).

After the signal processing of the waveforms is performed, it is possible to create a curve of the wave velocity as a function of the wavelength (Krstulovic-Opara et al.[13] and Nazarian and Desai[14]). A model of the system is created that matches the observed results. If no *a priori* information is available, it is challenging to guarantee that the proposed model is the one that gives the best results. This lack of uniqueness can be problematic. However, in many practical applications information is available that constrains the model. For instance, if the purpose of the study is to identify the thickness and properties of the materials existing in a concrete pavement and the subgrade, a natural model will consist of a series of layers parallel to each other. A computer program can vary the thickness and

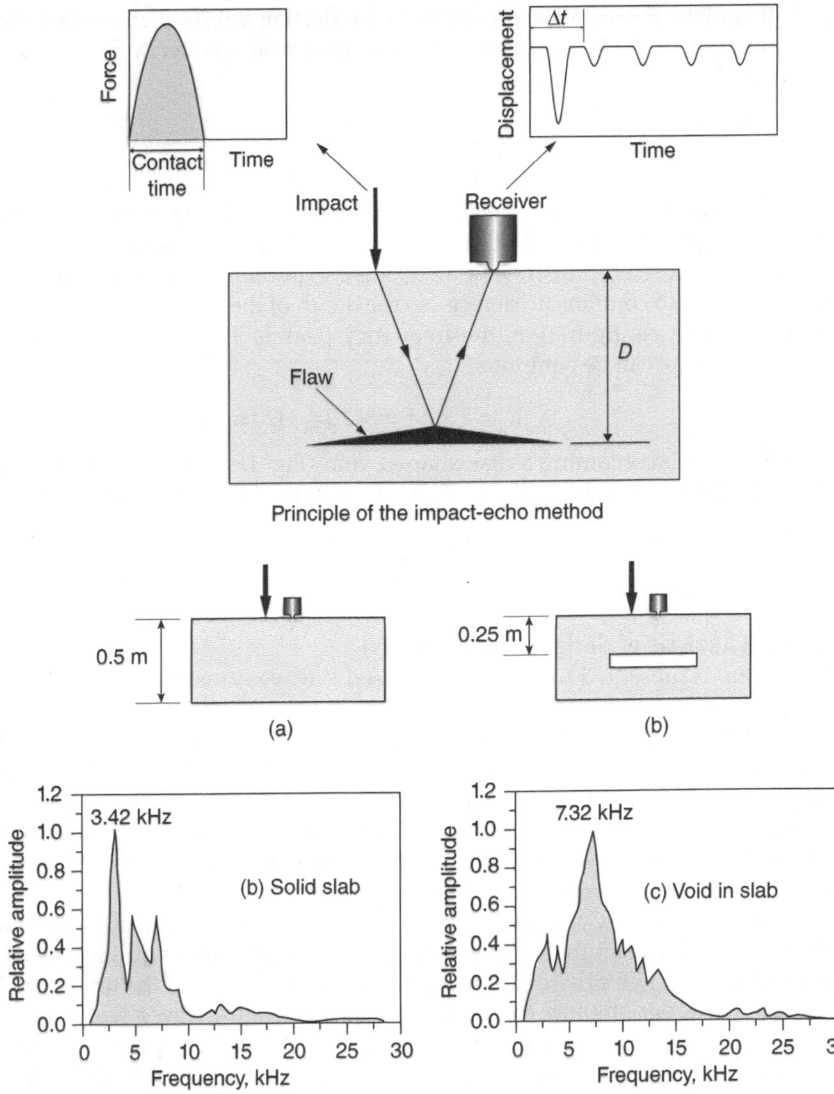

FIGURE 11-15 Application of the impact-echo method to identify a disk-shaped void in a slab. (After ACI 228.2R-98, *Nondestructive Test Methods for Evaluation of Concrete Structures.*)

material properties of each layer, assess the global response, and then compare with the experimental results to identify the configuration that best fits the observed data.

11.6.4 Acoustic Emission

Acoustic emission (AE) is a noninvasive, nondestructive method that analyzes the noises created when materials deform or fracture. Each acoustic emission event is a signature of an actual mechanism, a discrete event that reflects a given material response. As shown in Fig. 11-17, acoustic emission waves propagate through the

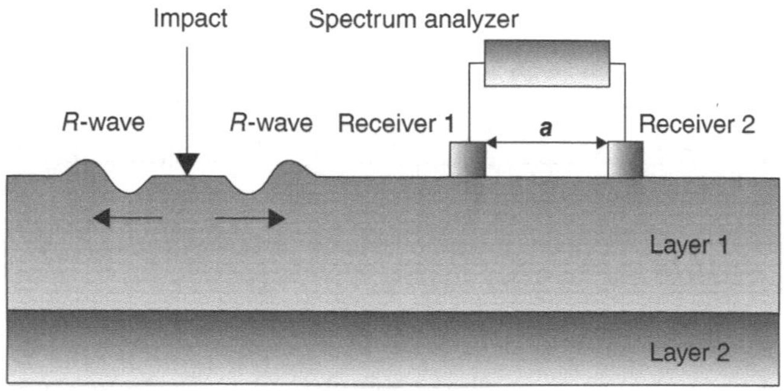

FIGURE 11-16 The set up for the SASW method. The impact is usually provided by hitting the surface with a hammer. The two receivers are used to measure the surface displacement caused by the surface wave created by the impact.

material and can be detected on the surface by a sensor, which turns the vibrations into electrical signals. The sound of fracture propagation was originally called acoustic emission since it is acoustic and audible, however, the frequency of these emissions can range from the audible range to many megahertz.

There is a critical difference between ultrasonic and acoustic emission methods. In the former, a known signal is imparted into a material and the material's response to the signal is studied, while in the latter the signal is generated by the material itself. Acoustic emission waves consist of *P*-waves (longitudinal waves) and *S*-waves (shear waves) and may include surface, reflected, and diffracted waves as well. These waves are originated by microcrack formation or propagation in concrete.

A material can generate acoustic emission of two basic waveforms: continuous and burst (see Fig. 11-18). Materials with high attenuation, such as concrete, quickly decrease the wave amplitude while materials with low attenuation, such as metals, maintain the wave amplitude. A schematic acoustic emission waveform obtained from concrete is shown in Fig. 11-18c. It is critical that the noise be minimized or it will interfere with the *P*-wave, making it hard to detect its arrival time. There are many methods available to count the occurrence of acoustic emission events. A simple method consists of measuring the number of times the amplitude of the acoustic emission wave is higher than a

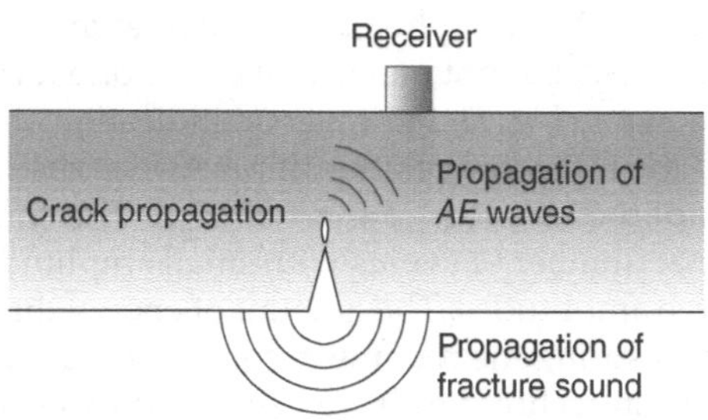

FIGURE 11-17 Generation, propagation, and detection of acoustic emission (AE). (After Ohtsu, M., The History and Development of Acoustic Emission in Concrete Engineering, *Concr. Res.*, Vol. 48, pp. 321–330, 1996.)

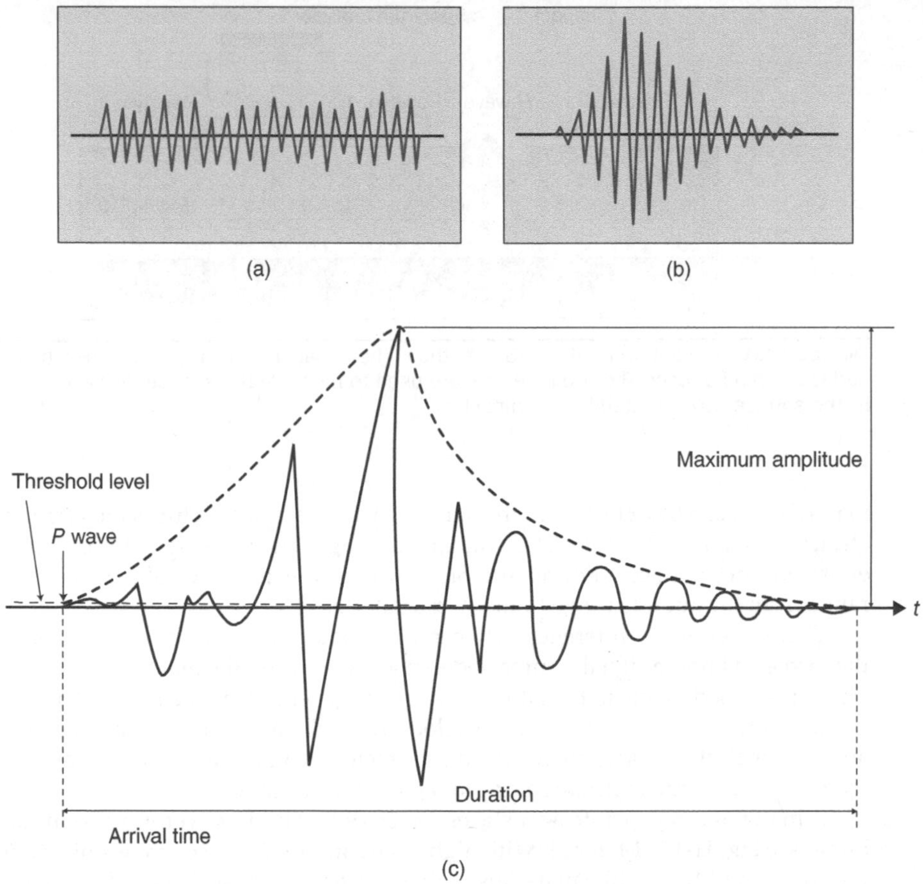

FIGURE 11-18 Basic types of acoustic emission waveforms: (*a*) continuous emission, (*b*) burst emission, and (*c*) concrete emission. (Figs. 11-18 *a* and *b* after Mindess, S., *Acoustic Emission Methods*, Handbook on Nondestructive Testing of Concrete, Malhotra, V.M., and N.J. Carino, eds., CRC Press, Boca Raton, FL, 1991, Fig. 11-18c from Ohtsu, M., The History and Development of Acoustic Emission in Concrete Engineering, *Concr. Res.*, Vol. 48, pp. 321–330, 1996.)

preset threshold value. More sophisticated schemes are available when the amplitude of the AE waves is small and not much above the noise level. The maximum amplitude of the AE wave shown in Fig. 11-18*c* is a good indication of the relative size of the event. Ohtsu[15] proposed the following relationship between the number of AE events, N, and the maximum amplitude, A: $\log_{10} N = \alpha - b\log_{10} A$. The equation, which has a negative slope, indicates that number of events with small amplitudes is larger than the number of events with large amplitudes. The amount of energy dissipated during the event can be estimated by measuring the root mean square of the wave.

Acoustic emission techniques have been used extensively to assess the nature of "the process zone," the region of discontinuous microcracking ahead of the continuous (visible) crack. Maji et al.[16] found that beyond the peak load most of the AE events occurred near the crack tip in a process zone extending about 25 mm ahead of the crack tip, and a longer distance behind it indicating ligament connections behind the visible crack tip. Berthelot and Robert[17] found that a damage zone appeared to grow in size as the crack progressed, reaching a length of up to 160 mm and a width of up to 120 mm. Suaris and Van Mier[18] compared crack propagation in tension (mode I) and in shear (mode II) in mortar. Li et al.[19] have shown that AE techniques are capable of detecting rebar corrosion in an early corrosion stage. Ohtsu showed examples of the possibilities of using AE to detect damages caused by alkali-silica reaction and freezing-thawing cycles. Yutama et al.[20] presented a case study from Japan where AE was applied in order to ensure the safety of an arch dam under construction in severe climate conditions.

Acoustic emission is a promising technique to study the fracture process in concrete, and to monitor concrete structures for their structural integrity. However, additional research is needed to resolve some of the following issues.

Concrete is a dispersive medium and many of the theoretical and analytical tools available for metals are not necessarily valid for AE signals from concrete.

The quantitative analysis of acoustic emission in concrete is difficult because the actual exact source mechanisms are not known or fully characterized beforehand, and the propagating medium is not a homogeneous, isotropic, and elastic solid. Material properties can change by an order of magnitude over short distances.

11.7 Electrical Methods

11.7.1 Resistivity

The resistivity of concrete is an important parameter in the corrosion of reinforced concrete structures. As presented in Chap. 5, high-resistivity concrete has little possibility of developing reinforcement corrosion. In the field, the electrical resistivity is determined by measuring the potential differences at the concrete surface caused by injecting a small current at the surface.

The relationship between current i and potential V is given by Ohm's law:

$$i = \frac{V}{R} \tag{11-23}$$

where R is the resistance of the system. Resistance is not a material property as it depends on the dimensions of the system. Just as ultimate load is normalized by the specimen dimensions to determine the strength of the material, the resistance is also normalized to establish *resistivity* ρ as a material property.

$$R = \rho \frac{L}{A} \tag{11-24}$$

where L is the length and A is the cross section.

Because electrical resistivity is determined by applying current at the concrete surface and measuring the changes of potential at specific points at the surface, it is appropriate to study the simple case of determining the potential at one point (P) when current is applied at one source (S), as shown in Fig. 11-19. If the current sink is placed far away, the current flows radially from the source and generates hemispherical equipotential surfaces.

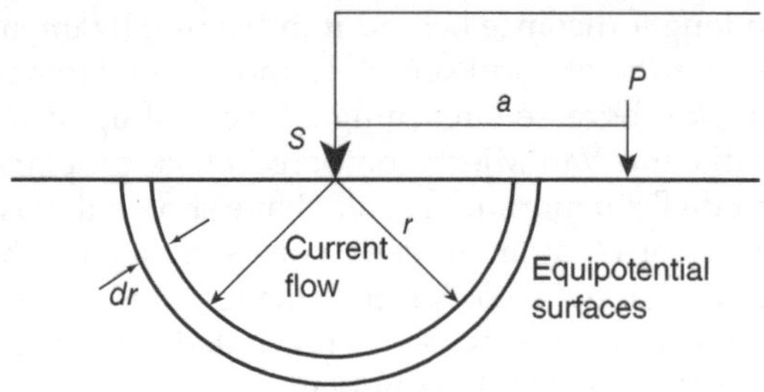

FIGURE 11-19 Determination of the potential at point P due to a point source of current S.

The difference of potential, dV, between two equipotential surfaces separated by dr is given by

$$dV = idR = i\rho \frac{dr}{2\pi r^2} \tag{11-25}$$

To obtain the potential at point P, we integrate the previous expression from distance "a" to infinity and use the usual convention that the potential at infinity is defined as zero. The following expression is obtained:

$$V = \frac{i\rho}{2\pi} \int_{r=a}^{\infty} \frac{dr}{r^2} = \frac{i\rho}{2\pi a} \tag{11-26}$$

In principle this equation can be used to map the potential at any point in the concrete. However, it is not practical to extend long cables to establish the "far away" condition required by integration up to infinite. A more practical configuration is shown in Fig. 11-20 where a small current is impressed on the concrete surface and removed at the sink placed within a finite distance from the source. The difference of potential is measured between two points P_1 and P_2. The potential at point P_1 can be obtained by using Eq. (11-26) and subtracting the contribution from the sink (note that the distance between P_1 and the sink is b).

$$V_1 = \frac{i\rho}{2\pi a} - \frac{i\rho}{2\pi b} \tag{11-27}$$

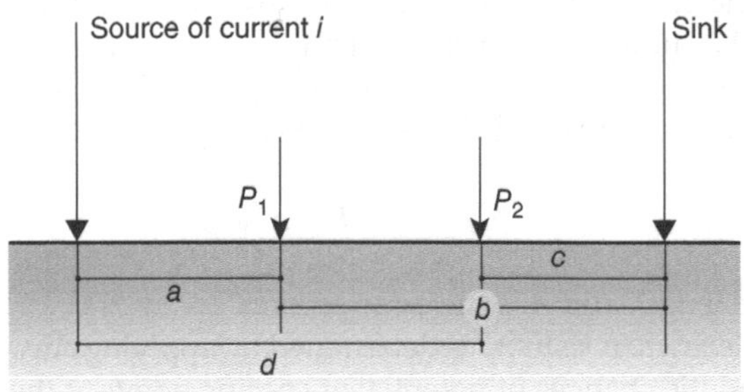

FIGURE 11-20 Determination of the resistivity of a material using two potential electrodes at P_1 and P_2.

Similarly the potential at point P_2:

$$V_2 = \frac{i\rho}{2\pi d} - \frac{i\rho}{2\pi c} \tag{11-28}$$

Therefore, the difference of potential is given by:

$$\Delta V = V_1 - V_2 = \frac{i\rho}{2\pi}\left[\left(\frac{1}{a} - \frac{1}{b}\right) - \left(\frac{1}{d} - \frac{1}{c}\right)\right] \tag{11-29}$$

and the following expression for resistivity is obtained:

$$\rho = \frac{2\pi\Delta V}{i}\left(\frac{1}{\frac{1}{a} - \frac{1}{b} - \frac{1}{d} + \frac{1}{c}}\right) \tag{11-30}$$

According to Ward,[21] a special case of this configuration, where the spacing between the sources, P_1, P_2, and the sink are equal to a, was developed by Wenner. The resistivity for this array is given by

$$\rho = \frac{2\pi a \Delta V}{i} \tag{11-31}$$

Across an inhomogeneous substructure, the pattern of current distribution in the test region is distorted and it is possible to create zonal maps of different resistivities. These maps can be constructed using arrays of electrodes arranged in various configurations. When the voltage is not in phase with the current, the resistivity becomes complex and it is referred to as electrical impedance, and will be discussed later. Monteiro et al.[22] showed that the reinforcing bars embedded in concrete can be located from surface measurements of resistivity and that the electrical impedance, also measured at the surface of the reinforced concrete, can assess the state of corrosion existing in the steel bars.

Because the flow of electric current in concrete is an electrolytic process, increasing ionic activity causes a decrease in the resistivity of concrete. Therefore, high water-cement ratio, high degree of saturation of concrete, and high concentration of dissolved salts in the pore solution led to lower concrete resistivity. Table 11-1 shows the CEB (Comite Euro-International du Beton)[23] recommendations for the likely corrosion rate.

Concrete Resistivity (Ω·m)	Likely Corrosion Rate
>200	Negligible
100–200	Low
50–100	High
<50	Very high

TABLE 11-1 CEB-192 Recommendation Based on Concrete Resistivity to Estimate the Likely Corrosion Rate

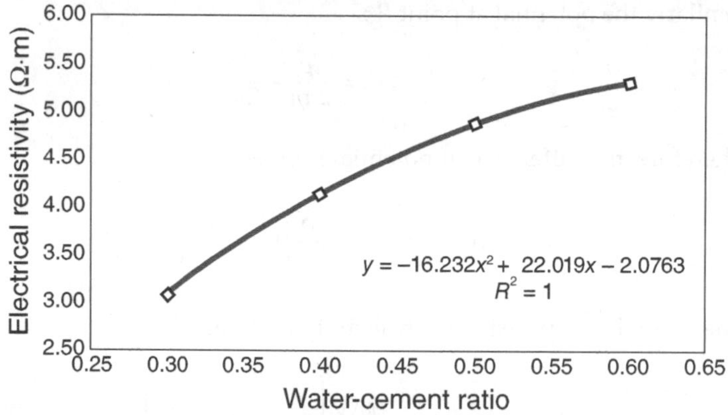

FIGURE 11-21 Relationship between electrical resistivity and w/c ratio of fresh concrete. (After Mancio, M., J.R. Moore, Z. Brooks, P.J.M. Monteiro, and S.D. Glaser, Instantaneous In-Situ Determination of Water-Cement Ratio of Fresh Concrete, *ACI Mat. J.*, Vol. 107, pp. 587–593, 2010.)

CEB-192 (Comite Euro-International du Beton) proposes that the likely corrosion rate is negligible for concrete with resistivity higher than 200 Ω·m, low for resistivity values in the range of 100 to 200 Ω·m, high for resistivity values in the range of 50 to 100 Ω·m, and very high for concrete with resistivity lower than 50 Ω·m.

Mancio et al.[24] developed a four-electrode probe using a Wenner array that can be immersed in fresh concrete to provide an instantaneous and accurate measure of its water-cement (w/c) ratio. In fresh cement pastes (a solution-saturated porous media), porosity (or water content) varies significantly with the w/c ratio and has a strong influence on the bulk resistivity. Figure 11-21 illustrates the relationship between the w/c ratio and electrical resistivity of fresh concrete. As expected, the electrical resistivity of fresh concrete increased with increasing w/c ratio (in contrast to hardened concrete). At a lower w/c ratio, the lower electrical resistivity (or higher conductivity) can be explained by a greater overall concentration of ions in solution, while for a higher lower w/c ratio, the increased resistivity can be attributed to a less concentrated pore solution. Worth noting is that the characteristic resistivity measured for each concrete mix is primarily a function of the resistivity of the pore fluid rather than the amount of pore fluid. All fresh concrete mixes tested experimentally had the same consistency and, therefore, approximately the same "porosity" (in the fresh, solution-saturated state, porosity equals water content plus entrapped air). In other words, the quantity of pore fluid in each mix was nearly constant. Therefore, any difference in the characteristic resistivity between samples must be attributed to the nature of the pore fluid rather than the amount of pore fluid.

A strong direct correlation was observed between electrical resistivity and the w/c ratio of fresh concrete. Resistivity readings from the probe were used as input in the fitted curve, and then the corresponding fitted quadratic equation (Fig. 11-21) was solved to obtain an estimate of the w/c ratio. (The equations are valid only between the upper and lower values of w/c ratio tested.) The estimated values are virtually identical to the actual w/c values for all mixes tested, with variations remaining smaller than 0.01.

11.8 Electrochemical Methods

11.8.1 Introduction to Electrochemistry of Reinforced Concrete

When a metal is inserted in an aqueous environment, a potential difference develops across the metal-aqueous solution interface. The rate at which the reactions and the movement of charges across the interface occurs is influenced by the magnitude of the potential difference. In the aqueous solution, the distribution of particles close to the metal surface is no longer homogeneous, nor are the forces isotropic. Water molecules are polar and therefore are attracted to the charged surface and orient themselves along the interface. An excess of ions of the opposite charge can accumulate near the interface. In analogy with a parallel-plate condenser, the system of the two oppositely charged planes is referred to as a double layer.

Small ions generate large Coloumbic forces, and therefore have a greater chance to become hydrated. This means that most cations are hydrated and most anions, being large, are not hydrated. This also explains why some anions cause more corrosion damage. The large anions are unhydrated and can get closer to the metal surface, even though they may not participate directly in the corrosion reactions. The ion can be adsorbed on the metal surface forming an inner-sphere complex when no water molecule is between the surface function group and the ion, or an outer-sphere complex when at least one water molecule exists between the surface and the ion. Ions can also be adsorbed in the diffuse swarm of the double layer in order to neutralize the surface charge (Fig. 11-22).

It is not possible to directly determine the potential difference across the interface, therefore it is necessary to establish a reference scale, usually the potential difference across a standard hydrogen electrode. There are two extreme cases of electrode behavior: *perfectly polarizable* and *reversible*. When a potential is applied to a system, a reversible electrode allows the current to flow across its interface until the system restores the equilibrium while a perfectly polarizable electrode does not allow the current to pass across its interface.

Corrosion of reinforcing bars in the electrolytic concrete pore solution involves electron or charge transfer through the chemical reactions at the interface. The difference

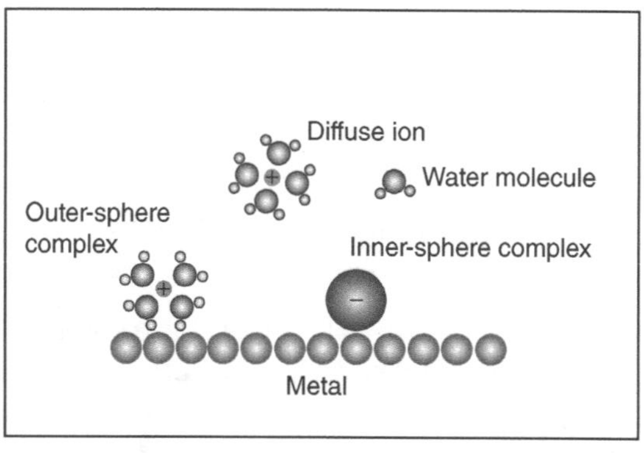

FIGURE 11-22 Three modes of adsorption of ions by a metal surface. (Modified from Sposito, G., *The Chemistry of Soils*, Chap. 12, Oxford University Press, New York, 1989.)

in chemical potential between the reinforcing bars and electrolyte is the driving force for the charge transfer to occur. The electrode potentials of the anodic and cathodic processes will change with the corrosion reaction rate until a stable, or steady state potential (E_{eq}) is achieved.

Consider the general electrochemical reaction:

$$aA + bB + Ze^- = cC + dD$$

The value of equilibrium potential (E_{eq}) is given by the Nerst equation:

$$E_{eq} = E^0 + \frac{RT}{ZF} \ln \frac{[A]^a [B]^b}{[C]^c [D]^d}$$

where F = Faraday's constant
E^0 = standard electrode potential
R = gas constant
T = absolute temperature
$[i]$ = activity of component i

At the equilibrium potential the anodic (i_a) and cathodic (i_c) currents are opposite and equal to i_0.

Deviation from the equilibrium condition can be expressed by the electrode polarization potential, also known as *overpotential* (η_a or η_c) where:

$$\eta_a = E - E_{eq} \tag{11-32}$$

$$\eta_c = E_{eq} - E \tag{11-33}$$

There is a strong incentive to experimentally determine the *polarization curves* of the metal in a certain electrolyte. These curves are in the form of the polarization potential and the log of the current density. A schematic polarization curve is shown in Fig. 11-23. The curve for reinforced concrete is more complex and should be obtained experimentally. However, the general principle is the same and the objective is to determine the existing

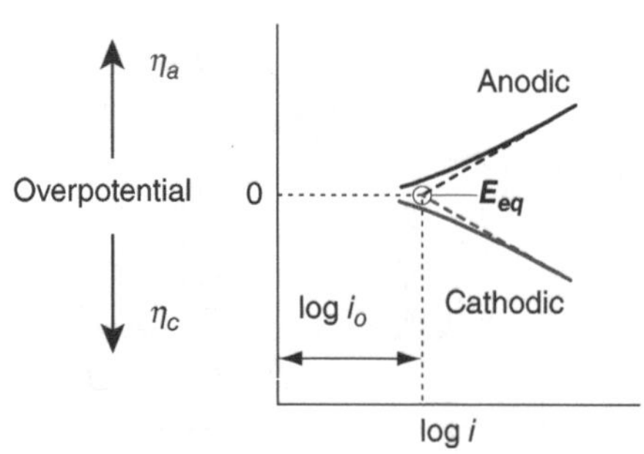

FIGURE 11-23 Polarization curves for an electrode process.

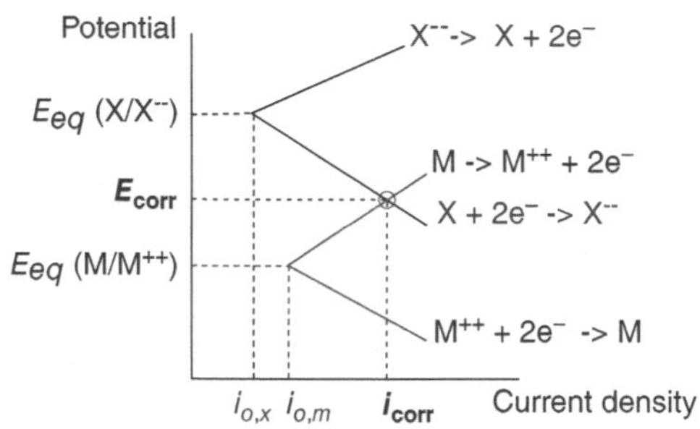

FIGURE 11-24 A simple corrosion process with one polarized anodic reaction ($M \rightarrow M^{++} + 2e^-$) and one polarized cathodic reaction ($X + 2e^- \rightarrow X^-$) that are coupled.

current density by using the following relationship between polarization potential and current density:

$$\eta_a = \beta_a \log\left(\frac{i_a}{i_0}\right) \tag{11-34}$$

$$\eta_c = \beta_c \log\left(\frac{i_c}{i_0}\right) \tag{11-35}$$

where β_a and β_c are the Tafel coefficients and i_0 is the exchange current density.

A corrosion process has at least one polarized anodic reaction and one polarized cathodic reaction that are coupled. Each reaction is characterized by its equilibrium potential and polarization curves. During corrosion, the electrical charges produced by the dissolution of the metal are consumed by the cathodic reactions. Due to conservation of electrical charge, the anodic current density has to be equal to the cathodic current density. To obtain this condition, the system must achieve an intermediate potential between the two equilibrium potentials. As shown in Fig. 11-24 this intermediate potential and its corresponding current density are referred to the corrosion potential (E_{corr}) and the corrosion current density (i_{corr}), respectively.

11.8.2 Corrosion Potential

The corrosion potential of the steel in reinforced concrete can be measured as the voltage difference between the steel and a reference electrode in contact with the surface of the concrete. Half-cell measurements may be made relatively easily, using only a high-impedance voltmeter and a standard reference electrode, such as a copper-copper sulfate electrode. As shown in Fig. 11-25, the voltmeter connects the steel with the reference electrode such that the steel is at the positive terminal of the voltmeter. It is important to maintain a good contact between the reference electrode and the concrete. A standard test procedure is described in ASTM C 876 and the ASTM criteria for corrosion of steel in concrete are shown in Table 11-2.

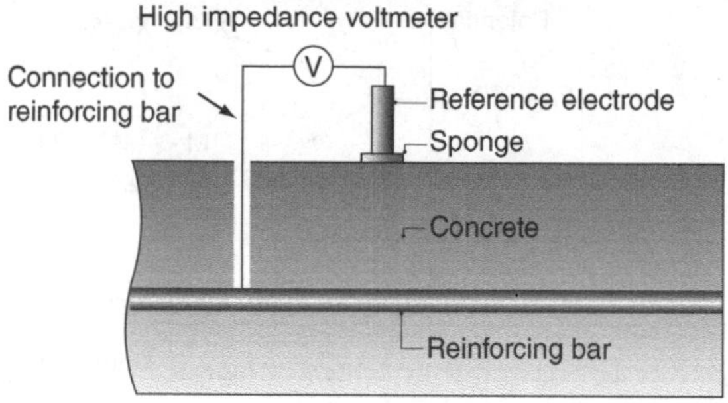

FIGURE 11-25 System for measuring the half-cell potential. The electrode is moved on the concrete surface to assess the risk of corrosion at various locations.

The potential recorded in the half-cell measurement can be used to indicate the probability of corrosion of the steel reinforcement, as shown in Table 11-2. The reference electrode can be moved over the concrete surface to develop a potential map that shows the possible sites of active corrosion in the structure. This is an attractive method when planning repairs or monitoring cathodic protection.

The half-cell potential measurement has been widely used in the field and the method provides a quick and inexpensive means of identifying zones that need further analysis or repair; however, the results of the test may be affected by the following:

Degree of humidity in concrete. The measurement is very sensitive to the humidity existing in the concrete. More negative potentials result for concrete with higher degree of saturation. Misleading conclusions regarding the risk of corrosion may result in a structure where segments, otherwise identical, are exposed to dry conditions while others remain moist.

Oxygen content near the reinforcement. The lack of oxygen near the reinforcement results in more negative potentials as compared to more aerated zones. It should be noted, however, that the lack of oxygen actually significantly reduces the corrosion rate, even though it causes more negative potential. Therefore, care should be taken when analyzing the risk of corrosion in structures where the lack of oxygen near the reinforcement is expected, such as submersed or underground construction.

Microcracks. Localized corrosion can be generated by microcracks that also modify the concrete resistivity, consequently affecting the corrosion potential measurement.

Stray currents. The presence of stray currents will significantly affect the measurements of the half-cell potential.

Measured Potential (mV vs. CSE)	Corrosion Probability
> −200	Low, less than 10% probability of corrosion
−200 ~ −350	Uncertain
< −350	High, greater than 90% probability of corrosion

TABLE 11-2 ASTM Criteria for Corrosion of Steel in Concrete (ASTM C 876)

For these reasons, ASTM C 876 has specified that the standard does not apply in certain environments, including concrete which is too dry, carbonated, or has a coated surface or when the reinforcing steel has a metallic coating. While the half-cell potential technique is popular, it must be recognized that it is not quantitative. First, unless the full polarization curve is known, the kinetic corrosion rate cannot be predicted. Second, due to contrast in size between the reinforcing steel or reinforcing cage and the reference electrode, the location along the reinforcement where potential is measured is uncertain.

11.8.3 Polarization Resistance

It is instructive to continue the analysis started in the introduction of electrochemistry of reinforced concrete corrosion where the concepts of polarization potential were introduced. Let us apply an overpotential to a corroding couple, and define the resulting difference between the cathodic and anodic current at the same potential as the applied cathodic current density, $i_{\mathrm{app},c}$:

$$i_{\mathrm{app},c} = i_c - i_a \tag{11-36}$$

Using Eqs. (11-35) and (11-36), Eq. (11-37) can be expressed as:

$$i_{\mathrm{app},c} = i_{\mathrm{corr}}(10^{\eta_c/\beta_c} - 10^{\eta_a/\beta_a}) \tag{11-36a}$$

The slope at the origin of the polarization curve is defined as the polarization resistance, R_p:

$$R_p = \left(\frac{d\eta}{d\,i_{\mathrm{app},c}}\right)_{\eta\to 0} = \frac{\beta_a\beta_c}{2.3i_{\mathrm{corr}}(\beta_a+\beta_c)} \tag{11-37}$$

or

$$R_p = \frac{B}{i_{\mathrm{corr}}} \tag{11-38}$$

where B is given by

$$B = \frac{\beta_a\beta_c}{2.3(\beta_a+\beta_c)} \tag{11-39}$$

Equation (11-38) was proposed by Stern and Geary[25] and it provides an important relationship to determine the corrosion rate of a metal from the polarization resistance. This lead to the development of *linear polarization resistance* methods that directly measure the polarization resistance R_p by applying a small DC voltage (dE_{app}) as a perturbation, and the response as a current flow (dI) is recorded, or vice versa. The application of linear polarization resistance by voltage perturbation is called potentiostatic, and when performed by current perturbation the technique is termed galvanostatic, as shown in Fig. 11-26. The potentiostatic measurement is used more often in practice. The inciting voltage must be small to hold the assumption that the system is linear, see Fig. 11-27.

Linear polarization resistance is widely used both in the field and in laboratory. Typical values of polarization resistance for steel embedded in concrete are shown in

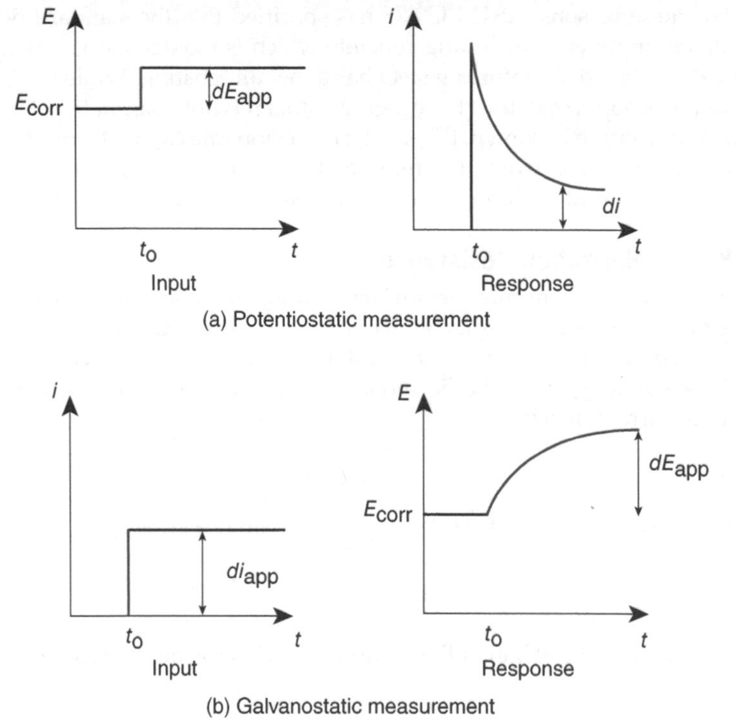

FIGURE 11-26 Linear polarization resistance measurement.

Table 11-3. If the parameter B is known for a reinforced concrete system, the corrosion rate can be quantitatively obtained by Eq. (11-38). For corrosion of reinforcing steel in concrete, the value of B is generally in range of 26 to 52 mV. In many practical applications, a value of 26 mV is selected when the steel is actively corroding in concrete and a value of 52 mV is selected when the steel is passive.

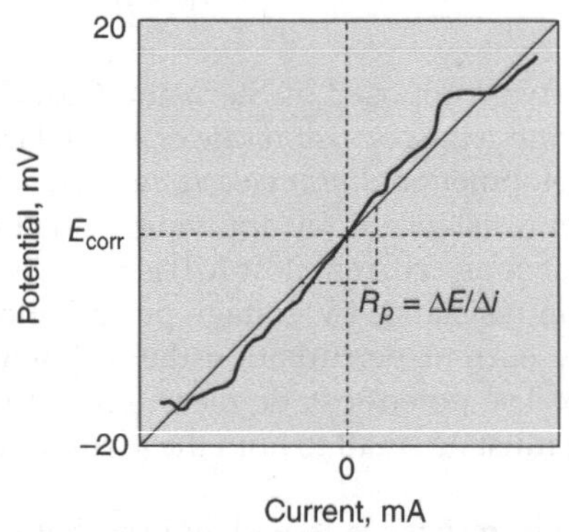

FIGURE 11-27 A representation of the polarization resistance curve.

Rate of Corrosion	Polarization Resistance, R_p (kΩ·cm²)	Corrosion Penetration, p (μm/year)
Very high	$0.25 < R_p < 2.5$	$100 < p < 1000$
High	$2.5 < R_p < 25$	$10 < p < 100$
Low/moderate	$25 < R_p < 250$	$1 < p < 10$
Passive	$250 < R_p$	$p < 1$

TABLE 11-3 Typical Polarization Resistance for Steel in Concrete

There are many advantages in using the linear polarization technique for reinforced concrete:

1. The determination of the corrosion rate is an important parameter in the assessment of the life cycle of the structure.
2. Commercial equipment is available to measure the polarization resistance in existing structures.
3. It has been verified experimentally that the magnitude of the corrosion rate measured by polarization resistance is similar to that measured by gravimetry.
4. The measurement times are small.
5. The method applies small pertubations that do not interfere with the existing electrochemical processes.

However, there are some limitations to this technique:

1. The whole reinforcing bar or the steel mat, along which uniform corrosion is rare, must be polarized. Also, the polarization is nonuniform due to much smaller size of the counter electrode relative to the reinforcing steel and due to the high resistivity of concrete. Therefore, the measured R_p is essentially an average result for all of the steel in the structures whereas the actual corrosion is nonuniform.
2. This method assumes that the concrete resistivity is low, when in reality the concrete resistivity is usually high. For instance, when concrete is dry, the uncompensated ohmic drop can be large, which may induce a significant error in the measured R_p.
3. Linear polarization is a DC method. Because of the electrical capacitance across the steel/concrete interface, it takes time to obtain a full response. As a result, the scanning rate plays an important role in obtaining accurate measurements.

Of the above limitations, the first has attracted the most research effort. A popular and commercialized technique is to use a "guardring" electrode to confine polarization or the current to the measured site. The setup and principle of the guard-ring electrode (GE) is shown in Fig. 11-28. It works with the same principle as the counter electrode (CE) but with an independent current source. In addition, the current from CE is confined within the guard ring. Therefore, the measured length of the steel reinforcement is known.

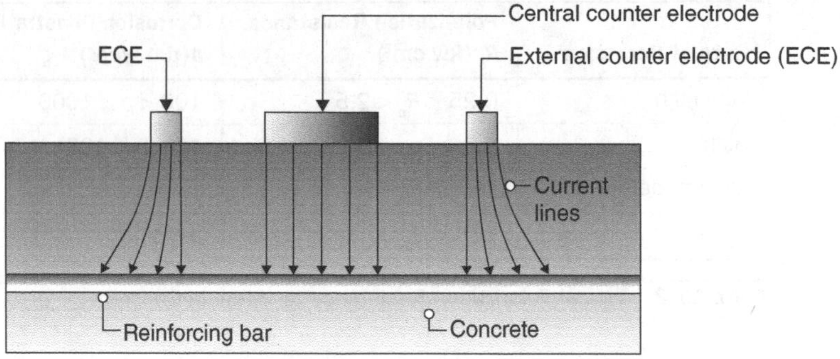

FIGURE 11-28 Confinement of current lines around the counter electrode ("guard ring") due to the presence of external counter electrode. (After Feliu, S., J.A. Gonzalez, M.S. Feliu, and M.C. Andrade, *ACI Mat. J.*, Vol. 87, p. 458, 1990.)

11.8.4 Electrochemical Impedance Spectroscopy

Electrochemical impedance spectroscopy, or AC impedance, is an informative method because not only is R_p measured, but also the physical processes in concrete and steel/concrete interface are assessed. Impedance measurement employs small-amplitude alternating (AC) signals in a wide range of frequency as a perturbation. Many processes occurring at the surfaces can absorb electrical energy originating a time lag and a phase angle between the perturbation and the response. In electrochemical impedance spectroscopy measurements, a sine or cosine wave of AC current with magnitude I_0 and frequency f is commonly used as the input. The output is recorded as a voltage response with the magnitude $V(f)$ and a phase angle $\phi(f)$ with respect to the current (see Fig. 11-29). The ratio between the voltage $V(f)$ and current $I(t)$ is called the frequency dependent impedance, $Z(f)$, which is a complex number and therefore has an in-phase real (Z_r) component and an out-of-phase imaginary (Z_i) component:

$$Z(f) = Z_r(f) + i\, Z_i(f) \tag{11-40}$$

The impedance modulus is defined as $|Z|$:

$$|Z|^2 = (Z_r)^2 + (Z_i)^2 \tag{11-41}$$

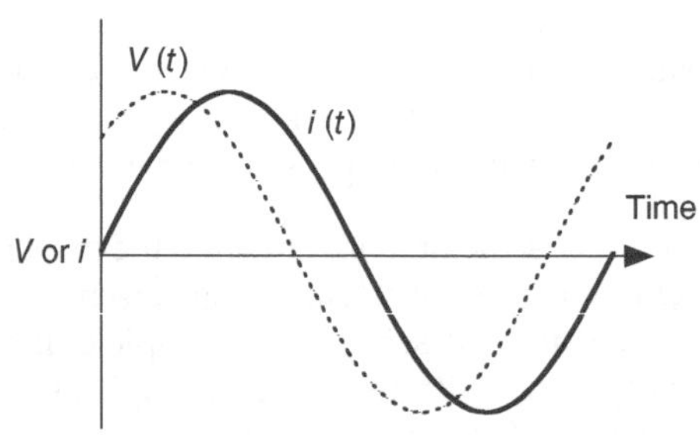

FIGURE 11-29 Voltage response *V* to sinusoidal current signal *i*.

Therefore, the in-phase and out-of-phase component can be expressed as:

$$Z_r(f) = |Z|\cos(\phi(f)) \tag{11-42}$$

$$Z_i(f) = |Z|\sin(\phi(f)) \tag{11-43}$$

Equivalent circuits have been used to model the impedance of complex systems. Pure resistance and pure capacitance represent two types of impedance to the charge transfer. Energy is dissipated through electrons or ions flowing through a resistance element, which constitutes a nonfrequency-dependent impedance that has only a real part. A capacitance element represents an energy storage process or charge separation under an electrical field. It creates an alternating electric current under an alternating electrical field, and the impedance will decrease with frequency. At the same time, the charge separation requires operational time, which causes a phase lag in the response with respect to the applied electrical field. The combination of the resistance and capacitance elements in an electrical circuit can produce a frequency-dependent complex impedance.

The simplest and most widely used equivalent electrical circuit is the concrete resistance R_c in series with a parallel double layer capacitance C_{dl} and charge transfer resistance R_p described in Fig. 11-30. This model is analogous to the standard solid model of viscoelasticity discussed in Chap. 13. In the same manner that by adding springs and dashpots created more sophisticated models, improved modeling can be obtained by combining resistance and capacitance in series or in parallel. However, the basic configuration shown in Fig. 11-30 captures important features of impedance spectroscopy.

The impedance of this circuit is given by:

$$Z = Z_r + iZ_i = R_c + \frac{R_p}{1 + i\omega R_p C_{dl}} \tag{11-44}$$

where $\omega = 2\pi f$. Multiplying the numerator and denominator of the last term of Eq. (11-44) by $1 - i\omega R_p C_{dl}$, gives

$$Z = R_c + \frac{R_p}{1 + \omega^2 R_p^2 C_{dl}^2} - \frac{i\omega R_p^2 C_{dl}}{1 + \omega^2 R_p^2 C_{dl}^2} \tag{11-45}$$

Eliminating ω leads to

$$\left[Z_r - \left(R_c + \frac{R_p}{2}\right)\right]^2 + Z_i^2 = \left(\frac{R_p}{2}\right)^2 \tag{11-46}$$

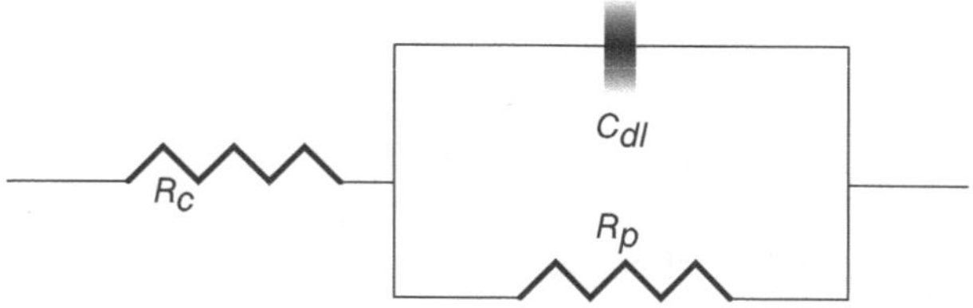

FIGURE 11-30 Model of equivalent circuit. R_c represents the electrolyte resistance or the concrete; R_p represents the polarization resistance; C_{dl} represents the double layer capacitance.

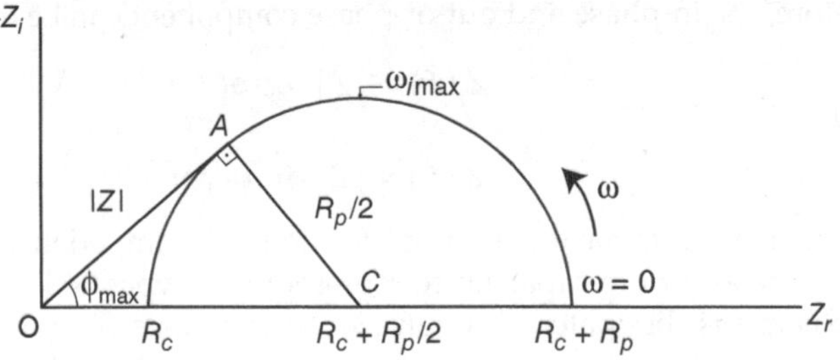

FIGURE 11-31 Nyquist plot for the impedance of the electric circuit shown in Fig. 11-30.

which is the equation of a circle with center $(R_c + R_p/2, 0)$ and radius $R_p/2$. This equation can be displayed in a graph with the in-phase real (Z_r) component in the horizontal axis and the out-of-phase imaginary (Z_i) in the vertical axis. This representation, shown in Fig. 11-31, is known as the Nyquist plot.

Comparing Eqs. (11-44) and (11-45), it is easy to identify the in-phase real (Z_r) component and the out-of-phase imaginary (Z_i) as

$$Z_r = R_c + \frac{R_p}{1+\omega^2 R_p^2 C_{dl}^2} \tag{11-47}$$

and

$$Z_i = -\frac{\omega R_p^2 C_{dl}}{1+\omega^2 R_p^2 C_{dl}^2} \tag{11-48}$$

Equations (11-47) and (11-48) show that the values of the in-phase real (Z_r) component are equal to ($R_c + R_p$) as the frequency approaches zero, and R_c as the frequency approaches infinity. Once the value of the polarization resistance is determined, it is possible to estimate the corrosion rate, i_{corr}, using Eq. (11-38). Semicircles with a large diameter in the Nyquist plot indicate that the polarization resistance, which is numerically equal to the diameter of the circle, is high and therefore the corrosion rate becomes small. As corrosion increases, the values of the charge transfer resistance decrease and, consequently, the diameter of semicircles in the Nyquist plot also decreases.

The frequency $\omega_{i\max}$ when the out-of-phase component (Z_i) reaches its maximum, can be found by differentiating Eq. (11-48) with respect to frequency:

$$\frac{dZ_i}{d\omega} = 0 \tag{11-49}$$

which gives

$$\omega_{i\max} = \frac{1}{R_p C_{dl}} \tag{11-50}$$

The phase angle ϕ is defined as

$$\phi = \arctan\left(\frac{Z_i}{Z_r}\right) \tag{11-51}$$

Using the Eqs. (11-47) and (11-48), Eq. (11-51) becomes

$$\phi = \arctan\left(\frac{-\omega R_p^2 C_{dl}}{R_c + R_p + R_c \omega^2 R_p^2 C_{dl}^2}\right) \tag{11-52}$$

It is possible to determine the maximum phase angle ϕ_{max} by observing the triangle OAC in the Nyquist plot shown in Fig. 11-31:

$$\sin\phi_{max} = \frac{R_p/2}{R_c + R_p/2} \tag{11-53}$$

The frequency $\omega_{p\,max}$ where the maximum phase angle, ϕ_{max}, occurs can be found by differentiating Eq. (11-52):

$$\frac{d(\tan\phi)}{d\omega} = 0 \tag{11-54}$$

which gives

$$\omega_{p\max} = \sqrt{\frac{R_c + R_p}{R_c}}\,\frac{1}{R_p C_{dl}} \tag{11-55}$$

This is nicely represented in a Bode plot (Fig. 11-32) where the impedance modulus, |Z|, and the phase angle are plotted as a function of the frequency. The modulus of the impedance decreases with frequency. At low frequencies, the imaginary component of the impedance disappears, and the value of the impedance becomes the sum of R_p and R_c. Similarly, at high frequencies, the imaginary component of the impedance disappears and the value of the impedance becomes equal to R_c. The figure also indicates the dependence of the frequency on the phase angle.

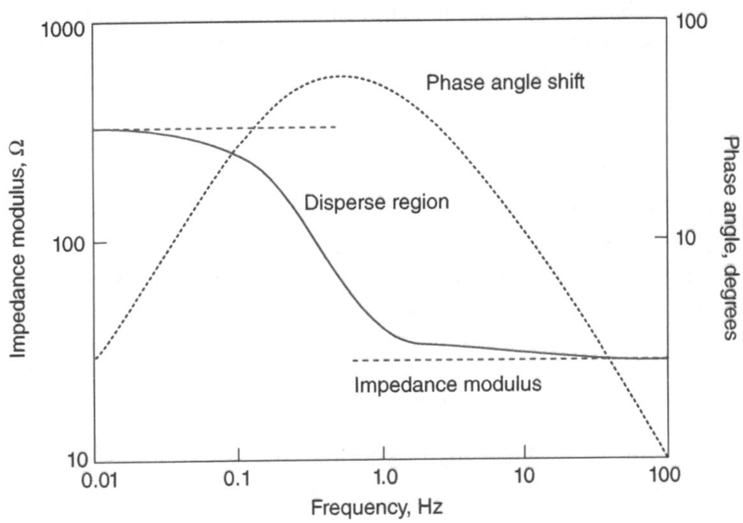

FIGURE 11-32 A plot of impedance spectrum plot with frequency for the equivalent circuit described in Fig. 11-30 using the following values: R_e = 30 Ω, R_p = 300 Ω, and C_d = 0.003 F.

The formulation presented here is adequate for simple corrosion systems with one step electron transfer. For many cases of practical applications, this simplification is inadequate and it is necessary to define a charge transfer resistance R_t. Gabrielli and Keddam[26] developed theoretical relationships between corrosion current and charge transfer and polarization resistance for corrosion involving complex stages, including passive dissolution and diffusional control.

For such conditions, a simple electrical circuit may not adequately describe the behavior. As a result, more elements are introduced in the equivalent circuits in order to accurately simulate the complex electrochemical processes. For instance, a Warburg impedance, whose magnitude varies inversely with the square root of the frequency, should be included if the process is diffusion controlled.

A reinforced concrete element under an electrical field has at least three electrical impedances: the concrete, the steel/concrete interface, and the steel. Given a broad enough frequency range, each component displays frequency dependence, although their response to electrical signals will occur at different rates. According to the underlying principle different physical processes can be distinguished by their response to perturbation in different frequency ranges. The charge transfer modes in a measured system depend on the material's composition, microstructure, and interfacial reactions. The underlying charge transfer mechanisms are:

Interface between the reinforcing bar and concrete pore solution. There are basically two paths for current to pass across the metal-electrolyte interface. In the first path, often called the *Faradaic* path, the charges are carried across the interface by chemical reactions. The reaction impedance is the change transfer resistance R_t whose magnitude is inversely proportional to the corrosion rate. This path may also involve the diffusion of ions toward the interface and it can be represented by a frequency-dependent Warburg impedance. In the second path, the charges do not cross the interface and the current is carried by the charging and discharging of the electrical double layer between steel and concrete. As discussed before, the double layer is basically a charge separation which expands and contracts under an impressed electric field, and behaves electrically as a capacitance.

Film and corrosion products. Impedance spectroscopy is a powerful tool for detecting the presence of films and corrosion products on the steel surface. Interesting work in this area has been done by John et al.[27] and Ford et al.,[28] however, there is still no complete understanding of how they affect the overall corrosion of reinforced concrete.

Concrete matrix. Current may be conducted through concrete by the ions present in its pore solution. A capacitive effect also exists, but the factors responsible for it are not well understood. The capacitance of a concrete matrix, however, is so small (around 10^{-11}F) that its frequency response regime is in MHz. Christensen et al.[29] have summarized the theory, measurement, and applications of impedance spectroscopy as applied to cement-based materials.

AC impedance spectroscopy has been used in the laboratory; however, field applications have been limited for several reasons:

1. The equipment used is quite bulky and complex. The required equipment includes a signal generator, a potentiostat, and a frequency analyzer.

2. As with linear polarization, the entire reinforcing bar or network must be polarized. Also, because uniform corrosion is rare, measurements are essentially an average value along the steel reinforcement.
3. Similar to the linear polarization, this method also requires a physical connection to the steel embedded in concrete. To provide a direct electrical connection to the reinforcement, cover-concrete must be removed and later patched. This process is time-consuming and labor intensive, and may adversely affect the corrosion rate of the reinforced member.
4. AC impedance measurements can have lengthy data acquisition times, especially for low frequency measurements.

Often, this method is still too complicated for practical use, however, EIS is a powerful method due to its ability to separate the individual processes, therefore the technique is often used in corrosion research. Corrosion rate can be estimated from the impedance spectrum and in addition, the methodology has been used to study the effects of corrosion inhibitors, coatings, and pitting corrosion.

11.9 Electromagnetic Methods

11.9.1 Covermeter

Covermeter is the generic term for equipment used to locate steel reinforcing bar in concrete and to estimate the thickness of the concrete cover over the reinforcement. Unlike concrete, steel bars interact strongly with low-frequency electromagnetic waves applied at the surface of the concrete, making it easy to identify their location. Before describing the commercial equipment available, the next paragraphs give a very brief review of magnetic fields generated when current passes through electrical coils as these fields are used to detect the reinforcing bars.

The magnetic field B_r at a point P generated by a current $I_0 \cos \omega t$ through a small electrical coil can be decomposed in a component $B_{//}$ parallel to the axial direction and a component B_p perpendicular to it (see Fig. 11-33):

$$B_{//} = \frac{\mu_0 m_0 \cos \omega t}{4\pi r^3}(3\cos^2\theta - 1) \tag{11-56}$$

$$B_p = \frac{\mu_0 m_0 \cos \omega t}{4\pi r^3} 3\sin\theta \cos\theta \tag{11-57}$$

where μ_0 is the permeability constant of air and $m_0 \cos \omega t$ is the magnetic moment (area of the coil × the current). These equations show that along the dipole ($\theta = 0$ or π) there is no field perpendicular to the dipole ($B_p = 0$).

Another important property of a coil exposed to a time-dependent magnetic field is that an electromotive force ε is generated and its magnitude can be determined using Faraday's law of induction:

$$\varepsilon = -N \frac{d\Phi_B}{dt} \tag{11-58}$$

where Φ_B is the magnetic flux and N is the number of turns in the coil.

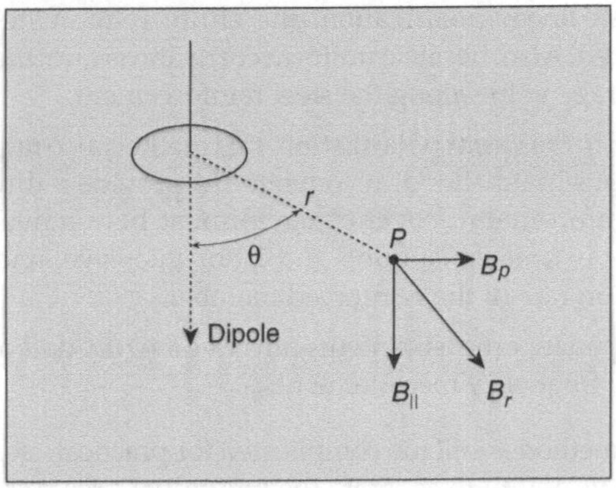

FIGURE 11-33 Electromagnetic field caused by impressing a current into an electrical coil.

Let us now discuss the two main methods used in commercial covermeters. The first method uses an excitation coil to generate a magnetic flux. The flux travels through the concrete and its intensity is measured by a sensing coil. The whole circuit is closed by the presence of a ferromagnetic core. Concrete is not a good conductor of magnetic flux, that is, it has a high *magnetic reluctance*. When the detector moves to a position near the reinforcement, the sensing coil starts to indicate a large increase in the magnetic flux as steel is a very good conductor (Fig. 11-34). The intensity of the current measured at the sensing coil is greatly influenced by the depth of the cover, therefore, with a proper calibration, it is possible to estimate the depth of the concrete cover in the field.

The second method consists in finding the location of the reinforcing bar by scanning the concrete surface with an electrical coil attached to an AC source and a current indicator (see Fig. 11-35). When the detector scans areas with no reinforcement, the current indicator remains in the same position. However, as the detector gets closer to the reinforcement, the indicator starts showing a decrease in current until it reaches a minimum

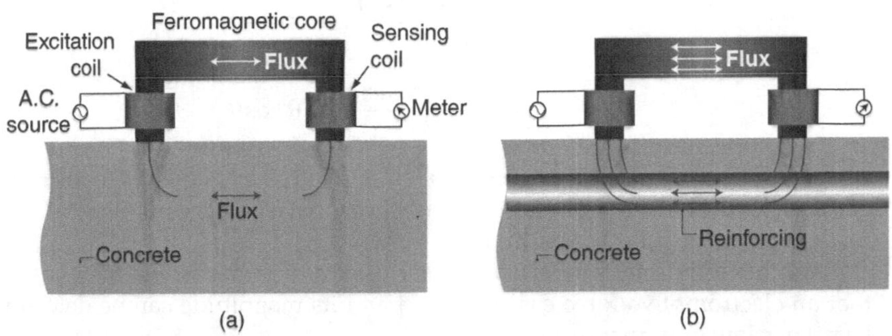

FIGURE 11-34 (*a*) When the covermeter is placed on a nonreinforced area, the sensing coil will register a small flux since concrete has high reluctance. (*b*) When the covermeter is placed on top of the reinforcing steel, the sensing coil will measure a high flux because the bar has low reluctance. (After ACI 228.2R-98, *Nondestructive Test Methods for Evaluation of Concrete Structures*.)

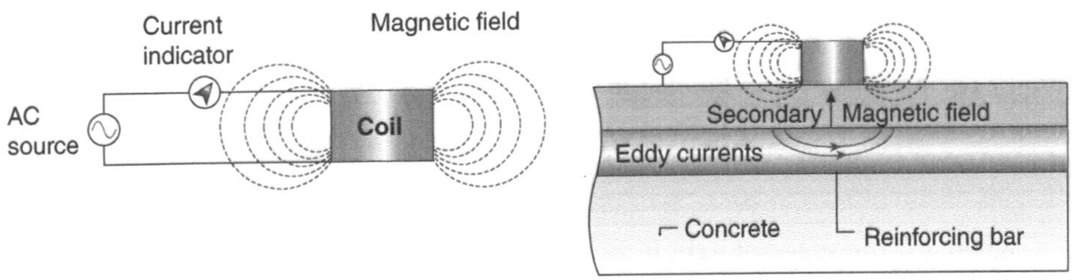

FIGURE 11-35 Eddy currents are induced in a conductor when the magnetic flux going through it changes. These currents can be a nuisance when they appear in structural systems because the currents increase the temperature of the material. Eddy currents can be used to find the steel reinforcement embedded in concrete. Figure 11-35 shows the basic configuration: a current indicator and electrical coil that generates magnetic field because of the AC source. When the detector is close to the reinforcing bar, eddy currents are generated in the bar which creates a secondary field whose presence is sensed by the current indicator. (After ACI 228.2R-98, *Nondestructive Test Methods for Evaluation of Concrete Structures*.)

value at the moment that the detector is on top of the reinforcement. The magnetic field generated by the coil induces eddy currents in the reinforcing steel bar, which in turn produces a secondary magnetic field. This secondary field creates a current in the opposite direction, explaining the reason for the decrease in current observed in the indicator when the steel bar is present.

11.9.2 Ground Penetrating Radar

Ground penetrating radar (GPR) methods use electromagnetic energy, typically at frequencies of 50 to 1500 MHz, to probe the subsurface. This method has been used in concrete structures to detect voids and delaminations, locate the reinforcing bars, measure the pavement thickness, and monitor structural changes. Recently, field research has been performed on the material characterization of concrete, such as water content, degree of cement hydration, and presence of chlorides. Since the field applications of radar to concrete structures go beyond the characterization of the subsurface, Clemena[30] has suggested that the term short-pulse radar would be more appropriate than ground penetrating radar.

The propagation velocity and amplitude of the recorded electromagnetic waveform can be analyzed to infer subsurface properties or changes in material properties. The general form of a propagating electromagnetic wave in one dimension is:

$$A_R = A_0 \exp(-\alpha x) \exp i\omega(t - x/v) \tag{11-59}$$

where A_R = recorded amplitude
t = time
x = distance traveled
v = phase velocity
A_0 = amplitude at $x = 0$ and $t = 0$
α = attenuation coefficient
ω = angular frequency
$i^2 = -1$

Neglecting dispersion, for nonmagnetic materials such as concrete and for the high frequencies that are typically used in the assessment of concrete structures (~1 GHz), the phase velocity v primarily depends on the relative dielectric constant (ε) of the material:

$$v \approx \frac{c}{\sqrt{\varepsilon}} \tag{11-60}$$

where c is the speed of light in vacuum (3×10^8 m/s). The relative dielectric constant of air is 1, of water is approximately 80, and of dry natural or structural materials is in the range of 4 to 10. Thus, relative dielectric constants of both natural and structural materials vary primarily as a function of water content, but are also affected by variations in porosity, mix proportions, temperature, pore fluid solution, and particle or pore shape.

At interfaces between materials of different electromagnetic properties, part of the signal travels through the interface to the next layer, and the rest of the signal returns to the surface and is recorded by the receiving antenna. The magnitude of the amplitude that is returned to the surface is a function of the electromagnetic impedance contrast of the two materials; the greater the electromagnetic impedance, the stronger is the GPR reflection. For low loss materials at high GPR frequencies, the reflection coefficient (R) quantifies the magnitude of the returned signal amplitude of the reflected wave as a function of the values of the upper (ε_1) and lower (ε_2) material relative dielectric constants:

$$R = \frac{\sqrt{\varepsilon_1} - \sqrt{\varepsilon_2}}{\sqrt{\varepsilon_1} + \sqrt{\varepsilon_2}} \tag{11-61}$$

Using Eq. (11-61), the amplitude of the recorded event can be analyzed to obtain information about boundaries when there are changes in dielectric constant, such as at the concrete-steel interface.

A GPR system consists of an impulse generator, which repeatedly sends a particular voltage and frequency source to a transmitting antenna (T_x). A signal propagates from the transmitting antenna through the concrete and is reflected, scattered, and attenuated; the receiving antenna (R_x) subsequently records the modified signal. The reflected (R) signal returned to the receiving antenna typically is displayed as a waveform of voltage variations that is a function of time. The most commonly used GPR acquisition mode is common-offset reflection, which involves collecting one trace per surface location from a transmitter-receiver antenna pair that has a fixed separation distance (S) as the antenna is moved along the ground/concrete surface or from an elevated position above the ground surface (Fig. 11-36). Data collected in this mode are typically displayed as wiggle-trace profiles, with the distance that the radar antenna pair has traversed on the horizontal axis and two-way signal travel-time on the vertical axis. In addition to reflected energy, other events arrive at the receiving antenna from the transmitting antenna, including the airwave (A) and the ground wave (G), as shown in Fig. 11-36. Using a standard display, GPR data often provide a type of "cross section" of the near subsurface along the GPR traverse that illustrates boundaries having different electromagnetic properties. The depth of penetration of the GPR signal is a function of the radar system parameters, target parameters, and the electromagnetic properties of the materials being investigated, and can be calculated using a radar range equation.

GPR is attractive as a field tool because it is nondestructive, has high spatial resolution, and the data can be acquired rapidly and even under traffic conditions.

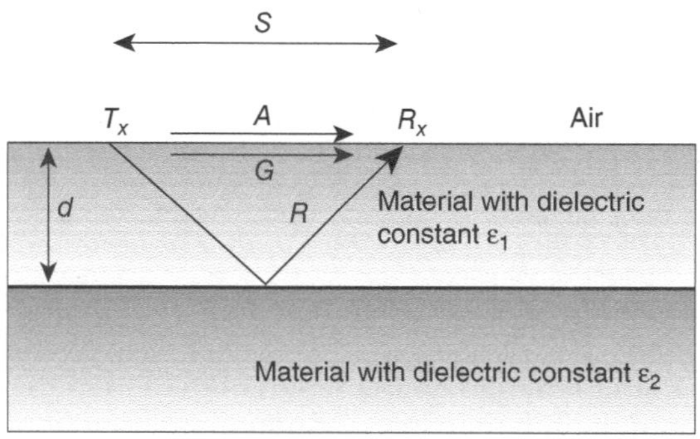

FIGURE 11-36 Schematic showing surface GPR common offset arrivals between GPR transmitting antenna (T_x) and receiving antenna (R_x), which are separated by a distance S. Arrivals include the airwave (*A*), ground wave (G), and reflected events (*R*) that occur when there is a contrast of dielectric constants (κ) across a boundary, such as at the rebar-concrete interface. (After ACI 228.2R-98, *Nondestructive Test Methods for Evaluation of Concrete Structures.*)

The large difference between the electromagnetic properties of the concrete and the steel reinforcing bar generates a large reflection coefficient at the concrete/steel interface as described by Eq. (11-61). As the antenna pair is directly over the axis of the reinforcing bar, the travel time of the signal to the bar is shortest; as the antenna pair moves away from the rebar axis, the travel time increases. This allows the determination of the location of the reinforcing bars since the movement of the antenna pair over the reflector yields a reflection hyperbola on the radar profile image, as illustrated in Fig. 11-37. The air and ground wave arrivals (refer to Fig. 11-36) have been suppressed by processing, and the radar profile predominantly displays the hyperbolic response of the signal travel time as the transmitter/receiver pair traverse over the embedded reinforcing bars.

Besides being able to locate the reinforcement, GPR can detect delaminations that may occur in a concrete structure due to corrosion of the steel reinforcement. As discussed before, delaminations can be located by sounding methods, usually by dragging chains over the surface or by hitting the surface with a hammer. These methods are adequate when the area to be studied is limited and it does not suffer by closure while taking the measurements. In concrete pavements, for instance, closing a lane often causes undesirable consequences to the traffic flow. Under such circumstances, GPR becomes an attractive tool because the measurements are fast and often cause little or no disruption to traffic. The method used in the inspection of a concrete deck for the presence of delamination is similar to the one described for the location of the reinforcing bars. The delamination will manifest itself by generating an additional echo as indicated schematically in Fig. 11-38.

The relative dielectric constant of water is 80; however, when it reacts with portland cement to produce hydration products and becomes chemically bound its dielectric constant drops down to approximately 5. The evolution of cement hydration produces significant changes in the dielectric constant of concrete; therefore, radar can be used as a powerful method of monitoring the degree of hydration in concrete structures. Using the same principle, radar can also be used to determine the water content of fresh concrete.

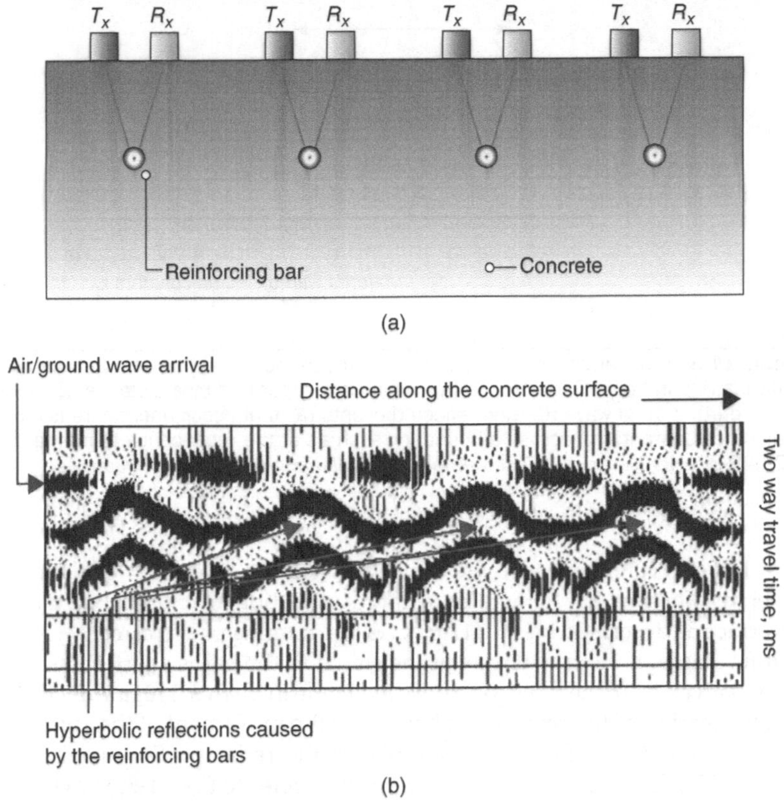

FIGURE 11-37 (*a*) Schematic showing collection of surface GPR data over four embedded reinforcing bars. T_x is transmitter and R_x is receiving antenna. (*b*) Example GPR profile collected over the four embedded reinforcing bars using a 1200-MHz GPR system. (From Hubbard, S.S., J.Y. Zhang, P.J.M. Monteiro, J.E. Peterson, and Y. Rubin, Experimental Detection of Reinforcing Bar Corrosion Using Nondestructive Geophysical Techniques, *ACI Mat. J.*, Vol. 100, No. 6, pp. 501–510, 2003.)

11.9.3 Infrared Thermography

Imperfections and localized zones of high porosity have different thermal properties than the rest of the concrete, so under heat flux they will produce zones with different temperatures than the surrounding concrete. Chapter 13 discusses how to compute the temperature distribution in a solid using finite element models. It also includes a discussion on the three fundamental modes of heat transfer: conduction, convection, and radiation. In conduction, a heat flux develops if there is a temperature difference between two areas of the body; heat will flow from the hotter to the cooler zone. Voids and imperfections have low thermal conductivity, therefore their presence creates a thermal anomaly in the temperature distribution of the concrete structure under heat flux, as shown in Fig. 11-39. The condition of heat entering the concrete structure is typical of a warm day and the condition of heat flowing from the concrete structure happens during a cold night, for instance. Weil[31] suggests that the best thermal contrast is obtained two or three hours after sunrise or sunset. In cases where the concrete structure does not experience

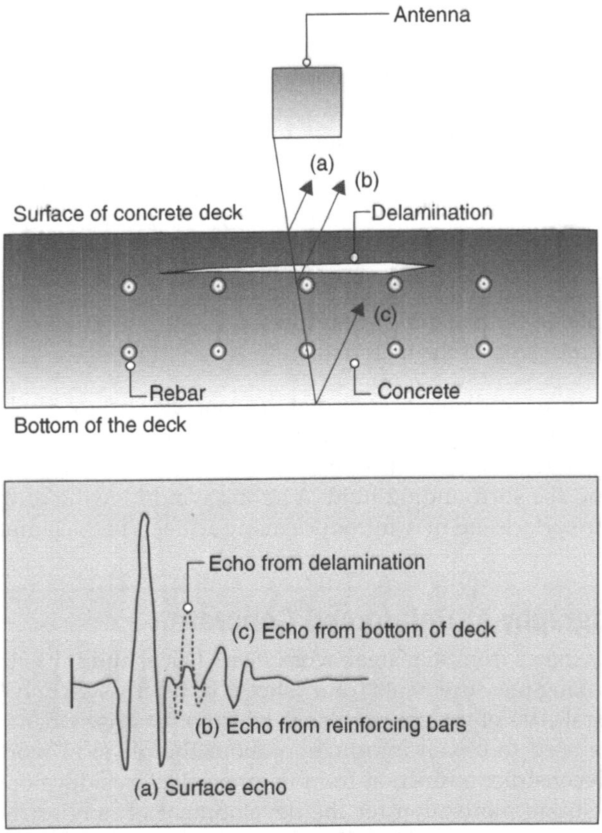

FIGURE 11-38 Schematic showing GPR echoes from a reinforced concrete deck. The delamination present in the top mat of reinforcing bars originates additional reflection. (From Clemena, G.G., V.M. Malhotra, and N.J. Carino, eds., Short-Pulse Radar Methods, in *Handbook on Nondestructive Testing of Concrete*, p. 260, CRC Press, Boca Raton, FL, 1991.)

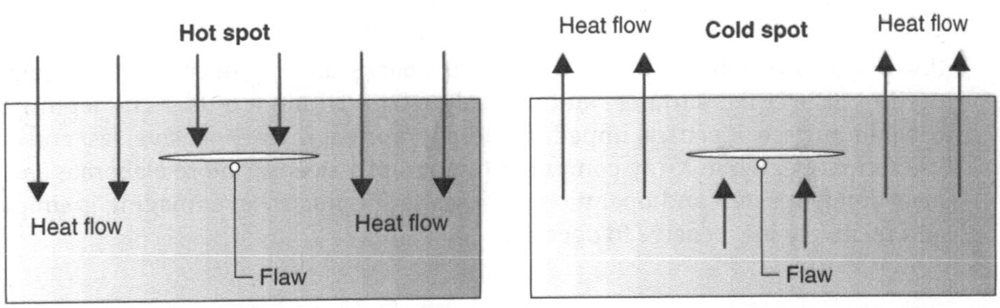

FIGURE 11-39 Effect of a flaw or imperfection on the temperature distribution of a concrete slab. When the heat flows inward, such as in a hot day, and since the flaw does not conduct heat well the temperatures above the flaw will be higher. The opposite happens when the heat flow is outward, such as during a cold night when the concrete is hot, and the surface above the flaw will remain colder than the surrounding concrete. (After ACI 228.2R-98, *Nondestructive Test Methods for Evaluation of Concrete Structures.*)

heat flux because it is not exposed to sunlight, it becomes necessary to artificially heat the structure. The influence of the flaws in affecting the temperature at the surface will be negligible if they are too small compared to the depth of the structure or if they are located too deep inside the concrete.

Due to thermal radiation, materials emit electromagnetic waves according to the Stefan-Boltzmann law:

$$\phi = \varepsilon \sigma T^4 \tag{11-62}$$

where ϕ = rate of energy radiation per unit area
σ = Stefan-Boltzmann constant
ε = emissivity of the material

The temperature gradients observed at the concrete surface can be imaged using infrared thermographic scanning systems.

Convection is the heat transfer mechanism that develops between the surface of the material and the surrounding fluid. A strong wind blowing at the surface of concrete will cool the surface and may introduce many artifacts in the temperature measurement.

11.10 Tomography of Reinforced Concrete

Tomography comes from the Greek word *tomos* (slice) and it has the goal of imaging an object by taking measurements from "slices" of its cross section. Tomography belongs to the general class of inverse problems, which often arise when inferences from observations are used to obtain information about the physical world. The mathematical theory to reconstruct a function from its projection was developed in 1917 by Radon. There is a strong motivation for the development of nondestructive mapping of the internal structure of a body, including intrinsic complexity and anisotropy of the phases, imperfections, cracks, inhomogeneties, and anisotropies. There are new, emerging technologies that have the potential to assess adequately concrete structures: computed tomography with X-rays or γ-rays, electrical impedance tomography, and backscattering microwave imaging. X-ray computed tomography shows promise in high-resolution (<5 μm) detection of cracks, and in locating rebars in moderate size structures. Electrical impedance tomography has the potential for locating rebars, water-filled fractures, and similar zones of high conductivity in concrete with the possibility of gaining information about corrosion around reinforcing bars. Backscattering microwave imaging can be used to inspect flat surfaces, and to locate rebars within a 6- to 7-cm depth from the concrete surface. Electrical impedance tomography is an inexpensive, fast and easy-to-use technique, while X-ray computed tomography needs a more elaborate and costly measurement setup and operation. Backscattering microwave imaging is simple, fast, and relatively inexpensive to operate.

11.10.1 X-Ray Computed Tomography

In computed tomography, the image of an object is reconstructed from projections of the object. Most commonly the projections are obtained by using penetrating X-rays, although other modalities for measuring projection data are also available. A radiation source (X-rays or γ-rays) is rotated a full 360° around the structure under inspection,

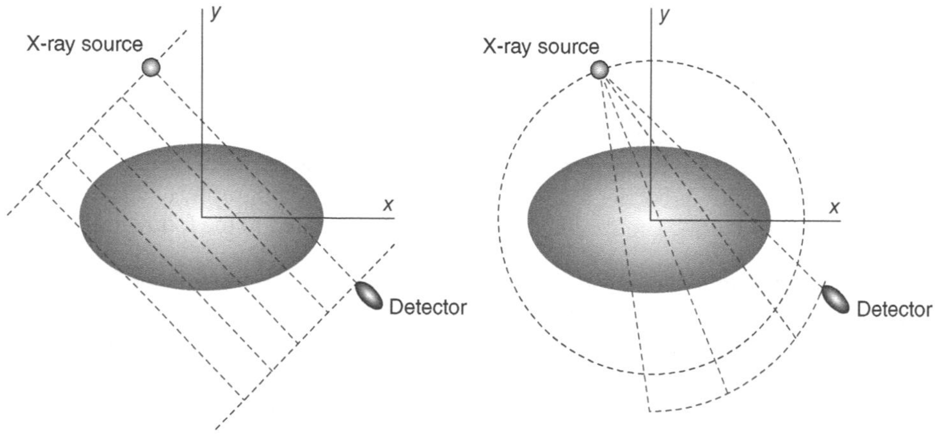

FIGURE 11-40 Parallel-beam and fan-beam geometry in X-ray CT.

The attenuation along the straight line in the parallel-beam geometry is mathematically described by the Radon transform, and in the fan-beam geometry by the X-ray transform. Each source-detector pair measures a line integral over the attenuation distribution.

and at each source position, the attenuation of the radiation penetrating through the material is measured with a linear array, or a two-dimensional array of detectors at the opposite side. For each source position, the projection of the attenuation is the measured parameter. Computerized reconstruction algorithms are then used to reconstruct a cross sectional image of the density distribution (or the attenuation coefficient distribution) of the structure at the detector-source plane.

There are two basic configurations: parallel beam and fan beam geometry (Fig. 11-40), the former having little practical use. All medical scanners, for example, are based on fan beam geometry where an X-ray source is approximated as a point source, with an array of detectors, all of which are simultaneously exposed to the radiation. The attenuation of the X-rays between the X-ray source and each detector is expressed as a line integral over the penetrated material, and a projection consists of all the line integrals from a specific location of the X-ray source. Unfortunately, fan beam geometry is described by the X-ray transform, which has cumbersome mathematical properties, while the parallel beam geometry is described by the Radon transform, which has elegant mathematical properties.

The objective of X- or γ-ray CT is to reconstruct object absorption cross sections from projections through the object. The beam intensity is attenuated as it transverses the object. The amount of attenuation is related to the atomic number of the phases distributed in the object as well as their density distribution. For a well-collimated monoenergetic beam, the observed transmitted intensity I at the detector is given by:

$$I = I_0 e^{-\int_0^t f(t)dt} \tag{11-63}$$

where t indicates the thickness of the attenuating material that has been penetrated and I_0 is the incident radiation intensity. The attenuation properties of the material are described by the product $f(t) = \mu(t)\ r(t)$, where $\mu(t)$ is the mass-attenuation coefficient (m^2/kg) and $\rho(t)$ is the density (kg/m^3).

Laboratory tests show that X-ray tomography can be a powerful tool in detecting the location of reinforcing bars, the presence of voids, and the existence of discontinuities. The tomographic image reconstruction requires the object to be X-rayed from every angle around it. In many potential applications it is impossible to have the X-ray source and the array of X-ray detectors rotate a full circle around the object. However, due to significant developments in reconstruction algorithms limited angle geometry does not need to be a serious barrier to the application of X-ray tomography to inspection of certain concrete structures. The special resolutions are associated to the following parameters[32]:

Finite detector width and radiation source spot size. The attenuation takes place along a radiation beam of finite width, instead of, ideally, a line.

Data sampling rate. Improvement in the projection data sampling requires an increase in the number of line integrals measured per projection, and an increase in the number of projections per 360° around the object.

Uniformity of the angular sampling. Gaps in the angular coverage of projection data cause visible artifacts in the reconstructed image.

Scattering of radiation into an off-beam detector, giving rise to measurement noise. Collimators in front of the detectors can be used to reduce false measurements, but at the expense of a reduced number of detectors per projection.

Total photon count. Increased radiation source strength produces an increased intensity of the transmitted radiation, and thus better counting statistics, and signal-to-noise ratio. X-ray tubes generate a higher intensity beam than isotope sources.

Energy of the radiation. The energy of the radiation should be matched to the attenuation properties, and the thickness of the object. X-ray tubes generate a spectrum of X-ray energies, while isotope sources emit essentially monoenergetic γ-radiation when beam hardening (higher attenuation of low-energy vs. higher-energy radiation) shifts toward higher energies the energy spectrum of the radiation penetrating through the medium may cause a deterioration of the image quality, especially when imaging large objects.

11.10.2 Collapsing a Three-Dimensional World into a Flat Two-Dimensional Image

Visual perceptions can be misleading. If you have not seen a rhino before, you may describe the animal shown in Fig. 11-41*a* as having two opposing horns. However, if you are a zoo enthusiast you will automatically realize that two rhinos are in the picture and that the peculiar image is a consequence of representing the three-dimensional world into a flat two-dimensional image. Nevertheless, this automatic mental processing can lead to erroneous conclusions as shown in Fig. 11-41*b* where instead of the obvious conclusion that two turtles should account for the presence of two heads, this particular turtle had indeed two heads. To capture the three dimensionality of the observation, microscopists use stereo photographs and medical doctors developed tomography.

Similar situations can rise when X-rays are used. X-ray radiography was one of the most important diagnostic tools for medical diagnosis developed in the last century. In the radiographic images, however, it is difficult to distinguish small changes in a complex structure because this technique compresses a three-dimensional object into a

(a) (b)

FIGURE 11-41 The human mind has the capacity of identifying complex information from two-dimensional images without problems, however, the lack of depth in the images can lead to false interpretations. Two rhinos are present in Fig. 11-41*a*, however, only one turtle born with two heads is present in Fig. 11-41*b*. (*Photograph courtesy of Casey Lazik.*)

two-dimensional image. A major breakthrough was achieved in 1972, when the first X-ray computed tomography equipment became available in a clinical environment. The importance of X-ray tomography was recognized by the 1979 Nobel Prize given to Hounsfeld and Cormack for their experimental and theoretical work. To show the difference between radiography and tomography applied to reinforced concrete, a $13.5 \times 15 \times 9$-cm concrete block was cast with several reinforcing bars ranging from 0.4 to 1.5 cm in diameter embedded mostly in the vertical direction. Figure 11-42 compares the radiographic and CT images of the concrete block with multiple reinforcing bars. In both images the reinforcing bars are clearly visible. However, the CT image reveals the spatial location and diameters of the various rebars. The gray level distribution in the tomograms reveals different linear attenuation values representative of the background (dark gray), concrete (medium gray), and rebar (light gray to white).

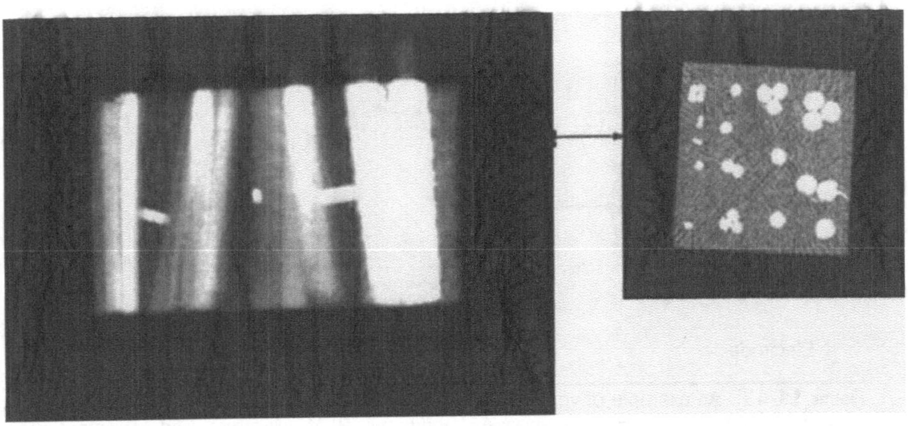

FIGURE 11-42 Radiographic (left) and CT (right) image of a concrete block with reinforcing bars.

11.10.3 Electrical Resistance Tomography

Electrical resistance tomography (ERT) is an imaging modality used to estimate the internal conductivity distribution whereby current is injected into an object and voltage measurements are then acquired from the object boundary. The use of ERT is considered a diffusive modality because if a current is injected into the target volume between any two points, the electric current (density) spreads into the whole volume. As a consequence, reconstruction of the image is more complicated than, for example, when using X-ray tomography, because X-rays travel in straight line, and one measurement carries information from a very restricted subset of the target. With ERT, its reconstruction problem is an ill-posed inverse problem. The practical definition of inverse ill-posed problems is that they are extremely sensitive, even to moderately small measurement and modeling errors.

In ERT, alternating currents, I_ℓ, are injected through electrodes attached on the surface of the object, and the resulting voltages, V_k, are measured between different pairs of electrodes (Fig. 11-43).

The most accurate model for ERT is the complete electrode model, where the target volume under investigation is denoted by Ω, its exterior boundary by $\partial\Omega$, and the patch under the ℓth electrode by e_ℓ. The complete electrode model follows the Poisson equation:

$$\nabla\cdot(\sigma\nabla u)=0, \qquad \bar{r}\in\Omega \tag{11-64}$$

and the following boundary conditions:

$$\int_{e_\ell}\sigma\frac{\partial u}{\partial \bar{n}}dS=I_\ell, \quad \bar{r}\in e_\ell,\ \ell=1,\ldots,L \tag{11-65}$$

$$\sigma\frac{\partial u}{\partial \bar{n}}=0, \quad \bar{r}\in\partial\Omega\setminus\cup_{\ell=1}^{L}e_\ell \tag{11-66}$$

$$u+z_\ell\sigma\frac{\partial u}{\partial \bar{n}}=U_\ell, \quad \bar{r}\in e_\ell,\ \ell=1,\ldots,L \tag{11-67}$$

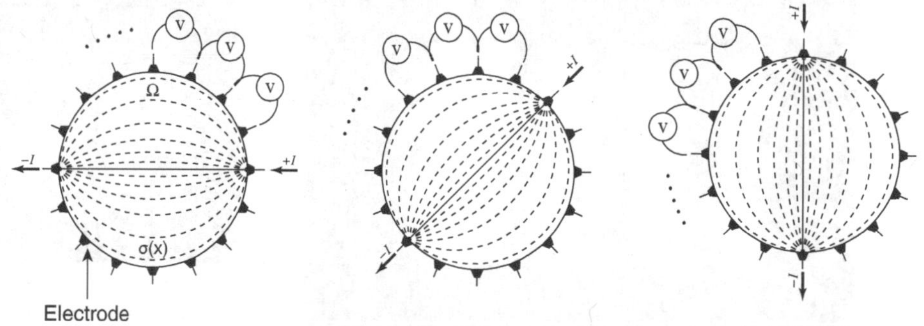

FIGURE 11-43 An example of voltage measurements corresponding to three current injections in ERT. (After Karhunen, K., A. Seppänen, A. Lehikoinen, P.J.M. Monteiro, and J.P. Kaipio, Electrical Resistance Tomography Imaging of Concrete, *Cem. Conc. Res.*, Vol. 40, pp. 137–145, 2010.)

where $\sigma = \sigma(\bar{r})$ is the conductivity, $\bar{r} \in \Re^3$ is the spatial coordinate, $u = u(\bar{r})$ is the electric potential in the body, U_ℓ is the potential on the ℓth electrode, z_ℓ is the contact impedance between the ℓth electrode and the object, and $\bar{n}$ is an outward unit normal. The boundary conditions [Eqs. (11-65) to (11-67)] can be interpreted in the following way: $\sigma \frac{\partial u}{\partial \bar{n}}$ is the current density through the boundary; therefore, Eq. (11-65) indicates that the integral of the current density over the ℓth electrode is equal to total current I_ℓ through the electrode. Furthermore, the boundary condition defined in Eq. (11-66) implies that there are no currents through the electrode-free boundary. Equation (11-67) represents the boundary condition for the electric potential. If the electric contact between an electrode and the target surface were perfect, the electric potential u under each electrode e_ℓ would be constant and equal to U_ℓ. The contact impedance, z_ℓ, however, causes a potential drop equal to $z_\ell \sigma \frac{\partial u}{\partial \bar{n}}$ between the electrode and the target surface.

In addition, currents I_ℓ need to satisfy the charge conservation condition

$$\sum_{\ell=1}^{L} I_\ell = 0 \qquad (11\text{-}68)$$

and the potential reference level has to be fixed, for example, by writing

$$\sum_{\ell=1}^{L} U_\ell = 0 \qquad (11\text{-}69)$$

As stated earlier, the inverse problem of ERT, that is, reconstructing the internal conductivity distribution and the contact impedances given the boundary voltage measurements, is an ill-posed inverse problem. This implies that the solution of the inverse problem is not unique, and it is sensitive to modeling errors and measurement noise. Hence, the solution of the inverse problem requires special methods.

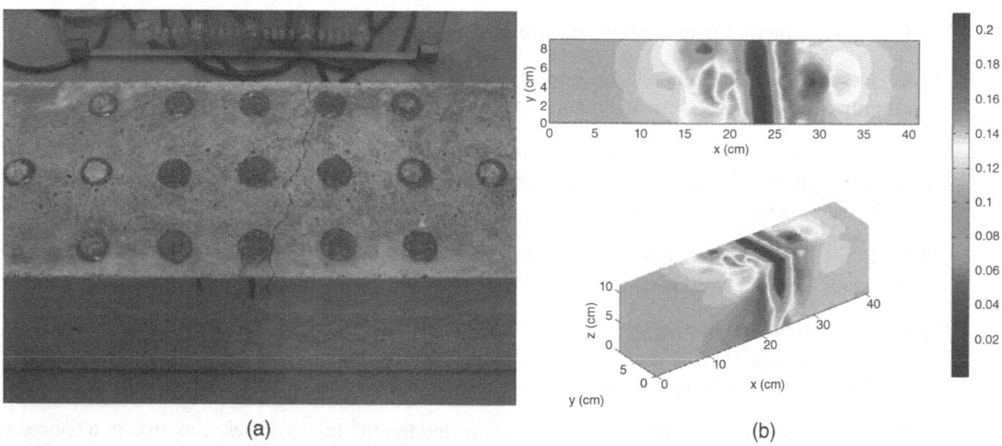

FIGURE 11-44 (*a*) Photo of a specimen taken after ERT measurements. The dark circles on the top surface are the remaining gel used to attach the electrodes; (*b*) surface plots of the reconstructed conductivity distribution of the specimen. (After Karhunen, K., A. Seppänen, A. Lehikoinen, J. Blunt, J.P. Kaipio, and P.J.M. Monteiro, Electrical Resistance Tomography for Assessment of Cracks in Concrete, *ACI Mater. J.*, Vol. 107, No. 5, Sep.–Oct. 2010.)

The reconstruction methods in ERT can be divided in deterministic and statistical methods. The deterministic approach replaces the ill-posed inverse problem by a well-defined problem that is stable and has a unique solution. This procedure is called *regularization*. The statistical approach requires developed an explicit prior model for the target, and the solution of the inverse problem is sought via the a *posteriori* distribution, which combines the prior and measurement models and the information given by the measurements.

Karhunenet et al.[33] studied the capability of ERT for crack detection and characterization in more realistic geometries in which the sensor array was attached to one planar surface of concrete. As a demonstration, they analyzed a specimen where an apparent crack—induced by three-point bending—propagated vertically at least through the half of the beam. The ERT reconstruction was in agreement with the visual inspection of the specimen; in the reconstruction, a nonconductive region reaching from the top surface to middle of the beam was detected. The conductivity under this region was also lower than the background conductivity. This was most likely due to microcracking that occurred in this domain. Note that the nonconductive area on the top surface is slightly slanted, which is also in agreement with the photo of the specimen (Fig. 11-44).

Test Your Knowledge

11.1 List four nondestructive test method for the evaluation of concrete quality. Which one can be used as a direct substitute for determining the compressive strength of concrete?

11.2 What are the principles behind the following test procedures: Schmidt hammer test, Windsor probe test, pullout test, pulse velocity test? Explain which you would recommend for deciding the formwork removal time.

11.3 Your company has been hired to perform the assessment of the damage of a building that had been exposed to a fire for 1 h. Write a memo describing the protocol of the site investigation including what nondestructive tests should be used to determine the best repair strategy for the reinforced concrete structure.

11.4 Explain the maturity concept, its application, and its limitations.

11.5 Describe the difference between longitudinal and transverse waves. Which one is more commonly used to detect damage in concrete? Why?

11.6 Unlike steel, concrete has large inhomogeneities such as aggregates and flaws. What will be the consequences of such inhomogeneities in the selection of the wave frequency to be used for a nondestructive test on concrete?

11.7 In Sec. 11.2, we developed an Eq. (11-19) to determine the thickness of a low-velocity horizontal layer. Derive a similar equation for a low-velocity layer with dipping angle i.

11.8 Can you use the method described in Sec. 11.2 to determine the thickness of the frozen layer in a concrete exposed to low temperature? What about the thickness of the dry layer in a concrete exposed to very low relative humidity? Explain your answer.

11.9 Explain how the impact method can determine the thickness of delamination in a concrete pavement.

11.10 Devise a scheme to determine the location of a crack using acoustic emission.

11.11 Why do resistivity measurements in a structure provide useful information on the potential corrosion resistance of reinforced concrete? List some limitations of this method.

11.12 Explain why the determination of the corrosion rate in reinforced concrete is a critical parameter in all modern models for the prediction of life cycle.

11.13 Describe the difference between X-ray radiography and X-ray tomography.

11.14 In medical tomography, the sensors completely surround the patient. In many civil engineering applications it is not feasible to completely surround the structure. What type of problems will that generate?

References

1. Bungey, J.H., *The Testing of Concrete in Structures*, Chapman and Hall, New York, 1982.
2. Malhotra, V.M., *In Situ/Nondestructive Testing of Concrete*, ACI SP-82, American Concrete Institute, Detroit, pp. 1–16, 1984.
3. Hansen, F., and E.J. Pedersen, Maturity Computer for Controlled Curing and Hardening of Concrete, *Nordisk Betong*, Vol. 1, p. 19, 1977.
4. Carino, N.J., The Maturity Method, in *Handbook on Nondestructive Testing of Concrete*, Malhotra, V.M., and N.J. Carino, eds., Chap. 5, pp. 101–146, CRC Press, Boca Raton, FL, 1991.
5. Ollivier, J.P., M. Massat, and L. Parrot, Parameters Influencing Transport Characteristics, *Performance Criteria for Concrete Durability*, Kropp, J., and H.K. Hilsdorf, eds., Rilem Report 12, pp. 33–96, E & FN Spon, London, 1995.
6. Geiker, M., H. Grube, T. Luping, L.O. Nielsen, and C. Andrade, Laboratory Test Methods, *Performance Criteria for Concrete Durability*, Kropp, J., and H.K. Hilsdorf, eds., Rilem Report 12, pp. 213–279, E & FN Spon, London, 1995.
7. Paulmann, K., and C. Molin, On-Site Test Methods, *Performance Criteria for Concrete Durability*, Kropp, J., and H.K. Hilsdorf, eds., Rilem Report 12, pp. 258–279, E & FN Spon, London, 1995.
8. Figg, J.W., Methods of Measuring the Air and Water Permeability of Concrete, *Concr. Res.*, Vol. 25, pp. 213–219, 1973.
9. Bolt, B.A., Earthquakes and Geological Discovery, *Scientific American Library*, New York, 1993.
10. Burger, H.R., Exploration Geophysics of the Shallow Subsurface, Prentice-Hall, Englewood Ciffs, NJ, 1992.
11. Sansalone, M., Impact-Echo: The Complete Story, *ACI Struc.*, Vol. 94, No. 6, pp. 777–786, 1997.
12. Sansalone, M., and N.J. Carino, Laboratory and Field Study of the Impact-Echo Method for Flaw Detection in Concrete, in *Nondestructive Testing of Concrete*, Lew, H.S., ed., SP-112, American Concrete Institute, Detroit, March 1989.
13. Krstulovic-Opara, N., R.D. Woods, and N. Al-Shayea, Nondestructive Testing of Concrete Structures Using the Rayleigh Wave Dispersion Method, *ACI Mat. J.*, Vol. 93, pp. 75–86, 1996.
14. Nazarian, S., and M.R. Desai, Automated Surface Wave Method: Field Testing, *J. Geotech. Eng.*, Vol. 119, pp. 1094–1111, 1993.
15. Ohtsu, M., The History and Development of Acoustic Emission in Concrete Engineering, *Concr. Res.*, Vol. 48, pp. 321–330, 1996.
16. Maji, A.K., C. Ouyang, and S.P. Shah, Fracture Mechanisms of Quasi-Brittle Materials Based on Acoustic Emission, *J. Mat. Res.*, Vol. 5, pp. 206–217, 1990.

17. Berthelot, J.M., and J.L. Robert, Modeling Concrete Damage by Acoustic Emission, *J. Acoust. Emission*, Vol. 6, pp. 43–60, 1987.
18. Suaris, W., and J.G.M. Van Mier, Acoustic Emission and Source Location in Concrete Subjected to Mixed-Mode Loading, in *Fracture and Damage of Concrete and Rock, Proceedings of the Second International Conference on Fracture and Damage of Concrete and Rock*, Vienna, Rossmanith, H.P., ed., E & FN Spon, London, pp. 157–165, 1993.
19. Li, Z., F. Li., A. Zdunek, E. Landis, and S.P. Shah, Application of Acoustic Emission Technique to Detection of Reinforcing Steel Corrosion in Concrete, *ACI Mat. J.*, Vol. 95, pp. 68–76, 1998.
20. Yuyama, S., T. Okamoto, O. Minemura, N. Sakata, and K. Maruyama, Acoustic Emission Applications to an Arch Dam Under Construction, *J. Acoust. Emission*, Vol. 16, pp. S75–S84, 1998.
21. Ward, S.H., Resistivity and Induced Polarization Methods, *Geotechnical and Environmental Geophysics*, Ward, S.H., ed., Vol. 1, SEG, Tulsa, OK, pp. 147–190, 1990.
22. Monteiro, P.J.M., F. Morrison, and W. Frangos, Non-Destructive Measurement of Corrosion State of Reinforcing Steel in Concrete, *ACI Mat. J.*, Vol. 95, No. 6, pp. 704–709, Nov.–Dec. 1998.
23. CEB-192, Diagnosis and Assessment of Concrete Structures—State-of-the-Art Report. Bulletin D'Information, Case Postale 88, CH-1015 Lausanne, 1989.
24. Mancio, M., J.R. Moore, Z. Brooks, P.J.M. Monteiro, and S.D. Glaser, Instantaneous In-Situ Determination of Water-Cement Ratio of Fresh Concrete, *ACI Mat. J.*, Vol. 107, pp. 587–593, 2010.
25. Stern, M., and A.L. Geary, Electrochemical Polarization, I. A Theoretical Analysis of the Shape of Polarization Curves, *J. Electrochem. Soc.*, pp. 56–63, 1957.
26. Gabrielli, G., and M. Keddam, Review of Applications of Impedance and Noise Analysis to Uniform and Localized Corrosion, *Corrosion*, Vol. 48, pp. 794–810, 1992.
27. John, D.G., P.C. Searson, and J.L Dawson, Use of AC Impedance Technique in Studies on Steel in Concrete in Immersed Conditions, *Br. Corros. J.*, Vol. 16, No. 2, pp. 102–106, 1981.
28. Ford, S.J., and T.O. Mason, Monitoring the Corrosion of Reinforcing Steel in Cement-Based Systems Using Impedance Spectroscopy, in *Mechanics of Chemical Degradation of Cement-Based Systems*, Scrivener, K.L., and J.F. Young, eds., pp. 125–133, 1997.
29. Christensen, B.J., R.T. Coverdle, R.A Olson, S.J. Ford, H.M. Jennings, and T.O. Mason, Impedance Spectroscopy of Hydrating Cement-Based Materials: Measurement, Interpretation, and Application, *J. Am. Ceram. Soc.*, Vol. 77, No. 11, pp. 2789–2891, 1994.
30. Clemena, G.G., Short-Pulse Radar Methods, in *Handbook on Nondestructive Testing of Concrete*, Malhotra, V.M., and N.J. Carino, eds., pp. 253–274, CRC Press, Boca Raton, FL, 1991.
31. Weil, G.J., Infrared Thermographic Techniques, *Handbook on Nondestructive Testing of Concrete*, Malhotra, V.M., and N.J. Carino, eds., CRC Press, Boca Raton, FL, 1991.
32. Monteiro, P.J.M., C.Y. Pichot, and K. Belkebir, Computer Tomography of Reinforced Concrete, Chap. 12, *Materials Science of Concrete*, American Ceramics Society, Westerville, OH, 1998.
33. Karhunen, K., A. Seppänen, A. Lehikoinen, J. Blunt, J.P. Kaipio, and P.J.M. Monteiro, Electrical Resistance Tomography for Assessment of Cracks in Concrete, *ACI Mat. J.*, Vol. 107, No. 5, Sep.–Oct. 2010.

Suggestions for Further Readings

General

ACI 228.1R-03, In-Place Methods to Estimate Concrete Strength, *ACI Manual of Concrete Practice*, American Concrete Institute, Farmington Hills, MI, 2003.

ACI 228.2R-98, Nondestructive Test Methods for Evaluation of Concrete Structures, *ACI Manual of Concrete Practice*, American Concrete Institute, Farmington Hills, MI, 2005.

Bungey, J.H., and S.G. Millard, *Testing of Concrete in Structures*, 3d ed., Blackie Academic & Professional, Glasgow, 1996.

Carino, N.J., in *Concrete Technology: Past, Present, and Future*, ACI SP-144,1, American Concrete Institute, Detroit, MI, 1994.

Carino, N.J., Nondestructive Test Methods, *Concrete Construction Engineering Handbook*, Nawy, E.G., ed., Chap. 19, CRC Press, Boca Raton, FL, 1997.

Galan, A., Combined Ultrasound Methods of Concrete Testing, *Developments in Civil Engineering*, Elsevier, New York, 1990.

Maierhofer, C., H.W. Reinhardt, and G. Dobmann, eds., *Non-Destructive Evaluation of Reinforced Concrete*, Vols. 1 and 2, Woodhead Publishing Limited, Cambridge, 2010.

Malhotra, V.M., and N.J. Carino, eds., *Handbook on Nondestructive of Concrete*, CRC Press, Boca Raton, FL, 2004.

Newman, J., and B.S. Choo, eds., *Advanced Concrete Technology: Testing and Quality*, Butterworth-Heinemann, Oxford, 2003.

Pessiki, S., and L. Olson, eds., *Innovations in Nondestructive Testing of Concrete*, SP-168, American Concrete Institute,1, Farmington Hills, MI, 1997.

Acoustic Emission

Mindess, S., Acoustic Emission Methods, *Handbook on Nondestructive Testing of Concrete*, Malhotra, V.M., and N.J. Carino, eds., CRC Press, Boca Raton, FL, 1991.

Electrochemical Methods

Andrade, C., Monitoring Techniques, in *Corrosion of Steel in Concrete*, Schiessl, P., ed., Report of the Technical Committee 60-CSC Rilem, Chapman and Hall, London, pp. 79–95, 1988.

Bentur, A., S. Diamond, and N.S. Berke, *Steel Corrosion in Concrete: Fundamentals and Civil Engineering Practice*, E & FN Spon, London, 1997.

Bertolini, L., B. Elsener, P. Pedeferri, and R. Polder, *Corrosion of Steel in Concrete: Prevention, Diagnosis, Repair*, Wiley-VCH, Weinheim, 2004.

Tomography

Garboczi, E.J., Three-dimensional mathematical analysis of particle shape using X-ray tomography and spherical harmonics: Application to aggregates used in concrete, *Cement and Concrete Research*, 32, 1621–1638, 2002.

Monteiro, P.J.M., C.Y. Pichot, and K. Belkebir, Computer Tomography of Reinforced Concrete, Chap. 12, *Materials Science of Concrete*, American Ceramics Society, Westerville, OH, 1998.

PART III

Advances in Concrete Technology

CHAPTER 12

Special Types of Concrete

Preview

Conventional portland-cement concrete mixtures suffer from certain deficiencies. Attempts to overcome these deficiencies have resulted in the development of special concrete types that are described in this chapter. For each concrete type, general considerations that led to its development, materials and mix-proportions, typical properties, and applications are discussed.

Ordinary concrete, made with natural aggregate, has a low strength-weight ratio compared to steel. This places concrete at an economic disadvantage when designing structural members for tall buildings, long-span bridges, and floating structures. There are three ways to address this problem, and all have found commercial application. According to the first approach, the density or the unit weight of concrete can be reduced by substituting lightweight aggregate in place of conventional aggregate. Lightweight aggregate made by calcination of clay or shale is commonly used to produce structural lightweight concrete that has about one-third less unit weight than conventional concrete. In accordance with the second approach, the strength of concrete can be raised substantially. High-strength concrete with compressive strengths ranging from 60 to 120 MPa is now commercially available in many metropolitan areas. Such high strengths have been made possible since 1970s with the advent of superplasticizers or high-range water reducing admixtures. The third approach, which is a relatively recent development, combines the first two approaches. It involves the use of high-strength lightweight aggregate particles in superplasticized mixtures to produce high-strength, lightweight concrete.

Superplasticized concrete mixtures, on account of their low water-to-cementitious ratio, generally perform well on exposure to some aggressive environmental conditions that are known to cause early deterioration of ordinary concrete. Durable concrete mixtures are now being used in the construction of marine concrete structures for a service life of 100 to 150 years, compared to 40 to 50 years with conventional concrete. The use of superplasticizing admixtures is not limited to high-strength and high-durability applications. As structures become larger and more complex, massive placements of highly workable concrete mixtures are required for the fabrication of heavily reinforced structural members. In fact, superplasticized concrete mixtures in combination with large proportion of fine mineral particles and viscosity-modifying chemical admixtures have been developed to produce self-consolidating concrete for special applications. The term high-performance concrete is used in some literature to refer to the family of special concrete types, which are superior to conventional concrete in one or more properties such as workability, strength, and durability.

Restrained shrinkage on drying is frequently the cause of concrete cracking. This has long been recognized in the design and construction practice of relatively thin structural elements such as floor and pavement slabs. To counteract this problem, shrinkage-compensating concrete containing expansive cements or cement additives were developed and are being successfully used.

Poor impact resistance is yet another deficiency from which concrete suffers as a building material. This characteristic has been substantially improved by using the concept of microlevel reinforcement. Fiber-reinforced concrete mixtures containing steel, glass, or polypropylene fibers are being employed successfully in the structures where resistance to impact is important.

Imperviousness is an important materials' characteristic for durability to strong chemical solutions. Concrete mixtures containing polymers have been developed which show very low permeability and excellent chemical resistance. Overlays composed of such concrete mixtures are suitable for protection of reinforcing steel from corrosion in industrial floors and bridge decks.

Heavyweight concrete made with high-density minerals is about 50 percent heavier than normal concrete containing conventional aggregate. This type of concrete is being used for radiation shielding in nuclear power plants when limitations of usable space require a reduction in the thickness of the shield.

The large amount of voids in pervious concrete makes it highly permeable and allows the water to drain quickly through concrete rather than allowing the water to drain over the pavement. Other important environmental benefits of pervious concrete includes the possible reduction of urban heat island effect due to its permeable characteristics and relatively lighter appearance and reduction of irrigation water in urban areas by allowing the water to be stored in soils so that vegetation can be grown in adjacent areas.

Mass concrete for dams and other large structures has been around for some time, but methods selected to control the temperature rise have had a considerable influence on the construction technology during the last 70 years. Precooling of concrete materials has virtually eliminated the need for expensive postcooling operations and has made faster construction schedules possible. Dams are also being built with roller-compacted concrete, using ordinary earth-moving equipment, at high speeds and low costs.

12.1 Structural Lightweight Concrete

12.1.1 Definition and Specifications

Structural lightweight concrete is a structural concrete in every respect except that, for reasons of overall cost economy, the concrete is made with a cellular lightweight aggregate so that its unit weight is approximately two-thirds of the unit weight of concrete made with typical natural aggregate. Because the density of concrete, not the strength, is the primary objective, the specifications limit the maximum permissible unit weight of concrete. Also, because highly porous aggregates tend to reduce the concrete strength greatly, the specifications require a minimum 28-day compressive strength to ensure that the concrete is of structural quality.

ACI 213R-03, *Guide for Structural Lightweight-Aggregate Concrete*[1] defines *structural lightweight aggregate concretes* as those having a 28-day compressive strength in excess of

17 MPa (2500 psi) and a 28-day, an equilibrium density between 1120 and 1920 kg/m^3 (70 and 120 lb/ft^3). The concrete may consist entirely of lightweight aggregates or, a combination of lightweight and normal-weight aggregates. From the standpoint of workability and other properties, it is a common practice to use normal-weight sand as fine aggregate, and to limit the nominal size of the lightweight coarse aggregate to a maximum of 19 mm. According to ASTM C 330, fine lightweight and coarse lightweight aggregates are required to have a dry-loose weight not exceeding 1120 kg/m^3 (70 lb/ft^3) and 880 kg/m^3 (55 lb/ft^3), respectively. The specification also contains requirements with respect to grading, deleterious substances, and concrete-making properties of aggregate, such as strength, unit weight, drying shrinkage, and durability of concrete containing the aggregate. The ASTM C 330 Standard Specifications requirements for compressive and tensile strengths and unit weight of structural lightweight concrete are shown in Table 12-1.

12.1.2 Mix-Proportioning Criteria

For various reasons the absolute volume method, which is the basis of the ACI method of proportioning normal-weight concrete mixtures (Chap. 9), is not useful for designing lightweight concrete mixtures. First, the relation between strength and water-cement ratio cannot be effectively employed because it is difficult to determine how much of the mixing water in concrete will be absorbed by the aggregate. The difficulty is caused not only by the large amounts (10 to 20 percent) of water absorption by porous aggregate, but also by the fact that some aggregates continue to absorb water for several weeks. Therefore, reliable estimates of moisture deviation from the saturated-surface dry (SSD) condition and of the SSD bulk specific gravity are very difficult. Also, unlike normal-weight aggregates, the bulk specific gravity of lightweight aggregates can vary widely with grading.

Air-Dried, 28-Day Unit Weight, Max. [kg/m^3 (lb/ft^3)]	28-Day Splitting Tensile Strength, min. [MPa (psi)]	28-Day Compressive Strength, min. [MPa (psi)]
All Lightweight Aggregates		
1760 (110)	2.2 (320)	28 (4000)
1680 (105)	2.1 (300)	21 (3000)
1600 (100)	2.0 (290)	17 (2500)
Combination of Normal Sand and Lightweight Aggregate		
1840 (115)	2.3 (330)	28 (4000)
1760 (110)	2.1 (310)	21 (3000)
1680 (105)	2.1 (300)	17 (2500)

Note: The compressive strength and unit weight shall be the average of three specimens, and the splitting tensile strength shall be the average of eight specimens.

Source: Reprinted with permission from the *Annual Book of ASTM Standards*, Sec. 4, Vol. 04.02 ASTM, Philadelphia, PA.

Table 12-1 Requirements for Structural Lightweight Concrete

Workability considerations in freshly made lightweight-aggregate concrete require special attention because, with high-consistency mixtures, the aggregate tends to segregate and float on the surface. To control this tendency, it is often necessary to limit the maximum slump and to entrain air, irrespective of whether durability of concrete to frost action is needed. Approximately 5 to 7 percent air entrainment is generally required to lower the mixing water requirement while maintaining the desired slump and reducing the tendency for bleeding and segregation. Consequently, structural engineers' specifications for lightweight concrete usually include minimum permissible values for compressive strength, maximum values for unit weight and slump, and both minimum and maximum values for air content.

For the purpose of mix design, the compressive strength of lightweight aggregate concrete is usually related to cement content at a given slump rather than to the water-cement ratio. In most cases, the compressive strength at a given cement and water content can be increased by reducing the maximum size of coarse aggregate and/or partial replacement of lightweight fine aggregate with a good-quality natural sand.

With some lightweight aggregates, it may be possible to use the ACI 211.1 volumetric method for proportioning normal-weight concrete mixtures and adjust the proportions by trial and error until the requirements of workability of fresh concrete and physical properties of hardened concrete are satisfactorily met. It is often convenient to start with equal volumes of fine and coarse aggregate and to make adjustments as necessary for achieving the desired slump with minimum segregation. The problem of mix proportioning with lightweight aggregates and the methods to get around them are described in ACI Standard 211.2 (*Standard Practice for Selecting Proportions for Structural Lightweight Concrete*), which should be consulted when a more accurate procedure is needed.

12.1.3 Properties

Workability

The properties of fresh concrete made with lightweight aggregate and the factors affecting them are essentially the same as with normal-weight concrete. Due to the low density and rough texture of porous aggregate, especially in the crushed state, the workability of concrete needs special attention. In general, placing, compacting, and finishing lightweight aggregate concrete requires relatively less effort; therefore, even 50 to 75 mm slump may be sufficient to obtain workability that is similar to a 100- to 125-mm slump, normal-weight concrete.

With lightweight-aggregate concrete mixtures, high slump and overvibration are the two causes that are generally responsible for drawing the heavier mortar away from the surface where it is needed for finishing. This phenomenon, called *floating of the coarse aggregate*, is the reverse of what is experienced with normal-weight concrete (e.g., segregation resulting into an excess of mortar at the surface). ACI 213R-03 recommends a maximum of 125 mm slump for achieving a good surface in floors made with lightweight-aggregate concrete. Slump loss can be a serious problem when aggregate continues to absorb considerable amount of water after mixing. This problem can be controlled by batching the aggregate in a damp condition.

Unit Weight

Next to workability, unit weight and strength are the two properties generally specified for structural lightweight concrete. With given materials, it is generally desired to have the highest possible strength–unit weight ratio with the lowest cost of concrete. Specifications limit the air-dried unit weight of concrete to a maximum of 1840 kg/m^3 (115 lb/ft^3), but there is no minimum limit. However, it is a common experience that when a highly porous aggregate larger than 19 mm maximum size is used, the unit weight of concrete can be reduced to less than 1440 kg/m^3 (90 lb/ft^3), but the product may not be able to meet the minimum 17-MPa (2500-psi) compressive strength at 28 days that is required for structural quality lightweight concrete.

The use of normal sand to control the properties of hardened concrete tends to increase the unit weight, although this tendency is partially offset by the opposite effect of entrained air, which is invariably prescribed for improving the workability. Most structural lightweight concretes weigh between 1600 and 1760 kg/m^3; however, in special cases, job specifications may allow higher than 1840 kg/m^3 unit weight.

Strength

Design strengths of 20 to 35 MPa 28-day compressive strength are common, although by using a high cement content and good-quality lightweight aggregate of small size (i.e., 9 or 13 mm maximum) it has been possible in some precast and prestressing plants to produce 40 to 50 MPa concrete. Lightweight aggregates with controlled microporosity have been developed to produce 70 to 75 MPa concrete with 1840 to 2000 kg/m^3 unit weight.

The splitting tensile strength of concrete cylinders (ASTM C 496) is a relative measure of tensile strength. The data in Table 12-1 show that, like normal-weight concrete, the ratio between the splitting tensile strength and the compressive strength decreases significantly with the increasing strength of lightweight concrete. The modulus of rupture of continuously moist-cured lightweight concrete also behaves in the same manner; tests on dried specimens show that the data are extremely sensitive to the moisture state. Examination of fractured specimens of lightweight aggregate concrete after the splitting tension test clearly reveals that, unlike normal-weight concrete, it is the aggregate and not the interfacial transition zone that is generally the weakest component of the system (Fig. 12-1). Scanning electron micrographic studies (Fig. 12-2) show that due to the pozzolanic reaction, the bond strength between the aggregate and the cement paste is generally stronger than the individual aggregate particles.

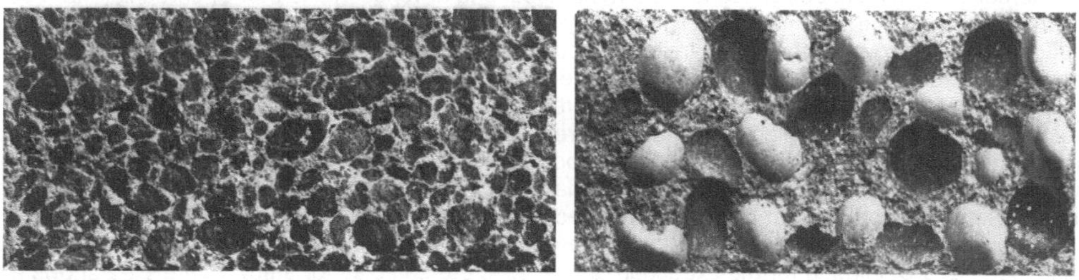

(a) (b)

FIGURE 12-1 Fracture surface from concrete cylinders after splitting tension test: (*a*) concrete made with lightweight aggregate; (*b*) concrete made with rounded flint aggregate. (*Photographs courtesy of P. Nepper-Christensen, Aalborg Portland Co., Aalborg, Denmark.*)

In lightweight aggregate concrete, the fracture passes through the cellular aggregate particles because both the interfacial transition zone and the cement paste are generally stronger. On the other hand, with normal-weight concrete the aggregate particles are dense and strong, and fracture is usually in the interfacial transition zone or the bulk cement paste, not through the aggregate.

FIGURE 12-2 The lightweight aggregate concrete in the upper deck of the San Francisco-Oakland Bay Bridge has delivered a superb performance for over 77 years. (*Figure 12.2a courtesy of Maryanne Mc Darby; Fig. 12.2b concrete sample from the Bay Bridge given by Rick Maggenti.*)

Built in 1936, the San Francisco-Oakland Bay Bridge carries about approximately 280,000 vehicles per day on its two decks. Originally, cars traveled on the upper deck and trucks and trains on the lower deck; subsequently the lower deck was also dedicated to road traffic. The bridge introduced groundbreaking structural design, construction methods, and advanced concrete. Of particular interest was the pioneering utilization of lightweight aggregate (LWA) concrete in the upper deck. Thomas E. Stanton, Jr., the discoverer of alkali-aggregate reaction in concrete (see Chap. 5), was the Caltrans materials and research engineer for the project. He wrote the following summary: "Considerable saving in the cost of the bridge was secured through the reduction in weight of members in portions of the structure by the use of lightweight aggregate concrete in the upper roadway deck. The design was based on concrete weighing not over 100 lbs per cubic foot [1600 kg/m^3] with a strength not less than 3000 pounds per square inch [21 MPa] at twenty-eight days with not over seven sacks of cement per cubic yard [390 kg/m^3]...Bottom form have been stripped and show perfect concrete surface without honey comb of any kind."* The recent construction of the new span connecting Oakland to the Yerba Buena Island (see Fig. 1-9) allowed for the extraction of LWA concrete samples for study. Figure 12-2 shows an optical micrograph of the concrete confirming the excellent interfacial transition zone between the LWA and the cement paste.

(*From Thomas E. Stanton, Jr., Report on "Cement & Concrete Control, San Francisco-Oakland Bay Bridge," March 1935.)

Dimensional Stability

In the ACI Building Code 318, the elastic modulus for normal-weight or structural lightweight concrete is computed using the equation $E_c = 0.043\,W_c^{1.5}\sqrt{f_c'}$ MPa (Chap. 4). The values of the elastic modulus thus obtained may deviate from those obtained experimentally (ASTM C 469) by ±15 to 20 percent. From a large number of test specimens Schideler[2] found that the static elastic moduli for 20 and 40 MPa lightweight concrete containing expanded-clay aggregate were 10 and 14 GPa, respectively. Concrete mixtures with complete replacement of lightweight sand by normal sand generally give 15 to 30 percent higher elastic modulus. Experimental studies indicate that the ultimate compressive strain of most lightweight concretes may be somewhat greater than the 0.003 value assumed for design purposes.

Compared to normal-weight concrete, the concrete mixtures made with lightweight aggregate exhibit higher moisture movement that results in high ultimate drying shrinkage (typically 800×10^{-6} m/m), and a considerably higher creep (typically 1600×10^{-6} m/m). It seems that low strength and low elastic modulus have a more pronounced effect on the creep than on the drying shrinkage.

Effects of replacement of fine lightweight aggregate with normal sand on the creep and the drying shrinkage are shown in Fig. 12-3. In view of the comparatively low tensile strength of lightweight concrete (2 to 3 MPa), a pronounced tendency for the drying-shrinkage cracking is to be expected. This tendency is offset to some degree by the lower modulus of elasticity and higher creep, which accounts for the greater extensibility

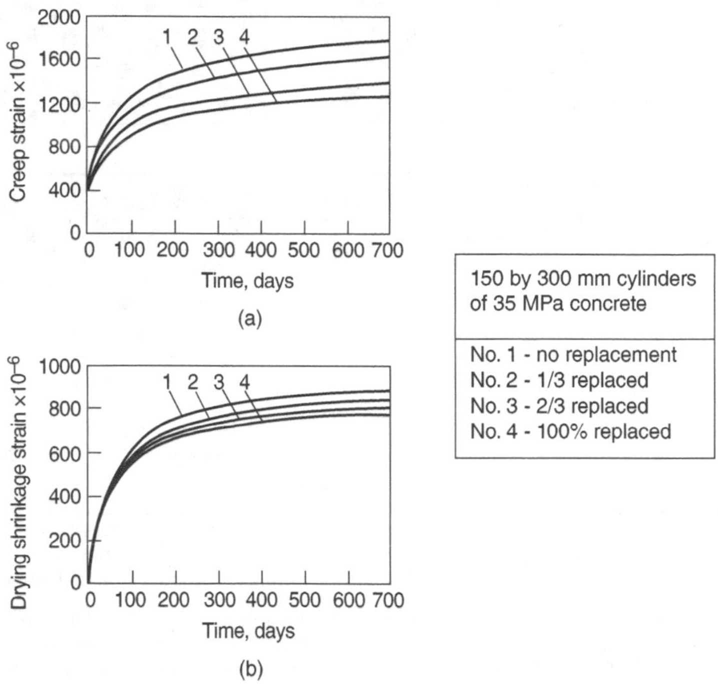

Figure 12-3 Effect of replacing lightweight fines by natural sand on (*a*) creep strain; (*b*) drying shrinkage strain. (From Orchard, D.F., *Concrete Technology*, Vol. 1, Elsevier, Barking, Essex, U.K., 1979.)

of lightweight concrete. Furthermore, Kulka and Polivka[3] point out that the creep and shrinkage of concrete, both lightweight and normal-weight, in real structures is much smaller than the values obtained with laboratory specimens. For example, when designing a lightweight-concrete bridge structure, the shrinkage and creep values were reduced from the laboratory test data by about 15 to 20 percent due to the size effect of the member, 10 to 20 percent due to the humidity of the environment, and 10 to 15 percent due to the reinforcement, which amounts to about 50 percent overall reduction.

Internal Curing

The use of pre-wetted lightweight aggregate can provide *internal curing* of the concrete as these particles act as reservoirs that can release water for hydration of the cementitious matrix and for replacement of water lost by evaporation or self-desiccation (Fig. 12-4). Experimental tests confirm that internal curing reduces the autogenous shrinkage and delays drying shrinkage.[4] The use of internal curing is becoming more popular as guidelines for mixture proportions are available. For instance, Bentz et al.[5] developed a simple and yet powerful mixture proportioning method. Internal curing

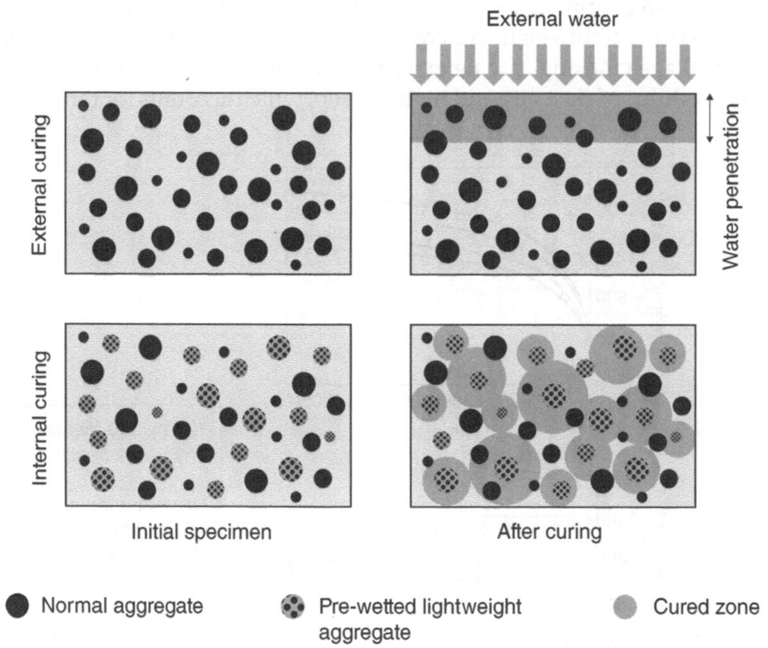

FIGURE 12-4 Illustration of external (conventional) and internal curing of concrete using pre-wetted lightweight aggregate. (After Bentz, D.P., and W.J. Weiss, Internal Curing: A 2010 State-of-the-Art Review, Report NISTIR 7765, National Institute of Standards and Technology, 2011.)

For external curing, the concrete surface is often exposed to external water, which can be absorbed. Curing compounds and evaporation retarders are also frequently used to minimize water loss. For internal curing, the pre-wetted lightweight aggregate particles act as reservoirs that release water for continuing the cement hydration process or for replacing the moisture lost by evaporation or self-desiccation. Effective internal curing requires that the pre-wetted lightweight aggregate be uniformly distributed and close to each other so that there is an overlapping of the cured zone, ensuring that an efficient curing of the cement paste matrix occurs.

can be particularly useful when supplementary cementitious materials (SCMs), such as fly ash and slag, are used for cement replacement. Because SCMs require longer curing time, the internal curing permits the continuation of the chemical reactions over a longer period of time.

Internal curing has the potential to significantly reduce the high autogenous volume in high-strength concrete. Because the strength of the concrete could be lowered if lightweight aggregate replaces normal-weight aggregate, often only a partial replacement is used. Cusson and Hoogeveen[6] studied internal curing generated by pre-wetted fine lightweight aggregate as partial replacements to regular sand. The results indicated that 178 kg/m^3 of saturated lightweight aggregate in a high-performance concrete mixture eliminated the tensile stress associated with autogenous shrinkage without jeopardizing the early-age strength and elastic modulus of the concrete.

Durability

The freeze-thaw resistance of air-entrained lightweight concrete is similar to that of normal-weight concrete. Air entrainment is especially helpful when aggregates are close to saturation and concrete will be subject to frost action. The hydraulic pressure caused by the expulsion of water from the aggregate can be accommodated by the entrained air present in the cement mortar, thus preventing damage to the concrete.

Although air-dried lightweight aggregate concrete tends to show a greater degree of moisture absorption, this does not mean high permeability. In fact, the permeability of lightweight concrete is low, and therefore its durability to aggressive chemical solutions is generally quite satisfactory. Holm et al.[7] reported that concrete cores taken from a 5.2-km-long lightweight concrete deck (which was then 20 years old) revealed little evidence of microcracking, whereas, from a parallel location, cores of concrete of the same age but containing a quartz gravel as aggregate were found to contain major cracks. Similarly, cores of lightweight concrete from a World War I freighter, the USS *Selma*, which had been exposed to a marine environment for 60 years, also revealed little microcracking. The principal reason for low permeability and excellent durability of lightweight concrete is the general absence of microcracks in the aggregate-cement paste interfacial transition zone, which, according to Holm et al., is due to similarities of elastic moduli between the lightweight aggregate and the mortar fraction. Also, due to the pozzolanic reaction between the thermally activated clay minerals in lightweight aggregate particles and calcium hydroxide present in the hydrated cement paste, the interfacial transition zone is normally dense and strong.

Lightweight aggregates are porous and therefore easily friable when compared to normal-weight rocks and minerals. Consequently, concrete containing lightweight aggregates generally shows *poor resistance to heavy abrasion*. Replacement of lightweight fines with natural sand improves the abrasion resistance of concrete.

The *coefficient of thermal expansion* of expanded-shale lightweight aggregate concrete is of the order of 6 to 8×10^{-1} per °C. In regard to resistance to heat flow, although insulating lightweight concretes (not described here) are better than structural lightweight concrete because of lower aggregate density, the *thermal conductivity* of the latter is still about half as much as that of normal-weight concrete, and therefore the fire endurance is considerably better. Typical values of thermal conductivity for concrete mixtures containing all expanded-clay aggregate vs. those containing expanded-clay coarse aggregate with natural sand are 0.91×10^{-3} and 1.36×10^{-3} cal/(s·cm²·°C/cm), respectively.

12.1.4 Applications

The use of lightweight aggregate concrete in a structure is usually predicated on a lower overall cost of the structure. While lightweight concrete will cost more than normal-weight concrete per cubic meter, the structure may cost less as a result of the reduced dead weight and lower foundation cost. Wilson[8] cites several examples, including the following, to demonstrate that application of lightweight concrete can result in lower costs for foundations and reinforcing steel.

In 1936, the construction of the lightweight concrete bridge deck for the San Francisco-Oakland Bay Bridge resulted in a $3 million saving in steel (see Fig. 12-2). Since then, numerous lightweight concrete bridge decks have been built throughout the world. Strength is not a major consideration in floor slabs; therefore, a large amount of lightweight aggregate concrete is used to reduce the dead weight of concrete in floors of high-rise buildings. An example of this application is the 71-stories Lake Point Tower in Chicago Illinois. The floor slabs from the second to the 70th level were made of cast-in-place concrete with 1730 kg/m^3 unit weight and 20 to 22 MPa 7-day compressive strength. Similarly, for the construction of the Australian Square building in Sydney, a 50-stories circular tower, which is 184 m high by 42.5 m diameter, a 13 percent saving in the construction cost was achieved through the use of 31,000-m^3 lightweight-aggregate concrete beams, columns, and floors above the seventh-floor level. The concrete had an average compressive strength of 34.3 MPa at 28 days and a density of 1792 kg/m^3. One Shell Plaza, Houston, Texas, is an all-lightweight concrete structure of 52 stories, containing a 70 by 52 by 2.5 m lightweight concrete pad, 18 m below grade. A concrete mixture of 1840 kg/m^3 density, and 41.2-MPa compressive strength was used for shear walls, columns, and mat foundation. If normal concrete had been used, only a 35-story structure could have been safely designed due to the limited bearing capacity of the soil.

Kulka and Polivka[9] state that the basic economy of lightweight-aggregate concrete can be demonstrated by the savings in reinforcement. With ordinary reinforced concrete the economic advantage is not as pronounced as with prestressed concrete. The prestressing force in most cases is computed strictly from the dead load of the structure; consequently, a weight reduction of 25 percent results in a substantial reduction in the weight of prestressing tendons. Among other advantages of reduction in the weight of concrete is the higher resistance of shear elements to earthquake loading since seismic forces are largely a direct function of the dead weight of the structure. However, it should be noted that major applications of lightweight concrete throughout the world continue to remain in the production of precast concrete elements and prefabricated panels. Due to lower handling, transportation, and construction costs the lightweight-aggregate concrete products are ideally suited for this type of construction.

Although expanded clay and shale aggregates are most suitable for the production of structural-quality lightweight concrete, the escalation of fuel costs has priced these aggregates out of many markets. Consequently, there is a renewed interest in finding natural lightweight aggregate of good quality. According to Bryan,[10] there are ample reserves of a suitable volcanic rock, called rhyolite, in Arizona, Colorado, California, Nevada, New Mexico, Oregon, and Utah. The unit weight of air-dried rhyolite concrete at 28 days is typically between 1760 kg/m^3 (110 lb/ft^3) and 2000 kg/m^3 (125 lb/ft^3), with compressive strength well over 28 MPa.

12.2 High-Strength Concrete

12.2.1 A Brief History of Development

In the late 1960s, high-range water reducing admixtures (superplasticizers) composed of salts of naphthalene sulfonate and melanine sulfonate were developed in Japan and Germany, respectively. The first applications in Japan were confined to *high-strength* precast products and cast-in-place girders and beams for bridges. In Germany, initially the objective was to develop underwater concrete mixtures possessing high fluidity without segregation. As it is possible to achieve both high strength and high workability simultaneously with the superplasticized concrete mixtures, they were found to be very suitable for use in the production of cast-in-place structural components for tall buildings Figs. 12-5 and 12-6.

12.2.2 Definition

Concrete is defined as "high-strength" solely on the basis of compressive strength at a given age. In the 1970s, before the advent of superplasticizers, concrete mixtures that showed 40 MPa or more compressive strength at 28 days were called high-strength concrete. Later, when 60 to120 MPa concrete mixtures became commercially available, *in 2002 the ACI Committee on High Strength Concrete revised the definition to cover mixtures with a specified design strength of 55 MPa or more.*

Although conventional practice is to specify concrete strength based on the 28-day test result, there is a growing movement to specify the 56- or 90-day strength because many structural elements are not fully loaded for periods as long as two to three months or even longer. When high strength is not needed at an early age, it is best not to specify it so as to achieve a number of benefits such as cement saving, ability to use relatively large proportions of mineral admixtures, and a more durable product.

12.2.3 Significance

Thirty years ago, tall buildings in Manhattan (New York City) were almost all steel-frame. Today, perhaps one-third of tall commercial buildings are concrete frame. It is believed that the choice of steel frame vs. reinforced concrete frame was decided in favor of the latter primarily on account of the high construction speed. Also, commercial availability of high-strength concrete provided an economical alternative to bulky columns of conventional concrete for the lower floors of high-rise buildings. According to one report, the load-carrying capacity of columns in multistory buildings increased 4.7 times for a threefold increase in price. For the construction of reinforced-concrete-frame buildings, 30 stories and higher, normal-size columns can be made in the upper third of the building with conventional 30 to 35 MPa concrete; however, the use of high-strength concrete is justified for slender columns in the lower two-thirds of the building.

12.2.4 Materials

Cement

Ordinary portland cement of any type (e.g., meeting ASTM C 150 Standard Specification) can be used to obtain concrete mixtures with compressive strengths up to 50 MPa. To obtain a higher strength while maintaining good workability, it is necessary to use

(a) January 1994 (b) September 1994

(c) May 1995 (d) September 1995 (e) February 1996

FIGURE 12-5 Construction sequence of the Petronas Twin Towers. (*Photographs courtesy from the Thornton Tomasetti Group.*)

The Petronas Towers in Malaysia's capital city, Kuala Lumpur, is the third tallest building in the world. The 452-m high structure composed of two, 88-story buildings and their pinnacles, optimized the use of steel and reinforced concrete. Steel was used primarily in the long-span floor beams, while reinforced concrete was used in the central core, in the perimeter columns, and in the tower perimeter ring beams. The strength of the concrete used in the building and foundation ranged from 35 to 80 MPa. The concrete mixture for the 80 MPa concrete contained 260 kg/m^3 portland cement, 260 kg/m^3 of a cementitious and pozzolanic blending material with 30 kg/m^3 silica fume, and 10 L/m^3 high-range water reducer to obtain a water-cement ratio of 0.27. The strength test was performed at 56 days to allow the slower reacting materials, such as fly ash, to contribute to the strength gain. High-strength mixtures were used in the lower level columns, core walls, and ring beams. Compared to a steel structure, an added benefit of using reinforced concrete was efficient damping of vibrations, an important consideration for the building's occupants in light of the structure's potential exposure to moderate and high winds.

FIGURE 12-6 Taipei 101 building in Taiwan. (*Photograph courtesy of Samsung C&T Corporation.*)

The Taipei 101 building at 509 m high is the second tallest building in world and is located in center of financial power in Taiwan. The design of 101 floors was a deliberate choice. One hundred is seen as a classical number of perfection and celebratory, and the addition of one more floor is perfection plus one. Taipei 101 was the first building to break the half-kilometer record in height, and it received a LEED Platinum certification, making Taipei 101 the tallest green building in the world. The construction and design of tall buildings in Taiwan is challenging because it is necessary to take into account typhoon winds, large earthquakes, and weak soil conditions. To provide sufficient structural strength and stiffness, the super-columns in Taipei 101 were filled with high-strength concrete at level 62 and below. To reduce gravity loads, steel was used in columns above level 62. The concrete had a designed compressive strength of 69 MPa at 90 days and had the following mix proportion (in kg/m^3): 340 of cement, 150 of slag, 30 of silica fume, 760 of coarse aggregate, 957 of fine aggregate and 6.24 of admixture. Autogenous shrinkage of the concrete was as low as 300×10^{-6}.

chemical and mineral admixtures in combination with cement. In such cases, cement-admixture compatibility becomes an important issue. Experience has shown that, with naphthalene or melamine sulfonate type superplasticizers, low-C_3A and low-alkali portland cements generally produce concrete mixtures which do not show high slump loss with time. This situation has changed because it is reported that polyacrylate copolymers, a new generation of superplasticizers, do not cause excessive slump loss with most types of portland, or blended portland cements.

Aggregate

With normal-strength concrete, *the type and amount of aggregate* plays an important role in the volume stability of concrete, but it has a limited influence on the strength. In high-strength concrete, the aggregate remains important for volume stability, but it also plays an important role on the strength and stiffness of concrete. The low water-cement ratio used in high-strength concrete mixtures causes densification of both the matrix and the interfacial transition zone. Moreover, some aggregate types such as granite and quartzite may develop microcracks in the transition zone due to differential thermal shrinkage and, thus hinder the development of high mechanical strength. Therefore, proper care must be taken in the selection of aggregates for high-strength concrete. Based on the results of an experimental study, Aitcin and Mehta[11] recommend that hard and strong aggregate types with high modulus of elasticity and a low coefficient of thermal expansion are better for producing very high-strength concrete mixtures (Fig. 12-7).[12]

As discussed in Chap. 3, with a given water-cement ratio, the strength of a concrete mixture can be increased significantly by simply reducing the *maximum size of the coarse aggregate* because this has a beneficial effect on the strength of the interfacial transition zone. According to Aitcin[12] the higher the targeted strength, the smaller should be the maximum size of coarse aggregate. Up to 70 MPa compressive strength concrete can be

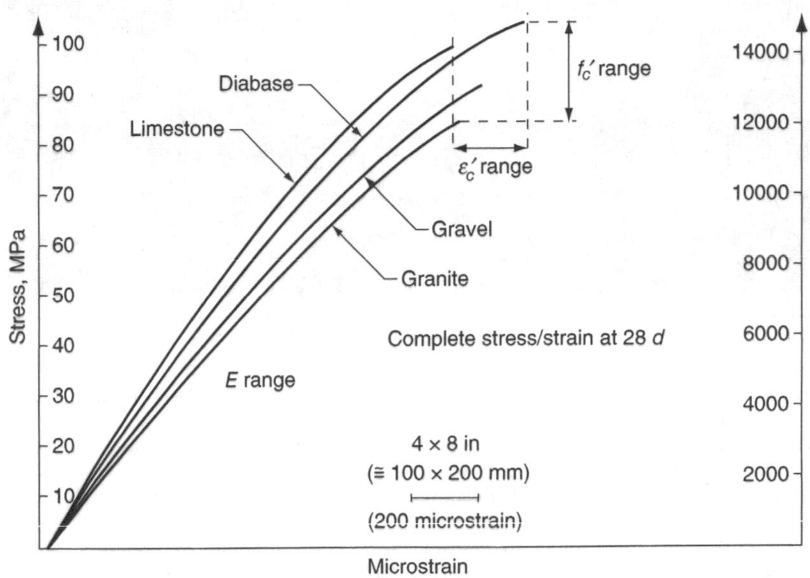

FIGURE 12-7 Stress-strain curves for high-strength concrete at 28 days. (Adapted from Aitcin, P.C., *High-Performance Concrete*, E & FN Spon, London, p. 591, 1998.)

produced with a good-quality coarse aggregate of 20 to 25 mm maximum size. To produce 100 MPa compressive strength, aggregates with 14 to 20 mm maximum size should be used. Commercial concretes with compressive strengths of over 125 MPa have been produced with 10 to 14 mm maximum size coarse aggregate.

In regard to fine aggregate, any material with the particle size distribution meeting the ASTM Standard Specification C 39 (see Chap. 7) is adequate for high-strength concrete mixtures. Aitcin recommends using fine aggregates with higher fineness modulus (approx. 3.0) for the following reasons: (a) high-strength concrete mixtures already have large amounts of small particles of cement and pozzolan, therefore, the presence of very small particles in the fine aggregate is not needed to improve the workability, (b) the use of a coarser fine aggregate requires less water to obtain the same workability, and (c) during the mixing process, the coarser particles will generate higher shearing stresses that help to prevent the flocculation of cement particles.

Admixtures

Depending on the desired properties, a high-strength concrete may contain one or more types of chemical admixtures, such as plasticizing, set-controlling, and air-entraining (see Chap. 8). Also, for reasons discussed in the next paragraph, it is common to use mineral admixtures such as fly ash, slag, and silica fume.

Poorly crystalline calcium silicate hydrate (C-S-H) is the major phase present in hydrated portland cement pastes; crystalline phases such as calcium hydroxide and capillary pores may be considered as microstructural inhomogeneities in the system. If microstructural inhomogeneities in the hydrated portland cement paste are strength limiting, the obvious solution is to modify the microstructure so that the components causing the inhomogeneities are either completely eliminated or reduced. In the case of cement products, an inexpensive and effective way to achieve this objective is through the *incorporation of a pozzolanic material*. As described in Chap. 8, pozzolanic admixtures such as fly ash react with calcium hydroxide to form a reaction product that is similar in composition and properties to C-S-H. The pozzolanic reaction is also accompanied by a reduction in the total volume and size of capillary pores—an effect that is equally important for the enhancement of strength.

In addition to cost reduction and a more homogeneous end product when a part of the portland cement in concrete is replaced by a pozzolan, another major benefit accrues in the form of lower temperature rise due to less heat of hydration. Because of the high cement content, the users of high-strength concrete mixtures frequently experience thermal cracking in large structural elements. Thus, in certain cases, reducing the risk of thermal cracking is by itself enough of a justification for partial cement replacement with a pozzolan.

At a given water-cement ratio, when Class F fly ash is used as partial replacement for portland cement, the early strengths (3 and 7 days) of concrete cured at normal temperatures are reduced almost in direct proportion to the amount of fly ash in the total cementitious material (cement + fly ash). However, concrete mixtures containing Class C fly ash or ground blast-furnace slag tend to show some strength gain at 7 days. Highly reactive pozzolans such as silica fume, metakaolin, and rice husk ash usually begin to make a strength contribution at about 3 days. Under conditions of acceleration of cement hydration, such as those present in the production of steam-cured concrete products, small differences in the reactivity or the pozzolanic activity index will not have much effect on the strength.

Substantial strength gain at early age can be attained by partial substitution of fly ash or ground blast-furnace slag for fine aggregate provided this is not accompanied by an increase in the water requirement of the concrete mixture. It should be noted that this approach will neither offer an economic advantage nor reduce the risk of thermal cracking with massive structural elements, because no reduction in the cement content has been made.

12.2.5 Mixture Proportioning

The requirements of high strength, and small aggregate size mean that the cementitious materials content of the concrete mixture would be high, typically above 400 kg/m^3. Cementitious contents of 600 kg/m^3 and even higher have been investigated but are undesirable for reasons of high cost and excessive thermal and drying shrinkage. Furthermore, with increasing proportion of cement in concrete, a strength plateau is reached, that is, there is hardly much strength gain above a certain cement content. As explained above, this is probably due to the inherent inhomogeneity of the hydrated portland cement paste that contains randomly distributed areas of crystalline calcium hydroxide within the principal phase (i.e., poorly crystalline calcium silicate hydrate). These areas present weak regions that are vulnerable to microcracking under tensile stress.

Published literature contains several methods of proportioning high-strength concrete mixtures such as those discussed by ACI Committee 363,[13] de Larrad,[14] and Mehta and Aitcin.[15] Due to its simplicity, only the latter method is described here. According to this method, high-strength concrete mixtures are classified into five strength grades, with 28-day compressive strengths 65, 75, 90, 105, and 120 MPa. To keep the drying shrinkage and creep low, the cement paste- aggregate volume ratio is fixed at 35:65. The strength enhancement is achieved by reducing the water content with the help of a superplasticizing admixture, and by partial substitution of cement with mineral admixtures. Also, as a first approximation, the fine-coarse aggregate volume ratio is kept at 2:3. Depending on the desired rate of strength development and the local availability of fly ash (FA), ground granulated blast furnace slag (BFS), and condensed silica fume (CSF), this procedure offers the concrete producer three options, that is, use only portland cement (PC), use portland cement in combination with fly ash or slag, or use portland cement in combination with silica fume and fly ash or slag.

Based on the assumptions discussed above, Table 12-2[16] shows the calculated trial batch mix proportions that seem to compare reasonably well with the laboratory and field experience from different regions of North America. The results of an inter-laboratory study on some commercially available concrete mixtures, with 28-day compressive strength ranging from about 80 to 120 MPa, are reported by Burg and Ost.[16] The mix proportions are shown in Table 12-3. Table 12-4 shows mix proportions of 60 to 120 MPa concrete used in the construction of some well-known high-rise buildings in four different metropolitan areas of North America. The strength and other properties of the five hardened concrete mixtures of Table 12-3 are shown in Table 12-5.[17]

12.2.6 Microstructure

The description of the interfacial transition of normal strength concrete is presented in Chap. 2. This porous zone, where cracks often originate, prevents efficient load transfer between the coarse aggregate and the cement mortar. From the materials and mix designs used for making high-strength concrete mixtures, it is to be expected that the

Strength Grade	Avg. Strength (MPa)	Option	Cementitious Materials			Total* Water	Coarse Agg.	Fine Agg.	Total Batch	W/C
			PC	FA or BFS	CSF					
A	65	1	534	–	–	160	1050	690	2434	0.30
		2	400	106	–	160	1050	690	2406	0.32
		3	400	64	36	160	1050	690	2400	0.32
B	75	1	565	–	–	150	1070	670	2455	0.27
		2	423	113	–	150	1070	670	2426	0.28
		3	423	68	38	150	1070	670	2419	0.28
C	90	1	597	–	–	140	1090	650	2477	0.23
		2	447	119	–	140	1090	650	2446	0.25
		3	447	71	40	140	1090	650	2438	0.25
D	105	–	–	–	–	–	–	–	–	–
		2	471	125	–	130	1110	630	2466	0.22
		3	471	75	42	130	1110	630	2458	0.22
E	120	–	–	–	–	–	–	–	–	–
		2	495	131	–	120	1120	620	2486	0.19
		3	495	79	44	120	1120	620	2478	0.19

*The mix proportions are for non-air-entrained concrete, although 2 percent entrapped air is assumed. Total water includes the water in the superplasticizing admixture, the dosage of which may range from 3 to 10 L/m^3, depending on the consistency and strength requirements, cementitious content, and superplasticizer type.

TABLE 12-2 Calculated Mixture Proportions for the First Trial Batch according to the Mehta and Aitcin's Procedure, kg/m^3

kg/m^3	Mix 1	Mix 2	Mix 3	Mix 4	Mix 5
Portland cement, Type I	564	475	487	564	475
Silica fume	–	24	47	89	74
Fly ash	–	59	–	–	104
Coarse aggregate (10 mm max. size)	1068	1068	1068	1068	1068
Fine aggregate (Crushed dolomite)	647	659	676	593	593
Total water	158	160	155	144	152
Water-cementitious ratio	0.28	0.29	0.29	0.22	0.23

Source: Adapted from Burg, R.G., and B.W. Ost, Research and Development, Bull. RD104, *Portland Cement Association*, p. 62, 1994.

TABLE 12-3 Mix Proportions of High-Strength Superplasticized Concrete Mixtures Commercially Produced in Chicago Area

kg/m³	Mix 1*	Mix 2†	Mix 3‡	Mix 4§
Portland cement, ASTM Type I	400	360	315	564
Fly ash	100	150	–	–
Granulated BF slag	–	–	135	–
Silica fume	–	–	36	44
Coarse aggregate	1150	1157	1130	1100
Fine aggregate	590	603	645	682
Superplasticizer	–	3	6	21
Total water	162	148	145	124
w/cm	0.33	0.29	0.30	0.20
Avg. 28-d, f_c: MPa	56	80	82	138

*Mix 1 was used for the cast-in-place concrete columns, spandrel beams, and shear walls in the lower stories of a 75-story (Texas Commerce Tower) building in Houston.

†Mix 2 was used for the composite steel and concrete columns of a 72-story building (Interfirst Plaza) in Dallas.

‡Mix 3 was used for the concrete columns of the Nova Scotia Plaza Tower in Toronto, Canada.

§Mix 4 was used to provide a high-stiffness concrete (50 GPa Modulus of Elasticity) for 18 columns of the Two Union Square Building in Seattle.

Table 12-4 Mix Proportions of High-Strength Concrete Used in Four Different Metropolitan Areas of North America

	Mix 1	Mix 2	Mix 3	Mix 4	Mix 5
Compressive Strength of Moist-Cured Cylinders, MPa (ASTM C39)					
Age, days					
3	58.3	57.6	55.3	74.8	55.6
7	65.9	71.9	72.9	95.6	79.5
28	78.6	88.5	91.9	118.9	107.0
56	81.4	97.3	94.2	121.2	112.0
91	86.5	100.4	96.0	131.8	119.3
272	98.5	107.6	98.1	134.7	129.8
Tensile Strength of Moist-Cured Cylinders, MPa (ASTM C496)					
91 days	6.4	5.9	5.0	7.2	6.6
Modulus of Rupture of Moist-Cured Prisms, MPa (ASTM C78)					
91 days	9.4	10.1	9.9	13.7	9.6
Modulus of Elasticity of Moist-Cured Cylinders, GPa (ASTM C78)					
91 days	45.8	47.0	46.4	51.4	48.5
Drying Shrinkage of Prisms, Millionth (ASTM C157)					
90 days	573	447	383	320	340
400 days	703	593	530	473	483
Rapid Chloride Permeability of Concrete, Coulombs AASHTO, T277 (ASTM C157)					
287 days	2570	650	580	108	n.a.

Table 12-5 Properties of Commercially Available High-Strength Concrete

microstructure of the product is relatively free from inhomogeneities. With aggregate particles that are hard and strong, and with a strong interfacial transition zone due to the pore refinement and grain refinement (see Fig. 6-16), the microstructure of high-strength concrete permits efficient load transfer between the cement mortar and the coarse aggregate. Thus, the high elastic modulus and other mechanical properties of the material are directly attributable to the microstructure.

12.2.7 Properties of Fresh and Hardened Concrete

Consistency

By using a heavy dosage of the superplastizer, it is possible to obtain slump values on the order of 200 to 250 mm with concrete mixtures containing a very low water content, and a high content of cement and mineral admixtures. The presence of large amounts of fines helps to reduce the tendency for bleeding and segregation in these high-consistency mixtures. The construction of heavily reinforced, cast-in-place, shear walls, columns, floors, and other high-strength structural members in high-rise buildings would have been difficult without the availability of concrete mixtures possessing high workability.

Chemical and Autogenous Shrinkage

Using the stoichiometric amount of water needed for complete hydration of the cement paste in a closed system, it can be shown that the volume of the hydration products, V_h, would be less than the sum of the volume of water and the volume of cement that is hydrated.

The initial volume of the cement paste is given by

$$V_i = V_{ci} + V_{wi} \tag{12-1}$$

where V_{ci} and V_{wi} are the initial volume of cement and water, respectively.

During hydration, the current volume V_t of the paste can be expressed as

$$V_t = V_c + V_w + V_h \tag{12-2}$$

where V_c and V_w are the current volume of cement and water, respectively. The volume of the hydration products, V_h, includes their own (nano) porosity (26 to 28 percent); and the current volume of water, V_w, can be defined as the volume of *capillary water* (it can be also defined as the volume of *evaporable water*).

Chemical shrinkage ε_{ch} is the relative volume reduction

$$\varepsilon_{ch} = 1 - \frac{V_c + V_w + V_h}{V_{ci} + V_{wi}} \tag{12-3}$$

which starts from 0 and continuously increases. In the literature, chemical shrinkage is also referred to as the *Le Chatelier contraction*, in honor of the French scientist who first observed the phenomenon.

Table 12-6 gives an example of the volume change during the hydration of C_3S. As explained in Chap. 6, there is still some uncertainty in the composition of C-S-H, therefore two probable reactions are analyzed.

Before setting, the chemical shrinkage is not constrained and, therefore, it will induce shrinkage of the same magnitude in the cement paste. As a rigid network of

	Hydration								Chemical Shrinkage (%)
	C_3S	+	5.3 H	→	$C_{1.7}SH_4$	+	1.3 CH		7.2
Molar mass (g/mol)	228.32		18.02		227.2		74.09		
Mass (g)	1.00		0.418		0.995		0.422		
Density (g/mL)	3.15		0.998		2.01		2.24		
Volume (mL)	0.317		0.419		0.495		0.188		
	2 C_3S	+	6.5 H	→	$C_3S_2H_{3.5}$	+	3 CH		8.5
Molar mass (g/mol)	228.32		18.02		351.49		74.09		
Mass (g)	1.00		0.257		0.770		0.487		
Density (g/mL)	3.15		0.998		2.50		2.24		
Volume (mL)	0.317		0.257		0.308		0.217		

Source: Adapted from Justnes, H., E. Sellevold, B. Reyniers, D. Van Loo, A.V. Gemert, F. Verboven, and D.V. Gemert, *Autogenous Shrinkage of Concrete*, Tazawa, E., ed., E & FN Spon, London, pp. 71–80, 1998.

TABLE 12-6 Calculation of Chemical Shrinkage of C_3S

hydration products starts to develop, the values of the chemical shrinkage and that of the measured shrinkage in the cement paste start to diverge, since the rigidity of the paste restrains the volume change. This is analogous to the shrinkage of the cement paste being higher than the shrinkage of concrete because of the higher restraint present in concrete.

The measured deformation of cement paste in a closed system is called *autogenous shrinkage*. The Japanese Concrete Institute defines autogenous shrinkage as *the macroscopic volume reduction of cementitious materials when cement hydrates after initial setting. Autogenous shrinkage does not include volume change due to loss or ingress of substances, temperature variation, and application of an external force and restraint.*[18] This shrinkage develops internally in the whole volume of concrete and is sometimes referred to as self-desiccation shrinkage.

Some autogenous shrinkage occurs in all types of concrete; however, it is too high to ignore in the case of high-strength concrete that typically contains a high cement content and low water-cement ratio. The magnitude of this shrinkage is further increased if silica fume has been used in the concrete mixture.

To understand the stress development due to drying and autogenous shrinkage, it is necessary to introduce the concepts of surface chemistry. The presence of a surface breaks the molecular symmetry that exists inside a material. The molecules at the surface have different energy than the molecules inside the bulk material. Surface energy, U_{surf}, is the difference between the energy of the molecules at the surface and the energy that they would have within the body. It can be expressed as

$$U_{surf} = \alpha S \tag{12-4}$$

where S is the area of the interface and α is the surface tension. Dimensional analysis indicates that the units of the surface tension can be expressed either by energy per unit area or force per unit length.

Nature brings a body to its minimum energy. When considering surface energy, there is a tendency to reduce the interface between two phases. Ignoring the effects of gravity, small spherical drops of liquid and gas bubbles are good examples of surface minimization for a given volume. The decrease in surface induces a contraction of the drop, increasing its internal pressure and making it higher than the external pressure. The difference between internal and external pressure is called surface pressure (p_{surf}). A balance of energy requires that the reduction in surface energy (αdS) be equal to the work done by the surface forces in reducing the surface. The work done can be expressed as $p_{\text{surf}}\, dV$ where dV is the volume change. Therefore

$$\alpha dS = p_{\text{surf}}\, dV \tag{12-5}$$

This expression can be used to obtain useful expressions depending on the geometry:

Sphere of radius r $\left(S = 4\pi r^2,\ V = \frac{4\pi r^3}{3}\right)$

$$p_{\text{surf}} = \frac{2\alpha}{r} \tag{12-6}$$

Cylinder of radius r and height h ($S = 2\pi rh$, $V = \pi r^2 h$)

$$p_{\text{surf}} = \frac{\alpha}{r} \tag{12-7}$$

These relationships, often referred to as *Young-Laplace equations*, clearly show that as the size of the sphere or the cylinder decreases, the magnitude of the surface pressure increases greatly. Radius r can be positive or negative depending on whether the surface is concave or convex. In the example of the liquid drop, the surface is convex so radius r is positive, and the surface tension tends to compress the liquid increasing its internal pressure. For concave curvatures, such as in capillary tubes or air bubbles in liquids, the pressure in the liquid is lower than in the air.

Consider a thin layer of liquid between two parallel plates shown in Fig. 12-8. If the meniscus is concave, the pressure in the liquid is less than the outside pressure, making the plates to approach each other. The opposite would happen when the meniscus is convex. From the geometry shown in Fig. 12-8, the following equation can be obtained:

$$p_{\text{surf}} = \frac{\alpha}{r} = \frac{2\alpha\cos\theta}{d} \tag{12-8}$$

where d is the distance between the plates and θ is the *contact angle* between the liquid and the solid, measured in the liquid. Force F, which brings the plates together, can be determined by multiplying the surface pressure p_{surf} by the area of contact S,

$$F = \frac{2\alpha S\cos\theta}{d} \tag{12-9}$$

For simplicity, this derivation was performed considering two plates, but similar results can be established for small capillaries that exist in a porous medium, such as

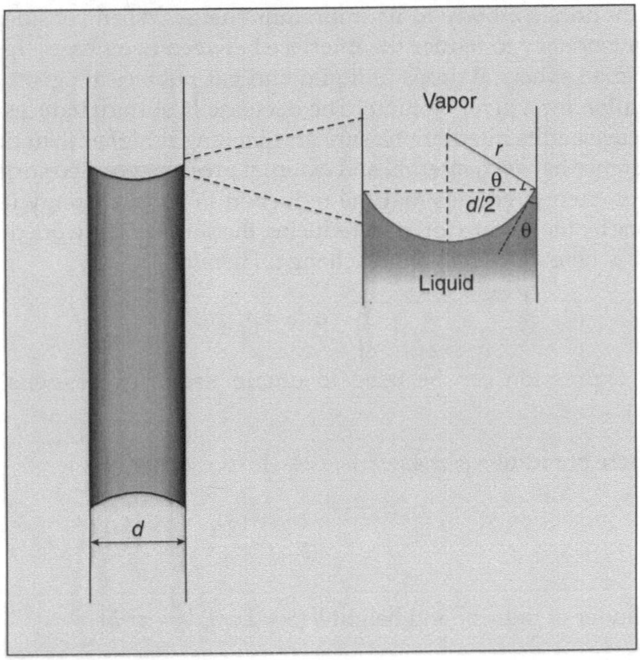

FIGURE 12-8 Formation of a meniscus between two plates. (After Landau, L., A. Ajjiezer, and E. Lifshitz, *General Physics: Mechanics and Molecular Physics*, Pergamon Press, New York, 1967.)

cement paste. For a capillary of radius r, the surface pressure is given by $p_{surf} = 2\alpha\cos\theta/r$. There the development of a concave meniscus in a small capillary also generates lower pressure in the liquid than in the vapor and similarly originates forces that bring the capillary walls together. This process of pulling the walls closer is analogous to shrinkage of the capillary walls; it can be used to estimate the autogenous volume change. In a porous material, the internal stress σ_c in the matrix caused by the capillary pressure can be estimated by

$$\sigma_c(1-V) + p_{surf}V = 0 \tag{12-10}$$

where V is the relative volume of pores. In normal strength concrete, the volume of pores ranges from 15 to 20 percent, whereas for high-strength concrete the value is lower than 15 percent. The capillary pressure can be estimated from the relative humidity existing inside the concrete, either by mathematical equations or tabulated data from surface chemistry textbooks. For instance, a relative humidity of 90 percent inside the concrete generates a capillary pressure of 14.6 MPa. Assuming a porosity of 15 percent in the concrete and using the above equation leads to a compressive stress σ_c of 2.6 MPa. This stress will cause a viscoelastic strain ε_c in the concrete given by

$$\varepsilon_c = \frac{\sigma}{E_c}(1+\varphi) \tag{12-11}$$

where E_c is the elastic modulus and φ the creep coefficient. In the example where the compressive stress was 2.6 MPa, assuming an elastic modulus of 35 GPa and a creep coefficient of 2, we obtain a strain of 222×10^{-6}. This value is of the same order of magnitude as reported by experimental results. More sophisticated analysis can be found in the work of Hua et al.[19]

Thermal Shrinkage

In addition to autogenous shrinkage, massive structural members made of high-strength concrete mixtures are generally vulnerable to early-age cracking from thermal shrinkage that can be very high when freshly cast concrete is exposed to a cool ambient temperature within a few days after casting. Due to the high cement content, the adiabatic temperature rise is considerable in the case of high-strength concrete mixtures. For example, with Concrete Mix No. 1-5 (Table 12-3), Burg and Ost[16] reported 50 to 58°C temperature rise in thermocouples installed at the center of 1220 mm cube specimens within a period of 30 to 50 h after casting.

Drying Shrinkage and Creep

The data in Table 12-5[17] show that, in spite of low water content and high-quality aggregate, the long-term (400 days) drying shrinkage values of typical high-strength concrete mixtures (Mix 1, 2, and 3) ranged between 500 and 700 microstrain. Such high-drying shrinkage values are attributed to the high cementitious content of high-strength concrete mixtures. A slight reduction in drying shrinkage was observed when the water/cementitious material (w/cm) was reduced from 0.29 to 0.22 or 0.23 (Mix 4 and 5).

Burg and Ost[16] also reported the specific creep data for the high strength concrete mixtures of Table 12-3, under sustained loading for 430 days. The applied stress was equal to 39 percent of the compressive strength of concrete. As expected, the specific creep was lowest for the concrete mixtures with highest strength. The specific creep was approximately 64 millionth/MPa for Mix 1, and 25 millionth/MPa for Mix 4. These values are ½ to 1/3 of the specific creep typically reported for ordinary concrete. Note that a combination of the high drying shrinkage and low creep is also responsible for the high cracking potential of cast-in-place, high-strength concrete mixtures.

Strength

Several conclusions can be drawn from the data in Table 12-5, which is valuable because it represents typical behavior of modern high-strength concrete mixtures. First, with compressive strengths ranging from 90 to 130 MPa at 91 days, note that 50 to 75 MPa strength was attained at the early age of 3 days. Therefore, the benefit from the use of such high-strength mixtures is obvious for high-speed construction and for precast and prestressed concrete products. Second, compared to the reference concrete (Mix 1), 40 to 50 percent enhancement in the 91-day compressive strengths was achieved with a combination of lower w/cm and partial substitution of cement by either silica fume (Mix 4), or by silica fume and fly ash (Mix 5). However, at the early age of 3 days, only Mix 4, with 14 percent silica fume and 0.22 w/cm, gave higher strength than the reference concrete mixture.

From the 91-day tensile and flexural strength data, tensile-compressive strength ratio in high strength concrete was approximately 5.5 to 6 percent, whereas the flexural-compressive strength ratio was about 10 percent. Due to the relatively more brittle microstructure, these values are significantly lower than the typical values with conventional concrete.

With high-strength concrete, due to a decrease in microcracking in the interfacial transition zone, the ascending branch of the stress-strain curve is steeper and more linear to a higher percentage of the peak strength than with normal-strength concrete. Iravani and MacGregor[20] reported linearity of the stress-strain diagram at 65 to 70, 75 to 80, and above 85 percent of the peak load for concrete with compressive strengths of 65, 95, and 105 MPa. From their experimental study, the authors suggested the following *strength values for sustained loading:*

> 70 to 75 percent (of the short-time loading strength) for 65 MPa concrete, 75 to 80 percent for 95 MPa concrete without silica fume, and 85 to 90 percent for 105 MPa concrete with silica fume.

These results indicate that the reduction in porosity, which is necessary for strength, helps to delay the formation and propagation of microcracks, thus increasing the allowable sustained load. Silica fume contributes to the densification and grain refinement of the cement paste and the interfacial transition zone, therefore, it is not surprising that its use increases the allowable sustained load.

Elastic Modulus

The data in Table 12-5 show that, with 91-day-old high-strength concrete mixtures of approximately 0.29 w/cm, there was an upper limit of about 47 GPa modulus of elasticity in compression. Note that all five concrete mixtures contained a high quality coarse aggregate (crushed dolomite, 10 mm max. size). A slight increase in the modulus was obtained only when the w/cm was reduced to 0.22 and a significant amount of cement was replaced with silica fume and fly ash (Mix 4 and 5).

Equations developed for normal-strength concrete to estimate the elastic modulus from the compressive strength are not applicable to high-strength concrete. Extrapolation beyond the limits of validity of these equations often leads to overestimation of the elastic modulus. According to Tomosawa and Noguchi[21] the following formula is found to be more applicable:

$$E_c = 3.35 \times 10^4 k_1 k_2 (w/2400)^2 \sqrt[3]{f_{cm}/10} \text{ MPa}$$

where k_1 is equal to 1.20 for limestone, 0.95 for basalt and quartzite, and 1.0 for other aggregates; w is the unit weight of concrete in kg/m^3.

Durability

Among others, the two common causes of deterioration of reinforced and prestressed concrete structures are exposure to: (a) cycles of freezing and thawing and (b) seawater or deicing chemicals. As with ordinary concrete, proper air entrainment (Chap. 5) is necessary to protect high-strength concrete from freezing and thawing cycles. Some researchers believe that concrete mixtures with less than 0.25 w/cm do not require air entrainment for frost resistance because there would be hardly any free or freezable water.

The penetration of seawater or deicing chemicals is a frequent source of chloride ions that have the ability to destroy the protective coating normally present on the surface of reinforcing steel, which is one of the prerequisites for initiation of the corrosion process. Another prerequisite is the electrical resistivity of concrete, which depends on the concrete permeability and the ionic concentration of pore fluid.

The ASTM C 1202, rapid chloride penetration test (based on AASHTO T 277) gives an indirect estimate of both the chloride permeability and the electrical resistivity of concrete, expressed as number of coulombs passed through a test specimen under prescribed standard conditions. With more than 4000 coulombs charge passed, ASTM C 1202 rates the chloride permeability as *high*; with 2000 to 4000 coulombs, as *moderate*; with 1000 to 2000 coulombs, as *low*; and with less than 1000 coulombs, as *very low*. From the chloride penetration data on five high-strength concrete mixtures (Table 12-5), it should be noted that the reference concrete without any mineral admixtures (Mix 1) permitted almost four times the coulombs charge, when compared to the concrete mixtures containing mineral admixtures (e.g., Mix 2 and Mix 3) in spite of the fact that all three concretes were made with similar cementitious content and water content.

The dramatic effect of mineral admixtures on the chloride penetration values obtained in the ASTM C 1202 test is probably due to the combined effect of a decrease in the permeability of concrete and a decrease in the ionic concentration of the pore fluid. The latter will also have the effect of increasing the electrical resistivity of concrete. Due to a direct relation between the resistivity of concrete and the corrosion of steel reinforcement, the finding that the incorporation of mineral admixtures considerably increases the electrical resistivity of concrete is of great significance in the development of high-performance concrete mixtures. Note that with a further decrease of the w/cm ratio and increase of the proportion of mineral admixtures (Mix 4), the electrical charge passed, dropped to a mere 108 coulombs, which is indicative of very high electrical resistivity. In fact, from direct tests on the resistivity of concrete and the corrosion rates of reinforcing steel, Burg and Ost[16] confirmed that the incorporation of mineral admixtures greatly improved the corrosion resistance of concrete mixtures.

Whereas the laboratory experience with high strength concrete mixtures is generally satisfactory, the field experience is limited to a short time period of 30 years, and the results are rather mixed. Using laboratory size specimens, it could be easily shown that high-strength concrete possesses excellent resistance to permeation of water and aggressive solutions. Therefore, its use was extended from high-rise buildings to long-span bridges, undersea tunnels, offshore oil platforms, and cast-in-place concrete bridge decks and parking garages that are frequently exposed to deicing chemicals.

For indoor applications, for example, foundation, floor slabs, columns and beams of high-rise buildings, high-strength concrete mixtures have continued to perform satisfactorily. In outdoor applications under severe environmental conditions such as exposure to aggressive water, cycles of wetting and drying, heating and cooling, and freezing and thawing, the concrete tends to crack easily due to the low creep and high shrinkage strain characteristics described earlier. Once the watertightness of the concrete is lost, the material would start deteriorating from a number of causes. This explains the reason for unsatisfactory experience in the U.S. with a number of *cast-in-place* concrete bridge decks made of high-strength concrete. However, some long-span bridges, massive beams and piers of precast and prestressed *high-strength concrete structures* are being designed for a 100-year service life. This is because prestressing can control cracking by preventing the development of high tensile stress from thermal, autogenous, and drying shrinkage strains. Some of these structures are described in the "high-performance concrete" section.

12.2.8 High-Strength, Lightweight Aggregate Concrete

Lightweight aggregate (LWA) concrete, with compressive strengths up to 50 MPa, can be commercially produced with high-quality lightweight aggregates. According to Hoff,[22] in 1980s independent studies in North America and Norway resulted in the development of high-strength LWA concrete. The need for offshore concrete structures in the

Arctic and sub-Arctic regions was the driving force in North America. These structures were built in a temperate climate and then floated to their final location. To provide sufficient resistance to ice abrasion and substantial buoyancy while carrying heavy load, it was necessary to use high-strength LWA concrete. The impetus in Norway was the discovery of oil and gas in deeper waters (250 to 350 m) and poor sub-sea soil conditions. To accommodate these conditions, relatively large structures are needed. The structures are built on-shore and floated from the construction site to the point of installation hundreds of kilometers offshore. The use of high-strength LWA concrete in the upper regions of the structure allows it to float and to be towed with an adequate margin of safety.

It has been known for a long time that both the density and the strength of LWA concrete are heavily dependent on the density and the strength of the lightweight aggregate particles. From the results of a North American joint-industry study, Hoff concluded that LWA concrete of 60 MPa or higher strength can be produced by using a high-quality lightweight aggregate, silica fume, and superplasticizer. Malhotra[23] was able to achieve about 70 MPa compressive strength and approximately 1900 kg/m^3 density with superplasticized and air-entrained concrete mixtures containing an expanded-shale lightweight aggregate of Canadian origin, 400 kg/m^3 ASTM Type III portland cement, 80 kg/m^3 fly ash, and 28 kg/m^3 silica fume.

Zhang and Gjorv[24] reported the results of a Norwegian investigation using 400 to 600 kg/m^3 of high early strength portland cement, 9 percent silica fume (as cement replacement), a naphthalene-based superplasticizer, and five high-quality lightweight aggregates of varying microporosity and strength. An expanded clay aggregate, with a dense outer crust, produced concrete with 1800 kg/m^3 density and approximately 100 MPa compressive strength at 28 days. The concrete mixture contained 500 kg/m^3 cement, 50 kg/m^3 silica fume, and 0.36 water-cementitious ratio. Test data showed that the compressive strength of concrete mixtures was not significantly affected by either an increase of the cement content or by partial replacement of the cement with silica fume. Also, it was not significantly affected by partial replacement of natural sand with lightweight sand. Zhang and Gjorv's study clearly shows that the lightweight aggregate microstructure is the primary factor in controlling the compressive strength of high-strength LWA concrete.

Examples of successful use of high-strength LWA concrete include two oil exploration concrete structures placed in the Arctic Ocean, namely, the Tarsuit and the Glomar Beaufort Sea 1. The latter, built in Southern Honshu, Japan, and towed to the Alaskan Beaufort Sea, contains about 9200 m^3 of LWA concrete that has a unit weight of 1790 kg/m^3 and a 56-day compressive strength of 62 MPa. High-strength LWA concrete has also been used in several bridges in Norway. Two of these are floating bridges with pontoons made of LWA concrete. The superstructure of several conventional bridges of double cantilever type is built wholly or partly of LWA concrete.

The selection of high-strength LWA concrete for building the Heidrun tension-leg platform hull and supporting beams provided an important breakthrough for the development of materials and construction practice for high-strength LWA concrete. According to Moksnes and Sandvik,[25] approximately 66,000 m^3 of Grade LC 60 LWA concrete (60 MPa specified strength at 28 days) was produced on shore over a 1-year period. The superplasticized, air-entrained concrete mixture contained 420 kg/m^3 high-strength portland cement, 40 kg/m^3 silica fume slurry (50 percent solids), 590 kg/m^3 of LWA (16 mm max. size), 700 kg/m^3 normal sand, and 0.36 water-cementitious ratio. The typical unit weight of concrete was 1940 kg/m^3, and the 28-day compressive strength of cubes averaged 78 MPa. For quality control purposes, the most critical parameter was judged to be the uniform and predictable moisture content of LWA (preferable in the range 0 to 5 percent). Before delivery, all LWA concrete was retempered with an additional dosage of superplasticizer to restore the workability lost during transportation.

In regard to the thermal cracking potential with high-strength LWA concrete structures, the following observation by Moksnes and Sandvik[25] is noteworthy:

> LWA concrete has lower thermal conductivity and lower heat storage capacity than normal-density concrete. Experience has shown that the cement content needed for high-strength LWA concrete is equal to or slightly higher than normal concrete of the same strength. Consequently, the temperature rise due to heat of hydration will be higher in LWA concrete than in normal concrete. Maximum temperatures of more than 80°C may be generated in large sections unless cooling measures are employed. By replacing a part of the mixing water with crushed ice, temperature reductions on the order of 10°C was achieved.

According to Harmon,[26] in the Arctic Circle, the 711-m-long Raftsundet Bridge, linking the Lofoten Islands with the mainland Norway, used a 60-MPa LWA concrete of 1975 kg/m^3 unit weight in the central span which is 298 m long. The superplasticized concrete mixture is composed of 430 kg/m^3 high-strength portland cement, 23 kg/m^3 silica fume, 745 kg/m^3 normal sand, and 550 kg/m^3 of 12 mm maximum size light-weight aggregate imported from North Carolina, U.S.A. Due to the high quality of the aggregate, placement of concrete by pumping did not have any adverse effect on its quality.

Furthermore, Harmon[26] reported that the Rugsund Bridge linking the Norwegian mainland with the Rugsund Island contains a 60-MPa LWA concrete of 1975 kg/m^3 unit weight in the central part of the main span. This allowed an increase in the length of the main span from 172 to 190 m with the result that the foundations could be moved to shallow water depths, which simplified the construction and reduced the cost by 15 percent. As with the Raftsundet and the Rugsund Bridges, a similar LWA concrete was used for the construction of the 224 m long middle span of this three-span post-tensioned, cast-in-place box girder concrete bridge. The placement of the LWA concrete by pumping significantly reduced the time of construction.

Using high-strength LWA concrete for bridges allows for longer spans and lighter girders. Liles and Holland[27] reported that the Georgia Department of Transportation successfully used high-strength LWA concrete for two spans of the four-span I-85 Ramp crossing State Route 34 in Newnan. The LWA concrete was composed of expanded slate for coarse aggregate and manufactured granite sand; during construction the light-weight aggregates were soaked before batching. The designed compressive strength of the concrete was 69 MPa (10,000 psi) and the girds had a span of 33.5 m (110 ft). At 56 days, field tests showed that the compressive strength exceeded the specified design strength and that the values of the chloride permeability test were very low, in the range of 284 to 360 coulombs.

12.3 Self-Consolidating Concrete

12.3.1 Definition and Significance

The use of high-strength concrete mixtures, with dense steel reinforcement, has successfully met the need of the construction industry for stronger and more ductile concrete structures. However, the *constructability* of highly congested reinforced concrete elements requires the fresh concrete mixtures to be very fluid. The advent of superplasticizers made it possible to achieve slump values on the order of 200 to 250 mm, without the use of too much water. Even then the use of *flowing concrete mixtures* presents the risk of bleeding, segregation, and settlement, which, by weakening the interfacial transition zone between the cement paste and aggregate (also reinforcing steel), would have an adverse effect on the mechanical properties as well as on the durability of concrete.

The risk of materials separation in concrete becomes especially great in heavily reinforced structures with high placement heights and with excessive use of vibrators during the consolidation. So, here is the problem: how can we enhance the stability of the composite system which, when exposed to a high shear rate, behaves as a non-stable dispersion of aggregate particles suspended in a fluidized cement paste? In other words, how to produce concrete mixtures possessing a *high workability*, that is, high fluidity and high cohesiveness simultaneously? In late 1970s and early 1980s pioneering work by German, Italian, and Japanese researchers led to the development of high-workability concrete mixtures that are commercially known by various names such as self-consolidating concrete, self-compacting concrete, self-leveling concrete, or rheoplastic concrete. The *self-consolidating concrete* (SCC) may be defined as a flowing concrete that can be cast into place without the use of vibrators to form a product free of honeycombs (i.e., no unfilled space within the formwork) and bug holes (i.e., no entrapped air voids).

12.3.2 Brief History of Development

The San Marco dry dock at Trieste (Italy) was constructed in 1980, using underwater placement of 40,000 m^3 concrete. It was decided to develop a superplasticized concrete mixture possessing both high fluidity and high cohesiveness. With large dosage (7 kg/m^3) of sulfonated naphthalene-condensate type superplasticizer, high proportion of sand (65 percent sand-to-coarse aggregate ratio), and a relatively high content of cement and mineral admixtures (400 kg/m^3), Collepardi et al.,[28] were able to produce a *rheoplastic* concrete mixture that was self-leveling and highly cohesive. The authors also reported that, in 1983–84, during the placement of superplasticized concrete for a thick foundation for Mass Transit Railway of Hong Kong, the engineers discovered that *the use of poker vibrators was unnecessary*. This is because the concrete mixture was able to flow, under its own weight, around the reinforcing mat.

Although superplasticizing admixtures are expensive, the ability to place concrete rapidly and to consolidate it with little or no cost presents savings that may equal or even exceed the cost of the superplasticizer, while producing a better quality end product. Some of the early work in Japan for development of SCC mixtures is described by Miura et al.[29] and Tanaka et al.[30] The early approaches to enhance the stability of high-fluidity concrete mixtures under conditions of high shear consisted of controlling the volume and the maximum size of the coarse aggregate. According to the result of an investigation by Grube and Rickert[31] most concrete was able to pass easily through the reinforcing bars when the total volume of the coarse and fine aggregates remained below 60 percent, and the volume of the coarse aggregate (12 mm maximum size) did not exceed 25 percent. This means that the paste volume, consisting of water, cement, and mineral admixtures (e.g., fly ash, pulverized limestone, and slag) was at least 40 percent.

In Germany and the United States, there was a considerable interest in the development of antiwashout chemical admixtures for increasing the viscosity of fresh concrete to make it suitable for repairing submerged concrete structures without dewatering. The viscosity-modifying admixtures (VMA) are generally composed of two basic types of chemicals, namely, hydrolyzed starches and biopolymers such as *welum gum*. The latter is relatively expensive but is also more effective for producing self-compacting concrete. Incorporation of a VMA into a concrete mixture enables it to become cohesive even without the use of high content of fine particles of cement and mineral admixtures. This is because the molecular structure of a typical VMA facilitates the removal of large amounts of water by physical adsorption. Consequently, the VMA-containing concrete mixtures exhibit thixotropic (pseudo-plastic) behavior. They appear solidified but, due to the release of the physically adsorbed water, flow readily when agitated, pumped, or vibrated. However, once the shearing action stops, the viscosity increases again.

12.3.3 Materials and Mixture Proportions

Besides sand, coarse aggregate (19 or 25 mm maximum size), and ordinary or blended portland cement, the necessary ingredients for making SCC include superplasicizers, viscosity-modifying admixtures, and mineral admixtures of fine particle size. In general, depending on the approach taken for controlling the bleeding and segregation, SCC mixtures may be classified into two categories: (a) those which contain a high powder content (e.g., more than 400 kg/m^3 cement + fly ash + pulverized slag or limestone), (b) those which contain a viscosity-modifying chemical admixture such as welum gum, hydrolyzed starch, silica fume, and ultrafine amorphous colloidal silica, also called *nanosilica*. The SCC mixtures of the second category do not require a high cementitious content for controlling the viscosity. Typical proportions of some SCC mixtures are shown in Table 12-7 and are briefly discussed next.

According to Collepardi[32] (Table 12-7) Mix A, a close precursor of modern SCC mixtures, was used at the end of 1970s for underwater placement of a self-compacting, nonsegregating concrete. The high consistency (260 mm slump) was achieved by using a relatively large dosage of naphthalene sulfonate type superplasticizer. Excellent cohesiveness was achieved by limiting the maximum aggregate size to 15 mm and by incorporation of 180 kg of very fine sand (0.075 to 0.6 mm) into 1170 kg of total sand. Collepardi et al. also gave mix proportions for high-strength SCC and mass SCC. Mix B represents a high-strength SCC (over 90 MPa compressive strength) that was used in the construction of World Trade Center in San Marino. A high-efficiency acrylic superplasticizer was used to obtain high fluidity. At the same time, high dosage (65 kg/m^3) of silica fume was used to prevent bleeding and segregation. Mix C represents a SCC suitable for massive structures that require control of the heat of hydration. Due to the low portland cement content (120 kg/m^3), the temperature rise in the concrete was below 20°C, which is considered safe for preventing thermal cracking.

Mix D is from an experimental study by Bouzouba and Lachemi,[33] which showed that relatively inexpensive SCC mixtures can be produced by incorporation of high volume of ASTM Class F fly ash. SCC mixtures with 0.35 w/cm to 0.45 were made using ASTM Type I normal portland cement, 40 to 60 mass percent cement replacement by fly ash, a local natural sand, crushed limestone aggregate of maximum 19 mm size, an air-entraining admixture, and small dosage of a sulfonated naphthalene-formaldehyde superplasticizer. Compared to the control 35 MPa concrete with no superplasticizer, the high-volume fly ash SCC was flowable, resistant to segregation and to thermal cracking, and did not result in any increase in cost. Only the setting time was longer by 3 to 4 h.

Mix No.	A	B	C	D	E	F
Mix proportions, kg/m³						
Ordinary portland cement	400	465	120	207	330	360
Fly ash	–	–	127	207	–	–
Slag	–	–	184	–	–	–
Standard sand	1170*	710	964	845	960	960
Coarse aggregate	630	915	822	843	900	900
Water	190	175	176	188	180	180
Superplasticizer	7.0	4.6	5.0	0.5	3.0	3.5
Viscosity modifying agents						
Welum gum	–	–	–	–	0.13	0.25
Silica fume	–	65	–	–	30	–
Ultrafine colloidal silica	–	–	8	–	–	–
Properties						
Slump, mm	260	–	–	240	220	220
Slump flow,† mm	–	730	800	–	–	–
w/cm	0.47	0.33	0.40	0.45	0.50	0.50
28-day compressive strength, MPa	–	95	40	33	46	43

*Mix A contains 990 kg/m³ standard sand and 180 kg/m³ of very fine sand.
†Diameter of the concrete pat after removal of the slump cone.
Mix A, B, C: Collepardi, M., et al., *ACI*, SP-222, pp. 1–18, 2004.
Mix D: Bouzubaa, N., and M. Lachemi, *Cem. Concr. Res.*, Vol. 31, pp. 413–420, 2001.
Mix E & F: Khayat, K.H., and Z. Guizani, *ACI Mat. J.*, Vol. 94, No. 4, pp. 332–340, 1997.

TABLE 12-7 Composition and Properties of Typical Self-Consolidating Concrete Mixtures

Mix E and F, from an experimental investigation by Khayat and Guizani,[34] contain welum gum powder as a viscosity-modifying admixture (VMA) in combination with a sulfonated naphthalene-based superplasticizer, a well graded sand (2.5 fineness modulus) and a coarse aggregate of 10 mm maximum size. Mix F, with 0.25 kg/m³ welum gum, produced a SCC that showed greater stability to segregation than Mix E, which contained 0.13 kg/m³ welum gum together with 30 kg/m³ silica fume.

12.3.4 Properties of SCC

The principal difference between ordinary concrete and SCC is the superior rheological characteristics of the latter. A typical SCC is a flowing concrete which gives a slump value in excess of 200 mm and a slump-flow value in excess of 600 mm (diameter of the concrete pat after removal of the slump cone) is highly cohesive, and can be placed and compacted without the help of vibrators. These properties are usually achieved by a higher cement paste-to-aggregate ratio than is normally used with ordinary concrete mixtures. Consequently, the drying shrinkage of SCC tends to be high, and the thermal shrinkage will also be high when the concrete contains a relatively large proportion of portland cement and other reactive powders (e.g., finely ground granulated blast-furnace slag or ASTM Class C fly ash).

ASTM C1712-09[35] covers the rapid assessment of static segregation resistance of normal-weight self-consolidating concrete. The test gives an assessment of whether static segregation is likely to occur.

Ordinary SCC mixtures are designed for 0.45 to 0.50 w/cm and they generally acquire 28-day compressive strength on the order of 40 MPa (Table 12-7). By adjusting the w/cm ratio and the component of the cementitious materials, it is possible to produce both high-strength SCC (Mix B), and low-heat SCC (Mix C and D).

12.3.5 Applications

In Europe and Japan, SCC has been used for underwater concreting and for the construction of heavily reinforced structures. According to Spiratos et al.,[36] in North America at present SCC is being used primarily in precast concrete plants where there is a high degree of quality control. Some ready-mixed concrete plants are reluctant to produce SCC due to the high cost and additional requirements for quality control that are necessary when using viscosity-controlling admixtures.

12.4 High-Performance Concrete

12.4.1 A Brief History of Development

Conventionally, products that last longer are called *high-performance products*. From laboratory investigations many researchers have reported that properly proportioned and cured mixtures of superplasticized concrete, with 0.4 or less w/cm, show a little or no permeability, which is the most desired property for long-term durability of structures exposed to corrosive environments. In fact, superplasticized concrete mixtures made with blended portland cements containing mineral additives exhibit unusually low permeability ratings in the ASTM C 1202, rapid chloride penetration test.

In a 1990 publication, Mehta and Aitcin[15] suggested the term *high-performance concrete* (HPC) for concrete mixtures that possess the following three properties: high-workability, high-strength, and high durability. However, as described next, ACI has taken a different approach which has resulted in a situation that many of the concrete mixtures that are being marketed today as HPC may not prove to be durable under demanding environmental conditions.

12.4.2 ACI Definition and Commentary on High-Performance Concrete

In 1991, ACI's Technical Activities Committee formed a task-group on high-performance concrete to develop a definition and description of the performance characteristics of HPC. The definition and commentary that was approved by the ACI in 1998 are as follows:

> HPC is defined as a concrete meeting special combination of performance and uniformity requirements that cannot always be achieved routinely using conventional constituents and normal mixing, placing, and curing practices.

According to the commentary on the definition, a high-performance concrete is a concrete in which certain characteristics are developed for a particular application and environment. Examples of characteristics that may be considered critical for particular application are

- Ease of placement
- Compaction without segregation
- Early age strength
- Long-term strength and mechanical properties
- Permeability
- Density
- Heat of hydration
- Toughness
- Volume stability
- Long life in severe environments

Note that according to the ACI definition, durability under severe environmental conditions is an optional, not a mandatory requirement for HPC. This can be problematic for those who assume that the term high-performance automatically implies a long service life. As described next, field experience with cast-in-place HPC mixtures, meeting the ACI definition, shows that those possessing a very-high early strength are prone to cracking at early-age from high autogenous shrinkage and high thermal shrinkage.

12.4.3 Field Experience

The Strategic Highway Research Program (SHRP) in the United States defined HPC for highway structures by three requirements, namely a maximum w/cm, a minimum durability factor to cycles of freezing and thawing (ASTM C 666, Method A), and a minimum early-age or ultimate compressive strength. Under SHRP, four types of HPC were subsequently developed: very early strength (14 MPa in 6 h), high early strength (34 MPa in 24 h), very high strength (69 MPa in 28 days), and high early strength with fiber-reinforcement.

Based on SHRP recommendations, the U.S. Federal Highway Administration (FHWA) sponsored a national program of field testing HPC bridge decks. The assumption that "*stronger concrete mixtures would be more durable*" did not turn out to be true in the case of many *cast-in-place* and exposed concrete structures, therefore FHWA has revised the definition of HPC for highway structures. A recent publication[37] contains the following statement:

> HPC is a concrete that has been designed to be more durable and *if necessary*, stronger than conventional concrete. HPC mixtures are essentially composed of the same materials as conventional concrete mixtures. But the proportions are designed or engineered to provide the strength and durability needed for the structural and environmental requirements of the project.

Unlike the ACI definition of HPC, which is broad in one sense (e.g., an attempt to cover all types of concrete) and restrictive in another sense (e.g., use of special materials and construction practices), the FHWA guideline for HPC is more practical and useful. First, by bringing the issue of concrete durability in the forefront, it emphasizes that strength is not synonymous with durability and, *for high-performance*, high durability is more important than high strength. Second, it encourages the use of local materials, which is a step in the right direction from standpoint of cost, and materials and energy conservation.

12.4.4 Applications

Many sophisticated concrete structures are now being designed for longer service life, for example, 100 to 120 years. When exposed to aggressive fluids and demanding environmental exposure conditions, it is essential that the concrete should remain crack-free and impermeable for a long period. HPC mixtures are being used for the construction of structural components of offshore oil-drilling platforms, long-span bridges, and highway bridge decks. Examples of HPC mixtures and relevant features of the construction practice are described as follows:

Off-Shore, Oil Drilling Platforms

Since 1970s some 20 concrete platforms have been installed in the British and Norwegian sector of the North Sea. The construction process involves the fabrication of massive oil-storage cells and prestressed concrete shafts on a dry dock under rigorous quality control conditions, and then towing the structural components to the job site. The concrete is required to withstand the corrosive action of seawater, and impacts and erosion from high tidal waves. According to Moksnes and Sandvik,[22] earlier North Sea platforms were designed for a service life of approximately 30 years and have performed satisfactorily. For the platforms completed in 1993–95, Troll A and Heidrun, the design life has been extended to 50 to 70 years. Based on previous experience, extra precautions were taken to prevent thermal cracking. The precautions included cooling the concrete constituents, partly replacing the mixing water with crushed ice, increasing the silica fume content from 3 to 9 percent, and decreasing the w/cm ratio from about 0.40 to 0.36. The mix proportions of the lightweight (1940 kg/m^3), high-strength (70 MPa specified cube strength at 28 days), and pumpable (220 to 250 mm slump) concrete for the Heidrun offshore platform, which contains a unique tension-leg floating in 345 m deep seawater, are described in Sec. 12.2.8.

Long Span Bridges

HPC is being extensively used now for the fabrication of precast pylons, piers, and girders of many long span bridges in the world. Materials and construction practices used for the following three bridges are briefly described here: The Normandie Bridge in France, the East Bridge of the Great Belt Link in Denmark, and the Confederation Bridge in Canada.

According to Aitcin,[12] the Port de Normandie cable-stay bridge was the longest in the world when it was built in 1993. It has 2141 m overall length and a center span of 856 m. Approximately 35,000 m^3 of 60-Grade HPC (60 MPa specified strength at 28 days) was used in the construction of pylons and cantilever beams. The concrete mixture was composed of 425 kg/m^3 blended portland cement containing 8 percent silica fume, 770 kg/m^3 fine aggregate, 1065 kg/m^3 coarse aggregate of 20 mm maximum size, 153 kg/m^3 water (w/cm = 0.36), and 11 L/m^3 of melamine-type superplasticizer. Measures to prevent thermal cracking included the use of insulated forms, and protecting the deck slab with hot air for 18 h followed by a 6-h waiting period before exposing the concrete to ambient temperature.

The Great Belt Link in Denmark shown in Fig. 12-9 and the Confederation Bridge in Eastern Canada provide examples of HPC components that are designed to last at least

(a) (b)

FIGURE 12-9 Construction of the Great Belt Link. (*Photograph, courtesy of Ben Gerwick.*)

At a cost of 4 billion USD, the Great Belt fixed link in Denmark provided a major improvement to the Northern European transportation system. The island of Sprogo divides the 18-km Great Belt into two parts. The Great Belt link has a railway tunnel, a high-level motorway bridge across the East Channel and a low-level bridge for rail and motorway across the West Channel. High-quality precast concrete segments were fabricated on dry docks under controlled environment. Even for the 50,000-ton precast concrete units, construction tolerances were within a few centimeters.

100 years in severe environmental conditions. In both cases, the bridge pier foundation elements, the main span, and the cantilever girders of the segmentally constructed bridges were prefabricated on land and then transported to the site by a marine-crane vessel. The concrete is required to withstand not only the corrosive action of seawater but also the impact of ice floes. Furthermore, exposure to many freezing and thawing cycles each winter made it necessary to use air-entrained in concrete mixtures. Mix-proportions and properties of concrete used for the construction of the Confederation Bridge are shown in Table 12-8. Note that prestressing and the use of precooled concrete were important factors in controlling the thermal cracking.

According to Langley et al.,[38] the 12.9-km-long Confederation Bridge between Prince Edward Island and the mainland of Canada (across the Northumberland Strait), consists of 44 main spans of 250 m length each and massive main pier shaft and foundation elements fabricated on land with Class A concrete (Table 12-8). Approach pier foundations and some mass concrete sections, requiring control of thermal cracking, were built with Class C concrete that contained approximately 32 percent fly ash as a cement replacement material. Both concrete mixtures contained 7.5 percent silica fume by mass of the total cementitious materials. The specifications for Class A concrete included a minimum of 55 MPa compressive strength and a maximum of 1000 coulombs chloride permeability (ASTM C 1202 test) at 28 days. Some piers, with an abrasion resistant ice shield, were made with 80 MPa concrete. The requirements for Class C concrete were 30 and 40 MPa minimum compressive strength at 28 and 90 days, respectively. The requirements for strength, chloride permeability, workability, and air entrainment were easily met. Mix proportions and properties of the HPC concrete used to fabricate the piers and T-beams for the eastern segment of the Great Belt Link Bridge were similar to the Class C concrete of the Confederation Bridge (Table 12-8), except that the cement content was somewhat higher and the fly ash content was lower.

Mix Proportions, kg/m³	Class A Concrete for Main Piers and T-Beams	Class C Concrete for Massive Foundations	Abrasion-Resistant Ice Shield Concrete
Portland cement	416	285	478
Silica fume	34	22	42
Fly ash, Class F	–	133	60
Fine aggregate	737	744	650
Coarse aggregate	1030	1054	980
Water	153	159	142
Superplasticizer	3	2	6
W/cm	0.34	0.37	0.25
Properties			
Entrained air, %	6.1	7.0	–
Slump, mm	200	185	–
Compressive Strength, MPa			
1-day	35	9.7	–
3 days	52	27.4	–
28 days	82	50.0	100
91 days	–	76.0	–
Rapid Chloride Permeability, Coulombs (AASHTO T277)			
28 days	300	420	–
90 days	–	–	–

Source: Langley, W.S., R. Gilmour, and E. Trompsch, ACI SP-154, 1995.

Table 12-8 Confederation Bridge HPC Mix Proportions and Properties

Bridge-Decks, Pavements, and Parking Structures

Concrete mixtures containing low cementitious materials content (<300 kg/m^3) and w/cm > 0.5 are prone to show premature deterioration when exposed to corrosive conditions, such as seawater or deicing salts. Published literature contains numerous reports of early deterioration of concrete in bridge decks, pavements, and parking structures (Chap. 5). The advent of superplasticizers provided an impetus for the development of very high-strength concrete mixtures that found their way quickly into *cast-in-place* structures designed for long-term durability under severe environmental conditions.

Burrows[39] has summarized the early field experience with cast-in-place HPC bridge decks in Virginia, Kansas, Texas, and Colorado. The superplasticized concrete mixtures used for the construction of the bridge decks typically contained a high cement content (400 to 500 kg/m^3), a low w/cm (0.30 to 0.35), and usually 9 to10 percent silica fume by mass of the cementitious material. In Denver, Colorado, a bridge deck was made with this type of concrete that showed, within a short period, severe early-age cracking attributable to high thermal and autogenous shrinkage which was attributed to the use of a high-early strength concrete containing too much of a reactive portland cement,* silica fume, and a low w/cm (0.31). Consequently, several state and public transportation agencies in the United States have now revised their concrete specifications for bridge decks, with special attention to cracking at early-age and durability issues. For instance, according to Bognacky et al.,[40] the Port Authority of New York and New Jersey, which maintains many major public transportation facilities in New York City metropolitan area, is now specifying permeability as the primary concrete property to determine payment to contractors. Their test results showed that concrete mixtures containing less than 400 kg/m^3 cementitious content and 0.4 w/cm ratio, with 30 percent cement replacement by fly ash or 40 percent cement replacement by slag, gave considerably low coulomb values in the AASHTO T-277 (ASTM C 1202) rapid chloride permeability test than would have been possible with mixtures without fly ash or slag.

A review of the typical mixture proportions and properties of concrete currently being used for long-term durability of cast-in-place bridge decks, pavements and other infrastructure in New York, New Jersey, New Mexico, Texas, and Virginia shows that, in general:

- The total cementitious material content does not exceed 400 kg/m^3 and, typically, 30 to 40 percent portland cement is being replaced by fly ash or granulated blast-furnace slag. The use of pozzolans and slag is considered mandatory for achieving low permeability and for controlling the thermal cracking.
- A low chloride permeability rating, for example, maximum 1500 or 2000 coulombs at 56 days (ASTM C1202 Test Method) is specified because this property is more desirable than high-early strength for concrete structures exposed to deicing chemicals and seawater. A very low chloride permeability (<1000 coulombs) can be readily achieved even at the early age of 28 days by incorporation of 7 to 10 percent silica fume, metakaolin, or rice husk ash in the concrete mixture.

Good construction practice and adequate curing are critical for achieving a long service life in the field. According to Gjorv,[41] lack of good construction practice and quality control were the principal causes for premature deterioration of concrete due to the corrosion of reinforcement in bridge decks exposed to Norwegian marine environment. He pointed out that similar materials and mix proportions had been used for the construction of precast concrete elements for offshore drilling platforms which are in excellent condition after 30 years of service in the North Sea. Note that precast concrete products are generally made under more rigid quality control than cast-in-place concrete, and often they are prestressed.

In conclusion, concrete mixtures with high volume of fly ash or ground granulated blast-furnace slag, qualify for being classified as high-performance concrete because they are not only economical but also more durable. Furthermore, the performance of cements and concrete mixtures in the future will be judged from standpoint of sustainability of the concrete industry. An emerging technology of high-volume fly ash concrete mixtures that are sustainable, durable, and almost self-consolidating is described next.

---

*Note that modern portland cements are very reactive. The ASTM Type I/II cement used for the Denver viaduct concrete had approximately 400 m^2/kg Blaine fineness and over 70 percent $C_3S + C_3A$.

12.4.5 High-Performance, High-Volume Fly Ash Concrete

The high-volume fly ash (HVFA) concrete system, originally developed by Malhotra and his associates[42–45] has emerged as a powerful tool by which we can build concrete structures in the future that would be far more durable and resource-efficient than those made of conventional portland-cement concrete. Whether used as a component of blended cements or as a mineral admixture added to concrete batch during mixing, the fly ash content of concrete must be above 50 percent by mass of the total cementitious material in order to meet Malhotra and Mehta's definition for HVFA concrete.[46]

In the past, the HVFA concrete mixtures generally did not perform well with respect to strength development, drying shrinkage, and durability. This is because the fly ash produced by old thermal power plants was coarser and usually contained high carbon. Laboratory and field experience have shown that fly ash from the modern thermal power plants, generally characterized by low carbon content and high fineness, *when used in a large volume, is able to impart excellent workability to concrete at a water content that is 15 to 20 percent lower than without the fly ash.* Further reductions in the mixing water content can be achieved with better aggregate grading and with the help of a superplasticizing admixture. As described next, a large reduction in the mixing water content has a highly beneficial effect on the cracking resistance and durability characteristics of concrete.

To understand why *HVFA concrete is more crack resistant* than a conventional portland cement concrete, typical proportions of two concrete mixtures are shown in Table 12-9.[47] The mixtures are designed to achieve 25 MPa strength, and 125 to 150 mm slump. Compared to the conventional concrete mixture, the HVFA system contains one-third less mixing water. As a result, not only the w/cm ratio of the HVFA concrete is lower but also the total volume of the cement paste is nearly 16 percent less. Consequently, the drying shrinkage, which is directly related both to the w/cm and the proportion of cement paste present in the concrete, is much reduced. At the same time, due to a drastic reduction in the amount of portland cement, the HVFA concrete

Materials	Conventional Concrete		HVFA Concrete	
	By Mass (kg/m^3)	By Volume (m^3/m^3)	By Mass (kg/m^3)	By Volume (m^3/m^3)
Cement	307	0.098	154	0.048
Fly ash	–	–	154	0.064
Water	178	0.178	120	0.120
Entrapped air (2%)	–	0.020	–	0.020
Coarse aggregate	1040	0.385	1210	0.450
Fine aggregate	825	0.319	775	0.298
Total	2350	1.000	2413	1.000
w/cm	0.58	–	0.38	–
Paste volume:	–	(0.296)	–	(0.252)
Percent:		29.6%		25.0%

TABLE 12-9 Comparison of Mixture Proportions for 25-MPa Concrete

generates nearly 40 percent less heat of hydration at early age and, therefore, in massive structural members the potential for thermal shrinkage and cracking is also greatly reduced.

What about strength and durability characteristics? Typically, the water content of the HVFA concrete mixtures is maintained within the narrow limits of 100 and 120 kg/m^3 (170 to 200 lb/yd^3). Depending on the desired heat of hydration characteristic and compressive strength level, which may range between 20 and 40 MPa at 28-days for ordinary concrete, the portland cement content may vary between 100 and 200 kg/m^3 (170 to 340 lb/yd^3) and the superplasticizer dosage between 1 and 4 L/m^3. For frost resistance, the use of an air-entraining admixture is mandatory. The setting time is somewhat longer and the early strength development rate of HVFA concrete is slower. However, under warm weather conditions, the strength at 24 to 48 h is generally adequate for formwork removal. When possible the forms may be kept in place for 7 days or even longer to maintain the moist-curing environment. The long-term strength and impermeability of HVFA concrete is, generally, far superior to ordinary concrete with no fly ash or a small quantity of fly ash. Also, from laboratory test data on the corrosion of steel reinforcement, alkali-aggregate expansion, and sulfate attack, Malhotra[45] has confirmed the excellent durability characteristics of HVFA concrete mixtures.

Mixture proportions, construction practice, and properties of HVFA concrete used for building structures in North America are described in various publications.[46–49] Mehta and Langley[49] described the construction of a large HVFA concrete foundation designed to endure for at least one thousand years (Fig 12-10). The foundation consists of a massive, unreinforced, monolith structure composed of two parallel slabs, each measuring 36 × 17 × 0.62 m. There was one-week interval between the two castings. Each slab was cast with a 20-MPa, HVFA concrete mixture containing 106 kg/m^3 ASTM Type I cement, and 142 kg/m^3 Class F fly ash that registered only 13°C maximum temperature rise. When inspected last, 5 years after the construction, the exposed surfaces did not reveal a single crack in the concrete, and the strength was nearly double the specified value. In addition to this structure that is still under construction, photographs of two other structures having similar HVFA concrete foundations are shown in Figs. 12-11*a* and *b*.

Representing the other end of the spectrum, Fig. 12-11*c* shows a photograph of another structure that contains heavily reinforced shear walls and a massive, post-tensioned, reinforced concrete foundation, 3.4 m deep and 1.8 m wide. The primary goal of the owner and structural designer was to build a crack-free, *green concrete*, structure. The HVFA mixtures for the shear walls as well as for the foundation were designed to have 150 mm slump and 35 MPa strength at 28 days. Additionally, to facilitate the re-use of formwork, the wall concrete mixture was designed to obtain 20 MPa strength in 7 days. In the case of foundation concrete, the primary concern was thermal cracking, therefore, the goal was to keep the temperature difference between the interior and the surface of the concrete within the desired limit of 25°C. Superplasticized HVFA concrete mixtures, with 0.32 w/cm, 195 kg/m^3 Type I portland cement, and 195 kg/m^3 Class F fly ash met the shear wall concrete specifications satisfactorily. However, for the foundation concrete mixture the cement content had to be reduced to 160 kg/m^3 in order to control the maximum temperature rise. Due to the excellent workability of HVFA concrete, the finished surface of the walls and the foundation showed no

(a)

(b)

Figure 12-10 Hindu Temple, Kauai Island, Hawaii. (*a*) Placement and finishing of HVFA concrete; (*b*) monolith, HVFA concrete foundation supporting an assembly of carved column and beam sections, designed for a service life of 1000 years.

honeycombing and no bug holes. Also, inspection of the walls, some two years after the construction, revealed no cracks.

In developing countries like India, HVFA concrete is being used now for the construction of low-cost housing and pavements. With growing interest in durability and sustainability issues, other countries are also expected to adopt this technology.

(a)

(b)

FIGURE 12-11 (*a*) The BAPS temple and cultural complex at Bartlett, near Chicago. The entire superstructure made of carved segments of white marble, is supported by drilled piers and massive unreinforced slabs composed of HVFA, (*b*) Placement and consolidation of HVFA concrete showed high workability. (*Source: Sachi Construction Company, Chicago, Il.*)

(c)

Figure 12-11 (c) Heavily reinforced shear halls for the seismic retrofit of Barker Hall at the University of California's Berkeley campus are built with HVFA concrete. (*Source: D. Manmohan and P.K. Mehta, Concrete Intl., Vol. 24, No. 8, 2002, pp. 64–70.*) (*Continued*)

Compared to ordinary portland cement concrete, HVFA concrete shows superior performance characteristics in so many respects (Table 12-10)[46] that it deserves to be classified as a high-performance concrete.

12.5 Shrinkage-Compensating Concrete

12.5.1 Definition and the Concept

Shrinkage-compensating concrete[50] is an expansive cement concrete which, when properly restrained by reinforcement of other means, will expand an amount equal to or slightly greater than the anticipated drying shrinkage. Because of the restraint, compressive

Flowability/pumpability	Easier
Workability/compactibility	Easier
Bleeding	None or negligible
Finishing	Quicker
Setting time	Slower up to 2 h
Early strength (up to 7 days)	Lower but can be accelerated
Ultimate strength (90 days +)	Higher
Crack resistance	Higher
• Plastic shrinkage	Higher if unprotected
• Thermal shrinkage	Lower
• Drying shrinkage	Lower
Resistance to penetration of chloride ions	Very high after 3 months
Electrical resistivity	Very high after 3 months
Durability	
• Resistance to sulfate attack	Very high
• Resistance to alkali-silica expansion	Very high
• Resistance to reinforcement corrosion	High
Cost	
• Materials	Lower
• Labor	Similar
• Life cycle	Very low
Environmental benefits (reduced CO_2 emission)	Very high

Table 12-10 HVFA Concrete Compared to Conventional Portland-Cement Concrete

stresses will be induced in the concrete during expansion. Subsequent drying shrinkage will reduce these stresses. Ideally, a residual compression will remain in the concrete, eliminating the risk of shrinkage cracking.

A graphical representation of the concept in Fig. 12-12 compares the behavior of portland cement concrete with a Type K expansive cement concrete* during the early moist-curing and subsequent air-drying periods. Briefly speaking, this is how the concept underlying the use of shrinkage-compensating concrete works to reduce or eliminate the drying-shrinkage cracking in a reinforced concrete element. As the Type K cement hydrates, large amount of ettringite are formed. When the concrete sets and develops strength, it will bond to the reinforcement and at the same time start expanding if sufficient quantities of curing water are present. Since the concrete is bonded to steel, its expansion under the restraining influence of the steel will induce tension in the latter while the *concrete itself goes into compression*. At the end of moist curing, when the

*The properties and applications of shrinkage-compensating concretes described in this book are base on Type K expansive cements (Chap. 6), which is the only type of expansive cement being produced in the United States.

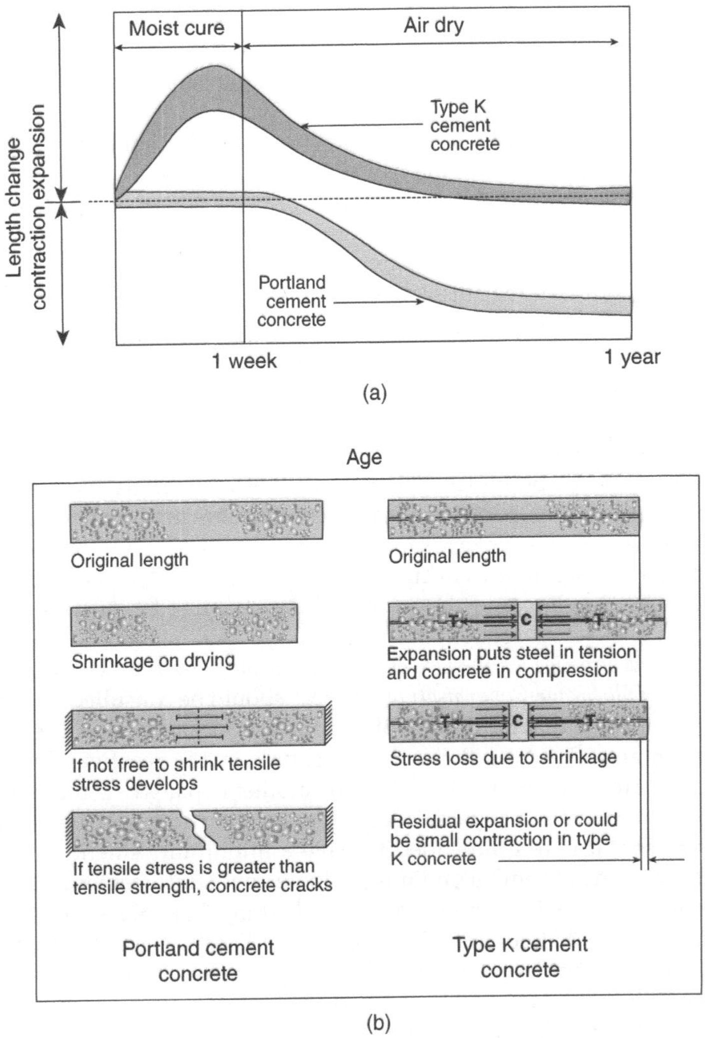

FIGURE 12-12 (*a*) Comparison of length change characteristics between portland cement and Type K cement concrete; (*b*) illustration showing why Type K cement concrete is resistant to cracking from drying shrinkage. (From Williams, J.V., Jr., *Concr. Int.*, Vol. 3, No. 4, p. 58, 1981.)

element is exposed to drying conditions, it will shrink like a normal portland cement concrete. However, the shrinkage will first relieve the *precompression* before inducting tensile stress in the concrete due to drying shrinkage is thus reduced.

12.5.2 Significance

Shrinkage on drying is a fundamental property of the calcium silicate hydrate (see Chap. 2), which is the principal hydration product of portland cements. Therefore, as described in Chap. 4, small and medium-size concrete elements are prone to undergo considerable drying shrinkage within a few months, and they usually crack to relieve the tensile stress induced under restraint. Cracks in concrete are unsighted; by increasing the permeability of concrete the cracks also become instrumental in reducing durability of structures exposed to aggressive waters. In some cases, they may even threaten the safety of a concrete structure.

Current design and construction practices assume that concrete will crack, and try to get around the problem in many ways, such as selection of materials and proportioning of concrete mixtures that will shrink less, provision of adequate joints in floor slabs or pavements, and reinforcement of concrete elements with steel. The advent of expansive cement has offered an alternative and cost-effective approach. As a result, since the first industrial use of the shrinkage-compensating concrete in the United States about 50 years ago, the material has found successful application in the construction of a variety of elements, such as floor slabs, pavements, roofs, water storage tanks, and sewage digesters.

12.5.3 Materials and Mix Proportions

According to ACI Committee 223, the same basic materials and methods necessary to produce high-quality portland cement concrete are required to produce satisfactory results in the use of shrinkage-compensating concrete. Additional care is necessary to provide continuous moist curing for at least 7 days after placement in order for the expansion to develop, and the structural design must be such as to ensure adequate expansion to offset subsequent drying shrinkage. For details, ACI Committee 223, 223-98 *Recommended Practice for the Use of Shrinkage-Compensating Cements* and 233R-10 *Guide for the Use of Shrinkage-Compensating Concrete*, should be consulted. Expansion in concrete can be achieved either by the use of a modified portland cement (e.g., Type K cement, which contains $C_4A_3\bar{S}$ as the principal source of the reactive aluminate needed for ettringite formation) or by the addition of another expansive cement or additive to portland cement concrete. To ensure adequate expansion and restraint when Type K cement is being used, it is recommended to have a minimum cement content of 515 lb/yd^3 (305 kg/m^3) concrete with a minimum 0.15 percent reinforcement. The composition and properties of expansive cements are described in Chap. 6. Although Type K cement is used in the United States and China, generally Type O cement is used in Japan.

In regard to admixtures, the air-entraining admixtures are as effective with shrinkage-compensating concrete as with portland cement concrete in providing freeze-thaw and deicer salt durability. Calcium chloride, excessive amounts of fly ash and other pozzolans, and some water-reducing agents tend to reduce expansion by causing an imbalance between the rate of ettringite formation and the rate of strength development in the concrete.

For the determination of mix proportions, it is suggested that ACI 211.1 (Chap. 9) procedures be used except that compared to portland cement concrete, a slightly higher water-cement ratio may be employed in shrinkage-compensation concrete for achieving the same strength level. Owing to the relatively large amount of water needed for ettringite formation, and the water-imbibing property of ettringite, approximately 10 percent additional water may be used with a Type K expansive cement concrete, without strength impairment, in order to produce a consistency similar to that of a Type I portland cement concrete having the same cement content. In addition to the report of ACI 223, several other reports, such as the one by Williams,[51] contain useful information for mix proportioning and design and construction procedures for shrinkage-compensating concrete structures. Typical mix proportions and properties

Mix Proportions (No Admixtures Were Used)	
	kg/m³ (lb/yd³)
Type K cement	350 (588)
Coarse aggregate (1 in., max.)	1070 (1790)
Fine aggregate	770 (1287)
Water	185 (312)
Water-cement ratio	0.53
Properties	
Slump	4¾ in. (120 mm)
Specified compressive strength	4000 psi (27 MPa)
Average 28-day strength achieved	6034 psi (41 MPa)

Source: Based on Hoffman, M.W., and E.G. Opbroek, *ACI Concr. Int.*, Vol. 1, No. 3, pp. 19–25, 1979.

TABLE 12-11 Typical Mix Proportions and Shrinkage-Compensating Concrete Used in the Construction of a Floor Slab

of shrinkage-compensating concrete used in a floor slab described by Hoffman and Opbroek[52] are given in Table 12-11.

12.5.4 Properties

Workability

Because of the water-imbibing characteristic of ettringite, which forms in relatively large quantities during very early stages of hydration, the concrete mixtures tend to be stiff but highly cohesive. Compared to normal portland cement, the use of a somewhat higher water-cement ratio (without the possibility of strength impairment) than recommended by the standard w/c-strength relationships for normal portland cement concrete, is therefore permitted with expansive cements for achieving a reasonable consistency. Slumps in the range of 100 to 150 mm are recommended for most structural members, such as slabs, beams, reinforced walls, and columns. Because it is more cohesive than portland cement concrete and has less tendency to segregate, the Type K shrinkage-compensating concrete is reported to be especially suitable for placement by pumping.

Slump Loss

Slump loss under hot (concrete temperatures 32°C or higher) and dry conditions is more serious a problem in shrinkage-compensating concrete than with normal portland cement concrete. As a result of slump loss, excessive retempering of concrete on the job site not only will reduce the strength but also the expansion, which defeats the purpose for which the concrete is used. At higher than 17 to 29°C ambient temperatures, unless the concrete is cooled, both the amount of ettringite formed and the rate of its formation may be large enough to cause severe slump loss and quick setting. During the construction of Houston's wastewater and sludge treatment plant,[53] the contractor experienced extended periods of 38°C ambient temperatures, which necessitated the use of 50 to 125 lb/yd^3 (30 to 75 kg/m^3) of ice in Type K concrete in order to lower the fresh concrete temperature to between 24 and 29°C at the job site.

Plastic Shrinkage

Because of lack of bleeding and quicker stiffening and setting of concrete under hot, dry, and windy conditions, plastic shrinkage cracking is another problem for which extra precautions must be taken when using the shrinkage-compensating concrete. When fresh concrete is likely to be in contact with an absorptive surface such as a dry soil or an old concrete, the base should be thoroughly saturated by soaking it the evening before placement. Special precautions should be taken to avoid placement delays at the job site when using ready-mixed concrete. For slabs, fog spraying or covering the surface with wet blankets soon after placement is desirable in order to prevent rapid moisture loss.

Strength

The development of compressive, tensile, and flexural strength in shrinkage-compensating concrete is generally influenced by the same factors as portland cement concrete. Polivka and Willson[54] found that for a given water-cement ratio (in the range 0.4 to 0.65), the compressive strengths of Type K cement concrete were significantly higher than that of Type I portland cement concrete (Fig. 12-13*a*). In the case of shrinkage-compensating concrete, a denser cement paste matrix and a stronger interfacial transition zone between the cement paste and the coarse aggregate are the factors responsible for strengths higher than those of a portland cement concrete made with an equivalent water-cement ratio.

Volume Changes

The drying-shrinkage characteristics of a shrinkage-compensating concrete are comparable to those of a corresponding portland cement concrete; the rate and the magnitude of shrinkage in both the cases are affected by the same factors, such as aggregate content and type, and water content. However, in the case of shrinkage-compensating concrete, the influence of the water-cement ratio on expansion during the early moist-curing period is quite important. Polivka and Willson's data (Fig. 12-13*b*) showed that with a water-cement ratio of 0.53 or less, the magnitude of the initial expansions was such that there was always some residual expansion even after 2 months of drying shrinkage. The magnitude of expansion decreased substantially with the increase in water-cement ratio (e.g., 0.76). Since the degree of needed precompression in shrinkage-compensating concretes may reduce considerably with water-cement ratios above 0.6, it is recommended that low water-cement ratios be used even when this is not needed from the standpoint of strength.

Everything else remaining the same, curing conditions are equally important for determining the overall volume change in shrinkage-compensating concrete. Probably for the reason cited in Chaps. 5 and 6 (i.e., the role of water in ettringite expansion), Kesler[55] found (Fig 12-13*c*) that the expansion of shrinkage-compensating concrete made with a Type M cement was significantly reduced when wet burlap or water curing was replaced by a curing procedure based on moisture retention (polyethylene sheet covering); only a little expansion occurred when the sealed-in-mixing water was the sole source of curing water. Available data on modulus of elasticity, creep, and Poisson's ratio show that for shrinkage-compensating concretes these properties are within the same range as those of portland cement concretes of comparable quality.

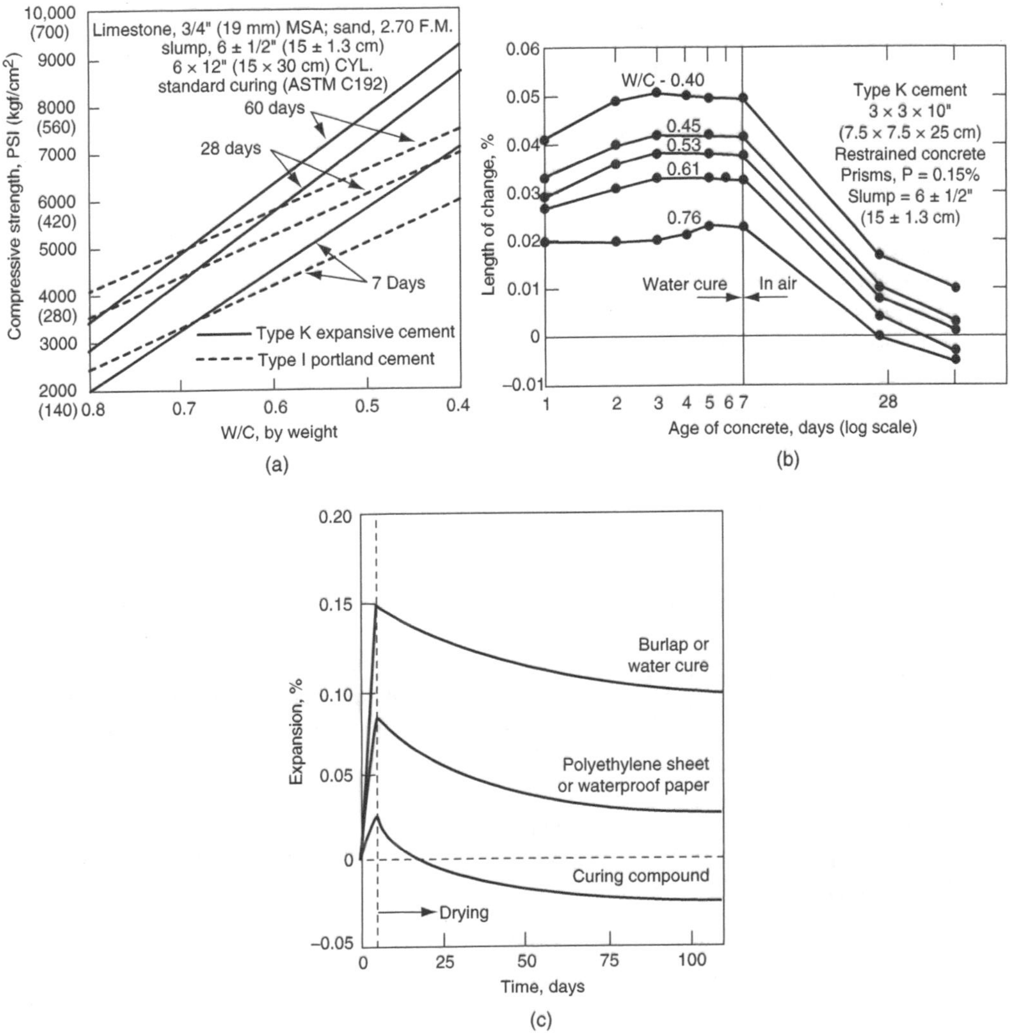

Figure 12-13 Factors affecting the properties of expansive cement concrete: (*a*) effect of water-cement ratio on strength; (*b*) effect of water-cement ratio on expansion; (*c*) effect of curing conditions on expansion. [(*a*), (*b*) From Polivka, M., and C.W. Wilson, *Expansive Cement*, ACI SP-38, 1973; (*c*) From Kesler, C.E., *Proc. ASCE J. Const. Div.*, Vol. 102, No. CO1, 1976.]

Durability

For several reasons, such as the restrained expansion of concrete, lack of bleeding, and little or no microcracking by drying shrinkage, the shrinkage-compensating concrete provides a more dense and essentially impermeable mass than does portland cement concrete of an equivalent water-cement ratio in the range of 0.4 to 0.6. Consequently, laboratory and field experience has shown that, in general, Type K cement concretes possess a significantly higher resistance to abrasion, erosion, and chemical attack by aggressive solutions. Tests have also shown that when Type K expansive cements are made with an ASTM Type II or V portland cement clinker, the concrete shows a durability to sulfate attack that is similar to concrete made with a sulfate-resisting portland cement. This has been an important consideration in the use of Type K cement concrete for sanitary structures.

Properly designed, placed, and cured shrinkage-compensating concrete mixtures offer as much resistance to freezing and thawing, and deicer scaling, as portland cement concretes of comparable quality. Generally, a given amount of an air-entraining admixture will produce a volume of air and void spacing in a shrinkage-compensating concrete which is comparable to that of portland cement concrete, all other conditions being the same. For quick reference, a general comparison of the properties of shrinkage-compensating concrete with portland cement concrete of the same water-cement ratio is presented in Table 12-12.

12.5.5 Applications

Since the 1960s, expansive cements have been used in several countries for the purpose of producing both shrinkage-compensating and self-stressing concretes. It appears that most of the applications have been in structural elements, such as slabs, pavements, prestressed beams, and roofs. For obvious reasons, several hundred applications are reported for the construction of water and sewage-handling structures, such as water storage tanks, digesters, filtration plants, equalization basins, sludge removal

Type of Property	Characteristics of Shrinkage-Compensating Concrete Relative to Portland Cement Concrete of Similar Water-Cement Ratio
Workability	
Consistency	Stiffer
Cohesiveness	Better
Time of set	Quicker
Strength	Better
Impermeability	Better
Drying shrinkage	Similar
Creep	Similar
Elastic modulus	Similar
Overall dimensional stability	Better
Durability	
Resistance to abrasion	Better
Resistance to erosion	Better
Resistance to sulfate attack	Similar to Type V portland cement concrete*
Resistance to frost action	Similar when equivalent air entrainment present

*This is applicable to Type K cements made with ASTM Type II or V portland cement.

TABLE 12-12 Comparison of Properties of Shrinkage-Compensating Concrete with Portland Cement Concrete facilities, spillways, cooling tower basins, and swimming pools.[51] Some applications are described below.

Construction requirements for a 150-mm, 186,000-m^2 floor slab on grade for a warehouse in the Midwest included minimal joints and cracks and a carrying capacity for heavy loads both fixed and moving.[52] Normal portland cement concrete would have required small placements, checkerboard casting, 50 percent of slab reinforcement cut at saw joints, and saw cuts performed 12 h after casting. The concrete is generally cast in a 12 by 12 m (40 by 40 ft) section and then saw-cut into 6 by 6 m (20 by 20 ft) sections (Fig. 12-14*a*). Immediately after cutting, the cuts are cleaned and sealed until the slab has fully cured; then the joints are uncovered and filled with a joint compound. Following this practice the amount of construction and saw-cut joints required for the warehouse would be 54 km (33.4 mi). Instead with a Type K shrinkage-compensating cement concrete, 24 by 36 m (80 by 120 ft) sections were cast without any saw cuts, and the amount of joints required totaled only 10.5 km or 6.6 mi (Fig. 12-14*b*). A reduction in the number of joints by about 80 percent by the use of shrinkage-compensating concrete was considered to be a big advantage from the standpoint of appearance, and construction and maintenance costs of the floor.

Over the last half century, several countries have used Type K expansive cements to produce both shrinkage-compensating and self-stressing concretes. Most of the applications have been in structural elements, such as slabs, pavements, prestressed beams, and roofs. Keith et al.[56] reported successful application of pavements using shrinkage-compensating concrete. By compensating for the shrinkage, the joints could be set as far apart as 32 m (105 ft) without any significant cracks or curling. The pavement was 15 cm (6 in.) thick with a single layer of No. 4 bars at 46 cm (18 in.) on center with a 3.8-cm (1.5-in.) concrete cover from the top of the pavement. The final design specified over 9300 m^2 (100,000 ft^2) of expansive concrete paving, with a 28-day strength of 28 MPa (4000 psi). A more-uniform-than-normal aggregate size distribution was used, plus 3.8-cm (1.5-in.) maximum-size aggregate rather than the standard 1.3 to 2.5 cm (0.5 to 1 in.) aggregate, to keep the mix "stony" while keeping a good workability. Natural sand (rather than manufactured) was used for enhanced pumpability, workability, and finishability. The maximum concrete temperature at placement was 29°C (85°F), which increased the window of finishability. Although the cost was higher than the ordinary

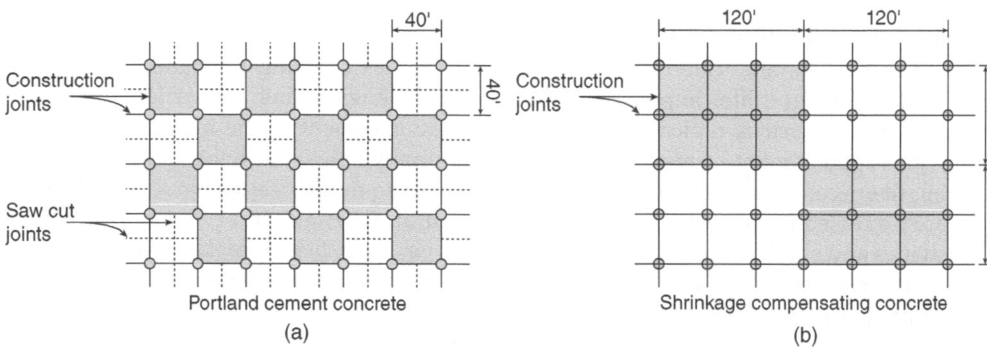

FIGURE 12-14 Relative proportions of constructions joints in concrete slabs made with (*a*) portland cement; (*b*) shrinkage-compensation cement. (From Hoffman, M.W., and E.G. Opbroek, *Concr. Int.*, Vol. 1, No. 3, pp. 19–25, 1979.)

concrete pavement, it was offset to a large degree by the elimination of most of the pavement joints, especially when considering life-cycle costs.

Ramey et al.[57] conducted laboratory testing to identify good shrinkage-compensating concrete mix proportions and assessed their sensitivity to construction quality in bridge deck applications. In conjunction with the Alabama Department of Transportation, they designed a test bridge using shrinkage-compensating concrete for the deck for one span and a comparison span using portland cement. The laboratory results indicate that the hardened strength of shrinkage-compensating concrete was comparable to the portland cement–based mix. Addition of silica fume improves the strength of Type K mixes when properly cured, but it also increases the risk of reducing the expansion when mix does not undergo ideal curing conditions. The conclusion of field study was that a moist curing condition is absolutely critical for using shrinkage-compensating concretes for application to bridge decks.

The distribution center of COTO superstores located near Buenos Aires, Argentina, is another good example of the application of shrinkage-compensating concrete.[58] To minimize the joints number as well as avoiding cracking and curling, the building's 60,000 m^2 (646,000 ft^2) slabs were constructed with shrinkage-compensating concrete rather than conventional concrete. Instead of using Type K cement in the mixture, expansive additives (Expan-K) were used together with Type I portland cement to create expansive concrete. Typical dimensions of the slabs were 40×25 m (130×82 ft). The joints were armor-plated with two steel pieces (one on each side) that were 50×19 mm (2×0.75 in.). Dowels were not used to allow the free expansion of the slabs in all directions. Considering the slab thickness and reinforcement ratio, the targeted expansion was set within the range of 0.03 to 0.06 percent after 14 days of moist curing. Laboratory tests indicate that with an additive dosage of 12 percent (by weight of cement), an expansion value of 0.04 percent could be adopted after 14 days of curing under water. The expansion developed mainly during the first 3 to 4 days. After 14 days the samples were air cured, resulting in a reduction of expansion to 0.03 percent at 28 days due to the compensation from the dry shrinkage. The project was considered a success as no appreciable curling and no visible cracks appeared.

12.6 Fiber-Reinforced Concrete*

12.6.1 Definition and Significance

Discrete fibers are added to concrete to improve its cracking resistance, toughness, strength, fatigue life, impact resistance, or other engineering properties. This *fiber-reinforced concrete* is typically made with hydraulic cement as the main binder, along with supplementary cementitious materials in many cases. However, benefits of adding fibers can also be obtained for concretes utilizing nonhydraulic binders. The fibers are produced from steel, plastic, glass, and natural materials, and come in a variety of shapes and sizes. Amongst the different fiber types, steel fibers are the most commonly used for structural applications and account for many of the nonstructural applications. Aggregates are used in fiber-reinforced concrete, but the size distribution of aggregates should be chosen according to fiber type and the performance objectives.

---

*The authors thank Professor John Bolander for his contribution to this section.

The properties of fiber-reinforced concrete depend on the type, amount, and dimensions of the fibers added, as well as the properties of the cement-based matrix and the fiber-matrix interface. The fiber-matrix interface and the geometry of the fiber control the bond of the fiber to the matrix. In addition, the spatial and orientation distributions of the fibers within the concrete affect the cracking performance and toughness of the composite material. For many applications, it is desirable to have the fibers uniformly distributed at random within the concrete. However, preferential alignment of the fibers can be advantageous for some applications, particularly when the direction of loading is well defined.

In common practice, the amount of fibers added to the concrete mixture depends on the properties of the cement-based matrix and the performance objectives. Typically, lower amounts of fibers (e.g., less than about 1 percent volume fraction) are used to control shrinkage cracking. Prime application areas are slabs and pavements, which have large exposed surfaces that are highly susceptible to shrinkage cracking. Disperse fibers offer various advantages over steel bars and wire mesh to reduce shrinkage cracking: (a) the fibers are distributed throughout the concrete volume, more effectively resisting local crack formation and growth; (b) the fibers are less sensitive to corrosion than the reinforcing steel bars; and (c) fiber use can reduce labor costs, since they partially or completely replace the bars or wire mesh commonly used for crack control. Higher amounts of fibers are typically used to achieve significant improvements in modulus of rupture, fracture toughness, and impact resistance. These composites are used in construction methods such as shotcrete and in structures that require energy absorption capability, improved resistance to delamination, spalling, and fatigue.

Lessons in the History of Building Materials

When was the concept of fiber reinforcement first used in building material? According to Exodus 5:6, Egyptians used straw to reinforce mud bricks. There is evidence that asbestos fiber was used to reinforce clay pots about 5000 years ago.

Nature offers examples of fiber-reinforced building materials. Professor Alberto Fava of the University of La Plata in Argentina points to the Hornero, which is a tiny bird native to Argentina, Chile, Bolivia, and other South American countries. The bird has been painstakingly building straw-reinforced clay nests on treetops since before the advent of humans (see photograph).

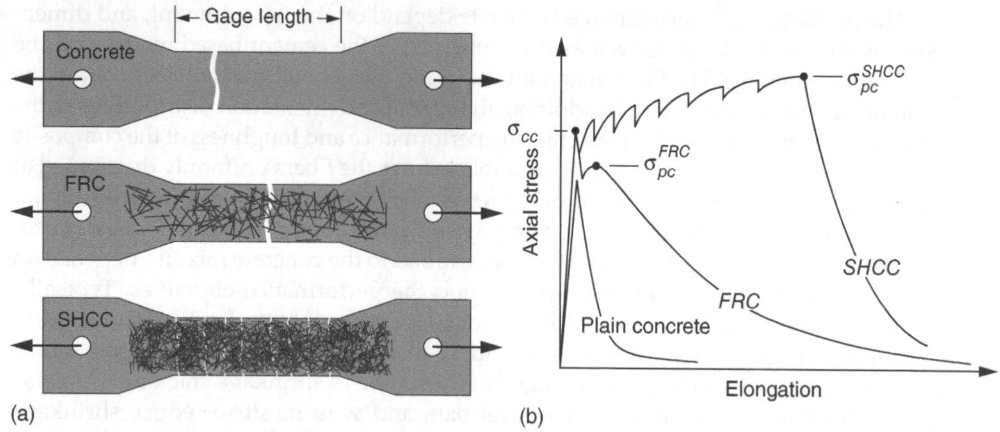

FIGURE 12-15 Schematic representation of the tensile behavior of plain and fiber-reinforced concrete: (*a*) tensile specimens subjected to equivalent axial displacements; and (*b*) axial stress versus elongation over the gage length.

These new materials offer interesting possibilities for the improved structural design and for the retrofit of existing structures. Even though the market for fiber-reinforced concrete is still small compared to the overall production of concrete, in North America there has been a yearly growth rate of about 20 percent and Li[59] reports that the worldwide yearly consumption of fibers used in concrete is 300,000 tons.

12.6.2 Toughening and Strengthening Mechanisms

The strength and toughness properties of fiber-reinforced concrete can be measured through the load testing of specimens in tension, as shown in Fig. 12-15. Both plain (unreinforced) and fiber-reinforced concretes respond almost linearly up until the point of first cracking. The addition of fibers increases or decreases the elastic modulus of the composite material, depending on the type of fiber added. A similar effect is made upon tensile strength. Due to the small amounts of fibers added to concrete, typically ranging up to 2 percent by volume, the elastic modulus and strength of fiber-reinforced concrete are not largely different from those of plain concrete.

After first cracking, the load-carrying capacity of plain concrete diminishes rapidly (Fig. 12-15*b*). It is for this reason that, in ordinary structural design, concrete is assumed to carry no tension after cracking. In contrast, fiber-reinforced concrete exhibits significant postcracking strength and toughness, since the fibers bridge (and thus transfer load) across cracks. Only one crack forms if the postcracking strength, σ_{pc}, is less than the first cracking strength, σ_{cc}. The gradual reduction in load with elongation is due to the pullout of fibers bridging the crack. In some cases, fiber pullout can be precluded by fiber rupture, which is a less desirable failure mechanism due to its more brittle nature. The large majority of applications fit into this general class of fiber-reinforced concrete in which failure evolves through the opening of a single crack.

Through engineering efforts at the material scale, a newer class of *strain-hardening cementitious composite* (SHCC) has been developed. Strain hardening is achieved through the formation of multiple cracks. For this to occur, postcracking strength must be greater than the first cracking strength (i.e., $\sigma_{pc} > \sigma_{cc}$), as shown in Fig. 12-15*b*.

The small drops in load that appear along the hardening branch are associated with the formation of individual cracks. This process of multiple crack formation continues until the load on the composite reaches the collective pullout resistance of the fibers bridging the critical crack. Upon reaching the postcracking strength, fracture localizes in the form of fiber pullout at the critical crack location. Dramatic improvements in strength and toughness can be achieved through such strain-hardening behavior. Furthermore, for the same deformation over the gage length, the formation of multiple fine cracks is generally better from a durability perspective, since cracking opening strongly affects the permeability of the composite material. These and other attributes of strain-hardening fiber-reinforced concretes are described later in this section.

Toughness can be measured as the area under the stress-elongation curve produced by a uniaxial tensile test, such as that illustrated in Fig. 12-15. Due to the difficulties in performing uniaxial tension tests, however, toughness is more commonly measured using flexural tests. The load-displacement behavior in flexure resembles that in tension: linear behavior occurs up to near peak load for both plain and fiber-reinforced concrete; plain concrete exhibits sudden loss of strength after peak load, whereas fiber-reinforced concrete provides significant postcracking strength and toughness (Fig. 12-16). Toughness can be estimated as the area under the load-displacement curve divided by the area of the fracture surface. Hardening behavior, accompanied by multiple cracking, can occur during flexural testing, but it does not necessarily imply strain-hardening behavior, as it would for the tensile test.

According to the report by ACI Committee 554,[60] the total energy absorbed in fiber debonding as measured by the area under the load-deflection curve before complete separation of a beam is at least 10 to 40 times higher for fiber-reinforced concrete than for plain concrete. The magnitude of improvement in toughness is strongly influenced by fiber concentration and resistance of fibers to pull out, which is influenced by other factors such as shape and surface texture of the fibers.

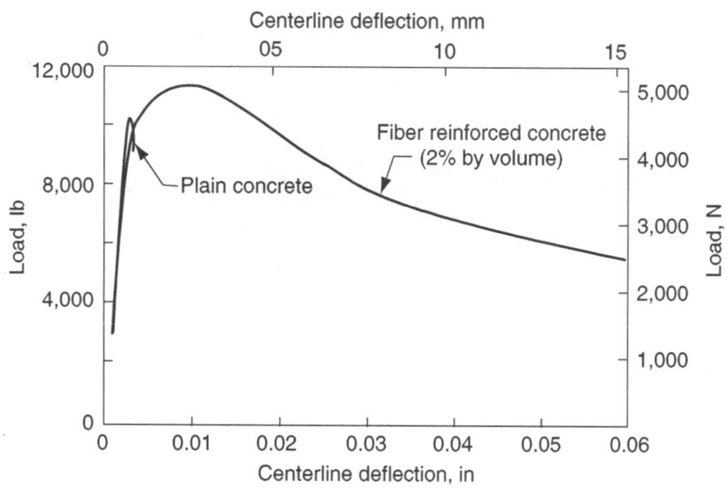

FIGURE 12-16 Load-displacement behavior of plain and fiber-reinforced concrete in flexure. (From Hanna, A.C., *Report RD049.01P*, Portland Cement Associate, Skokie, IL, 1977.)

The energy absorbed G_c by the fiber pullout can be computed[61] as

$$G_c = \frac{1}{2} V_f g \tau \frac{L_f^2}{d_f} \tag{12-12}$$

where V_f, L_f, and d_f are the fiber volume fraction, length, and diameter, respectively, and g and τ are interface parameters. The increase in the energy will increase the characteristic length ($l_{ch} = EG_c/\sigma_t^2$) of the material (see Chap. 13) leading to more ductile systems.

From a material and structural point of view, there is a delicate balance in optimizing the bond between the fiber and the matrix. If the fiber-matrix bond is weak, the ability of fibers to transfer load across cracks (and thus the postcracking strength) is diminished. In this situation, the fibers do not significantly increase the toughness of the system. On the other hand, if the bond with the matrix is too strong, many of the fibers may break before they dissipate energy by pulling out. In this case, the fibers behave as nonactive inclusions leading to only marginal improvement in the mechanical properties.

Concrete contains a large number of defects that exist prior to loading or appear during loading in the form of microcracks. Figure 12-17*a* shows a model of a material volume containing a disk-shaped defect, defined by its center coordinates, radius r, and surface-normal direction **n**. Consider one of the hybrid-fiber composites tested by Lawler et al.[62] that contains: steel macrofibers of length l_{macro} = 30 mm and diameter d_{macro} = 0.5 mm; and steel microfibers of length l_{micro} = 6 mm and diameter d_{micro} = 0.022 mm. (Note that the microfiber diameter is on the order of the cement grain sizes.) The macro- and microfibers

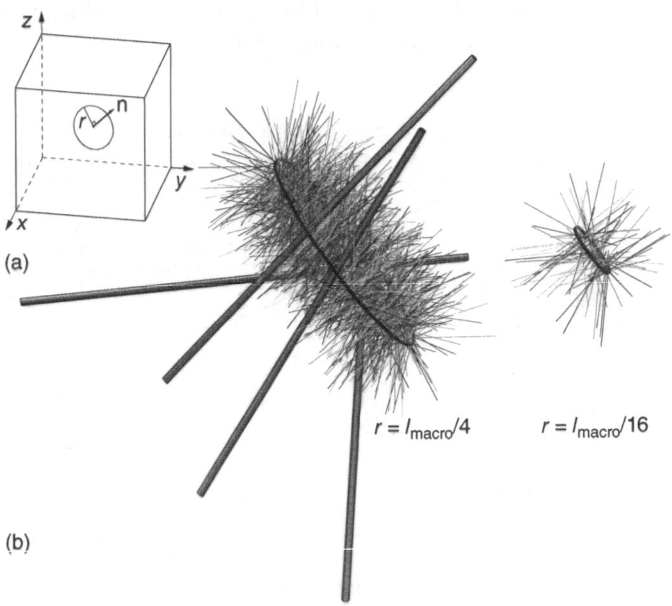

Figure 12-17 (*a*) Disk-shaped defect in a material volume; and (*b*) micro-/macrofiber combinations that bridge defects of differing size. (*Figures courtesy of John Bolander.*)

are randomly introduced throughout the material volume at dosages of 0.5 and 0.32 percent, respectively, by volume. For the larger defect size shown in Fig. 12-17*b*, both micro- and macrofibers bridge the defect surfaces. For the smaller defect size, however, the defect surfaces are bridged only by microfibers (for a large majority of random realizations of the fiber distribution).

The comparison shown in Fig. 12-17 is useful in illustrating, for the roughly same volume fraction of each fiber type, the dramatic differences in the number of micro- and macrofibers and their dispersions within the material volume. The greater number of microfibers allows them to effectively restrain the formation and growth of microcracks. In addition, the collection of microfibers crossing a fracture plane possesses a much larger bonded surface area relative to that of the same volume fraction of macrofibers. These two characteristics of microfibers enhance composite strength, so that it can become much higher than that of either the plain matrix or macrofiber-reinforced composite. In contrast, macrofibers are not efficient in resisting microcrack growth and coalescence into traction-free matrix cracks. Due to the longer length of most macrofibers, however, larger crack openings can be accommodated prior to fiber pullout. This results in more ductile behavior compared to microfiber composites.

Implicit within these discussions of strength and toughness is the assumption that the fibers have been uniformly distributed within the material volume. Nonuniformity of the fiber distribution (e.g., regions with significantly lower fiber content) can compromise composite performance. The ability to disperse the fibers and avoid fiber segregation, and workability of the fiber-reinforced concrete mixture, are primary considerations in the design of the composites to meet performance objectives. Factors affecting the workability of fiber-reinforced concrete are discussed later in this section.

The idea of *hybrid-fiber composites*, in which two or more fibers types are combined, offers solutions to some material design problems. For example, combinations of microfibers and macrofibers can provide significant improvements in both strength and toughness, as shown conceptually in Fig. 12-18. In the experimental study of Cyr et al.,[63] $100 \times 25 \times 4\text{-mm}^3$ panels of extruded fiber-reinforced mortar were tested in flexure. The study considered polypropylene (PP) fibers, glass fibers, polyvinyl alcohol (PVA) fibers,

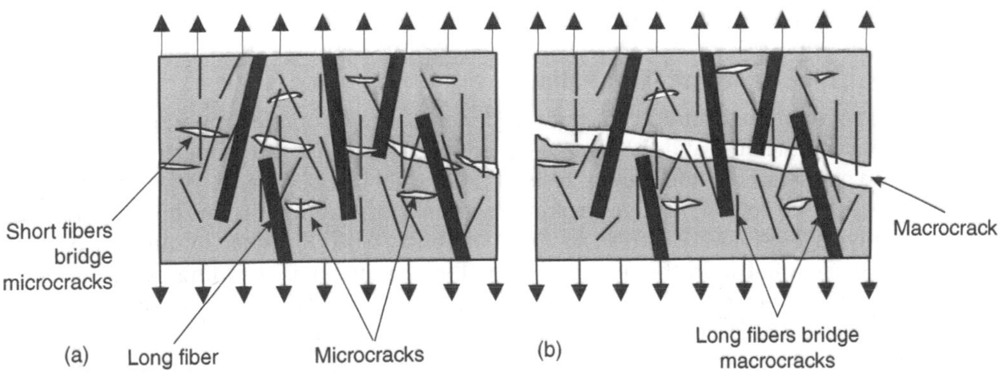

FIGURE 12-18 Schematic illustration of hybrid fiber concrete: (*a*) microfiber pinning of small defects increases tensile strength; and (*b*) macrofiber bridging of macrocracks improves composite toughness. (Adapted from Marković, I., *High-performance hybrid-fibre concrete: development and utilization*, Ph.D. dissertation, Technical University of Delft, The Netherlands, 2006.)

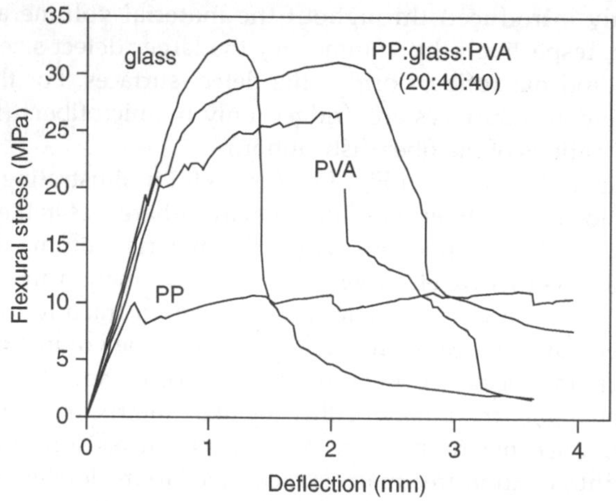

Figure 12-19 Flexural testing of hybrid-fiber composites. (Adapted from Cyr, M.F., A. Peled, and S.P. Shah, Extruded hybrid fiber-reinforced cementitious composites, *7th Int. Sym. Ferrocement and Thin Reinforced Cem. Comp. (Ferro-7)*, Centre for Construction Materials and Technology, National University of Singapore, pp. 199–207, 2001.)

and their hybrid combination as indicated in Fig. 12-19. The polypropylene fiber composite exhibits the most ductile behavior, but at lower strength. The glass fiber composite exhibits the highest strength, but fails in a brittle manner. The PVA fiber composites provide strength and ductility, but are expensive due to the cost of PVA. The hybrid combination of these fibers provides the best solution in terms of strength, ductility, and economy.

In other studies, it has been shown that hybrid-fiber combinations can act synergistically, providing improvements beyond the sum contribution of each fiber type acting alone. One such mechanism is the improved confinement offered by microfibers in the anchorage zones of hooked macrofibers (i.e., the presence of microfibers can improve the pullout performance of macrofibers). These new materials offer interesting possibilities for improved structural design and for the retrofit of existing structures.

12.6.3 Mechanics of Fiber Pullout

The behavior of fiber-reinforced concrete depends largely on the properties of the interface between the fiber and the cement-based matrix. Mechanical properties of the interface can be assessed using pullout tests of individual fibers. A variety of configurations have been used for such pullout testing; two ordinary configurations are shown schematically in Figs. 12-20*a* and *b*. In the most basic case, the fiber is pulled from the cement-based matrix in the direction of its longitudinal axis. Loading can be applied directly to the free end of the fiber or by casting both ends of the fiber within the cement-based matrix. The latter approach is useful when gripping the fiber is problematic.

For straight fibers, it is often assumed the bond properties are uniform along the embedded length, l_e, and can be represented by a constitutive relation in terms of bond shear stress and bond slip (Fig. 12-21). The ascending branch of the bond stress-slip curve corresponds to elastic deformation over the interphase thickness and therefore does not represent true slip between the fiber and matrix. Debonding occurs upon

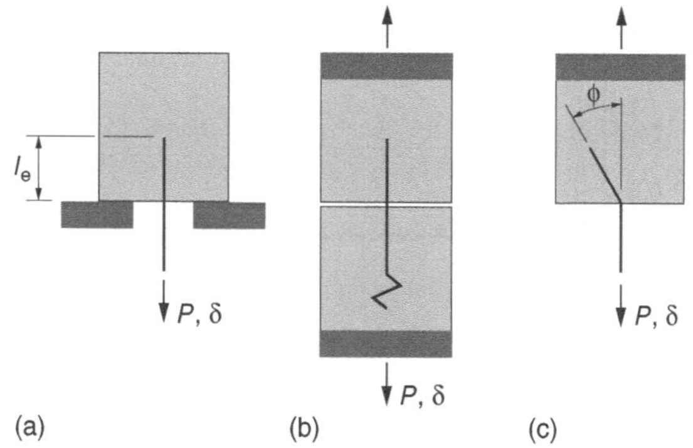

FIGURE 12-20 Pullout testing of individual fibers from a cement-based matrix: (*a*) one-sided loading, (*b*) two-sided loading, and (*c*) entry angle of the fiber is skew to the direction of loading.

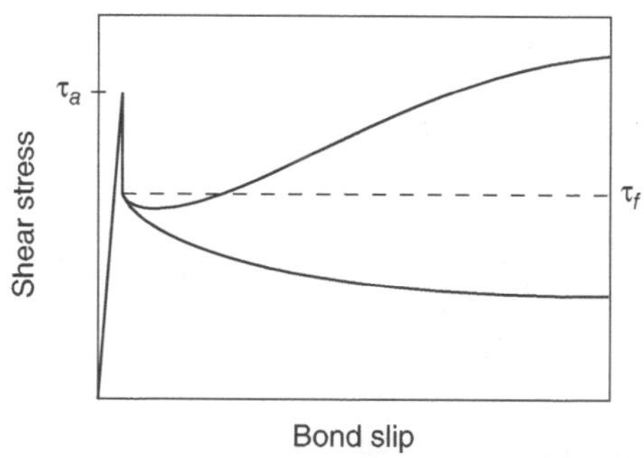

FIGURE 12-21 Constitutive modeling of fiber-matrix interface. (*Figure courtesy of John Bolander.*)

reaching adhesion strength, τ_a, and is followed by frictional pullout. Frictional resistance typically decays with increasing bond slip, but slip-hardening is possible for some material combinations.

Typical results for the pullout of straight, smooth steel fibers are presented in Fig. 12-22. Once adhesion bond is broken over the embedded length, frictional resistance to pullout decays rapidly for small slip values and then becomes approximately constant. Constant frictional resistance leads to linear decrease in the pullout force with increasing slip. Straight, smooth fibers do not effectively carry load at higher slip values and, therefore, most steel macrofibers have surface or end geometries that promote load transfer through mechanical interlocking. For example, the hooked-end steel fibers shown in Fig. 12-23 provide substantially increased pullout strength and toughness relative to straight, smooth fibers. Looking at the results for the hooked-end fiber, the undulation in the softening branch is associated with the straightening of the hooked-end within a high-strength matrix.

The preceding discussion of fiber pullout has assumed the fiber is aligned with the direction of tensile loading, as pictured in Figs. 12-20*a* and *b*. In fiber-reinforced concrete, however, the fibers are typically dispersed in a random manner, such that fiber entry from a crack face would be skew to the crack opening direction (Fig. 12-20*c*). Pullout testing

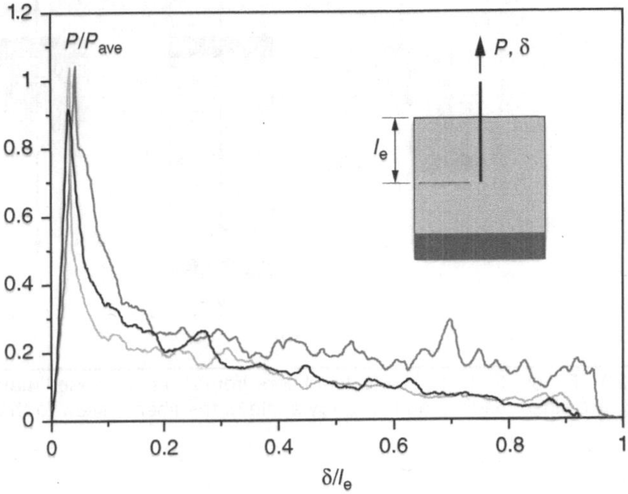

FIGURE 12-22 Normalized load versus pullout displacement curves for straight, smooth steel fibers embedded in a mortar matrix. For the three test results, $(l_e)_{ave}$ = 13.9 mm and P_{ave} = 77.9 N. (*Courtesy of Stillmaker, K., and J.E. Bolander.*)

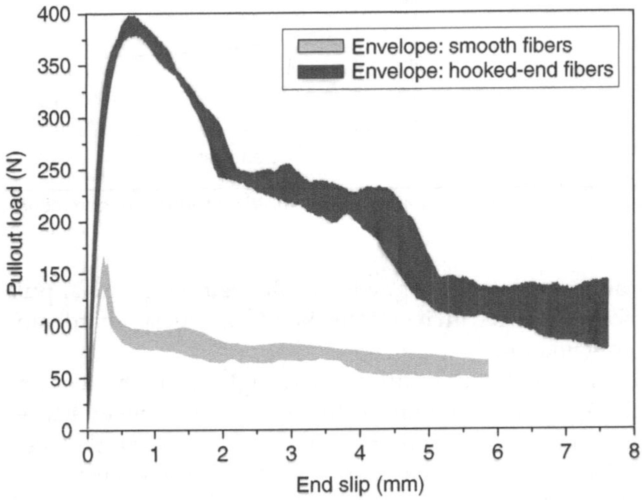

FIGURE 12-23 Pullout response of straight and hooked-end steel fibers from a self-consolidating concrete. (Adapted from Cunha, V.M.C.F., J.A.O. Barros, and J.M. Sena-Cruz, Pullout behavior of steel fibers in self-compacting concrete, *J. Mat. Civil Eng., ASCE*, Vol. 22, No. 1, pp. 1–9, 2010.)

has shown that fiber entry angle ϕ affects strength and toughness in several ways: (1) with nonzero entry angle, frictional resistance to pullout is increased due to fiber bearing against the matrix (so-called snubbing effect); (2) nonzero entry angle implies the fiber will experience flexural deformation, which can lead to plastic hinging in ductile fibers or increased susceptibility to rupture of brittle fibers; and (3) for severe entry angles, there is a tendency for spalling of the matrix near the fiber entry point. In general,

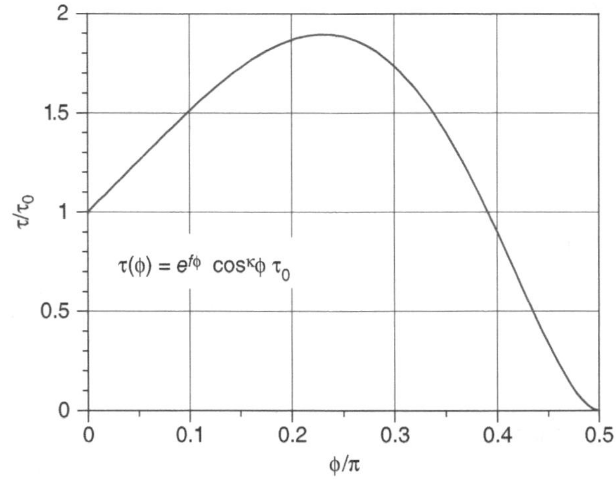

Figure 12-24 Apparent bond strength as a function of fiber inclination to the loading direction. Parameters f and κ depend on matrix composition and fiber type. (Adapted from Lee, Y., S.T. Kang, and J.K. Kim, Pullout behavior of inclined steel fiber in an ultrahigh strength cementitious matrix, *Cons. Build. Mater.*, Vol. 24, pp. 2030–2041, 2010.)

relative to the aligned fiber case, moderate entry angles improve pullout strength and toughness whereas large entry angles have the opposite effect. This is illustrated by the apparent bond strength relation in Fig. 12-24, which was developed through the pullout testing of straight steel fibers. The first term in the relation accounts for the snubbing effect nearby the fiber entry point (Li et al.[64]), whereas the second term accounts for spalling of the matrix due to large entry angles of the fiber. The dependency of apparent bond strength on inclination angle varies according to the type of fiber. For low-modulus synthetic fibers, Ouyang et al.[65] found that

$$\tau(\phi) = \tau_0/\cos\phi \tag{12-13}$$

up to fiber entry angles ϕ of about 45°, after which apparent strength diminished. In this relation, τ_0 is the bond strength for $\phi = 0°$.

The results of fiber pullout testing have been useful in designing for improved performance of the fiber-reinforced concrete. In general, improvements in the pullout curve translate to improvements of the composite material. However, that observation is contingent upon the ability to disperse fibers. Closely spaced fibers exhibit a group effect, in which the efficiency of the group of fibers can be significantly less than the sum of the efficiencies of the individual fibers.

12.6.4 Materials and Mix Proportioning

Fibers

Typical properties of various kinds of fibers are given in Table 12-13, and some typical fibers are shown in Fig. 12-25. Round steel fibers have diameters in the range 0.25 to 0.75 mm. Flat steel fibers have cross sections ranging from 0.15 to 0.4 mm thickness by 0.25 to 0.9 mm width. Crimped and deformed steel fibers are available both in full

Type of Fiber	Tensile Strength (ksi)	Young's Modulus (10^3 ksi)	Ultimate Elongation (%)	Specific Gravity
Acrylic	30–60	0.3	25–45	1.1
Asbestos	80–140	12–20	~0.6	3.2
Cotton	60–100	0.7	3–10	1.5
Glass	150–550	10	1.5–3.5	2.5
Nylon (high tenacity)	110–120	0.6	16–20	1.1
Polyester (high tenacity)	105–125	1.2	11–13	1.4
Polyethylene	~100	0.02–0.06	~10	0.95
Polypropylene	80–100	0.5	~25	0.90
Rayon (high tenacity)	60–90	1.0	10–25	1.5
Rock wool (Scandinavian)	70–110	10–17	~0.6	2.7
Steel	40–400	29	0.5–35	7.8

Source: ACI Committee 544, Report 544.IR-82, *Concr. Int.*, Vol. 4, No. 5, p. 11, 1982.

TABLE 12-13 Typical Properties of Fibers

length or crimped at the ends only. To facilitate handling and mixing, fibers collated into bundles of 10 to 30 with water-soluble glue are also available. Typical aspect ratios range from about 30 to 150. Typical glass fibers have diameters of 0.005 to 0.015 mm, but these fibers may be bonded together to produce glass fiber elements with diameters of 0.013 to 1.3 mm. Since ordinary glass is not durable to chemical attack by portland cement paste, alkali-resistant glass fibers with better durability have been developed.[66] Fibrillated and woven polypropylene fibers are also being used.

The type of fiber, its length, volume fraction, bond with the matrix will have a significant influence on the response of the composite. The bridging of the tensile cracks by the fibers is reflected in stress (σ)-crack opening (δ) behavior. For randomly oriented short straight fibers, this relationship can be expressed as

$$\sigma = \frac{1}{2} V_f g \tau \frac{L_f}{d_f} \left(1 - \frac{\delta}{\frac{L_f}{2}} \right)^2 \tag{12-14}$$

where V_f, L_f, and d_f are the fiber volume fraction, length, and diameter, respectively, and g and τ are interface parameters. This equation was developed for short fibers where the pullout load is lower that the fiber strength. If this assumption is not valid, it becomes necessary to incorporate the influence of the fiber strength in the formulation. Table 12-14 shows some typical values for fibers used in structural elements.

The peak load, σ_o can be obtained from the previous equation:

$$\sigma_o = \frac{1}{2} V_f g \tau \frac{L_f}{d_f} \tag{12-15}$$

This relationship is a good indication of the efficiency of the fibers in the composite. Observing the variation of the properties in Table 12-14, it is clear that the aspect ratio controls the magnitude of the peak stress.

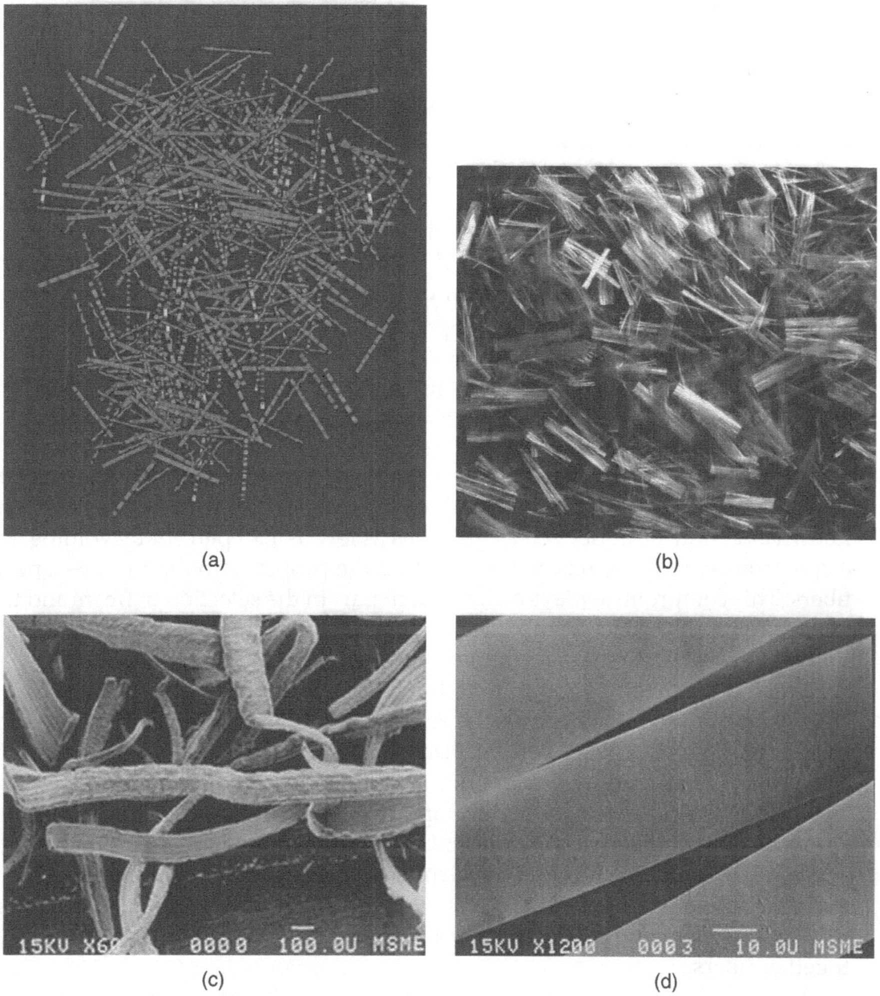

Figure 12-25 Some fiber types used in mortar and concrete: (*a*) cold drawn deformed wire steel fibers, (*b*) fibrillated polypropylene fibers, (*c*) steel microfibers, (*d*) polypropylene microfiber. [(*a*) and (*b*) *courtesy of SI Corporation*; (*c*) and (*d*) *courtesy of Claudia Ostertag*.]

Fiber Type	Aspect Ratio L_f/d_f	τ (MPa)	g
Steel	100	4	2
Olefin	75	2	1–2
Carbon	300	2–5	1
Polyethylene	340	0.5–5	2

Source: From Li, V.C., Large volume, high-performance applications of fibers in civil engineering, *J Appl. Polym. Sci.*, Vol. 83, pp. 660–686, 2002.

Table 12-14 Effect of Fiber Type on the (σ)-Crack Opening (δ) Behavior

General Considerations

It is well known that the addition of any type of fibers to plain concrete *reduces the workability*. Regardless of the fiber type, the loss of workability is roughly proportional to the volume concentration of the fibers in concrete. Since fibers impart considerable stability to a fresh concrete mass, the slump cone test is not a good index of workability. For example, introduction of 1.5 volume percent steel or glass fibers to a concrete with 200 mm of slump is likely to reduce the slump of the mixture to about 25 mm, but the placeability of the concrete and its compactibility under vibration may still be satisfactory. Therefore, the Vebe test (Chap. 10) is considered more appropriate for evaluating the workability of fiber-reinforced concrete mixtures.

The effects of fiber content and aspect ratio on Vebe time, as found by researchers at the British Building Research Establishment,[67] are shown in Fig. 12-26*a*. From the standpoint of strengthening and toughening the concrete, it is desirable to have a large fiber aspect ratio as well as a high fiber concentration in concrete; on the other hand, the data in the figure clearly show that increases in these two variables would have the effect of reducing the workability. In fact, data from the work of Swamy and Mangat[68] show that steel fibers with aspect ratios greater than about 100 tended to produce the phenomenon known as *curling up* at fiber concentrations as low as 1.13 percent by volume. Evidently, a compromise must be reached in selecting the proper amount and the aspect ratio of fibers. This compromise plays an important part in the selection of fibers and the design of fiber-reinforced concrete mixtures. In general, the amount of steel fibers in concrete is limited to about 2 percent by volume, with a maximum aspect ratio of 100.

Even at 2 percent concentration, the workability of concrete or mortar containing steel fibers is reduced sharply as the aggregate size goes up (Fig. 12-26*b*). According to the *ACI Guide for Specifying, Mixing, Placing, and Finishing Steel Fiber Reinforced Concrete*,[69] aggregates larger than 19 mm are not recommended for use in steel-fiber concrete. In fact, during the earlier stages of development of fiber-reinforced concrete, some investigators recommended that it was desirable to use no more than 25 percent by mass coarse aggregate of 9 mm maximum size. Generally, the requirement of proper workability in mixtures containing fibers can be met by the use of air entrainment, plasticizing admixtures, higher cement paste content (with or without a pozzolan), and the use of glued-together fibers.

It should be noted that by using modern acrylic-based superplasticizers, advanced viscosity-modifying agents and mineral admixtures it has been possible to develop *fiber-reinforced self-compacting concrete*.[70,71] Greenough and Nehdi[72] reported that, due to its high workability, self-consolidating concrete was very efficient in accommodating the addition of fibers and that fiber-reinforced self-compacting concrete beams performed better under shear loading than comparable conventional fiber-reinforced concrete.

Mix proportions, properties, and applications of steel-fiber-reinforced concrete are described below because concretes containing other types of fibers are less commonly used for structural concrete applications.

Typical Concrete Mixtures

ACI Committee 544[73] states that steel-fiber-reinforced concrete is usually specified by strength and fiber content. Normally, the flexural strength is specified for paving applications, and compressive strength for structural applications. Typically, either a flexural strength of 700 to 1000 psi (5 to 7 MPa) at 28 days, or a compressive strength of 5000 to 7000 psi (34 to 48 MPa), is specified. For normal-weight concrete, fiber contents from as

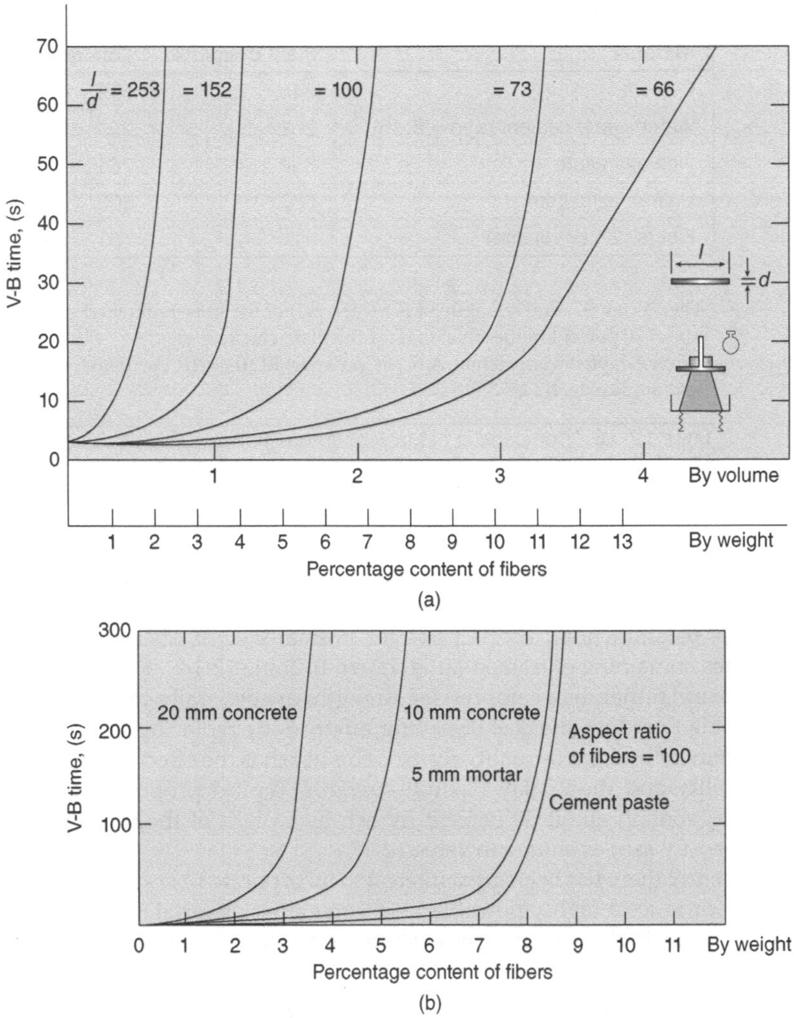

FIGURE 12-26 (*a*) Effect of fiber aspect ratio on workability of mortar; (*b*) effect of aggregate size and fiber content on workability. (From Edington, J., D.J. Hannant, and R.I.T. Williams, *BRE Current Paper 69/74*, July 1974. *By courtesy of Building Research Establishment, U.K., and reproduced by permission of Her Majesty's Stationery Office. Crown Copyright*.)

low as 30 kg/m^3 or 50 lb/yd^3 (0.38 percent by volume) to as high as 220 kg/m^3 or 365 lb/yd^3 (2 percent by volume) have been specified, although the usual upper limit is 95 to 120 kg/m^3 or 160 to 200 lb/yd^3 (1.2 to 1.5 percent by volume).

For steel fibers with dimensions of $0.25 \times 0.55 \times 25$ mm^3 ($0.01 \times 0.022 \times 1$ in^3), mix proportions and properties of fiber-reinforced concrete mixture suitable for highways and airport pavements and overlays were investigated by the U.S. Portland Cement Association.[74] A chart was developed to determine the increase in the cement content and the decrease in aggregate proportions for the fiber additions in the range 0.5 to 2 percent by volume. Using this chart, the mix proportions in Table 12-15 show how at

Material	Plain Concrete	Fiber-Reinforced Concrete*
Cement	446	519
Water (water-cement ratio = 0.45)	201	234
Fine aggregate	854	761
Coarse aggregate	682	608
Fibers (2% by volume)	—	157

*The 14-day flexural strength 8 MPa (1150 psi) of the fiber-reinforced concrete was about 20 percent higher than that for the plain concrete.

Source: Adapted from Hanna, A.N., PCA Report RD 049.01P, Portland Cement Association, Skokie, IL, 1977.

Table 12-15 Comparison of Mix Proportions between Plain Concrete and Fiber-Reinforced Concrete (kg/m³)

a given water-cement ratio, the cement paste content has to be increased with a corresponding decrease in the proportion of aggregates to maintain adequate workability when 2 percent steel fibers are added to the plain concrete mixture. The range of proportions recommended by ACI 544 for normal-weight fiber-reinforced mortars and concretes containing entrained air is shown in Table 12-16.

Flexural rather than compressive strengths are generally specified for pavements. It is possible to reduce the cost of paving mixtures by replacing a substantial proportion of the cement by a good-quality fly ash. This also has the effect of improving the needed workability and the 28-day flexural strength. Typical proportions of airfield paving mixtures contain about 30 percent fly ash by weight of the total cementing material (cement + fly ash) as shown in Table 12-17.

Recently there has been great interest to incorporate fibers into self-compacting concrete to produce a highly flowable, nonsegregating material for use in precast applications and in high seismic zones where concrete structures are heavily reinforced.

	Mortar	10-mm Maximum Aggregate	19-mm Maximum Aggregate
Cement (lb/yd³)	415–712	356–593	297–534
Water-cement ratio	0.30–0.45	0.35–0.45	0.40–0.50
Percent of fine to coarse aggregate	100	45–60	45–55
Entrained air content (%)	7–10	4–7	4–6
Fiber content (vol. %)			
Deformed steel fiber	0.5–1.0	0.4–0.9	0.3–0.8
Smooth steel fiber	1.0–2.0	0.9–1.8	0.8–1.6
Glass fiber	2–5	0.3–1.2	—

Source: ACI Committee 544, Report 544.IR-82, *Concr. Int.*, Vol. 4, No. 5, p. 16, 1982.

Table 12-16 Range of Mix Proportions for Normal-Weight Fiber-Reinforced Concrete

Material	Mixture 1* 10-mm Maximum Aggregate	Mixture 2* 19-mm Maximum Aggregate
Cement	297	311
Fly ash	139	148
Fine aggregate	813	854
Coarse aggregate	872	789
Water	151	168
Steel fibers (0.6–1.0% by volume)†	49–83	49–83

*Suitable amounts of water-reducing and/or air-entraining admixtures may be included. Flexural strengths at 28 days ranged from 7.2 to 7.6 MPa.

†The lower volume was found to be adequate when crimped-end fibers were used instead of flat steel fibers.

Source: J. ACI., Proc., Vol. 81, No. 2, pp. 140–148, 1984.

TABLE 12-17 Typical Proportions for Airfield Paving Mixtures Containing Fly Ash (kg/m³)

Trono et al.[75] studied confinement and tension stiffening effects on self-consolidated, hybrid fiber-reinforced concrete composites and compared to plain self-consolidated concrete specimens without fibers. Their compression test results of the self-consolidated, hybrid fiber-reinforced concrete specimens with transverse confinement indicated that even at very low reinforcement ratios, the specimens had superior softening behavior beyond the peak load as compared to unreinforced self-consolidated specimens. Table 12-18 summarizes the mixture proportions used in the study. Ferrara et al.[76]

Material	Plain Self-Consolidating Concrete	Self-Consolidating Hybrid Fiber-Reinforced Concrete
Cement	398	398
Fly ash	131	131
Fine aggregate	1074	1046
Coarse aggregate	430	419
Water	238	238
Superplasticizer	0.46	0.46
Viscosity modifying admixture	2.22	2.22
30-mm long steel macrofibers	—	1.3
8-mm long polyvinyl alcohol microfibers	—	0.2

Water, cement, fly ash, and fine and coarse aggregate are given in kg/m³. Superplasticizer and viscosity modifying admixture are given as a wt. % of binder materials (cement + fly ash). Fibers are given in volume %.

Source: Trono, W., G. Jen, D. Moreno, S. Billington, and C.P. Ostertag, Confinement and tension stiffening effects in high performance self-consolidated hybrid fiber-reinforced concrete composites, in *High Performance Fiber Reinforced Cement Composites 6*, Parra-Montesinos, G.J., H.W. Reinhardt, A.E. Naaman, eds., Springer, 2012.

TABLE 12-18 Material Proportions for Self-Consolidating Concrete Mixtures developed a robust methodology for the mix proportions of fiber-reinforced self-compacting concrete by extending the formulation of the *rheological paste model*[77] originally developed for self-consolidating mixtures.

12.6.5 Properties

It is difficult to evaluate some of the properties of fiber-reinforced concrete. Guidance on the subject is provided by ACI Report 544.2R.[78] A summary of the essential properties is given below.

Workability

The marked effect of fiber additions on workability of fresh concrete mixtures has already been described. For most applications, typical mortar or concrete mixtures containing fibers possess very low consistencies; however, the placeability and compactibility of the concrete is much better than reflected by the low consistency.

Strength

For low and moderate fiber content, the most important contribution of fiber reinforcement in concrete is not to strength but to the flexural toughness of the material. This can be observed from the data in Fig. 12-27, which is taken from the report of an experimental study by Shah and Rangan. The test results in the tabulated data are from concrete specimens containing a constant amount of fibers (1 percent by volume) of given cross

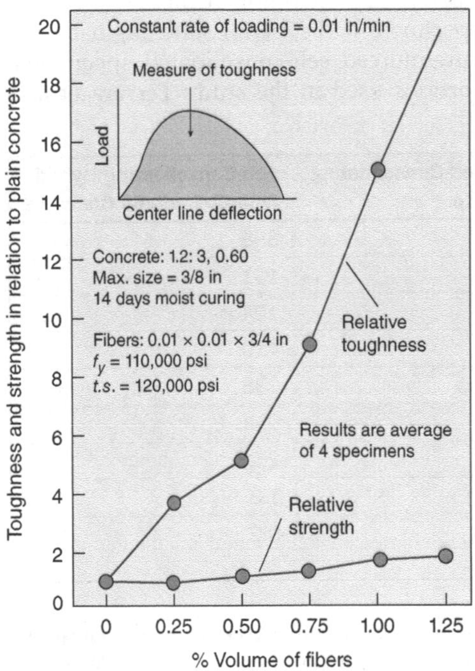

(a)

Effect of aspect ratio			
Type reinforcement	Aspect ratio L/d	Relative strength	Relative toughness
Plain concrete	0	1.00	1.0
	25	1.50	2.0
Random	50	1.60	8.0
fiber	75	1.70	10.5
	100	1.50	8.5
Effect of type reinforcement			
Conventional tensile bar	—	3.15	—
Random fibers	75	1.00	—

(b)

Figure 12-27 Factors affecting properties of fiber-reinforced concrete: (*a*) influence of increasing fiber volume; (*b*) influence of increasing the aspect ratio. (From Shah, S.P., and B.V. Rangan, *J. ACI, Proc.*, Vol. 68, No. 2, p. 128, 1971.)

section (0.25 by 0.25 mm) but with aspect ratio varied by increase in fiber length from 6.5 to 25 mm. The authors concluded:

- It can be seen that increasing the length of fibers up to a point increases strength as well as toughness. The increase in toughness is an order of magnitude while that in strength is only mild. Comparison between conventional reinforcement and fiber reinforcement is also shown in the table. The percentage volume of steel was the same in both cases (1 percent). Conventional reinforcement was a deformed bar of 0.233 in. (6 mm) diameter and was placed with 0.5 in. (12 mm) cover from the tension face of the beam. It can be seen that the conventional reinforcement gave a maximum flexural load that was more than three times that of fiber-reinforced concrete.[79]

Clearly when flexural strength is the main consideration, fiber reinforcement of concrete is not a substitute for conventional reinforcement. From the curve in Fig. 12-27, which is on the effect of volume of fibers on flexural strength and toughness of beams of concrete containing 0.25 × 0.25 × 19-mm (0.01 × 0.01 × 0.75-in) steel fibers, it is obvious that increasing the volume fraction of fibers enhanced both the flexural strength and toughness; whereas the increase in toughness was as much as 20 times for 1.25 percent volume of fibers, the increase in strength was less than twofold.

It seems that the first crack strength (Fig. 12-28*a*) and the ultimate flexural strength of fiber-reinforced systems (i.e., regardless of whether the matrix is hardened cement paste, mortar, or concrete) can be predicted using a composite material approach:

$$\sigma_c = A\sigma_m(1-V_f)+BV_f\frac{l}{d} \tag{12-16}$$

where σ_c and σ_m are strengths of the composite (containing the fiber) and the matrix, respectively; V_f is the volume fraction and l/d the aspect ratio of fibers; and A and B are constants.

Swamy and his associates at Sheffield University[80] made regression analyses of test data from several investigations of fiber-reinforced pastes, mortars, and concretes with a wide range of mix proportions and fiber geometries to determine the values of the constants. The respective values of A and B were found to be 0.843 and 2.93 for first crack strength, and 0.97 and 3.41 for the ultimate strength of the composite. The coefficient of correlation from the regression analysis was reported to be 0.98.

Toughness and Impact Resistance

As shown by the data in Fig. 12-27, the greatest advantage in fiber reinforcement of concrete is the improvement in flexural toughness (total energy absorbed in breaking a specimen in flexure). Related to flexural toughness are the impact and fatigue resistance of concrete, which are also increased considerably. Unfortunately, for want of satisfactory tests for impact resistance of fiber-reinforced concrete, it has been difficult for researchers to assess the exact magnitude of improvement. Typical data showing the relative improvement achieved by *the substitution of crimped steel fibers for indented fibers* are shown in Fig. 12-28*b*.

Even *low-modulus fibers* such as nylon and polypropylene have been found to be very effective in producing precast concrete elements subject to severe impact. According to Johnston,[81] a construction company in the United Kingdom annually produces

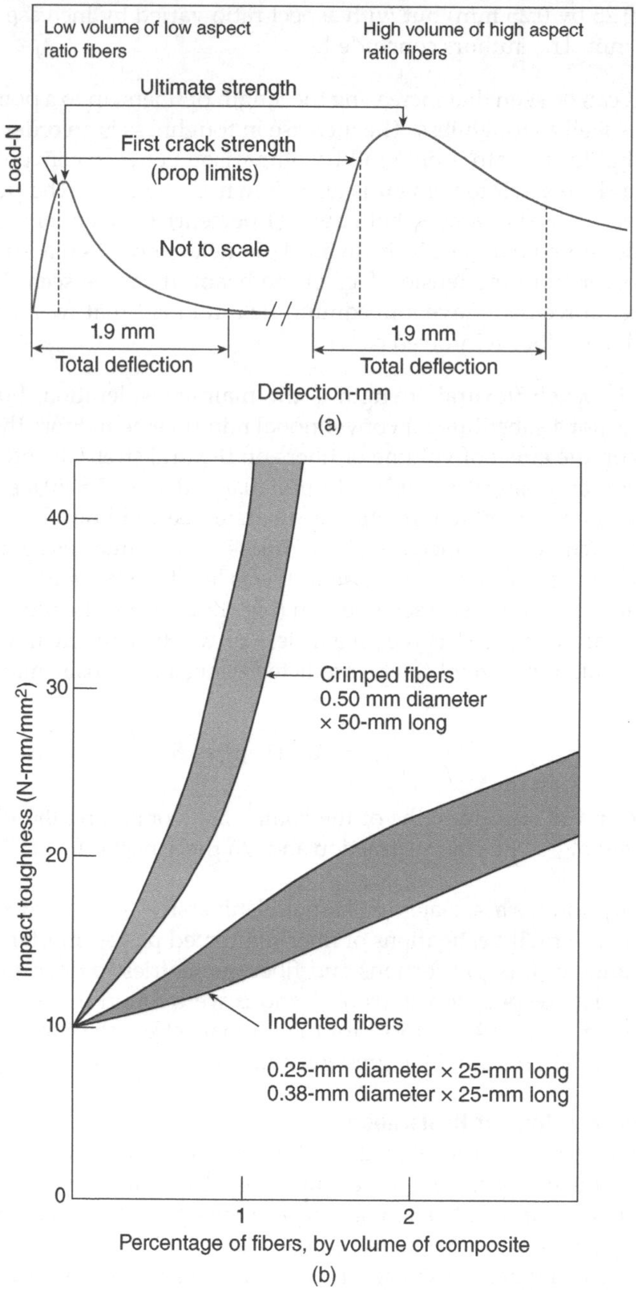

FIGURE 12-28 (*a*) Influence of fiber reinforcement on the first crack strength; (*b*) influence of crimped versus indented fibers on the impact strength. [(*a*) From Johnston, C.D., in *Progress in Concrete Technology*, V.M. Malhotra, ed., CANMET, Ottawa, 1980; (*b*) From Edington, J., D.J. Hannant, and R.I.T. Williams, *BRE Current Paper 69/74*, 1974. *By courtesy of Building Research Establishment, U.K., and reproduced by permission of Her Majesty's Stationery Office. Crown Copyright*.]

about 500,000 hollow segmental precast pile units. These units, which are 915 mm long by 280 mm in diameter with a wall thickness of about 50 mm, form the outer casing for conventional reinforced concrete piles subsequently cast within. By employing a concrete with 10-mm aggregate and 0.5 percent by volume of 40-mm-long fibrillated twine of polypropylene, resistance to breakage during pile driving was reduced by 40 percent compared to the steel mesh reinforced units used previously.

It has been shown[82] that the addition of fiber to conventionally reinforced beams increased *the fatigue life* and decreased the crack width under fatigue loading. In general, a properly designed fiber-reinforced concrete will have a fatigue strength of about 90 percent of the static strength at 2×10^6 cycles when nonreversed loading is used, and about 70 percent when full reversal of load is used.

Elastic Modulus, Creep, and Drying Shrinkage

Inclusion of steel fibers in concrete has little effect on the modulus of elasticity, drying shrinkage, and compressive creep. Tensile creep is reduced slightly, but flexural creep can be substantially reduced when very stiff carbon fibers are used. However, in most studies, because of the low volume the fibers simply acted as rigid inclusions in the matrix, without producing much effect on the dimensional stability of the composite.

Durability

Fiber-reinforced concrete is generally made with a high cement content and a low water-cement ratio. When well compacted and cured, concretes containing steel fibers seem to possess excellent durability as long as fibers remain protected by the cement paste. In most environments, especially those containing chloride, surface rusting is inevitable but the fibers in the interior usually remain uncorroded. Long-term tests of steel-fiber concrete durability at the Battelle Laboratories in Columbus, Ohio, showed minimal corrosion of fibers and no adverse effect after 7 years of exposure to deicing salt.[83]

As stated earlier, ordinary glass fiber cannot be used in portland cement mortars or concretes because of chemical attack by the alkaline cement paste. Zirconia and other alkali-resistant glass fibers possess better durability to alkaline environments, but even these are reported to show a gradual deterioration with time. Similarly, most natural fibers, such as cotton and wool, and many synthetic polymers suffer from lack of durability to the alkaline environment of the portland cement paste.

According to ACI 544,[84] steel-fiber-reinforced concretes have been investigated for durability to surface corrosion and cavitation. Tests by the Corps of Engineers suggest that erosion resistance against scour from debris in flowing water isn't improved. These tests show that when erosion is due to the gradual wearing away of concrete due to relatively small particles of debris rolling over the surface at low velocity, the quality of aggregate and hardness of the surface determine the rate of erosion (Chap. 5). Inclusion of fibers in concrete did not cause any improvement; in fact, if the use of fibers results in a higher water-cement ratio and a larger paste volume, an increase in wear would occur. However, when the loss of material is due to the cavitation impact of large debris, fibrous concrete has proved durable. The largest project of this kind yet undertaken was the repair of the stilling basin of the Tarbella Dam in Pakistan, with a 500-mm overlay of concrete containing 444 kg/m^3 cement and 73 kg/m^3 fibers.[85]

For the purpose of easy reference, a chart prepared by Johnston on the relative improvements in various properties of plain concrete, when 25 to 38 mm straight steel fibers are incorporated, is shown in Fig. 12-29.

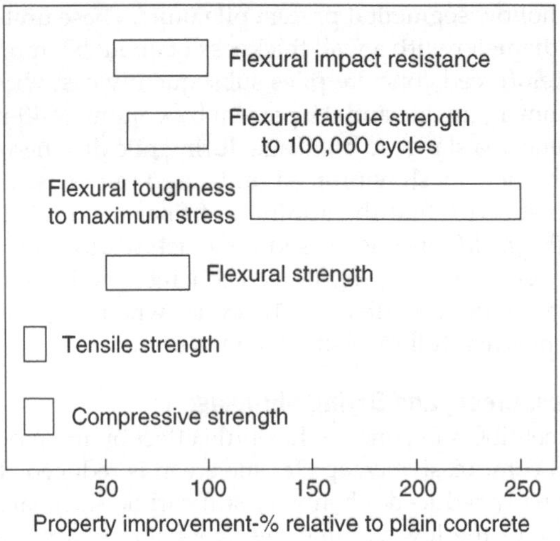

FIGURE 12-29 Relative improvement in various properties of concrete by fiber reinforcement. (From Johnston, C.D., in *Progress in Concrete Technology*, V.M. Malhotra, ed., CANMET, Ottawa, 1980, p. 502.)

12.6.6 Development of Ultra-High-Performance Fiber-Reinforced Composites

There is a new generation of high performance fiber-reinforced composites. In many of these materials the strength, toughness, and durability are significantly improved. While there are still competing approaches to obtain an optimized mechanical behavior, the following composites have already demonstrated the potential for new applications:

Compact Reinforced Composites

Researchers[86] in Denmark created Compact Reinforced Composites using metal fibers, 6 mm long and 0.15 mm in diameter, and volume fractions in the range of 5 to 10 percent. High frequency vibration is needed to obtain adequate compaction. These short fibers increase the tensile strength and toughness of the material. The increase of strength is greater than the increase in ductility; therefore, the structural design of large beams and slabs requires that a higher amount of reinforcing bars be used to take advantage of the composite. The short fibers are an efficient mechanism of crack control around the reinforcing bars. The final cost of the structure will be much higher than if the structure would be made by traditional methods; therefore, the use of compact reinforced composites is mainly justified when the structure requires special behavior, such as high impact resistance or very high mechanical properties.

Reactive Powder Concrete

Researchers in France[87] developed reactive powder concrete (RPC) by optimizing the following parameters in the mix proportions and curing conditions[88]: (a) elimination of the coarse aggregate, (b) improvement of the packing of the granular material, (c) application of pressure to the fresh concrete, (d) exposure of the material to a heat treatment, and (e) use of steel fibers, typically 13 mm long and 0.15 mm in diameter,

Fiber-Reinforced Composite	σ_o(MPa)	σ_t(MPa)	ε_t(%)	E(GPa)	G_c(kJ/m^2)	l_{ch}(m)
Plain Concrete	—	2–5	0.01	15–30	0.1–0.2	0.25–0.4
Steel FRC (V_f = 1% hooked end)	4	4.5	0.05–0.5	32.5	5	8
Carbon FRC (V_f = 2%)	5.5	5	0.1–0.2	5.6	1–3	0.2–0.7
Polymer FRC (V_f = 1% olefin)	0.75–1.5	4.5	0.1	30	1–4	1.5–6
SIFCON (V_f = 4–20% steel)	20–35	6–32	0.5	30–70	20–30	2–17
CRC (V_f = 6% steel, together with steel bars)	40	120	1–2	100	1200	8.3
RPC 200 (V_f = 2.4% steel)	9.6	10–24	0.5–0.7	54–60	15–40	4.2–8.1
ECC (V_f = 2% PE)	4.5–8	2.5	3–6	22–35	27	95–150

Source: From Li, V.C., Large volume, high-performance applications of fibers in civil engineering, *J. Appl. Polym. Sci.*, Vol. 83, pp. 660–686, 2002.

TABLE 12-19 Mechanical Properties of Some Fiber-Reinforced Concrete

with a maximum volume fraction of 2.5 percent. This composite uses fibers that are twice as long as the compact reinforced composites.

Therefore, because of workability limitations, it cannot incorporate the same volume fraction of fibers. The material can obtain a compressive strength of 200 MPa when cured in hot water at 90°C for three days and a compressive strength of 800 MPa when cured at 400°C. Commercial versions of this product have further improved the tensile strength of the matrix, chemically treated the surface of the fiber, added microfibers, and eliminated the need of thermal treatment. Typical properties of RPC are given in Table 12-19 and Fig. 12-30 shows a commercial application of this material.

Slurry-Infiltrated-Fibered Concrete

The processing of this composite consists of placing the fibers in a formwork and then infiltrating a high w/c ratio mortar slurry to coat the fibers. Fiber with high surface area are used. Compressive and tensile strengths up to 120 MPa and 40 MPa, respectively, have been obtained. Modulus of rupture up 90 MPa and shear strength up to 28 MPa have been also reported. The manufacturing process makes the product to be anisotropic. In direct tension along the direction of the fibers, the material shows a very ductile response (see Fig. 12-31). This composite has been used in pavement slabs and repair.

Engineered Cementitious Composite

The ultrahigh ductility of this composite, characterized by a tensile strain capacity of 3 to 7 percent, was obtained by optimizing the interactions between fiber, matrix, and their interface. Mathematical models[89] were developed so that a small volume fraction

Figure 12-30 With a 120-m span and a 130-cm arch depth the Cheong footbridge in Korea has the world record of slenderness (i.e., of the ratio: depth divided by span length). Even though the engineering applications of ultra-high-performance fiber-reinforced composites at still at their infancy, the material has unique properties and has the potential of changing the structural design philosophy. (*Courtesy of Paul Acker, Lafarge Company.*)

of 2 percent could attain the large ductility. Figure 12-32*a* shows a uniaxial tensile stress-strain curve for an engineered cementitious composite (ECC) sample. The material has a very high strain capacity and toughness and controlled crack propagation (Fig. 12-32*b*). Typical properties of ECC are given in Table 12-19.

The manufacturing of ECC can be done by normal casting or by extrusion. By using an optimum amount of superplasticizer and a non-ionic polymer with steric action, it was possible to obtain self-compacting casting. Experimental results with extruded pipes indicate that the system has a plastic yield behavior instead of the typical brittle fracture exhibited when plain concrete is used. This composite has been studied to repair pavements and bridge decks, to retrofit structures to withstand seismic loads, and to be an important component in the structural design of new structures.

Multiscale Fiber-Reinforced Concrete

Researchers of the Laboratoire Central des Ponts et Chaussees (France)[90] proposed to combine short and long fibers to increase the tensile strength, the bearing capacity, and the ductility of the composite. With this blend, good workability was achieved with fiber volume fractions up to 7 percent. One typical combination of fibers is 5 percent straight drawn steel fibers, 5-mm long and 0.25 mm in diameter, and 2 percent hooked-end drawn steel fibers, 25-mm long and 0.3 mm in diameter. Typical properties of multiscale fiber-reinforced concrete (MSFRC) are given in Table 12-19.

(a) Specimen failure under compression

(b) Sifcon

(c) Reinforced concrete

FIGURE 12-31 (*a*) The failure mode of *slurry-infiltrated-fibered concrete* (SIFCON) is different than that of conventional concrete. Figures 12-31*b* and *c* show the impressive deformation capacity of SIFCON compared to a reinforced concrete element subjected to shear even though both specimens had similar matrix properties. (*Courtesy of Tony Naaman.*)

12.6.7 Classifications of Fiber-Reinforced Cement Composites

Fiber-reinforced cement composites (FRCC) may be classified according to their performance under loading, as shown in Fig. 12-33. The diagram presents three classifications with respect to toughness of the material. For the lowest toughness classification of fiber-reinforced concrete (FRC), within which the large majority of applications reside, both the material and associated structural element soften after cracking occurs. However, the toughness of the material is still several orders of magnitude greater than that of ordinary, unreinforced concrete. For the intermediate toughness classification of ductile fiber-reinforced cement composites (DFRCC), the material exhibits strain softening after cracking, but the structural element (tested in bending, for example) exhibits displacement hardening. For the highest toughness classification of strain hardening cementitious composites (SHCC), both the material and associated structural element harden after cracking occurs. The literature pertaining to this subject area contains

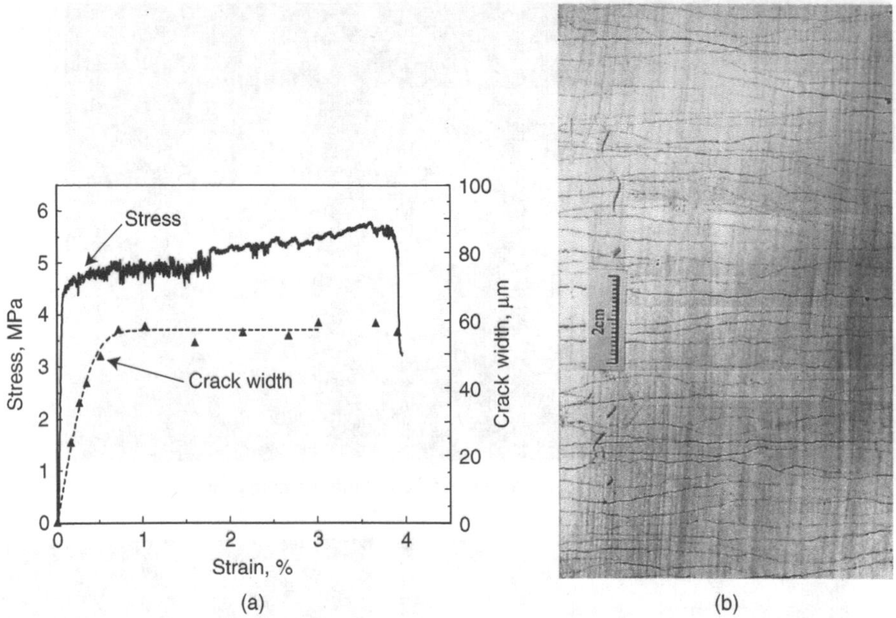

Figure 12-32 (*a*) Tensile stress-strain and crack-width-strain relations for a pseudo strain-hardening ECC; (*b*) controlled damage of engineered cementitious composite. (*Courtesy of V.C. Li.*)

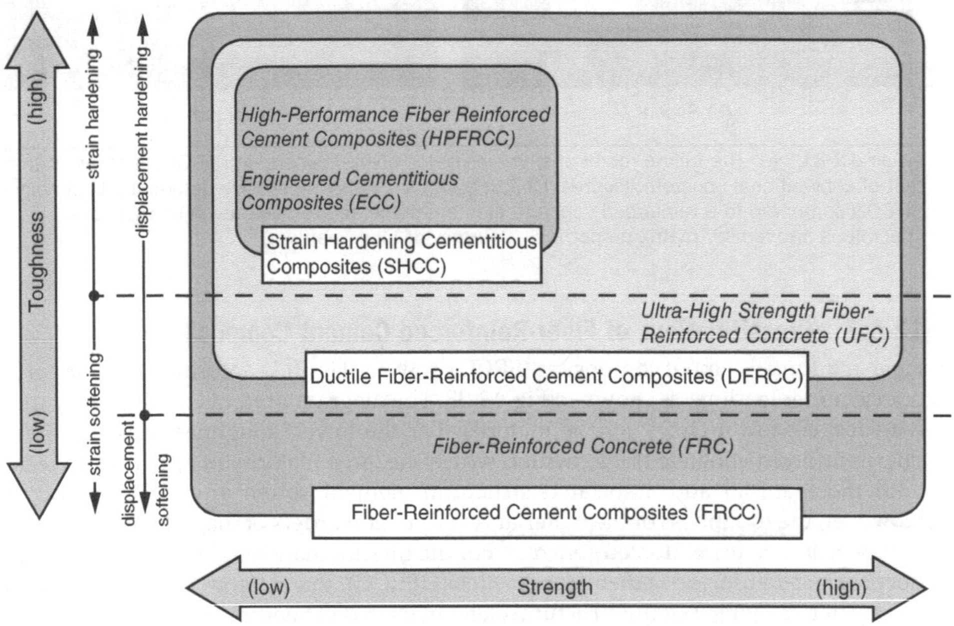

Figure 12-33 General classification of the performance of fiber-reinforced cement composites under loading. (Adapted from JCI-C82, Symposium on New Uses of Fiber-Reinforced Cement Composites, *Japan Concrete Institute*, 2012, in Japanese.)

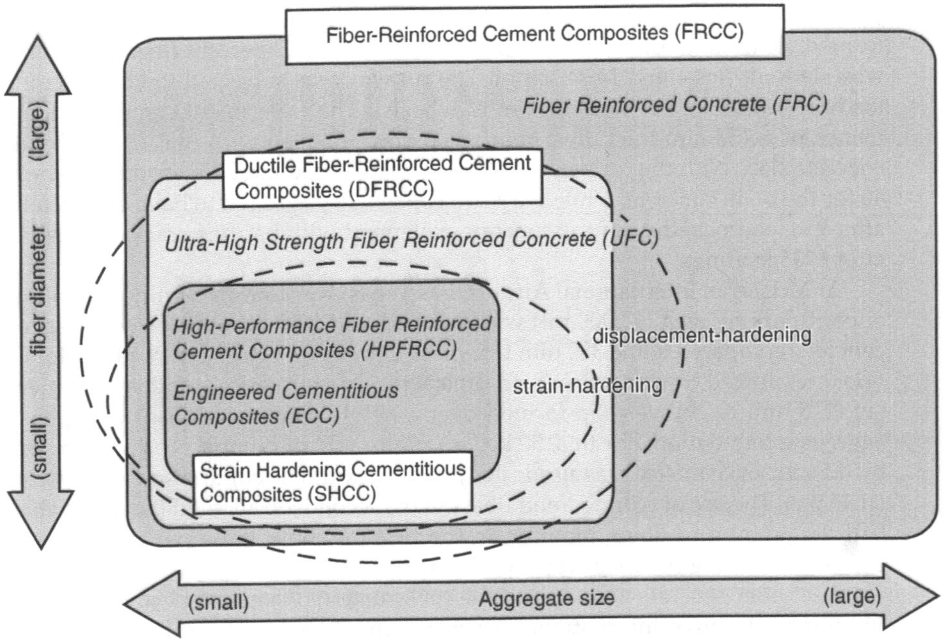

Figure 12-34 Dependence of toughness classification on fiber diameter and aggregate size. (Adapted from JCI-C82, Symposium on new uses of fiber-reinforced cement composites, *Japan Concrete Institute*, 2012, in Japanese.)

alternative naming conventions for these three categories of toughness. There are needs for developing high-strength, high-toughness materials, as indicated by the lack of available technologies appearing in the upper-right region of Fig. 12-33.

High toughness is mainly the result of fine multiple cracking, which is more readily achieved using small diameter fibers dispersed within fine-grained matrices. All else being equal, smaller fiber diameter leads to higher bonded area and more uniform distribution of reinforcement. The presence of coarser aggregates is problematic for several reasons, including their large size relative to the desired crack spacing and the introduction of large flaws associated with debonding of the aggregate-matrix interface. General dependencies of FRCC toughness on fiber diameter and aggregate size are illustrated in Fig. 12-34.

12.6.8 Applications

A report of worldwide applications involving fibrous concrete was compiled by ACI Committee 544; Henager[91] summarized some of the large and more interesting projects. Highlights from typical applications included in this summary and in the report by ACI Committee 544[92] are given here.

The first structural use of steel-fiber-reinforced concrete was in 1971 for the production of demountable 3250 mm^2 by 65 mm-thick panels for a parking garage at London's Heathrow Airport. The concrete contained 3 percent by weight of 0.25 mm diameter by 25-mm-long cold-drawn steel fibers. At the time of the last reported inspection, after 5 years of use, the slabs showed no signs of cracking.

Also in 1971, the U.S. Army Construction Engineering Research Laboratory performed controlled testing of fiber-reinforced concrete (2 percent fiber by volume) runway slabs at Vicksburg, Mississippi. The runway was subjected to C5A aircraft wheel loading (13,600 kg per wheel, over 12 wheels). The experiment compared the performance of a 150-mm-thick fiber-reinforced concrete slab with that of a 250-mm plain concrete slab. With the former, the first crack appeared after 350 loadings compared to 40 for the plain concrete; while the plain concrete was judged to have completely failed after 950 loadings, the fibrous concrete pavement with hairline cracks was serviceable after 8735 loadings.

At McCarran International Airport, Las Vegas, Nevada, an existing asphaltic-paved aircraft parking area (53,000 m^2) was overlaid with 150-mm-thick steel-fiber-reinforce concrete, compared to the 380 mm thickness that would have been required for conventional reinforced concrete. Based on satisfactory performance, another pavement overlay (175 mm thick) was slip-formed over a newly constructed base of asphalt on an aggregate foundation. For this, 50 kg/m^3 (85 lb/yd^3) of crimped-end fiber 50 mm long by 0.5 mm in diameter was used; the 28-day flexural strength of concrete was 7 MPa (1045 psi). The use of crimped-end fibers in the second job allowed the use of about one-half the amount of fibers required in the first job. Both these parking areas support DC-10 and Boeing 747 loads (up to 775,000 lb).

Steel-fiber concrete was used in the construction of a new taxiway (850 m × 23 m × 170 mm) at Cannon International Airport, Reno, Nevada. A total of 5600 m^3 of steel-fiber concrete containing 12 mm maximum aggregate and 52 kg/m^3 (87 lb/yd^3) of crimped fibers of the type employed at Las Vegas were used. The taxiway was constructed in three 7.5-m-wide strips with 12 m between sawed transverse joints; keyed longitudinal construction joints were placed at 7.5-m intervals.

Two large areas of rock slopes were stabilized with steel fibrous shotcrete, one at a refinery in Sweden, the other at a railroad cut along the Snake River in the state of Washington. In Sweden, 4100 m^3 of shotcrete containing 55 kg/m^3 of fibers was used. On the Snake River job, where 5300 m^3 of shotcrete containing 120 kg/m^3 (200 lb/yd^3) to 150 kg/m^3 (250 lb/yd^3) fibers (0.25 × 12 mm) was utilized, the slope was first rock-bolted, then covered with 60 to 75 mm of shotcrete. An estimated $50,000 was saved over the alternate plan of using plain shotcrete reinforced with wire mesh. Several tunnel linings and other lining repair projects in Japan and the United States using steel fibrous concrete have been reported.

Finally, with regard to typical uses of steel-fiber-reinforced concrete, the following statement in the report of ACI 544.1R-96 aptly summarizes the position:

> *Generally, for flexural structural components, steel fibers should be used in conjunction with properly designed continuous reinforcement. Steel fibers can reliably confine cracking and improve resistance to material deterioration as a result of fatigue, impact, and shrinkage or thermal loads. In applications where the presence of continuous tensile reinforcement is not essential to the safety and integrity of the structure, such as floors on grade, pavements, overlays, and shotcreting linings, the improvements in flexural strength, impact resistance, toughness, and fatigue performance associated with the fibers can be used to reduce section thickness, improve performance, or both.*[93]

If it is assumed that the cost of mixing, transporting, and placing concrete does not change by fiber incorporation, the difference in cost between in-place fibrous concrete and plain concrete will not be large. Also, compared to plain concrete, since the thickness of fibrous concrete slabs designed for a given load can be substantially reduced, the overall difference in the first cost may turn out to be negligible. Considering the service life, therefore, fibrous concrete would appear to be cost-effective.

12.7 Concrete Containing Polymers

12.7.1 Nomenclature and Significance

Concretes containing polymers can be classified into three categories: *polymer concrete* (PC) is formed by polymerizing a mixture of a monomer and aggregate—there is no other bonding material present; *latex-modified concrete* (LMC), which is also known as *polymer portland cement concrete* (PPCC), is a conventional portland cement concrete which is usually made by replacing a part of the mixing water with latex (polymer emulsion); and *polymer-impregnated concrete* (PIC), which is produced by impregnating or infiltrating a hardened portland cement concrete with a monomer and subsequent polymerizing the monomer in situ.

Both PC and LMC have been in commercial use since the 1950s; PIC was developed and has been in use since the 1970s. Depending on the materials employed, PC can develop compressive strengths of the order of 140 MPa (20,000 psi) within hours or even minutes and is therefore suitable for emergency concreting jobs in mines, tunnels, and highways. LMC possess excellent bonding ability to old concrete, and high durability to aggressive solutions; it has therefore been used mainly for overlays in industrial floors, and for rehabilitation of deteriorated bridge decks. In the case of PIC, by effectively sealing the microcracks and capillary pores, it is possible to produce a virtually impermeable product which gives an ultimate strength of the same order as that of PC. PIC has been used for the production of high-strength precast products and for improving the durability of bridge deck surfaces.

In his review on the use of polymers in concrete, Fowler[94] states that "*it should be understood that one of the primary limitations of concrete-polymer materials is cost. The cost of polymers can range from 10 to 100 times that of portland cement, and even considering that the specific gravity of cement is about 2.5 times that of polymer, the cost per unit volume of polymer composites is still considerably higher than portland cement concrete. Due to this higher cost, the volume of concrete polymer per unit area needs to be minimized which makes its use for high volume applications, e.g., pavements, foundations, sidewalks, and hydraulic structures impractical except in very unusual cases where durability renders concrete unusable.*" A brief summary of materials, production technology, and properties is given next. For details, other publications should be consulted.[95]

12.7.2 Polymer Concrete

Polymer concrete (PC) is a mixture of aggregates with a polymer as the sole binder. To minimize the amount of the expensive binder, it is very important to achieve the maximum possible dry-packed density of the aggregate. For example, using two different size fractions of 19 mm maximum coarse aggregate and five different size fractions of sand in an investigation carried out at the Building Research Institute of Japan, Ohama[96] attempted to match Fuller's curve for maximum density of the aggregate mixture; the voids in the range of 20 to 25 percent were filled with a 1:1 mixture of unsaturated polymer resin and finely ground limestone. It was important to use dry aggregate because the presence of moisture caused a serious deterioration in the properties of the polymer concrete.

In the Japanese study described earlier, a peroxide catalyst and an accelerator were included with the monomer for promoting subsequent polymerization in concrete. Two different curing procedures were adopted; thermal curing at 50 to 70°C, or room temperature curing at 20°C. The heat-cured specimens gave about 140 MPa (20,000 psi) compressive strength in 5 h, whereas the normal-cured specimens gave about 105 MPa (15,000 psi) in 7 days. The polyester resins were attractive because of their relatively lower cost compared to other products. Commercial products are available with a variety of formulations, some capable of hardening to 105 MPa (15,000 psi) within a few minutes without thermal treatment. Epoxy resins are higher in cost but offer advantages such as adhesion to wet surfaces. The use of styrene monomer, and methyl methacrylate (MMA) with benzoyl peroxide catalyst and an amine promoter, seems to be increasing in PC formulations. Products with increased strength have been obtained by adding to the PC monomer system a silane coupling agent, which increases the interfacial bond between the polymer and aggregate.

The properties of PC are largely dependent on the amount and properties of polymer in the concrete; for example, PC made with MMA is a brittle material that shows a nearly linear stress-strain relationship with high ultimate strength, but the addition of butyl acrylate produces a more ductile material (Fig. 12-35*a*). Typical mechanical properties of a polyester PC and a polymethyl methacrylate PC are shown in Table 12-20. By-product sulfur from oil refineries has a low viscosity at 120°C and has been used successfully for making PC.

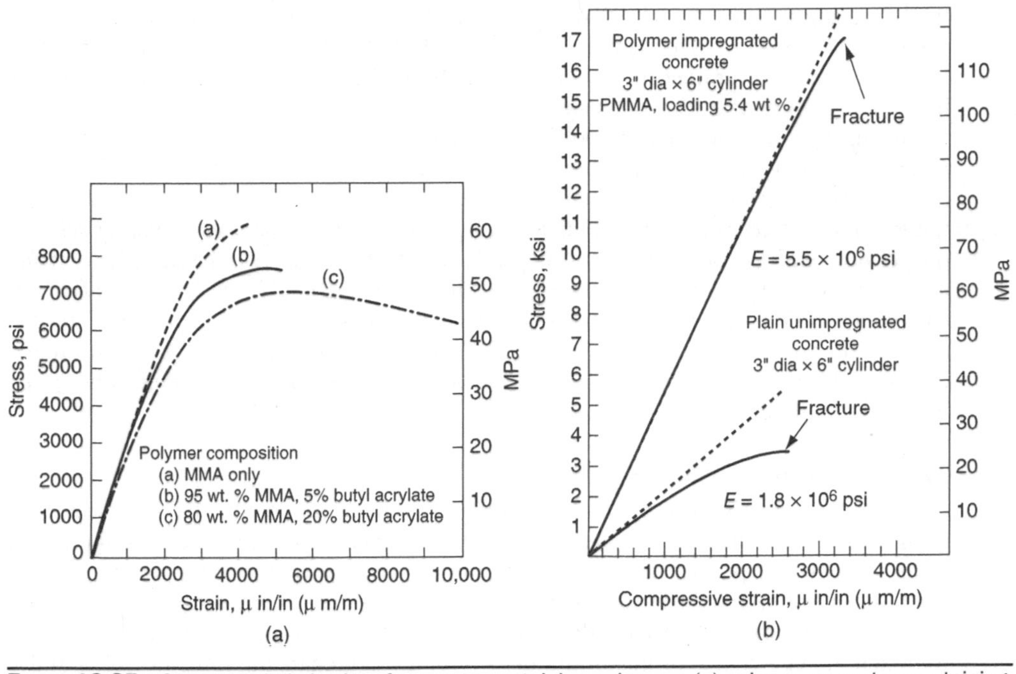

FIGURE 12-35 Stress-strain behavior of concrete containing polymers: (*a*) polymer concretes containing different polymer types; (*b*) polymer-impregnated concrete containing poly-methylmethacrylate. [(*a*) From Dikeou, J.T., in *Progress in Concrete Technology*, ed., Malhotra, V.M., CANMET, Ottawa, 1980; (*b*) From Steinberg, M., *Polymers in Concrete*, ACI SP-40, p. 12, 1973.]

	PC			PIC			
	Polyester 1:10 Polymer/ Aggregate Ratio	LMC			LMC Containing Styrene Butadiene Air-Cured		MMA Impregnated Thermal-Catalytical Polymerization
		Polymerized MMA 1:15	Control Moist-Cured	Air-Cured		Control Unimpregnated	
Compressive strength	18,000	20,000	5800	4500	4800	5300	18,000
Tensile strength	2000	1500	535	310	620	420	1500
Flexural strength	5000	3000	1070	610	1430	740	2300
Elastic modulus, $\times 10^6$	5	5.5	3.4	—	1.56	3.5	6.2

Source: Polymer-impregnated concrete from ACI SP-40, LMC from ACI Committee 548, and PC from Dikeau, J.T., in *Progress in Concrete Technology*, Malhotra, V.M., ed.,CANMET, Ottawa, 1980.

TABLE 12-20 Typical Mechanical Properties of Concretes Containing Polymers (PSI)

Due to good chemical resistance and high initial strength and modulus of elasticity, industrial use of PC has been mainly in overlays and repair jobs. Thermal and creep characteristics of the material are usually not favorable for structural applications of PC. Polyester concretes are viscoelastic and will fail under a sustained compressive loading at stress levels greater than 50 percent of the ultimate strength. Sustained loadings at a stress level of 25 percent did not reduce ultimate strength capacity for a loading period of 1000 h. The researchers recommend, therefore, that polyester concrete be considered for structures with a high ratio of live to dead load and for composite structures in which the polyester concrete may relax during long-term loading.

12.7.3 Latex-Modified Concrete

The materials and the production technology for latex-modified concrete (LMC) are the same as those used in normal portland cement concrete except that latex, which is a colloidal suspension of polymer in water, is used as an admixture. Earlier latexes were based on polyvinyl acetate or polyvinylidene chloride, but these are seldom used now because of the risk of corrosion of steel in concrete in the latter case, and low wet strengths in the former. Elastomeric or rubberlike polymers based on styrenebutadiene and polyacrylate copolymers are more commonly used now.

A latex generally contains about 50 percent by weight of spherical and very small (0.01 to 1 μm in diameter) polymer particles held in suspension in water by surface-active agents. The presence of surface-active agents in the latex tends to incorporate large amounts of entrained air in concrete; therefore, air detraining agents are usually added to commercial latexes. Since 10 to 25 percent polymer (solid basis) by weight of cement is used in typical LMC formulations, the addition of latex provides a large quantity of the needed mixing water in concrete. For reasons given later, the application of LMC is limited to overlays where durability to serve environmental conditions is of primary concern. Therefore, LMC is made with as low an addition of extra mixing water as possible; the spherical polymer molecules and the entrained air associated with the latex usually provide excellent workability. Typically, water-cement ratios are in the range 0.40 to 0.45, and cement contents are on the order of 390 to 420 kg/m^3 (650 to 700 lb/yd^3).

It should be noted that unlike polymerization of monomers by additives and thermal activation, the hardening of a latex takes place by drying or loss of water. There is some internal moisture loss in a fresh LMC mixture because water is needed for the hydration of portland cement; however, this is not sufficient to develop adequate strength. Consequently, dry curing is mandatory for LMC; the material cured in air is believed to form a continuous and coherent polymer film which coats the cement hydration products, aggregate particles, and even the capillary pores.

The mechanical properties of a typical latex-modified mortar (sand/cement ratio of 3) containing 20 percent polymer solids by weight and dry-cured at 50 percent RH for 28 days, are compared to those of control portland cement mortars in Table 12-20. It is evident from the data that LMC is better than the control material in tensile and flexural strengths. However, the strength gains are not impressive enough to justify the use of expensive latexes in making LMC products. The most impressive characteristics of LMC are its ability to bond strongly with old concrete, and to resist the entry of water and aggressive solutions. It is believed that the polymer film lining the capillary pores and microcracks does an excellent job in impeding the fluid flow in LMC. These characteristics have made the LMC a popular material for rehabilitation of deteriorated floors, pavements, and bridge decks.

12.7.4 Polymer-Impregnated Concrete

As stated earlier, the concept underlying polymer-impregnated concrete (PIC) is simply that if voids are responsible for low strength as well as poor durability of concrete in severe environments, then eliminating them by filling with a polymer should improve the characteristics of the material. In hardened concrete, the void system, consisting of capillary pores and microcracks, is very tortuous. It is difficult for a liquid to penetrate it if the viscosity of the liquid is high and the voids in concrete are not empty (they contain water and air). Therefore, for producing PIC, it is essential not only to select a low-viscosity liquid for penetration but also to dry and evacuate the concrete before subjecting it to the penetration process.

Monomers such as methyl methacrylate (MMA) and styrene are commonly used for penetration because of relatively low viscosity, high boiling point (less loss due to volatilization), and low cost. After penetration, the monomer has to be polymerized in situ. This can be accomplished in one of three ways. A combination of promoter chemical and catalysts can be used for room-temperature polymerization; but it is not favored because the process is slow and less controllable. Gamma radiation can also induce polymerization at room temperature, but the health hazard associated with it discourages the wide acceptance of this process in field practice. The third method, which is generally employed, consists of using a monomer-catalyst mixture for penetration, and subsequently polymerizing the monomer by heating the concrete to 70 to 90°C with steam, hot water, or infrared heaters.

From the above it should be apparent that the technology of producing PIC will be far more complex than that of conventional concrete. Although techniques have been developed to penetrate hardened concrete in the field, for obvious reasons PIC elements are more conveniently made as precast products in a factory. Typically, the sequence of operation consists of the following steps:

1. *Casting conventional concrete elements.* As will be discussed later, since the quality of concrete before penetration is not important from the standpoint of properties of the end product, no special care is needed in the selection of materials and proportioning of concrete mixtures. Section thickness is generally limited to a maximum of about 150 mm, since it is difficult to fully penetrate thick sections.
2. *Curing the elements.* Following the removal of elements from forms, at ambient temperatures conventional moist curing for 28 days or even 7 days is adequate because the ultimate properties of PIC do not depend on the pre-penetration concrete quality. For fast production schedules, thermal curing techniques may be adopted.
3. *Drying and evacuation.* The time and temperature needed for removal of free water from the capillary pores of moist-cured products depend on the thickness of the elements. At the drying temperatures ordinarily used (i.e., 105 to 110 °C), it may require 3 to 7 days before free water has been completely removed from a 150 by 300 mm concrete cylinder. Temperatures on the order of 150 to 175 °C can accelerate the drying process so that it is complete in 1 to 2 days; the risk of thermal cracking due to rapid heating and cooling may not have much significance if all pores and microcracks are eliminated subsequently by the monomer penetration.

   The dried elements should be evacuated before immersion in the monomer if relatively rapid (within 1 h) and complete penetration is desired. This may not be essential in durability applications, in which case overnight monomer soaking of the dried concrete, without prior evacuation, will result in one-half to three-fourths depth of penetration from the exposed surface.
4. *Soaking the dried concrete in a monomer.* The in situ penetration of concrete in the field may be achieved by surface ponding, but precast elements are directly immersed in the monomer-catalyst mixture. Commercial monomers contain inhibitors that prevent premature polymerization during storage; the catalyst serves to overcome the effect of the inhibitor. In the case of MMA, 3 percent by weight of benzoyl peroxide may be used as a catalyst.

   As stated earlier, due to the tortuous void system of hardened concrete, complete penetration of the dried specimens by soaking or infiltration is very difficult. In a study by Sopler et al., 10-cm cubes of a 0.56-water-cement ratio concrete were moist-cured for 7 days, dried at 150°C for 4 days, and soaked in MMA for periods ranging from 5 min to 48 h. Inspection of polymerized specimens soaked for 48 h showed that the total polymer loading was about 4½ percent by weight of concrete, and the middle 3-cm core was still unpenetrated (i.e., only 3½-cm-deep penetration was achieved from each surface). In the first 5 min of exposure about 38 percent of the 48-h loading was realized; thereafter the rate of penetration slowed down; penetration depths at 100 min, 4 h, and 8 h were 2, 2½, and 3 cm, respectively. Consequently, when full penetration is desired, rather than soaking it is necessary to impregnate the monomer under pressure.

5. *Sealing the monomer.* To prevent loss of monomer by evaporation during handling and polymerization, the impregnated elements must be effectively sealed in steel containers or several layers of aluminum foil; in the rehabilitation of bridge decks this has been achieved by covering the surface with sand.
6. *Polymerizing the monomer.* As discussed earlier, thermal-catalytical polymerization is the preferred technique. The time for complete polymerization of the monomer in the sealed elements exposed to steam, hot water or air, or infrared heat at 70 to 80°C may vary from a few to several hours. In the case of a MMA-benzoyl peroxide mixture, no differences in strength were found between specimens polymerized at 70°C with hot air for 16 h or with hot water for 4 h. Polymerization in hot water simplifies the precast PIC production process by eliminating the need for sealing.

The strain-stress behavior of PIC are shown in Fig. 12-35. Various studies have shown that the properties of PIC are not affected by the initial quality of the unimpregnated concrete. For example, Sopler et al. subjected concretes of three different qualities, 20 MPa or 3000 psi (0.83 water-cement ratio), 38 MPa or 5500 psi (0.56 water-cement ratio), and 59 MPa or 8500 psi (0.38 water-cement ratio), to the same polymer-impregnation process; all concretes achieved similar strength characteristics, although a higher polymer loading was required in the 0.83-water-cement ratio concrete.

As expected, the sealing of microcracks and pores renders the material brittle. The stress-strain curve in compression remains linear up to about 75 percent of ultimate load, and the deviation from linearity is only 10 to 15 percent at failure. Due to the absence of adsorbed water, shrinkage and creep are insignificant.

In spite of high strength, PIC elements will be of little interest for structural use due to the size limitation. There has been considerable interest in applications of PIC where the main consideration is the excellent durability of the material to abrasion, frost action, and attack by strong chemical solutions. The durability characteristics of PIC are due to impermeability and the absence of freezable water. PIC also shows promise for substrate treatment in the rehabilitation of concrete bridge decks. There is evidence to suggest that the overlay systems most commonly used for rehabilitation of deteriorated pavements may not be wholly compatible with untreated concrete substrate. Cady et al.[97] investigated combinations of several substrate treatments and overlays and concluded that only the MMA (soaked or impregnated) substrate did not suffer from debonding, and showed excellent durability to frost action.

12.8 Shotcrete

12.8.1 Definitions and Brief History

The ACI Committee 506R[98] defines *shotcrete* as a mortar or concrete pneumatically projected onto a surface at high velocity. Thus the name of the product is derived from the special method of concrete placement. The concrete mix is composed of the same ingredients that are used for making conventional concrete structures, namely water, aggregates, and cementing materials such as portland cement, coal fly ash, natural pozzolans, ground granulated blast-furnace slag, and silica fume. In Europe, shotcrete is generally referred to as "*Sprayed Concrete.*" Shotcrete technology is well-suited for the repair of damaged or deteriorated concrete, and in new construction for providing mining and tunneling ground support.

In a 1910 patent, Carl Akeley's invention of a double-chamber gun for shooting sand-cement mixture was given the proprietary name *"Gunite."* Shotcrete technology has evolved to the point that it is now used in various sectors of the construction industry. In 1930s, the American Railway Engineering Association introduced the term *shotcrete*, for products using the Gunite process. In 1951, the American Concrete Institute approved the term shotcrete to describe both mortar and concrete products made by the dry-mix process. Since 1966, according to the report of ACI Committee 506, the concrete cast by the wet-mix process is also covered under the term shotcrete.

12.8.2 The Shotcreting Process

Shotcreting can be done either by the dry-mix process or the wet-mix process. In the *dry-mix process*, all ingredients except water are thoroughly mixed before they are fed into the shotcreting gun, which then conveys the material into the delivery hose by a metering device. The material is then carried by compressed air to a nozzle assembly. A water-ring, which is fitted inside the nozzle assembly, is capable of injecting mixing water under pressure. The water is thoroughly mixed with the dry ingredients and this mixture is then jetted from the nozzle at high velocity onto the surface to be shotcreted.

In the *wet-mix process*, the batching and mixing operations are similar to conventional ready-mixed concrete, whereby all the ingredients including the mixing water are accurately measured and thoroughly mixed in a central mixer or truck mixers and transported for delivery to the shotcreting equipment at the job site. Compressed air is injected at the nozzle to increase the velocity and improve the shooting pattern.

A comparison of between the wet-mix versus the dry-mix shotcreting processes (see Table 12-21) demonstrates the following:

- The wet-mix process provides for better control in incorporating the water and a more thorough mixing the ingredients.
- The dry-mix process does have one advantage in that an experienced nozzleman can exercise instantaneous control over the mixing water and the consistency of shotcrete to meet variable field conditions.

Dry-Mix Process	Wet-Mix Process
1. Instantaneous control over mixing water and consistency of the mixture at the nozzle to meet variable field conditions.	1. Mixing water is controlled at the mixing equipment and can be accurately measured.
2. Better suited for placing mixtures containing lightweight aggregates or refractory materials.	2. Better assurance that the mixing water is thoroughly mixed with other ingredients.
3. Capable of being transported longer distance	3. Less dust and cementitious materials lost during the shooting operation.
4. Delivery hoses are easier to move.	4. Normally has lower rebound, resulting in less waste.
5. Lower volume per hose size.	5. Higher volume per hose size.

Source: From *Guide to Shotcrete*, ACI Committee 506, American Concrete Institute, Farmington Hills, Michigan, 2005.

TABLE 12-21 Comparison of Dry-Mix and Wet Mix Processes

- The wet process generally has a lower rebound waste, and a higher productivity per hose size.
- Dry-mix shotcrete has the advantage of being transportable over longer distances. Since the 1980s, there is an increasing trend worldwide to replace the dry-mix shotcreting process with wet-mix shotcreting.

12.8.3 Mixture Proportioning

According to the ACI *Guide to Shotcrete*, many different types of concrete with different properties are being produced by shotcreting; these include high-strength concrete, fiber-reinforced concrete, and underground shotcrete. As a general rule, the composition of the concrete mix affects the properties of hardened shotcrete in the same way as normal concrete. For instance, if we compare the effect of adding coarse aggregate to a shotcrete mortar mix, the reduced surface area of coarse aggregate at the cost of fine aggregate permits a lower water content but with the same consistency. This has many beneficial effects on the properties of hardened concrete—such as higher strength and lower permeability. In addition, replacement of fine aggregate with coarse aggregate enables considerable reduction in the drying shrinkage and creep. The impact of coarse aggregate particles on the plastic mortar during the shotcreting process improves the in-place density of the product. For both dry-mix and wet-mix processes, the ACI *Guide to Shotcrete* (506R, 2005) cautions that the coarse-aggregate shotcrete with more than 30 percent coarse aggregate has greater rebound loss, is more difficult to finish, and cannot be used for thin layers. Based on current practice in the United States, mix proportions and typical properties of hardened shotcrete are summarized below.

There are some deleterious as well as beneficial effects associated with the placement of concrete by shotcreting. As a result, the properties of in-place shotcrete may differ significantly from the properties of as-batched, ready-mixed concrete. This should be taken into consideration in selecting materials and mix proportions for shotcrete. The ACI 506.4 Report[99] serves as a guide for evaluating the in-place dry-mix and wet-mix shotcrete. Based on the latest reports of various subcommittees of ACI 506R, useful information on proportioning of shotcrete mixtures, is presented next.

Because rebound is low in the wet-mix shotcreting process, the in-place proportions should not vary significantly from the as-batched concrete mix. In the dry-mix shotcreting process, however, the rebound material tends to contain a higher proportion of aggregate compared with cementing material. Such variation in the composition of in-place shotcrete not only influences strength and other properties, but also sand streaks and porous zones. Note that this type of inhomogeneity in the composition of in-place concrete normally does not occur in the case of conventionally placed concrete.

12.8.4 Properties

Compressive Strength

To a great extent the compressive strength of dry-mix shotcrete is governed by the cement-aggregate ratio; in the case of wet-mix shotcrete it is governed by the water-cementitious materials (w/cm) ratio. For ordinary applications, the 28-day specified compressive strength of 30 to 50 MPa is readily achieved. By reducing the w/cm by adding a superplasticizer and using silica fume, the 28-day strength of 85 to 100 MPa has been reported for both the dry-mix and the wet-mix shotcreting process.

Early-age strength development is more important than ultimate strength in the repair and rehabilitation of structures, and for installation of tunnel lining and underground support. In these cases, chemical accelerating admixtures are often used to achieve early setting and hardening. Specifications can be made for 8 h, 1 day, 3 days, 7 days, and 28 days minimum strength requirement. Commonly used set accelerators are sodium and potassium carbonate, calcium aluminate cement, and triethanol amine. Note that the early-age strength acceleration is generally accompanied by a reduction in the long-term strength and durability.

Flexural Strength and Bond Strength

According to ACI 506R, in the past traditionally welded wire fabric was used in shotcrete tunnel linings to provide ductility. Presently, welded-wire reinforcement is increasingly being replaced by steel or synthetic polymer fibers. Fiber reinforcement imparts toughness and postcracking load-bearing capacity to the lining. It also helps control restrained shrinkage cracking and improves impact resistance.

Bond strength is usually measured by shear or direct tension using a pull-off test. Because shotcrete is physically driven onto the receiving surface, it usually develops good bond with concrete, steel, rock, and masonry. It is recommended that shotcrete should develop a minimum bond strength of 0.7 MPa; properly applied shotcrete on a well-prepared substrate usually develops a bond strength of more than 1 MPa.

Drying Shrinkage

As described by the ASTM Standard Test C157, the 3-month drying shrinkage of shotcrete varies between 0.06 and 0.10 percent, which is greater than conventional concrete because of less coarse aggregate, and increased cement and water content as is typically found in a shotcrete mix. Shrinkage is an important property with respect to potential cracking and bond failure, especially when shotcrete is used to repair structures. Note that set acceleration tends to increase the drying shrinkage.

Resistance to Freezing and Thawing

With respect to freezing and thawing resistance, shotcrete behaves similarly to conventional concrete; therefore, its behavior is strongly dependent on the w/cm and the quality of the air-void system. According to ASTM C666, critically saturated wet-mix shotcrete requires an in-place air-void system with a minimum air content of 4 percent and a maximum air-void spacing factor of 0.3 mm to resist rapid freezing and thawing cycles. Because a considerable amount of air is lost during the shooting process or wet-mix shotcrete, ACI Committee 506 recommends a minimum air content of 6 percent before shooting of the concrete mix.

Other Properties

An overall indication of the quality of shotcrete can be obtained by determining the moisture absorption capacity and volume of permeable voids in specimens of hardened shotcrete (ASTM C642). According to ACI 506R, the acceptable values of a permeable void system is 14 to 17 percent; the typical absorption ranges between 6 and 9 percent. These values are greatly influenced by the w/cm of the concrete mix; a low w/cm will yield a relatively low volume of permeable voids and low absorption. In general, high values of permeable voids or absorption are indicative of poor quality and potential durability. For given constituent materials and w/cm, both shotcrete and conventional concrete would have similar coefficients of permeability and a coefficient of thermal expansion similar to steel. The density of high-quality shotcrete (2230 to 2390 kg/m^3) and the modulus of elasticity (17 to 40 GPa) are also similar to conventional concrete.

12.8.5 Field Applications

Shotcrete is the preferred method over conventional concrete in a variety of repair applications and in the case of new construction where normal casting techniques cannot be employed. Typical examples of where the use of shotcrete is appropriate include jobs where formwork is too complicated, access to work area is difficult, or thin layers or concrete are required. According to ACI 506R *Guideline to Shotcrete,* normal shotcrete without any special admixtures is the most commonly used material in the following applications:

- *New structures.* Roofs, thin shells and walls, prestressed tanks, canals, foundation shoring, and artificial rocks.
- *Lining and coating.* For brick and concrete block masonry, underground support for tunnels, slope protection, erosion control, fireproofing of structural steel members, and steel pipelines.
- *Repair-work.* For deteriorated concrete in bridges, culverts, sewers, dams, tunnels, buildings, sea walls, and piers.
- *Strengthening and reinforcing.* To strengthen and reinforce concrete beams, columns, slabs, masonry walls, and for seismic rehabilitation of shear walls.
- *Refractory shotcrete.* Because of the high speed of installation, shotcrete has become the method of choice for concrete casting in new construction and for repairs of all types of refractory linings in industrial furnaces.
- *Ground support.* Shotcrete is extensively used for temporary or permanent ground support and, in fact, it has become the primary method of choice for ground support in tunnels and mines. As described below, to achieve better performance, the shotcrete can be reinforced with steel or polypropylene fibers.

12.8.6 Fiber-Reinforced Shotcrete

Fiber-reinforced shotcrete (FRS) is conventional shotcrete with added steel or synthetic fibers. Recent research has shown that performance of shotcrete can be improved by adding a small quantity of discrete fibers. These fibers are generally divided into two classes based on their diameter: less than 0.3-mm-diameter fibers are known as *microfibers*; greater than 0.3-mm-diameter fibers are called *macrofibers*. Commercially used macrofibers of steel are generally 19 to 35 mm long, and 0.4 to 0.8 mm in diameter. Polypropylene-based fibers—of less than 33 μm in diameter and less than 12 mm long—are generally used as microfibers. Steel macrofibers in shotcrete usually range from 12 to 60 kg/m^3. Synthetic fibers are generally used in the range of 5 to 9 kg/m^3 for macrofibers and 0.6 to 2.4 kg/m^3 for microfibers.

Significant improvements in toughness and impact resistance occur with the addition of fiber reinforcement. These improvements are accompanied by several benefits in field performance when compared to normal shotcrete, such as reduction in shrinkage cracking and fire-related explosive spalling. Banthia and Campbell[100] have shown that both steel and synthetic *macrofibers* are effective. Fiber reinforcement not only delays crack formation, but also reduces the crack width and total area where cracks occur.

Padron and Zollo[101] have reported that the addition of as low as 0.1 percent volume percentage of *microfibers* can provide *resistance to plastic shrinkage cracking* if the rate of moisture loss is high at early ages. The addition as low as 0.25 percent by volume *macrofibers* provide *resistance to drying shrinkage cracking*, according to a report by Grzybowski and Shah.[102] Field applications of FRS have been discussed in numerous publications; a few examples are described next.

The recently completed renovation of the 125-years old Weehawken tunnel for the Hudson-Bergen rapid-transit rail system in New Jersey is a good example of shotcrete that was reinforced with steel macrofibers and synthetic microfibers. According to Garrett[103] and Tatnall,[104] this project required enlargement of the tunnel diameter and installation of a new railway system about halfway through the 12.66-km-long tunnel. To provide ground support for the 12.8 shaft of the tunnel and for station elevators, a shotcrete reinforced with steel macrofibers was used. To provide resistance to explosive spalling in the case of fire, a low dosage, 1 to 3 kg/m^3, of polypropylene microfibers were incorporated into shotcrete used in the tunnel lining between the 8.2-m-wide tunnel and 19.8-m-wide station. This was done in response to the fire-related explosive spalling that occurred in the 157-km-long Gotthardt Tunnel fire in the Swiss Alps.

Shotcrete-Lined Tunnels

In introductory remarks to his recently published book,[105] Alun Thomas says that the design and construction of both hard rock and soft ground tunnels have been revolutionized by the advent of the shotcreting method. By enabling the installation of permanent concrete lining, shotcreting has unlocked the true potential of this method of construction to minimize construction cost and time. According to the author, as a result of the extensive use of shotcreting to line the chalk caverns in the construction of the Channel Tunnel (1989), shotcreting is rapidly growing in popularity in the United Kingdom.

Shotcreting for Repair and Rehabilitation of Structures

Worldwide, there are numerous field applications of the shotcreting method of concrete construction for repair and rehabilitation of structures. Gilbride et al.[106] reported that steel macrofibers incorporated into shotcrete were used to repair the deteriorated concrete piles and decks of concrete berths in the Gulf of Funde at the Port of Saint-John, Canada; similar structures exposed to the marine environment in the Caribbean have been repaired with shotcrete containing synthetic microfibers. Seismic retrofit of Little Rock Dam in Southern California required stiffening of arches with a 100-mm-thick bonded overlay of steel fiber-reinforced shotcrete is described by Forrest et al.[107]

12.9 Heavyweight Concrete for Radiation Shielding

12.9.1 Significance

Concrete is commonly used for biological shielding in nuclear power plants, medical units, and atomic research and testing facilities. Other materials can be employed for this purpose, but concrete is usually the most economical and has several other advantages. Massive walls of conventional concrete are being used for shielding purposes. However, where usable space is limited, the reduction in the thickness of the shield is accomplished by the use of heavyweight concrete. Heavyweight concretes are produced generally by using natural heavyweight aggregates. The concrete unit weights are in the range 210 to 240 lb/ft^3 (3360 to 3840 kg/m^3), which is about 50 percent higher than the unit weight of concrete containing normal-weight aggregates.

12.9.2 Concrete as a Shielding Material

Two types of radiation have to be considered in the design of biological shields. First are X-rays and gamma rays, which have a high power of penetration but can be absorbed adequately by an appropriate mass of any material. Most materials attenuate these high-energy, high-frequency electromagnetic waves primarily according to the Compton scattering effect, the attenuation efficiency being approximately proportional to the mass of the material in the path of radiation. Since attenuation is not influenced by the type of material, different materials of the same mass have the ability to offer equal protection against X-rays and gamma rays.

The second type of radiation involves *neutrons*, which are heavy particles of atomic nuclei and do not carry an electrical charge. Thus neutrons are not affected by the electric field of the surroundings and therefore slow down only on collision with atomic nuclei. According to Polivka and Davis, for an efficient neutron shield it is often desirable to incorporate the following three classes of materials:

> The shield should contain some heavy material, such as iron, whose atomic mass is 56, or elements of higher atomic number. These heavy elements slow down the fast neutrons by inelastic collisions. It is desirable to have light elements, preferably hydrogen, to further slow down the moderately fast neutrons through elastic collisions. Hydrogen is particularly effective because it weighs about the same as a neutron. Finally, it is necessary to remove the slow-thermal neutrons by absorption. Hydrogen is effective in this action but in the process emits a 2.2 million volt gamma ray that requires considerable shielding itself. Boron, on the other hand, has not only a very high cross section (absorbing power) for neutrons but emits gamma rays of only 0.0478 million volts. For these reasons boron-containing materials are a factor in the required shield thickness.[108]

Commenting on the shielding ability of concrete, Polivka and Davis state that concrete is an excellent shielding material that possesses the needed characteristics for both neutron and gamma ray attenuation, has satisfactory mechanical properties, and has a relatively low initial as well as maintenance costs. Also the ease of construction makes concrete an especially suitable material for radiation shielding.

Since concrete is a mixture of hydrogen and other light nuclei, and nuclei of higher atomic number, and can be produced within a relatively wide range of density, it is efficient in absorbing gamma rays, slowing down fast neutrons, and absorbing resonance and slow neutrons. The hydrogen and oxygen, contained in chemically combined form in the hydrated cement, moderate the neutron flux satisfactorily. The oxygen may also be present in concrete in another form in addition to water; concrete with siliceous aggregate is about one half of oxygen.

The 1989 classical book by Kaplan[109] on concrete radiation shielding contains the most extensive information on the concrete properties required for use in nuclear power plants. More recent data on the attenuation coefficients for gamma rays can be found in Gurler and Tarim[110] and Basher et al.[111]

12.9.3 Materials and Mix Proportions

Except for the heavyweight aggregates (Table 7-3) and some hydrous ores as well as boron minerals, the same materials and proportioning methods are used for producing heavyweight concrete mixtures as are used for conventional normal-weight concrete. For details pertaining to concrete-making materials for biological shielding, the standard specifications should be consulted: ASTM C 637-09[112] (specification for aggregates for radiation shielding concrete) and ASTM C 638-09[113] (nomenclature of constituents of aggregates for radiation shielding concrete). ASTM C 637 describes two classes for aggregates for radiation-shielding concrete. Class I consists of minerals that have high specific density and contain substantial proportions of atoms of high atomic weight. Class II consists of minerals of substantial boron content that are effective in absorbing neutrons without generating highly penetrating gamma rays. Guidelines for petrographic analysis of high-density and boron-bearing aggregates can be found in a recent publication by Lee et al.[114]

Because of the high density of aggregate particles, segregation of fresh concrete is one of the principal concerns in mix proportioning. From the standpoints of high unit weight and a lower tendency for segregation, it is desirable that both fine and coarse aggregate be produced from high-density rocks and minerals. Due to the rough shape and texture of crushed aggregate particles, heavyweight concrete mixtures tend to be harsh. To overcome this problem it is customary to use a finer sand, a greater proportion of sand in aggregate than with conventional concrete, and cement contents higher than 360 kg/m^3 (600 lb/yd^3). It should be noted that to get around the problem of segregation, sometimes other than conventional methods, such as preplaced aggregate concreting, may be employed. In this method, after filling the forms with compacted aggregate coarser than 6 mm, the voids in the aggregate are filled by pumping in a grout mix containing cement, fine sand, pozzolans, and other pumpability aids.

12.9.4 Important Properties

As stated above, the workability of fresh concrete can be a problem. Heavyweight concrete can be pumped or placed by chutes over short distances only, because of the tendency of coarse aggregate to segregate. Concretes containing borate ores, such as colemanite and borocalcite, may suffer from slow setting and hardening problems because these minerals are somewhat soluble, and borate solutions are strong retarders of cement hydration. Unit weights of concrete containing barite, magnetite, or ilmenite aggregate are in the range of 3450 to 3760 kg/m^3 (215 to 235 lb/ft^3); when hydrous and boron ores (which are not of high density) are used as partial replacement for heavyweight aggregate, the unit weight of concrete may come down to about 3200 to 3450 kg/m^3 (200 to 215 lb/ft^3).

Sakr[115] studied the effects of replacing portland cement with silica fume or rice husk ash in concrete mixtures containing ilmenite and barite. The experimental results indicated that concrete samples cast with 15% silica fume had the highest density; compressive, tensile, flexural, and bond strengths; modulus of elasticity; and attenuation coefficient values of all the mixtures studied. Rice husk ash replacements produced concrete with higher mechanical properties than the reference concrete. The gamma attenuation coefficient of the concrete mixtures was not affected by replacing cement with silica fume or rice husk ask.

Massive shielding walls need not be designed for more than 14 MPa (2000-psi) compressive strength; for structural concrete, strengths of the order of 20 to 35 MPa (3000 to 5000 psi) are sufficient and not difficult to achieve with the high cement contents normally used. Strength is, however, of principal concern in the design of heavyweight concrete mixtures suitable for use in prestressed concrete reactor vessels (PCRV). These are pressure vessels that operate at higher stress levels and temperatures than conventional structures, and concrete is subject to appreciable thermal and moisture gradients. In such cases, inelastic deformations such as creep and thermal shrinkage should be minimized because they can cause microcracking and loss of prestress. Obviously, the elastic modulus of aggregate and compatibility of coefficients of thermal expansion between aggregate and cement paste should be considered to minimize microcracking.

The reactor vessels are usually designed to operate with concrete temperatures up to 71°C, but higher accidental temperatures and some thermal cycling is expected during the service life. Considerable strength loss can occur when concrete is subjected to wide and frequent fluctuations in temperature; hence PCRV concrete is designed not only for high density but also for high strength. In a study at the Corps of Engineers, Waterways Experiment Station,[116] using 430 to 575 kg/m^3 (720 to 970 lb/yd^3) Type I portland cement, 12 mm or 38 mm-maximum magnetite or ilmenite aggregate, and an 0.30 to 0.35 water-cement ratio, heavyweight concretes (3680 kg/m^3 unit weight) were produced which gave 52 to 65 MPa (7600 to 9400-psi) compressive strength at 7 days, and 62 to 76 MPa (9000 to 11,000 psi) at 28 days.

12.10 Pervious Concrete

12.10.1 Definition and Specification

Pervious concrete is a special concrete that has been designed to contain a relatively high proportion of large pores together with a large volume fraction of such pores. According to ACI Committee 522R,[117] typically the pore sizes are in the 2 to 8 mm range, and the total volume fraction of such pores varies between 10 and 35 percent. Gap-grading the coarse aggregate—either by eliminating or minimizing the volume of fine aggregate in the concrete mixture—creates a large volume of coarse pores in the hardened concrete. This type of concrete is also known as enhanced-porosity concrete or no-fines concrete.[118] Because of its highly porous microstructure, this concrete is pervious to fluids and relatively low in strength when compared to conventional concrete mixtures.

Pervious concretes can achieve compressive strengths on the order of 20 MPa (3000 psi); higher strengths can be obtained by using special structural design and construction practice, and chemical admixtures. The durability of pervious concretes can also be enhanced by using supplementary cementitious materials such as silica fume, coal fly ash, and ground granulated blast furnace slag. The load-bearing capacity of these concretes can be increased by installing a well-compacted subgrade of coarse aggregates beneath the pavement. With proper control of subgrade preparation, concrete compaction, and curing, pervious concrete pavements can last 20 to 40 years with little or no maintenance.

The Environmental Protection Agency (EPA) now requires public and private agencies to provide sustainable solutions to various problems associated with the mismanagement of storm water runoff and considers the use of pervious concrete among the best management practices in its manual of technologies for storm water management. Pervious concretes are also being promoted by the Green Building Council's LEED (Leadership in Energy and Environmental Design) rating system.[119]

12.10.2 Benefits of Pervious Concrete

The benefits of using pervious concrete as a paving material compared to conventional concrete can be classified into three broad categories: (1) safety, (2) environmental benefits, and (3) economic advantages.

FIGURE 12-36 Surface texture of a pervious concrete pavement. (*Courtesy of Omkar Deo.*)

Safety

The considerable number of voids in pervious concrete makes it highly permeable, which allows the water to drain quickly through concrete instead of excess water draining over the pavement. This eliminates the spraying and pooling of water, and prevents hazards such as hydroplaning. Also, due to the lack of fine aggregates, the concrete has a unique surface texture (shown in Fig. 12-36) made up of rounded and angular aggregates, such as limestone, gravel, or crushed stone. The exposed coarse aggregate of pervious concrete provides enhanced traction for vehicles. This prevents hazardous driving conditions during severe weather such as heavy rain and snow.

Environmental Benefits

The use of pervious concrete pavement helps in recharging the groundwater table that would remain mostly depleted if a conventional or rather impervious concrete pavement is used. Other important environmental benefits of pervious concrete include the possible reduction of urban heat island effect due to its permeable characteristics and relatively lighter appearance, and a reduction of irrigation water in urban areas by allowing the water to be stored in soils, thereby encouraging the growth of vegetation in adjacent areas.[120] Research[121] has also determined that pervious pavements used for parking lots have the potential to trap approximately 98 percent or more of oil introduced from vehicular spills. Recently there has been considerable interest in using pervious concrete to mitigate the tire-pavement interaction noise found in concrete pavements.[122–124]

Economical Advantages

The use of pervious concrete pavements can eliminate the need for expensive measures for storm water runoff control, such as retention ponds, swales, and lagoons. By reducing the storm water impact fees charged by several government agencies, they minimize the demands for sewer systems. Pervious concrete is also recognized as a pavement material having a low life-cycle cost.[119]

Materials	Proportions kg/m³ (lb/yd³)
Cementitious materials	270–415 (450–600)
Aggregate	1190–1480 (2000–2500)
Water-cement ratio	0.27–0.34
Aggregate-cement ratio	4–5.5
Fine-coarse aggregate ratio	0–1

TABLE 12-22 Typical Ranges of Materials Proportions in Pervious Concrete

12.10.3 Mixture Proportioning

Mixture proportioning for pervious concrete has traditionally relied on trial-and-error procedures. A series of trial batches are developed in the laboratory and then tested in the field to ensure expected behavior and performance.[117] The goal of mixture proportioning is to achieve a good balance between void content, paste content, workability, and strength. Trial batches serve as preliminary checks to achieve the intended results in the final production of the mix. The typical range of mixture proportions are shown in Table 12-22.[117]

The aggregate-cement ratio and water-cementitious materials ratio (w/cm) are the two most important factors in the mixture that affect the mechanical characteristics. In general, a water-cementitious materials ratio between 0.30 and 0.40 is used to produce pervious concrete mixtures.[117,120] Chemical admixtures, which can influence workability and setting time, are also used.

Because the fine aggregate content is nearly or entirely eliminated in pervious concrete mixtures, the aggregate grading used is typically either single-sized coarse aggregates or grading between ¾ and 3/8 in. (19.0 and 9.5 mm).[117] Commonly used gradations of coarse aggregate in ASTM C-33 include No. 67 (¾ in. to No. 4), No. 8 (3/8 in. to No. 16), or No. 89 (3/8 in. to No. 50) sieves (in metric units: No. 67 [19.0 to 4.75 mm], No. 8 [9.5 to 2.36 mm], or No. 89 [9.5 to 1.18 mm]), respectively.[120] The use of larger aggregate sizes (20 mm maximum size) in the mixture proportioning is recommended to prevent clogging when applying pervious concrete pavements.[118]

Chemical admixtures are used in pervious concrete for the same reasons as in conventional concrete. Generally due to the quick-setting tendency of pervious concrete mixtures, set-retarding admixtures are commonly used.[120] High-range or medium-range water reducing admixtures are frequently used depending on the water-cementitious materials ratio. Air-entraining admixtures can also be used for environments susceptible to freezing and thawing.[120]

Deo and Neithalath's[125] novel mixture proportioning methodology for pervious concretes achieves the desired porosities for both fresh and hardened state by using a set of materials and compaction methods. ASTM C 1688, *Test Method for Determining the Fresh Density and Void Content of Pervious Concretes*, is used to determine the fresh porosity of pervious concrete mixtures; the hardened state porosities are evaluated using a volumetric method.[122] Initially, the dry-rodded unit weight for each of the different aggregates is measured in accordance with ASTM C 29, and the aggregate void content is calculated from the measured dry-rodded unit weight. If a pervious concrete mixture is to be designed to achieve a specific or certain porosity (ϕ_{design}), the minimum paste volume needed ($v_{p\text{-min}}$) can be assumed as follows:

$$(v_{\text{void-agg}} - \phi_{\text{design}})$$

where $v_{p\text{-min}}$ is based on the assumption that the aggregates in the mixture are compacted in a manner similar to the ASTM C 29 procedure, and the paste is relatively incompressible. Since the compaction effort based on ASTM C 1688 procedure is found to be lower than that based on ASTM C 29 procedure, the paste volume fraction is varied to obtain the design porosity. This variation or difference in the porosity is adjusted by either increasing the paste volume fraction or by increasing the compaction effort in the total volume fraction of the concrete mixture.

12.10.4 Applications

Although pervious concrete does not have a significant load-bearing capacity when compared to conventional concrete, it is adequate for a wide variety of applications, including: low-volume pavements, driveways, parking lots, sidewalks, pathways, tennis courts, patios, swimming pool decks, noise-barriers, and load-bearing walls.[117] Figure 12-37 shows common applications of pervious concrete.

12.10.5 Properties of Pervious Concretes

Compressive Strength

The 28-day compressive strength of pervious concrete ranges from 3.5 to 28 MPa (500 to 4000 psi), with 17 MPa (2500 psi) as a typical value.[120] Since no standards for fabrication of specimens have been developed yet, drilled cores are considered to be reliable for measurement of in situ concrete compressive strength.[117] Typically, the compressive strength of pervious concrete is not used for quality control purposes.

Flexural Strength and Dynamic Modulus of Elasticity

The flexural strength and dynamic modulus of elasticity are important parameters in considering the structural behavior of pavements. Flexural strength of pervious concrete generally ranges from 1 to 3.8 MPa (150 to 550 psi)[120] and is influenced by many factors, especially the degree of compaction and porosity. Dynamic modulus of elasticity can be measured by nondestructive testing of a beam described in ASTM C 215 or by the use of a portable ultrasonic pulse velocity measuring instrument (see Chap. 11).

Permeability

The interconnected voids present in a pervious concrete specimen are responsible for the permeability of the specimen, which is directly dependent on the porosity, pore sizes, and pore roughness. The conventional method of measuring the permeability of a concrete specimen is done by driving the fluid under pressure through its miniature pores; however, since pervious concrete consists of larger voids and thereby occupying more volume, a different technique is applied to measure its hydraulic conductivity. Darcy's law is used to describe the hydraulic conductivity of the pervious concrete specimen because the flow regime inside the pervious concrete for most of the practical applications is laminar. Calculation of water penetration coefficient (coefficient of

(a) (b) (c) (d)

FIGURE 12-37 Common applications of pervious concrete: (*a*) a roadway, (*b*) a sidewalk, (*c*) a swimming pool deck, and (*d*) a driveway. (*Courtesy of Omkar Deo.*)

permeability) has also been quantified using an apparatus in which water was allowed to permeate through the pervious concrete specimen to a predetermined level. The use of a falling head permeability apparatus has been found to be a relatively uncomplicated procedure, yielding results that are easy to interpret.[124]

Typical flow rates through pervious concrete are from 740 cm/h (290 in./h) to 1960 cm/h (770 in./h); values higher than 4190 cm/h (1650 in./h) have been measured in the laboratory.[120] As with a variety of both natural and synthetic porous materials, pervious concretes also exhibit a porosity-permeability relationship that is exponential in nature; see Fig. 12-38. The measured permeabilities are in the range of $4.0 \times 10^{-10} - 2.5 \times 10^{-9}$ m^2.

Porosity and Pore Structure Characteristics

Porosity is probably the most important property because it dictates the mechanical and functional behavior of pervious concrete. Pervious concretes with a value of less than 15 percent are considered as having low porosity, while those with 30 percent are considered as having high porosity. A reasonable average value for preliminary structural

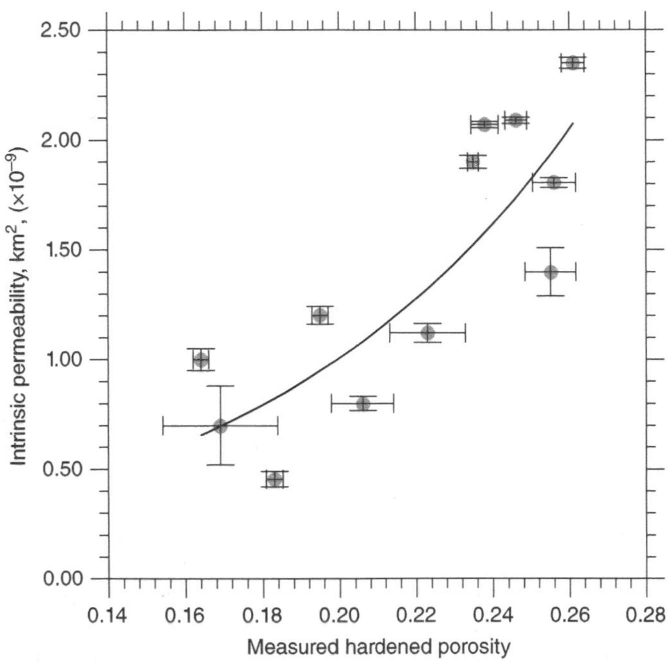

FIGURE 12-38 Variation of intrinsic permeability with porosity for various pervious concrete mixtures.

and hydrological design is 20 percent.[120] Porosity is also highly dependent on several factors such as aggregate gradation, cementitious material content, w/cm, and compactive effort.[117] The pervious concrete specimen is made up of complex interconnections of pores and apertures (throats connecting the pores) of varying degree of shapes and sizes. Many researchers[122–124] have used different methods and techniques to quantify the pore structure features in different types of macroporous concrete.

Density

The density of pervious concrete is a function of the mixture proportion and the consolidation of the mixture. In-place densities of pervious concretes are commonly found to be in the range of 1600 to 2000 kg/m^3 (100 to 125 lb/ft^3).[120] ASTM C 1688 describes a method to determine the fresh density of pervious concrete mixtures by consolidating using a standard Proctor hammer (2.5 kg [5.5 lb] in weight and 30 cm [12 in.] drop height) at 20 blows per lift for 2 equal lifts. Previous methods of consolidation included ASTM C 1176 and ASTM C 1435. Note that the density of pervious concretes is likely to be higher in field applications.

Acoustic Absorption

Because pervious concrete has a large volume of interconnected pores of considerable sizes, they are highly effective in acoustic absorption. The material can be employed as a means to reduce noise generated by tire-pavement interaction on concrete pavements. This noise reduction occurs from a combination of reduced noise generation and increased sound absorption. Pervious pavements alter the generation of noise by minimizing the air pumping between tire and road surface. In addition, the pores absorb sound through internal friction between the moving air molecules and the pore walls.

Evaluation of the sound absorption characteristics of pervious concrete using an impedance tube has been studied.[122] The absorption coefficient, which is a measure of a material's ability to absorb sound, was measured. A material with an absorption coefficient of 1.0 indicates a purely absorbing material, whereas a material with an absorption coefficient of 0 indicates that the material is purely reflective. Normal concrete, for example, typically has an absorption coefficient of 0.03 to 0.05; pervious concrete typically has an absorption range from 0.1 (for poorly performing mixtures) to nearly 0.8 (for mixtures with optimal pore volume and sizes).

Freeze-Thaw Resistance

Pervious concrete pavements exposed to freezing and thawing cycles in cold climates must be frost resistant. The cement paste must contain adequate entrained air, and the effects of saturation of the large pores must be considered. The open structure of pervious concrete allows the ingress of water into the specimen, and subsequent freezing of the saturated pervious concrete pavements could lead to rapid deterioration. Freeze-thaw damage of pervious concretes has been reported, which is primarily due to deterioration of the cement paste.[126] Schaefer et al.[127] reported failure due to splitting and deterioration of aggregate, as well as paste deterioration (which appeared to be dependent on the aggregate type and size). Pervious concrete beams with dimensions of 76 by 76 by 381 mm (3 by 3 by 15 in.) and cylinders with dimensions of 102 by 203 mm (4 by 8 in.) subjected to freezing in air and thawing in air, or freezing in air and thawing in water, or freezing in water and thawing sustained over 160 freeze-thaw cycles when damp or in air, failed within 45 freeze-thaw cycles when submerged in water.

Neithalath et al.[123] evaluated the frost resistance of pervious concrete specimens subjected to freezing and thawing resistance using either rapid freezing and thawing in water, in accordance with ASTM C 666 at 5 to 6 cycles per day, or slow freezing and thawing in water in a controlled temperature chamber where the specimens were subjected to one cycle per day. Specimens without air entraining admixtures (AEA) subjected to ASTM C 666 exposures had 50 to 70 percent loss of dynamic modulus after 65 cycles. Specimens with AEA suffered a dynamic modulus loss of 20 to 60 percent after 70 cycles. Specimens subjected to slow freezing and thawing in water experienced a dynamic modulus loss of 5 to 7 percent after 80 cycles.

Although techniques to protect pervious concrete from the effects of freezing and deicing salts currently exist,[117] the combined effects of clogging on potential saturation, reduced infiltration of the subsoil, changes in the depth of freezing, and seasonal differences in storm intensity, including snow melt, have not been fully determined.

12.10.6 Potential Drawbacks and Challenges of Pervious Concretes

Despite the benefits of using pervious concrete pavements, certain drawbacks and challenges still need to be addressed, including the following: (1) due to its low strength, it cannot be used in heavily trafficked areas; (2) pervious concretes require special construction practices and extended curing time; (3) the fresh concrete mixture is highly sensitive to water content; (4) special attention and care is required in the design of the concrete mixture when expansive soils are used; (5) currently, there is a lack of standardized test methods; and (6) they require providing a permeable base and edge drains.[117]

12.11 Mass Concrete

12.11.1 Definition and Significance

ACI Committee 116 has defined *mass concrete* as concrete in a massive structure, for example, a beam, columns, pier, lock, or dam where its volume is of such magnitude as to require special means of coping with the generation of heat and subsequent volume change. There is a popular assumption that the composition and properties of mass concrete are of interest only to those who are involved in the design and construction of dams; this definition attempts to correct that erroneous impression because many construction practices developed over a long period in building large concrete dams are applicable to structures far less massive.

Designers and builders of large concrete dams were the first to recognize the significance of temperature rise in concrete due to heat of hydration, and subsequent shrinkage and cracking that occurred on cooling. Cracks parallel to the axis of the dam endanger its structural stability; a monolithic structure (i.e., essentially free from cracking) will remain in intimate contact with the foundation and abutments and will behave as predicted by the design stress distributions. Concrete piers, columns, beams, walls, and foundations for large structures are much smaller than a typical concrete gravity dam. If they are several meters thick and are made of high-strength concrete mixtures (high cement content), the problem of thermal cracking can be as serious as in dams.

ACI Committee 207 has authored comprehensive reports[128] on concrete for dams and other massive structures, which should be consulted for details. Based mainly on these reports, only a brief summary of the general principles and their applications to lean (low-strength) concrete are presented here.

12.11.2 General Considerations

As discussed in Chap. 4,* the tensile stress on cooling concrete can, at first, be assumed to be the product of four quantities: $K_r E\alpha\Delta T$, where ΔT is the temperature drop, α the coefficient of thermal expansion, E the elastic modulus, and K_r the degree of restraint. Since the temperature drop and the resulting stress do not occur simultaneously, a correction in the calculated stress is necessary to take into account the stress relief due to creep. Thus the product of the quantities listed in the equation, minus the stress relaxed by creep, would determine the actual stress; concrete will crack when the magnitude of stress exceeds the tensile strength of the material (Fig. 4-1). Methods of determination of the tensile strength of massive unreinforced concrete structures are discussed in Chap. 3. The temperature drop is the easiest to control and has received the most attention in dam building. At the heart of various construction practices for reducing the temperature drop is the recognition that a cost-effective strategy is to restrict the heat of hydration, which is the source of temperature rise in the first place. Basic principles, derived primarily from experience governing the selection of materials, mix proportions, and construction practices for controlling the temperature rise or drop in mass concrete, are discussed later.

---

*For an advanced treatment of the subject, see Chap. 13.

12.11.3 Materials and Mix Proportions

Cement

As discussed earlier (Chap. 6), the heat of hydration of a cement is a function of its compound composition and fineness. Portland cements which contain relatively more C_3A and C_3S show higher heats of hydration than do coarser cements with less C_3A and C_3S. For instance, from the adiabatic-temperature-rise curves for a mass concrete containing 223 kg/m^3 of any one of the five types of portland cements (Fig. 4-20), it can be seen that between a normal cement (Type I) and a low-heat cement (Type IV) the difference in temperature rise was 23°C in 7 days and 9°C in 90 days. It should be noted that at this cement content (223 kg/m^3), the total temperature rise was above 30°C even with the low-heat cement.

In the event that temperature rise and the subsequent temperature drop of the order of 30°C is judged too high from the standpoint of thermal cracking, one way to lower it would be by reducing the cement content of the concrete provided that this can be done without compromising the minimum strength and workability requirements needed for the job. By using several methods, which are described below, it is possible to achieve cement contents as low as 100 kg/m^3 in mass concrete suitable for the interior structure of gravity dams. With such low cement contents, even ASTM Type II portland cement is considered adequate; substitution of 20 percent pozzolan by volume of portland cement produces a further drop in the adiabatic temperature rise (Fig. 4-24).

Admixtures

With cement contents as low as 100 kg/m^3, it is essential to use a low water content to achieve the designed 1-year compressive strength (in the range 13 to 17 MPa), which is normally specified for interior concrete of large gravity structures. Approximately 4 to 8 percent entrained air is routinely incorporated into the concrete mixtures for the purpose of reducing the water content while maintaining the desired workability. Increasingly, water-reducing admixtures are simultaneously being employed for the same purpose. While pozzolans are used primarily as a partial replacement for portland cement to reduce the heat of hydration, most fly ashes when used as pozzolans have the ability to improve the workability of concrete and reduce the water content by 5 to 8 percent.

Aggregate

With concrete mixtures for dams, every possible method of reducing the water content that would permit a corresponding reduction in the cement content (i.e., maintaining a constant water-cement ratio) has to be explored. In this regard, the two cost-effective methods are the choice of the largest possible size of coarse aggregate, and the selection of two or more individual size groups of coarse aggregate that should be combined to produce a gradation approaching maximum density on compaction (minimum void content). Typical coarse aggregate gradation limits for mass concrete and idealized combined grading for 150-mm and 75-mm maximum aggregate are given in the report by ACI Committee 211.[129]

Figure 12-39*a*, based on the U.S. Bureau of Reclamation's investigations on mass concrete for Grand Coulee Dam, shows the extent of reduction in water content by the use of entrained air and the largest possible size of aggregate. The same data are illustrated in Fig. 12-39*b* which shows that at a given water-cement ratio and consistency, as the maximum aggregate size is increased, both the water and the cement contents are reduced.

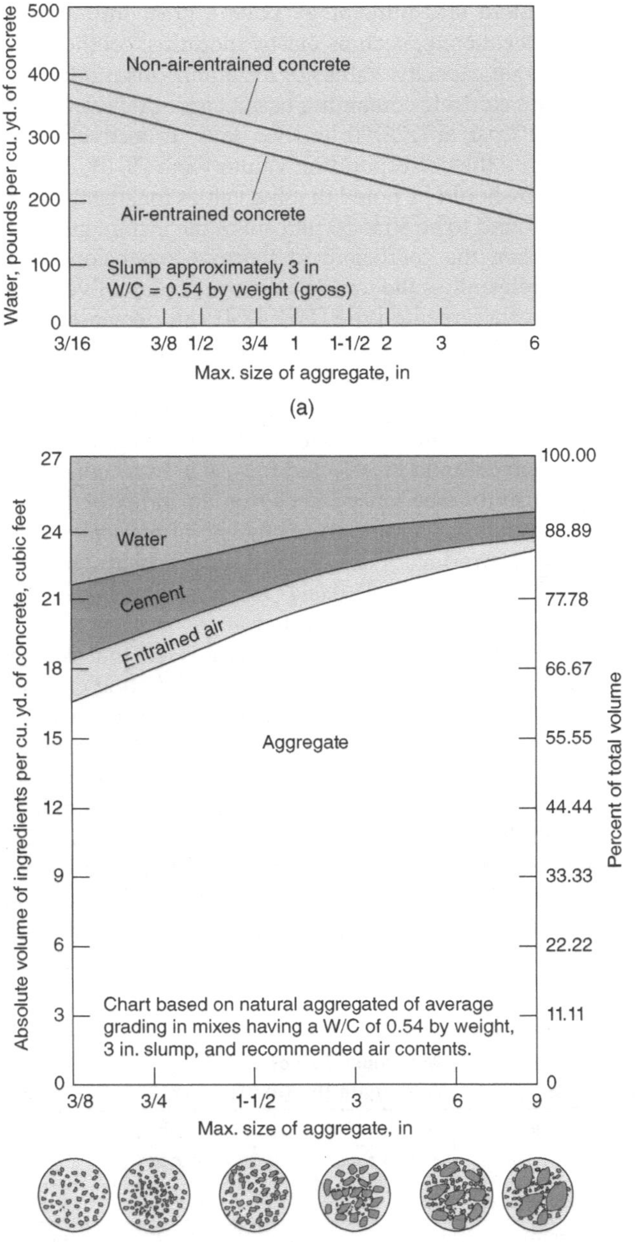

FIGURE 12-39 Effect of maximum size of aggregate on (*a*) water content of concrete, (*b*) concrete mix proportions. (From *Concrete Manual*, 8th ed., U.S. Bureau of Reclamation, Denver, Colorado, 1981.)

Aggregate content and mineralogy have a great influence on properties that are important to mass concrete, such as elastic modulus, coefficient of thermal expansion, diffusivity, and strain capacity. Values of instantaneous modulus of elasticity at various ages a typical mass concrete containing basalt aggregate were 16, 24, 28 and 34 GPa (2.3, 3.5, 4.1, and 5.0×10^6 psi) at 7, 28, 90, and 365 days, respectively; at the same ages but with sandstone aggregate the corresponding values were 29, 31, 36 and 39 GPa (4.2, 4.5, 5.2, and 5.7×10^6 psi). It should be noted that the values for sustained modulus after 365 days under load were found to be 50 to 60 percent of the instantaneous modulus.

As stated earlier, the coefficient of thermal expansion of concrete is one of the parameters that determines the tensile stress on cooling. Everything else remaining the same, the choice of aggregate type (Fig. 4-24) can decrease the coefficient of thermal expansion by a factor of more than 2. With typical mass concrete mixtures (237 kg/m^3 cement content, 30:70 fine to coarse aggregate ratio, and high degree of saturation), the coefficients of thermal expansion (in millionths/°C) were 5.4 to 8.6 for limestone, 8.3 for basalt, and 13.5 for quartzite aggregate.[130] Obviously, aggregates with a low coefficient of thermal expansion should be selected for use in mass concrete whenever feasible.

Thermal diffusivity (see Chap. 4 also) is an index of the ease or difficulty with which concrete undergoes temperature change; numerically, it is the thermal conductivity divided by the product of density and specific heat. Thermal diffusivity values of 0.003 m^2/h (0.032 ft^2/h) for basalt, 0.0047 m^2/h for limestone, and 0.0054 to 0.006 m^2/h for quartzite aggregate have been reported.

As suggested before (Chap. 4), some designers feel that designs based on maximum tensile strain rather than stress are simpler for predicting cracking behavior when the forces can be expressed in terms of linear or volumetric changes. Some of the factors controlling the tensile strain capacity are given by the data in Table 12-23. The data show that compared to mortar and concretes, the pure cement paste of the same water-cement ratio has a considerably higher tensile strain capacity. In general, the tensile strain capacity increased with the period of hydration and decreased with the size of coarse aggregate. For example, in the case of natural quartzite aggregate, the values of 28-day tensile strain capacity were 165, 95, and 71×10^{-6} with 4.75, 37.5, and 75 mm maximum aggregate size, respectively. With aggregate of the same type and maximum size, the strain capacity increased by 50 percent (from 95 to 139×10^{-6}) when smooth-textured aggregate

Mix	Aggregate	Maximum Size of Aggregate [in. (mm)]	W/(C+F)*	Tensile Strain Capacity 10^{-6}	
				7 Days	28 Days
1	Quartzite, natural	3 (75)	0.68	45	71
2	Quartzite, natural	1½ (37.5)	0.68	76	95
3	Quartzite, natural	No. 4 (4.75)	0.68	138	165
4	None (paste)		0.68	310	357
5	Quartzite, crushed	1½ (37.5)	0.68	119	139
6	Quartzite, natural	1½ (37.5)	0.40*	151	145

*All mixes contained 30% fly ash by absolute volume, except mix 6, which had no fly ash.
Source: Carlson, R.W., D.L. Houghton, and M. Polivka, *J. ACI, Proc.*, Vol. 76, No. 7, p. 834, 1979.

Table 12-23 Effect of Maximum Size of Aggregate on Tensile Strain Capacity of Concrete was replaced by the rough-textured aggregate (crushed rock). An increase of a similar order of magnitude was recorded in concrete containing 38 mm natural quartzite when the water-cement ratio was reduced from 0.68 to 0.40. Thus the use of crushed rock and a low water-cement ratio in mass concrete provide effective ways to enhance the tensile strain capacity.

Mix Design

Reports by ACI Committee 211[131] and Scanlon[132] contain a detailed description of procedures for designing mass concrete mixtures. The procedure is the same as used for determining the concrete mix proportions for normal-weight concrete (Chap. 9). Some of the points from Appendix 5 of ACI Committee 211 Recommended Practice[129] are discussed later.

In addition to the largest size of aggregate, determination of the water content should be based on the stiffest possible consistency of fresh concrete that can be adequately mixed, placed, and compacted. Typically, mass concrete slumps in unreinforced structures are of the order of 25 ± 12 mm. If the job-site equipment is inadequate for handling concrete with a stiff consistency, alternative equipment should be sought rather than increasing the water and the cement contents of the concrete mixture. In the case of precooled concrete, the laboratory trial mixtures should also be made at low temperature because less water will be needed to achieve the given consistency at 5 to 10°C than at normal ambient temperatures (20 to 30°C), due to the slower hydration of cement at low temperatures.

Determination of the cement content of mass concrete is guided by the relation between water-cement ratio and strength, which seems to be significantly affected by the aggregate texture (Table 12-24). In a moderate or mild climate, generally a maximum of 0.8-water-cement ratio concrete is permitted for interior of dam and lock walls, and 0.6 for exterior surfaces exposed to water. The maximum compressive stress in gravity dams that are properly designed against overturning and sliding is fairly low; in MPa units it is usually 0.025 to 0.03 times the height of dam in meters.[133] Thus a 100-m-high dam would have a maximum stress of about 3 MPa (450 psi). Stress in arch dams may be as high as 8 MPa (1200 psi). For safety it is recommended that the concrete strength should be four times the maximum stress at 1 year after construction.

Water/Cement Ratio by Weight	Approximate 28-Day Compressive Strength (f_c'), psi (MPa)	
	Natural Aggregate	Crushed Aggregate
0.40	4500 (31.0)	5000 (34.5)
0.50	3400 (23.4)	3800 (26.2)
0.60	2700 (18.6)	3100 (21.4)
0.70	2100 (14.5)	2500 (17.2)
0.80	1600 (11.0)	1900 (13.1)

Note: When pozzolan (P) is used, the strength should be determined at 90 days, and the water/cement ratios may be converted to $W/(C+P)$ ratios by the use of special equations.
Source: ACI 211, *Concr. Int.*, Vol. 2, No. 12, p. 76, 1980.

TABLE 12-24 Approximate Compressive Strengths of Air-Entrained Mass Concrete for Various Water-Cement Ratios

Gravity dams completed before 1940 (i.e., before the use of air-entraining, water-reducing, and pozzolanic admixtures) contained concretes made with approximately 376 lb/yd^3 (223 kg/m^3) cement content.

Comparison of laboratory specimens under standard curing conditions with cores drilled from high dams containing 223 kg/m^3 cement showed that the actual strengths in the structure were considerably above that required: for example, Hoover Dam, 29 MPa (4260 psi); Grande Coulee Dam, 55 MPa (7950 psi); and Shasta Dam, 35 MPa (5100 psi). Even more impressive strength gains were observed in concretes containing pozzolans. Hungry Horse Dam (172 m high and 98 m thick at the base) was the first large dam built by the U.S. Bureau of Reclamation in which less than 223 kg/m^3 cement was used. Concrete mixtures for the Hungry Horse Dam, which was completed in 1952, and subsequent large dams such as Flaming Gorge and Glen Canyon, contained 111 kg/m^3 Type II portland cement, 56 kg/m^3 pozzolan, and air-entraining admixtures.

Generally, a minimum of 3 to 4 percent air is always specified for mass concrete, although in practice 6 to 8 percent air is at times incorporated without any strength loss because the water-cement ratio is substantially reduced. Typically, 35 percent fly ash by volume of total cementing material is used for interior concretes, and 25 percent for exposed concrete. For sands with average fineness (2.6 to 2.8 fineness modulus), approximate coarse aggregate content is 78 to 80 percent by absolute volume of the total aggregate; thus the fine aggregate content is only 20 to 22 percent.

Construction Practices for Controlling Temperature Rise

In addition to the reduction of cement content in concrete mixtures, certain construction practices are used to control the temperature rise in massive concrete structures. Again, only a brief description is given below; an excellent and detailed report on cooling and insulating systems for mass concrete has been prepared by ACI 207.[134]

Postcooling

The first major use of postcooling of in-place concrete was in the construction of Hoover Dam in the early 1930s. In addition to control of temperature rise, a primary objective of postcooling was to shrink the columns of concrete composing the dam to a stable volume so that the construction joints could be filled with grout to ensure monolithic action of the dam. Due to the low diffusivity of concrete (0.065 to 0.084 m^2 or 0.7 to 0.9 ft^2 per day), it would have taken more than 100 years for dissipation of 90 percent of the temperature rise if left to natural processes. This is in spite of the fact that a low-heat cement (Type IV) was used in making the concrete mixture, which contained 150 mm (6-in.)-maximum aggregate; also a special method of block construction for efficient heat dissipation was employed.

The cooling was achieved by circulating cold water through thin-wall steel pipes (typically 25 mm in nominal diameter, 1.5 mm in wall thickness) embedded in the concrete. In Hoover Dam, the circulation of cold water was started after the concrete temperature had reached 65°C (i.e., several weeks after the concrete had been placed). Subsequently, for the construction of several large dams the U.S. Bureau of Reclamation followed essentially the same practice, except that circulation of cooling water was started simultaneously with the placement of concrete. Also, pipe spacing and lift thickness are varied to limit the maximum temperature to a predesigned level in all seasons.

According to the ACI 207 Recommended Practice,[134] during the first few days following placement the rate of cooling or heat removal can be as high as possible because the elastic modulus of concrete is relatively low. The strength and the elastic modulus generally increase rapidly until after the initial peak in concrete temperature has been experienced, which may be some time during the first 15 days following placement. Thereafter cooling should be continued at a rate such that the concrete temperature drop generally does not exceed 0.6°C (1°F) per day. Experience has shown that most mass concretes having average elastic and thermal expansion properties can sustain a temperature drop from 11 to 17°C (20 to 30°F) over approximately a 30-day period following the initial peak. When concrete has become elastic, it is important to have the temperature drop as slowly as possible to allow for stress relaxation; under the slow cooling conditions, concrete can stand a 20°C drop in temperature without cracking.

Precooling

The first use of precooling of concrete materials to reduce the maximum temperature of mass concrete was by the Corps of Engineers during the construction of Norfork Dam in the early 1940s. A part of the mixing water was introduced into the concrete mixture as crushed ice so that the temperature of in-place fresh concrete was limited to about 6°C. Subsequently, combinations of crushed ice, cold mixing water, and cooled aggregates were utilized by Corps of Engineers in the construction of several large concrete gravity dams (60 to 150 m high) to achieve placing temperatures as low as 4.5°C.

According to ACI 207 Recommended Practice, one of the strongest influences on the avoidance of thermal cracking in mass concrete is the control of placing temperature. Generally, the lower the temperature of the concrete when it passes from a plastic state to an elastic state, the less will be the tendency toward cracking. In massive structures, each 10°F (6°C) lowering of the placing temperature below the average air temperature will result in a lowering by about 6°F (3°C) of the maximum temperature the concrete will reach.

To raise the temperature by 1°F, water absorbs 1 Btu/lb heat, whereas cement and aggregates absorb only 0.22 Btu/lb. Therefore, pound for pound, it is more efficient to use chilled water in reducing the temperature of concrete. Of course, the use of ice is most efficient, because ice absorbs 144 Btu/lb heat when it changes to water. For concrete homogeneity, it is important that all the ice in the concrete mixture has melted before the conclusion of mixing. Therefore, flake ice of biscuit-shaped extruded ice is preferable to crushed ice blocks. Cooling the coarse aggregate while en route to the batch bins by spraying with chilled water may be necessary to supplement the use of ice and cooled mixing water.

Surface Insulation

The purpose of surface insulation is not to restrict the temperature rise, but to regulate the rate of temperature drop so that stress differences due to steep temperature gradients between the concrete surface and the interior are reduced. After the concrete has hardened and acquired considerable elasticity, decreasing ambient temperatures and rising internal temperature work together to steepen the temperature gradient and the stress differential. Especially in cold climates, it may be desirable to moderate the rate of heat loss from the surfaces by covering with pads of expanded polystyrene or urethane (k factor of the order of 0.2 to 0.3 Btu-in./h-ft^2°F).

12.11.4 Application of the Principles

Price reviewed the last 50 years of construction practices in the United States to show the development of strategies for control of cracking in concrete dams. In the construction of Hoover (1935), Grande Coulee (1942), and Shasta (1945) Dams, which contain

2.4, 8.0, and 4.5 million cubic meters of concrete, respectively, ASTM Type IV low-heat portland cement (223 kg/m^3 cement content) was used, and concrete was postcooled by circulating cold water through the embedded pipes. The heights and scheduling of placements were controlled, and special block construction procedures were devised for more efficient heat dissipation. All three dams remained free of objectionable cracks and leakage. Also in the early 1940s, the Tennessee Valley Authority utilized post-cooling in the construction of Fontana Dam. In all the cases, postcooling not only reduced the temperatures rise, particularly in the base of the dam, which was more vulnerable to cracking due to the restraining effect of the foundation, but also stabilized the columns within the construction period of the dam so that the construction joints between columns could be filled with grout to ensure monolithic action.

Beginning with Norfork Dam (1945), precooling of concrete materials was successfully used to control cracking in Detroit Dam. By limiting the cement content (Type II) to 134 kg/m^3 and the concrete placement temperature to 6 to 10°C, the temperature rise was restricted to 17°C above the mean ambient temperature. No postcooling of concrete was necessary. Also, the exposed surfaces of concrete were protected from rapid cooling by covering with an insulating material. The complete absence of objectionable cracking in the Detroit Dam blocks, as long as 102 m, pointed the way to future application of these techniques. Thus the trend toward lower peak concrete temperatures began when it was discovered that this allowed the use of larger monolith lengths without harmful consequences. Precooling and postcooling were used in combination in the construction of several large dams, notably Glen Canyon Dam, Dworshak Dam, and Libby Dam. In every case the temperature rise was limited to 14°C.

Limiting the temperature drop to less than 20°C by precooling concrete-making materials is possible only when, at the same time, the cement content is reduced substantially. Compared to 223 kg/m^3 cement for the Hoover Dam concrete, only 111 kg/m^3 cement and 56 kg/m^3 pozzolan (pumicite) were used for the Glen Canyon Dam concrete, which gave about 20 MPa (3000 psi) compressive strength at 28 days, and 48 MPa (7000 psi) at 360 days. Such a low cement content could be used because a considerable reduction in the water content was achieved by using properly graded aggregates, an air-entraining admixture, and a water-reducing admixture. The data in Table 12-25 illustrate how water content varies with the maximum size of aggregate, fine-coarse aggregate ratio, and the use of admixtures. For example, instead of a 210-kg/m^3 water requirement with a 10-mm-maximum aggregate size and no admixtures present, only 92 kg/m^3 water is needed to achieve the same consistency when 150-mm-maximum aggregate and both water-reducing and air-entraining admixtures were used. According to Price, for a given aggregate and consistency of concrete, about 35 percent reduction in the water content can be accomplished by using entrained air, a water-reducing admixture, a lower sand content, and a lower placing temperature.

Laboratory investigations on mass concrete for Itaipu Dam (Fig. 1-1) showed that the permissible <20°C rise in adiabatic temperature of concrete could be achieved when the cement content was held to about 108 kg/m^3. However, the designed 1-year compressive strength of 14 MPa was not achieved with this cement content even with 13 kg/m^3 pozzolan present, when 38 or 75 mm-maximum aggregate size was used. This is because the water-cement ratio required for the desired consistency was too high. A mixture of 150-, 75-, and 38-mm-maximum aggregates, and incorporation of an air-entraining admixture, caused enough lowering of the water-cement ratio to produce 25 to 50 mm-slump concrete with a 17.5-MPa compressive strength at 90 days. This concrete mixture

Maximum Size of Coarse Aggregate [mm (in.)]	Recommended Air Content (% for full mix)	Sand, Percent of Total Aggregate by Solid Volume	Non-Air- Entrained Concrete: Average Water Content [kg/m³ (lb/yd³)]	Air-Entrained Concrete: Average Water Content [kg/m³ (lb/yd³)]	Air-Entrained Concrete with WRA:* Average Water Content [kg/m³ (lb/yd³)]
10 (3/8)	8	60	210 (352)	190 (320)	180 (300)
12.5 (½)	7	50	200 (336)	180 (305)	170 (285)
20 (¾)	6	42	185 (316)	165 (280)	157 (265)
25 (1)	5	37	178 (300)	158 (265)	148 (250)
40 (11/2)	4.5	34	165 (280)	145 (245)	135 (230)
50 (2)	4	30	158 (266)	136 (230)	127 (215)
80 (3)	3.5	28	143 (242)	120 (205)	112 (190)
150 (6)	3	24	125 (210)	98 (165)	92 (150)

* WRA stands for water-reducing admixture.
Source: Price, W.H., *Concr. Int.*, Vol. 4, No. 10, p. 43, 1982.

TABLE 12-25 Approximate Air and Water Contents Per Cubic Meter for Concretes Containing Different Maximum Size Aggregate

contained 108 kg/m³ cement, 13 kg/m³ fly ash, 85 kg/m³ water, 580 kg/m³ sand, and a mixture of 729, 465, and 643 kg/m³ coarse aggregates with 38, 75, and 150 mm maximum size, respectively. The concrete placement temperature was restricted to 6°C by precooling all coarse aggregate with chilled water, followed by cold air, using some of the mixing water at 5°C and most of it in the form of flake ice.

12.12 Roller-Compacted Concrete

Concept and Significance

The development of roller compacted concrete (RCC) caused a major shift in the construction practice of mass concrete dams and locks. The traditional method of placing, compacting, and consolidating mass concrete is at best a slow process. Improvements in earth-moving equipment made the construction of earth and rock-filled dams speedier and, therefore, more cost-effective. In 1970, in an effort to stop their decreasing share of the market, the builders and the designers of concrete dams held a conference in Asilomar, California, at which, Professor Jerome Raphael,[135] from the University of California at Berkeley, said

> Could it be that concrete gravity dams with their small volume of very expensive concrete, and earthfill dams with their large volume of very cheap fill material are simply the end points or the extremes of a whole class of gravity dams varying continuously from one to the other?

Raphael also pointed out that the fill material has only 10 percent of the cohesion of mass concrete. It would be feasible, therefore, to design an optimum gravity dam that took advantage of the excellent cohesion of mass concrete, with the advantage of rapid transportation and consolidation of fill material.

The first successful application of RCC technology was demonstrated in 1974. The repair of the collapsed intake tunnel of Tarbela Dam proved that the material had more than adequate strength and durability. The maximum placement of 18,000 m^3 of RCC in one day,[136] which is still a world record, was a clear evidence of the potential of this new construction method. Presently, this technology has been incorporated in most hydroelectric projects. For instance, RCC will be used extensively in cofferdams of the Three-Gorges Dam, the largest concrete dam in the world (Fig. 12-40).

ACI 207.5R defines *roller compacted concrete* (RCC) as concrete compacted by roller compaction. The concrete mixture in its unhardened state must support a roller while being compacted. Thus RCC differs from conventional concrete principally in its consistency requirement. For effective consolidation, the concrete mixture must be dry enough to prevent sinking of the vibratory roller equipment but wet enough to permit adequate distribution of the binder mortar in concrete during the mixing and vibratory compaction operations. Figure 12-41 shows the typical sequence of transportation, placement, and compaction of RCC.

FIGURE 12-40 Three-Gorges Dam, People's Republic of China. (*Courtesy of Kamran Nemati.*)

The Three Gorges Dam, estimated to cost $30 billion, is the largest concrete dam in the world. The 6000-km-long Yangtze River, the third longest in the world, was responsible for large floods that caused major damage to the public safety and to the economy of the region. The Three Gorges Dam was designed to control the flood of the river by creating a lake with a total storage capacity of 40 billion m^3. Naturally, such a massive volume of water extending over a large area raised environmental concerns.

The hydroelectric project will generate 18,200 MW of electrical power, representing 15 percent of China's total electrical power. The 185-m-high concrete gravity dam has a total length of 2300 m along its axis and the project will use 28 million m^3 of concrete. To minimize the thermal stresses in the mass concrete, low-heat cement and forty percent of fly-ash were used. Roller-compacted concrete was used in the tallest and most voluminous cofferdams ever built.

The Chinese engineers developed new creative methodologies, such as the use of grout-enriched RCC for upstream faces of main dams and the sloping-layer RCC placing method. These innovations are attracting world-wide attention from engineers eager to further enhance the efficiency of RCC.

Transportation

Placement

Compaction

FIGURE 12-41 The transportation, placement, and compaction of roller-compacted concrete are fast and efficient. This method has become increasingly popular in the construction of recent concrete dams.

The U.S. Army Corps of Engineers[137] list the following advantages of using RCC:

- Costs: Depending on the complexity of the structure, RCC costs 25 to 50 percent less than conventional concrete.
- Rapid construction: For large projects, RCC dams can be finished 1 to 2 years earlier compared to regular mass concrete dams.

- Spillways: Compared to embankment dams, which normally require that spillways be constructed in an abutment, RCC dams offer the attractive and cost-effective alternative of constructing the spillway in the main structure of the dam.
- Diversion and cofferdam: The costs of river diversion and damages caused by cofferdam overtopping are smaller for RCC dams than for embankment dam.

Hirose and Yanagida[138] list several additional advantages from the Japanese experience with RCC dams when compared with conventional concrete dam construction:

- Cement consumption is lower because much leaner concrete mixtures can be used.
- Formwork costs are lower because of the layer placement method.
- Pipe cooling is unnecessary because of the low temperature rise.
- Cost of transporting, placement, and compaction of concrete is lower, because concrete can be hauled by end dump trucks; spread by bulldozers and compacted by vibratory rollers.
- Rates of equipment and labor utilization are high because of the higher speeds of concrete placement.

12.12.1 Materials and Mix Proportions

Cement

The consolidation by a roller does not require special cements; however, when RCC is to be used in mass concrete, the recommendation of selecting cements with lower heat generation should be followed.

Admixtures

Mineral admixtures are used extensively in RCC mixtures. The use of large amounts of mineral admixtures reduces both the adiabatic temperature rise of concrete and costs, and improves durability. In the United States, Class F fly ash is the most common mineral admixture used in dams, however, in other parts of the world Class C fly ash, slag, and natural pozzolan have also been used.

Air-entraining and water-reducing admixtures are used in RCC compositions that contain a higher volume of paste. Set-retarding admixtures can extend the time up to which the concrete lift should remain unhardened, reducing the risk of cold joints with the subsequence lift. In RCC mixtures of dry consistency, however, chemical admixtures show rather a limited effectiveness.

Aggregates

Aggregates greater than 76 mm in diameter (3 in.) are seldom used in RCC because they can cause problems in spreading and compacting the layer. Usually the maximum aggregate size is limited to 38 mm (1.5 in.). The size of coarse aggregate has a significant influence on the degree of compaction in small layers. This influence is less marked in relatively thicker layers especially when large vibratory rollers are employed. Aggregate gradation, which will be discussed below, has an important influence on the mixture proportions. The use of material finer than 75 μm (No. 200 mesh sieve) produces a more cohesive mixture by reducing the volume of voids.

In many RCC dams, the stress level in the structure is rather low, making it possible to use lower-quality aggregate, particularly in the interior of the structure. In this case, care should be taken to assess the long-term durability, particularly if the structure is exposed to a cold climate.

Concrete Mixture Proportioning

There are two main approaches for proportioning RCC mixtures: the first approach uses the principles of soil compaction to produce a lean RCC, where the optimum water content of the concrete is the one that produces the maximum dry density of the mixture. This method does not utilizes the conventional concept of minimizing the water-cement ratio to maximize the concrete strength; the best compaction gives the best strength, and the best compaction occurs at the most wet mix that will support the operating vibrating roller. This method was used in Willow Creek, Copperfield, Middle Fork, and Monksville Dams. The overriding criteria for these mixtures are the compressive and shear strength since the dam using this type of concrete typically will have an impermeable upstream face made either by traditional mass concrete or precast panels.

The second approach uses traditional concrete technology methods to produce high-paste RCC mixtures. Upper Stillwater and Elk Creek Dams are examples of dams that were built using this approach. The overriding criteria for these mixtures are the shear strength between the lifts and low permeability of concrete since no protective, impermeable face is used upstream.

Table 12-26[140] summarizes the fundamental differences in RCC mixtures proportioned by the soil compaction approach and the concrete technology approach. ACI 207.5R identifies five methods of RCC mixture proportioning: (1) the method based on specific limits of consistency, (2) relying on trial mixture tests to select the most economical aggregate-cementitious combination, (3) the method using soils compaction concepts, (4) the method developed by the Japanese Ministry of Construction,[139] and (5) the method developed by the U.S. Army Corps of Engineers.[137] Details of the first three methods can be found in ACI 207.5R, and in Hansen and Reinhardt.[140]

12.12.2 Laboratory Testing

RCC is a zero-slump concrete whose properties are strongly dependent on the mixture proportions and on the quality of compaction. Concrete is consolidated in the field using vibrating rollers. Despite extensive research on this subject, there is as yet no unanimously accepted methodology to simulate the field condition in preparing laboratory samples. Casting samples for dry-consistency mixtures often requires impact compaction, which can be accomplished using equipment similar to the modified Proctor test for soils. Great care should be taken in selecting the adequate volumetric compacting energy: if too low the sample may develop undesirable layers, if too high the aggregate can break changing the granulometric curve. For RCC mixtures containing a large amount of paste, the test samples can be prepared with a Vebe vibrating table. A third method of sample preparation requires tamping either by a vibrating rammer or by a pneumatic pole tamper. Marques Filho et al.[141] developed a laboratory setup that uses a novel device that consolidates concrete layers, realistically simulating site conditions, including several curing conditions. The laboratory tests results were successfully compared to field results obtained at the Dona Francisca Hydro Power Plant, Brazil.

	Soil Compaction Approach	Concrete Technology Approach
Mix proportioning basis	Optimum moisture/maximum dry density	Good consolidation and a low water-cement ratio
Characteristics of voids	Not all voids are filled with paste Particle-to-particle contact in aggregate	Voids filled with paste
Theoretical air-free density (%)	Less than 98	Greater than 98
Consistency (Vebe or VC time)	45 s or more	45 s or less
Aggregate Grading	Graded or pit-run	Very well graded to minimize voids
Amount of Fines (<75 μm)	Up to 10% by total weight	Less fines, especially if high fly-ash content is used
Cement + pozzolan content	Less than 120 kg/m^3	Greater than 120 kg/m^3
Cohesion	Low value—less than 1.4 MPa	High value—greater than 1.4 MPa
Compressive strength	High per unit weight of cement; greater at top of the lift	Decreases with more water; greater at bottom of the lift
Permeability	Depends on the void content of the mix and degree of compaction and segregation	Depends on properties and amount of paste
Seepage control	Upstream "membrane"	Entire gravity dam
Placement	Segregation is a problem	Segregation is less of a problem
Compaction	Vibratory or heavy rubber-tired roller	Vibratory roller
Primary action of roller	Dry Compaction	Consolidation
Compacted lift thickness	Generally 0.3 m. Possibility of voids at the bottom	0.3–0.75 m Possibility of paste at surface and slight chance of voids in the bottom

Source: Adapted from Hansen, K.D., and W.G. Reinhardt, *Roller-Compacted Concrete Dams*, McGraw-Hill, New York, pp. 39–47, 1991.

Table 12-26 Comparison between the Two Methods for Proportioning RCC Mixtures

12.12.3 Properties

Strength

For RCC mixtures made according to the concrete technology approach, where the volume of the paste exceeds the volume of the voids between the aggregate, the compressive strength follows the dependence on the water-cement ratio as predicted by Abram's rule. For RCC mixtures made according to the soil mechanics approach, where the cement paste may not fill the voids between the aggregate, Abram's rule does not apply, and strength is often plotted as a function of the moisture content.

It is not surprising that Abram's rule does not work for these lean RCC mixes. The relationship was developed for fully compacted concrete where the porosity is in the continuous cement paste matrix. Ferret's rule,[142] which was proposed in 1892, is more appropriate under this condition because it accounts for all the porosity due to the hydration process and to the lack of vibration. His expression is given by

$$f_c = k\left(\frac{c}{c+w+a}\right)^2 \tag{12-17}$$

where f_c = compressive strength
k = constant
c = volume of cement
w = volume of water
a = volume of air

Alternatively, Feret's equation can be expressed as

$$f_c = \frac{k}{(1+w/c+a/c)^2} \tag{12-18}$$

For plastic concrete the air-cement ratio (a/c) is small and it can be neglected; however, for dry mixes that can entrap larger amounts of air, the a/c ratio may become significant.

Table 12-27 shows compressive strength results for some RCC mixtures. RCC concrete has about the same ratio between the compressive and tensile strength as normal concrete, ranging from 7 to 13 percent. As with normal concrete, the ratio depends on the strength level, age, aggregate type, and cement content. Naturally, RCC containing very low cement content or marginal aggregate will have much lower tensile strength.

The shear stress developed between the layers of the RCC is a critical parameter in the design of the dam. The shear stresses τ is related to the normal stress σ, the cohesion c, and the angle of internal friction ϕ, by Coulomb's law: $\tau = c + \sigma\phi$. For Willow Creek Dam, cohesion and angle of internal friction between layers ranged from 0.7 to 1.5 MPa (100 to 215 psi) and 0.25 to 0.42 MPa (37 to 62 psi), respectively. For Upper Stillwater these values ranged from 0.5 to 4 MPa (74 to 575 psi) and 0.3 to 0.5 MPa (42 and 71 psi), respectively.

Elastic Modulus and Poisson's Ratio

As described in the section on mass concrete, the thermal stresses generated by heat of hydration are proportional to the elastic modulus of concrete. Therefore, lean RCC mixtures, which produce concrete with low elastic modulus, are attractive to designers. As with regular concrete, the elastic modulus of RCC depends on the degree of hydration, volume and type of aggregate, and water-cement ratio. Poisson's ratio for CCR typically ranges from 0.15 to 0.20. The elastic properties of RCC used in dams are shown in Table 12-27.

Creep

The long-term deformation of RCC depends on the amount and the type of aggregate, the water-cement ratio, the age of loading, and the duration of loading. RCC with lower compressive strength and lower elastic modulus will normally show high creep, which

Dam	Cement (kg/m³)	Pozzolan (kg/m³)	Water (kg/m³)	W/(C+P)	Age Days	Compressive Strength (MPa)	Elastic Modulus (GPa)	Poisson's Ratio
Willow Creek (USA) Interior Mix	47	19	107	1.61	3	2.88		
					7	3.98	8.27	
					28	8.08	10.96	0.14
					90	11.94	13.17	0.17
					365	18.08		
Willow Creek (upstream face)	104	0	110	1.06	3	4.52		
					7	6.87	15.17	
					28	12.7	18.41	0.19
					90	18.2	19.17	0.18
					365	26.06		
Willow Creek (downstream face)	104	47	110	0.72	3	5.44		
					7	7.91	16.55	0.21
					28	14.2	20.06	0.21
					90	27.31	22.41	
					365	28.59		
Upper Stillwater (USA)	108	125	109	0.47	7	9.37		
					28	14.69	7.10	0.13
					90	24.20	9.10	0.14
					365	35.99	11.79	0.17
Upper Stillwater	72	160	103	0.45	7	5.31		
					28	8.41	5.65	0.13
					90	14.82		
					365	32.96	10.96	0.2
Upper Stillwater	77	140	107	0.43	7	7.65		
					28	11.17	6.34	0.13
					90	19.10		
					365	34.20	12.14	0.18
Serra da Mesa (Brazil)	60	140	128	0.64	7	13.4	13.8	0.19
					28	16.5	17.9	0.13
					90	21.6	20.4	0.17
					365	22.7	24.3	0.20
Cana-Brava (Brazil)	60	100	101	0.631	7	3.3	2.0	
					28	5.6	6.5	
					90	9.6	17.0	0.31
					180	11.3	21.9	

TABLE 12-27 Composition, Strength, and Elastic Properties of Some RCC Mixtures

is a critical factor in determining the stress relaxation when thermal strain is restrained. Lean concrete with large amounts of fines also shows high creep.

Thermal Properties

The adiabatic temperature rise of RCC is similar to conventional mass concrete mixtures and depends on the amount and type of cementitious material used in the mixture. The specific heat, conductivity, and coefficient of thermal expansion are a function of the type and amount of aggregate used in the mixture.

Durability

The coefficient of permeability of RCC is a critical parameter for long-term performance of dams, particularly if no impermeable membrane has been used at the upstream face of the dam. The construction process of RCC generates porous zones between the lifts where water can percolate. Depending on the mixture proportions and construction process, the coefficient of permeability can vary over 8 orders of magnitude. For instance, the lean concrete at Willow Creek dam had a coefficient of permeability of 2×10^{-4} m/s, while the coefficient of permeability at Upper Stillwater Dam was 4×10^{-12} m/s. Willow Creek Dam, however, has an impermeable membrane at its upstream face.

The parameters that affect permeability of concrete are discussed in Chap. 5. It is important to point out that besides optimizing the mixture proportions, it is equally crucial to provide proper compaction throughout the structure. If the moisture content in concrete goes beyond the critical saturation point, the performance of non-air entrained RCC to cycles of freezing and thawing will be poor; however, if the structure does not become saturated, the frost resistance of RCC is satisfactory. Air-entrainment of very lean RCC mixtures has not been very successful. As discussed in Chap. 5, the erosion resistance of concrete is a function of the strength of the concrete and the quality of the aggregate. Laboratory and field tests have demonstrated excellent erosion resistance of RCC mixtures including the ones used for repairs at Tarbela dam, Toutle dam spillway, and the Kerrville dam.

12.12.4 Construction Practice

The overall planning of a RCC dam is conceptually different from a gravity dam. To minimize thermal stresses, traditional mass concrete is built in separate, monolith blocks. This process is slow but allows great flexibility; if a problem develops in one of the blocks, the construction front moves to another block. RCC dams do not have such luxury. The operation is continuous, building one horizontal lift at a time. There are no special requirements for batching and mixing of RCC, which can be produced using the same equipment as for conventional mass concrete. Ready-mixed concrete trucks cannot be used to transport RCC because the zero-slump concrete is too dry and cannot be discharged. To obtain significant economical benefits, special care must be taken in the selection of equipment and construction methods for fast placement and consolidation of RCC. Conveyor systems can be an efficient method of transporting RCC. Elk Creek Dam used conveyor belts to deliver concrete from the mixer to the job site, and from the discharge point end-dump trucks were used. This delivery system allowed a maximum placement rate of 765 m^3/h, with an average of 460 m^3/h. Transporting and placing RCC with end-dump trucks followed by dozers that remix and spread the RCC mixture is a fast and economical method. Alternatively, scrapers and bottom-dump trucks can be used.

The success of a RCC dam is often contingent on the correct selection of lift thickness, which depends on the mixture proportions and on the equipment available. If the lift is too thin, the placement rates will be small, thereby reducing the advantages of using RCC. If the lift is too thick, the compaction will not be adequate, creating horizontal layers of higher porosity, thereby compromising the strength and durability of the structure. Normally, the thickness of the lifts ranges from 0.15 to 0.90 m; in the United States a lift thickness of 0.3 m is often used. Compaction of the lift is achieved by using a vibrating steel-wheel roller. The selection of the roller depends on the desired compaction force, drum size, frequency, and operating speed. Compaction of the lift should be performed as soon as possible, typically within 10 min after spreading and no more than 40 min after mixing. Once adequate compaction is achieved, good curing conditions for the finished surface are essential; the surface should be kept in a moistened condition until the next lift is placed. The dry consistency of RCC results in difficulty in bonding fresh concrete to hardened concrete. This bond can be improved between the lifts by reducing the time of casting the lifts or by increasing the paste content in the mixture. Investigations have shown that using special high-consistency bedding mixtures for starting the new concrete placement is helpful in reducing the cold joints. Typically, bedding mixtures contain 360 to 460 kg/m^3 of cement, 170 to 220 kg/m^3 of fly ash, and 4.75-mm maximum size aggregate. Figure 12-42 shows the sequence of placement and consolidation of RCC lifts.

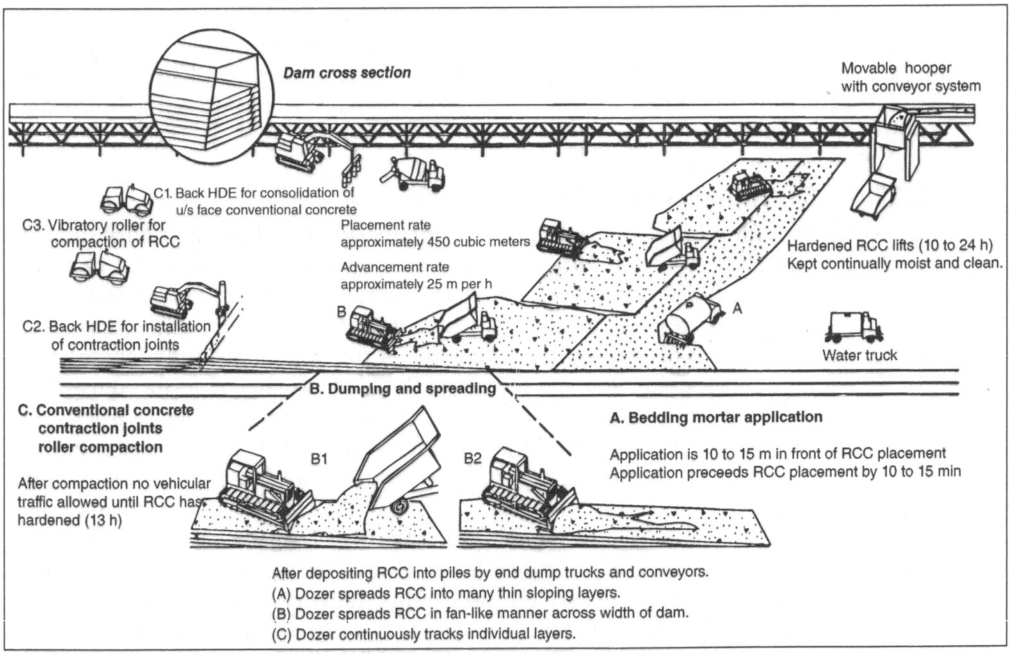

FIGURE 12-42 Sequence of RCC construction. (After Roller-Compacted Concrete. Technical Engineering and Design Guides as Adapted from the *U.S. Army Corps of Engineers*, ASCE Press, New York, p. 100, 1994.)

12.12.5 Applications

Willow Creek Dam

In 1982, construction of the world's first all RCC structure, the Willow Creek Dam (Fig. 12-43), was completed one year after the work was opened for bidding. A saving of \$13 million resulted over the alternative rock-filled design, which called for a 3-year schedule. If conventional concrete had been used as a reference, the savings would be considerably more. The in-place RCC cost averaged about \$26 per cubic meter, compared to \$85 per cubic meter estimated for conventional mass concrete. The aggregates for the Willow Creek Dam concrete contained approximately 25 percent of a silty, sandy gravel overburden. There was no aggregate washing; in fact, the presence of material passing a No. 200 sieve improved the compactibility and increased the strength of the compacted concrete. Although a maximum aggregate size of 230 mm had been used for the Tarbela RCC Dam in Pakistan, a 75-mm maximum size was preferred for the Willow Creek Dam in order to minimize the tendency for segregation. Most likely because of the dry consistency and low amount of cement paste, the use of either water-reducing or air-entraining admixtures did not provide any benefit with the concrete mixtures. Durability to frost action was not required because most of the concrete will never be subjected to wet-dry and freeze-thaw cycles. The exposed upstream face of the structure was made of conventional air-entrained concrete. Mix proportions and material properties for the range of RCC concretes used at the Willow Creek Dam are shown in Table 12-28.

Conventional design and construction of concrete dams typically locates the vertical joints every 12 to 18 m. According to Schrader,[143] the Willow Creek Dam is constructed

FIGURE 12-43 Willow Creek Dam, Oregon, the world's first all-roller-compacted concrete structure, shown about 18 weeks after the start of placement. (Photograph courtesy of Schrader, E., *Concr. Int.*, Vol. 6, No. 5, 1984.)

	Spillway Face	Downstream Face	Upstream Face	Interior Mix
Cement (Type II)	187	104	104	47
Fly ash	80	47	0	19
Water	119	110	110	108
Max aggregate size (mm)	38	76	76	76
Compressive strength (MPa)				
90 days	30.7	19.7	15.9	7.3
1 year	42.4	31.4	25.2	11.6

Source: Schrader, E., *Concr. Int.*, Vol. 4, No. 10, p. 17, 1982.

TABLE 12-28 Mix Proportions (kg/m^3) and Material Properties of Roller-Compacted Concretes

as one monolithic mass with no vertical joints. No postcooling, aggregate chilling, or ice were used. Due to the very low cement content of the interior concrete, the adiabatic temperature rise was only 11°C in 4 weeks. Seepage was observed downstream at the face of the dam during the filling of the reservoir. To reduce this seepage, the U.S. Corps of Engineers grouted the dam and the foundation.

Tamagawa Dam

The Tamagawa Dam in Japan was built on the Tama River for the purposes of flood control, water supply, and power generation. The site is in the mountains close to Lake Tazawa, the deepest lake in Japan. The conditions in the winter are so harsh that construction stops for 5 months due to snowfall. The use of RCC was selected after studies indicated that compared to traditional mass concrete it would save approximately one year of construction time. The dam is 100 m high and has a riverbed width of approximately 200 m, making the concrete volume for the dam, 1.14 × million m^3. The daily maximum placement of concrete was 5800 m^3.

The Tama river water is acidic with a pH 3.7. Special specifications were developed for the design of the upstream face of the dam. Also, alluvial rocks could not be used as aggregate which were obtained from approximately 4 km from the dam site. Maximum aggregate size of 150 mm was selected for mix proportioning. To improve the durability of concrete to acid attack and to reduce the thermal stresses, a moderate-heat portland cement with 30 percent fly ash were used in the concrete mixture (see Table 12-29).

The fundamental design criterion for this dam was to ensure that the concrete exposed to the environment be as watertight and durable as a conventional mass concrete dam. Therefore, rich mix proportions and slipforms were used for the upstream and downstream faces. Two inclines were used as the main facility for conveying concrete to the body of the dam.

The horizontal construction joints were treated by employing road sweepers and water jets. Subsequently, a bedding layer of mortar 1.5 cm thick was placed before the next lift was cast. Concrete was spread in four, 25-cm, layers to form one lift. Usually, one day of concrete placement was followed by 2 days of curing. The average strength of the RCC cores was 20 MPa, which is significantly higher than the design strength of 13 MPa. Field tests indicated that the coefficient of permeability was 2.5×10^{-8} m/s.

Content (kg/m³)						
Water	Cement + Fly-Ash	Sand	Coarse Aggregate	Admixture	Quantity ($10^3 \times m^3$)	Location That Was Used
115	240	440	1572	0.60	100	Upstream exterior
112	220	446	1592	0.55	90	Downstream exterior
108	180	497	1587	0.45	40	Near the foundation
95	130	657	1544	0.33	750	Interior
106	160	523	1582	0.40	40	Interior
138	270	513	1397	0.68	90	Reinforced concrete
129	240	527	1436	0.60	30	Reinforced concrete

Source: From Yamauchi,T., J. Harada, T. Okada, and S. Shimada; Construction of Tamagawa Dam by the RCD Method, Commission Internationale des Grands Barrages, Quinzieme Congres, Lausanne, 1985.

TABLE 12-29 Mix Proportions Used in the Tamagawa Dam

A careful thermal stress analysis was performed to determine: (a) the magnitude of the stress originating from the difference of the peak and equilibrium temperature; (b) the magnitude of the stress caused by temperature gradients within the concrete mass; and (c) the magnitude of the stress caused during winter months when construction was interrupted. The following procedure was used to reduce the thermal stresses[144]:

Cement: Low cement content was used in the interior structure of the dam. Fly-ash was used to reduce the adiabatic temperature rise.

Lift thickness: The thermal analysis indicated that close to the foundation, where the restraint is higher, the thickness of the lifts should be 0.5 m. Away from the foundation, the thickness of the layers can be increased to 0.75 m (location EI 348 m), and finally to 1.0 m (above location EI 348 m).

Cooling: Mixing water was cooled to 4°C in the summer, and the lift surfaces were sprayed with water after the placement.

Insulation: The lift surfaces were insulated when it was necessary to halt construction because of weather related interruptions.

Field-work indicated that this procedure was successful in reducing the thermal stresses, and no thermal-induced cracks were found in the dam.

Test Your Knowledge

12.1 (a) Slump loss and floating of coarse aggregates can be major problems with fresh, lightweight concrete mixtures. How are they controlled? (b) The compressive strength of conventional lightweight concrete is limited to about 48 MPa (7000 psi). How can this be increased?

12.2 (a) Compared to normal-weight concrete of the same water-cement ratio, a structural lightweight concrete would show higher drying shrinkage but less tendency to crack. Can you explain why? (b) In spite of the cellular structure of aggregate, lightweight concretes show less microcracking and excellent durability. Why?

12.3 (a) In high-rise buildings, what are the advantages of constructing shear walls and columns with high-strength concrete? (b) Discuss why the use of water-reducing and pozzolanic admixtures is essential for producing ultra-high-strength concrete. (c) Superplasticized concrete is, in general, prone to slump loss. How can this problem be overcome in construction practice?

12.4 What is the significance of superplasticized flowing concrete to the construction industry? What changes may have to be made in the mix proportions for flowing concrete?

12.5 Explain how the concept works for eliminating drying shrinkage cracking by the use of shrinkage-compensating concrete.

12.6 Compare normal portland cement concrete to shrinkage compensating concrete. Assume they have the same water-cement ratio (e.g., 0.6).

12.7 What is the principal advantage of using fiber-reinforced concrete? Explain how the concrete acquires this property.

12.8 Write a short note on the selection of materials and proportioning of mixtures suitable for use in fiber-reinforced concrete, with special attention to how the requirements of toughness and workability are harmonized.

12.9 What do you understand by the terms *polymer concrete, latex-modified concrete,* and *polymer-impregnated concrete*? What is the principal consideration in the design of polymer concrete mixtures?

12.10 Compare the technologies of producing latex-modified and polymer-impregnated concretes. Also, compare the typical mechanical properties and durability characteristics of the two concrete types.

12.11 (*a*) Discuss the two principal problems in radiation shielding. What are the benefits of using concrete as a shielding material? (*b*) Explain why both high strength and high density are sought in concrete for PCRV (prestressed concrete reactor vessels).

12.12 Define *mass concrete.* What type of cements and admixtures are used for making mass concrete mixtures? From the standpoint of tensile strain capacity, which factors are important in mix proportioning?

12.13 Write a brief note on the pros and cons of construction practices for controlling temperature rise in concrete.

12.14 What is roller-compacted concrete? How do the materials and construction technology for building a dam with roller-compacted concrete differ from conventional mass concrete practice?

References

1. ACI Committee 213, Report 213R-03, Guide for Structural Lightweight Concrete', *ACI Manual of Concrete Practice*, 2012.
2. Schideler, J.J., *J. ACI*, Vol. 54, pp. 299–329, 1957.
3. Kulka, F., and M. Polivka, *Consult. Eng.*, Vol. 42, No. 12, 1978.
4. Schlitter, J., R. Henkensiefken, J. Catro, K. Raoufi, J. Weiss, and T. Nantung, Development of Internally Cured Concrete for Increased Service Life, *Publication FHWA/IN/JTRP-2010*, West Lafayette, Indiana, 2010.
5. Bentz, D.P., P. Lura, and J.W. Roberts, Mixture Proportioning for Internal Curing, *Concr. Int.*, Vol. 27, pp. 35–40, 2005.

6. Cusson, D., and T. Hoogeveen, Internal Curing of High-Performance Concrete with Pre-Soaked Fine Lightweight Aggregate for Prevention of Autogenous Shrinkage Cracking, *Cem. Concr. Res.*, Vol. 38, pp. 757–765, 2008.
7. Holm, T.A., T.W. Bremner, and J.B. Newman, *Concr. Int.*, Vol. 6, No. 6, pp. 49–54, 1984.
8. Wilson, H.S., in *Progress in Concrete Technology*, Malhotra,V.M., ed., CANMET, Ottawa, pp. 141–187, 1981.
9. Kulka, F., and M. Polivka, *Consult. Eng.*, Vol. 42, No. 12, 1978.
10. Bryan, D.L., *A. ACI, Proc.*, Vol. 74, No. 11, p. N23, 1977.
11. Aitcin, P.C., and P.K. Mehta, *ACI Mat. J.*, Vol. 87, No. 2, pp. 103–107, 1990.
12. Aitcin, P.C., *High-Performance Concrete,* E & FN Spon, London, p. 591, 1998.
13. ACI 363 R-92, State-of-the-Art Report on High Strength Concrete, *ACI Manual of Construction Practice*, p. 55, 1993.
14. de Larrad, F., *Cem. Concr. Aggregates*, Vol. 12, No.1, pp. 47–52, 1990, American Society for Testing Materials, Philadelphia, PA.
15. Mehta, P.K., and P.C. Aitcin, *Cem., Concr. Aggregates, J.*, Vol. 12, No. 2, pp. 70–78, 1990, American Society for Testing Materials, Philadelphia, PA.
16. Burg, R.G., and B.W. Ost, Research and Development, Bull. RD104, *Portland Cement Association*, p. 62, 1994.
17. Justnes, H., E. Sellevold, B. Reyniers, D. Van Loo, A.V. Gemert, F. Verboven, and D.V. Gemert, *Autogenous Shrinkage of Concrete*, Tazawa, E., ed., E & FN Spon. London, pp. 71–80, 1998.
18. Technical Committee Report on Autogenous Shrinkage of Concrete, Japan Concrete Institute, *Autogenous Shrinkage of Concrete*, Tazawa, E., ed., E & FN Spon, London, pp. 3–62, 1998.
19. Hua, C. P., Acker, and A. Ehrlacher, Analyses and Models of the Autogenous Shrinkage of Hardening Cement Paste, *Cem. Concr. Res.*, Vol. 25, No.7, pp. 1457–1468, 1995.
20. Iravani, S., and J.G. MacGregor, *ACI Mat. J.*, Vol. 95, pp. 636–647, 1998.
21. Tomosawa, F., and T. Noguchi, *Proceedings of 3rd International Symposium on Utilization of High-Strength Concrete*, Vol. 2, Lillehammer, Norway, pp. 1209–1216, 1993.
22. Hoff, G.C., *Proceedings of 2nd International Symposium on High Strength Concrete*, Hester, W.T., ed., ACI SP-121, Detroit, MI, pp. 619–644, 1990.
23. Malhotra, V.M., *Proceedings of 2nd International Symposium on High Strength Concrete*, Hester, W.T., ed., ACI SP-121, pp. 645–666, 1990.
24. Zhang, M., and O.E. Gjorv, *ibid*, pp. 667–681.
25. Moksnes, J., and M. Sandvik, Paper Presented at a Symposium Honoring O.E. Gjorv, *3rd CANMET/ACI Conference on Marine Structures*, 1996.
26. Harmon, K.S., *High-Performance Structural Lightweight Concrete*, ACI Special Publication 218, Ries, J.P., and T.A. Holm, eds., pp. 189–197, 2004.
27. Liles, P., and R.B. Holland, High Strength Lightweight Concrete for Use in Precast, Prestressed Concrete Bridge Girders in Georgia, HPC Bridge Views, 61, 2010.
28. Collepardi, M., R. Khurana, and M. Valenti, *ACI SP-119*, pp. 471–492, 1989.
29. Miura, N., N. Takeda, R. Chikamatsu, and S. Sogo, *ACI SP-140*, pp. 163–186, 1993.
30. Tanaka, K., K. Sato, S. Watanabe, I. Arima, and K. Sueraga, *ACI SP-140*, pp. 25–51, 1993.
31. Grube, H., and J. Rickert, Concrete Technology Report 1998, 2000, Vorschungsinstitut der Zementindustrie, pp. 39–47, 2001.
32. Collepardi, M., A. Borosi, S. Collepardi, and R. Troli, *ACI SP-222*, pp. 1–18, 2004.
33. Bouzubaa, N., and M. Lachemi, *Cem. Concr. Res.*, Vol. 31, pp. 413–420, 2001.

34. Khayat, K.H., and Z. Guizani, *ACI Mat. J.*, Vol. 94, No. 4, pp. 332–340, 1997.
35. ASTM International C1712–09 Standard Test Method for Rapid Assessment of Static Segregation Resistance of Self-Consolidating Concrete Using Penetration Test, West Conshohocken, PA, pp. 19428–2959, 2009.
36. Spiratos, N., M. Page, N.P. Mailvaganam, V.M. Malhotra, and C. Jolicoeur, *Superplasticizers for Concrete*, Supplementary Cementing Materials for Sustainable Development, Ottowa, Canada, p. 322, 2003.
37. HPC Bridges, Building Bridges for the 21st Century, *U.S. Department of Transportation*, Publication No. FHWA-SA-98-084, Federal Highway Administration, Washington D.C., 1998.
38. Langley, W.S., R. Gilmour, and E. Trompsch, *ACI Special Publication*, No. 154, pp. 543–564, 1995.
39. Burrows, R.W., *ACI Monograph No. 11*, American Concrete Institute, p. 78, 1998; Mehta, P.K., and R.W. Burrows, *Concr. Int.*, Vol. 23, No. 3, pp. 57–63, 2001.
40. Bognacky, C.J., J. Marsano, and W.C. Bauman, *Concr. Int.*, Vol. 22, No. 9, pp. 50–56, 2000.
41. Gjorv, O.E., *Concr. Int.*, Vol. 16, No. 4, pp. 35–39, 1994.
42. Malhotra,V.M., *Concr. Int.*, Vol. 8, No.12, pp. 28–31, 1986.
43. Bisaillon, A., M. Livest, and V.M. Malhotra, *ACI Mat. J.*, Vol. 19, No. 2, pp. 178–187, 1994.
44. Bouzoubaa, A., M.L. Zhang, A. Bilodeau, and V.M. Malhotra, *ACI, SP-178*, pp. 575–603, 1998.
45. Malhotra, V.M., *Concr. Int.*, Vol. 24, No. 7, pp. 30–34, 2002.
46. Malhotra,V.M., and P.K. Mehta, *High-Performance, High-Volume Fly Ash Concrete*, 2d ed., Supplementary Cementing Materials for Sustainable Development, Ottawa, Canada, 2005.
47. Mehta, P.K., *Concr. Int.*, Vol. 24, No. 7, pp. 23–28, 2002.
48. Langley, W.S., and G.H. Leaman, *ACI SP-178*, pp. 575–603, 1998.
49. Mehta, P.K., and W.S. Langley, *Concr. Int.*, Vol. 22, No. 7, pp. 27–32, 2000.
50. ACI Committee 223, Report 223R-10, Guide for the Use of Shrinkage-Compensating Concrete *Manual of Concrete Practice*, Part 1, 2012.
51. Williams, J.V., Jr., *Concr. Int.*, Vol. 3, No. 4, pp. 57–61, 1981.
52. Hoffman, M.W., and E.G. Opbroeck, *Concr. Int.*, Vol. 1, No. 3, pp. 19–25, 1979.
53. Houston Plant Uses Type K Concrete, *Concr. Int.*, Vol. 3, No. 4, pp. 42–47, 1981.
54. Polivka, M., and C. Willson, *ACI SP-38*, pp. 227–237, 1973.
55. Kesler, C.E., *ASCE, Proc. J. Constr. Div.*, Vol. 102, No. C01, pp. 41–49, 1976.
56. Keith, F.R., W.W. Walker, and J.A. Holland, HYPERLINK "http://www.ssiteam.com/files/file/A%20Peach%20Of%20A%20Pavement%20Using%20Shrinkage-Compensating%20Concrete.pdf" \t "_blank"A Peach of a Pavement Using Shrinkage-Compensating Concrete, *Concr. Int.*, pp. 27–32, May 1996.
57. Ramey, G.E., D.W. Pittman, and G.K. Webster, Shrinkage-Compensating Concrete for Bridge Decks, *Concr. Int.*, Vol. 21, pp. 29–34, April 1999.
58. Luco, L.F., R. Pombo, and R. Torrent, Shrinkage-Compensating Concrete in Argentina, *Concr. Int.*, Vol. 25, pp. 91–101, 2003.
59. Li, V.C., Large Volume, High-Performance Applications of Fibers in Civil Engineering, *J. Appl. Polym. Sci.*, Vol. 83, pp. 660–686, 2002.
60. Report ACI 544.1R-82, *Concr. Int.*, Vol. 4, No. 5, pp. 9–30, 1982.
61. Li, V.C., *ASCE J. Mater. Civil Eng.*, Vol. 4, No. 1, pp. 41–57, 1992.
62. Lawler, J.S., D. Zampini, and S.P. Shah, Microfiber and Macrofiber Hybrid Fiber-Reinforced Concrete, *J. Mater. Civil Eng. ASCE*, Vol. 17, No. 5, pp. 595–604, 2005.

63. Cyr, M.F., A. Peled, and S.P. Shah, Extruded Hybrid Fiber Reinforced Cementitious Composites, *7th Int. Symp. Ferrocem. Thin Reinforced Cem. Composites (Ferro-7)*, Centre for Construction Materials and Technology, National University of Singapore, pp. 199–207, 2001.
64. Li, V.C., Y. Wang, and S. Backer, Effect of Inclining Angle, Bundling, and Surface Treatment on Synthetic Fiber Pull-Out from a Cement Matrix, *Composites*, Vol. 21, No. 2, pp. 132–140, 1990.
65. Ouyang, C., A. Pacios, and S.P. Shah, Pullout of Inclined Fibers from Cementitious Matrix, *J. Eng. Mechanics, ASCE*, Vol. 120, No. 12, pp. 2641–2659, 1994.
66. Majumdar, A.J., Glass Fiber Reinforced Cement and Gypsum Products, *Proc. R. Soc., London*, Vol. A319, pp. 69–78, 1970.
67. Edington, J.E., P.J. Hannant, and R.I.T. Williams, Building Research Establishment, *BRE Current Paper CP 69/74*, 1974.
68. Swamy, R.N., and P.S. Mangat, *Cem. Concr. Res.*, Vol. 4, No. 3, pp. 451–465, 1974.
69. Report ACI 544.3R-84, *J. ACI, Proc.*, Vol. 81, No. 2, pp. 140–48, 1984.
70. Bouzoubaâ, N., and M. Lachemib, Self-Compacting Concrete Incorporating High Volumes of Class F Fly Ash: Preliminary Results, *Cem. Concr. Res.*, Vol. 31, No. 3, pp. 413–420, 2001.
71. Aydin, A.C., Construc. *Building Mat.*, Vol. 21, pp. 1149–1154, 2007.
72. Greenough, T., and M. Nehdi, Shear Behavior of Fiber-Reinforced Self-Consolidating Concrete Slender Beams, *ACI Mat. J.*, Vol. 105, pp. 468–477, 2008.
73. Ibid.
74. Hanna, A.N., *Steel Fiber Reinforced Concrete Properties and Resurfacing Applications*, Portland Cement Association, Skokie, Ill., Report RD049.01P, 1977.
75. Trono, W., G. Jen, D. Moreno, S. Billington, and C.P. Ostertag, Confinement and Tension Stiffening Effects in High Performance Self-Consolidated Hybrid Fiber Reinforced Concrete Composites, *in High Performance Fiber Reinforced Cement Composites 6*, Parra-Montesinos, G.J., H.W. Reinhardt, A.E. Naaman, eds., Springer, 2012.
76. Ferrara, L., Y.-D. Park, and S.P. Shah, A Method for Mix-Design of Fiber-Reinforced Self-Compacting Concrete, *Cem. Concr. Res.*, Vol. 37, pp. 957–971, 2007.
77. Saak, A.W., H.M. Jennings, and S.P. Shah, New Methodology for Designing Self-Compacting Concrete, *ACI Mat. J.*, Vol. 98, No. 6, pp. 429–439, 2001.
78. ACI Report 544.2R, Measurement of Properties of Fiber Reinforced Concrete, *ACI Man. Concr. Prac.*, 2012.
79. Shah, S.P., and B.V. Rangan, *J. ACI, Proc.*, Vol. 68, No. 2, pp. 126–135, 1971.
80. Swamy, R.N., P.S. Mangat, and C.V.S.K. Rao, *Fiber Reinforced Concrete*, ACI SP-44, pp. 1–28, 1974.
81. Johnston, C.D., in *Progress in Concrete Technology*, V.M. Malhotra, ed., CANMET, Ottawa, pp. 452–504, 1980.
82. Report ACI 544.1R-96, *ACI Man. Concr. Prac.*, ACI International, 2001.
83. Ibid.
84. Ibid.
85. Johnston, C.D., *Progress in Concrete Technology*.
86. Bache, H.H., *Fracture Mechanics of Concrete Structures: From Theory to Applications;* Elfgren, L., ed., Chapman Hall, London/New York, pp. 382–398, 1989.
87. Richard, P., and M.H. Cheyrezy, *ACI Spring Convention*, San Francisco, CA, 1994.
88. Bonneau, O., C. Poulin, J. Dugat, P. Richard, and P.C. Aitcin, *Concrete International*, pp. 47–49, 1996.

89. Li, V.C., Engineered Cementitious Composites-Tailored Composites through Micromechanical Modeling, in *Fiber Reinforced Concrete: Present and the Future*, N. Banthia et al., eds., CSCE, Montreal, pp. 64–97, 1998.
90. Rossi, P., Ultra-High-Performance Fiber-Reinforced Concretes, *Concr. Int.*, pp. 46–52, 2001.
91. Henager, C.H., *Concr. Int.*, Vol. 3, No. 9, pp. 13–18, 1981.
92. Report ACI 544.1R-96, *ACI Man. Concr. Prac.*, ACI International, 2001.
93. Report ACI 544.1R-96, *ACI Manual of Concrete Practice*, ACI International, 2001.
94. Fowler, D.W., Polymers in Concrete: A Vision for the 21st Century, *Cem. Concr. Composites*, pp. 449–452, 1999.
95. Chandra, S., and Y. Ohama, *Polymers in Concrete*, CRC Press, Boca Raton, FL, 1994; *Polymers in Concrete: The First Thirty Years*, Kaeding, A.O., and Prusinsk, R.C., eds., SP 214, 2003, ACI Committee 548, Report 548.1R-09, Guide for the Use of Polymers in Concrete, *ACI Manual of Concrete Practice*, American Concrete Institute, Farmington Hills, MI, 2012.
96. *Polymers in Concrete, ACI Special Publication SP-40*, American Concrete Institute, Farmington Hills, MI, 1973.
97. Cady, P.D., R.E. Weyers, and D.T. Wilson, *Concr. Int.*, Vol. 6, No. 6, pp. 36–44, 1984.
98. *Guide to Shotcrete*, ACI Committee 506, American Concrete Institute, Farmington Hills, Michigan, 2005.
99. *Guide for the Evaluation of Shotcrete*, ACI Committee 506.4R-94, American Concrete Institute, Farmington Hills, Michigan, 2004.
100. Banthia, N., and K. Campbell, Restrained Shrinkage Cracking in Bonded Fiber Reinforced Shotcrete, The Interfacial Transition Zone in Cementitious Composites, RILEM Proceedings, V. 35, E&FNSpon, pp. 216–223, 1998.
101. Padron, I., and R.F. Zollo, Effect of Synthetic Fibers on Volume Stability and Cracking or Portland Cement Concrete and Mortar, *ACI Mater. J.*, Vol. 87, No. 4, July-Aug., pp. 327–332, 1990.
102. Grzybowski, M., and S.P. Shah, Shrinkage Cracking in Fiber Reinforced Concrete, *ACI Materials Journal*, Vol. 87, No. 2, Mar.-Apr., pp. 138–148, 1990.
103. Garrett, R., Construction Chemistry at Weehawken, *Tunnelling & Trenchless Construction*, Dec., pp. 21–25, 2004.
104. Tatnall, P.C., Shotcrete in Fires: Effects of Fibers on Explosive Spalling, *Shotcrete*, Vol. 4, No. 4, pp. 10–12, 2002.
105. Thomas, A., *Sprayed Concrete Lined Tunnels*, Taylor and Francis, New York, 2008.
106. Gilbride, P., D.R. Morgan, and T.W. Bremner, Deterioration and Rehabilitation of Berth Faces in Tidal Zones at the Port of Saint John, *Shotcrete*, Vol. 4, No. 4, Fall, pp. 32–38, 2002.
107. Forrest, M.P., D.R. Morgan, J.R. Obermeyer, P.L. Parker, and D.D. LaMoreaux, Seismic Retrofit of Littlerock Dam, *Shotcrete*, Vol. 6, No. 1, pp. 20–26, 2004.
108. Polivka, M., and H.S. Davis, ASTM STP 169B, Chap. 26, pp. 420–434, 1979.
109. Kaplan, M.F. *Concrete Radiation Shielding, Nuclear Physics, Concrete Properties Design and Construction*, Longman Press, 1989.
110. Gurler, O., and U.A. Tarim, An Investigation on Determination of Attenuation Coefficients for Gamma-Rays by Monte Carlo Method, *J. Radioanal Nucl Chem*, Vol. 293, pp. 397–401, 2012.
111. Bashter, I.I., A.S. Makarious, and A.E.S. Abdo, Investigation of Hematite-Serpentine and Ilmenite-Limonite Concretes for Reactor Radiation Shielding, *Ann. Nucl. Energy*, 23, 65–71, 1996.

112. ASTM C 637-09 Standard Specification for Aggregates for Radiation-Shielding Concrete, *ASTM Int.*, 100 Barr Harbor Drive, West Conshohocken, PA, United States, 2010.
113. ASTM C 638-09 Standard Descriptive Nomenclature of Constituents of Aggregates for Radiation-Shielding Concrete, *ASTM Int.*, 100 Barr Harbor Drive, West Conshohocken, PA, United States, 2010.
114. Lee, S.Y., A.M. Daugherty, and D.J. Broton, Assessing Aggregates for Radiation-Shielding Concrete, *Concr. Int.*, pp. 31–38, 2013.
115. Sakr, K., Effects of Silica Fume and Rice Husk Ash on the Properties of Heavy Weight Concrete, *J. Mater. Civil Eng.*, pp. 367–376, 2006.
116. Mather, K., *J. ACI, Proc.*, Vol. 62, pp. 951–962, 1965.
117. ACI 522R, Pervious Concrete, *American Concrete Institute*, Farmington Hills, Michigan, 2010.
118. Meininger, R.C., No-Fines Pervious Concrete for Paving *Concr. Int.*, Vol. 10, No. 8, pp. 20–27, 1988.
119. U.S. Green Building Council, Green Building Rating System for New Construction and Major Renovations, *Leadership in Energy and Environmental Design (LEED)*, 2002.
120. Tennis, P.D., M.L. Leming, and D.J. Akers, Pervious Concrete Pavements, *Portland Concrete Association: National Ready Mixed Concrete Association*, Silver Spring, MD, 2004.
121. Pratt, C.J., A.P. Newman, and P.C. Bond, Mineral Oil Bio-Degradation within a Permeable Pavement: Long Term Observations, *Water Sci. Technol.*, Vol. 39, pp. 103–109, 1999.
122. Marolf, A., N. Neithalath, E. Sell, and J. Weiss, The Influence of Aggregate Gradation on the Acoustic Absorption of Enhanced Porosity Concrete, *ACI Mat. J.*, Vol. 101, No.1, pp. 82–91, 2004.
123. Neithalath, N., J. Weiss, and J. Olek, Modeling the Influence of Pore Structure on the Acoustic Absorption of Enhanced Porosity Concrete, *J. Advanced Concr. Technol.*, Vol. 3, pp. 29–40, 2005.
124. Neithalath, N., J. Weiss, and J. Olek, Characterizing Enhanced Porosity Concrete Using Electrical Impedance to Predict Acoustic and Hydraulic Performance, *Cem. Concr. Res.*, Vol. 36, No. 11, pp. 2074–2085, 2006.
125. Nelson, P.M., Designing Porous Road Surfaces to Reduce Traffic Noise, *TRL Annual Rev.*, Transportation Research Laboratories, U.K., 1994.
126. Yang, J., and G. Jiang, Experimental Study on Properties of Pervious Concrete Pavement Materials, *Cem. Concr. Res.*, Vol. 33, No. 3, pp. 381–386, 2003.
127. Schaefer, V.R., K. Wang, Suleiman, J. Kavern, Mix Design Development for Pervious Concrete in Cold Weather Climates, Final Report, *National Concr. Pavement Technol. Center*, Iowa State University, Ames, IA, 2006.
128. Report of ACI Committee 207 on Mass Concrete, *ACI Manual of Concrete Practice*, Part 1, 2012.
129. ACI Committee 211, *Concr. Int.*, Vol. 2, No. 12, pp. 67–80, 1980.
130. Report ACI 207.1R-96, Mass Concrete, *ACI Manual of Concrete Practice*, American Concrete Institute, Farmington Hills, MI, 2002.
131. ACI Committee 211, *ACI Mat. J.*, Vol. 89, No. 5, 1991.
132. Scanlon, J.M., Mix Proportioning for Mass Concrete, *ACI SP-46*, pp. 74–96, 1974.
133. Price, W.H., *Concr. Int.*, Vol. 4, No. 10, pp. 36–44, 1982.
134. ACI Committee 207, Cooling and Insulating Systems for Mass Concrete, *Concr. Int.*, Vol. 2, No. 5, pp. 45–64, 1980.

135. Raphael, J.M., The Optimum Gravity Dam, *Proceedings, Conference on Rapid Construction of Concrete Dams*, American Society of Civil Engineers, New York, 1970.
136. Hansen, K.D., Roller Compacted Concrete: A Civil Engineering innovation, *Concr. Int.*, Vol. 18, p. 49, 1996.
137. Roller-Compacted Concrete, Technical Engineering and Design Guides as Adapted from the U.S. Army Corps of Engineers, ASCE Press, New York, p. 100, 1994.
138. Hirose,T., and T. Yanagida, *Concr. Int.*, Vol. 6, No. 5, pp.14–19, 1984.
139. *Technical Guide to RCD Construction Method*, Technical Center for National Land Development, Japan, July 1981.
140. Hansen, K.D., and W.G. Reinhardt, *Roller-Compacted Concrete Dams*, McGraw-Hill, New York, pp. 39–47, 1991.
141. Marques Filho, J., V.A. Paulon, P. Monteiro, W.P. Andrade, and D.D. Molin, Laboratory Device to Simulate Roller-Compacted Concrete Placement, *ACI Mat. J.*, Vol. 105, No. 2, pp. 125–130, Mar-Apr. 2008.
142. R. Féret, Sur la compacité desmortiers hydrauliques. Mémoires et documents relatifs à l'art des constructions et au service de l'ingénieur. Annales des Ponts et ChaussÇes, Vol. 4, 1892.
143. Schrader, E., *Concr. Int.*, Vol. 4, No. 10, pp. 15–25, 1982.
144. Yamauchi,T., J. Harada, T. Okada, and S. Shimada, Construction Work and Quality and Temperature Control for Tamagawa RCD Dam, Commission Internationale des Grands Barrages, Seizieme Congres, San Francisco, 1988.

Suggestions for Further Study

Lightweight Concrete

ACI Committee 213, Report 213R-03, Guide for Structural Lightweight Concrete', *ACI Manual of Concrete Practice*, Part 1, 2012.

Bentz D.P., and W. J. Weiss, Internal Curing: A 2010 State-of-the-Art Review, *Report NISTIR* 7765, National Institute of Standards and Technology, 2011.

Chandra S., and L. Berntsson, *Lightweight Aggregate Concrete*, Noyes Publications, Norwich, NY, 2002.

Chandra, S., and L. Berntsson, *Lightweight Aggregate Concrete: Science, Technology, and Applications*, Noyes Publications, New York, 2002.

Kovler, K., and O. Jensen, eds., *Internal Curing of Concrete*, RILEM Publications S.A.R.L., Paris, 2007.

Short, A., and W. Kinniburgh, *Lightweight Concrete*, Applied Science Publishers, Barking, Essex, 1978.

High-Strength Concrete

Bickley, J.A., *High-Strength Concrete: An International Perspective*; American Concrete Institute, Farmington Hills, MI, SP-167, p. 322, 1996.

Caldarone M.A. High-Strength Concrete: a practical guide, Taylor & Francis, New York, p. 252, 2009.

Hester, W., ed., *Proceedings of High Strength Concrete Conference* ACI SP-212, 1990.

Malier, Y., *High Performance Concrete : From Material to Structure*, E & FN Spon, London, p. 542, 1992.

Nawy, E.G., *Fundamentals of High-Performance Concrete*, 2d ed., Wiley, New York, 2001.

Shah, S.P., and S.H., Ahmad, *High Performance Concrete: Properties and Applications*, McGraw-Hill, New York, p. 403, 1994.

Self-Consolidating Concrete

Daczko, J.A., *Self-Consolidating Concrete: Applying What We Know*, Spon Press, London and New York, 2012.

Collepardi, M., Self-Compacting Concrete: What is New? *Seventh CANMET/ACI International Conference on Superplasticizers and Other Chemical Admixtures in Concrete*, Farmington Hills, MI, American Concrete Institute, pp. 1–16, 2003.

Khayat, K.H., Workability, Testing, and Performance of Self-Consolidating, *ACI Mat. J.*, Vol. 96, No. 3, pp. 346–353, 1999.

Loukili, A., *Self-Compacting Concrete*, John Wiley & Sons, Inc., Hoboken, NJ, 2011.

Okamura, H., Self-Compacting High-Performance Concrete, *Concr. Int.*, Vol. 19, No. 7, pp. 50–54, 1997.

Expansive Cement Concrete

ACI Committee 223, Report 223-10, Guide for the Use of Shrinkage-Compensating Concrete, *ACI Manual of Concrete Practice*, 2012.

Cedric Willson Symposium on Expansive Cement, *ACI SP-64*, American Concrete Institute, Detroit, MI, 1980.

Klein Symposium on Expansive Cement Concretes, *ACI SP-38*, American Concrete Institute, Detroit, MI, 1973.

Fiber-Reinforced Concrete

ACI Committee 544, Report 544.1R-96, State-of-the-Art Report on Fiber Reinforced Concrete, 544.5R-10, Report on the Physical Properties and Durability of Fiber-Reinforced Concrete, *ACI Manual of Concrete Practice*, American Concrete Institute, Farmington Hills, MI, 2012.

Balaguru, P.N., and S.P. Shah, *Fiber-Reinforced Cement Composites*, McGraw-Hill, New York, p. 530, 1992.

Banthia N., M. Criswell, P. Tatnall, K. Folliard, eds., *Innovations in Fiber-Reinforced Concrete for Value*, SP-216, American Concrete Institute, Farmington Hills, MI, 2003.

Bentur, A., and S. Mindess, *Fibre Reinforced Cementitious Composites*, Elsevier, New York, p. 449, 1990.

Blunt, J.D., and C.P. Ostertag, Deflection Hardening and Workability of Hybrid Fiber Composites, *ACI Mat. J.*, Vol. 106, pp. 265–272, 2009.

Bolander, J.E., S. Choi, and S.R. Duddukuri, Fracture of Fiber-Reinforced Cement Composites: Effects of Fiber Dispersion, *Int. J. Frac.*, Vol. 154, pp. 73–86, 2008.

Thakur, R., S.K. Das, and S. Ghosh, *Fiber Reinforced Self Compacting Concrete: Rheology and Compatibility of Ingredients*, VDM Verlag, 2011.

Concrete Containing Polymers

ACI Committee 548, Report 548.1R-09, Guide for the Use of Polymers in Concrete, 548.3R-09, Report on Polymer-Modified Concrete. *ACI Manual of Concrete Practice*, American Concrete Institute, Farmington Hills, MI, 2012.

Fowler, D.W., Polymers in Concrete: A Vision for the 21st Century, *Cem. Concr. Composites*, pp. 449–452, 1999.

Ohama, Y., *Handbook of Polymer-Modified Concrete and Mortars: Properties and Process Technology*, Noyes Publications, Park Ridge, NJ, p. 236, 1995.

Ohama, Y., M. Kawakami and Y. Fukuzawa, eds., *Polymers in Concrete*, E&FN Spon, 1997.

Walters, D.G., ed., *Polymer Concrete*, American Concrete Institute, SP -137, Detroit, MI, p. 130, 1993.

Heavyweight Concrete

ACI Committee ACI 349-06 Report on Code Requirements for Nuclear Safety Related Concrete Structures, 221R-96: Guide for Use of Normal Weight and Heavyweight Aggregates in Concrete, 304.3R-96: High Density Concrete: Measuring, Mixing, Transporting, and Placing, *ACI Manual of Concrete Practice*, 2006.

ACI, *Concrete for Nuclear Reactors*, SP-34 (3 vols.), 1973.

ASTM, *Radiation Effects and Shielding*, STP 169B, pp. 420–434, 1979.

ASTM C 638-09, Standard Descriptive Nomenclature of Constituents of Aggregates for Radiation-Shielding Concrete, *ASTM International*, 100 Barr Harbor Drive, West Conshohocken, PA, United States, 2010.

ASTM C 637-09, Standard Specification for Aggregates for Radiation-Shielding Concrete, *ASTM International*, 100 Barr Harbor Drive, West Conshohocken, PA, United States, 2010.

Mass Concrete and Roller Compacted Concrete

ACI Committee 207,1R-05 Report on Guide to Mass Concrete, 207.4R-05, Report on Cooling and Insulating Systems for Mass Concrete, 207.5R-11 Report on Roller Compacted Concrete, *ACI Manual of Concrete Practice*, 2012.

Berga, L., J.M. Buil, C. Jofre, and S. Chonggang, eds., *Roler Compacted Concrete Dams*, Swets & Zeitlinger, Lisse, 2003.

Berga, L., J.M. Buil, C. Jofré, S. Chonggang, eds., *RCC Dams–Roller Compacted Concrete Dams: Proceedings of the IV International Symposium on Roller Compacted Concrete Dams*, Madrid, Spain, Taylor & Francis, 17–19 Nov. 2003.

Berga, L., J. M. Buil, C. Jofre, S. Chonggang, eds., *Roller Compacted Concrete Dams: Proceedings of the IV International Symposium on Roller Compacted Concrete Dams*, A.A. Balkema, Rotterdam, 2003.

Hansen, K., The Many Faces of RCC, *Civil Engineering—ASCE*, Vol. 71, No. 4, pp. 48–53, April 2001.

Roller-Compacted Concrete, *Engineer Manual*, No. 1110-2-2006, U.S. Army Corps of Engineers, Washington, D.C., 2000.

Shotcrete

ACI Committee, 506R-05: Guide to Shotcrete, 506.5R-09: Guide for Specifying Underground Shotcrete, 506.1R-08: Guide to Fiber-Reinforced Shotcrete, *ACI Manual of Concrete Practice*, American Concrete Institute, Farmington Hills, MI, 2013.

Bernard E.S., Shotcrete: Engineering Developments, Swets & Zeitlinger, Lisse, 2001.

Morgan, D.R., Shotcrete: *A Compilation of Papers*, American Concrete Institute, Farmington Hills, MI, 2009.

Pervious Concrete

ACI Committee 522, 522R-10 Report on Pervious Concrete, 522.1-08 Specification for Pervious Concrete Pavements, *ACI Manual of Concrete Practice*, American Concrete Institute, Farmington Hills, MI, 2013.

Garber, G., *Paving with Pervious Concrete*, Schiffer Publishing, Ltd., 2011.

CHAPTER 13

Concrete Mechanics

Preview

The theory of composite materials has been extensively used to model the elastic behavior of advanced ceramics, rocks, soils, and so forth. In this chapter, the theory will be applied to estimate the elastic moduli of concrete, when the elastic moduli and the volume fractions of cement paste and aggregate are known. Concrete is a porous material with a number of microcracks even before load is applied. The *differential scheme* and *Mori-Tanaka method* allow the computation of the effect of these imperfections on the elastic moduli of concrete. The concept of upper and lower bounds for elastic moduli is discussed, and the *Hashin-Shtrikman bounds* are presented.

In Chap. 4, the concepts of creep and stress relaxation in viscoelastic materials were introduced. In this chapter, a number of rheological models that aid understanding of the underlying mechanisms are presented. Changes in properties of concrete, with time, makes the task of mathematical modeling more complex. Methods for incorporating the age of concrete into mathematical models are discussed. Some viscoelastic formulations, such as the principle of superposition, are also discussed.

The concept of thermal shrinkage was introduced in Chap. 4, whereas the technology to reduce cracking due to thermal stresses was discussed in Chap. 12. In this chapter, the finite element method for computing the *temperature distribution in mass concrete* is introduced. Examples of application of this method to a number of practical situations in concrete are also given.

Fracture mechanics of concrete has become a powerful method for studying the behavior of plain and reinforced concrete members in tension. The traditional concept of linear fracture mechanics and its limitations when applied to concrete are discussed in this chapter. The use of the finite element method for cracks in structures is demonstrated. Also presented are nonlinear fracture mechanics models, such as the *fictitious crack model, and the smeared crack model*, together with fracture resistance curves. The importance of size effect is illustrated, using the smeared crack band model.

There are many competing theories to model *concrete exposed to freezing temperatures* but few of them provide comprehensive quantitative predictions of the mechanical behavior, while accounting for the multiscale physics of the confined crystallization of ice. When a part of the liquid in the pores solidifies, a pressure build up is generated, and excess liquid is expelled from the freezing sites toward the remaining part of the porous network. In turn, with increasing cooling a cryo-suction process drives the liquid toward the frozen sites. The chapter introduces *unsaturated poroelasticity theory* that provides the formulation to compute the stresses and strains developed when concrete is exposed to low temperatures.

The *alkali-silica reaction* involves reactive siliceous aggregates and highly alkaline concrete pore solution that react to produce an alkali-silicate gel, which can imbibe water and expand. Many dams, reinforced concrete structures, and pavements require continuous repair or replacement. The chapter reviews recent advancements in the mechanical modeling of this deleterious reaction.

13.1 Elastic Behavior

This section introduces some fundamental approaches to describe composite materials. These methods can be used to estimate the elastic moduli of concrete if the elastic moduli of the concrete phases and their volume fractions are known. Chapter 4 presented the relationship between elastic modulus and compressive strength as recommended by ACI and FIP. These relationships are empirical and, by and large, were developed from experimental results on normal structural concretes with compressive strength ranging from 21 up to 120 MPa. The difference between predicted and measured values using this type of formulation can be significant, and special care should be taken when estimating the elastic modulus for different types of concrete (i.e., ultra high-strength concrete, light-weight aggregate concrete, and mass concrete). For instance, the maximum size aggregate and the volume fractions of aggregate and cement paste in mass concrete are quite different from structural concrete, therefore, predicting the elastic modulus using the ACI equation would not be reliable. Assume you are the designer of a large concrete dam and you want to perform a preliminary thermal stress analysis. Unfortunately, in most cases, experimental results are not available at this stage of the project, yet an estimation of the elastic modulus of concrete is crucial for predicting thermal stresses. To solve this problem you must obtain an estimate of the elastic properties using a composite materials formulation that incorporates the elastic moduli and the volume fractions of the cement paste and aggregate.

The two simplest models used to simulate a composite material are shown in Figs. 13-1*a* and *b*. In the first model the phases are arranged in a parallel configuration, imposing a condition of uniform strain. This arrangement is often referred to as the *Voigt model*. In the second model, the phases are arranged in a series configuration imposing a condition of uniform stress; this geometry is known as the *Reuss model*.

We solve the Voigt model by using a simple strength of materials approach. For a first approximation, lateral deformations are neglected. The following equations are obtained:

$$\text{Equilibrium equation} \qquad \sigma A = \sigma_1 A_1 + \sigma_2 A_2 \tag{13-1}$$

$$\text{Compatibility equation} \qquad \varepsilon = \varepsilon_1 = \varepsilon_2 \tag{13-2}$$

$$\text{Constitutive relationship} \qquad \sigma = E\varepsilon \tag{13-3}$$

Substituting Eq. (13-3) into (13-1) we obtain: $E\varepsilon A = E_1\varepsilon_1 A_1 + E_2\varepsilon_2 A_2$ (13-4)

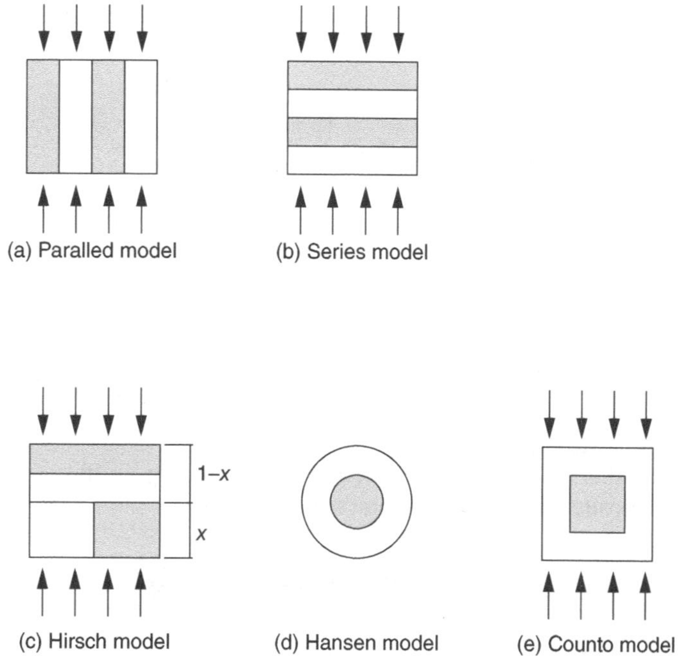

FIGURE 13-1 Traditional two-phase models for concrete.

Using the compatibility Eq. (13-2), Eq. (13-4) reduces to

$$EA = E_1A_1 + E_2A_2$$

For composite materials it is more convenient to deal with volume than area, therefore, for unit length:

$$EV = E_1V_1 + E_2V_2$$

or

$$E = E_1c_1 + E_2c_2 \tag{13-5}$$

where $c_i = V_i/V$ is the volume fraction of the ith phase. Using the same approach to solve for the series (Reuss) model:

$$\frac{1}{E} = \frac{c_1}{E_1} + \frac{c_2}{E_2} \tag{13-6}$$

To obtain further insight into these models, let us rederive the parallel and series models to include lateral deformations. Because the structural models shown in Fig. 13-1 do not allow the introduction of Poisson's ratio (ν), let us consider a homogeneous body of volume V and bulk modulus K subjected to a uniform hydrostatic compression P. The total stored strain energy W is given by

$$W = \frac{P^2V}{2K} \tag{13-7}$$

or

$$W = \frac{\varepsilon^2 KV}{2} \tag{13-8}$$

where

$$\varepsilon = \frac{dV}{V} = -\frac{P}{K} \tag{13-9}$$

is the volumetric strain.

The parallel model assumes that each phase undergoes the same strain in the two-phase composite. Therefore, the total stored energy is

$$W = W_1 + W_2 = \frac{\varepsilon^2 K_1 V_1}{2} + \frac{\varepsilon^2 K_2 V_2}{2} \tag{13-10}$$

where the subscripts 1 and 2 identify the phases. Equating the strain energy in the composite, Eq. (13-8), to the equivalent homogeneous medium, Eq. (13-10), leads to the following expression for the effective bulk modulus:

$$K = c_1 K_1 + c_2 K_2 \tag{13-11}$$

A similar expression can be obtained for the effective shear modulus G. The effective modulus of elasticity can be calculated from Eq. (13-11) in combination with the following relations from the theory of elasticity:

$$E = \frac{9KG}{3K+G} = 2G(1+v) = 3K(1-2v) \tag{13-12}$$

Using Eqs. (13-11) and (13-12), the effective modulus of elasticity for the parallel model can be given by[1]

$$E = c_1 E_1 + c_2 E_2 + \frac{27 c_1 c_2 (G_1 K_2 - G_2 K_1)^2}{(3K_v + G_v)(3K_1 + G_1)(3K_2 + G_2)} \tag{13-13}$$

where K_v and G_v refer to the values obtained using the Voigt model. For the special case where the two phases have the same Poisson's ratio, Eq. (13-13) reduces to Eq. (13-5), $E = E_1 c_1 + E_2 c_2$, which was obtained neglecting deformations.

The series model assumes that the stress state in each phase will be a uniform hydrostatic compression of magnitude P. The total stored energy for the composite is given by

$$W = W_1 + W_2 = \frac{P^2 V_1}{2K_1} + \frac{P^2 V_2}{2K_2} = \frac{P^2}{2}\left(\frac{V_1}{K_1} + \frac{V_2}{K_2}\right) \tag{13-14}$$

The effective bulk modulus can be obtained by equating Eqs. (13-7) and (13-14):

$$\frac{1}{K} = \frac{c_1}{K_1} + \frac{c_2}{K_2} \tag{13-15}$$

Using the relationships for elastic modulus given by Eq. (13-12), Eq. (13-15) can be rewritten as

$$\frac{1}{E} = \frac{c_1}{E_1} + \frac{c_2}{E_2} \tag{13-16}$$

Note that Eq. (13-16) is the same as Eq. (13-6), which was obtained when we neglected lateral deformations.

Neither the Voigt nor the Reuss models are precise, except in the special case where the moduli of the two materials are equal. This is because the equal-stress assumption satisfies the stress equations of equilibrium, but, in general, gives rise to displacements that are discontinuous at the interfaces between the two phases. Similarly, the equal-strain assumption leads to an admissible strain field, but the resulting stresses are discontinuous.

Hill,[2] who used various energy considerations from the theory of elasticity, showed that the parallel and series assumptions lead to upper and lower bounds on G and K. This result is significant because, given the elastic moduli of the phases and their volume fractions, it allows the determination of the maximum and minimum allowable value for the concrete elastic moduli. If the maximum and minimum values are close, the problem is solved from an engineering point of view. When hard inclusions are dispersed in a softer matrix, however, the maximum and minimum values are far apart, as shown in Fig. 13-2. Therefore, it is recommended to establish stricter upper and lower bounds, such as the Hashin-Shtrikman bounds to be discussed later after a brief review of more elaborate models.

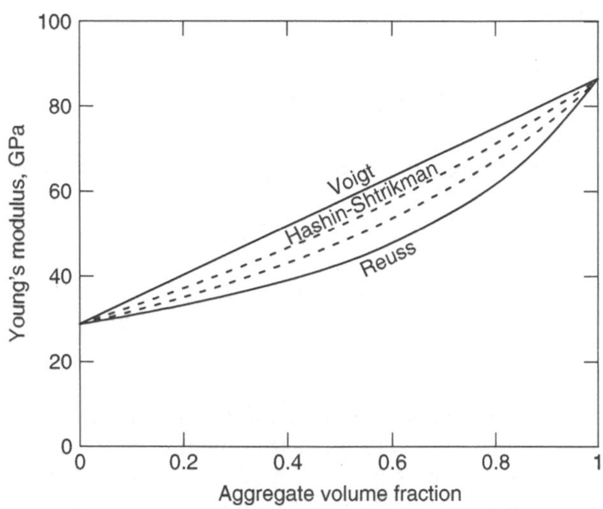

FIGURE 13-2 Bounds for Young's modulus (elastic moduli of the matrix E_m = 28.7 GPa, K_m = 20.8 GPa, and elastic moduli of the aggregate: E_a = 86.7 GPa, K_a = 44 GPa).

Hirsch[3] proposed a *model* (Fig. 13-1*c*) that relates the modulus of elasticity of concrete to the moduli of the two phases (aggregate and matrix), their volume fractions, and an empirical constant, x.

$$\frac{1}{E_c} = (1-x)\left(\frac{c}{E_a} + \frac{1-c}{E_m}\right) + x\left(\frac{1}{cE_a + (1-c)E_m}\right) \tag{13-17}$$

where

$$c = \frac{V_a}{V_c}$$

Here x and $(1-x)$ indicate the relative contributions of the parallel (uniform strain) and series (uniform stress) models. Researchers have used the Hirsch model to estimate the degree of bonding between the cement paste matrix and aggregate in concrete. It is assumed that perfect bonding would lead to the uniform strain situation, whereas a complete lack of bonding would lead to the uniform stress situation. This assumption is incorrect because the uniform stress model ($x = 0$) does not imply *no bond* between the phases, nor does the uniform strain model ($x = 1$) indicate a *perfect bond*. It is perhaps more appropriate to view Eq. (13-17) as merely providing a result that lies somewhere between the parallel and the series predictions. For practical application, the value 0.5 for x is often recommended, which gives the arithmetic average of the parallel and series moduli.

Hansen[4] proposed a *model* (Fig. 13-1*d*) that consists of spherical aggregate located at the center of a spherical mass of matrix. This model was based on a general formulation by Hashin,[5] which, for the particular case when Poisson's ratio of both phases is equal to 0.2, yielding

$$E = \left(\frac{c_1E_1 + (1+c_2)E_2}{(1+c_2)E_1 + c_1E_2}\right)E_1 \tag{13-18}$$

where phase 1 corresponds to the matrix and phase 2 to the aggregate.

Counto[6] considered the case (Fig. 13-1*e*) where an aggregate prism is placed at the center of a prism of concrete, both with the same ratio of height to area of cross section. By using a simple strength of materials approach, the modulus of elasticity for the concrete can be given by

$$\frac{1}{E} = \frac{1-\sqrt{c_2}}{E_1} + \frac{1}{\left(\frac{1-\sqrt{c_2}}{\sqrt{c_2}}\right)E_1 + E_2} \tag{13-19}$$

Again, phase 1 corresponds to the matrix and phase 2 to the aggregate.

Let us now return to the original assignment of computing the elastic modulus of concrete. The task has become quite complicated because of the large number of models from which to choose from. Selecting the most appropriate model to use may be challenging. Fortunately, for mass concrete, where a large amount of aggregate is present, the predictions are not far apart. For example, a typical lean concrete mixture with

75 percent of aggregate and 25 percent of cement paste by volume. For a given age, assuming the elastic modulus of the cement paste and aggregate (say a quartzite) to be 20 and 45 GPa, respectively, the predicted value of the elastic modulus of the concrete obtained using the Voigt model is 38.8 GPa and from the Reuss model 34.3 GPa. No matter how sophisticated or simple a two-phase model may be, any prediction should lie inside these bounds. The Hirsch, Hansen, and Counto models predict elastic modulus as 36.5, 36.4, and 36.2 GPa, so for practical purposes, the three models estimate the same elastic modulus for this particular example. In these examples, it has been relatively simple to estimate the elastic modulus. In other cases of estimating elastic moduli may be more problematic.

The models presented so far are limited in computing the effect of voids, cracks, and phase changes (such as water to ice during freezing of the cement paste). Another shortcoming of these methods is that they do not take into account any of the specific geometrical features of the phases or how the pores and aggregate particles interact with one another under various loading conditions. As a general rule for two-component materials, the effect of the shape of the inclusion is more important when the two components have vastly different moduli, but is of minor importance when the two components have roughly equal moduli. Hence, we can use models that ignore aggregate shape when trying to estimate the moduli of a mixture of cement paste and aggregate.

Nielsen and Monteiro[7] studied the limitations of two-phase models for concrete and proposed that concrete be modeled as a three-phase material consisting of aggregate particles, surrounded by a transition zone, embedded in the cement paste matrix. Hashin and Monteiro[8] developed a mathematical model based on the following assumption: concrete is a composite consisting of a matrix in which are embedded spherical particles, each of which is surrounded by a concentric spherical shell, which will be called the *interphase*. Matrix, particle, and interphase materials were considered elastic isotropic and the entire composite was assumed to be statistically homogeneous and isotropic. First the authors considered the classical problem of analytical determination of the effective elastic properties of the composite in terms of constituent properties and internal geometry. Then the model was used to determine of interphase properties in terms of particle and matrix properties and effective properties. Using experimental data, their analysis indicated that the shear and Young's moduli of the interphase are about 50 percent of those of the original bulk cement paste, while the bulk modulus is on the order of 70 percent of the original bulk cement paste.

When studying the effect of pores and cracks, sophisticated models are needed that explicitly consider the shape of the "inclusions." Two of the most accurate models available for estimating the effect of pores and/or cracks on the elastic moduli are the *differential scheme* and the *Mori-Tanaka method*. Discussion on the rationale of these methods is beyond the scope of this chapter. For two critical idealized pore shapes, namely, the sphere and the "penny-shaped" crack, the results have a relatively simple form which are described below.

If a solid body of modulus E_0 and Poisson's ratio v_0 contains a volume fraction c of spherical pores, its overall moduli E will be[9]

$$\text{Differential method:} \quad E = E_0(1 - c)^2 \tag{13-20}$$

$$\text{Mori-Tanaka method:} \quad E = E_0(1 - c)/(1 + \alpha c) \tag{13-21}$$

where $\alpha = 3(1 + \nu_0)(13 - 15\nu_0)/6(7 - 5\nu_0)$. Parameter α is nearly independent of ν_0, and is approximately equal to 1. Therefore, the Mori-Tanaka method essentially predicts $E = E_0(1 - c)/(1 + c)$. The differential method also predicts a slight, but algebraically complicated, dependence of E on ν_0. This dependence is negligible for practical purposes, and has been omitted from Eq. (13-20).

Since a small volume fraction of very thin cracks can cause an appreciable degradation of the moduli, it is not convenient to quantify their concentration using volume fractions. If the body is filled with circular cracks, instead, we use the crack-density parameter Γ, which is defined by $\Gamma = Na^3/V$, where a is the radius of the crack in its plane, and N/V is the number of cracks per unit volume. The effective moduli of a body containing a density Γ of circular cracks are as follows for the two different models

$$\text{Differential method:} \qquad E = E_0 e^{-16\Gamma/9} \tag{13-22}$$

$$\text{Mori-Tanaka method:} \qquad E = E_0/(1+\beta\Gamma) \tag{13-23}$$

where $\beta = 16(10 - 3\nu_0)(1 - \nu_0^2)/45(2 - \nu_0)$. For typical values of ν_0, β is essentially equal to $16/9 = 1.78$. Again, the differential method predicts a negligible dependence of E on ν_0, which we have omitted from Eq. (13-20).

More general treatments of the effect of pores on the elastic moduli assume that the pores are oblate spheroids of a certain aspect ratio. The sphere (aspect ratio = 1) and the crack (aspect ratio = 0) are the two extreme cases. Figure 13-3 shows the elastic moduli as a function of porosity, for various pore aspect ratios, as calculated using the Mori-Tanaka model. The detailed equations used to generate this figure can be found in the references given at the end of the chapter.

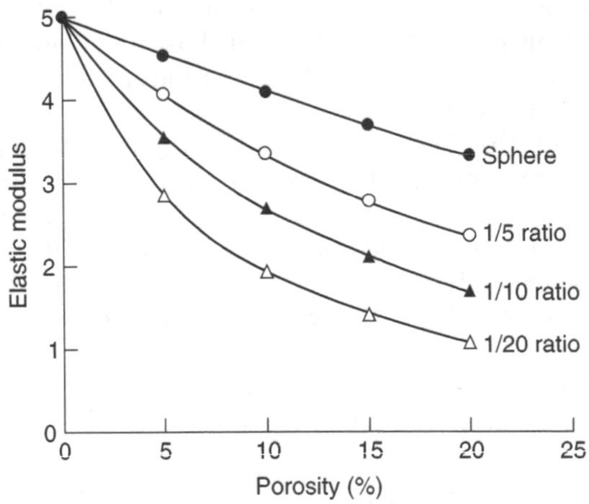

Figure 13-3 Effect of porosity and shape of pores on the elastic modulus of a material.

Transport Properties

This section has concentrated on various methods for estimating elastic modulus, however, other important properties can also be predicted using the theorems of composite materials. Consider the following relationships that have the same mathematical structure:

Electrical conduction: $j = \sigma E$

Thermal conduction: $Q = -\kappa \nabla T$

Dielectric displacement: $D = \varepsilon E$

Magnetic induction: $B = \mu H$

Diffusion: $Q = -D \nabla c$

For each of these five transport relationships, the flux vector is related to the driving force vector by a second-order physical property tensor, that is, a 3×3 matrix (σ, κ, ε, μ, D). For isotropic materials, the electrical conductivity σ, the thermal conductivity κ, the dielectric constant ε, the magnetic susceptibility μ, and the diffusion constant D reduce to a single constant. It should be noted that the elastic moduli is a fourth order tensor and, even for isotropic materials, contains two independent constants. Any model that can predict, say, diffusion constant D from the individual phases properties, will also be able to predict σ, κ, ε, and μ.

Hashin and Shtrikman derived the following bounds for transport constants. For thermal conductivity ($\kappa_2 > \kappa_1$), in the three-dimensional case we have for the upper bound:

Upper bound:

$$\kappa_u = \kappa_2 + \frac{c_1}{\dfrac{1}{\kappa_1 - \kappa_2} + \dfrac{c_2}{3\kappa_2}}$$

and for the lower bound

$$\kappa_1 = \kappa_1 + \frac{c_2}{\dfrac{1}{\kappa_2 - \kappa_1} + \dfrac{c_1}{3\kappa_1}}$$

The number 3 in the denominator should be replaced by 2 and 1 for two-dimensional or one-dimensional cases, respectively. Similar equations apply for the other transport constants.

13.1.1 Hashin-Shtrikman (H-S) Bounds

Although the Voigt and Reuss models produce an upper and lower bounds for the elastic moduli, as shown in Fig. 13-2, the bounds are often far apart, in which case they are of little use for certain specific cases. For instance, if we assume a volume fraction of 0.6 in Fig. 13-2, the upper and lower bounds are 63.9 and 47.9 MPa, respectively. The spread is large and, therefore, of limited use for engineering applications. Fortunately, Hashin and Shtrikman (H-S) developed more stringent bounds for a composite material, which in a statistical sense, is both isotropic and homogeneous. The H-S bounds were derived using variational principles of the linear theory of elasticity for multiphase materials of arbitrary phase geometry. For two-phase composites the expressions take the form:

$$K_{low} = K_1 + \frac{c_2}{\frac{1}{K_2 - K_1} + \frac{3c_1}{3K_1 + 4G_1}} \qquad K_{up} = K_2 + \frac{c_1}{\frac{1}{K_1 - K_2} + \frac{3c_2}{3K_2 + 4G_2}} \tag{13-24}$$

$$G_{low} = G_1 + \frac{c_2}{\frac{1}{G_2 - G_1} + \frac{6(K_1 + 2G_1)c_1}{5G_1(3K_1 + 4G_1)}} \qquad G_{up} = G_2 + \frac{c_1}{\frac{1}{G_1 - G_2} + \frac{6(K_2 + 2G_2)c_2}{5G_2(3K_2 + 4G_2)}} \tag{13-25}$$

where K and G are the bulk and shear moduli, respectively. Here $K_2 > K_1$; $G_2 > G_1$. K_{up} and G_{up} refer to the upper bounds and K_{low} and G_{low} to the lower bounds.

Figure 13-2 shows that the H-S bounds are inside the Voigt-Reuss bounds. Using the previous example for a volume fraction of 0.6, the H-S bounds give 58.4 and 54.0 MPa. The range is significantly narrower than that obtained using the Voigt-Reuss bounds.

13.2 Viscoelasticity

There are two methods used to study the one-dimensional viscoelastic behavior of concrete: (*a*) the creep test, where the stress is kept constant and the increase in strain over time is recorded, and (*b*) the relaxation test, where the strain is kept constant and the decrease in stress over time is recorded. Experimental results from both creep and relaxation tests are shown in Fig. 13-4, where the creep response is a function of the duration of loading and the age of concrete when the load was applied. The longer the concrete is under load, the greater the deformation, and the greater the age of loading, the lower the deformation. This behavior classifies concrete as an aging viscoelastic material. In fact, most of the mechanical properties of concrete are age-dependent. The mathematical formulation for aging materials is more complex than for nonaging materials; this section presents basic expressions for aging materials.

Creep and relaxation experiments are time-consuming, but worthwhile as they yield significant information about the viscoelasticity of the material. Contrary to elastic behavior where two constants are used to describe a homogeneous isotropic elastic material for viscoelastic behavior an evolution law is necessary to describe how the stress or strain changes over time. In this section, rheological models are presented that produce such evolution laws, in addition to some practical equations used in design codes. Rheological models will be used to provide some insight into the viscoelastic behavior of concrete, explaining for instance why the rate of stress decrease in the relaxation test is faster than the rate of strain increase in the creep test.

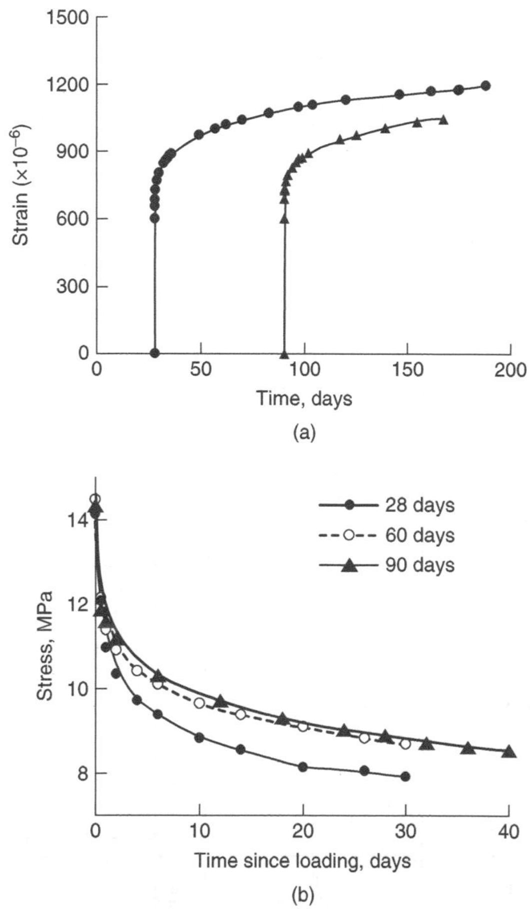

FIGURE 13-4 (*a*) Creep test; (*b*) relaxation test of concrete.

(*a*) Creep tests with a constant stress of 14.5 MPa loaded at 28 and 90 days; (*b*) relaxation tests performed at 28, 60, and 90 days. All the specimens had the same composition. The original data are from Thomas, K., D. Pirtz, and P. J. M. Monteiro, Proceedings of the ACI Journal, Vol. 83, p. 433, 1986.

Unfortunately, in real concrete structures the state of stress or strain is unlikely to be constant over time. To model more complex loading conditions, the principle of superposition and integral representations are presented. These methods allow to compute the strain if the creep function and stress history are known or to compute the stress if the relaxation function and strain history are known.

If no experimental data are available (i.e., creep or relaxation test results), the recommendations of a code or a model are used: FIB model code 2010, ACI-209, and the Bazant-Panula model. For technological aspects of the viscoelastic behavior of concrete, refer to Chap. 4.

13.2.1 Basic Rheological Models

The behavior of viscoelastic materials can be successfully estimated by the creation of rheological models based on two fundamental elements: the linear spring and the linear viscous dashpot. For the linear spring (see Table 13-1*a*) the relationship between stress and strain is given by Hooke's law:

$$\sigma(t) = E\varepsilon(t) \tag{13-26}$$

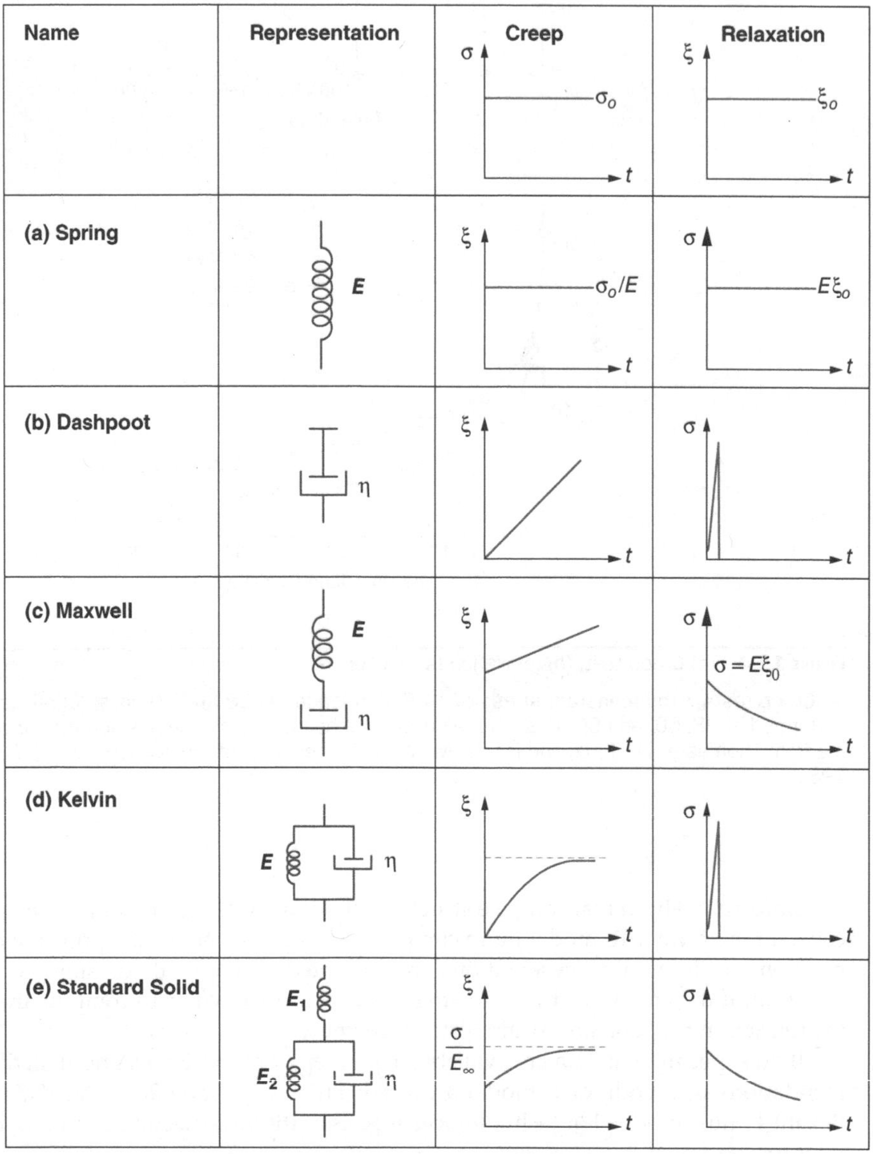

TABLE 13-1 Simple Rheological Models and Their Creep and Relaxation Response

The response of the spring to the stress is immediate. During a creep test, where the stress σ_0 is kept constant, the stain will be σ_0/E, constant over time. Similarly, for a relaxation test, where the strain ε_0 is kept constant the stress will be $\varepsilon_0 E$, constant over time.

The viscous dashpot can be visualized as a piston displacing a viscous fluid in a cylinder with a perforated bottom. Newton's law of viscosity:

$$\dot{\varepsilon}(t) = \frac{\sigma(t)}{\eta} \tag{13-27}$$

where $\dot{\varepsilon} = \dfrac{d\varepsilon}{dt}$ = the strain rate

η = the viscosity coefficient

states that the strain rate is proportional to the stress. Therefore, for the creep experiment, the dashpot will deform at a constant rate, as shown in Table 13-1*b*. For a relaxation experiment with the application of an instantaneous constant strain, the stress becomes instantaneously infinite, as indicated in Table 13-1*b*.

Complex formulations can be obtained by combining springs and dashpots in different configurations. One of the simplest combinations consists of assembling one spring and one dashpot in series or in parallel. The *Maxwell model* comprises a linear spring and a linear viscous dashpot connected in series, as shown in Table 13-1*c*. The following equations apply:

$$\text{Equilibrium equation} \qquad \sigma_E(t) = \sigma_\eta(t) = \sigma(t) \tag{13-28}$$

$$\text{Compatibility equation} \qquad \varepsilon(t) = \varepsilon_E(t) + \varepsilon_\eta(t) \tag{13-29}$$

$$\text{Constitutive relationship (spring)} \qquad \sigma_E(t) = E\varepsilon_E(t) \tag{13-30}$$

$$\text{(dashpot)} \qquad \sigma_\eta(t) = \eta\dot{\varepsilon}_\eta(t) \tag{13-31}$$

Differentiating Eqs. (13-29) and (13-30) with respect to time t and using Eqs. (13-28) and (13-31):

$$\dot{\varepsilon}(t) = \frac{\dot{\sigma}(t)}{E} + \frac{\sigma(t)}{\eta} \tag{13-32}$$

Note that for a rigid spring ($E = \infty$), the model reduces to a Newtonian fluid; likewise, if the dashpot becomes rigid ($\eta = \infty$), the model reduces to a Hookean spring. The response of the Maxwell model to various kinds of time-dependent stress or strain patterns can be determined by solving Eq. (13-32). For instance, consider again a creep test, with the initial conditions $\sigma = \sigma_0$ at $t = 0$. Integrating Eq. (13-32), we obtain:

$$\varepsilon(t) = \frac{\sigma_0}{E} + \frac{\sigma_0 t}{\eta} \tag{13-33}$$

The model predicts that the strain increases without bounds. This is characteristic of many fluids; therefore, a material described by Eq. (13-32) is known as a "Maxwell" fluid. When the system is unloaded at time t_1 the elastic strain σ_0/E in the spring recovers instantaneously, while a permanent strain $(\sigma_0/\eta)t_1$ remains in the dashpot.

In a relaxation experiment, where the strain ε_0 is constant, the model predicts:

$$\sigma(t) = E\varepsilon_0 e^{-Et/\eta} \tag{13-34}$$

The ratio $T = \eta/E$ is called the *relaxation time*, and it helps characterize the viscoelastic response of the material. A small relaxation time indicates that the relaxation process will be fast. In the extreme case of a purely viscous fluid, $E = \infty$, Eq. (13-34) would indicate an infinitely fast stress relaxation, $T = 0$; while for an elastic spring, $\eta = \infty$, the stress would not relax at all, since $T = \infty$.

The *Kelvin model* combines a linear spring and a dashpot in parallel as shown in Table 13-1*d*. The following equations apply:

Equilibrium equation $$\sigma(t) = \sigma_E(t) + \sigma_\eta(t) \tag{13-35}$$

Compatibility equation $$\varepsilon(t) = \varepsilon_E(t) = \varepsilon_\eta(t) \tag{13-36}$$

Constitutive relationship (spring) $$\sigma_E(t) = E\varepsilon_E(t) \tag{13-37}$$

(dashpot) $$\sigma_\eta(t) = \eta\dot{\varepsilon}_\eta(t) \tag{13-38}$$

Resulting in the differential equation $$\sigma(t) = E\varepsilon(t) + \eta\dot{\varepsilon}(t) \tag{13-39}$$

Note that the model reduces to a Hookean spring if $\eta = 0$, and to a Newtonian fluid if $E = 0$. Equation (13-39) may be used to predict strain if the stress history is given or to predict stress if the strain history is given. For instance, for the creep experiment, integrating Eq. (13-39) with the boundary condition $\sigma = \sigma_0$ at time $t_0 = 0$ yields:

$$\varepsilon(t) = \frac{\sigma_0}{E}(1 - e^{-Et/\eta}) \tag{13-40}$$

In Eq. (13-40), the strain increases at a decreasing rate and has an asymptotic value of σ_0/E, as shown in Table 13-1*d*. During the creep test the stress is initially carried by the dashpot and, as time goes by, the stress is transferred to the spring. Analogous to the relaxation time, we define the *retardation time* as $\theta = \eta/E$. A small retardation time indicates that the creep process will be fast. In the extreme case of an elastic spring ($\eta = 0$), the final strains would be obtained instantaneously since $\theta = 0$.

The Kelvin model requires an infinite stress to produce the instantaneous strain necessary for the relaxation test, which makes it physically impossible to perform.

The Maxwell and Kelvin models have significant limitations in representing the behavior of most viscoelastic materials. As discussed before, the Maxwell model shows a constant strain rate under constant stress, which may be adequate for fluids, but not for solids. The Kelvin model cannot predict a time-dependent relaxation and does not show a permanent deformation upon unloading.

A more complex, representative model is the *standard solid* model, where a spring is connected in series with a Kelvin element as shown in Table 13-1*e*.

Assuming ε_1 and ε_2 to be the strain in the spring and Kelvin elements, respectively, the total strain, for the standard solid, is given by

$$\varepsilon = \varepsilon_1 + \varepsilon_2 \tag{13-41}$$

Since the stress in the spring and the Kelvin element is the same, the stress can be determined using Eq. (13-39):

$$\sigma(t) = E_2\varepsilon_2(t) + \eta\frac{\partial\varepsilon_2(t)}{\partial t} \tag{13-42}$$

where $\partial/\partial t$ is a differential operator that may be handled as an algebraic entity,

$$\sigma(t) = \varepsilon_2(t)\left(E_2 + \eta\frac{\partial}{\partial t}\right) \tag{13-43}$$

leading to

$$\varepsilon_2(t) = \frac{\sigma(t)}{\left(E_2 + \eta\frac{\partial}{\partial t}\right)} \tag{13-44}$$

Therefore, we can obtain the strain for the standard solid by using Eq. (13-41)

$$\varepsilon(t) = \frac{\sigma(t)}{E_1} + \frac{\sigma(t)}{\left(E_2 + \eta\frac{\partial}{\partial t}\right)} \tag{13-45}$$

or

$$E_1\varepsilon(t)\left(E_2 + \eta\frac{\partial}{\partial t}\right) = E_1\sigma(t) + \sigma(t)\left(E_2 + \eta\frac{\partial}{\partial t}\right) \tag{13-46}$$

which leads to the differential equation

$$\eta E_1\dot{\varepsilon}(t) + E_1E_2\varepsilon(t) = \eta\dot{\sigma}(t) + (E_1 + E_2)\sigma(t) \tag{13-47}$$

Equation (13-47) can be integrated for an arbitrary stress history,

$$\varepsilon(t) = \frac{\sigma(t)}{E_1} + \frac{1}{\eta}\int_0^t \sigma(\tau)e^{-E_2(t-\tau)/\eta}\,d\tau \tag{13-48}$$

For the particular case of the creep experiment, Eq. (13-48) reduces to

$$\varepsilon(t) = \frac{\sigma_0}{E_1} + \frac{\sigma_0}{E_2}[1 - e^{-E_2t/\eta}] \tag{13-49}$$

which can be rewritten as

$$\varepsilon(t)=\sigma_0\left(\frac{E_1+E_2}{E_1E_2}-\frac{1}{E_2}e^{-E_2t/\eta}\right) \tag{13-50}$$

Equation (13-50) indicates that the strain is proportional to σ_0, changing from σ_0/E_1 at $t=0$ to σ_0/E_∞ at $t=\infty$. E_∞ is called the asymptotic modulus and is given by

$$E_\infty=\frac{E_1E_2}{E_1+E_2} \tag{13-51}$$

During the creep test, the elastic modulus of the standard solid model, $E_c(t)$, reduces from the initial value E_1 to its asymptotic value E_∞, according to the following law:

$$\frac{1}{E_c(t)}=\frac{\varepsilon(t)}{\sigma_0}=\frac{E_1+E_2}{E_1E_2}-\frac{1}{E_2}e^{-E_2t/\eta} \tag{13-52}$$

We now integrate Eq. (13-47) for an arbitrary strain history

$$\sigma(t)=\varepsilon(t)E_\infty+(E_1-E_\infty)\int_0^t e^{-(E_1+E_2)(t-\tau)/\eta}\,\dot{\varepsilon}(\tau)d\tau \tag{13-53}$$

In the particular case of relaxation experiment the stress evolution is given by

$$\sigma(t)=\varepsilon_0\left[E_\infty+(E_1-E_\infty)e^{-(E_1+E_2)t/\eta}\right] \tag{13-54}$$

Equation (13-54) indicates that the stress is proportional to ε_0 changing from $E_1\varepsilon_0$ at $t=0$, up to $E_\infty\varepsilon_0$ at $t=\infty$. Therefore, during the relaxation test the elastic modulus $E_r(t)$, reduces from the initial value E_1, to its asymptotic value E_∞, according to the following law:

$$E_r(t)=\left[E_\infty+(E_1-E_\infty)e^{-(E_1+E_2)t/\eta}\right] \tag{13-55}$$

Even though both creep and relaxation may be understood as a decrease in elastic modulus over time from E_1 to its asymptotic value E_∞, Eqs. (13-50) and (13-53) have different rates of decrease. In a relaxation test, the decrease in the elastic modulus occurs at significantly faster rate than in the creep test. As an example, let us take the following values for concrete: $E_1=35$ GPa, $E_2=18$ GPa, $T(E_2/\eta)=1/300$ days. Figure 13-5 illustrates the faster reduction of elastic modulus during relaxation than for the creep test.

Example 13-1 The testing of materials is usually performed either by controlling the stress or strain rate. Study the response of a standard solid model loaded under these conditions. Solve the problem analytically and then expand the discussion for instantaneous, slow, and medium stress and strain rates; assume the following properties for the standard solid: $E_1=35$ GPa, $E_2=18$ GPa, $T=1$ min.

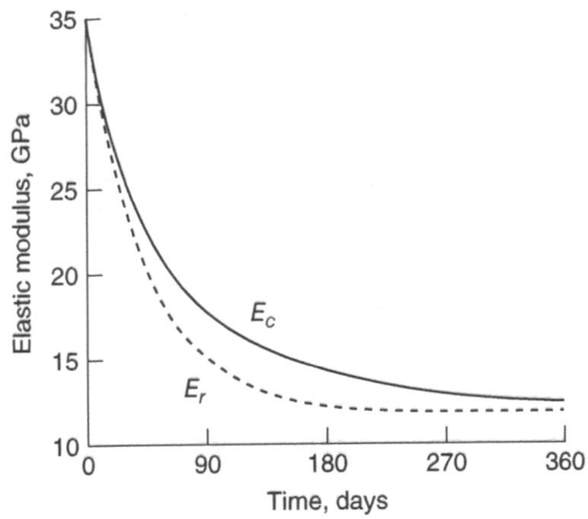

FIGURE 13-5 Decrease of elastic modulus for relaxation and creep.

(A) Test with a constant stress rate (v): The stress increases linearly with time, according to

$$\sigma(t) = vt \tag{13-56}$$

The strain in the standard solid model is obtained by combining Eqs. (13-48) and (13-56),

$$\varepsilon(t) = \frac{vt}{E_1} + \frac{v}{\eta}\int_0^t \tau e^{-E_2(t-\tau)/\eta}\, d\tau \tag{13-57}$$

which leads to

$$\varepsilon(t) = \frac{vt}{E_\infty} - \frac{v\eta}{E_2^2}(1 - e^{-E_2 t/\eta}) \tag{13-58}$$

Figure 13-6 presents the stress [Eq. (13-56)] as a function of strain [Eq. (13-58)] using the given material properties, showing that the stress-strain diagram is strongly influenced by the rate of loading.

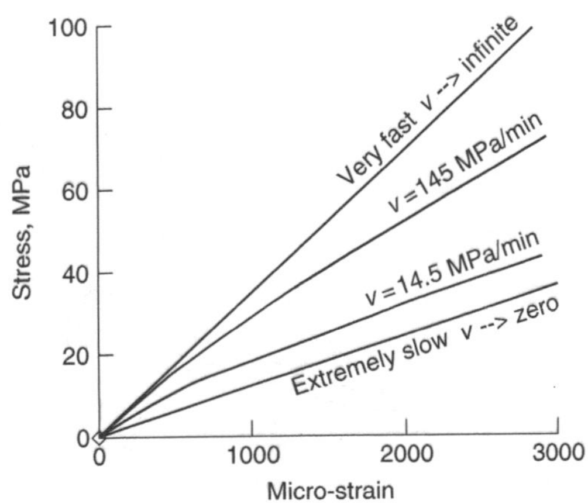

FIGURE 13-6 Effect of stress rate on the stress-strain diagram.

Note that the stress-strain diagram may be nonlinear, a common feature for viscoelastic materials where the strain at a given time is influenced by the entire stress history. This phenomenon will be presented in more detail in the following sections.

The stress-strain relationships shown in Fig. 13-6 are bounded by very slow and very fast rates. The latter gives the upper bound and physically corresponds to the linear spring (E_1) absorbing all the stress, as the Kelvin element has no time to deform. For very slow rates, the standard solid model responds with the asymptotic modulus E_∞, and physically corresponds to the linear spring E_1 in series with the spring from the Kelvin element E_2 the dashpot does not contribute to the stiffness of the system.

(B) Test with constant strain rate: The strain increases with time, according to

$$\varepsilon(t) = vt \tag{13-59}$$

The stress in the model is obtained by combining Eqs. (13-53) and (13-59),

$$\sigma(t) = vtE_\infty + (E_1 - E_\infty)\int_0^t ve^{-(E_1+E_2)(t-\tau)/\eta}\,d\tau \tag{13-60}$$

which leads to

$$\sigma(t) = vtE_\infty + (E_1 - E_\infty)v\frac{\eta}{(E_1+E_2)}(1-e^{-(E_1+E_2)t/\eta}) \tag{13-61}$$

Figure 13-7 shows the stress [Eq. (13-61)] in function of strain [Eq. (13-59)] with the specified concrete properties.

Example 13-2 Study the response of a viscoelastic material subjected to a cyclic strain $\varepsilon(t) = \varepsilon_0 \cos wt$, where ε_0 is the strain amplitude and w the frequency. Write explicit equations for the Maxwell and Kelvin models.

For a linear elastic spring, the stress will be in phase with the cyclic strain that is

$$\sigma(t) = E\varepsilon(t) = E\varepsilon_0 \cos wt \tag{13-62}$$

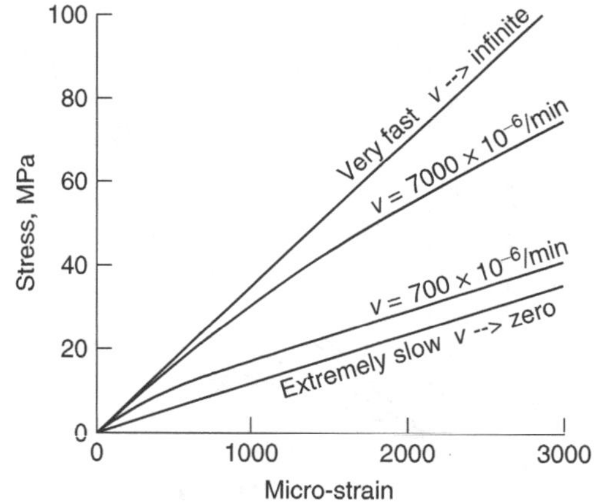

FIGURE 13-7 Effect of strain rate on the stress-strain diagram.

For a Newtonian fluid the stress will lead the strain by $\pi/2$:

$$\sigma(t) = \eta\dot{\varepsilon}(t) = -\eta w\varepsilon_0 \sin wt = \eta w\varepsilon_0 \cos(wt + \delta) \tag{13-63}$$

where

$$\delta = \frac{\pi}{2}$$

For a viscoelastic material the phase difference between stress and strain ranges from 0 to $\pi/2$. A convenient way of representing oscillatory strain is by using the expression:

$$e^{iwt} = \cos wt + i \sin wt \tag{13-64}$$

Taking the real part of the expression, the strain equation can be rewritten as

$$\varepsilon(t) = \varepsilon_0 \, e^{iwt} \tag{13-65}$$

The stress oscillates with the same frequency w, but leads the strain by a phase angle δ, where

$$\sigma(t) = \sigma_0 \, e^{i(wt+\delta)} \tag{13-66}$$

which can be rewritten as

$$\sigma(t) = \sigma_0 \, e^{i\delta} \, e^{iwt} = \sigma^* e^{iwt} \tag{13-67}$$

where σ^* is the complex stress amplitude given by

$$\sigma^* = \sigma_0 \, e^{i\delta} = \sigma_0(\cos\delta + i \sin\delta) \tag{13-68}$$

A complex modulus E^* can be defined as

$$E^* = \frac{\sigma^*}{\varepsilon_0} = \frac{\sigma_0(\cos\delta + i\sin\delta)}{\varepsilon_0} = E_1 + iE_2 \tag{13-69}$$

where E_1, the storage modulus, is in phase with the strain, and is given by

$$E_1 = \frac{\sigma_0}{\varepsilon_0}\cos\delta \tag{13-70}$$

E_2, the loss modulus, is the imaginary part, and is given by

$$E_2 = \frac{\sigma_0}{\varepsilon_0}\sin\delta \tag{13-71}$$

and the magnitude of the complex modulus is given by

$$|E^*| = \sqrt{E_1^2 + E_2^2} \tag{13-72}$$

It should be noted that

$$\tan\delta = \frac{E_2}{E_1} \tag{13-73}$$

represents the mechanical loss per cycle of strain.

For the Maxwell model: The constitutive equation for the Maxwell model is given by Eq. (13-32)

$$\sigma + \frac{\eta}{E}\dot{\sigma} = \eta\dot{\varepsilon} \tag{13-74}$$

Using Eqs. (13-65) and (13-67) we obtain

$$\sigma_0 e^{i\delta}\left(1 + iw\frac{\eta}{E}\right) = iw\varepsilon_0\eta \tag{13-75}$$

or

$$\sigma^*\left(1 + iw\frac{\eta}{E}\right) = iw\varepsilon_0\eta \tag{13-76}$$

Therefore, the complex modulus can be expressed by

$$E^* = \frac{\sigma^*}{\varepsilon_0} = \frac{iw\eta}{1 + \frac{iw\eta}{E}} \tag{13-77}$$

Separating the real and imaginary parts we find

$$E^* = \frac{\eta^2 w^2 / E}{1 + \eta^2 w^2 / E^2} + i\frac{\eta w}{1 + \eta^2 w^2 / E^2} \tag{13-78}$$

Hence the magnitude of the complex modulus is given by

$$|E^*| = w\eta\left(1 + \frac{\eta^2 w^2}{E^2}\right)^{-1/2} \tag{13-79}$$

and

$$\tan\delta = \frac{E_2}{E_1} = \frac{E}{w\eta} \tag{13-80}$$

Taking the material constants from the previous example, the magnitude of complex modulus can be plotted against the angular frequency, as shown in Fig. 13-8. Note that for very high frequencies

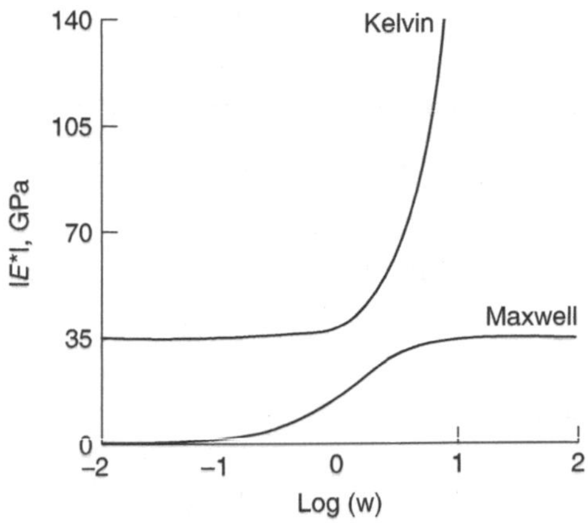

FIGURE 13-8 Complex elastic modulus in function of frequency.

the dynamic modulus approaches the spring constant E and for the low frequencies the dynamic modulus approaches zero.

For the Kelvin model: The constitutive equation for the Kelvin model is given by Eq. (13-39)

$$\sigma(t) = E\varepsilon(t) + \eta\dot{\varepsilon}(t) \tag{13-81}$$

Using Eqs. (13-66) and (13-81), we obtain

$$\sigma_0 e^{i\delta} = \varepsilon_0(E + iw\eta) \tag{13-82}$$

Therefore, the complex modulus is expressed by

$$E^* = \frac{\sigma^*}{\varepsilon_0} = E + iw\eta \tag{13-83}$$

and the magnitude of the complex modulus by

$$|E^*| = (E^2 + w^2\eta^2)^{1/2} \tag{13-84}$$

The mechanical loss for the model is

$$\tan\delta = \frac{\eta}{E}w \tag{13-85}$$

Again, if we take the material constants from the previous example, the results for the Kelvin model can be plotted, as shown in Fig. 13-8. Note that for low.frequencies the dynamic modulus is given by the spring constant E, while for high frequencies the stiffness increases.

The significantly different responses for the Maxwell and Kelvin models under oscillatory stress points to the advantage of performing such a test to assess which model is most representative for a specific material.

13.2.2 Generalized Rheological Models

The modeling of viscoelastic behavior can be improved by combining a large number of springs and dashpots in series or in parallel. By adding many elements, several relaxation times can be obtained, which is characteristic of complex materials such as concrete.

When generalizing the Maxwell model, we must choose to connect the units either in series or in parallel. Let us start by studying the response when the units are connected in series, as shown in Fig. 13-9. The constitutive equation has the form:

$$\dot{\varepsilon}(t) = \dot{\sigma}(t)\sum_{i=1}^{n}\frac{1}{E_i} + \sigma(t)\sum_{i=1}^{n}\frac{1}{\eta_i} \tag{13-86}$$

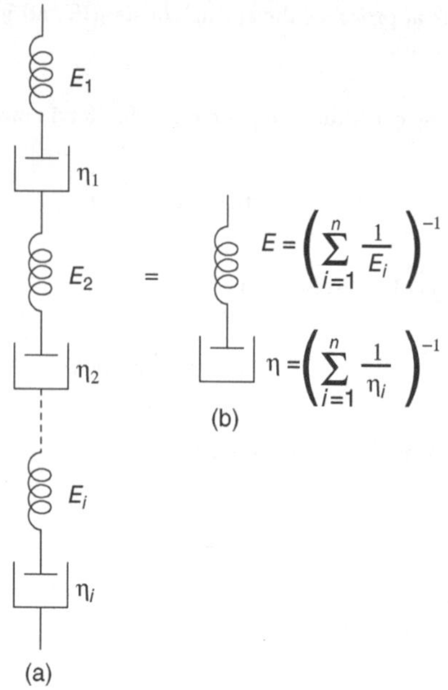

FIGURE 13-9 Generalized Maxwell model in series.

where n is the number of elements. Because the equation is equivalent to Eq. (13-32), the chain of elements is identical to a single Maxwell element, as shown in Fig 13-9*b*, therefore not much was accomplished by connecting the units in series.

Let us now analyze the response when the units are connected in parallel, as shown in Fig. 13-10*b*.

The strain in each unit of a generalized Maxwell model in parallel is given by

$$\frac{\partial}{\partial t}\varepsilon_i(t) = \left\{\frac{1}{E_i}\frac{\partial}{\partial t} + \frac{1}{\eta_i}\right\}\sigma_i(t) \tag{13-87}$$

The stress for the generalized model is given by

$$\sigma(t) = \left\{\sum_{i=1}^{n} \frac{\partial/\partial t}{\frac{1}{E_i}\frac{\partial}{\partial t} + \frac{1}{\eta_i}}\right\}\varepsilon(t) \tag{13-88}$$

and the relaxation function for the generalized Maxwell model is

$$E(t-\tau) = \sum_{i=1}^{n} E_i\{\exp-(t-\tau)/T_i\} \tag{13-89}$$

indicating that the response of the material depends on a distribution of relaxation times. This formulation is useful in modeling complex viscoelastic materials.

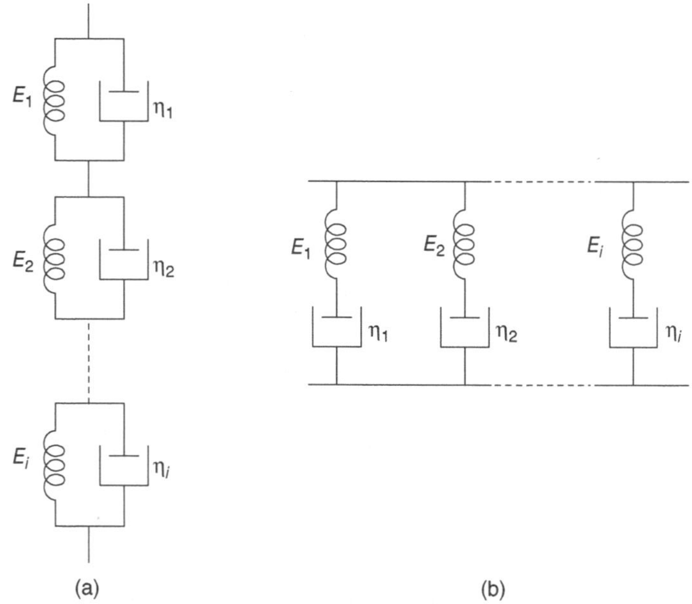

FIGURE 13-10 (*a*) Generalized Kelvin model in series and (*b*) generalized Maxwell model in parallel.

When generalizing the Kelvin model the same question arises: Should we connect the units in series or parallel? We start with the units connected in parallel, as shown in Fig. 13-11. The constitutive equation for the model has the form

$$\sigma(t) = \varepsilon(t)\sum_{i=1}^{n} E_i + \dot{\varepsilon}(t)\sum_{i=1}^{n} \eta_i \tag{13-90}$$

which has the same form as a Kelvin element shown in Fig. 13-11.

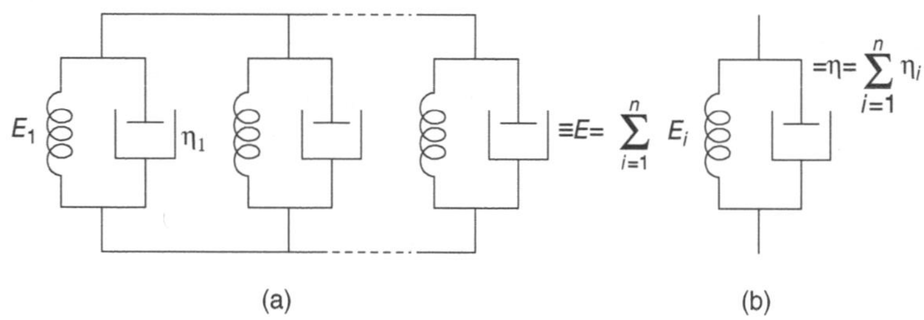

FIGURE 13-11 Generalized Kelvin model in parallel.

Consider a generalized Kelvin model in series (see Fig. 13-10*a*). The stress in each unit is given by

$$\sigma_i(t) = \left(E_i + \eta_i \frac{\partial}{\partial t}\right)\varepsilon_i(t) \tag{13-91}$$

The strain for the generalized model is given by

$$\varepsilon(t) = \sum_{i=1}^{n} \left\{ \frac{1}{E_i + \eta_i \frac{\partial}{\partial t}} \right\} \sigma(t) \tag{13-92}$$

Equations (13-87) and (13-92) are differential equations of the general form

$$\sum_{i=1}^{h} p_i \frac{d^i\sigma}{dt^i} = \sum_{i=1}^{l} q_i \frac{d^i\varepsilon}{dt^i} \tag{13-93}$$

The specific creep function for the generalized Kelvin model in series is

$$\Phi(t-\tau) = \sum_{i=1}^{n} \frac{1}{E_i}\{1 - \exp(-t-\tau)/\theta_i\} \tag{13-94}$$

In order to model the material's response adequately, the spring constants E_i and the dashpot constants η_i should vary over a large range. Sometimes when modeling a fluid or a solid, it is convenient to take some limiting value for the spring or dashpot constant. It should be noted that a Maxwell model with infinite spring constant or a Kelvin model with zero spring constant becomes a dashpot. Conversely a Maxwell model with infinite viscosity or a Kelvin model with zero viscosity results in a spring.

13.2.3 Time-Variable Rheological Models

Concrete changes its mechanical properties with time due to hydration reaction. In the models presented so far, however, the elastic modulus E and the viscosity coefficient η are constant over time. Consequently they have limited success in modeling the complex response of concrete. To include aging of concrete, we will now study how the differential equations for the basic elements—the spring and dashpot—change when their mechanical properties change with time.

Consider a linear spring with elastic modulus varying in time. Hooke's law can be expressed in two forms:

$$\sigma(t) = E(t)\varepsilon(t) \tag{13-95}$$

and

$$\dot{\sigma}(t) = E(t)\dot{\varepsilon}(t) \tag{13-96}$$

The equations are not equivalent. Solid mechanics literature defines a body following Eq. (13-95) to be elastic, whereas a body following Eq. (13-96) to be hypoelastic.

A linear viscous dashpot with viscosity coefficient varying in time is expressed unequivocally by

$$\sigma(t) = \eta(t)\dot{\varepsilon}(t) \tag{13-97}$$

If we reconstruct the previous models (Maxwell, Kelvin, standard-solid, generalized) for aging materials such as concrete, the equations for a Maxwell element with a hypoelastic spring or with an elastic spring are given by

$$\dot{\varepsilon}(t) = \frac{\dot{\sigma}(t)}{E(t)} + \frac{\sigma(t)}{\eta(t)} \tag{13-98}$$

$$\dot{\varepsilon}(t) = \frac{\dot{\sigma}(t)}{E(t)} + \left(\frac{1}{\eta(t)} + \frac{d}{dt}\frac{1}{E(t)}\right)\sigma(t) \tag{13-99}$$

Note that Eqs. (13-98) and (13-99) may be expressed as

$$\dot{\varepsilon}(t) = q_0(t)\dot{\sigma}(t) + q_1(t)\sigma(t) \tag{13-100}$$

where $q_0(t)$ and $q_1(t)$ are independent functions of time. Equation (13-100) represents the constitutive law for the Maxwell model with either an elastic or a hypoelastic spring.

Dischinger[10] used the aging Maxwell element, Eq. (13-98), to derive the so-called rate-of-creep method. The specific creep function $\Phi(t, \tau)$, that is the strain per unit stress at time t for the stress applied at age τ, is given by

$$\Phi(t,\tau) = \frac{1}{E(\tau)} + \int_\tau^t \frac{d\tau'}{\eta(\tau')} \tag{13-101}$$

The creep coefficient $\varphi(t, \tau)$ representing the ratio between the creep strain and the initial elastic deformation is

$$\varphi(t,\tau) = E(\tau)\int_\tau^t \frac{d\tau'}{\eta(\tau')} \tag{13-102}$$

Equation (13-98) can be expressed in function of the creep coefficient as

$$\frac{\partial\varepsilon}{\partial\varphi} = \frac{1}{E(t)}\frac{\partial\sigma}{\partial\varphi} + \frac{\sigma}{E(\tau)} \tag{13-103}$$

The Dischinger formulation implies that the creep curves are parallel for all ages. Experimental results do not indicate that the assumption is valid, as evident in Fig. 13-4*a*, where the creep curves are not parallel. Usually this method substantially underestimates the creep for stresses applied at ages greater than τ.

A Kelvin element with an elastic spring is described by

$$\sigma(t) = E(t)\varepsilon(t) + \eta(t)\dot{\varepsilon}(t) \tag{13-104}$$

and for a hypoelastic spring by

$$\dot{\sigma}(t) = [E(t) + \dot{\eta}(t)]\dot{\varepsilon}(t) + \eta(t)\ddot{\varepsilon}(t) \tag{13-105}$$

Equations (13-104) and (13-105) are not equivalent.

(a) $E - K_e$	(b) $E - K_h$
(c) $H - K_h$	(d) $H - K_e$
(e) $M // E$	(f) $M // H$

TABLE 13-2 Possible combinations for the standard solid to model an aging material.

Let us now consider the standard solid. Previously for nonaging materials, we solved the model for the Kelvin element in series with a spring. The same differential equation would have been obtained for a Maxwell element in parallel with a spring. For aging materials, the number of combinations for the standard solid greatly increases, as indicated in Table 13-2.[11]

With the notation:

E = elastic spring
H = hypoelastic spring
K_e, K_h = Kelvin element with elastic and hypoelastic spring, recpectively
M = Maxwell element [Note: Eq. (13-100) satisfies both springs]
–, // = series and parallel configurations

As an example, let us solve case (a) which was previously analyzed for a nonaging material. In this model we have an elastic spring with an elastic modulus $E_1(t)$ in series with a Kelvin model with an elastic modulus $E_2(t)$ and a dashpot of viscosity $\eta(t)$. Let ε_1 and ε_2 denote the strains of the spring and of the Kelvin element, respectively. Therefore,

$$\varepsilon(t) = \varepsilon_1(t) + \varepsilon_2(t) \qquad \varepsilon_1(t) = \sigma(t)/E_1(t) \tag{13-106}$$

$$E_2(t)\varepsilon_2(t) + \eta(t)\dot{\varepsilon}_2(t) = \sigma(t) \tag{13-107}$$

Eliminating ε_1 and ε_2, we obtain:

$$E_2(t)\varepsilon(t) + \eta(t)\dot{\varepsilon}(t) = \left(1 + \frac{E_2(t)}{E_1(t)} + \eta \frac{d}{dt}\frac{1}{E_1(t)}\right)\sigma(t) + \frac{\eta(t)}{E_1(t)}\dot{\sigma}(t) \tag{13-108}$$

These aging models can be generalized to obtain the following differential constitutive equation:

$$\left(\frac{d^n}{dt^n} + p_1(t)\frac{d^{n-1}}{dt^{n-1}} + \cdots + p_n(t)\right)$$
$$\varepsilon(t) = \left(q_0(t)\frac{d^n}{dt^n} + q_1(t)\frac{d^{n-1}}{dt^{n-1}} + \cdots + q_n(t)\right)\sigma(t) \tag{13-109}$$

It should be mentioned that models having two or more Maxwell elements in parallel or Kelvin elements in series will not, in general, lead to a differential equation, but rather to an integro-differential equation.

13.2.4 Superposition Principle and Integral Representation

In the lifetime of a concrete structure it is unlikely that the load will be kept constant as in a creep test nor will the strain be kept constant, as in a relaxation test. In order to estimate the strain at a given time from a known stress history further assumptions are necessary. McHenry made a significant contribution by postulating the following *Principle of Superposition:*[12]

"The strains produced in concrete at any time t by a stress increment at any time t_0 are independent of the effects of any stress applied either earlier or later than t_0. The stresses that approach the ultimate strength are excluded."

Experimental results indicate that the principle of superposition works well for sealed concrete specimens, that are for basic creep. When creep is associated with drying shrinkage other methods should be used.

The principle of superposition may also be formulated as follows "the effect of sum of causes is equal to sum of effects of each of these causes."[13] Consider $\varepsilon_1(\tau)$ and $\varepsilon_2(\tau)$, the strains resulting from the stress history $\sigma_1(\tau)$ and $\sigma_2(\tau)$, respectively. For a *linear* viscoelastic material we simply add the two stress histories

$$\sigma(\tau) = \sigma_1(\tau) + \sigma_2(\tau) \tag{13-110}$$

Using the principle of superposition, the following strain history is obtained:

$$\varepsilon(\tau) = \varepsilon_1(\tau) + \varepsilon_2(\tau) \tag{13-111}$$

Next, by using the principle of superposition and a known creep function, we can determine at any time the strain for a given stress history. For a creep test we may write the strain $\varepsilon(t)$ as a function of the stress σ_0, time t, and age of loading τ,

$$\varepsilon(t) = \Phi(\sigma_0, t, \tau) \tag{13-112}$$

In the linear range Eq. (13-112) may be written as

$$\varepsilon(t) = \sigma_0\,\Phi(t, \tau) \tag{13-113}$$

where $\Phi(t, \tau)$ is the specific creep function.

Figure 13-12 shows an arbitrary stress changing with time. Breaking the stress history up into small intervals, we have

$$\sigma(t) \cong \sum_{i=0}^{n} \Delta\sigma(\tau_i), \qquad \tau_n = t \tag{13-114}$$

Using Eq. (13-113), the strain history is given by

$$\varepsilon(t) \cong \sum_{i=0}^{n} \Delta\sigma(\tau_i)\Phi(t, \tau) \tag{13-115}$$

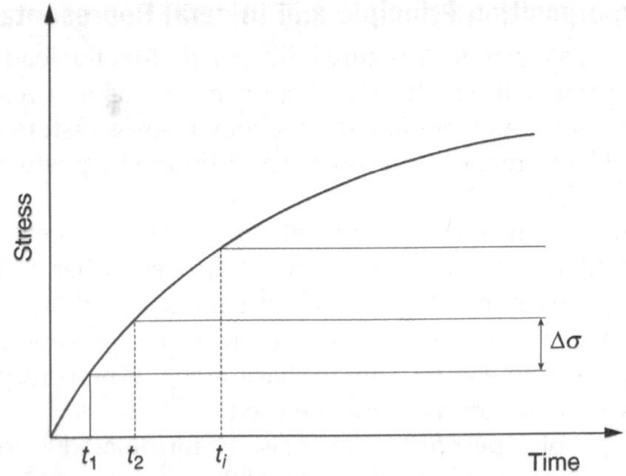

FIGURE 13-12 Incremental application of load over time.

and in the limit

$$\varepsilon(t) = \int_{\tau_0}^{t} \Phi(t,\tau)d\sigma(\tau) \tag{13-116}$$

Equation (13-116) is often referred to as the hereditary or Volterra integral. It shows that at time t the strain $\varepsilon(t)$ not only depends on the stress $\sigma(t)$ but rather on the whole stress history. Integrating Eq. (13-116) by parts we obtain

$$\varepsilon(t) = \frac{\sigma(t)}{E(t)} - \int_{\tau_0}^{t} \sigma(\tau)\frac{\partial\Phi(t,\tau)}{\partial\tau}d\tau \tag{13-117}$$

where $E(t) = 1/\Phi(t,t)$.

Our next objective is to compute the stress for a given strain history and relaxation function $E(t, \tau)$. Equations analogous to Eqs. (13-115) and (13-116) can be formulated.

$$\sigma(t) = \int_{\tau_0}^{t} E(t,\tau)\dot{\varepsilon}(\tau)d\tau \tag{13-118}$$

$$\sigma(t) = E(t)\varepsilon(t) - \int_{\tau_0}^{t} \varepsilon(\tau)\frac{\partial E(t,\tau)}{\partial\tau}d\tau \tag{13-119}$$

where $E(t) = E(t,t)$.

13.2.5 Mathematical Expressions for Creep

As we mentioned before, creep tests are time-consuming and special care needs to be taken to select a creep function that best fits the experimental results. In addition, the relatively short (time-wise) creep experiments, the selected creep function also must predict the long-term deformation. Previously, the curve fitting was done manually; researchers had to use intuition and experience to select simple and well-behaved functions.

Today, because curve-fitting can be performed on personal computers, the number and degree of sophistication of the functions has increased significantly. Before presenting some functions for creep of concrete commonly used in structural analysis, we will make the following general statements regarding the specific creep function $\Phi(t, \tau)$. Consider it as a guideline in case you feel the need to introduce a new creep function.

1. For a given age of loading τ, the creep function is a monotonic increasing function of time t;

$$\frac{\partial \Phi(t, \tau)}{\partial t} \geq 0 \tag{13-120}$$

2. However, the rate of creep increment is always negative;

$$\frac{\partial^2 \Phi(t, \tau)}{\partial t^2} \leq 0 \tag{13-121}$$

3. The aging of concrete causes a decrease in creep as the age of loading τ increases. For a given value of load duration $(t - \tau)$ due to aging of concrete;

$$\left(\frac{\partial \Phi(t, \tau)}{\partial \tau}\right)_{(t-\tau)} \leq 0 \tag{13-122}$$

4. Creep has an asymptotic value

$$\lim_{t \to \infty} \Phi(t, \tau) \leq M \tag{13-123}$$

In many structural models, the function $\Phi(t,\tau)$ is separated into instantaneous and delayed components.

$$\Phi(t, \tau) = \frac{1}{E(\tau)} + C(t, \tau) \tag{13-124}$$

if we take aging of the concrete into account, the specific creep function $C(t, \tau)$ is further separated into:

$$C(t, \tau) = F(\tau) f(t - \tau) \tag{13-125}$$

By writing $C(t, \tau)$ in this fashion, we indicate that at a given time concrete should recall not only the actions to which it was subjected since time τ, given by the function $f(t - \tau)$, but also its own material state at time τ, given by the function $F(\tau)$. Therefore, function $F(\tau)$ characterizes the aging of concrete. The following expressions for $F(\tau)$ and $f(t - \tau)$ have been traditionally used for fitting short term experimental data, with the objective of predicting the long-term deformation.

Expressions for $f(t - \tau)$

1. *Logarithmic expression*: The U.S. Bureau of Reclamation[14] proposed using the following logarithmic expression for its projects dealing with mass concrete. When the stress-strength ratio does not exceed 0.40, the following equation is used:

$$f(t - \tau) = a + b \log[1 + (t - \tau)] \tag{13-126}$$

Constants a and b are easily obtained when the creep data are plotted semilogarithmically. The equation was originally developed for modeling basic creep of large dams, and the duration of load $(t-\tau)$ is measured in days. The expression is unbounded and usually overestimates the later creep.

2. *Power expression*: The general expression is given by

$$f(t-\tau) = a(t-\tau)^m \tag{13-127}$$

Constants a and m can be easily obtained on a log-log plot, where the power expression gives a straight line. The expression captures the early creep well but overestimates the later creep with unbounded results.

3. *Hyperbolic expression*: Ross[15] proposed the following hyperbolic expression:

$$f(t-\tau) = \frac{(t-\tau)}{a+b(t-\tau)} \tag{13-128}$$

This expression provides a limiting value for creep, $1/b$. It usually underestimates early creep but provides good agreement for late creep. ACI code uses this formulation for creep evolution.

4. *Exponential expression*: The exponential expression provides a limiting value for creep. In its simplest formulation it is given by

$$f(t-\tau) = a(1-e^{-b(t-\tau)}) \tag{13-129}$$

It does not provide a good fit for experimental values. For numerical analysis more terms are usually incorporated.

Expressions for $F(\tau)$

$F(\tau)$ takes into account the aging of concrete, therefore it should be monotonically decreasing. While expressions for $f(t-\tau)$ have been developed during the last 70 years, expressions for $F(\tau)$ are much more recent. Among the expressions, we cite:

1. Power law:

$$F(\tau) = a + b\tau^{-c} \tag{13-130}$$

2. Exponential:

$$F(\tau) = a + be^{-c\tau} \tag{13-131}$$

13.3 Temperature Distribution in Mass Concrete

Chapter 4 presented the problems created by the temperature rise in mass concrete due to the exothermic reactions of cement. A simple equation was introduced that related the tensile stress in concrete to its coefficient of thermal expansion, elastic modulus, creep coefficient, degree of restraint, and temperature change. One major difficulty remains: how to compute the temperature distribution in complex geometries, and how to incorporate the incremental construction into the analysis.

One of the challenges in mass concrete design is to maximize the thickness of the concrete layers without causing thermally induced cracks and to minimize the time between the placement of the next layer. The designer is under pressure from the contractor, who wants large layers to be placed in rapid succession in order to speed up construction. For large projects, major economic savings can be achieved when the size and placement of the lifts are perfectly orchestrated; the penalty for not coordinating construction of the layers is to incur large labor costs while the construction crew waits for the next placement or repairing, or even demolishing an overly thick layer that cracked as a result of thermal stresses.

This section introduces the finite element method, the most powerful tool available to compute temperature distributions in solid materials. After deriving the fundamental equations, a series of simulations of concrete constructions will be analyzed using this method. The technology of selecting materials, mix proportions, and construction practices has been presented in Chap. 12.

13.3.1 Heat Transfer Analysis

The fundamental equation governing the distribution of temperature in a solid subjected to internal heat generation was developed by Fourier. Consider a parallelepiped representing a volumetric element of a material, with conductivity coefficient k (Fig. 13-13). The change in heat flux in the x-direction is given by the equation:

$$\frac{\partial}{\partial x}\left(k\frac{\partial T}{\partial x}\right)dx\,dy\,dz \tag{13-132}$$

where T is the temperature.

Similarly for the y and z directions:

$$\frac{\partial}{\partial y}\left(k\frac{\partial T}{\partial y}\right)dy\,dz\,dx \tag{13-133}$$

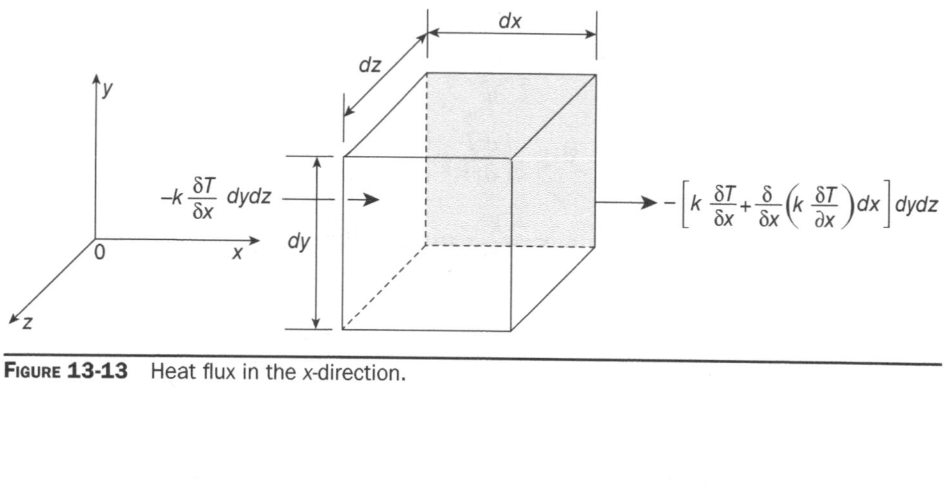

FIGURE 13-13 Heat flux in the x-direction.

and

$$\frac{\partial}{\partial z}\left(k\frac{\partial T}{\partial z}\right)dz\,dx\,dy \tag{13-134}$$

Addition of the flux variation in the three directions, Eqs. (13-132) to (13-134) determines the amount of heat introduced in the interior of the element per unit time:

$$\left\{\frac{\partial}{\partial x}\left(k\frac{\partial T}{\partial x}\right)+\frac{\partial}{\partial y}\left(k\frac{\partial T}{\partial y}\right)+\frac{\partial}{\partial z}\left(k\frac{\partial T}{\partial z}\right)\right\}dx\,dy\,dz \tag{13-135}$$

In the above derivation the material was considered isotropic. Considering it also homogeneous, Eq. (13-135) becomes

$$k\left(\frac{\partial^2 T}{\partial x^2}+\frac{\partial^2 T}{\partial y^2}+\frac{\partial^2 T}{\partial z^2}\right)dx\,dy\,dz \tag{13-136}$$

For a material with mass density ρ and specific heat c, the increase of internal energy in the element is given by

$$\rho c\,dx\,dy\,dz\frac{\partial T}{\partial t} \tag{13-137}$$

where t is the time.

When the material does not generate any heat, we equate Eqs. (13-136) and (13-137), obtaining

$$k\left(\frac{\partial^2 T}{\partial x^2}+\frac{\partial^2 T}{\partial y^2}+\frac{\partial^2 T}{\partial z^2}\right)=\rho c\frac{\partial T}{\partial t} \tag{13-138}$$

and then Eq. (13-138) is rewritten as

$$\kappa\nabla^2 T=\dot{T} \tag{13-139}$$

where

$$\dot{T}=\frac{\partial T}{\partial t}$$

$$\nabla^2 T=\left(\frac{\partial^2 T}{\partial x^2}+\frac{\partial^2 T}{\partial y^2}+\frac{\partial^2 T}{\partial z^2}\right)$$

$$\kappa=\frac{k}{c\rho}=\text{thermal diffusivity}.$$

Consider the case when heat generation occurs inside the material. Equation (13-136) when added to the quantity of heat generated in the interior of the element per unit of time, $w\,dx\,dy\,dz$, can be equated with the increase of internal energy in the element. Therefore, the Fourier equation is obtained

$$k\left(\frac{\partial^2 T}{\partial x^2}+\frac{\partial^2 T}{\partial y^2}+\frac{\partial^2 T}{\partial z^2}\right)+w=\rho c\frac{\partial T}{\partial t} \tag{13-140}$$

or

$$k\nabla^2 T + w = \rho c\dot{T} \tag{13-141}$$

In the steady-state, T and w are not function of time, therefore Eq. (13-141) becomes:

$$k\nabla^2 T + w = 0 \tag{13-142}$$

Note, Eq. (13-139) is applicable to any isotropic homogeneous material. We will concentrate on the problem of determining temperature distribution in mass concrete. In this case, the heat generation rate w is associated with the adiabatic temperature rise. In Chap. 4, the various factors affecting the temperature rise were presented; here we will show how to incorporate it into the Fourier equation. For a concrete with a density ρ and a cement content β (kg/m^3), the relationship between the adiabatic temperature rise T_a and the heat of hydration Q_h is given by

$$T_a = \frac{\beta}{c\rho}Q_h \tag{13-143}$$

The heat of hydration Q_h is obtained per unit mass of cement, therefore the factor β/ρ must be used to calculate the heat of hydration per unit mass of concrete. The heat generation rate w is related to the heat of hydration by the following equation:

$$w = \beta\frac{dQ_h}{dt} \tag{13-144}$$

Using Eq. (13-143) we obtain

$$w = \rho c\frac{dT_a}{dt} \tag{13-145}$$

In order to determine a unique solution to the Fourier Eq. (13-141), adequate initial and boundary conditions must be given. They should be compatible with the physical conditions of the particular problem.

13.3.2 Initial Condition

The initial condition must be defined by prescribing the temperature distribution throughout the body at time zero as a known function of x, y, and z.

$$T(x,y,z,t=0) = f(x,y,z) \tag{13-146}$$

13.3.3 Boundary Conditions

I. Prescribed temperature boundary. The temperature existing on a portion of the boundary of the body Γ_t is given as

$$T(x, y, z, t) = f(x, y, z, t) \qquad x, y, z \text{ on } \Gamma_t \tag{13-147}$$

This condition is also known as Dirichlet or essential boundary condition. In mass concrete, this condition may exist in the concrete-water contact, where the convection is small, making the temperature of the concrete that is in contact with water the same as that of the water.

II. Prescribed heat flow boundary. A prescribed heat flow boundary condition can be expressed as

$$k\frac{\partial T}{\partial n}(x, y, z, t) = q_n(x, y, z, t); \quad x, y, z \text{ on } \Gamma_q \tag{13-148}$$

where q_n is the given amount of heat flow at point (x, y, z), and n is the outward normal to the surface.

III. Convection boundary condition. The rate of heat transfer across a boundary layer is given by

$$k\frac{\partial T}{\partial n}(x, y, z, t) = h(T_e - T_s)^N \qquad x, y, z \text{ on } \Gamma_h \tag{13-149}$$

where h = heat transfer coefficient
T_e = known temperature of the external environment
T_s = surface temperature of the solid
Γ_h = portion of the boundary surface undergoing convective heat transfer

For a linear convection boundary condition, $N = 1$, and Eq. (13-149) becomes

$$k\frac{\partial T}{\partial n}(x, y, z, t) = h(T_e - T_s) = g(x, y, z, t) - hT_s \tag{13-150}$$

where $g(x, y, z, t) = hT_e$.

IV. Radiation boundary condition. Heat transfer by radiation between boundary condition surface Γ_r and its surroundings can be expressed by

$$q_r(x, y, z, t) = V\sigma\left(\frac{1}{\frac{1}{\varepsilon_r} + \frac{1}{\varepsilon_s} - 1}\right)\left[T_r^4 - T_s^4\right]; \qquad x,y,z \text{ on } \Gamma_r \tag{13-151}$$

where V = radiation view factor
σ = Stefan-Boltzmann constant
ε_r = emissivity of the external radiation source
ε_s = emissivity of the surface
T_r and T_s = absolute temperature of the radiation source and the surface, respectively

13.3.4 Finite Element Formulation

The finite element method is a powerful tool to solve thermal problems. The method is completely general with respect to geometry, material properties, and arbitrary boundary conditions. Complex bodies of arbitrary shape, including several different anisotropic materials, can be easily represented.

For mass concrete structures, the significant boundary conditions that apply are Cases I and III, that is, the prescribed temperature and the convection boundary conditions. Many approaches exist that present a finite element formulation for temperature distribution in mass concrete. Below, we will follow the approach suggested by Souza Lima et al.[16] First, we will start with the steady-state case, then move to the transient-state case. The objective is to solve the Fourier equation given the necessary initial and boundary conditions.

Consider a body with the two different boundary conditions: Γ_t where the temperature is prescribed and Γ_h where there is a convection boundary condition (see Fig. 13-14).

For a point P in Γ_t (steady-state case)

$$T = f(P) \tag{13-152}$$

and for a point P in Γ_h (steady-state case)

$$k\frac{\partial T}{\partial n} = g(P) - hT \tag{13-153}$$

Consider a continuous and differentiable function Φ in the domain shown in Fig. 13-14 with the condition $\Phi = 0$ along Γ_t. No condition on Φ is imposed along Γ_h. Φ

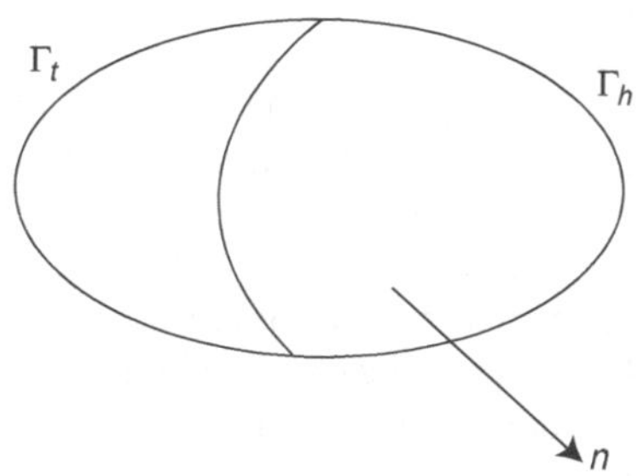

FIGURE 13-14 Boundary conditions in which the temperature is prescribed at Γ_t and convection is prescribed at Γ_h.

is often referred to as "weighting function" and is relevant to note that it is, and will remain, arbitrary.

Multiplying both sides of Eq. (13-142) by Φ, we obtain

$$k\Phi\nabla^2 T = -\Phi w \tag{13-154}$$

Integrating the above equation in domain V,

$$k\int_V \Phi\nabla^2 T\,dV = -\int_V \Phi w\,dV \tag{13-155}$$

Using the divergence theorem in the left side of Eq. (13-155):

$$k\int_V \Phi\nabla^2 T\,dV = -k\int_V \left(\frac{\partial\Phi}{\partial x}\frac{\partial T}{\partial x} + \frac{\partial\Phi}{\partial y}\frac{\partial T}{\partial y} + \frac{\partial\Phi}{\partial z}\frac{\partial T}{\partial z}\right)dV + k\int_V \Phi\frac{\partial T}{\partial n}dS \tag{13-156}$$

Since $\Phi = 0$ along Γ_t and using the boundary conditions defined above, Eq. (13-150) becomes

$$k\int_\Gamma \Phi\frac{\partial T}{\partial n}dS = k\int_{\Gamma_h} \Phi\frac{\partial T}{\partial n}dS = \int_{\Gamma_h} \Phi g\,dS - h\int_{\Gamma_h} \Phi T\,dS \tag{13-157}$$

Introducing Eqs. (13-156) and (13-157) into Eq. (13-154):

$$k\int_V \left(\frac{\partial\Phi}{\partial x}\frac{\partial T}{\partial x} + \frac{\partial\Phi}{\partial y}\frac{\partial T}{\partial y} + \frac{\partial\Phi}{\partial z}\frac{\partial T}{\partial z}\right)dV + h\int_{\Gamma_h} \Phi T\,dS = \int_V \Phi w\,dV + \int_{\Gamma_h} \Phi g\,dS \tag{13-158}$$

Equation (13-158) can be used to solve the steady-state Fourier heat equation. Consider a set of n functions Φi with $\Phi_i = 0$ on Γ_t. Thus any temperature field satisfying the boundary condition on Γ_h also satisfies Eq. (13-158) for each of the functions Φ_i.

Equation (13-158) may be used instead of Eqs. (13-142), (13-152), and (13-153) to solve approximately for T in the following manner:

$$T = \Phi_0 + \Sigma c_i \Phi_i \tag{13-159}$$

where c_i are unknown constants and Φ_0 is any smooth function satisfying the boundary condition on Φ_t. Of course the above may not satisfy Eq. (13-142) at every point in the body. However, substituting T given by Eq. (13-159) into Eq. (13-158) a system of linear equations is obtained that allows the determination of the coefficients c_i. By increasing the number of coefficients in Eq. (13-159) a better approximation to the solution is obtained. Φ_i are referred to as *interpolation functions* and they are almost invariably polynomials.

Finite element analysis idealizes the continuum by an assemblage of discrete elements or subregions. These elements may be of variable size and shape and are interconnected by a finite number of nodal points P_i. The interpolation functions Φ_i, should be chosen so that coefficients c_i are numerically equal to temperature T, at n nodal points P_i previously chosen in the domain. In order for the equality $c_i = T(P_i)$ to be true at the nodal points P_i the following conditions should be obeyed: $\Phi_i(P_i) = 1$, $\Phi_j(P_i) = 0 (j \neq i)$, and $\Phi_0(P_i) = 0$.

It is convenient to introduce a matrix formulation: $\{T\}$ is a vector of n elements with values $T(P_i)$ and $\{w\}$ is a vector of n elements with the values

$$\int_V \Phi_i w\,dV + \int_{\Gamma_h} \Phi_i g\,dS - w_{oi} \tag{13-160}$$

where w_{oi} is the value of the first term of Eq. (13-155), for $\Phi = \Phi_i$ and T is replaced by the function Φ_0.

For the steady-state case, this notation leads to

$$[K]\{T\} = \{w\} \tag{13-161}$$

where $[K]$ is the conductivity matrix $(n \times n)$ with values

$$[K] = K_{ij} = \int_V \nabla^T \Phi_i k \nabla^T \Phi_j\,dV \tag{13-162}$$

Determining the heat transfer in mass concrete is additionally complicated, because it involves the solution of the transient case and the continuous change of boundaries as construction progresses. To solve this problem, an incremental calculation of the linear transient problem is introduced. In the transient case, the Fourier equation is given by Eq. (13-141), which differs from the steady-state case by the term $\rho c\dot{T}$ and because w is a function of time. Using the divergence theorem:

$$\begin{aligned} &k\int_V \left(\frac{\partial \Phi}{\partial x}\frac{\partial T}{\partial x} + \frac{\partial \Phi}{\partial y}\frac{\partial T}{\partial y} + \frac{\partial \Phi}{\partial z}\frac{\partial T}{\partial z}\right) dV + h\int_{\Gamma_h} \Phi T\,dS \\ &= \int_V \Phi w\,dV + \int_{\Gamma_h} \Phi g\,dS - \rho c\int_V \Phi \dot{T}\,dv \end{aligned} \tag{13-163}$$

Or, if the matrix notation is used:

$$[k]\{T\} = \{w\} - [c]\{\dot{T}\} \tag{13-164}$$

where $[c]$ is the capacity matrix $(n \times n)$ with values

$$c_{ij} = \rho c\int_V \Phi_i \Phi_j\,dV \tag{13-165}$$

To integrate Eq. (13-164), an incremental method is usually employed. Taking small interval Δt

$$\{\dot{T}\} = \frac{1}{\Delta T}[\{T(t)\} - \{T(t-\Delta t)\}] \tag{13-166}$$

and incorporating Eq. (13-166) into Eq. (13-164)

$$\left([k] + \frac{1}{\Delta T}[c]\right)\{T(t)\} = \{w\} + \frac{1}{\Delta T}[c]\{T(t-\Delta t)\} \tag{13-167}$$

Starting from a known initial temperature distribution, we proceed stepwise. Equation (13-167) allows the determination of $\{T(\Delta t)\}$ for the first step. Once the new temperature is known, we proceed to the next step, giving a new increment Δt and continuing the process until the distribution of temperatures over the period of time of interest is known.

13.3.5 Examples of Application

Typical problems that a concrete technologist faces when studying thermal stresses in mass concrete include the type of aggregate, amount of pozzolan, size of the concrete lift, and temperature of fresh concrete that might affect the maximum temperature rise in concrete. To study these parameters a finite element model of a concrete block placed on a foundation rock can be developed, as shown in Fig. 13-15. The finite element mesh is made of 385 nodal points and 344 elements. Note that the size of elements in the concrete block is much smaller than in the foundation because we are mainly interested in temperature distribution inside the concrete block. The material properties for different types of concrete and for the foundation rock are shown in Table 13-3. An important parameter in thermal analysis is the adiabatic temperature rise. Figure 13-16 shows the assumed values for different levels of pozzolan replacements.

To illustrate the importance of lift thickness, consider the mesh shown in Fig. 13-15. Assume that the concrete was placed either: (a) in two lifts of 1.50 m placed 3 days apart, or (b) in one lift of 3.00 m. Given that the temperature distribution changes with time, however, the designer is usually concerned with the maximum temperature distribution that occurs within the concrete block. Figure 13-17 shows the maximum temperature distribution in the concrete block for both cases. The maximum temperature with two

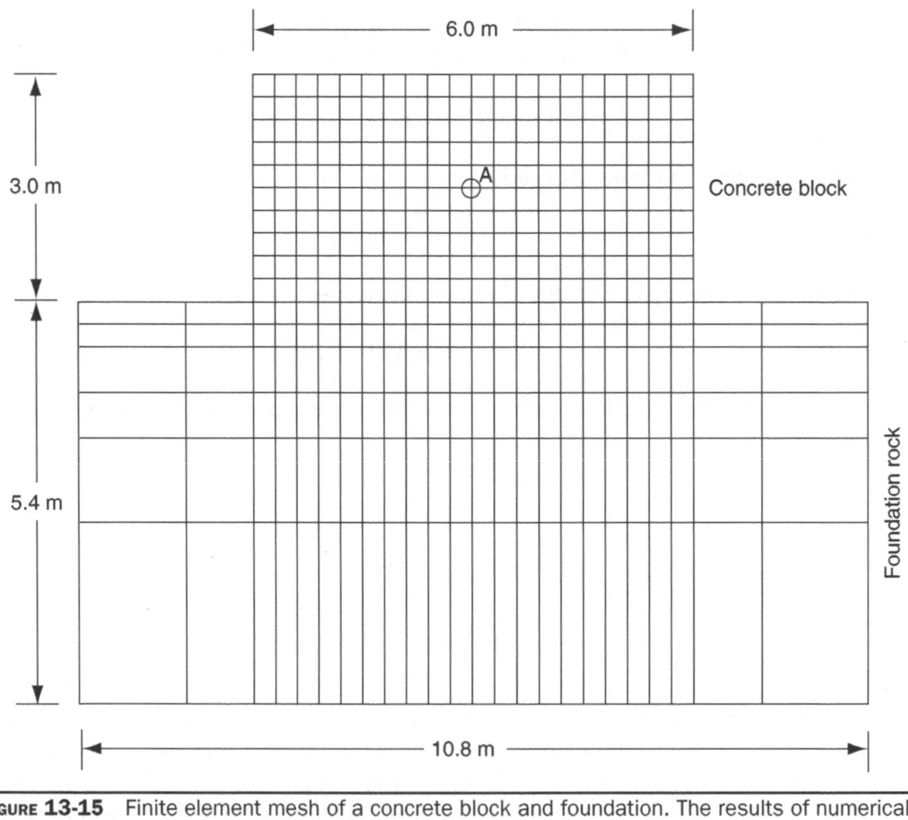

FIGURE 13-15 Finite element mesh of a concrete block and foundation. The results of numerical simulations are given in Figs. 13-17 and 13-20.

	Properties of Concrete Made with Different Aggregates			Foundation Rock
	Basalt	Granite	Gravel	
Thermal conductivity (kcal/m·h·°C)	1.740	2.367	3.690	2.800
Specific heat (kcal/kg °C)	0.24	0.23	0.22	0.20
Thermal diffusivity (m^2/h)	0.0029	0.0042	0.0075	0.0050
Density (kg/m^3)	2500	2450	2400	2800
Cement consumption (kg/m^3)	315	315	315	—

TABLE 13-3 Properties of Concrete and Foundation Rock

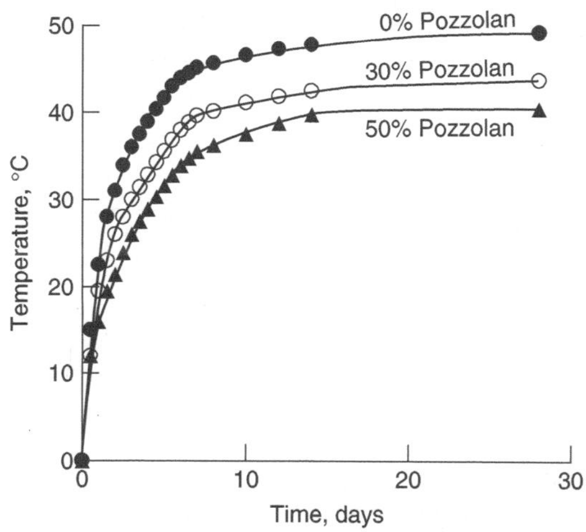

FIGURE 13-16 Effect of percentage of pozzolan on the adiabatic temperature rise.

1.50-m lifts was 46°C, which is likely to cause fewer problems than the 56°C reached using the one 3.00-m lift.

The thermal diffusivity of concrete is controlled mainly by the aggregate. To analyze the effect of thermal diffusivity on the temperature distribution, consider three types of aggregates: basalt, gravel, and granite. The temperature evolution for the point *A* (indicated in Fig. 13-15) is shown in Fig. 13-18. Concrete made with gravel has the highest thermal diffusivity, therefore, it dissipates heat faster and, consequently, shows the smallest temperature rise.

The use of pozzolans is an efficient method of controlling the temperature rise in concrete. Information on various types of pozzolans is presented in Chap. 8. The performance of three concrete mixtures is compared, and the adiabatic temperature rise for each type of concrete is shown in Fig. 13-16. The advantage of including pozzolans is illustrated in Fig. 13-19, where the maximum temperature rise is significantly reduced when such replacements are used.

Refrigeration is another powerful method of controlling temperature rise in mass concrete. Precooling can be achieved by replacing the mixing water by ice or by cooling

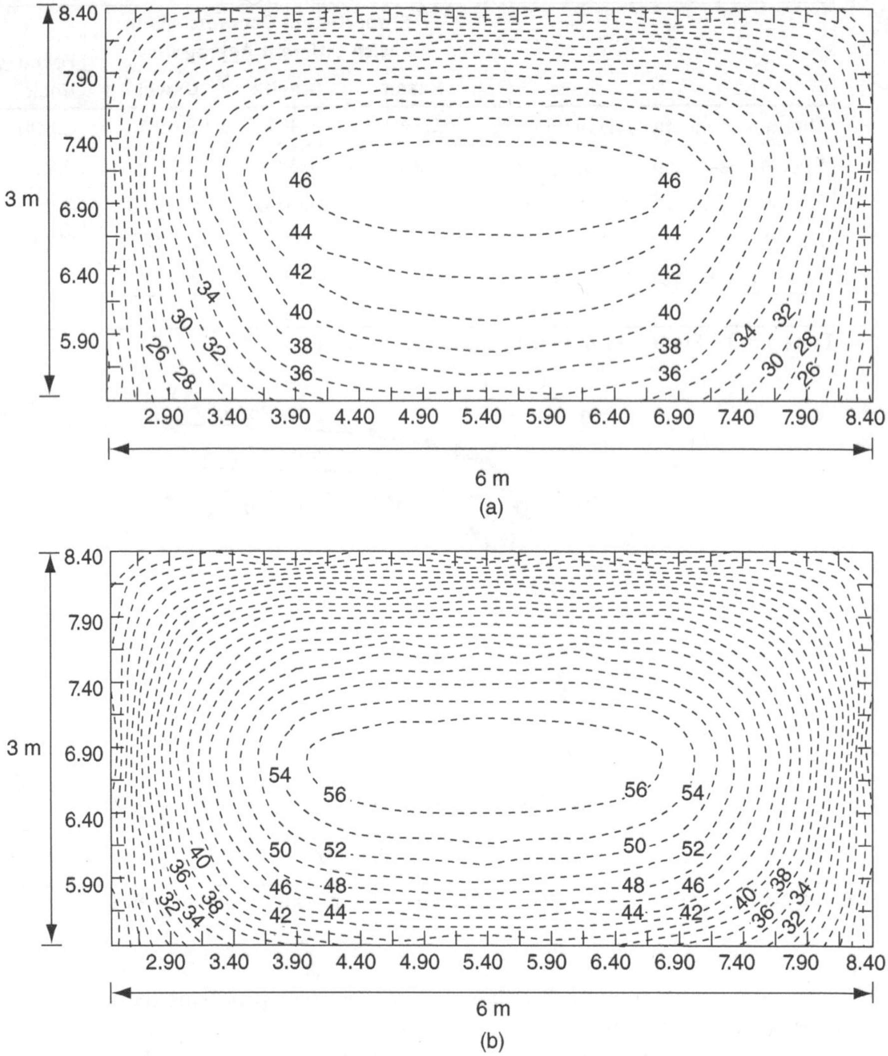

FIGURE 13-17 Maximum temperature distribution in the concrete block.

The size of the concrete lift is an important parameter in the temperature distribution in mass concrete. Thick lifts are attractive for fast construction, but high temperatures are usually generated in the concrete. Smaller lifts generate much lower temperatures, however, they may create problems with the construction scheduling. It is the responsibility of the engineer to establish the optimal thickness of the lifts. For this, a thermal analysis is usually performed using the finite element method. As an illustration, a thermal analysis was conducted for the finite element mesh shown in Fig. 13-15, with two conditions: (a) two concrete lifts of 1.50 m placed 3 days apart, and (b) one lift of 3.00 m. The temperature distribution in the concrete is shown above. For case (a) the maximum temperature in the concrete is 46°C that is much lower than the 56°C for case (b). For this analysis, the temperature of the fresh concrete was assumed to be 17°C, the aggregate was assumed to be granite, and no pozzolan was used.

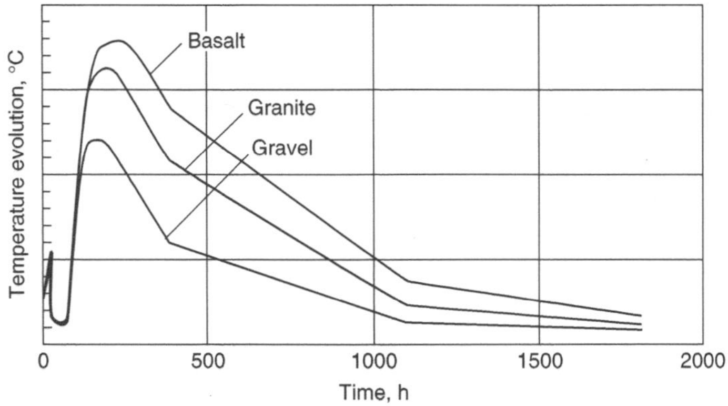

FIGURE 13-18 Temperature evolution for concrete with different thermal diffusivities.

Thermal diffusivity of concrete is a property which greatly influences the temperature distribution within the mass. Higher thermal diffusivity leads to faster heat loss which results in a lower maximum temperature. It may not always be desirable to have a rapid dissipation of heat, because the concrete may not have enough tensile strength at earlier ages. For this study two lifts of 1.50 m each was considered. The reason for the first temperature drop is given in the caption of Fig. 13-19.

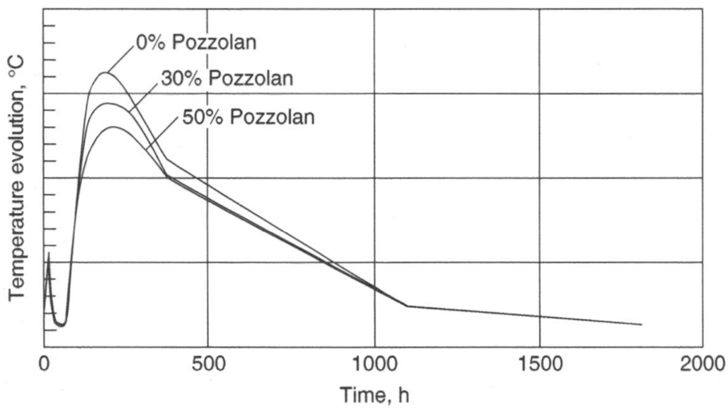

FIGURE 13-19 Influence of pozzolan on the temperature of concrete.

Pozzolans can significantly reduce the temperature inside mass concrete. The plot above shows the temperature evolution for point A of Fig. 13-15. Placement consisted of two concrete lifts of 1.50 m 3 days apart. Point A is at the top of the first lift, so there is an initial temperature increase, followed by a quick heat loss to the ambient temperature (17°C) until the next lift is placed. The temperature then increases up to a maximum, after which the block starts to cool.

the coarse aggregate. Postcooling can be achieved by circulation of cold water through pipes embedded in concrete. Usually precooling is preferred because it is more economical and does not involve extra labor, such as the embedding of pipes, pumping cold water, and eventually regrouting the pipes. The importance of temperature of fresh concrete is shown in Fig. 13-20. When the concrete is placed at 25°C, the maximum temperature is 52°C, compared to 42°C when the concrete is placed at 10°C.

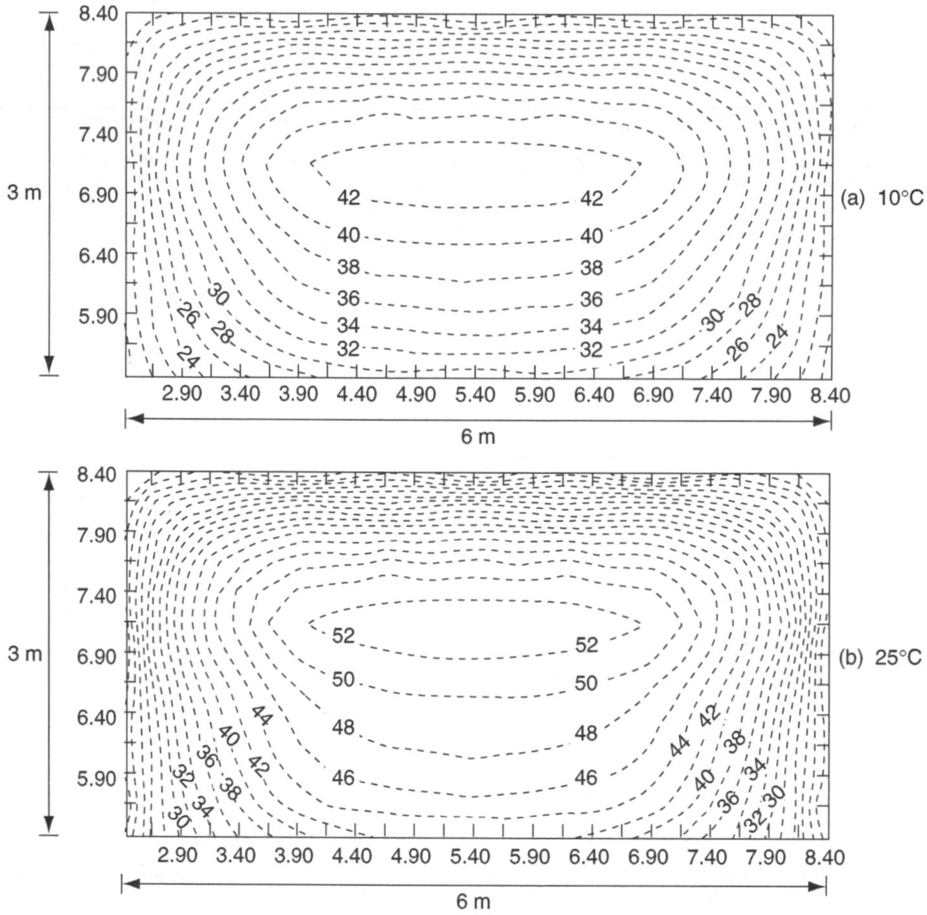

FIGURE 13-20 Effect of fresh concrete on the maximum temperature distribution.

One of the most effective methods of controlling the temperature rise in mass concrete is by lowering the temperature of fresh concrete. A simple method is to use ice instead of mixing water or precooling the aggregate. Unlike change in the size of lift, modifications in the temperature of fresh concrete do not affect construction scheduling. This finite element analysis assumed three different temperatures of fresh concrete: 10, 17, and 25°C. The temperature distribution for fresh concrete placed at a temperature of 17°C is shown in Fig. 13-17a. The concrete was placed in two 1.50 m lifts each, with granite as the coarse aggregate and an ambient temperature of 17°C.

13.3.6 Case Study: Construction of the Cathedral of Our Lady of the Angels in California, USA

Mass concrete is usually associated with the construction of large dams, office platforms, and massive foundations. This perception can be misleading because even fairly small members can develop fairly high thermal stresses, particularly when high-strength concrete is used. The present case study analyses an unsuspected application of mass concrete in a structure where normally thermal stresses are not a problem, however, due to architectural and construction requirements precautions against thermal cracking became necessary.

Substantially damaged by the 1994 Northridge, California earthquake, the aging St. Vibiana's Cathedral in Los Angeles was condemned by the City of Los Angeles. The Catholic Archdiocese decided to build a landmark Cathedral that reflects the rich diversity of Los Angeles. The $163 million complex consists of a new Cathedral, a conference center, a 47.5 m (156 ft) campanile tower, carillon, a residence hall, a $2^1/_2$ acre grand plaza, and a subterranean parking structure.

The architect chose lightly colored, sandblasted architectural concrete for the exterior and interior walls of the cathedral, as well as for other structures in the complex. While the architectural concrete work at the parking structure was being completing, the contractor was refining the forming techniques for the cathedral's intricate elements at an off-site mock-up. To create the 850 unique corner conditions called for in the cathedral's design, the contractor selected a wood-forming system, which were cost-effective and allowed for relatively flexible field adjustment. Water curing of the walls resulted in an unacceptable mottling effect and the only procedure for curing the walls that did not leave residue or unsightly mottling or discoloration waste leave the forms in place for several days. To achieve optimal color distribution, ensure concrete placement through congested seismic reinforcing, and account for possible traffic delays of ready-mix delivery, a 23-cm (9-in) slump superplasticized concrete mixture was selected. For durability requirements, a 0.45 water-cement ratio was selected (Fig. 13-21).

Unfortunately, the mock-up surfaces exhibited a high incidence of thermal-induced cracking, as well as interior to exterior thermal gradient cracking. Internal surface temperature probes were then installed in subsequent castings and recorded internal hydration temperatures of 110°C (230°F). The unique combination of the aforementioned practices, materials, and methods produced an unacceptable thermal outcome for two reasons: (1) the insulating effect of the double-sheeted wood forms created a nearly adiabatic condition (little loss of heat to the ambient), making the cathedral's even relatively thin walls behave like a mass concrete structure; (2) the Type III cement (chosen because of its light color), the durability driven water-cement ratio, and the choice of a rich concrete mix.

With just seven weeks until the first cathedral walls were to be cast, engineers embarked on a concurrent, iterative theoretical, and field test program. Finite element thermal stress models were developed and 19 large-scale formed concrete mock-ups were ultimately cast. The research was broken down into the following three studies. The first study focused on a method to limit the temperature rise of the concrete mix but maintain the color and adhere to the 0.45 water-cement ratio durability requirement. The second study focused on a way to lower the placement temperature of the concrete but, again, maintain the color uniformity of the concrete. The third study employed finite element models to simulate the evolution of temperature/stress/strength in the concrete walls in order to determine the optimum time to remove the formwork relative to the concrete stresses.

FIGURE 13-21 Use of architectural mass concrete in the Los Angeles Cathedral. (From Selna, D., and P.J.M. Monteiro, Cathedral of Our Lady of The Angels, Concr. Int., November, 2001.)

The Cathedral of Our Lady of The Angels was an interesting and challenging construction with unique architectural and technological requirements. Because of the concrete mixture proportion and construction practices, the internal temperature in the columns would crack the concrete unless proper measures were taken. The engineers had the challenge of optimizing the concrete mixture proportions and, at the same time, satisfy tough architectural design demands.

The first study developed a mix proportion that decreased the temperature but maintained the desired color, texture, and durability requirements. Heat rise reduction was easily achieved by altering cementitious content and eliminating the use of metakaolin. The next step was to find a suitable cement: Type II or V cement has low C_3A content, but all locally produced Type II and V cements had an unacceptable gray color. A white cement manufactured in Aalborg Denmark with a C_3A content of just 4 percent was located and a slightly gray Class F fly ash was selected. The new mix proportions generated maximum temperatures below 71°C (160°F), but the concrete cooled very slowly approximately 0.12°C/h (0.22°F/h). Limiting the interior to ambient gradients was important to control cracking. Further reduction in the ultimate temperature became critical as cooling rates were slow and would significantly delay the schedule.

In study two, the challenge was to find the most economical combination that reduced placement temperature but maintained consistency and color. Field tests containing 50 percent ice replacement was unsuccessful because it resulted in poor color distribution and mottling; however, chilled water reduced very effectively the temperature of fresh concrete without averse effect on the concrete color.

Study three used finite element analysis to predict temperature distribution and tensile stresses in the cathedral blocks. Stripping periods of the formwork in excess of three days had significant schedule impact. Computer analysis indicated that the temperature of the concrete reached its peak hydration temperature 24 h after casting. Based upon splitting tensile test data, simulations indicated that the concrete was strong enough in the time window between 9 and 12 h after casting, but not so warm yet that formwork could be removed without subjecting the concrete to excessive thermal shock. Simulations also indicated that while the window of 9 to 12 h was beneficial in terms of thermal stresses, there was a 24- to 29-h window after casting when thermal stresses were least favorable relative to concrete tensile strengths. Although the 9- to 12-h form removal was very favorable in terms of construction schedule and cost, same day stripping caused mottling and shingle detailing when forms were removed.

Numerical and field results showed that concrete could be stripped at 2½ days with an ambient to differential of 26°C (46°F), producing an excellent finish with no cracking and little impact to the construction schedule. The following batching, mix design, placing, and formwork removal procedures were successfully implemented:

1. Early morning placements, beginning at 5:00 A.M. and finishing no later than 9:00 A.M.
2. Ready mix made with low C_3A content white cement (318 kg/m^3).
3. Fly ash (56 kg/m^3) substitution—strict color consistency control procedures implemented.
4. Concrete formulated with chilled water (167 kg/m^3) and ice supplements as required to keep the concrete temperature below 24°C (75°F) at placement.
5. Trucks cooled with chilled water spray prior to loading.
6. Forms left in place a minimum of $2^1/_2$ days after placing, and longer as required to be within 25°C (45°F) of ambient. Internal temperatures measured using thermocouples.
7. Aggregates cooled using chilled water misters.

13.4 Fracture Mechanics

Applying fracture mechanics to concrete design can provide much insight on how the size of a structural element may affect the ultimate load capacity. It can also be a useful tool in predicting crack propagation. Consider a case where you are responsible for determining if a given crack in a large structure such as a concrete dam will propagate catastrophically under certain loading conditions. You can adopt a strength criterion that predicts that a crack will propagate when the stresses reach the ultimate tensile strength of the material. For sharp cracks, however, the theory of linear elasticity predicts that the stresses at the tip of the crack go to infinity, thereby assuming that the crack will propagate no matter how small the applied stress, an unlikely scenario.

Fracture mechanics, on the other hand, provides an energy criterion that does not have such drawbacks and allows for more precise predictions of the stability of the crack. The application of this energy criterion can be particularly useful when using traditional finite element methods to study cracks where mesh sensitivity becomes a problem. Figure 13-22 shows an example where the result is greatly affected by the size of the mesh when a strength criterion is used, however, little mesh sensitivity is observed when an energy criterion based on fracture mechanics is employed.

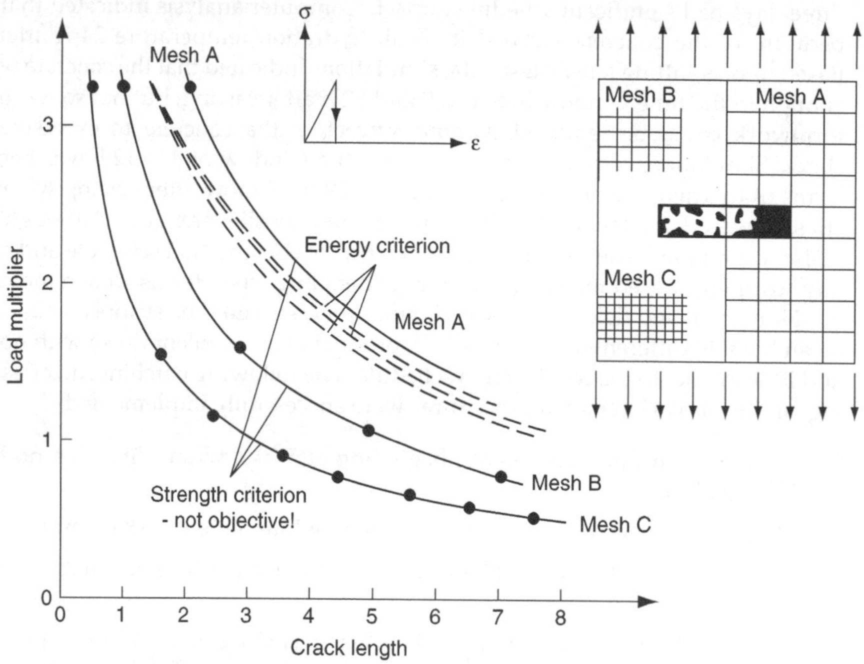

FIGURE 13-22 Example of mesh sensitivity. (From Bazant, Z.P., and L. Cedolin, *Stability of Structures*, Oxford University Press, New York, p. 917, 1991.)

Considering the advantages of using fracture mechanics for concrete, it is surprising that this is a relatively new area of research. The development of fracture mechanics for concrete was slow as compared to other structural materials.

Linear elastic fracture mechanics theory was developed in 1920, but it was not until 1961 that the first experimental research in concrete was conducted. Fracture mechanics had been applied successfully to design metallic and brittle materials for many years; however, comparatively few applications were found for concrete. This trend continued up until the middle 1970s when major advances were finally made. The contributions were based on the development of nonlinear fracture mechanics models, taking into account the behavior of the structure and behavior of concrete. Throughout the last two decades, intensive research has been performed and applications of fracture mechanics in the design of beams, anchorage, and large dams are becoming more common. That said, when compared to previous continuum theories covered in this chapter (elasticity, viscoelasticity, and thermal problems), fracture mechanics is not yet as mature a theory and this will be reflected in its presentation. A fair, but simplified, exposition of some of the existing fracture mechanics models for concrete are presented later at an introductory level.

13.4.1 Linear Elastic Fracture Mechanics

The presentation begins by highlighting the work of A.A. Griffith,[17] who is often regarded as the founder of fracture mechanics. His original interest focused on the effect of surface treatment in the strength of solids. It had been observed experimentally that small imperfections have a much less damaging effect on the material properties than the large imperfections. This was theoretically puzzling because the fracture criteria used at that time predicted that if the imperfections were geometrically similar, the stress concentrations caused by the imperfections should be the same so that the effect on the strength is the same, no matter the size of the imperfection. Griffith solved this problem by developing a new criterion for fracture prediction. In contrast to the simplistic strength approach, he suggested an energy balance approach, based not only on the potential energy of the external loads and on the stored elastic strain energy but also on another energy term: the *surface energy*. This surface energy γ is associated with the creation of fresh surface during the fracture process. Griffith applied his method to a crack of length $2a$ in an infinite plate of unit thickness.

Figures 13-23*a* and *b* show that when the crack is extended *under constant load*, the change in potential energy of the external load due to crack growth is $P\Delta x$ and the increase in strain energy is $1/2\ P\Delta x$. In other words, the decrease in potential energy of the external load is twice the increase in strain energy. During crack extension there is an increase of surface energy $4a\gamma$ (remember, the crack length is $2a$, and both the upper and lower surface of the crack should be included). Griffith used a result obtained by Inglis,[18] that the change in strain energy due to an elliptical crack in a uniformly stressed plate is $\pi a^2\sigma^2/E$, and therefore, the change in potential energy of the external load is $2\pi a^2\sigma^2/E$. The energy change of the plate, due to the introduction of the crack, is given by

$$U_{\text{cracked}} - U_{\text{uncracked}} = -\frac{2\pi a^2\sigma^2}{E} + \frac{\pi a^2\sigma^2}{E} + 4a\gamma \tag{13-168}$$

Minimizing the energy in relation to the crack length,

$$\frac{\partial}{\partial a}\left(-\frac{\pi a^2\sigma^2}{E} + 4a\gamma\right) = 0 \tag{13-169}$$

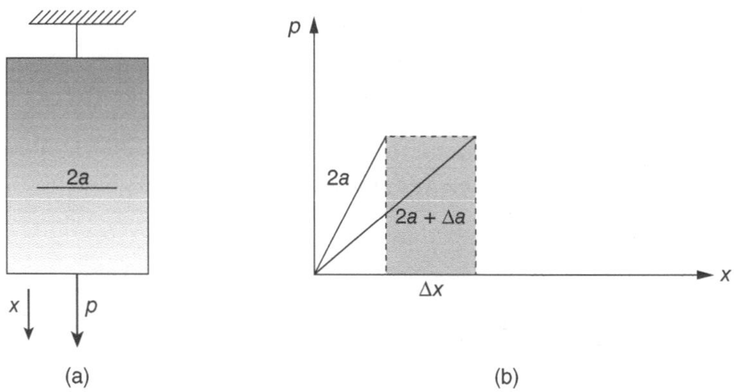

Figure 13-23 (*a*) Plate with crack 2a; (*b*) load-displacement diagram.

gives the critical stress (for plane stress)

$$\sigma = \sqrt{\frac{2E\gamma}{\pi a}} \tag{13-170}$$

This equation is significant because it relates the size of the imperfection ($2a$) to the tensile strength of the material. It predicts that small imperfections are less damaging than large imperfections, as observed experimentally. Even though Griffith's work was ignored for several years, it paved the way for the creation and development of the field of linear fracture mechanics.

One problem with Griffith's approach was that the surface energy obtained from his equation was found to be orders of magnitude higher than the one obtained using thermodynamical tests unrelated to fracture. The reason is that because the dissipative processes associated with the fracture propagation absorbs a significant amount of energy, the energy required for crack extension exceeds the thermodynamical value. Irwin[19] proposed that instead of using the thermodynamic surface energy, one should measure the characteristic surface energy of a material in a fracture test. He introduced the quantity G_c as the work required to produce a unit increase in crack area. G_c is also referred to as the *critical energy release rate*. Typically, G_c is determined experimentally, using simple specimen configuration. Once G_c for a given material is known, assuming that it is a material property, we have a powerful method for determining if a given crack will or will not propagate under any other loading condition. The process is quite simple: The energy release per unit increase crack area, G, is computed; if the energy release rate is lower than the critical energy release rate ($G < G_c$), the crack is stable. Conversely, if $G > G_c$ the crack propagates. In the particular case when the energy release is equal to the critical energy release rate ($G = G_c$), a metastable equilibrium is obtained.

The following analysis illustrates how to compute the value of G_c. Considering the plate, shown in Fig. 13-23, with thickness B, we can express the energy released by crack growth Δa as

$$GB\Delta a = P\Delta x - \Delta U_e \tag{13-171}$$

where ΔU_e is the change in elastic energy due to crack growth Δa. In the limit:

$$GB = P\frac{dx}{da} - \frac{dU_e}{da} \tag{13-172}$$

Introducing the compliance $c = x/P$, the strain energy U_e is given by

$$\Delta U_e = \frac{cP^2}{2} \tag{13-173}$$

Equation (13-172) becomes

$$GB = P\frac{d(cP)}{da} - \frac{d(cP^2/2)}{da} \tag{13-174}$$

or

$$G = \frac{P^2}{2B}\frac{dc}{da} \tag{13-175}$$

When, the compliance vs. crack length has been obtained for a given specimen configuration, the critical energy release rate G_c can be determined by recording the load at fracture.

Example 13-3 Compute the energy release rate for the double cantilever beam shown in Fig. 13-24. In addition, study the stability of the crack in its own plane under (*a*) load control and (*b*) displacement control. Shear deflections may be ignored.

The deflection of each cantilever can be easily found using simple beam theory:

$$\frac{\delta}{2} = \frac{Pa^3}{3EI} \tag{13-176}$$

where E is the elastic modulus and I is the moment of inertia,

$$I = \frac{1}{12}b\left(\frac{h}{2}\right)^3 \tag{13-177}$$

The compliance is given by

$$c = \frac{\delta}{P} = \frac{2a^3}{3EI} \tag{13-178}$$

Therefore, the energy release rate is given by

$$G = \frac{P^2}{2B}\frac{dc}{da} = \frac{P^2a^2}{BEI} \tag{13-179}$$

Stability criteria: A crack is stable if the derivative of the strain energy rate, with respect to crack length is negative. In other words,

$$\frac{1}{G}\frac{\partial G}{\partial a} < 0 \tag{13-180}$$

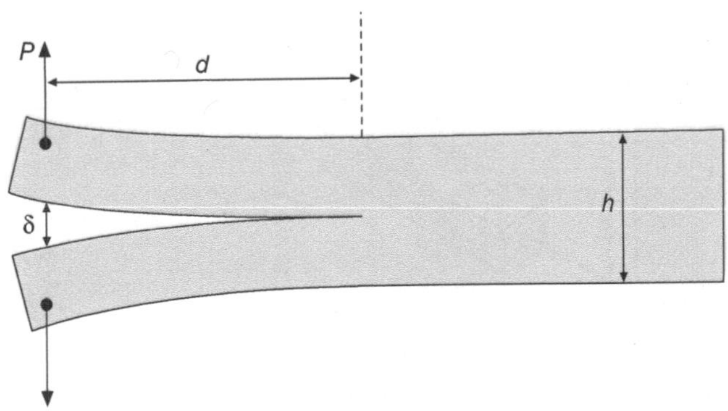

Figure 13-24 Double cantilever beam with thickness *B*.

1. For load control:

$$\frac{\partial G}{\partial a} = \frac{2P^2 a}{BEI} \tag{13-181}$$

(1/G)(MG/Ma) = 2a is a positive number, therefore the crack will propagate in an unstable way.

2. *For displacement control:* Combining Eqs. (13-179) and (13-178), the energy release rate can be expressed in terms of the deflection:

$$G = \frac{9EI\delta^2}{4a^4 B} \tag{13-182}$$

and

$$\frac{\partial G}{\partial a} = -\frac{9EI\delta^2}{a^5 B} \tag{13-183}$$

(1/G)(MG/Ma) = –4/a is a negative number, therefore, the crack will propagate in a stable manner.

Let us now analyze what happens to the stress field near the tip of a crack for the three configurations shown in Fig. 13-25. The three types of relative movements of two crack surfaces are classified as (*a*) *Mode I*: opening or tensile mode, (*b*) *Mode II*: sliding or in-plane shear mode, and (*c*) *Mode III*: tearing or antiplane shear mode.

Most practical design situations and failures are associated with Mode I. As shown in Fig. 13-26, the stresses at the tip of the crack for this mode are given by

$$\sigma_y = \frac{K_I}{\sqrt{2\pi r}} \cos\frac{\varphi}{2}\left(1 + \sin\frac{\varphi}{2}\sin\frac{3\varphi}{2}\right) \tag{13-184}$$

$$\sigma_x = \frac{K_I}{\sqrt{2\pi r}} \cos\frac{\varphi}{2}\left(1 - \sin\frac{\varphi}{2}\sin\frac{3\varphi}{2}\right) \tag{13-185}$$

$$\tau_{xy} = \frac{K_I}{\sqrt{2\pi r}} \left(\sin\frac{\varphi}{2}\cos\frac{\varphi}{2}\cos\frac{3\varphi}{2}\right) \tag{13-186}$$

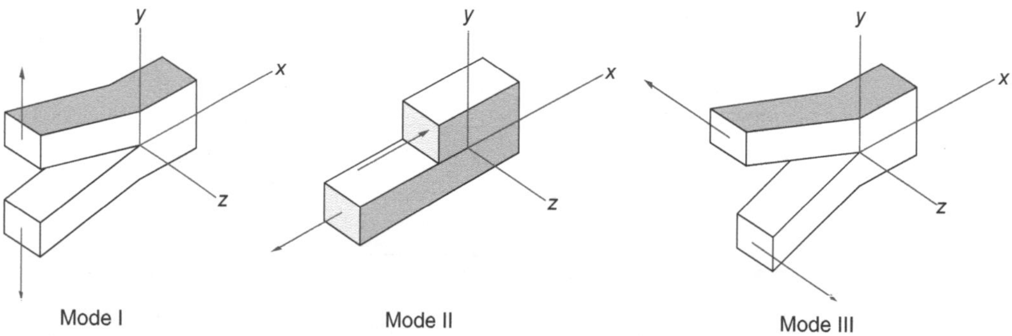

Figure 13-25 Basic modes of loading. (After Hertzberg, R.W., *Deformations and Fracture Mechanics of Engineering Materials*, Wiley, New York, p. 262, 1976.)

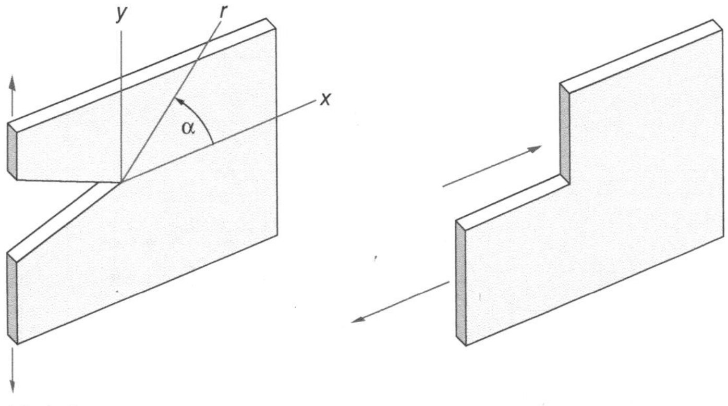

FIGURE 13-26 Coordinate system at the tip of the crack. (From Piltner, R., Spezielle finite element Lochen, Echen und Rissen, unter Vewendung von analytischen teillosunge, VDI-Verlag, Dusseldorf, No. 96, 1982).

K_I is called *stress-intensity* factor for Mode I. Dimensional analysis of Eqs. (13-184) to (13-186) indicates that the stress-intensity factor must be linearly related to stress and to the square root of a characteristic length. Assuming that this characteristic length is associated with the crack length, we have

$$K_I = \sigma\sqrt{a}\, f(g) \tag{13-187}$$

where $f(g)$ is a function that depends on the specimen and crack geometry.

The stress-intensity factor can be computed for a variety of crack shape configurations. Suppose we measure the value of the stress at fracture in a given test. Using Eq. (13-187) we determine the *critical stress intensity factor* K_c or, as it is usually called in the literature, the *fracture toughness*. If we make the assumption that K_c is a material property (as we did for critical energy rate G_c) we have another powerful tool of predicting critical combinations of stress and crack length for other configurations of Mode I.

Irwin showed that the energy release rate and the stress intensity factor approaches are equivalent. For linear elastic behavior, considering only Mode I and plane stress condition:

$$G_I = \frac{K_I^2}{E} \tag{13-188}$$

13.4.2 Concrete Fracture Mechanics

The first experimental research on fracture mechanics of concrete was performed by Kaplan in 1961. Subsequent research studied the effects of various parameters on K_c and G_c. Experimental studies indicated that the fracture toughness increases with increasing (*a*) aggregate volume, (*b*) maximum-size aggregate, and (*c*) roughness of the aggregate. As expected, the toughness decreases with increasing water-cement ratio and increasing air content. One of the problems encountered in the early stages of this research was that, instead of being a material property, the value of the fracture toughness K_c, was strongly

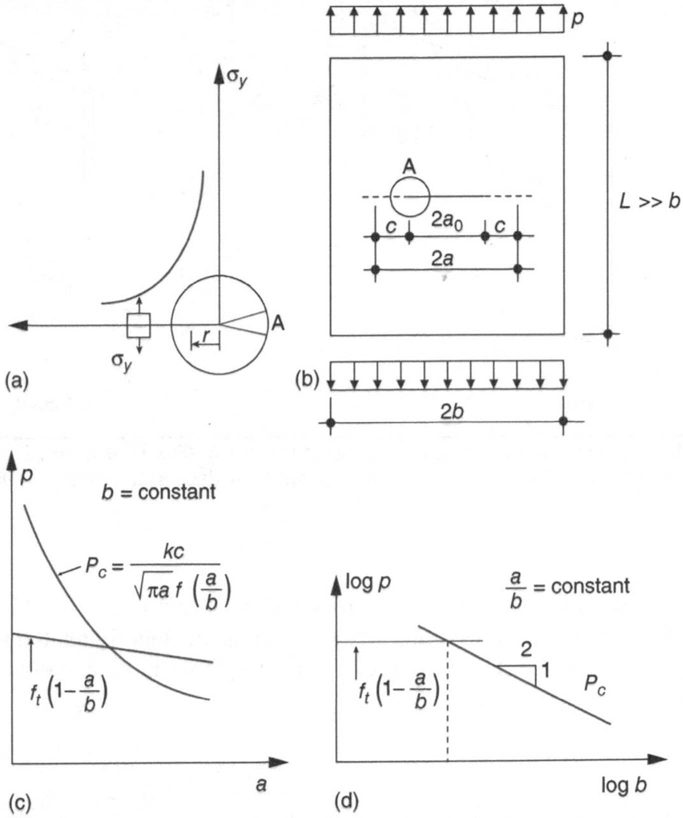

FIGURE 13-27 (*a*) Variation of σ_y at the crack tip in an elastic body; (*b*) cracked plate under tension; (*c*) comparison between ultimate values of applied tension, calculated according to fracture mechanics and tensile strength; (*d*) effect of plate width for geometrically similar plates. (From Cedolin, L., Introduction to Fracture Mechanics of Concrete, *El Cemento*, No. 4, p. 285, 1986.)

influenced by the size of the specimen tested. It soon became apparent that fracture mechanics measurements should not be made on small concrete specimens.

To analyze what happens to the ultimate stress when we change the dimensions of a cracked plate (see Fig. 13-27*b*) let us study the following case proposed by Cedolin.[20] The stress intensity for this configuration is given by $K = p\sqrt{\pi a}\, f(a/b)$ where $f(a/b)$ is a correction factor for the geometry.

The critical stress p_c associated with the fracture toughness K_c is given by

$$p_c = \frac{K_c}{\sqrt{\pi a}\, f(a/b)} \tag{13-189}$$

This relationship is shown in Fig. 13-27*c*. Instead of the fracture mechanics criteria, let us now analyze the strength criteria. The average tensile stress f_t in the plane that contains the crack will vary because the crack dimensions affect the net section of the specimen. This relationship is given by

$$p_t 2b = f_t(2b - 2a) \tag{13-190}$$

or

$$p_t = f_t\left(1 - \frac{a}{b}\right) \tag{13-191}$$

which is also shown in Fig. 13-27*c*. Therefore, as clearly demonstrated in Fig. 13-27*c*, for a small crack the strength criteria dominates, and we cannot infer fracture mechanics properties.

It is also fruitful to study the case of geometrically similar plates (a/b constant) and varying b. Equation (13-189) may be rewritten as

$$p_c = \frac{K_c}{\sqrt{b}\, f^*(a/b)} \tag{13-192}$$

where $f^*(a/b) = \sqrt{\pi a/b}\, f(a/b)$. Since (a/b) is constant, when Eq. (13-192) is plotted as function of b in a logarithmic scale it gives a straight line with slope $-1/2$ (Fig. 13-27*d*). Equation (13-191) is also plotted in Fig. 13-27*d*, and because a/b is constant it yields a straight line with zero slope. Again, we conclude that for small specimen sizes the strength criteria dominates and fracture mechanics properties cannot be inferred.

The ratio between the fracture mechanics criteria [Eq. (13-192)] and the strength criteria [Eq. (13-191)] is given by

$$\frac{p_c}{p_t} = \frac{K_c}{f_t\sqrt{b}(1 - a/b)\, f^*(a/b)} \tag{13-193}$$

It is convenient to define a *brittleness number*, $s = K_c/f_t\sqrt{b}$, to characterize the nature of the collapse; the lower the brittleness number the more brittle the behavior of the specimen. Fracture occurs in specimens with a small brittleness number, that is, for materials with a comparatively low fracture toughness, a high tensile strength, and in large specimens. The brittleness number characterizes the nature of the collapse for one-dimensional problems; for beams or slabs in flexure, additional information on the slenderness is necessary. It should be noted that the physical dimensions of the tensile strength [FL^{-2}] and fracture toughness [$FL^{-3/2}$] are different; however, the brittleness number is dimensionless.

The brittleness number can also be expressed as a function of elastic modulus E and energy release rate G, instead of the fracture toughness K_c: $s = \sqrt{EG}/(f_t\sqrt{b})$. This number helps to explain the experimental results where concretes made with high-strength silica fume cement paste usually have more fine microcracks than normal strength concrete (Fig. 13-28). In the high-strength matrix, the tensile strength can be two to five times greater than the normal-strength matrix; however, the increase in fracture energy or elastic modulus is not as much. Consequently, a high-strength matrix has a much lower brittleness number and is more susceptible to the development of cracks. A complete description of scaling laws for brittle materials is given by Bazant.[21]

13.4.3 Fracture Process Zone

Microcracks in concrete originate from strain localization and develop ahead of the crack tip, creating what is referred to as a *fracture process zone*. The characterization of this zone is of fundamental importance in the development of modern nonlinear

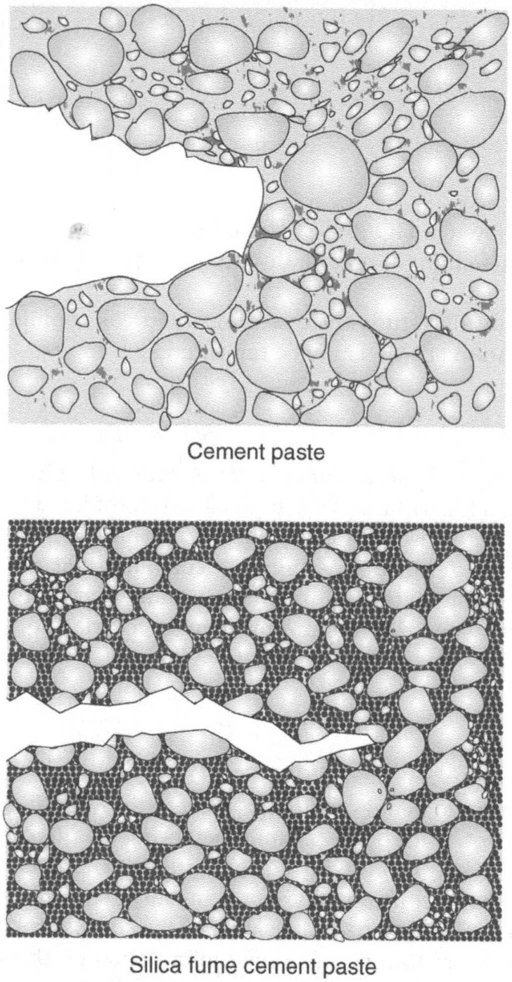

FIGURE 13-28 Structure of crack front in ordinary cement paste and in silica fume cement paste. (From Bache, H.H., Fracture Mechanics in Design of Concrete Structures, in *Fracture Toughness and Energy of Concrete*, Wittman, F.H., ed., Elsevier, Amsterdam, p. 582, 1986.)

fracture mechanics for concrete. Although the experimental characterization is challenging, recently new methods have been proposed.

In addition to other parameters, it is desirable to determine the position of the crack tip, the profile of the crack opening, and the overall state of microcracking ahead of the crack tip. Optical microscopy is one option, but the resolution is limited (in the order of 10 μm). Scanning electron microscopy has a much better resolution, but in traditional models the vacuum required for operation induces significant changes in the cracking pattern due to drying shrinkage. With the new generation of scanning electron microscopes, however, the study of saturated specimens and meaningful information of microcracking can be obtained.

Because of concrete's heterogeneity and a three-dimensional stress state along the crack front, the crack profile is not generally straight. Acoustic emissions (see Chap. 11) resulting from the sudden release of energy during the failure process provides useful information on the cracking mechanism. Acoustic emissions are transient elastic waves that can be detected at the surface by a transducer that converts an acoustic-pressure pulse into an electrical signal of very low amplitude.

Another powerful method of analyzing the fracture process zone is to use optical interferometry with laser light. In a study by Cedolin et al.[22] a reference grid (with a density of 1000 lines/mm) was created on the surface of the concrete specimen. When a load was applied to the specimen it produced a Morè fringe pattern, from which the extensional strain was determined.

Nemati at al.[23] were able to maintain the cracking pattern in concrete under load by impregnating the specimen with Wood's metal (Fig. 13-29). This metal is a fusible alloy and in the liquid phase it is nonwetting, with an effective surface tension of about 400 mN/m. It consists of 42.5 percent Bi, 37.7 percent Pb, 11.3 percent Sn, and 8.5 percent Cd. It has a melting range from 71.1 to 87.8°C (160 to 190°F) below the boiling point of water, and is solid at room temperature. While under load the concrete is intruded with Wood's metal under a pressure of 10.3 MPa (1500 psi), which should penetrate pores and cracks down to 0.08 mm. After intrusion the metal is allowed to solidify before unloading.

The additional elongation in the fracture zone can be estimated by introducing the additional strains ε_w over the length of the fracture zone, as shown in Fig. 13-30.

$$w = \int \varepsilon_w \, dx \qquad (13\text{-}194)$$

Unfortunately, the real strain distribution is often very hard to incorporate into an analytical model, and to date only simplified models have been proposed. Bazant and co-workers developed the *smeared crack band model*, where the entire fracture zone is represented by a band of microcracked material with width w_c. The model assumes a linear stress-strain relationship E_c up to the tensile strength f_t and a strain-softening relationship with slope E_t. The area enclosed by the diagram in Fig. 13-31 represents the fracture energy G_f given by

$$G_f = w_c \int_0^{\varepsilon_0} \sigma \, d\varepsilon_f = \frac{1}{2} w_c f_t^2 \left(\frac{1}{E_c} - \frac{1}{E_t} \right) \qquad (13\text{-}195)$$

This method proved to be very successful when used with the finite element method. Further simplification is obtained when the fracture process zone is modeled as a "tied crack" (Fig. 13-30), that is, a crack with a width w and a specified stress-elongation ($\sigma - w$) relationship. Because the aim of this model is to replace the real fracture process zone by an equivalent fictitious tied crack, this representation has been called *fictitious crack model*. The development of this model is presented in detail in the following section.

Fictitious Crack Model

The fictitious crack model was created and expanded upon by Hillerborg, Petersson, and co-workers. One of the objectives of the model is to capture the complex nature of

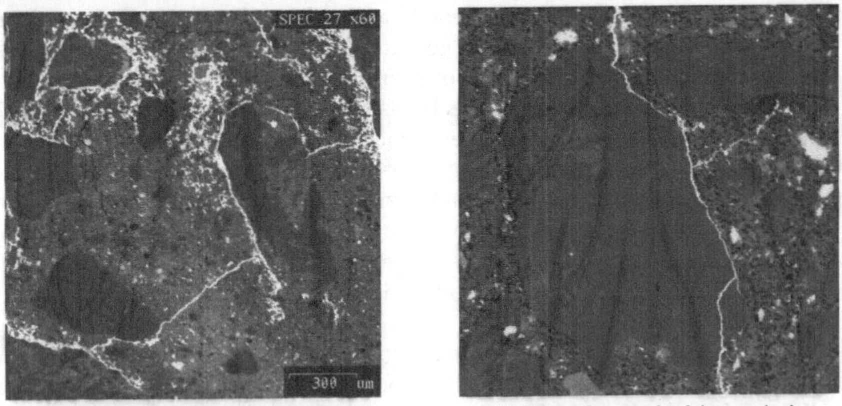

(a) SEM micrograph of the cracks in normal strength concrete loaded in uniaxial compression

(b) SEM micrograph of the cracks in high-strength concrete loaded in uniaxial compression

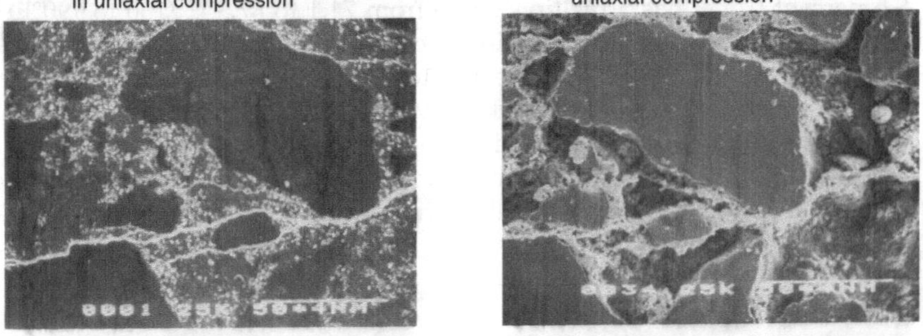

(c) Normal-strength concrete samples before (left) and after etching (right)

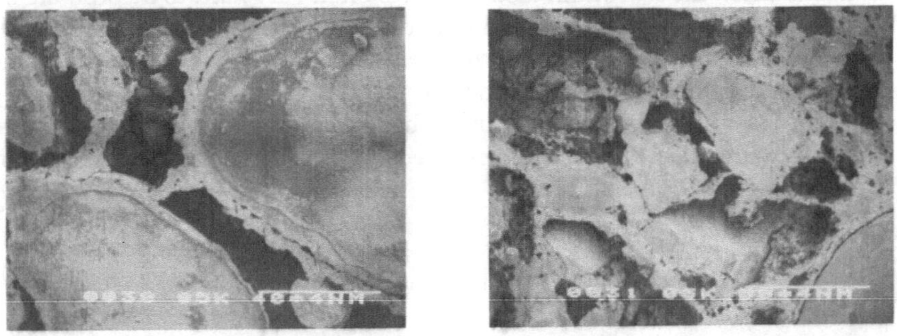

(d) Three-dimensional network of cracks

FIGURE 13-29 Preservation of cracks in concrete under load. [Images (*a*) and (*b*) from Nemati, K.M., P.J.M. Monteiro, and K.L. Scrivener, Analysis of Compressive Stress-Induced Cracks in Concrete, ACI Mat. J., Vol. 95, No. 5, pp. 617–631, 1998; images (*c*) and (*d*) from Nemati, K.M., and P.J.M. Monteiro, Cem. Concr. Res., Vol. 27, pp. 1333–1341, 1997.]

The equipment used for these experiments was specially designed and developed at the University of California at Berkeley to study the cracks in concrete samples as they exist under load. Cracks impregnated by Wood's metal are easy to locate using electron microcopy (Fig. 13-29a). The differences of cracking pattern in high strength concrete can be observed in Fig. 13-29b. After the alloy is solidified, the surface of the specimen can be etched with hydrochloric acid removing the cement paste and leaving a skeleton of alloy on top of the new surface (see Fig. 13-29c). Scanning electron microscopy can be used to analyze the complex three-dimensional network of cracks.

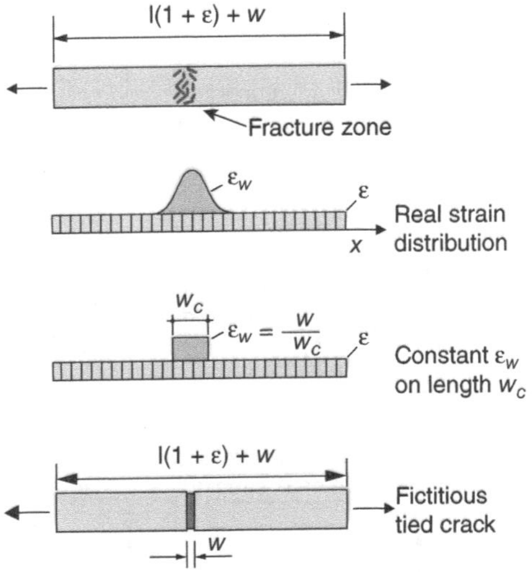

Figure 13-30 Strain distribution during fracture and two possible assumptions. (From Hillerborg, A., Numerical Methods to Simulate Softening and Fracture of Concrete, in *Fracture Mechanics of Concrete*, Sih, G.C., and A. Di Tommaso, eds., Martinus Nijhoff Publishers, Dordrecht, p. 148, 1985.)

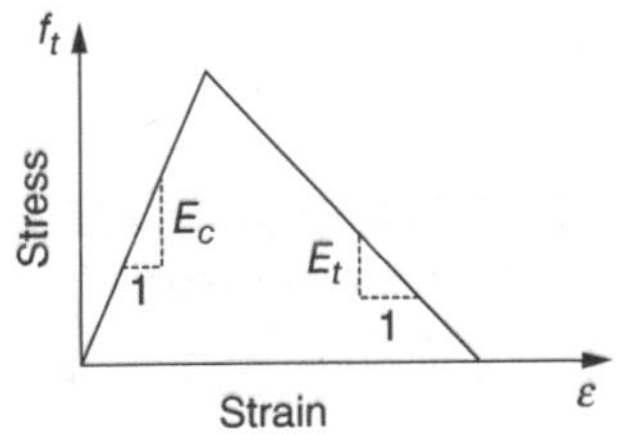

Figure 13-31 Stress-strain relationship for the smeared crack band model.

concrete in tension. The amount of microcracking in concrete, which is in tension, is small before the peak stress is reached, therefore, the deformation ε along the specimen can be assumed to be uniform, and the total elongation Δl of the specimen can be expressed in terms of the length of the specimen l (Fig. 13-32).

$$\Delta l = l\varepsilon \tag{13-196}$$

A localized fracture zone starts to develop just after the peak load is reached. In the model, this zone is assumed to form simultaneously across an entire cross section.

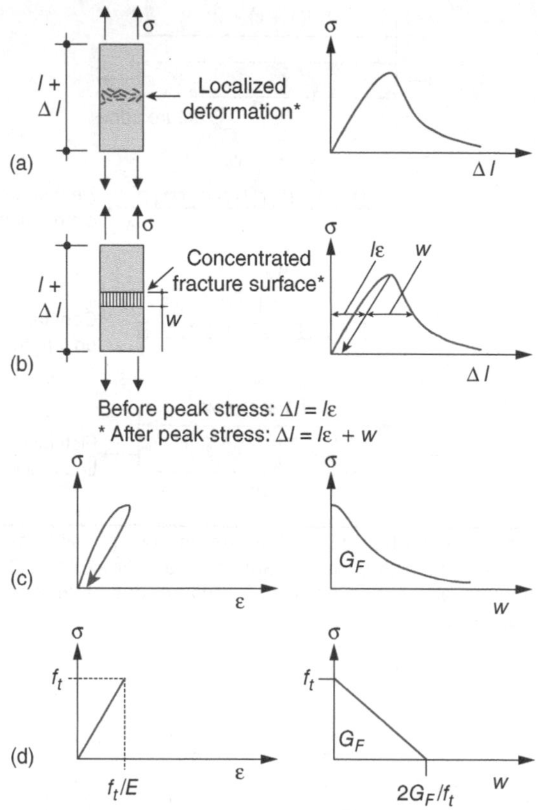

FIGURE 13-32 Fictitious crack model description of tensile fracture: (*a*) realistic structural behavior; (*b*) model of structural behavior; (*c*) model for description of properties of material; and (*d*) simplified properties of material. (From Gustafsson, P.J., and A. Hollerborg, Sensitivity in Shear Strength of Longitudinally Reinforced Concrete Beams to Fracture Energy of Concrete, *ACI Struc. J.*, p. 287, 1988.)

As the total elongation increases, the stress decreases and the region outside the fracture zone experiences an unloading, while inside the fracture zone, there is softening. The fracture zone remains localized and does not spread along the specimen, this is called *strain localization*, somewhat akin to that seen in plasticity. Beyond the peak stress, the total elongation of the specimen is the sum of the uniform deformation outside the fracture zone and the additional localized deformation w existing in the fracture zone, as shown in Fig. 13-32*b*.

$$\Delta l = l\varepsilon + w \tag{13-197}$$

As illustrated in Fig. 13-32*c*, two relationships are needed to characterize the mechanical behavior of concrete in tension: (1) a stress-strain ($\sigma - \varepsilon$) relationship for the region outside the fracture zone, and (2) a stress-elongation ($\sigma - w$) relationship for the fracture zone. Note that in the $\sigma - \varepsilon$ diagram, the horizontal axis is given by the strain, which is nondimensional, while for the $\sigma - w$ diagram, the horizontal axis is given by the elongation, which has units of length.

Although the curves shown in Fig. 13-32*c* may be influenced by the rate of loading and temperature, they are assumed to be independent of the shape and size of the specimen. Figure 13-32*d* shows simplified stress-strain and stress-elongation relationships. There is no fundamental reason to choose linear or bilinear relationships with the exception that they are numerically simple and seem to satisfy experimental results rather well. It should be mentioned that other researchers preferred to use a nonlinear stress-elongation ($\sigma - w$) relationship.

The fracture energy G_f is equal to the area under the stress-elongation curve.

$$G_f = \int_0^\infty \sigma(w)dw \tag{13-198}$$

Figure 13-33*a* shows typical experimental stress-elongation curves for different concrete mixture proportions. The results presented in Fig. 13-33*a* are redrawn in Fig. 13-33*b* to show that, even with different composition, the normalized stress-elongation curves have the same shape.

For very large specimens with deep preexisting cracks, the fracture energy G_f corresponds to the parameter G_c of the linear elastic fracture mechanics. While its measurement is fairly easy to make, the determination of the $\sigma - w$ relationship is not. Therefore, formulations, based on the fracture energy, such as the one indicated in Fig. 13-32, are usually preferred in analysis.

The fracture energy of concrete G_f is generally determined experimentally using a notched specimen loaded in flexure, according to RILEM Recommendation TC-50 FMC. The value for G_f is obtained by computing the area under the load-deflection relationship and dividing it by the net cross-section of the specimen above the notch. When experimental data are not available, FIP model code 2010 recommends the use of the following expression:

$$G_f = 73 f_{cm}^{0.18} \tag{13-199}$$

where f_{cm} is the mean compressive strength in MPa.

The stress-strain and stress-elongation curves are related in the following manner: the slope of the stress-strain diagram is E, and the slope of the stress-deformation curve is proportional to $f_t/(G_f/f_t)$. The ratio between the two slopes has units of length called the *characteristic length* (l_{ch}) of the material:

$$l_{ch} = \frac{EG_f}{f_t^2} \tag{13-200}$$

The characteristic length is often considered to be a material property, and it gives a measure of the brittleness of the material. Cement paste has a characteristic length in the range 5 to15 mm, mortar in the range 100 to 200 mm, and concrete 200 to 400 mm. Compared to normal-strength concrete, high-strength concretes and light-weight aggregate concrete have lower characteristic lengths.

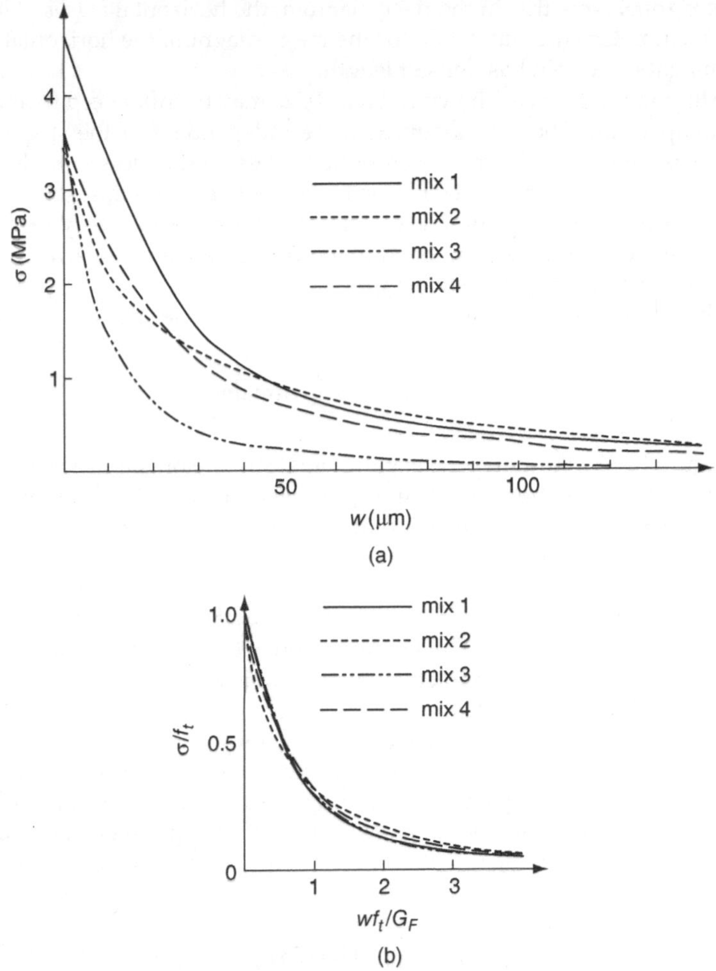

FIGURE 13-33 (*a*) σ-*w* curves for four concrete mixes (From Petersson, P., Crack Growth and Development of Fracture Zones in Plain Concrete and Similar Materials, Report TVBM-1006, Lund, Sweden, p. 167, 1981.); (*b*) the curves from (*a*) are redrawn to show that their shape is similar. (From Hillerborg, A., Numerical Methods to Simulate Softening and Fracture of Concrete, in *Fracture Mechanics of Concrete*, Shih, G.C., and A. DiTommaso, eds., Martinus Nijhoff Publishers, Dordrecht, p. 152, 1985).

The importance of the stress-strain and stress-elongation relationships in the design of concrete in tension must be stressed. The FIP model code 2010 recommends the following stress-strain relationships for uniaxial tension (Fig. 13-34).

$$\sigma_{ct} = E_c \varepsilon_{ct} \qquad \text{for} \qquad \sigma_{ct} \le 0.9 f_{ctm} \tag{13-201}$$

$$\sigma_{ct} = f_{ctm} - \frac{0.1 f_{ctm}}{0.00015 - (0.9 f_{ctm}/E_c)}(0.00015 - \varepsilon_{ct}) \qquad \text{for} \qquad 0.9 f_{ctm} \le \sigma_{ct} \le f_{ctm} \tag{13-202}$$

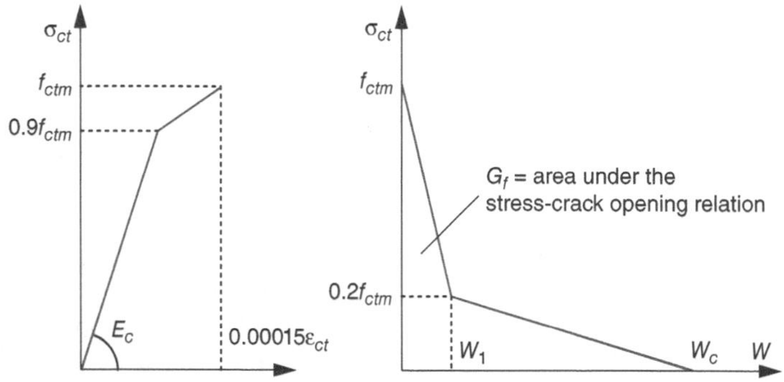

FIGURE 13-34 Stress-strain and stress-elongation for concrete in uniaxial tension. (After FIP Model Code 2010.)

where E_c = tangent modulus of elasticity in MPa
f_{ctm} = tensile stress in MPa
σ_{ct} = tensile stress in MPa
ε_{ct} = tensile strain

For the cracked section, the following bilinear stress-crack opening relation is recommended:

$$\sigma_{ct} = f_{ctm} \cdot \left(1 - 0.8 \cdot \frac{w}{w_1}\right) \quad \text{for} \quad w \leq w_1 \tag{13-203}$$

$$\sigma_{ct} = f_{ctm} \cdot \left(0.25 - 0.05 \cdot \frac{w}{w_1}\right) \quad \text{for} \quad w_1 < w \leq w_c \tag{13-204}$$

and

$$\begin{aligned} &w_1 - G_F/f_{ctm} \text{ in mm when } \sigma_{ct} - 0.20 \cdot f_{ctm} \\ &w_c = 5 \cdot G_F/f_{ctm} \text{ in mm when } \sigma_{ct} = 0 \end{aligned} \tag{13-205}$$

where w_1 = crack opening (mm)
w_c = crack opening (mm) for $\sigma_{ct} = 0$
G_f = fracture energy [Nm/m²]

13.4.4 Fracture at the Nanoscale*

In 1900 Max Planck introduced a fundamental idea that changed the subsequent development of physics: that the energy of radiation is transferred not continuously, but by portions (i.e., quantum) equal to $h\,\nu$, where ν is the frequency of the radiation and h is a fundamental constant.

*The authors wish to acknowledge the contributions of Prof. G.I. Barenblatt to the field of scaling. This section was based on the paper: Barenblatt G.I. and P.J.M. Monteiro, Scaling laws in nanomechanics, *Physical Mesomechanics*, vol. 13, pp. 245–248, 2010.

Naturally the quantum phenomena should enter the mechanisms of the deformation and fracture at a nanoscale. Therefore, the Planck constant h should enter the governing parameters at these scales. The other quantities influencing the deformation and fracture at these scales are Young's modulus E and density ρ, because the formation of defects at a small scale is always a dynamical process even if the global, macroscopic process of deformation is static.

Thus Planck's constant h, which is determined with high precision, is equal to 6.626 10^{-34} m^2kg/s. The density ρ for metals, concrete, etc., in ordinary conditions (materials that are not highly compressed, e.g., by converging shock waves) varies; for platinum ρ equals 21.0×10^3 kg/m^3, 19.0×10^3 kg/m^3 for tungsten and plutonium, and 1×10^3 kg/m^3 for water. The typical density of normal weight concrete is 2.4×10^3 kg/m^3.

Consider the quantity:

$$\frac{h}{(\rho E)^{1/2}} = \frac{h}{\rho (E/\rho)^{1/2}} \tag{13-206}$$

It is easy to see that this quantity is measured in m^4 because $(E/\rho)^{1/2}$ is measured in m/s and it is proportional to the velocity of elastic waves propagation in the material.

Therefore, the quantity

$$\lambda = \left(\frac{h}{\rho (E/\rho)^{1/2}} \right)^{1/4} \tag{13-207}$$

is measured in meters. Remarkably, this quantity is practically the same for a wide spectrum of materials from platinum to concrete and water (note that it is critical to observe the small exponent ¼). The speed of elastic wave propagation is in the order of 1.5×10^3 m/s for water and 4.0×10^3 m/s for concrete.

The calculation gives

$$\lambda = \left(\frac{h}{\rho (E/\rho)^{1/2}} \right)^{1/4} = A(10^{-40})^{1/4} \cong 10^{-10}\text{m} \tag{13-208}$$

which is the length of the "pixel" accepted in nanomechanics. Note that the length λ is a fundamental (not arbitrary assigned as 1 μm for micromechanics) physical quantity. In particular, the normal interatomic distances in crystal lattice have this order of magnitude.

This fact can have important consequences. Indeed, the parameter λ should enter the governing parameters that determine the properties of deformation and strength of structures. Therefore, the parameter Λ/λ —where Λ is the length scale of a specimen or of a structure—should be included in the laws of scaling. Generally speaking, this means that the scaling effect[24] could be observed in testing the macroscopic, large specimens or structures. It seems to be a plausible way to incorporate the nanomechanical (quantum mechanical) effects on fatigue, fracture, plasticity, creep, and other mechanical properties usually attributed to macroscopically large specimens and structures.

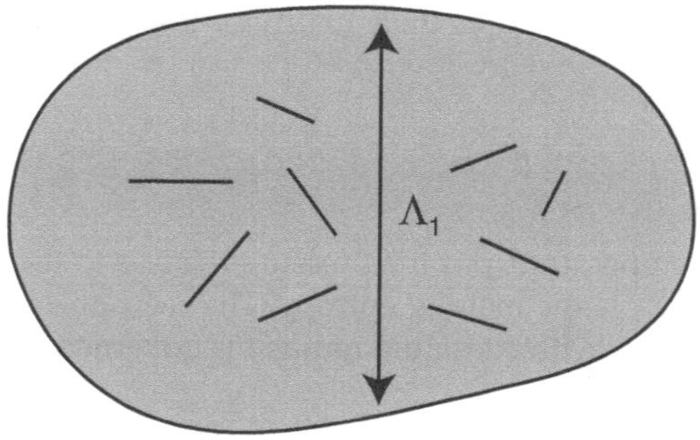

Figure 13-35 Interaction of many small cracks can prevent the development of complete similarity.

This formulation can provide new insights in the modeling of cracking for materials with complex nanostructures. Although the cohesive crack approach developed by Barenblatt[25] requires complete similarity ($\Lambda/\lambda >> 1$) and may be appropriate for a large number of materials, it was initially developed for very small cracks. The cracking in concrete involves multiscale physics where small cracks at the nano-level (with size on the order of λ) coalesce and interact with microcracks, thereby requiring different scaling laws. The first stage occurs in the domain $1 << \Lambda_1/\lambda << M_1$ (M is a large number), followed by a second stage over the domain $M_1 << \Lambda_2/\lambda << M_2$, and so on.

13.5 Poromechanics of Concrete Exposed to Freezing Temperatures

The damage caused by the formation of ice in concrete is described in Chapter 5 (Sec. 5.9). When concrete freezes, the resulting coupled and contrary actions cause it to deform due to the following mechanisms: (1) the difference of density between the liquid water and the ice crystal; (2) the drainage of the liquid water expelled from the freezing sites; (3) the interfacial effects between the different constituents, which govern the crystallization process in connection with the pore access radius distribution; (4) the cryosuction process, which drives liquid water toward the already frozen pores; and (5) the thermomechanical coupling between the deformable porous matrix, the liquid water, and the ice crystals. The phase transformation of a fluid inside a porous deformable body can be formulated in the context of poroelasticity originally developed by Biot[26] and expanded in the more general context of continuum thermodynamics by Coussy.[27,28] This section describes a poroelastic formulation for concrete exposed to low temperatures. The treatment of this model is based on the work of Coussy and Monteiro.[29,30]

13.5.1 Effect of the Pore Size Distribution on the Liquid Saturation Curve Below the Melting Point

The temperature of solidification of water in a given pore depends on the size of the throat that gives access to the pore. The chemical equilibrium between liquid water

(L) and ice crystal (C) requires the equality of the chemical potential of both phases, yielding

$$(1-\rho_C/\rho_L)\,\mathrm{d}p_L + \mathrm{d}p_C - \mathrm{d}p_L = \Sigma_m\,\mathrm{d}T \tag{13-209}$$

where T is the temperature, p is the pressure relative to the atmospheric pressure, ρ is the density, and Σ_m is the melting entropy. The mechanical equilibrium of a spherical solid-liquid interface with curvature radius r is governed by the Laplace equation:

$$p_C - p_L = \frac{2\gamma_{CL}}{r} \tag{13-210}$$

where γ_{CL} is the crystal-liquid interface energy. For common civil engineering applications, the cooling $\Delta T = T_m - T$ does not exceed 30°C, where $T_m = 0°C$ is the bulk melting point. In this case, the melting entropy Σ_m can be considered as constant (e.g., its variation with pressure and temperature is neglected). Combining Eqs. (13-209) and (13-210) gives

$$\frac{2\gamma_{CL}}{r} = \Sigma_m \Delta T - (1-\rho_C^0/\rho_L^0)p_L \tag{13-211}$$

Because of the small value of $1-\rho_C/\rho_L \le 1-\rho_C^0/\rho_L^0 \simeq 1-{}^{917}\!/_{997} \simeq 0.08$, where superscript $_0$ refers to the state at T_m, the last term in Eq. (13-211) is often neglected.* Therefore, Eq. (13-211) simplifies to the Gibbs-Thomson equation:

$$r = \frac{2\gamma_{CL}}{\Sigma_m (T_m - T)} \tag{13-212}$$

If the angle of contact between the liquid and the solid matrix is zero, then r becomes the minimum radius of the pore, where the current solid-liquid interface can locate. This implies that r is the radius of the smallest channels, where the ice crystals can propagate when temperature T is lower than melting point T_m.

In practice, the cumulative pore volume fraction $G(r)$ occupied by pores having a pore entry radius greater than r can be experimentally obtained through mercury porosity tests. As an example, using data extracted from the literature,[30] function $G(r)$ is given in Fig. 13-36. Cooling from an initially liquid-saturated mortar under drained condition $p_L = 0$, the volume fraction S_C of ice at temperature $T < T_m$ is given by

$$S_C = G\left(r = \frac{2\gamma_{CL}}{\Sigma_m (T_m - T)}\right) \tag{13-213}$$

Figure 13-37 shows ice saturation S_C as a function of the cooling, $\Delta T = T_m - T$, for the mortar given in Fig. 13-36. For this determination the following values were used: a zero

*This approximation is no longer valid when the value of the liquid pressure p_L is much greater than zero, which is the case in undrained condition (see Sec. 13.5.3).

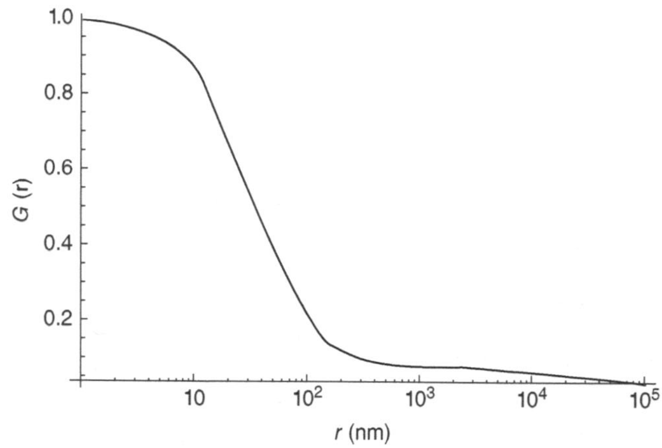

FIGURE 13-36 Cumulative pore volume fraction plotted against pore entry radius for mortar. (Adapted from Coussy, O., and P.J.M. Monteiro, Poroelastic model for concrete exposed to freezing temperatures, *Cem. Concr. Res.*, vol. 38, pp. 40–48, 2008.)

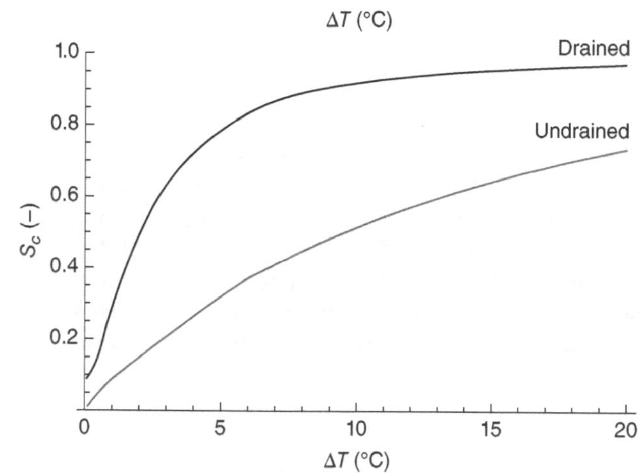

FIGURE 13-37 Ice saturation S_c plotted against cooling ΔT for the mortar of Fig. 13-36 under drained and undrained conditions.

contact angle and the values $\gamma_{CL} = 0.0409\ \mathrm{J \cdot m^{-2}}$, $\Sigma_m = 1.2\ \mathrm{MPa \cdot K^{-1}}$ at $T = T_m = 273$ K. Based upon the $S_C - \Delta T$ curve and through the Gibbs-Thomson law, a material characteristic cooling temperature $T_{\mathrm{ch}} = T_m - (\Delta T)_{\mathrm{ch}}$ can be introduced by weighting cooling ΔT with the number of pores whose entry radius r is associated with ΔT. This definition leads to express $(\Delta T)_{\mathrm{ch}}$ in the form

$$(\Delta T)_{\mathrm{ch}} = \int_{S=0}^{S=1} \Delta T(r) \times dS(r) \tag{13-214}$$

When using this definition, the value of the characteristic cooling temperature $T_{ch} = T_m - (\Delta T)_{ch}$ is found to be −6°C for the mortar shown in Figs. 13-36 and 13-37. For $\Delta T << (\Delta T)_{ch}$ a major fraction of water will remain in liquid form, although for $\Delta T >> (\Delta T)_{ch}$ almost all water will have solidified.

13.5.2 Unsaturated Poroelasticity for Freezing Materials

At low temperatures, the porous cement paste matrix deforms as a result of the pore pressure caused by the formation of ice. The equation of saturated poroelasticity can be extended to unsaturated conditions, where the porous space is filled by the liquid solution and the solid crystals. Before extending the equations of saturated poroelasticity, it is insightful to review their fundamental assumptions. Consider a representative volume $d\Omega_0$ extracted from a porous body. This representative volume has an initial porosity ϕ_0, and its initial temperature is melting temperature T_m. The porous body is then subjected to the cooling $\Delta T = T_m - T$. By adopting the atmospheric pressure as (zero) reference pore pressure, then the constitutive equations of a thermoelastic porous solid whose porous volume is subjected to uniform pore pressure p are

$$\begin{aligned} \sigma &= K\varepsilon - bp + 3aK\Delta T \\ s_{ij} &= 2Ge_{ij} \\ \varphi &= \phi - \phi_0 = b\varepsilon + p/N + 3\alpha_\phi \Delta T \end{aligned} \tag{13-215}$$

where σ and ε are the mean stress and the volumetric dilation, respectively; s_{ij} and e_{ij} the components of the deviatoric stress and the deviatoric strain tensors, respectively; ϕ is the current porosity; K, G, and 3α are the bulk modulus, shear modulus, and thermal volumetric dilation coefficient of the porous solid, respectively, which are the properties related to the empty porous material with a zero pore pressure; and b and N are the Biot coefficient and Biot modulus, respectively. These macroscopic properties are linked to bulk modulus k_s and thermal volumetric dilation coefficient $3\alpha_s$ of the solid matrix according to the relations:

$$b = 1 - \frac{K}{k_s}; \quad \frac{1}{N} = \frac{b - \phi_0}{k_s}; \; a = \alpha_s; \;\; \alpha_\phi = \alpha_s(b - \phi_0) \tag{13-216}$$

In the case of a freezing water-infiltrated porous material subjected to a temperature below 0°C, the pore pressure is no longer uniform since the porous space is partly occupied by ice crystals at pressure p_C, and partly occupied by the water remaining in liquid form at pressure p_L. Constitutive equations [Eq. (13-215)] are then extended in the form:

$$\begin{aligned} \sigma &= K\varepsilon - b_C p_C - b_L p_L + 3aK\Delta T \\ s_{ij} &= 2Ge_{ij} \\ \varphi_C &= b_C\varepsilon + p_C/N_{CC} + p_L/N_{CL} + 3a_C\Delta T \\ \varphi_L &= b_L\varepsilon + p_C/N_{CL} + p_L/N_{LL} + 3a_L\Delta T \end{aligned} \tag{13-217}$$

As represented in Eq. (13-215), $\varphi d\Omega_0$ is the change undergone by volume $\phi_0 d\Omega_0$ under thermomechanical loading; $\varphi_C d\Omega_0$ (resp. $\varphi_L d\Omega_0$) in Eq. (13-217) represents the change undergone by the volume $\phi_0 S_C d\Omega_0$ (resp. $\phi_0 S_L d\Omega_0$) under the combined action of ε, p_C, p_L, and ΔT. For its part, the overall volume currently occupied by ice crystals (resp. the liquid water) is $\phi_C d\Omega_0$ (resp. $\phi_L d\Omega_0$) with†

$$\phi_C = \phi_0 S_C + \varphi_C \quad (\text{resp. } \phi_L = \phi_0 S_L + \varphi_L) \tag{13-218}$$

where S_J is the current saturation related to phase J with

$$S_C + S_L = 1 \tag{13-219}$$

In Eq. (13-217), coefficients b_L and b_C are the generalized Biot coefficients related to the liquid and to the solid crystals respectively; N_{JK} are the generalized Biot coupling moduli; and $3a_L$ and $3a_C$ are the coefficients related to the thermal volumetric dilation of the pore volume occupied by the liquid and the ice crystals, respectively:

$$b_C + b_L = b = 1 - \frac{K}{k_s}; \quad \frac{1}{N_{JJ}} + \frac{1}{N_{LC}} = \frac{b_J - \phi_0 S_J}{k_s}; \quad a_J = \alpha_s (b_J - \phi_0 S_J) \tag{13-220}$$

Unsaturated poroelastic properties b_J and N_{JK} are current properties depending on the current saturation degree S_J. In contrast to general relations given in Eq. (13-216), knowledge of S_J, ϕ_0, and the solid matrix properties provide the missing relations that allow for the expression of b_J and N_{JK}. Assuming that all pores deform the same when they are subjected to the same pore pressure, Biot coefficient b_J can be expressed in the simple form[28,29]

$$b_J = b S_J \tag{13-221}$$

where b is the Biot coefficient related to the porous solid [as defined by Eqs. (13-215) and (13-216)].

13.5.3 Development of Strain during Freezing

Experimental studies of length changes of cement paste and concrete samples exposed to low temperatures are traditionally conducted without applying any external load. Under this stress-free condition, where $\sigma = 0$, the solid-liquid equilibrium condition [Eq. (13-209)],‡ the first of constitutive Eq. (13-217) and relations established in Eq. (13-220) combine to give the volumetric dilation ε in the form

$$\varepsilon = \frac{b p_L + (b_C \Sigma_m - 3\alpha_s K)\Delta T}{K} \tag{13-222}$$

†In poromechanics, this decomposition is known as the Lagrangian saturation and was originally developed by Olivier Coussy.

‡For simplicity, at this stage, assume that $1 - \rho_C^0 / \rho_L^0$ is small. Later on, the consequences of this assumption will be discussed in detail.

The determination of ε requires the determination of p_L. Therefore, it is assumed that the pores are fully saturated by liquid water before the porous material is exposed to low temperatures.

The current mass densities, ρ_C and ρ_L, of the crystal and liquid are linked to pressures p_C and p_L, and to the current cooling ΔT through the linearized constitutive equation:

$$\rho_C = \rho_C^0\left(1+\frac{p_C}{K_C}+3\alpha_C\Delta T\right) \tag{13-223}$$

and

$$\rho_L = \rho_L^0\left(1+\frac{p_L}{K_L}+3\alpha_L\Delta T\right) \tag{13-224}$$

where K_C and K_L, $3\alpha_C$ and $3\alpha_L$ are the bulk modulus and the thermal volumetric dilation[§] coefficient, respectively, of crystal and liquid. Consistent with Eq. (13-218), the total mass of water m_W currently contained in the porous material per unit of initial volume $d\Omega_0$ in both liquid and solid form is

$$m_W = \rho_C(\phi_0 S_C+\varphi_C)+\rho_L(\phi_0 S_L+\varphi_L) \tag{13-225}$$

Substituting the poroelastic constitutive equations both for crystal and liquid [see Eq. (13-217)], and combining Eqs. (13-223) and (13-224), Eq. (13-225) yields:

$$m_W = \rho_L^0\phi_0+\rho_L^0\left(\upsilon_{\Delta\rho}+\upsilon_{\varphi}\right) \tag{13-226}$$

where

$$\upsilon_{\Delta\rho} = \left(\frac{\rho_C^0}{\rho_L^0}-1\right)\phi_0 S_C \tag{13-227}$$

and

$$\upsilon_{\varphi} = b\varepsilon+\frac{p_L}{M_L}+\frac{p_C}{M_C}+3(\phi_0 S_L\alpha_L+\phi_0 S_C\alpha_C+a_L+a_C)\Delta T \tag{13-228}$$

while M_J is defined by

$$\frac{1}{M_J}=\frac{1}{N_{JJ}}+\frac{1}{N_{LC}}+\frac{\phi_0 S_J}{K_J} \tag{13-229}$$

The term $\upsilon_{\Delta\rho}$ captures the pore volume change due to the change in saturation S_J under the sole condition established in Eq. (13-219) and to the mass density difference between the two constituents. The term υ_{φ} expresses the pore volume change due to the

[§]These thermal dilation coefficients are not to be mistaken for $3a_L$ and $3a_C$.

thermomechanical loading in unsaturated conditions. Using the liquid-crystal equilibrium condition [see Eq. (13-209)]:

$$v_{\varphi} = b\varepsilon + \frac{p_L}{M} + \frac{\Sigma_m}{M_C}\Delta T + 3\sum_{J=L,C}\left(\phi_0 S_J \alpha_J + a_J\right)\Delta T \tag{13-230}$$

where $1/M = 1/M_C + 1/M_L$. Using Eqs. (13-219), (13-220), and (13-229) provides the expression:

$$\frac{1}{M} = \frac{1}{N} + \phi_0\left(\frac{S_C}{K_C} + \frac{S_L}{K_L}\right) \tag{13-231}$$

Because the liquid cannot escape from the sealed freezing sample, the total mass m_W remains constantly equal to the initial water mass $\rho_L^0\phi_0$; therefore

$$v_{\Delta\rho} + v_{\varphi} = 0 \tag{13-232}$$

Substituting Eqs. (13-227) and (13-230) into Eq. (13-232) provides an equation linking ε, p_L, and ΔT. Eliminating p_L between the latter and Eq. (13-222) while using Eq. (13-220) gives

$$\varepsilon = \varepsilon_{\text{th}} + \varepsilon_{\Delta\rho} + \varepsilon_{\Sigma_m} \tag{13-233}$$

where

$$\varepsilon_{\text{th}} = -3\left[\alpha_s + \frac{\phi_0 bM}{K_u}(S_C\alpha_C + S_L\alpha_L - \alpha_s)\right]\Delta T \tag{13-234}$$

$$\varepsilon_{\Delta\rho} = \frac{\phi_0 bM}{K_u} S_C\left(1 - \frac{\rho_C^0}{\rho_L^0}\right) \tag{13-235}$$

$$\varepsilon_{\Sigma_m} = \frac{bM}{K_u}\left(\frac{S_C}{M_L} - \frac{S_L}{M_C}\right)\Sigma_m\Delta T \tag{13-236}$$

where** $K_u = K + b^2M$. By exploiting the relations between the different poroelastic coefficients, Eq. (13-236) now becomes $\varepsilon_{\Sigma_m} = \frac{\phi_0 bM}{K_u} S_C S_L\left(\frac{1}{K_L} - \frac{1}{K_C}\right)\Sigma_m\Delta T$. Since a solid has a higher bulk modulus than a liquid, the liquid-solid equilibrium results in a volumetric dilation strain $\varepsilon_{\Sigma_m} \geq 0$, which would be the sole source of the swelling mechanism whenever $\rho_C^0 \geq \rho_L^0$: this is generally the case except for water and a few other substances.

**Note that because of Eq. (13-231), K_u represents the bulk modulus of a sealed porous material whose pores would be saturated by a fictitious fluid having $1/(S_L/K_L + S_C/K_C)$ as bulk modulus.

From Eqs. (13-233) to (13-236) the volumetric strain depends on the ice saturation, S_C. When a sealed sample is considered, cooling occurs under undrained condition where the liquid cannot escape from the sample [see Eq. (13-232)]. In this case the liquid is significantly pressurized. Opposite to the case studied in Sec. 13.5.1, in particular, Eq. (13-213), here it is necessary to take into account the influence of p_L on S_C, which means that the first term in the left-hand member of the liquid-solid equilibrium equation [see Eq. (13-209)] cannot now be discarded. In addition to the cumulative pore volume fraction $G(r)$ occupied by pores having a pore entry radius greater than r, which is already known (see Fig. 13-36), and the pore entry radius r above which the porous space is filled by ice—still governed by the Laplace equation [Eq. (13-210)]—the liquid pressure remains to be determined.

Considering Eqs. (13-222) and (13-232), where a poor compressibility of the solid matrix with regard to water and low values of thermal and entropic term exists when compared to the mass change density term $\phi_0 S_C\left(1-\frac{\rho_C^0}{\rho_L^0}\right)$, it is found that

$$p_L \simeq \left(1-\frac{\rho_C^0}{\rho_L^0}\right) S_C \frac{K_L K_C}{S_L K_C + S_C K_L} \tag{13-237}$$

In summary, the pore entry radius r, above which the porous space is filled by ice at a given cooling ΔT, is obtained by substituting Eq. (13-237) for p_L in Eq. (13-211) and solving the following equation:

$$\frac{2\gamma_{CL}}{r} + \left(1-\frac{\rho_C^0}{\rho_L^0}\right)^2 G(r)\frac{K_L K_C}{[1-G(r)]K_C + G(r)K_L} - \Sigma_m \Delta T = 0 \tag{13-238}$$

which then provides the ice saturation as a function of cooling $S_C = G[r(\Delta T)]$ for the undrained condition. For a given ice saturation, the cooling applied in the undrained condition should be greater than that applied in drained condition [see the first term of the left-hand side of Eq. (13-238)].

Figure 13-37 shows ice saturation S_C as a function of the cooling $\Delta T = T_m - T$ for the mortar given in Fig. 13-36 and tested in a sealed sample (the undrained condition). It clearly shows that less ice forms when compared to the drained condition. The used bulk moduli values are $K_L = 1.7$ GPa , $K_C = 7.81$ GPa. Figure 13-38 shows the volumetric dilation undergone during undrained freezing predicted for the mortar of Figure 13-36 from Eqs. (13-233) to (13-236). The following values were used: $k_s = 42.4$ GPa, $K = 17.7$ GPa, $\phi_0 = 0.34$, $3\alpha_s = 18 \cdot 10^{-6}$ K^{-1}, $3\alpha_L = -286.3 \cdot 10^{-6}$ K^{-1}, $3\alpha_C = 155 \cdot 10^{-6}$ K^{-1}. Figure 13-39 shows the resulting liquid pressure. Even with less-formed ice, the pressure intensity reaches very great values due solely to the water volumetric expansion of 9 percent that is associated with its solidification, according to Eq. (13-237).

13.5.4 Effect of Air-Entrained Voids

As above discussed, freezing of water in fully saturated porous material under undrained condition (like in a sealed sample) generates pressures so high that material

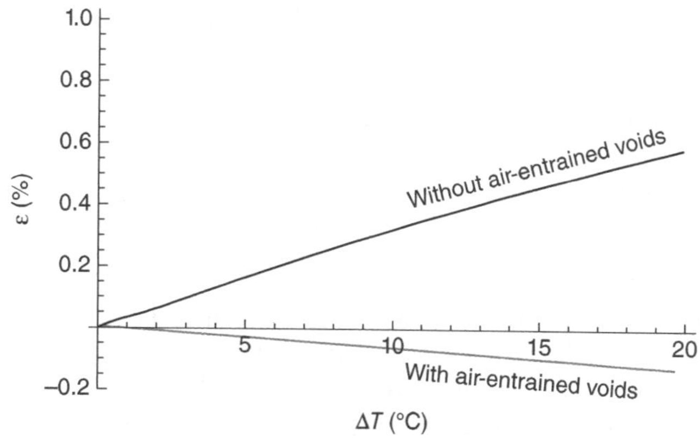

FIGURE 13-38 Volumetric dilation undergone during freezing predicted for the mortar of Figs. 13-36 and 13-37 in the absence of air-entrained voids (dilation, upper curve) and with air-entrained voids (shrinkage, lower curve). The latter curve accounts for the air void acting both as an expansion reservoir and as a cryopump (see Fig. 13-40).

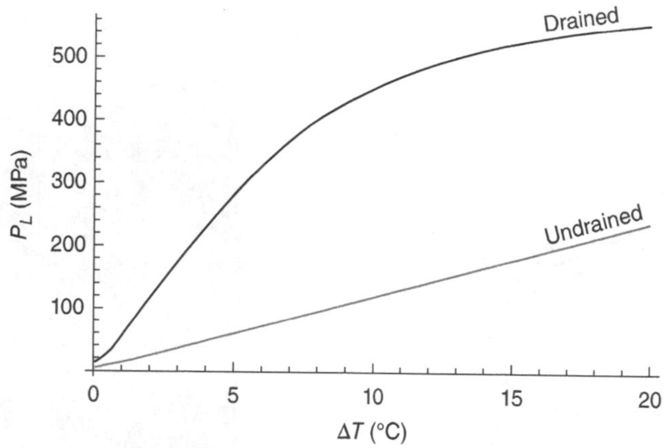

FIGURE 13-39 Liquid pressure generated during freezing predicted for the mortar of Figs. 13-36 and 13-37 in the absence of air-entrained voids.

In a sealed sample composed of a porous material fully saturated with water, the liquid pressurization limits the extension of the frozen zone and finally the liquid overpressure as the cooling increases. This is illustrated by comparing the curves calculated from the ice saturation versus cooling with and without considering the pressurization effect. In both cases, the intensity of the pressure becomes so high that damage is unavoidable.

damage is unavoidable. This problem can be avoided when micrometric air-voids are properly distributed inside the cement matrix. Powers in his seminal work[31] proposed the first comprehensive theory of concrete frost damage. For saturated concrete—as the temperature decreases below 0°C—some water is expelled from the capillary pores as a result of a volume increase caused by the ice formation, generating a hydraulic pressure as the transient water flows toward an air void. His model demonstrated the relevance of air voids in reducing hydraulic pressure; it also predicted analytically the value of a reasonable critical spacing factor between air voids to prevent cracking in the matrix.

Recently low-temperature SEM has revealed that the expelled water at the air-void entry is under the form of ice, as shown in Fig. 13-40. This observation allows for the completion of Powers' initial analysis: An air void can act both as an expansion reservoir and as a cryopump. Because the pressure in the air void is atmospheric, the liquid water expelled from the freezing sites freezes when entering the air void. As long as there is space for ice crystals to form in an air void (thus acting as an expansion reservoir), the pressure of these crystals remains atmospheric ($p_C = 0$). As a result, according to the liquid-solid equilibrium Eq. (13-209), the pressure of the liquid in contact with the crystals equals to $p_L = -\frac{\rho_L^0}{\rho_C^0}\Sigma_m \Delta T \leq 0$: the liquid depressurizes near the ice formed at the

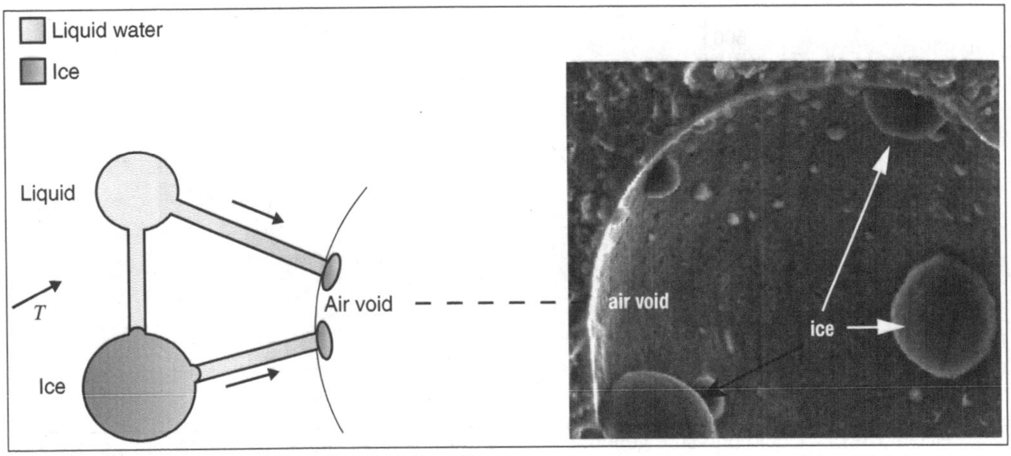

FIGURE 13-40 Air voids acting both as an expansion reservoir and as cryopumps.

The pressurized liquid water that is expelled from the freezing sites in the cement paste is drained toward the closest air void, where it freezes when entering and then remains at zero (atmospheric) pressure due to the large air-void extent. In order to further meet the liquid-crystal equilibrium condition, the liquid depressurizes within the porous matrix in the immediate vicinity of the air void. As a result, the air void acts as a cryopump, whose efficiency increases with increasing cooling. This process in the formation of ice has been directly observed in experiments performed on porous cement pastes. The diagram on left represents the cryosuction process, and the scanning electron microscopy image on the right shows the formation of ice inside the entrained air void.

air-void entry (thus acting as a cryopump). When equilibrium is restored (homogeneous pressure and temperature), substitution of the latter and $p_C = 0$ in the first of poroelastic constitutives Eq. (13-217) gives

$$\varepsilon\big|_{p_C(R)=0} = -\left(\frac{\rho_L^0}{\rho_C^0}\frac{b}{K}\Sigma_m S_C + 3\alpha_s\right)\Delta T \le 0 \tag{13-239}$$

where ice saturation S_C as a function of the cooling $\Delta T \le 20°\text{C}$ approximates to the curve of the undrained case shown in Fig. 13-37. As indicated in Eq. (13-239), the presence of air voids leads to volumetric shrinkage instead of dilation; compare the curves labeled "(with void)" and "(with no void)" in Fig. 13-38, as formerly observed by Powers and Helmuth,[32] in Fig. 13-37.

Predictions using Eq. (13-239) are equilibrium values. They correspond to infinitely slow cooling rates with regard to the rate of the liquid flow toward the air void achieving the equilibrium. This discussion is restricted to determining the validity range of assuming infinitely slow cooling rates. The characteristic time $\tau_{\Delta T}$, whereby the cooling rate is scaled with regard to the solidification within a porous material, can be assessed in the form:

$$\tau_{\Delta T} = (\Delta T)_{\text{ch}} / \overset{\circ}{\overline{\Delta T}} \tag{13-240}$$

where $(\Delta T)_{\text{ch}}$ is the characteristic cooling defined in Eq. (13-214), while $\overset{\circ}{\overline{\Delta T}}$ is the imposed cooling rate. Characteristic time τ_w, which scales the hydraulic diffusion of liquid water toward the void, can be expressed in the form:

$$\tau_w = \frac{L^2}{4D} \tag{13-241}$$

where L is the spacing factor, the distance between the centers of two neighboring spherical voids of radius R being equal to $2R + 2L$ (in other words, L is the half-distance between the surfaces of two adjacent air-voids). The hydraulic diffusivity, D, is expressed as follows:

$$D = \frac{\kappa}{\eta_L} \Big/ \left[\frac{b^2(1+\nu)}{3K(1-\nu)} + \frac{\phi_0}{K_L} + \frac{b-\phi_0}{k_s}\right] \tag{13-242}$$

where ν is the Poisson coefficient of the porous solid, and κ is its intrinsic permeability, while η_L ($\simeq 1.79 \times 10^{-3}\,\text{Pa} \cdot \text{s}$ at $273\,\text{K}$) is the liquid water viscosity. Cooling rates can be considered as infinitely slow with regard to the liquid flow if $\tau_w \ll \tau_{\Delta T}$, providing the condition:

$$L^2 \ll 4D \times (\Delta T)_{\text{ch}} / \overset{\circ}{\overline{\Delta T}} \tag{13-243}$$

To explore the latter condition, the value $\overset{\circ}{\overline{\Delta T}} = 3\,\text{K/h}$ is adopted, which is in the order of magnitude of in situ cooling rates. For the mortar here analyzed, $(\Delta T)_{\text{ch}} = 6\,\text{K}$

was found, cf. Eq. (13-214). If $\nu = 0.3$, $\kappa = 10^{-21}\,\mathrm{m}^2$, and the poroelastic properties reported above are taken, then Eqs. (13-242) and (13-243) yield $L \ll 8$ mm. This condition indicates that the prediction [see Eq. (13-239)] is relevant as regards in situ cooling rates because the spacing factor experimentally observed to prevent internal freezing damage are sub-millimeter in order (usually around 250 μm).

If this discussion is restricted to the mechanical effects, the right choice of spacing factor L ultimately has to meet the condition that the void must be large enough in order to avoid any in-void pressure build up. Therefore, the ability of unsaturated zones to accommodate the extra liquid water may be disregarded. The corresponding condition has then to express that the void volume $4\pi R^3/3$ is capable of welcoming the totality of the liquid water expelled from the adjacent freezing sites owing to the ice-liquid difference of density. The zone of influence of a void is the spherical shell, with inner radius R and outer radius $R+L$; its volume is $V_{\text{shell}} = (4\pi/3)[(R+L)^3 - R^3]$. At the equilibrium related to cooling ΔT, the volume of liquid that has frozen within this shell is $V_{L\to C} = \phi_0 S_C(\Delta T) \times V_{\text{shell}}$, where $S_C(\Delta T)$ is given by the $S_C - \Delta T$ curve. The volume change during solidification is expressed as $(\rho_L^0/\rho_C^0 - 1) \times V_{L\to C}$. Therefore, pressure build up within the void will not occur if the following relation is satisfied

$$\left(\frac{\rho_L^0}{\rho_C^0} - 1\right) \times \phi_0 S_C(\Delta T) \times \frac{4\pi}{3}[(R+L)^3 - R^3] < \frac{4\pi}{3} R^3 \tag{13-244}$$

from which a critical spacing factor $\bar{L}$ can be estimated as

$$L < \bar{L} = R\left(\left(1 + \frac{1}{\left(\frac{\rho_L^0}{\rho_C^0} - 1\right)\phi_0 S_C}\right)^{\frac{1}{3}} - 1\right) \tag{13-245}$$

As summarized in Fig. 13-41, Eq. (13-245) demonstrates that (1) the bigger the air void, the more water it can accommodate such that $\bar{L}$ increases with R; and (2) the higher the porosity or the quantity of ice formed or the change in mass density, the more pore water volume is displaced upon freezing: the air void shell must be thinner (a decrease in $\bar{L}$) so that the air void can still accommodate the expelled water. Figure 13-42 shows the normalized ratio $\bar{L}/R$ as a function of ΔT for the mortar of Fig. 13-36 whose $S_C - \Delta T$ is similar to the undrained curve in Fig. 13-37. The values of the ratio $\bar{L}/R$ reported in Fig. 13-42 agree with the order of magnitude of the experimental values[33] known to preserve the cement-based materials from bulk damage under freezing.

The analysis presented herein on the effect of air-entrained voids allows for estimating the critical spacing factor. This material property is the safe half-distance between the surfaces of neighboring air voids to ensure that they remain unfilled during freezing, thereby avoiding damaging mechanical pressurization typical to undrained loading on saturated porous materials. An air void can then act as expansion reservoir in addition to a cryopump, by which water is driven from the cement paste to the entry of

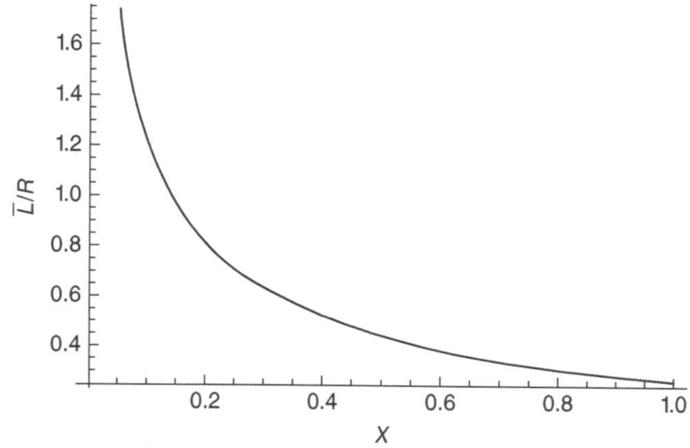

FIGURE 13-41 Critical spacing factor $\bar{L}$ normalized by an air-entrained void radius R plotted against the ice saturation, $X = \left(\rho_L^0 / \rho_C^0 - 1 \right) \phi_0 S_C$. In particular, more ice formation (at a cooling temperature) results in a lowering of the safe critical-spacing factor.

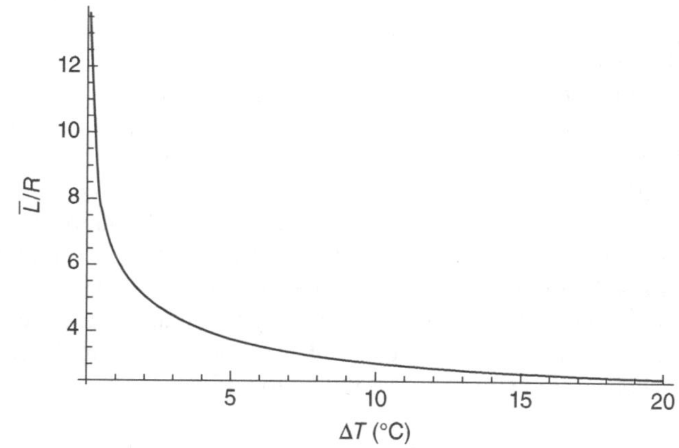

FIGURE 13-42 Critical spacing factor plotted against cooling for the mortar of Fig. 13-36

an air void where ice has formed. Such thermodynamics-based processes, like the so-called cryosuction, eventually lead to shrinkage.

13.6 Mechanics of Concrete Affected by the Alkali-Silica Reaction

Chapter 5 described the damage caused by the alkali-silica reaction (ASR). Mechanical models in the literature propose different approaches both for the kinetic of the reaction and the concrete constitutive behavior (e.g., damage theory, fracture mechanics, and plasticity). Modeling of the ASR and its effect on concrete structures is especially complicated by the influence of internal factors (i.e., the chemical composition of the concrete) and external factors (i.e., environmental conditions). Various approaches to this problem differ in the scale and depend on which phenomenon is being observed. The existing models can be grouped into two classes: (1) microscopic models, in which the reaction and its consequences are observed at the level of a reference volume element and usually consist of a reactive aggregate surrounded by the concrete matrix; and (2) macroscopic models, which consider the consequences of the reaction at the macroscopic level by studying internal variables. Models belonging to the first class focus on the mechanism of the reaction, but their applicability for simulating the behavior of real structures is limited. Although phenomenological models are very useful in predicting structural behavior, they neglect the mechanism of the reaction and concentrate only on its consequences. Details of these approaches can be found in the review articles by Pan et al.[34]

The micromechanical models developed by Bazant and Steffens[35] and Bazant et al.[36] considered the mechanical behavior of a reference volume element consisting in a unit cubic cell of concrete containing one spherical reactive particle. A spherical layer of ASR gel grows radially inward into the particle, which is controlled by the diffusion of water through the gel layer toward the reaction front. Periodical boundary conditions are then applied to the reference volume element in order to obtain the continuum model. In this model, diffusion laws are used to define the mechanism whereby water and ions penetrate the RVE until the reactive front develops around the aggregate. The differential equations are integrated numerically, and a parametric study of the problem clarifies the effects of particle size. Fracture mechanics can be applied at the microscopic level in order to model the effect of the ASR on the tensile strength and the volumetric expansion of concrete. Multon et al.[37] proposed a microscopic model to analyze the development of ASR expansion of mortars containing reactive aggregates of different sizes, taking into account the fixation of alkali. Figure 13-43 represents the relative elementary volume for several reactive aggregate sizes used by the authors. A damage model was introduced to assess the decrease in stiffness of the mortar as a result of cracking caused by the ASR. The authors then used this model to simulate free-expansion tests on different mortars specimens to assess the model's ability to represent real ASR expansions by reproducing the decrease of expansion with the size of the aggregate and the increase of the expansion with the alkali content. Comby-Peyrot at al.[38] and Dunant and Scrivener[39] studied the mechanical behavior of concrete affected by the ASR at the mesoscale. These models take into account the particle size distribution and the reactive aggregate location, which are explicitly presented with refined finite elements at an experimental scale. Based on the analysis of SEM images, Shin et al.[40] developed a microstructure-based finite element approach—including microstructure of aggregates, voids, paste, gels, and cracks—to analyze the mechanical response of concrete subjected to ASR. No structural applications of this approach are available in the literature.

Macroscopic model for concrete affected by the ASR are based on the combination of the salient aspects of the chemical reaction and its mechanical effects. The description of the kinetics of the reaction and the final concrete swelling is based on experimental results and in situ measurements; several laws have been proposed to account the influence of temperature, humidity, reactive sites distribution, and stress state. Originally the mechanical behavior of concrete was simulated using linear elastic constitutive models. Subsequently, several nonlinear models based on damage or fracture mechanics were developed; some of these included creep and shrinkage. The early models proposed for structural analysis in the presence of ASR interpreted the concrete expansion due to gel swelling as an equivalent thermal expansion. Here, the kinetics of the

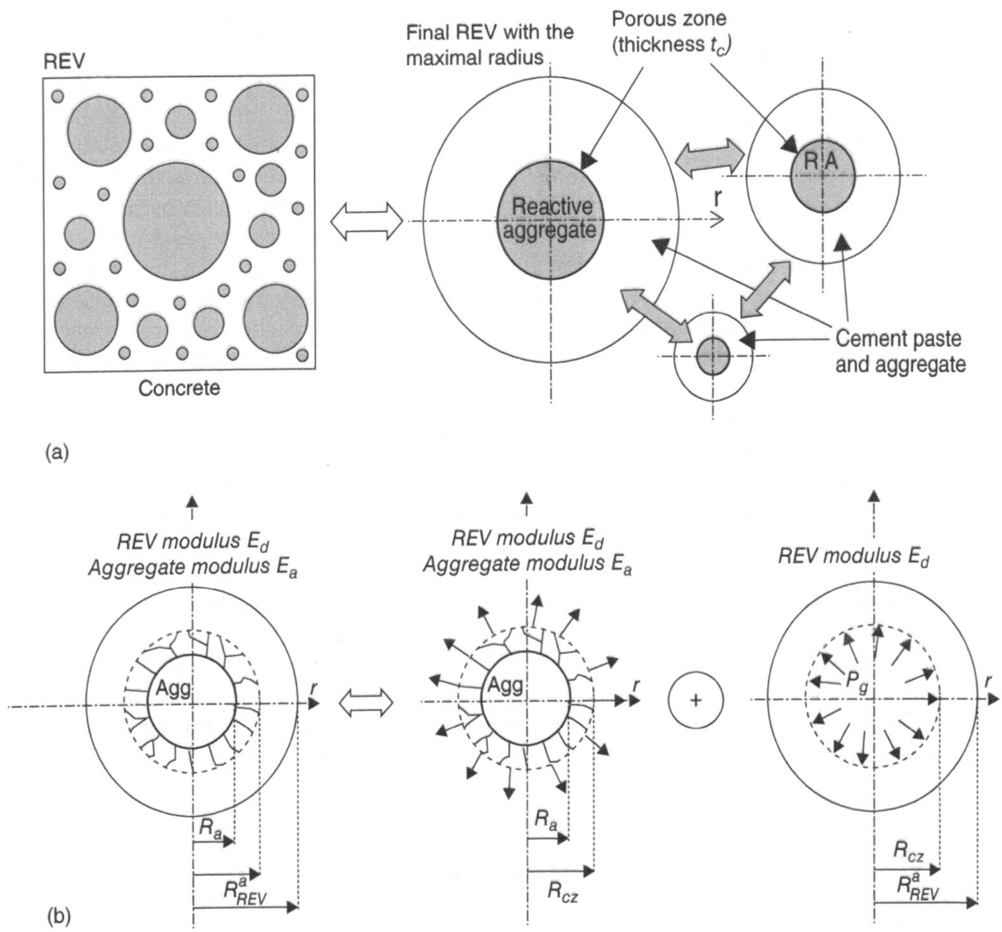

FIGURE 13-43 (*a*) Definition of the relative elementary volume for several reactive aggregate sizes and (*b*) mechanical equilibrium of the damaged relative elementary volume. [After Multon, S., A. Sellier, and C. Martin, Chemo-mechanical modeling for prediction of alkali silica reaction (ASR) expansion, *Cem. Concr. Res.*, vol. 39, pp. 490–500, 2009.]

chemical reaction was not considered, and the principal steps included a thermal analysis of the structure: accurate computation of the temperature distribution with time was determined by calibrating the diffusion constant first and then the thermal expansion coefficient, which were based on the displacements between experimental measures. Introduction of an anisotropic thermal expansion provides more reliable results as it considers indirectly the influence of the stresses on the gel expansion. Indeed, the application of a compressive stress in a given direction hinders the reaction and limits the gel expansion. In order to simulate a lower expansion in this direction, the value of the thermal coefficient α_i can be properly reduced. The presence of different constraints in different directions may cause nonisotropic compressive stress; for example, in arch dams the constraint in the direction of the abutments is much stronger than the constraints in the vertical and transversal directions. However, the introduction of an anisotropic thermal expansion factor involves an additional source of uncertainty regarding the quantity of expansion coefficients and their values. In their analysis of the Isola arch-gravity dam in Switzerland, Malla and Wieland[41] demonstrated that the expansive behavior causing the cracking of the dam could be satisfactorily explained by assuming an isotropic gradual increase of the volume of concrete due to the ASR. Macrocracks found on the upper part (downstream) and the lower part (upstream) of the upper gallery could result from the high values of stress in those zones. Models based on the equivalent thermal expansion allow a preliminary evaluation of the stresses in a structure affected by the ASR. Although this approach is advantageous because of its simplicity, it neglects the effect that the development of the reaction has on the mechanical strength and stiffness of the material. The unique link between the model and the effective stress state of the structure due to the ASR is the equality of the displacements measured in the most significant points of the structure.

Ulm et al.[42] developed a chemoelastic model within the framework of poromechanics. The theory of porous media, first proposed by Biot[43] with reference to soil mechanics, describes porous materials as the superposition of homogenized phases, the skeleton, and fluid filling the pores, and takes into account the change of the porosity and its consequences on the overall behavior. Used in different contexts (e.g., see the previous section on the application of poromechanics for ice formation), it was applied to the ASR for the first time by Ulm et al.[42]; other poromechanics models have been proposed since. To capture the expansive material behavior at a macrolevel, Ulm et al.[42] considered the concrete skeleton and the gel produced by the ASR as two phases acting in parallel, as shown in the one-dimensional sketch in Fig. 13-44. In this figure, σ is the macroscopic stress due to external forces, and ε is the corresponding overall strain.

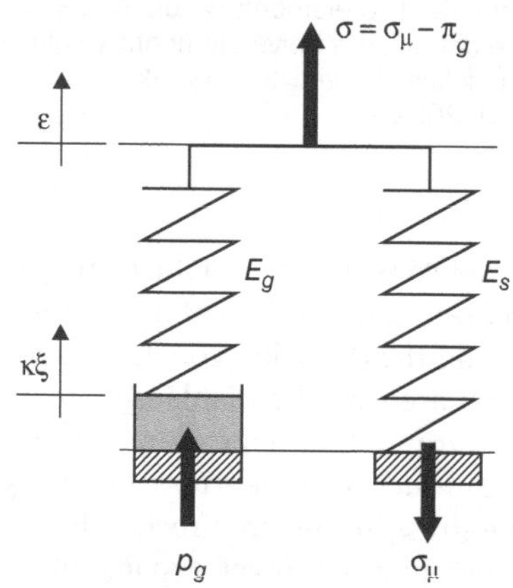

FIGURE 13-44 Chemoelastic pressure-spring device of ASR swelling. (From Ulm, F. J., O. Coussy, L. Kefei, and C. Larive, Thermo-chemo-mechanics of ASR expansion in concrete structures, *ASCE J. Eng. Mech.*, vol. 126, no. 3, pp. 233–242, 2000.)

Thus the solid skeleton is represented by the spring of stiffness E_s, and the corresponding stress is denoted by σ_μ; the gel is represented by an elastic spring of stiffness E_g and by a pressure cell with the swelling pressure denoted by p_g. The skeleton is assumed to be linearly elastic. The inelastic deformation of the cell is $k\xi$, with ξ being the reaction extent variable (ranging from 0 to 1) whose evolution follows the law proposed by Larive[44]; k is the intrinsic dilatation coefficient of the reaction products. Neglecting the free expansion of the reaction products in the concrete porosity, the stress equilibrium in this chemoelastic device becomes

$$\sigma = \sigma_\mu - p_g = E_s\varepsilon + E_g(\varepsilon - k\xi);\ \sigma_\mu = E_s\varepsilon;\ p_g = -E_g(\varepsilon - k\xi) \tag{13-246}$$

Worth noting is that under stress-free conditions [$\sigma = 0$ in Eq. (13-246)], the internal swelling pressure of the expansive products is self-balanced by the tension in the skeleton, causing the overall material to swell in response. In the absence of external loads, the total strain ε (measurable in laboratory test specimens) is related through a chemical dilatation coefficient to the reaction extent ξ, that is,

$$\varepsilon = \left(\frac{kE_g}{E_g + E_s}\right)\xi \tag{13-247}$$

This one-dimensional model can be extended to the three-dimensional case. The constitutive equations are obtained by partial derivation of a proper free energy. The development of the ASR is described by the phenomenological variable $\xi(t)$ $(0 \le \xi \le 1)$ as function of time t, the characteristic time τ_{ch}, and the latency time τ_{lat} of the reaction; both are dependent on temperature T. The expression of the function $\xi(t)$ was first proposed by Larive[44] as a result of a far-reaching experimental study at LCPC, and then used by Ulm et al.[42] in their model for the ASR. It assumes the form

$$\xi(t,T) = \frac{1 - e^{-\frac{t}{\tau_{ch}(T)}}}{1 + e^{-\frac{[t - \tau_{lat}(T)]}{\tau_{ch}(T)}}} \tag{13-248}$$

where T is the temperature in K, τ_{lat} is the latency time identifying the inflection point in the expansion curve, and τ_{ch} is the characteristic time describing the exponential decay of the second part of the expansion curve, the ratio τ_{lat}/τ_{ch}, experimentally determined, results between 2.36 and 2.48, and is 2.39 on average. Figure 13-45 shows the definition of the latency time and characteristic time. As shown experimentally, the kinetic energy of the ASR increases with temperature. The temperature influences the kinetics of the reaction through the latency time τ_{lat} and characteristic time τ_{ch}.

Typically, since the ASR is thermal activated, the influence of temperature on the characteristic and latency times is modeled by Arrhenius' law, which is commonly used for the description of thermal-activated phenome

$$\tau_{ch}(T) = \tau_{ch}(T_0)\exp\left[U_{ch}\left(\frac{1}{T} - \frac{1}{T_0}\right)\right];\ U_{ch} = 5400 \pm 500\ \text{K} \tag{13-249}$$

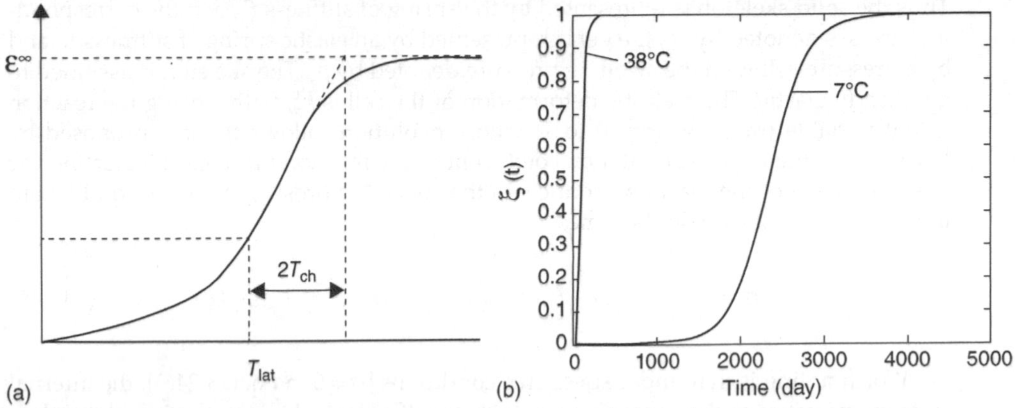

FIGURE 13-45 Physical interpretation of parameters describing swelling evolution of ASR-affected concrete. (From Larive, C., Apports combines de l'experimentationet de la modelisation a la comprehension de l'alcali-reaction et de ses effets mechaniques. Ph.D. thesis, LCPC, 1998.)

$$\tau_{\text{lat}}(T) = \tau_{\text{lat}}(T_0)\exp\left[U_{\text{lat}}\left(\frac{1}{T} - \frac{1}{T_0}\right)\right]; \; U_{\text{lat}} = 9400 \pm 500\ \text{K} \tag{13-250}$$

The model proposed by Ulm et al.[42] for the kinetic of the reaction has been adopted by several authors, as summarized next. The model assumes that temperature influences the kinetics of the reaction and not the final value of the volumetric expansion due to the ASR. After their work, others poromechanics models have been proposed. The points on which the subsequent literature has focused are as follows:

- Ulm et al.[42] did not take into account the volumetric fraction of the phases. This parameter has been introduced in several models (see Comi et al.[45]).
- The constitutive behavior of the concrete skeleton is assumed to be linearly elastic. Subsequent models have described the behavior of the concrete skeleton by different constitutive laws. Steffens et al.[46] considered an elastic-plastic behavior; Farage et al.[47] and Fairbairn et al.[48] applied a smeared crack approach; and Grimal et al.[49] and Comi et al.[45] proposed damage models. Grimal et al.[49] have also taken into account concrete creep.
- Some authors developed a model starting from the two-phase scheme proposed in Ulm et al.[42] (e.g., Farage et al.,[47] Fairbairn et al.,[48] and Comi et al.[45]), while others developed multiphase formulations (e.g., Grimal et al.[49] and Comi and Pignatelli[50]), where gel and water are considered as different phases.
- Usually for simplicity's sake, the gel formation and the gel expansion are considered as simultaneous. Steffens et al.,[46] however, adopted a two-stage mechanism to describe the formation and the expansion of the gel, taking into account the initial porosity filling process. The loss of the swelling capacity of the gel with time is taken into account through a characteristic time of aging.
- Dormieux et al.[51] and Lemarchand et al.[52] derived the homogenized equations typical of poromechanics that first appeared in Ulm et al.[42] for concrete affected

by the ASR starting from a micro-reference volume element and adopting a micromechanics approach.

Test Your Knowledge

13.1 Suppose the objective of a laboratory experiment is to measure the effect of freezing on the elastic modulus of concrete. You take two identical concrete samples from the fog room, and then test one sample in the saturated condition and the other, in a frozen condition (say –20°C). Assume that the freezing was done carefully and, therefore, did not generated microcracks. Which concrete will have a higher elastic modulus: saturated-concrete or frozen concrete?

13.2 A 33 MPa concrete is made with limestone aggregate. Suppose you replace 50 percent of the aggregate with solid steel balls (about the same size as the aggregates). Is the compressive strength going to increase? What about the elastic modulus? Please justify your answer.

13.3 A series of experiments on identical specimens of Maxwell material were performed such that in each experiment the strain rate was held constant. Sketch a family of stress-strain curves corresponding to three different strain rates: very slow, moderate, and very fast. For each case determine $E(0)$. Discuss the implications of results in practical applications.

13.4 Study the response of a standard-solid material subjected to a cyclic strain $\varepsilon(t) = \varepsilon_0 \cos wt$, where ε_0 is the strain amplitude and w is the frequency.

13.5 Using the principle of superposition, consider the following conditions for a 33 MPa compressive strength concrete (*justify* your answers):

(*a*) The basic creep of the concrete at 28 days is 300×10^{-6} under a compressive load of 7 MPa. Can you estimate the basic creep of the same concrete at 90 days under a compressive load of 7 MPa?

(*b*) The basic creep of the concrete at 28 days is 300×10^{-6} under a compressive load of 7 MPa. Can you estimate the basic creep of the same concrete at 28 days under a compressive load of 11 MPa?

(*c*) The basic creep of the concrete at 90 days is 1200×10^{-6} under a compressive load of 27 MPa. Can you estimate the basic creep of the same concrete at 90 days under a compressive load of 7 MPa?

(*d*) The basic creep of the concrete at 28 days is 300×10^{-6} under a compressive load of 7 MPa. The drying shrinkage (50 percent R.H.) at 28 days is 100×10^{-6}. Can you estimate the creep of the same concrete at 28 days under a compressive load of 11 MPa and exposed to 50 percent R.H.?

(*e*) The basic creep of the concrete at 28 days is 300×10^{-6} under a compressive load of 7 MPa. The drying shrinkage (50 percent R.H.) at 28 days is 100×10^{-6}. Can you estimate the basic creep of the same concrete at 28 days under a compressive load of 27 MPa?

13.6 A Burgers model is made by connecting a Maxwell and a Kelvin model in series. Suppose that a Burgers material is maintained under a constant stress until time t_1 and then unloaded. Draw the graph of strain vs. time.

13.7 Assume that a mass concrete structure should not have a temperature difference greater than 13°C. Given the following conditions: adiabatic temperature rise, 42°C; ambient temperature, 23°C; temperature losses, 15°C. Determine the maximum temperature of fresh concrete to avoid cracking.

13.8 Compute the energy release rate for the double cantilever beam when loaded by end moments.

13.9 Show for plane stress that the energy release rate G is equal to

$$G = \frac{K_I^2}{E} + \frac{K_{II}^2}{E} + \frac{L_{III}^2}{2S}$$

where S is the shear modulus.

13.10 Compare critically the advantages and limitations of the various techniques used for the determination of the fracture process zone in concrete.

References

1. Grimvall, G., *Thermophysical Properties of Materials*, North-Holland, Amsterdam, 1986.
2. Hill, R., *Proceedings of the Physical Society of London*, Vol. 65-A, p. 349, 1952.
3. Hirsch, T.J., *ACI J.*, Vol. 59, p. 427, 1962.
4. Hansen, T.C., *ACI J.*, Vol. 62, No. 2, pp. 193–216, Feb. 1965.
5. Hashin, Z., *J. Appl. Mech.*, Vol. 29, No. 1, pp. 143–150, March 1962.
6. Counto, U.J., *Mag. Concr. Res.*, Vol. 16, No. 48, pp.129–138, 1964.
7. Nilsen, A.U., and P.J.M. Monteiro, *Cem. Concr. Res.*, Vol. 23, pp. 147–151, 1993.
8. Hashin, Z., and P.J.M. Monteiro, *Cem. Concr. Res.*, Vol. 32, No. 8, pp. 1291–1300, 2002.
9. Zimmerman, R.W., *Mech. Mat.*, Vol. 12, pp. 17–24, 1991.
10. Dischinger, F., *Der Bauingenieur*, Vol. 18, pp. 487–520, 539–562, 595–621, 1937.
11. Lubliner, J., *Nucl. Eng. Design*, Vol. 4, p. 287, 1966.
12. McHenry, D.A., New Aspect of Creep in Concrete and Its Application to Design, *ASTM Proc.*, Vol. 43, pp. 1069–1084, 1943.
13. Sharma, M.G., *Viscoelasticity and Mechanical Properties of Polymers*, University Park, Pennsylvania, PA, 1964.
14. U.S. Bureau of Reclamation, Creep of Concrete Under High Intensity Loading, Concrete Laboratory Report No. C-820, Denver, Colorado, 1956.
15. Ross, A.D., *Struc. Eng.*, Vol. 15, No. 8, pp. 314–326, 1937.
16. Souza Lima, V.M., D. Zagottis, and J.C. André, *XI National Conference on Large Dams*, Ceará, Brazil, Theme I, 1, 1976.
17. Griffith, A.A., The Phenomena of Rupture and Flow in Solids, *Philosophical Transactions*, Royal Society of London, Series A 221, pp. 163–198, 1920.
18. Inglis, C.E., Stresses in a Plate due to the Presence of Cracks and Sharp Corners, *Trans. Inst.*, Naval Architects, Vol. 55, pp. 219–241, 1913.
19. Irwin, G.R., *Trans ASME, J. Appl. Mech.*, Vol. 24, pp. 361–364, 1957.
20. Cedolin, L., Introduction to Fracture Mechanics of Concrete, *Il Cemento*, p. 283, 1986.
21. Bazant, Z.P., *Proceedings of the Academy of Sciences*, Vol. 101, pp. 13400–13407, 2004.
22. Cedolin, L., S.D. Poli, and I. Iori, *J. Eng. Mech.*, Vol. 113, p. 431, 1987.
23. Nemati, K.M., P.J.M. Monteiro, and N.G.W. Cook, *ASCE J. Mat. Civ. Eng.*, Vol. 10, No. 3, pp. 128–134, 1998.
24. Barenblatt, G.I., *Scaling*, Cambridge University Press, 2003.
25. Barenblatt, G.I., On the equilibrium cracks formed in brittle fracture, *J. Appl. Mathematics Mechanics* (PMM), Vol. 16, No. 3, pp. 434–444; No. 4, pp. 706–721; No. 5, pp. 893–900, 1959.
26. Biot, M.A., General theory of three dimensional consolidation, *J. Appl. Phys.*, Vol. 12, pp. 155–164, 1941.
27. Coussy, O., *Poromechanics*, John Wiley and Sons, 2004.

28. Coussy, O., *Mechanics and Physics of Porous Solids*, John Wiley and Sons, 2010.
29. Coussy, O., and Monteiro, P.J.M., Poroelastic model for concrete exposed to freezing temperatures, *Cem. Concr. Res.*, Vol. 38, pp. 40–48, 2008; and Errata *Cem. Concr. Res.*, Vol. 39, pp. 371–372, 2009.
30. Coussy, O., and Monteiro, P.J.M., Unsaturated poroelasticity for crystallization in pores, *Computers and Geotechnics*, Vol. 34, pp. 279–290, 2007.
31. Powers, T.C., The air requirement of frost-resistant concrete, Highway Research Board, PCA, Bulletin 33, Vol. 29, pp. 184–211, 1949.
32. Powers, T., Helmuth, R.A., Theory of volume changes in hardened Portland cement paste during freezing, Proceedings, Highway Research Board, Bulletin n. 33, Portland Cement Association, Vol. 32, pp. 286–297, 1953.
33. Marchand, J., and Pigeon, M., Résistance du béton à l'écaillagedû au gel en présence de sels fondants: une revue des récentsdéveloppementsdans le domaine (strength of concrete to scaling upon freezing in presence of deicer salts: a review), in: J.P. Bournazel, Y. Malier (eds.), *Proc. of RILEM Conf. "Concrete: from material to structure,"* Arles, France, 1996.
34. Pan, J.W., Y.T. Feng, J.T. Wang, Q.C., Sun , C.H. Zhang, and D.R.J. Owen, Modeling of alkali-silica reaction in concrete: a review, *Front. Struct. Civ. Eng.*, Vol. 6, No. 1, pp. 1–18, 2012.
35. Bazant, Z.P., and A. Steffens, Mathematical model for kinetics of alkali-silica reaction in concrete. *Cem. Concr. Res.*, Vol. 30, pp. 419–428, 2000.
36. Bazant, Z.P., G. Zi, and C. Meyer, Fracture mechanics of ASR in concretes with waste glass particles of different sizes, *J. Eng. Mech., ASCE*, Vol. 126, No. 3, pp. 226–232, 2000.
37. Multon, S., A. Sellier, and C. Martin, Chemo-mechanical modeling for prediction of alkali silica reaction (ASR) expansion. *Cem. Concr. Res.*, Vol. 39, pp. 490–500, 2009.
38. Comby-Peyrot, I., F. Bernard, P.O. Bouchard, F. Bay, and E. Garcia-Diaz, Development and validation of a 3D computational tool to describe concrete behaviour at mesoscale. Application to the alkali-silica reaction. Comput. Mater. Sci., Vol. 46, No. 4, pp. 1163–1177, 2009.
39. Dunant, C.F., and K.L. Scrivener, Micro-mechanical modelling of alkali-silica-reaction-induced degradation using the AMIE framework. *Cem. Concr. Res.*, Vol. 40, No. 4, pp. 517–525, 2010.
40. Shin, J.H., L.J. Struble, and R.J. Kirkpatrick, Modeling alkali-silica reaction using image analysis and finite element analysis. Proc. of the 6th Int. Symp. on Cem. & Concr. and Canmet/ACI Int. Symp. on Concr. Tech. for Sustainable Development, Vol. 3, pp. 1669–1675, 2006.
41. Malla, S., and M. Wieland, Analysis of an arch-gravity dam with a horizontal crack. Comput. Struct., Vol. 72, pp. 267–278, 1999.
42. Ulm, F.J., O. Coussy, L. Kefei, and C. Larive, Thermo-chemo-mechanics of ASR expansion in concrete structures, *ASCE J. Eng. Mech.*, Vol. 126, No. 3, pp. 233–242, 2000.
43. Biot, M.A., General solutions of the equations of elasticity and consolidation for a porous material. *J. Appl. Mech.*, Vol. 55, No. (A-7), 1956.
44. Larive, C., Apports combines de l'experimentationet de la modelisation la comprehension de l'alcali-reaction et de ses effets mechaniques. Ph.D. thesis, LCPC, 1998.
45. Comi, C., B. Kirchmayr, and R. Pignatelli, Two-phase damage modeling of concrete affected by alkali-silica reaction under variable temperature and humidity conditions, *Int. J. Solids Struct.*, Vol. 49, pp. 3367–3380, 2012.
46. Steffens, A., K. Li, and O. Coussy, Aging approach to water effect on alkali-silica reaction degradation of structures, *J. Eng. Mech.*, Vol. 129, No. 1, pp. 50–59, 2003.

47. Farage, M.C.R, J.L.D. Alves, and E.M.R. Fairbain, Macroscopic modelling of concrete subjected to alkali-aggregate reaction. *Cem. Concr. Res.*, Vol. 34, pp. 495–505, 2004.
48. Fairbairn, E.M.R., F.L.B. Ribeiro, L.E. Lopes, R.D. Toledo-Filho, and M.M. Silvoso, Modelling the structural behaviour of a dam affected by alkali-silica reaction. *Commun. Numer. Methods Eng.*, Vol. 22, pp. 1–12, 2006.
49. Grimal, E., A. Sellier, Y. Le Pape, and E. Bourdarot, Creep, shrinkage, and anisotropic damage in alkali-aggregate reaction swelling mechanism—Part I: a constitutive model. *ACI Mater. J.*, Vol. 105, No. 3, pp. 227–235, 2008.
50. Comi, C., and R. Pignatelli, A three-phase model for damage induced by ASR in concrete structures. In IV International Conference on Computational Methods for Coupled Problems in Science and Engineering, 2011.
51. Dormieux, L., E. Lemarchand, D. Kondo, and E. Fairbairn, Elements of poro-micromechanics applied to concrete. *Mater. Struct.—Concr. Sci. Eng.*, Vol. 37, pp. 31–42, 2004.
52. Lemarchand, E., L. Dormieux, and F. Ulm, Micromechanics investigation of expansive reactions in chemoelastic concrete, *Phil. Trans. Royal Soc. A*, Vol. 363, pp. 2581–2602, 2005.

Suggestions for Further Study

Elastic Behavior

Christensen, R.M., A Critical Evaluation for a Class of Micromechanics Models, *J. Mech. Pays. Solids*, Vol. 18, No. 3, pp. 379–404, 1990.

Christensen, R.M., *Mechanics of Composite Materials*, Wiley, New York, 1976.

Hashin, Z., Analysis of Composite Materials—A Survey, *J. Appl. Mech.*, Vol. 50, pp. 481–505, 1983.

Hendriks, M.A.N., *Identification of Elastic Properties by a Numerical-Experimental Method, Heron*, Delft University of Technology, The Netherlands, Vol. 36, No. 2, 1991.

Torquato, S., *Random Heterogeneous Materials: Microstructure and Macroscopic Properties*, Springer-Verlag, New York, 2002.

Zohdi, T.I., and P. Wriggers, *An Introduction to Computational Micromechanics*, corrected second printing, Springer-Verlag Berlin Heidelberg, 2008.

Viscoelasticity

Bažant, Z.P., A. Hauggaard, S. Baweja, and F. Ulm, Microprestress-solidification theory for concrete creep. I: Aging and drying effects. *J. Eng. Mech.*, Vol. 123, No. 11, pp. 1188–1194, 1997.

Bažant, Z.P., and S. Baweja, Creep and shrinkage prediction model for analysis and design of concrete structures: Model B3. Adam Neville Symposium: Creep and shrinkage—structural design effects, *ACI SP-194*, A. Al-Manaseer, ed., American Concrete Institute, Farmington Hills, Michigan, 1–83, 2000.

Creus, G.J., *Viscoelasticity—Basic Theory and Applications to Concrete Structures*, Springer-Verlag, Berlin, 1986.

Flugge, W., *Viscoelasticity*, Springer-Verlag, New York, 1975.

Gilbert, R.I., *Time Effects in Concrete Structures*, Elsevier, New York, 1988.

Gilbert, R.I., and G. Ranzi, *Time-Dependent Behaviour of Concrete Structures*, Spon Press, Abingdon, Oxon, 2011.

Neville, A.M., W.H. Dilger, and J.J. Brooks, *Creep of Plain and Structural Concrete*, Longman, New York, 1983.

Rusch, H., D. Jungwirth, and H.K. Hilsdorf, *Creep and Shrinkage: Their Effect on the Behavior of Concrete Structures*, Springer-Verlag, New York, 1986.

Tanabe, T., K. Sakata, H. Mihashi, R. Sato, K. Maekawa, and H. Nakamura, *Creep, Shrinkage and Durability Mechanics of Concrete and Concrete Structures*, Taylor & Francis Group, London, U.K., 2009.

Ulm, F.-J., Z.P. Bazant, and F.H. Wittmann, eds., *Creep, Shrinkage, and Durability Mechanics of Concrete and Other Quasi-Brittle Materials: Proceedings of the Sixth International Conference*, Elsevier, Amsterdam, p. 811, 2001.

Thermal Stresses in Mass Concrete

Acker, P., and M. Regourd, Physicochemical Mechanisms of Concrete Cracking, in *Materials Science of Concrete II*, Skalny, J., and S. Mindess, eds., The American Ceramic Society, Westerville, OH, 1991.

Mihashi, H., and F.H. Wittmann, *Control of Cracking in Early Age Concrete*, Swets & Zeitlinger B.V. Lisse, The Netherlands, 2002.

Polivka, R.M., and E.Wilson, *DOT/DETECT: Finite Element Analysis of Nonlinear Heat Transfer Problems*, Report No. UCB/SESM-76/2. University of California, Berkeley, 1976.

Springenschmid, R. (ed.), *Thermal Cracking in Concrete at Early Ages*, E&FN Spong, London, 1995.

Springenschmid, R. (ed.), *Prevention of Thermal Cracking in Concrete at Early Ages*, E&FN Spong, London, 1998.

Wilson, E., *The Determination of Temperatures within Mass Concrete Structures*, Report No. UCB/SESM-68-17, University of California, Berkeley, 1968.

Fracture Mechanics of Concrete

Bazant, Z.P., *Scaling of Structural Strength*, Taylor & Francis Books, New York, NY, 2002.

Bazant, Z.P., and J. Planas, *Fracture and Size Effect in Concrete and Other Quasibrittle Materials*, CRC Press, Boca Raton, FL, 1998.

Bazant, Z.P., and L. Cedolin, *Stability of Structures: Elastic, Inelastic, Fracture, and Damage Theories*, World Scientific Publishing Co., 2010.

Elfgren, L., and S.P. Shah, eds., *Analysis of Concrete Structures by Fracture Mechanics*, Chapman and Hall, London, 1991.

Shah, S.P., S.E. Swartz, and C. Ouyang, *Fracture Mechanics of Concrete: Applications of Fracture Mechanics to Concrete, Rock and Other Quasi-Brittle Materials*, Wiley, New York, p. 552, 1995.

Shailendra, K., and V.B. Sudhirkumar, *Concrete Fracture Models and Applications*, Springer-Verlag, Berlin Heidelberg, 2011.

Sluys, L.J., and R. De Borst, *Rate-Dependent Modeling of Concrete Fracture, Heron*, Delft University of Technology, The Netherlands, Vol. 36, No. 2, 1991.

Van Mier, J.G.M., *Fracture Processes of Concrete: Assessment of Material Parameters for Fracture Models*, CRC Press, Boca Raton, FL, 1997.

Vipulanandan, C., and W.H. Gerstle, *Fracture Mechanics for Concrete Materials: Testing and Applications*, SP-201, ACI International, Farmington Hills, MI, 2001.

Whittman, F.H., ed., *Fracture Mechanics of Concrete*, Elsevier, Barking, Essex, U.K., 1983.

CHAPTER 14

Concrete in the Era of Global Warming and Sustainability

Preview

For a variety of reasons discussed in Chap. 1, concrete is the most widely used construction material today. What about the future? In his 1980 paper, *"Concrete for the Year 2000,"* Professor Clyde Kesler of the University of Illinois, said:

> Concrete, as a construction material, has been important in the past, is more useful now, and is confidently forecast to be indispensable in the future.[1]

This forecast was based on the time-honored rules of the marketplace, such as demand, supply, and technical and economic advantages of concrete over the alternative structural materials like lumber and steel. However, recently published reports[2–4] show that the choice of construction materials would depend on how well do they perform in the age of climate change and sustainability.

It seems that extreme weather cycles in the era of global warming will severely test the durability and sustainability of materials of construction. Field experience with structural members made of concrete shows that their service life can be greatly enhanced if the concrete mixture is designed to remain free from cracking under both structural and environmental loads.

In this chapter, an overview is presented of the primary forces that shaped the 20th century, and are continuing to shape today's world. These forces are population growth, urbanization, and technology choice for industrial and economic development. Concrete will continue to be the material of choice for general construction provided the concrete production follows basic principles of material science to make crack-resistant concrete mixtures. This is because we are entering an era where durability and sustainability of materials would be considered more important than high early strength and high speed of construction, which have been the main objectives of the industry for financial reasons.

14.1 The Forces Shaping Our World—An Overview

Change is inevitable. However, it is the rapid rate of change that often becomes disruptive. This is why, all of a sudden, we are confronted with the present situation when rapid speed of economic and industrial development has brought us to devastating cycles of weather changes on the planet Earth. Population growth, urbanization, technology choices and their environmental impact are, unquestionably, the key forces that are shaping the today's world.[5] These forces are interconnected as described below. The cumulative effect of these forces is global warming (climate change) that is threatening the survival of human life and society on the planet.

Population growth. It is generally believed that humans appeared on the planet Earth after the end of the ice age about 10,000 years ago. In the beginning, the population growth was at a slow rate, however it accelerated after the 19th century when the global population had reached 1.5 billion. In the beginning of the 20th century, the world population was 1.5 billion, and at the end it hit the 6 billion mark. A population growth rate of 4.5 billion people in a short period of 100 years was indeed explosive.

Urbanization. Statistics show a correlation between population growth and urbanization of the planet. At the beginning of the 20th century, approximately 10 percent of the people lived in cities; in the year 2001 nearly 3 of the 6 billion inhabitants lived in and around the cities. According to the United Nations' statistics,[6] the planet had 19 megacities,* 22 cities with 5 to 10 million, 370 cities with 1 to 5 million, and 430 cities with 0.5 to 1 million people. The number of moderately populated cities with population between 0.5 and 5 million has shown a tremendous increase in China. According to Thomas Friedman,[7] the scale and the scope of urbanization in China is mind-boggling. Thirty years ago, the urban population of China was about 20 percent, and now 40 percent of the 1.3 billion people live in cities. By 2020, the Chinese urban population is projected to grow to 60 percent—thus adding hundreds of satellite cities, and a skyrocketing demand for the energy-intensive materials like cement and steel that are needed for construction of new buildings, factories, highways, dams, and other structures.

Population statistics for the 20 largest cities of the world in the years 1980, 2000, and 2009 are presented in Table 14-1. It is evident from the data that there has been acceleration of urbanization in the emerging countries of the world. For instance, in India the population of Mumbai and Delhi has increased by 11 million and 10 million people, respectively, during the last 30 years. In China, during the same period, the population of Shanghai has grown by 8.5 million and, in Brazil, São Paulo city has added more than 7 million residents. From the 2009 world population statistics it is concluded that 17 of the largest 20 cities of the world are qualified for the "megacity" status. Many architects who are involved with urban planning believe that megacities cannot be sustainably developed.

Technology choices for economic development. During the 20th century, population growth and urbanization have played a great part in the enormous expansion of the energy sector, manufacturing sector, and transportation sector of the economy. Unfortunately, our technology choices have turned out to be wasteful and, at times, even harmful because decisions are based on short-term and narrow goals of the enterprise rather than a holistic view of the full range of consequences of the technology used. According to Hawken et al.,[8] *only 6 percent of the total global flow of materials, some 500 billion tonnes a year, actually ends up in consumer products, whereas much of the*

*A city with 10 million or more population is called a "megacity."

No*	Year 1980		Year 2000		Year 2009	
1	Tokyo	21.9	Tokyo	26.4	Tokyo	36.3
2	New York	15.6	Mexico City	18.1	Delhi	21.7
3	Mexico City	13.9	Mumbai	18.1	São Paulo	19.9
4	São Paulo	12.5	São Paulo	17.8	Mumbai	19.7
5	Shanghai	11.7	New York	16.6	Mexico City	19.3
6	Osaka	10.0	Los Angeles	13.1	New York	19.3
7	Buenos Aires	9.9	Kolkata	12.9	Shanghai	16.3
8	Los Angeles	9.5	Shanghai	12.8	Kolkata	15.2
9	Kolkata	9.0	Buenos Aires	12.6	Dhaka	14.2
10	Beijing	9.0	Dhaka	12.3	Buenos Aires	13.0
11	Paris	8.9	Karachi	11.8	Karachi	12.8
12	Rio de Janeiro	8.7	Delhi	11.7	Los Angeles	12.7
13	Seoul	8.3	Jakarta	11.0	Beijing	12.2
14	Moscow	8.1	Osaka	11.0	Rio de Janeiro	11.8
15	Mumbai	8.1	Manila	10.9	Manila	11.4
16	London	7.7	Beijing	10.9	Osaka	11.3
17	Tianjin	7.3	Rio de Janeiro	10.6	Lagos	10.2
18	Cairo	6.9	Cairo	10.5	Seoul	9.8
19	Chicago	6.8	Lagos	10.4	Jakarta	9.1
20	Essen	6.3	Seoul	9.9	Cairo	9.1

Source: The State of the World Cities Report, United Nations Center for Human Settlements, New York, June 2001. The 2009 World Urbanization Prospects, UNESCO Secretariat http://esa.un.org/wup2009/unup.

From the data in the table, it is clear that, in general, there has been an acceleration in the rate of urbanization of the world during the 30 years (from 1980 to 2009). For example, the population of Shanghai rose by 1.1 million during the 20 years period (1980–2000), however, more than three times as many people were added in the next 9 years period (2000–2009). Similarly, the population of New York City rose by 1 million during the 20 years (1980–2000) and by 2.7 million people during the next 9 years. Also, it may be noted from the 2009 population that 17 of the 20 largest cities have joined the category of megacities.

TABLE 14-1 Thirty Years of Population History of the World's 20 Largest Cities, millions

virgin materials are returned to the environment in the form of solid, liquid, and gaseous pollutants.

14.2 Concrete Requirements in the Near Future

There is no doubt that during the next 10 to 20 years, large volumes of structural materials will be needed for repair and replacement of deteriorating structures, and for building new housing and infrastructure to meet the basic needs of millions of poor migrants to towns and cities. Among the three primary materials of construction, namely portland-cement concrete, steel, and timber, concrete is the most adaptable structural material for use under changing climate conditions. The microstructure and properties of concrete can be easily altered by cost-effective methods, to improve the impermeability and durability on exposure to weather extremes. Examples of a few large projects that are being planned for execution in countries such as the United States and China are presented next.

America's infrastructure, built mostly during the World War II years of the 20th century, is confronting a crisis that threatens public safety, industrial development and economic growth. On March 19, 2013, the American Society for Civil Engineers (ASCE) released its Report Card for America's Infrastructure. Conducted by an Advisory Panel composed of the nation's leading civil engineers whose areas of expertise cover a broad spectrum, the fifth such assessment undertaken by the ASCE has conferred an overall grade of "D+," and estimated that an investment of 3.6 trillion U.S. dollars will be needed to raise the grade to a "B" level by 2020.

China is moving closer to completing the world's largest high-speed train network—a 10,000 miles honeycomb that will link major cities across an expanse similar in size to the United States. But that is only a small part of a trillion-dollar budget which also includes construction of a nationwide system of toll highways comparable to the U.S. Interstate Highway System; streamlined harbor port terminals capable of handling the country's vast export traffic, and several new state-of-the-art airports. Beijing, Shanghai, and Guangzhou have now some of the most sophisticated and integrated transport system in the world. China also wants to build high-speed rail lines across Asia, ultimately connecting to the European systems.

According to a recent report (News China Magazine, July 2013), new dams are being constructed nonstop on all major rivers and tributaries in the Yunnan Province and Sichuan Province. With a total installed capacity of 250 GW, China already is a top-ranking country in hydropower. The Three Gorges Dam, completed in 2006, added 21 GB, and a similar wattage will be added by eight smaller dams that are under construction on the Jinsha River. In Sichuan Province reportedly 7000 small dams are either under construction or completed.*

Developing countries, namely India, Brazil, and South Africa, have also many projects that would require large amounts of concrete in the near future. A question arises—do we have adequate resources of virgin raw materials for making portland-cement clinker, and aggregates for making concrete mixtures? Yes we do have large deposits of high-quality minerals and fossil fuels for making portland-cement clinker, and we also have abundance of rocks and sand deposits that are available worldwide, which are good sources of aggregates for concrete. In spite of such a bright future of the global concrete industry, this may not be realized because the carbon dioxide emissions attributable to human activities on the planet are increasing exponentially and are changing the global climate at a rate which threatens the survival of human life and built-environment on the Earth.

14.3 Environmental Impact of the Global Concrete Industry

Concrete is the largest manufactured product used by human society. The concrete industry has both negative and positive environmental impacts. Concrete mixtures for general construction, on the average, are composed of 12 percent portland cement,

*Note that the harmful environmental impacts of large dams are well known now.

8 percent mixing water, and 80 percent coarse and fine aggregate by mass. In 2010, approximately 33 billion tonnes yearly global concrete industry consumed nearly 3.7 billion tonnes portland-cement clinker and 27 billion tonnes of aggregate, in addition to 2.7 billion tonnes of mixing water and small amount of chemical admixtures. The mining, processing, and transport of such huge quantities of materials for making concrete, in addition to large amounts of limestone, clay, and fossil fuels for portland-cement clinker manufacture, require considerable energy, and adversely affect the ecology of virgin lands.

Portland cement, the principal hydraulic binder used in modern concrete mixtures, is a manufactured product which is not only energy-intensive (4 GJ/tonne of cement) but also responsible for large emissions of CO_2. The manufacture of 1 tonne of portland-cement clinker releases on the average 1 tonne CO_2 into the atmosphere. The world's yearly cement production of 4 billion tonnes of mostly portland-cement is responsible for nearly 7 percent of the total global CO_2 emissions.[9]

Typical ingredient of a general-use concrete mixtures are *a cementing material* composed of portland cement and one or more *supplementary cementing materials* (SCM) from the following list: fly ash from coal-fired power plants, granulated slag from iron blast-furnaces, pulverized limestone or limestone dust from quarries, and raw or calcined pozzolan, coarse and fine aggregates, mixing water, and one or more chemical admixtures.

There are positive environmental impacts of concrete which are not well-known. Industrial by-products such as coal fly ash and iron blast-furnace slag contain some toxic metals; and disposal of these by-products by dumping into landfills or ponds contaminates the drinking water supply. Instead, if these by-products are used as a part of the cementing materials in concrete mixtures, the toxic elements can get permanently bound in the hydration products of portland cement.

How can we reduce the adverse environmental impact of the concrete industry? The long-term approach to lower the environmental impact of any material lies in reducing its rate of consumption. It is estimated that 45 percent of the world's concrete is consumed for the construction of new buildings, 15 percent for new infrastructure, and 40 percent for repair and renovation of the built-environment. There are three ways to reduce the consumption of concrete. *First, the project footprint can be reduced by innovative architecture. Second, the thickness of foundations, columns, walls, and large beams can be reduced by smart structural design without compromising structural safety. Third, the durability of concrete mixtures can be improved by replacing large portion of portland cement with supplementary cementing materials.* Note that sustainability of the concrete industry also requires less consumption of the cementing materials in concrete mixtures, and less consumption of portland cement in the blended cements. Considerable cement-saving in concrete mixtures can also be achieved when the specified strength requirement is changed from 28 to 56 days. Significant cement savings are also achieved when the volume of cement paste in concrete is reduced by using aggregate mixtures of optimum size and grading. Finally, the slump of fresh concrete should be increased by using water-reducing chemical admixtures, rather than by the addition of more cement and more water. To pursue a holistic approach for sustainable industrial development, we must also practice industrial ecology. Simply stated, the practice of industrial ecology implies that waste products of one industry are recycled as substitutes for virgin raw materials of another industry, thereby reducing the environmental impact of both industries.

Reportedly, over a billion tonnes of *construction and demolition waste* are being disposed in road-bases and landfills every year, in spite of the fact that cost-effective technologies are available to recycle most of the waste as a partial replacement for coarse aggregate in concrete mixtures.[10] Similarly, most waste waters and undrinkable natural waters can be substituted for municipal water to mix concrete unless proven harmful by testing. Blended portland-cements containing high-volume fly ash from coal-fired power plants and granulated slag from blast-furnace iron industry offer other good example of industrial ecology because it offers a holistic solution to reduce the environmental impact of more than one industry.

Cementing materials containing either 20 to 30 percent coal fly ash or 30 to 50 percent blast-furnace slag by mass are already being used worldwide by the concrete construction industry. In Europe, concrete mixtures containing portland-slag cements containing 50 to 70 percent iron blast-furnace slag are well known for long-time durability to sulfate and seawater attack. Recent work in North America has shown that, with conventional materials and technology, it is possible to produce high performance concrete mixtures containing 50 to 60 percent fly ash by mass of the blended cementing material (Chap. 12). Fly ash is readily available in large amounts in China, India, and the U.S.[10]; therefore, in the near future fly ash would be the most powerful tool for reducing the environmental impact of the cement and concrete industry in those countries that require large quantities of cement for buildings and infrastructure.

Lord Shiva of the Industrial Materials World?

In the Hindu mythology, there are several gods. Shiva is an ascetic god who is always ready to help the needy. Once the gods collectively decided to dewater an ocean to obtain the pot of nectar of immortality which was lying at the ocean floor. However, in the dewatering process, a stream of poison was released and it started destroying the whole world. When no other god showed courage to handle the poison, Shiva came forward. He drank the whole stream of poison. This did not do any harm to him except changing the complexion of his skin to blue.

As a part of fuel for clinker burning, the cement industry is recycling hazardous organic wastes, thus recovering the energy value and conserving virgin fuel resources. Also, hydration products of portland-cement minerals are reported to serve as a safe sink for some toxic elements that are generally present in these hazardous wastes. Because portland-cement concrete is able to safely incorporate millions of tonnes of fly ash, slag, and other industrial by-products that contain toxic metals, it is like Lord Shiva of the industrial materials world.

14.4 Global Warming—An Introduction

By the end of the 20th century, it became clear that industrialization and urbanization of the world were happening at a speed which is not sustainable. In 2007, the U.N. sponsored Intergovernmental Panel on Climate Change (IPCC) issued a report on the physical science behind climate change. A summary of this report was published by Collins et al. in the Scientific American magazine.[11] According to the authors, there is inexorable evidence that recent global warming is associated with increasing atmospheric concentration of human-induced CO_2, because it is much higher than the preindustrial level of 280 ppm, and is accelerating rapidly. Independent studies by several national science academies have also confirmed the conclusions of the 2007 report of IPCC that the global climate is changing and presents a serious threat to human life and society on the planet Earth by increasing the frequency and intensity of extreme-weather events, like heat waves, rainstorms and flashfloods.

The term "global warming" refers to the small and steady increase in the average temperature of the Earth's surface due to the *greenhouse gas effect*. The greenhouse gases, namely carbon dioxide, methane, and nitrous oxide trap heat in the Earth's surface and the lower atmosphere, thus warming the entire planet. Karl et al. (Fig. 14-1) and other physicists have shown that a linear relationship exists between the average temperature of earth's surface and the atmospheric concentration of CO_2, which makes up 85 percent of the greenhouse gases. According to a recent report of the U.S. National Oceanic and Atmospheric Administration, the 2012 atmospheric CO_2 concentration (394 ppm), is the highest recorded in human history; it represents approximately 40 percent increase from the preindustrial concentration of 280 ppm CO_2.

From the ending of the last ice age, some 10,000 years ago, and the slow growth of industrialization in the 19th century A.D., the atmospheric concentration of CO_2 remained unchanged at the preindustrial level of 280 ppm until the beginning of the 20th century when, in response to accelerating industrialization on the earth, the atmospheric CO_2 started rising slowly at first and subsequently accelerated rapidly (Fig. 14-1). It can be deduced, therefore, that atmospheric CO_2 concentration above 280 ppm is attributable

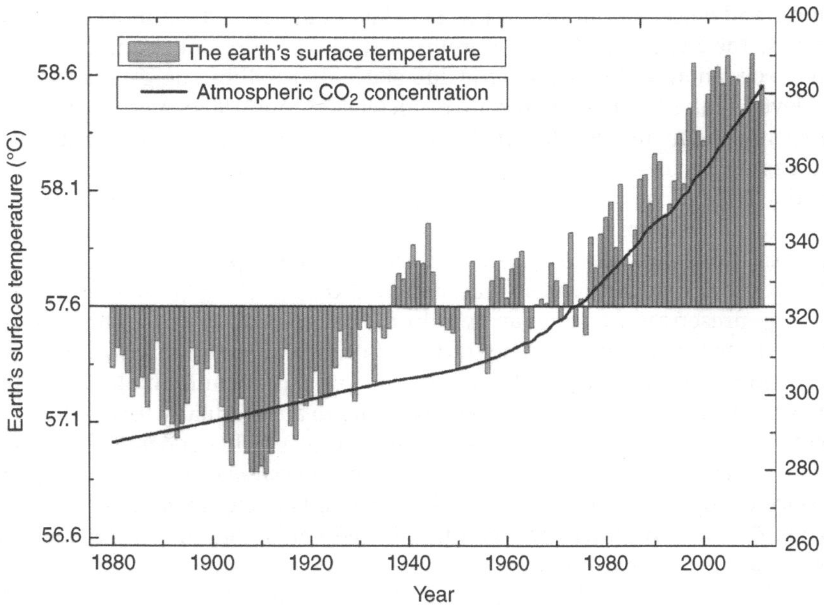

FIGURE 14-1 Relationship between global temperature and atmospheric CO_2 concentration. [Karl, T.R. et al., June 2009 (www.globalchange.gov)[12].]

to human activities on the planet Earth. In the following list some of the *observed effects* of global warming are stated:

- A sharp increase in the melting rates of glaciers, ice sheets, and polar caps.
- Threats to coastal populations from rising sea levels. The sea level has risen by 16 to 17 mm in the last 15 years.
- Unusual increase in the number and intensity of heat waves, droughts and wild fires. For example, in 2003 a heat wave killed more than a thousand people in Europe; North-Eastern Africa has been suffering from a severe famine; since the last 10 years nearly two-thirds of the land in the Midwestern United States is suffering from an unending drought.
- Unprecedented increase in the frequency and intensity of rainstorms and hurricanes all over the world (e.g., Hurricane Katrina ravaged the New Orleans in the United States once in 2005, and again in 2012).
- Impact of shifting rainfall patterns on the global resources of agriculture and water.
- Disruption of the earth's carbon cycle due to changes in botanical species on land and oceans.

14.5 Enhancement of Concrete Durability

In the concrete industry, the practice of industrial ecology provides only a *short-term* solution for sustainable development. In the *long term*, sustainable development will happen only if we make dramatic improvements in the resource productivity by significant advancement of durability. Hawken et al.[8] describe a movement launched in 1994 by the Factor Ten Club—a group of scientists, economists, and business people. The declaration of the Factor Ten Club states that within one generation, nations can achieve a tenfold increase in resource efficiency through 90 percent reduction in the use of energy and materials. Obviously, large savings in materials can result in the future if we begin to make products that would last much longer. For example, the resource productivity of concrete industry can leap by a factor of 5 if most of the structures built today would endure for 250 years instead of the conventional 50. Using a fundamental approach, it is possible to understand and control the primary causes of deterioration of concrete in field structures, as described next.

The presence of free water within the microstructure of a concrete member is the *first common denominator* behind most of the causes of deterioration, such as the corrosion of reinforcing steel, frost action, and sulfate and alkali attack. Essentially impermeable concrete structures, on exposure to air, gradually tend to lose watertightness. In addition to voids and some structural cracks, concrete contains invisible microcracks. The microcracks are the *second common denominator*. Depending on the service conditions, the microcracks in concrete become unstable and begin to grow. When the growing microcracks, interlink the voids and structural cracks, *the watertightness of the concrete is lost*.

Large capillary voids and massive calcium hydroxide crystals, formed during the hydration of portland cement create a plane of weakness in the transition zone between the cement paste and particles of aggregate (or embedded steel in reinforced concrete). A homogeneous microstructure resistant to microcracking can be obtained by using less mixing water, and a large amount of mineral admixtures with portland cement (e.g., 50 percent siliceous or a calcareous fly ash, or 60 to 70 percent ground blast-furnace slag). An alternative approach for use in the climate-change era is provided by the ternary cement compositions that are described in the next paragraph.

Ternary cements are multicomponent cements made by blending portland cement (clinker + gypsum) with two supplementary cementing materials, such as coal fly ash, natural pozzolans, granulated blast-furnace slag, limestone dust, silica fume, and reactive rice-husk ash. The properties of ternary-blended cements depend on the chemistry, mineralogy, particle size distribution, and surface area of the three components that make up the cementing material. Differences in physical and chemical characteristics of individual components and their proportions can be exploited for producing concrete mixtures with desired properties.

The early-age setting and hardening properties of concrete are controlled by portland cement (clinker + gypsum), which will continue to be one of the major components of the ternary-blended cements. From standpoint of sustainability, it is desirable that the proportion of portland clinker is limited to a maximum of 50 percent by mass of the ternary-blended cement.

Since the current rate of portland cement production is responsible for 7% carbon emissions annually, the industry must achieve a drastic cut in the global rate of portland cement and portland clinker production. In fact, as described in Section 14-6, the cement industry has been moving in the opposite direction.

The concrete manufacturing and construction practice of today is reductionistic because it is driven solely by the goal of high early strength of concrete and high speed of construction, which often leads to a product that is prone to cracking much before the end of design life. On the contrary, a holistic approach would have specified a blended portland cement containing large amount of industrial by-products, such as coal fly ash, granulated blast-furnace slag, and limestone dust. The hardened concrete, in spite of some retardation in the setting and hardening properties would have been a highly durable product made from low-cost and low-carbon cementing materials.

The question now is how can we accomplish a paradigm shift to a holistic approach from the currently prevailing reductionistic practices in the industry? According to Wilson[13] the process must begin at universities because this issue encompasses the entire field of engineering education today. Similarly, a holistic approach to address the climate change issue might save the human civilization from the threat of complete destruction if we take strong measures to get off the trajectory of rising carbon emissions as early as possible before reaching the point of no return (year 2040) on Fig. 14-3.

14.6 Accelerating Cement Production in the Era of Rapidly Changing Climate

According to Table 14-2, in the last decade of the 20th century (1990–1999), the global cement industry was growing at the average rate of 5 percent a year and it took 7 years (1990–1997) to increase the cement production by 500 million tonnes. Toward the end of

Year	Production	Year	Production
1990	1,040	2001	1,750
1991	1,180	2002	1,850
1992	1,120	2003	2,020
1993	1,290	2004	2,190
1994	1,370	2005	2,350
1995	1,440	2006	2,620
1996	1,490	2007	2,770
1997	1,540	2008	2,850
1998	1,540	2009	3,030
1999	1,600	2010	3,310
2000	1,660	2011	3,400

Source: U.S. Geological Survey.

TABLE 14-2 Global Cement Production, 1990–2010 (millions tonnes)

the previous decade, the global cement industry registered a similar increase in the production (from 2850 to 3310 million tonnes) in a much shorter period of 2 years (2008–2010). The plot of global cement production during the period 1990 to 2011 clearly shows an exponential increase in the yearly global production rate of cement (Fig. 14-2).

In the 20th century, the average surface temperature of the earth went up by 0.7°C. During the past 22 years, with only one exception, every year on the planet Earth has been warmer than the previous years. Since 1990s the frequency and intensity of extreme weather events has increased in many parts of the world. In order to bring

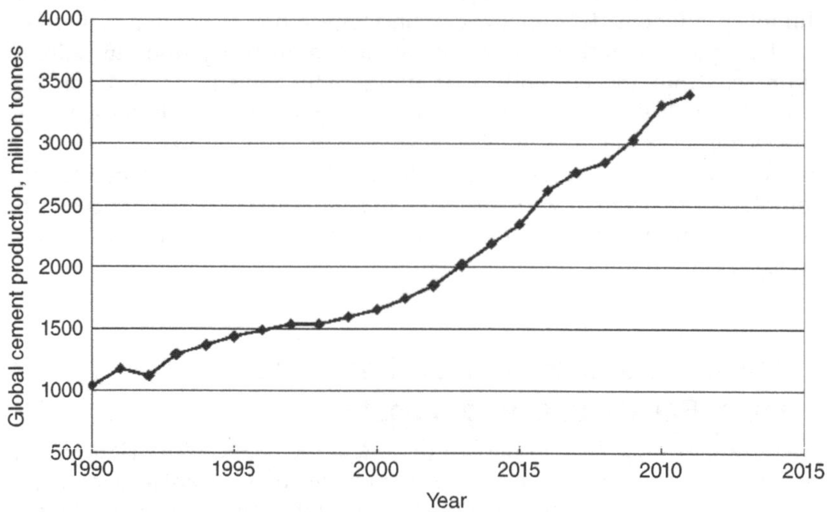

FIGURE 14-2 Global cement production.

under control the global climate disruption and its devastating effects on life on the earth, the carbon emission generated by human activities must be reduced, which is in direct conflict with exponentially rising global cement production.

14.7 Sustainability of the Concrete Industry—A Holistic Approach

In the rapidly changing climate era and accelerating cement production, how to define sustainability? In 1983, the U.N. set up the Brundtland Commission, which defined *Sustainable Development as the development that meets the needs of the present generation without compromising the ability of future generations to meet their own needs*. This definition was written in the era of population explosion when there was a growing threat of possible shortages of water, food, energy, and industrial materials. Now, we are confronting another threat which is climate change due to rising atmospheric CO_2 concentration from burning of fossil fuels to meet a variety of human needs. Because atmospheric CO_2 has proven to be a harmful pollutant, we need an amendment to the Brundtland definition of sustainability by including a reference to the earth's limit for safe disposal of harmful pollutants that may endanger human life and society on the planet Earth.

Figure 14-3 is an adapted version of the IPCC plot on historical and predicted atmospheric concentration of CO_2 during the years 800–2200. The horizontal and nearly straight line at 280 ppm CO_2 represents the concentration of Nature Controlled Carbon (NCC). In the preindustrial period until the year 2000, which marks the ending of 19th century and the beginning of the 20th century, the post-1900 industrial period total carbon emissions are attributable to both NCC and HCC (Human Controlled Carbon). As the value of total carbon emissions are known, we can subtract 280 ppm of NCC to obtain an estimate of the value of HCC. Based on this assumption, tabulated data is shown in Table 14-3.

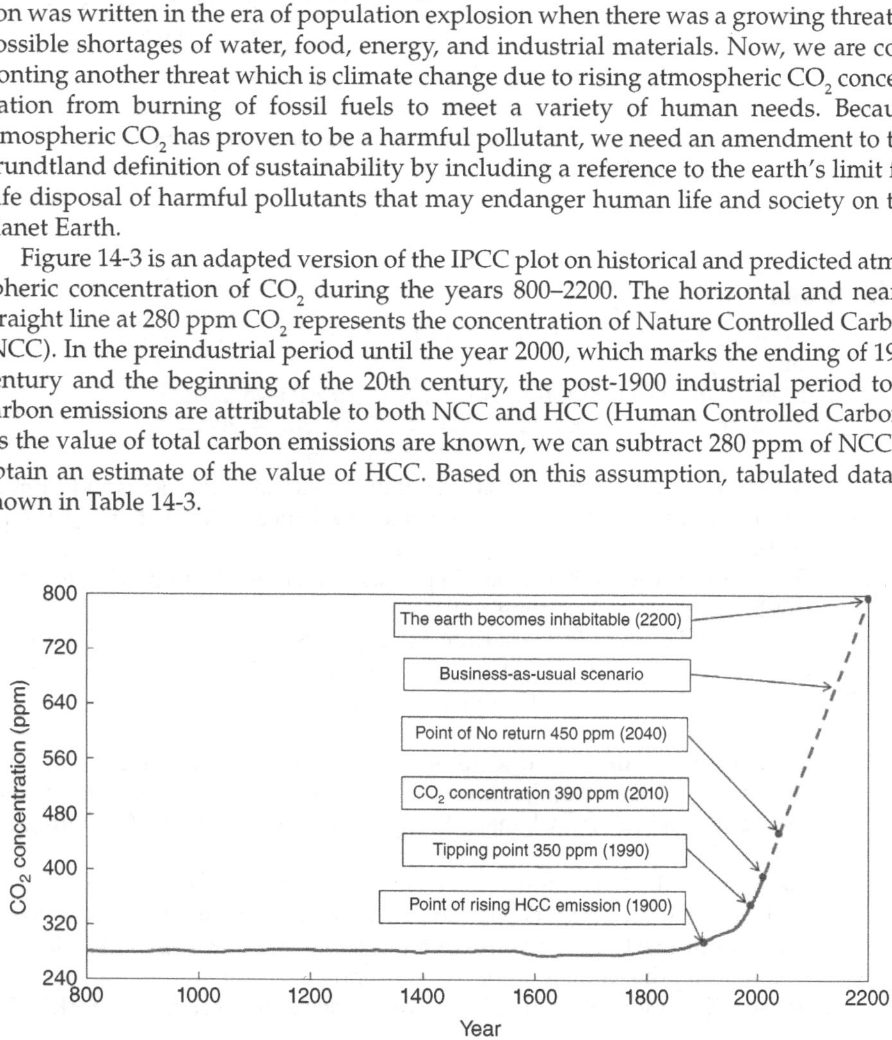

FIGURE 14-3 Historical and predicted CO_2 in the atmosphere. [Adapted from the IPCC Report, 2007, Solomon, S., D. Qin, M. Manning, Z. Chen, M. Marquis, K.B. Averyt, M. Tignor, and H.L. Miller (eds.). Cambridge University Press, Cambridge, United Kingdom and New York, NY, USA.]

Year	Total Atmospheric CO_2 Concentration	Preindustrial (NCC Emissions)	Industrial (HCC Emissions)	Observed Effect
1990	350	280	70	Beginning of extreme weather events attributable to HCC emissions
2010	390	280	110	Increasing frequency and intensity of heat waves, fires, droughts, rainstorms, and floods
2040	450*	280	170	Unprecedented losses to human life and property
2200	800*	280	520	—

*The data of the year 2040 and 2200 are computed with the prediction model of IPCC report.

Table 14-3 Atmospheric CO_2 Concentration (in ppm)

References

1. Kesler, C.E., *Progress in Concrete Technology*, Malhotra, V.M., ed., CANMET, Ottawa, pp. 1–23, 1980.
2. Mehta, P.K., *ACI Concr. Int.*, Vol. 24, No. 7, pp. 23–28, 2002.
3. Malhotra, V.M., *Concr. Int.*, Vol. 24, No. 7, pp. 30–34, 2002.
4. Holland, T.C., *Concr. Int.*, Vol. 24, No. 7, pp. 35–40, 2002.
5. Mehta, P.K., "Concrete Technology at the Cross Roads—Problems and Opportunities in the Past, Present, and Future Concrete," ACI Special Publication, SP-144, pp. 1–31, 1994.
6. The State of the World Cities Report 2001, United Nations Center for Human Settlements, New York, June 2001.
7. Friedman, T.L., *Hot, Flat, and Crowded*, Farrar, Straus and Giroux, New York, 2008.
8. Hawken, P., E. Lovins, and H. Lovins, *Natural Capitalism—Creating the Next Industrial Revolution*, Little Brown, Boston, p. 369, 1999.
9. Mehta, P.K., *ACI Concr. Int.*, Vol. 23, No. 10, pp. 61–66, 2001.
10. Corinaldesi, V., and G. Moriconi, *ACI SP-199*, pp. 869–884, 2001.
11. Collins, W., R. Coleman, J. Haywood, M.R. Manning, and P. Mote, "The Physical Science Behind Climate Change," Scientific American (www.SciAm.com), Aug. 2007, pp. 64–73.
12. Karl, T.R., J. M. Melillo, and T. C. Peterson, eds., 2009. Global climate change impacts in the United States. U.S. Global Change Research Program (USGCRP). Washington, DC: Cambridge University Press. Online at: http://downloads.globalchange.gov/usimpacts/pdfs/climate-impacts-report.pdf.
13. Wilson, E.O., *Consilience: The Unity of Knowledge*, Alfred Knof, New York, p. 325, 1998.

Index

Note: Page numbers followed by f denote figures; page numbers followed by t denote tables

B

G

H

Q

R

S

T